ZERO-WASTE TECHNOLOGIES IN AQUACULTURE

ZERO-WASTE TECHNOLOGIES IN AQUACULTURE

Global Perspectives and Methods toward a Circular Economy

Edited by

SOIBAM KHOGEN SINGH

Senior Scientist & Head, ICAR Research Complex for NEH Region, ICAR-Krishi Vigyan Kendra, Imphal, Manipur, India

DHARMENDRA KUMAR MEENA

Senior Scientist, ICAR-Central Inland Fisheries Research Institute, Barrackpore, Kolkata, West Bengal, India

MOHAMMAD HOSSEIN KHANJANI

Associate Professor, Department of Fisheries Sciences and Engineering, Faculty of Natural Resources, University of Jiroft, Jiroft, Iran

MOSLEM SHARIFINIA

Head, Aquaculture Scientific Department, Shrimp Research Center, Iranian Fisheries Sciences Research Institute, Agricultural Research, Education and Extension Organization (AREEO), Bushehr, Iran

ELSEVIER

ACADEMIC PRESS

An imprint of Elsevier

elsevier.com/books-and-journals

Academic Press is an imprint of Elsevier

125 London Wall, London EC2Y 5AS, United Kingdom
50 Hampshire Street, 5th Floor, Cambridge, MA 02139, United States

ISBN: 978-0-443-34689-7

For Information on all Academic Press publications visit our website at https://www.elsevier.com/books-and-journals

Publisher: Mara Conner
Acquisitions Editor: Vaishali Singh
Editorial Project Manager: Suman Deshwal
Production Project Manager: Abhishek Singh
Cover Designer: Arumugam Kothandan

Typeset by Aptara, New Delhi, India

Contents

16. FLOCponics as a sustainable and environmentally friendly system

Otávio Augusto Lacerda Ferreira Pimentel, Dariano Krummenauer, Wilson Wasielesky and Mohammad Hossein Khanjani

17. Integrated multitrophic aquaculture: Advancing sustainability and environmental stewardship

Pronob Das, Maibam Malemngamba Meitei, Muzammal Hoque, Syamala Karthireddy, Muralidhar P Ande and Basanta Kumar Das

18. Novel nutraceuticals from waste resources: Looking beyond the aqua-industry

Soibam Ngasotter, Sandhiya Venkatesh, Tao Kara, Madhulika, Sushil Lukram, David Waikhom, Domendra Dhruve and K. A. Martin Xavier

19. Zero-waste valorization of fibre crops: Towards a circular bioeconomy

Kanti Meena, A.K. Lavanya, Pandu Ranganath Diveti and Sonali Paul Mazumdar

20. Utilization of agroindustrial waste in ornamental fish culture

Mukesh Kumar Bairwa, Dharmendra Kumar Meena, Biswajit Dash and Saroj Kumar Swain

21. Nutritional interventions for sustainable aquaculture in a biobased circular economy

Nazeemashahul Shamna and Femi John Fawole

22. Introducing insects into a circular-based economy in aquaculture

Moslem Sharifinia, Mohammad Hossein Khanjani and Zahra Afshari Bahmanbeigloo

23. Novel flocculants from waste derivatives: A way to improve biofloc systems functioning

Reshmi Debbarma, Soibam Khogen Singh, Pronob Das, Sourabh Debbarma and Shrinath Vaijanath Gavhane

Contributors

Cristóbal Aguilera Aquaculture Program, Institute of Agrifood Research and Technology (IRTA), La Ràpita, Spain

Muralidhar P Ande ICAR-Central Institute of Fisheries Education, Kakinada Centre, Kakinada, Andhra Pradesh, India

Zahra Afshari Bahmanbeigloo Bushehr Regional Water Authority, Iran Water Resources Management Company, Bushehr, Iran

Mukesh Kumar Bairwa ICAR-Central Institute of Freshwater Aquaculture, Bhubaneswar, Odisha, India

Girija Saurabh Behere College of Fisheries, Datia, Rani Lakshmi Bai Central Agricultural University, Jhansi, Uttar Pradesh, India; ICAR-Central Inland Fisheries Research Institute, Barrackpore, Kolkata, West Bengal, India

Luis Otavio Brito Department of Fisheries and Aquaculture, Federal Rural University of Pernambuco (UFRPE), Recife, Pernambuco, Brazil

Koushik Chakroborty Department of Aquatic Environment and Resource Management, Sher-e-Bangla Agricultural University, Dhaka, Bangladesh

Balasaheb Ramdas Chavan ICAR-Central Inland Fisheries Research Institute, Prayagraj, Uttar Pradesh, India

Basanta Kumar Das ICAR-Central Inland Fisheries Research Institute, Barrackpore, Kolkata, West Bengal, India

Pronob Das ICAR-Central Inland Fisheries Research Institute, Regional Centre, Guwahati, Assam, India

Shyamal Chandra Sukla Das ICAR-Central Inland Fisheries Research Institute, Regional Centre, Guwahati, Assam, India

Biswajit Dash ICAR-Central Institute of Freshwater Aquaculture, Bhubaneswar, Odisha, India

Clarissa Vilela Figueiredo da Silva Campos Academic Unit of Serra Talhada, Federal Rural University of Pernambuco (UFRPE), Serra Talhada, Brazil

Danielle Alves da Silva Department of Fisheries and Aquaculture, Federal Rural University of Pernambuco (UFRPE), Recife, Pernambuco, Brazil

Jessika de Lima Abreu Department of Fisheries and Aquaculture, Federal Rural University of Pernambuco (UFRPE), Recife, Pernambuco, Brazil

Priscilla Celes Maciel de Lima Department of Fisheries and Aquaculture, Federal Rural University of Pernambuco (UFRPE), Recife, Pernambuco, Brazil

Carlos Yure Barbosa de Oliveira Department of Botany, Federal University of Santa Catarina (UFSC), Florianópolis, Santa Catarina, Brazil

Reshmi Debbarma College of Fisheries, Central Agricultural University, Agartala, Tripura, India

Sourabh Debbarma Fisheries College and Research Institute, Thoothukudi, Tamil Nadu Dr. J. Jayalalithaa Fisheries University, Tamil Nadu, India

Domendra Dhruve LSPN College of Fisheries, Dau Shri Vasudev Chandrakar Kamdhenu Vishwavidyalaya, Kawardha, Chhattisgarh, India

Pandu Ranganath Diveti ICAR-Central Research Institute for Jute and Allied Fibres, Kolkata, West Bengal, India

Maurício Gustavo Coelho Emerenciano Livestock & Aquaculture Program, Aquaculture Systems Team, Commonwealth Scientific and Industrial Research Organization, Bribie Island Research Centre, Woorim, QLD, Australia

Femi John Fawole Agricultural Research and Development Program, Central State University, Wilberforce, OH, United States

Baldassare Fronte Department of Veterinary Sciences, University of Pisa, Pisa, Italy

Shrinath Vaijanath Gavhane Dr. M.G.R. Fisheries College and Research Institute, Ponneri, Tamil Nadu Dr. J. Jayalalithaa Fisheries University, Tamil Nadu, India

Enric Gisbert Aquaculture Program, Institute of Agrifood Research and Technology (IRTA), La Ràpita, Spain

Dibakar Gogoi ICAR-Central Inland Fisheries Research Institute, Regional Centre, Guwahati, Assam, India

Alfredo Olivera Gálvez Department of Fisheries and Aquaculture, Federal Rural University of Pernambuco (UFRPE), Recife, Pernambuco, Brazil

Muzammal Hoque ICAR-Central Inland Fisheries Research Institute, Regional Centre, Guwahati, Assam, India

Md Shoebul Islam Shrimp Research Station, Bangladesh Fisheries Research Institute, Bagerhat, Bangladesh

Aruna Kamble College of Fisheries, Rangeilunda, Odisha University of Agriculture & Technology (OUAT), Bhubaneswar, Odisha, India

Manoj Tukaram Kamble Department of Agricultural and Fishery Science, Faculty of Science and Technology, Prince of Songkla University, Pattani, Thailand

Dibyendu Kamilya Agricultural and Food Engineering Department, Indian Institute of Technology Kharagpur, Kharagpur, West Bengal, India

Tao Kara ICAR-Central Institute of Fisheries Education, Mumbai, Maharashtra, India

Syamala Karthireddy ICAR-Central Institute of Fisheries Education, Kakinada Centre, Kakinada, Andhra Pradesh, India

Md. Idrish Raja Khan Department of Aquatic Animal Health Management, College of Fisheries Science, Chaudhary Charan Singh Haryana Agricultural University, Hisar, Haryana, India

Mohammad Hossein Khanjani Department of Fisheries Sciences and Engineering, Faculty of Natural Resources, University of Jiroft, Jiroft, Iran

Potluri Sai Kishore ICAR-Central Institute of Fisheries Education, Mumbai, Maharashtra, India

Dariano Krummenauer Aquaculture Marine Station, Institute of Oceanography, Federal University of Rio Grande, Rio Grande, Brazil

Ganesh Kumar College of Fisheries, Datia, Rani Lakshmi Bai Central Agricultural University, Jhansi, Uttar Pradesh, India

A.K. Lavanya ICAR-Central Research Institute for Jute and Allied Fibres, Kolkata, West Bengal, India

Sushil Lukram ICAR-Central Institute of Fisheries Education, Mumbai, Maharashtra, India

Dorothy MS National Fisheries Development Board-North Eastern Regional Center, Guwahati, Assam, India

Madhulika ICAR-Central Institute of Fisheries Education, Mumbai, Maharashtra, India

Sullip Kumar Majhi ICAR-Central Inland Fisheries Research Institute, Regional Centre, Guwahati, Assam, India

Prasenjit Mali Department of Aquatic Animal Health, Faculty of Fishery Sciences, West Bengal University of Animal and Fishery Sciences, Kolkata, West Bengal, India

Sonali Paul Mazumdar ICAR-Central Research Institute for Jute and Allied Fibres, Kolkata, West Bengal, India

Seema Vijay Medhe ASEAN Institute for Health Development, Mahidol University, Putthamonthon, Salaya, Nakhon Pathom, Thailand

Dharmendra Kumar Meena ICAR-Central Inland Fisheries Research Institute, Barrackpore, Kolkata, West Bengal, India

D.K. Meena College of Fisheries, Datia, Rani Lakshmi Bai Central Agricultural University, Jhansi, Uttar Pradesh, India

Kanti Meena ICAR-Central Research Institute for Jute and Allied Fibres, Kolkata, West Bengal, India

Maibam Malemngamba Meitei ICAR-Central Institute of Fisheries Education, Mumbai, Maharashtra, India

Tanuj Misra College of Fisheries, Datia, Rani Lakshmi Bai Central Agricultural University, Jhansi, Uttar Pradesh, India

Maitarambum Wartung Monsang College of Fisheries, Central Agricultural University (Imphal), Agartala, Tripura, India

Sukham Munilkumar ICAR-Central Institute of Fisheries Education, Mumbai, Maharashtra, India

Soibam Ngasotter LSPN College of Fisheries, Dau Shri Vasudev Chandrakar Kamdhenu Vishwavidyalaya, Kawardha, Chhattisgarh, India

Lokesh Pawar Faculty of Sciences, UGent Campus Sterre, University of Ghent, Krijgslaan, Gent, Belgium

Otávio Augusto Lacerda Ferreira Pimentel Department of Fisheries and Aquaculture, Federal Rural University of Pernambuco (UFRPE), Recife, Pernambuco, Brazil

S.M. Rafiquzzaman Department of Fisheries Biology and Aquatic Environment, Gazipur Agricultural University, Gazipur, Bangladesh

Sudhanshu Raman College of Fisheries, Datia, Rani Lakshmi Bai Central Agricultural University, Jhansi, Uttar Pradesh, India

Kalapala Satvik Kerala University of Fisheries and Ocean Studies, Kochi, Ernakulam, Kerala, India

Nazeemashahul Shamna Fish Nutrition, Biochemistry and Physiology Division, ICAR-Central Institute of Fisheries Education, Mumbai, Maharashtra, India

Moslem Sharifinia Aquaculture Department, Shrimp Research Center, Iranian Fisheries Sciences Research Institute, Agricultural Research, Education and Extension Organization (AREEO), Bushehr, Iran

Niti Sharma ICAR-Central Inland Fisheries Research Institute, Regional Centre, Guwahati, Assam, India

Angom Baleshwor Singh ICAR-Krishi Vigyan Kendra, Chandpur, Assam, India

Soibam Khogen Singh College of Fisheries, Central Agricultural University, Agartala, Tripura, India; ICAR Research Complex for NEH Region, Krishi Vigyan Kendra, Imphal, Manipur, India

Kaliyaperumal Sriranjani Institute of Fisheries Post Graduate Studies, Vaniyanchavadi, Tamil Nadu Dr. J. Jayalalithaa Fisheries University, Tamil Nadu, India

Saroj Kumar Swain ICAR-Central Institute of Freshwater Aquaculture, Bhubaneswar, Odisha, India

Karthireddy Syamala ICAR-Central Institute of Fisheries Education, Kakinada Centre, Kakinada, Andhra Pradesh, India

Sruti T.N. Dr. M.G.R. Fisheries College and Research Institute, Thalainayeru, Tamil Nadu, India

Sandhiya Venkatesh ICAR-Central Institute of Fisheries Education, Mumbai, Maharashtra, India

David Waikhom College of Fisheries, Central Agricultural University (Imphal), Lembucherra, Tripura, India

Wilson Wasielesky Aquaculture Marine Station, Institute of Oceanography, Federal University of Rio Grande, Rio Grande, Brazil

K. A. Martin Xavier ICAR-Central Institute of Fisheries Technology, Cochin, Kerala, India

Somnath Ramchandra Yadav College of Fishery Science, Maharashtra Animal & Fishery Sciences Unviersity (MAFSU), Nagpur, Maharashtra, India

Saeed Zahedi Department of Fisheries, Faculty of Natural Resources and Environment, Ferdowsi University of Mashhad, Mashhad, Iran

About the Editors

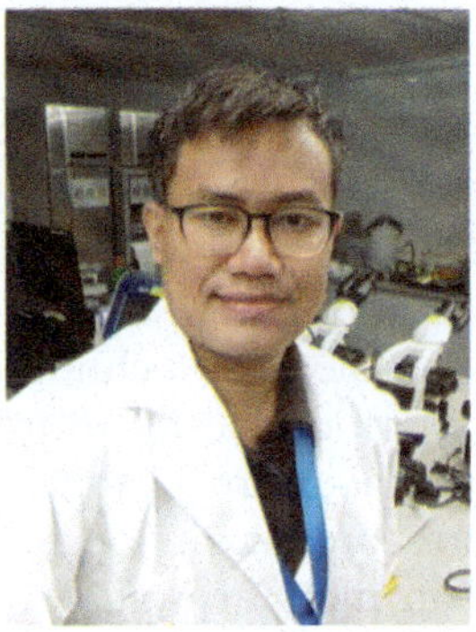

Dr. Soibam Khogen Singh is currently working as Senior Scientist cum Head of the Krishi Vigyan Kendra under the ICAR Research Complex for NEH Region, India. He completed his graduation from the Faculty of Fishery Science, West Bengal University of Animal and Fisheries Sciences, Kolkata, India, M.F.Sc. (Aquaculture) from the Fisheries College and Research Institute, Tamil Nadu Veterinary and Animal Sciences University, Chennai, India, and PhD (Aquaculture) from Central Institute of Fisheries Education, Mumbai, India. For over a decade, he has been dedicated to fostering excellence in fisheries education and research. As an Assistant Professor of Aquaculture at the College of Fisheries, CAU, Tripura, India, he played a pivotal role in shaping the next generation of aquaculture professionals. He has vast academic experience of teaching undergraduate, postgraduate, and PhD programs, reflecting his sincere commitment to nurturing talent at all levels. His research contributions are equally significant. He has guided 4 PG and 1 PhD students as a Chairman, and 25 others as committee members, fostering their research endeavors. His research, supported by external funding and collaborations, has yielded over 90 publications in reputed journals with high impact factor. His expertise is further reflected in the authorship of four books on fisheries. Recognizing his contributions to the field, he is serving as the Associate Editor of two prestigious journals (*Aquaculture Nutrition and Frontiers in Marine Science*), as well as editorial boards of journals such as *Scientific Reports, Discover Sustainability, Frontiers in Sustainable Food Systems, BMC Environmental Science and Frontiers in Aquaculture*. He has also undertaken faculty foreign training at James Cook University, Singapore. Beyond academia, he has been associated with five projects above 1 crore budgets, resulting in the development of six field-ready technologies. His commitment to outreach is evident in my involvement in vocational training, skill development programs, and national extension initiatives.

Dr. Dharmendra Kumar Meena, PhD (Fisheries Science), is a renowned Senior Scientist at ICAR-Central Inland Fisheries Research Institute (CIFRI), Kolkata, India (2006–present), with more than 20 years of pioneering research in sustainable aquaculture, freshwater ecology, and watershed management. At ICAR-CIFRI, he has pioneered a transformative "waste-to-wealth" paradigm, successfully commercializing innovative products including CIFRI CAGE GROW floating feed (Trademark 3625921) and the Reshmeen series feeds, which utilize agro-wastes and silkworm pupae as sustainable protein sources. With more than 107+ peer-reviewed publications (h-index 25, 3139 citations), his key works include macrozoobenthic indicators for Gangetic wetlands, Hilsa ranching protocols, and *Terminalia arjuna* as a fish health immunomodulator. A globally recognized editor, he serves as Associate Editor for *International Aquatic Research, Aquaculture Science and Management (BMC)*, and *Frontiers in Aquaculture*; Academic Editor for *PLOS ONE* and *Advances in Agriculture*; and Editorial Board Member for more than 20+ journals including *Aquaculture International, Scientific Reports, BMC Blue Biotechnology,* and *Fishes* (MDPI). He has also served as Lead Guest Editor for special issues on blue economy and sustainability. Dr. Meena is a Principal Member of the Bureau of Indian Standards (BIS) FAD12 Committee (Fish and Aquaculture), serves on the NFDB Project Monitoring Committee, and contributes as a NITI Aayog policy expert. His professional recognitions include Fellowships from the Zoological Society of India, Academy of Environmental Biology, Fisheries & Life Sciences, and Society of Biotic and Environmental Research, along with the Young Scientist Award (Academy of Environmental Biology) and multiple Best Presentation Awards. As a Master Trainer for ASCI and RKVY schemes, he has trained more than 10,000 stakeholders, and as a networking ambassador, he fosters international collaborations that bridge Indian fisheries innovation with global scientific communities for sustainable blue economy advancement.

Dr. Mohammad Hossein Khanjani is an Associate Professor in the Department of Fisheries Science and Engineering at the University of Jiroft, Iran. He obtained his MSc in Natural Resources Engineering and PhD in Aquaculture. His research interests include novel aquaculture technologies such as biofloc technology, integrated multi-trophic aquaculture, and recirculating aquaculture systems. His primary area of research is biofloc technology for sustainable shrimp farming. In addition, he serves on the editorial boards of Aquaculture, Fish and Fisheries, Aquaculture Studies, and the Indonesian Aquaculture Journal, and is an Associate Editor for the Journal of Fisheries. According to the Stanford University/Elsevier ranking of the world's top 2% of scientists (released in 2023, 2024, and 2025), he has been recognized as one of the most influential researchers globally. This ranking is based on citation metrics and scientific impact indicators from the Scopus database.

Dr. Moslem Sharifinia is a distinguished marine biologist and a research faculty member with extensive expertise in sustainable aquaculture. He holds a PhD in Marine Biology and currently serves as the Head of the Department of Marine Ecology at the Iran Shrimp Research Center. His research spans a wide range of topics, including nutrition, physiology, immunology, and biofloc technology in aquatic organisms. With a strong background in ecological fieldwork and data analysis, Dr. Sharifinia has made significant contributions to advancing our understanding of plankton and benthic communities and their roles in the health of shrimp and fish. His analytical approach and commitment to environmentally responsible aquaculture have earned him recognition as a respected voice in the field.

Preface

The book *Zero-Waste Technologies in Aquaculture: Global Perspectives and Methods toward a Circular Economy* serves as an important academic resource for improving sustainable food systems and their related sectors. The worldwide aquaculture sector presently encounters considerable constraints, characterized by a strong reliance on limited marine resources for aquafeed and the aggregate environmental deterioration due to untreated nitrogen-rich discharges. By employing circular bio-economy frameworks based on the principles of reduction, reuse, and recycling, the sector may significantly improve resource efficiency and reduce negative environmental impacts while boosting industrial performance. This book outlines the essential transition of traditional extractive models to advanced closed-loop systems. The discussion begins with an assessment of wastewater treatment through recirculatory principles, followed by a thorough examination of microbial-mediated zero-waste technology. Dedicated parts analyze symbiotic aquaculture and the notion of "Aquamimicry," emphasizing the simultaneous enhancement of water quality and aquatic animal health via biological management. The book synthesizes interdisciplinary expertise in biotechnology, environmental engineering, and aquatic nutrition, establishing the conceptual and practical framework for novel aquafeeds. This paradigm highlights the conversion of waste streams into high-value ingredients, including insect-based proteins. Moreover, the work offers a comprehensive analysis of Integrated Multitrophic Aquaculture and FLOCponics, illustrating how integrated bioremediation techniques can enhance systemic circularity by nearly 90%. The book culminates with a forward-looking evaluation of nutritional interventions in bio-based economies and the creation of innovative waste-derived flocculants, providing a crucial framework for aquaculturists, environmental scientists, and policymakers committed to sustainable resource management.

Soibam Khogen Singh
Dharmendra Kumar Meena
Mohammad Hossein Khanjani
Moslem Sharifinia

Acknowledgments

The authors would like to express their sincere gratitude to all the contributors of this book for their insightful presentations on the frontier areas of zero-waste technology in aquaculture. We would especially like to thank the Indian Council of Agricultural Research (ICAR), New Delhi, India, Central Agricultural University, Imphal, India, University of Jiroft, Iran, and Iran Shrimp Research Center (ISRC), Iran for their guidance and support. The publisher's consideration of the subject and invitation to edit the volume are appreciated by the authors.

Description of the Book

Zero-Waste Technologies in Aquaculture: Global Perspectives and Methods toward a Circular Economy: provides a definitive framework for transitioning the sector from traditional extractive models to advanced closed-loop systems through circular bio-economy principles. The book explores microbial-mediated zero-waste technologies and biological management strategies to optimize both water quality and aquatic animal health. Additional chapters on nutrient trapping and the use of insects as feed serve as a vital resource for practitioners and policymakers committed to sustainable resource valorization.

Key features:

- Shift from conventional methods to water-efficient practices such as biofloc technology and FLOCponics, which promote nutrient recycling and reduce environmental impact through advanced microbial management
- Aquamimicry and symbiotic-based aquaculture utilize natural biological processes to improve water quality and promote aquatic animal health without relying on intensive chemical interventions
- The upcycling of waste, including industrial wastes, agro-industrial waste, and fiber crops, into valuable resources for both commercial fish culture, as well as the extraction of novel nutraceuticals
- Introduction of insects and microbial-mediated proteins derived from bio-waste, offering a sustainable substitute for limited marine-based feed components
- Multitiered bioremediation utilizing the IMTA principle and innovative waste-derived flocculants to enhance overall system efficiency

Introduction

The book *Zero-Waste Technologies in Aquaculture: Global Perspectives and Methods toward a Circular Economy* serves as an important academic resource for improving sustainable food systems and their related sectors. The worldwide aquaculture sector presently encounters considerable constraints, characterized by a strong reliance on limited marine resources for aquafeed and the aggregate environmental deterioration due to untreated nitrogen-rich discharges. By employing circular bioeconomy frameworks based on the principles of reduction, reuse, and recycling, the sector may significantly improve resource efficiency and reduce negative environmental impacts while boosting industrial performance. This book outlines the essential transition of traditional extractive models to advanced closed-loop systems. The discussion begins with an assessment of wastewater treatment through recirculatory principles, followed by a thorough examination of microbial-mediated zero-waste technology. Dedicated parts analyze symbiotic aquaculture and the notion of "Aquamimicry," emphasizing the simultaneous enhancement of water quality and aquatic animal health via biological management. The book synthesizes interdisciplinary expertise in biotechnology, environmental engineering, and aquatic nutrition, establishing the conceptual and practical framework for "Aquafeed 3.0". This paradigm highlights the conversion of waste streams into high-value ingredients, including insect-based proteins. Moreover, the work offers a comprehensive analysis of Integrated Multitrophic Aquaculture and FLOCponics, illustrating how integrated bioremediation techniques can enhance systemic circularity by nearly 90%. The book culminates with a forward-looking evaluation of nutritional interventions in biobased economies and the creation of innovative waste-derived flocculants, providing a crucial framework for aquaculturists, environmental scientists, and policymakers committed to sustainable resource management.

CHAPTER

1

Global aquaculture growth: Challenges, opportunities, and key reforms

Somnath Ramchandra Yadav[a], *Balasaheb Ramdas Chavan*[b], *Seema Vijay Medhe*[c], *Manoj Tukaram Kamble*[d] *and Dharmendra Kumar Meena*[e]

[a]College of Fishery Science, Maharashtra Animal & Fishery Sciences Unviersity (MAFSU), Nagpur, Maharashtra, India [b]ICAR-Central Inland Fisheries Research Institute, Prayagraj, Uttar Pradesh, India [c]ASEAN Institute for Health Development, Mahidol University, Putthamonthon, Salaya, Nakhon Pathom, Thailand [d]Department of Agricultural and Fishery Science, Faculty of Science and Technology, Prince of Songkla University, Pattani, Thailand [e]ICAR-Central Inland Fisheries Research Institute, Barrackpore, Kolkata, West Bengal, India

1.1 Introduction

Aquaculture is an important source of food production globally, particularly in developing countries. Aquaculture is important for global food security, particularly in developing countries (Kamble et al., 2026a). However, disease remains a major limitation to sustainable production. Aquaculture has emerged as one of the most dynamic and rapidly expanding sectors of global food production, fundamentally reshaping the supply of aquatic foods worldwide. With capture fisheries output remaining largely static over the past several decades, aquaculture has assumed the primary role in meeting the growing global demand for fish and shellfish. Recent global assessments indicate that total fisheries and aquaculture production reached an unprecedented level of more than 223 million tons in 2022, with aquaculture accounting for more than half of the aquatic animals consumed directly by humans (FAO, 2024). This shift highlights aquaculture's central role in ensuring food and nutrition security, particularly in

Zero-Waste Technologies in Aquaculture: Global Perspectives and Methods toward a Circular Economy
DOI: https://doi.org/10.1016/B978-0-443-34689-7.00005-7

TABLE 1.1 Global aquaculture growth indicators (FAO, 2024).

Indicator	Global status
Total aquatic animal production	223.2 million tons
Contribution of aquaculture to human consumption	>57%
Global sector valuation	USD 472 billion
Aquatic food supply per capita	20.7 kg/year
Primary sector employment	62 million people

developing regions where fish serves as an affordable source of high-quality protein, essential fatty acids, and vital micronutrients (Chavan et al., 2015). Beyond food security, the sector supports millions of livelihoods (Table 1.1), contributes significantly to national economies, and plays a crucial role in the evolving blue economy framework.

The sustained growth of aquaculture has been underpinned by multiple drivers, including population growth, urbanization, rising incomes, changing dietary preferences, and increasing recognition of the health benefits of aquatic foods (Troell et al., 2023). Technological progress has further accelerated this expansion by advancing hatchery techniques, selective breeding, formulated feeds, water quality management, and health monitoring. Traditional extensive pond systems have gradually given way to semiintensive and intensive production models, including biofloc systems, biofilm/periphyton systems, recirculating aquaculture systems (RAS), offshore cages, and integrated multitrophic aquaculture (IMTA) (Avnimelech, 2009; Yadav et al., 2022; Alam et al., 2024; Halim et al., 2025). These transformations have enhanced production efficiency, land and water productivity, and year-round production potential, thereby strengthening aquaculture's position as a reliable and scalable food production system.

Despite these achievements, the rapid intensification and spatial expansion of aquaculture have generated a complex set of environmental, biological, and socioeconomic challenges (Mavraganis et al., 2020). Disease outbreaks and the growing threat of antimicrobial resistance (AMR) remain among the most serious constraints to productivity and profitability (Mohammed et al., 2025). Feed dependency—particularly on marine-derived fishmeal and fish oil—raises serious concerns about resource sustainability and ecological footprints (Majluf et al., 2024). Environmental issues such as nutrient pollution, habitat alteration, and biodiversity risks linked to exotic species introductions further complicate the sustainability landscape (Mavraganis et al., 2020). In addition, climate change has emerged as a pervasive stressor, influencing temperature regimes, water availability, salinity patterns, and the frequency of extreme weather events, thereby affecting both inland and coastal aquaculture systems (Yadav et al., 2024).

In response to these interlinked challenges, global policy discourse has increasingly shifted from a narrow focus on production maximization to a broader emphasis on sustainable, resilient, and responsible aquaculture development. The concept of "Blue Transformation," promoted by international agencies, represents a strategic framework for reorienting aquatic food systems toward sustainability (Stentiford et al., 2025). This approach emphasizes ecosystem-based production, efficient and circular use of resources, climate-resilient infrastructure,

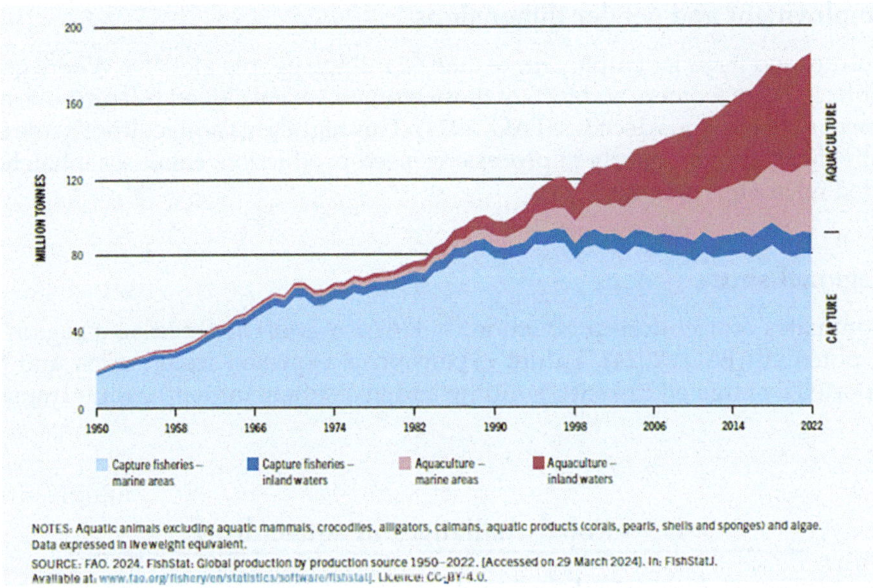

FIGURE 1.1 World fisheries and aquaculture production of aquatic animals.

strengthened biosecurity, inclusive value chains, and digital governance. It recognizes that future aquaculture growth must be achieved not through unchecked intensification but through innovation, regulation, and institutional reforms that balance productivity with environmental integrity and social equity.

Against this backdrop, this chapter examines the global growth of aquaculture through the dual lenses of emerging challenges and key reform imperatives. It synthesizes contemporary global trends, major biological and environmental constraints, technological transformations, circular economy-based production models, and evolving governance frameworks. By integrating scientific evidence, policy perspectives, and practical innovations, the chapter aims to provide a comprehensive understanding of how aquaculture can transition toward a climate-smart, resource-efficient, and socially inclusive sector capable of sustaining future global food and nutritional demands.

1.2 Global status and production trends

1.2.1 Production dynamics

Global aquaculture has grown at an exceptional pace since the 1990s, driven by technological innovation, expansion of commercially important species, adoption of intensified culture systems, and improved seed and feed quality (FAO, 2024). FAO figures (Fig. 1.1) show global aquatic food production rising steadily from 1950 to 2022, with aquaculture displaying the sharpest curve compared with wild harvests.

1.2.2 Employment and gender dimensions

Aquaculture and fisheries employ more than 62 million people globally. Where gender data are available, 24% are women, and 53% of these women are employed full time—a substantial improvement over previous decades (FAO, 2024). This highlights aquaculture's role in gender-inclusive livelihoods, particularly in processing, seed production, small-scale hatcheries, and value-added microenterprises.

1.2.3 Regional shifts

Asia dominates aquaculture production, but Africa is highlighted as a region with unexploited potential (FAO, 2024). Future expansion is expected from Africa and Southeast Asia, supported by targeted capacity-building and investment under the Blue Transformation roadmap.

1.3 Global challenges in aquaculture

1.3.1 Environmental sustainability and ecosystem integrity

Environmental sustainability represents one of the most pressing challenges to the long-term viability of aquaculture. Rapid intensification has significantly increased nutrient loading, leading to eutrophication, sedimentation, and deterioration of water quality in both inland and coastal farming systems (Kamble et al., 2025a; Mavraganis et al., 2020). The conversion of sensitive ecosystems, particularly mangroves and wetlands, for aquaculture development has resulted in habitat loss, biodiversity decline, and reduced coastal protection (Bao et al., 2013). In addition, the escape of exotic and genetically improved species from farms poses serious ecological risks through competition, hybridization, and disease transmission to native populations (Gozlan et al., 2024). In response to these growing concerns, global policy discourse increasingly emphasizes ecosystem-based aquaculture, carbon-neutral production, circular waste management, and nature-positive farming approaches aligned with international biodiversity and climate commitments.

1.3.2 Disease burden, biosecurity, and antimicrobial resistance

Disease outbreaks continue to be the single most devastating constraint to aquaculture growth, causing massive production losses and economic instability. Disease emergence and transmission are driven by a complex interaction of farming intensity, water quality, nutritional status, genetic susceptibility, and management practices (Kamble et al., 2024a; Lim et al., 2025). Indiscriminate use of antibiotics to combat these outbreaks has accelerated the global problem of AMR, raising serious concerns for aquatic animal health, environmental safety, and public health (Mohammed et al., 2025; Kamble et al., 2024b). AMR represents a growing concern in aquaculture, with significant implications for human and animal health. Multiple studies have documented the presence of antimicrobial-resistant bacteria and resistance genes in aquaculture environments worldwide.

Contemporary scientific consensus now strongly supports the One Health approach, which integrates aquatic animal health, environmental health, and human health into a unified disease management framework. Strengthening farm-level biosecurity, early disease diagnostics, vaccination strategies, and national surveillance systems are now recognized as essential foundations for resilient aquaculture development (Kamble et al., 2024c).

1.3.3 Managing microbes in sustainable aquaculture

In an aquaculture system, the major item of expenditure is feed, which is about 50% of the total cost involved. Only 20%–30% of the feed used in aquaculture is being assimilated by cultured animals, and the remainder is accumulated as waste. The accumulated waste includes NH_4^+–N and NO_2^-–N, which are toxic to the cultured animals in a closed system. These toxic wastes can be effectively converted into useful nutrients by producing microorganisms by manipulating the carbon and nitrogen ratio (C/N) with biofloc technology (Chavan et al., 2018; Wasave et al., 2020). Microbial management could support sustainable aquaculture development, with a focus on small-scale production systems. Traditional practices in microbial management have sustained production for decades; there are significant gaps in systematic data collection, scientific understanding of complex microbial interactions, and access to diagnostic tools. Development and management of microbial loops in culture systems are the main constraints in sustainable aquaculture.

1.3.4 Feed, resource dependency, and economic pressures

Feed and resource dependency constitute major structural constraints in global aquaculture. Feed alone accounts for nearly 60%–70% of production costs, while continued dependence on marine-derived fishmeal and fish oil exerts unsustainable pressure on wild fish stocks and marine food webs (Auchterlonie, 2024). At the same time, competition for freshwater, land, energy, and labor is intensifying across agricultural and industrial sectors (Boyd, 2009). These pressures disproportionately affect small-scale farmers, who often lack capital buffers to absorb fluctuating input costs. To overcome this challenge, global research has increasingly focused on alternative feed ingredients such as insect proteins, microalgae, yeast, biofloc microbial biomass, and agroindustrial byproducts (Debbarma et al., 2025; Kamble et al., 2025b; Medhe et al., 2025). Nutrient-recycling–based aquaculture systems offer major opportunities to reduce external feed dependency while enhancing farm-level efficiency, profitability, and environmental performance.

1.3.5 Climate change and governance constraints

Climate change has emerged as a dominant cross-cutting challenge shaping the future of aquaculture. Rising temperatures, salinity intrusion, hypoxic events, erratic rainfall patterns, cyclones, floods, and droughts are disrupting production cycles, weakening animal health, and damaging farm infrastructure (Maulu et al., 2021). These impacts are particularly severe in coastal and smallholder farming systems with limited adaptive capacity (Daunde et al., 2025). Climate-resilient aquaculture strategies such as IMTA, seaweed-based farming, offshore cage culture, RASs, and desert-zone aquaculture are now being promoted as adaptation and mitigation pathways (FAO, 2024). Alongside climatic stress, weak governance continues to restrict sustainable growth in many regions. Fragmented institutional authority, inadequate zoning, unclear licensing procedures, and poor regulatory enforcement often result in unplanned expansion and environmental conflict (Hishamunda et al., 2014). Regulatory modernization, spatial planning, and strengthened compliance mechanisms are therefore essential to align aquaculture growth with sustainability, climate action, and social equity goals.

1.4 Emerging nutrient-recycling–based aquaculture systems and their future prospects

Nutrient-recycling–based aquaculture systems such as biofloc technology (BFT), biofilm/periphyton systems, recirculating aquaculture systems (RAS), and IMTA represent a transformative shift toward resource-efficient, environmentally responsible, and climate-resilient aquaculture (Chavan et al., 2024). These systems are designed to convert organic wastes generated during fish and shrimp culture into biologically useful nutrients, thereby minimizing water exchange, reducing environmental pollution, improving feed utilization efficiency, and enhancing overall farm productivity. Biofloc and biofilm systems utilize indigenous microbial communities to assimilate nitrogenous wastes into microbial biomass that serves as a natural protein source, significantly lowering feed costs and improving water quality (Yadav et al., 2021). RAS enables near-zero water discharge through continuous mechanical and biological filtration, making it ideal for high-density farming in water-scarce and biosecure environments (Halim et al., 2025). IMTA integrates species from different trophic levels—such as finfish, shellfish, and seaweeds—where the waste of one species becomes the nutrient input for another, creating a self-regulating ecological production loop (Nederlof et al., 2022).

The future prospects of these nutrient-recycling systems are exceptionally promising, particularly under global pressures of climate change, land and water scarcity, feed insecurity, and stricter environmental regulations. These technologies support carbon-efficient production, reduced dependence on fishmeal and fish oil, improved biosecurity, and year-round intensive farming, even in urban, desert, and coastal saline environments. Their scalability—from small farmers to industrial operations—positions them as key drivers of the next generation of sustainable global aquaculture.

1.5 Digital and technological transformation

Digitalization and advanced technologies are rapidly reshaping the aquaculture sector, ushering in a new era of precision farming, automation, and data-driven decision-making. Recent technological reviews highlight the growing influence of robotics, drones, artificial intelligence (AI), biosensors, Internet of Things (IoT) telemetry, blockchain-based traceability, and augmented and virtual reality (AR/VR) platforms in modern aquaculture operations (Chandran et al., 2025). These innovations are transforming traditional farm management into intelligent production systems capable of optimizing resource use, minimizing operational risks, and improving productivity across different scales of aquaculture.

One of the most significant contributions of digital technologies lies in real-time environmental monitoring and farm surveillance. IoT-enabled sensors and biosensors now allow continuous tracking of critical water quality parameters such as dissolved oxygen, temperature, pH, salinity, and turbidity (Shete et al., 2024). Drones and underwater robotic systems enhance remote farm surveillance by enabling rapid detection of structural damage, stock behavior abnormalities, and unauthorized access in offshore and inland production systems (Amrita and Balasubramanian, 2025). AI-driven analytics further strengthen disease forecasting, stress detection, and behavioral assessment, allowing farmers to initiate early interventions and significantly reduce mortality losses (Yang et al., 2025).

Technological innovation has also revolutionized feeding strategies, animal welfare assessment, and harvest management. Intelligent feeding systems driven by AI and machine vision optimize feed delivery based on real-time fish appetite and swimming behavior, thereby improving feed conversion efficiency and reducing feed wastage (Kamble et al., 2026b; Zhang et al., 2023). Biometric monitoring tools enable noninvasive assessment of fish growth, stress levels, and health status, strengthening welfare compliance and productivity (Kamble et al., 2024d; Ramírez-Coronel et al., 2024). Blockchain platforms now ensure complete supply-chain transparency by enabling end-to-end traceability from hatchery to harvest and market, enhancing food safety assurance, regulatory compliance, and consumer confidence (Chandran et al., 2025).

At the institutional level, global agencies are complementing farm-level technological adoption with digital governance frameworks. FAO's AquaGRIS is a global platform for documenting and tracking aquatic genetic resources. It helps with broodstock management and genetic conservation (FAO, 2024). The aquaculture legal assessment and revision tool (ALART) supports digital modernization of aquaculture legislation and regulatory systems, improving governance efficiency and compliance (Hodgson, 2023). In parallel, the promotion of aquaculture park models as integrated sustainability hubs encourages the clustering of digital infrastructure, biosecurity systems, waste management technologies, and renewable energy solutions, collectively accelerating the transition toward smart, climate-resilient, and sustainable aquaculture systems (Tamilarasu et al., 2024).

1.6 Conclusion

Global aquaculture has firmly established itself as the backbone of future aquatic food supply, driven by its rapid growth, expanding technological base, and irreplaceable role in nutrition security, employment generation, and the blue economy. As demonstrated in this chapter, aquaculture now contributes more than half of the aquatic foods consumed by humans and supports more than 62 million livelihoods worldwide. Its continued expansion is both inevitable and essential in the face of stagnant capture fisheries and rising global demand for high-quality protein. However, this growth trajectory is increasingly constrained by interlinked challenges related to environmental degradation, disease outbreaks and AMR, feed and resource dependency, climate change impacts, and weak governance frameworks. These challenges underscore that further expansion of aquaculture cannot follow conventional input-intensive models without risking ecological instability and economic vulnerability.

The future of aquaculture therefore depends on a decisive transition toward sustainability-driven, climate-resilient, and technologically enabled production systems. Ecosystem-based management, nutrient-recycling technologies, alternative feed resources, One Health–based biosecurity strategies, and digital monitoring platforms must become central pillars of aquaculture development. At the same time, regulatory modernization, spatial planning, institutional coordination, and inclusive governance are essential to ensure environmentally responsible growth and equitable benefit sharing. When aligned with the principles of the Blue Transformation and circular economy, aquaculture holds exceptional potential to evolve into a resource-efficient, low-carbon, and socially inclusive sector. The successful implementation of these reforms will determine whether aquaculture emerges not only as the world's fastest-growing food production sector but also as one of its most sustainable and resilient.

References

Alam, M.M., Jørgensen, N.O.G., Bass, D., Santi, M., Nielsen, M., Rahman, M.A., Hasan, N.A., Bablee, A.L., Bashar, A., Hossain, M.I., Hansen, L.H., Haque, M.M., 2024. Potential of integrated multitrophic aquaculture to make prawn farming sustainable in Bangladesh. Front. Sustain. Food Syst. 8. doi:10.3389/fsufs.2024.1412919.

Amrita, C.M., Balasubramanian, K., 2025. Precise aquaculture development: overview of unmanned system technologies and its application to monitoring and management in aquaculture farms. J. Environ. Prot. Ecol. 26 (7), 2501–2513.

Auchterlonie, N.A., 2024. Fishmeal and fish oil in aquaculture feeds: global food security questions. Int. J. Environ. Stud. 81 (2), 896–913. doi:10.1080/00207233.2024.2307221.

Avnimelech, Y., 2009. Biofloc Technology: A Practical Guide Book. The World Aquaculture Society, Baton Rouge, LA.

Bao, H., Wu, Y., Unger, D., Du, J., Herbeck, L.S., Zhang, J., 2013. Impact of the conversion of mangroves into aquaculture ponds on the sedimentary organic matter composition in a tidal flat estuary (Hainan Island, China). Cont. Shelf Res. 57, 82–91. doi:10.1016/j.csr.2012.06.016.

Boyd, C.E., 2009. Aquaculture, Freshwater. In: Likens, G.E. (Ed.), Encyclopedia of Inland Waters. Academic Press (Elsevier), Oxford,, pp. 234–241.

Chandran, P.J.I., Khalil, H.A., Hashir, P.K., V., S., 2025. Smart technologies in aquaculture: an integrated IoT, AI, and blockchain framework for sustainable growth. Aquacult. Eng. 111. doi:10.1016/j.aquaeng.2025.102584.

Chavan, B.R., Kamble, M.T., Yadav, S.R., Medhe, S., Ponpornpisit, A., Rodhkum, C., Pirapat, N., 2024. Revolutionizing small scale aquaculture in India: sustainable nutrient recycling technologies and practices. Thai. J. Vet. Med. 54 (3), 77–83.

Chavan, B.R., Pawar, S.D., Kamble, M.T., Naik, S.D., 2018. Microorganisms in maintaining Carbon/Nitrogen (C/N) ratio in Intensive Aquaculture System: A Review. Int. J. Trop. Agric. 36 (4), 843–847.

Chavan, B.R., Yakupitiyage, A., Ataguba, G.A., Kamble, M.T., Medhe, S.V., 2015. Tilapia as Food Fish: Enhancement of Ω-3 polyunsaturated fatty acids in tilapia (*Oreochromis* Spp.). Int. J. Agric. Sci. 7 (9), 671–677.

Daunde, V., Kamble, M.T., Ponpornpisit, A., Chavan, B.R., Medhe, S.V., & Pirarat, N. (2025). Sustainable adaptation strategies for coastal aquaculture amidst climate change-induced temperature rise. Thai J. Vet. Med., 55(Suppl. 1), 197–199.

Debbarma, S., Deb, S., Yadav, N.K., Kashyap, S., Debbarma, P., Chrisolite, B., Yirang, Y., Meena, D.K., Singh, S.K., Choudhury, T.G., Patel, A.B., 2025. Waste not, want not: unlocking the innovative potential of organic and eco-friendly insect and algal resources for future aquaculture. Aquac. Int. 33 (2). doi:10.1007/s10499-024-01814-8.

FAO, 2024. FishStatJ data: Fisheries and Aquaculture - Global Statistical Collections. Food and Agriculture Organization (FAO). https://www.fao.org/fishery/en/statistics.

Gozlan, R.E., Bommarito, C., Caballero-Huertas, M., Givens, J., Mortillaro, J.M., Pepey, E., Ralaiarison, R.P., Senff, P., Combe, M., 2024. A one-health approach to non-native species, aquaculture, and food security. Water Biol. Secur. 3 (2). doi:10.1016/j.watbs.2024.100250.

Halim, M.A., Aziz, D., Arshad, A., Nur, N.L., Nabi, M.M., Islam, M.A., Syukri, F., 2025. A systematic analysis of recirculating aquaculture systems (RAS) and biofloc technology (BFT) for white leg shrimp (Litopenaeus vannamei) in the indoor farming system. Aquacult. Eng. 110. doi:10.1016/j.aquaeng.2025.102544.

Hishamunda, N., Ridler, N., Martone, E., 2014. Policy and Governance in Aquaculture: Lessons Learned and Way Forward. FAO, Rome, p. 1.

Hodgson, S., 2023. Aquaculture Legal Assessment and Revision Tool. Legal Guide. FAO, Rome. https://openknowledge.fao.org/handle/20.500.14283/cc5797en.

Kamble, M.T., Chaiyapechara, S., Salin, K.R., Bunphimpapha, P., Chavan, B.R., Bhujel, R.C., Medhe, S.V., Kettawan, A., Thiyajai, P., Thompson, K.D., Pirarat, N., 2024c. Guava and Star gooseberry leaf extracts improve growth performance, innate immunity, intestinal microbial community, and disease resistance in Nile tilapia (*Oreochromis niloticus*) against *Aeromonas hydrophila*. Aquaculture Reports 35, 101947. doi:10.1016/j.aqrep.2024.101947.

Kamble, M.T., Gallardo, W., Salin, K.R., Pumpuang, S., Chavan, B.R., Bhujel, R.C., Medhe, S.V., Kettawan, A., Thompson, K.D., Pirarat, N., 2024a. Effect of *Moringa oleifera* leaf extract on the growth performance, hematology, innate immunity, and disease resistance of Nile tilapia (*Oreochromis niloticus*) against *Streptococcus agalactiae* Biotype 2. Animals 14 (6), 953. doi:10.3390/ani14060953.

Kamble, M.T., Medhe, S.V., Chavan, B.R., Daunde, V.Y., Palekar, G.R., Tayade, S.H., Gabriel, N.N., Pirarat, N., 2025a. Ecotoxicological consequences of microplastics in aquatic life. In: Abubakr, A., Amin, A., Malik, M.M. (Eds.), *Microplastics: Ecological and food security implications*. Springer, Cham, pp. 187–220. doi:10.1007/978-3-031-94480-2_8.

Kamble, M.T., Medhe, S.V., Chavan, B.R., Gabriel, N.N., Hoseini, S.M., Meena, M.K., Pirarat, N., 2026b. Decarbonizing aquaculture: Integrating technology, policy, and nature for Net-Zero blue food systems. Aquac. Fish. doi:10.1016/j.aaf.2026.06.010.

Kamble, M.T., Pirarat, N., Chavan, B.R., Medhe, 2026a. *Phytobiotics for Sustainable Aquaculture: Innovations, Benefits, and Future Perspectives*. CRC Press, Boca Raton doi:10.1201/9781003646945.

Kamble, M.T., Salin, K.R., Chavan, B.R., Medhe, S.V., Thompson, K.D., Pirarat, N., 2024d. Length-weight relationship and condition factor of Nile tilapia (Oreochromis niloticus) fed diets supplemented with guava and star gooseberry leaf extract. F1000Research 13, 540. doi:10.12688/f1000research.145369.2.

Kamble, M.T., Wongprasert, K., Chavan, B.R., Daunde, V.V.Y., Palekar, G.K.R., Tayade, S.H., Thompson, K.D., Gabriel, N.N., Medhe, S.V., Pirarat, N., 2025b. Red Seaweeds in Aquaculture: Impacts on growth, immunity, antioxidant status, gene expression, and gut health. Annals of Animal Science doi:10.2478/aoas-2025-0092.

Kamble, M.T., Yostawonkul, J., Medhe, S.V., Chavan, B.R., Kumar, A., Palekar, G, R., Daunde, V.Y., Tayade, S., Gabriel, N.N., Ataguba, G.A., & Pirarat, N. (2024b). Innovative Feed Additives for Sustainable Aquaculture: Phytobiotics Encapsulated in Organic Nanoparticles. In N. N., Gabriel, K. P., Abasubong, V. N., Erasmus, & M. T., Kamble (Eds.), *Sustainable feed ingredients and additives for aquaculture farming: Perspectives from Africa and Asia*. Singapore: Springer Nature Singapore, pp. 501-520. doi:10.1007/978-981-97-4279-0_22.

Lim, K.C., Yusoff, F.M., Natrah, F.M.I., De Zoysa, M., Md Yasin, I.S., Yaminudin, J., Karim, M., 2025. Biological strategies in aquaculture disease management: towards a sustainable blue revolution. Aquac. Fish. 10 (5), 743–763. doi:10.1016/j.aaf.2025.04.002.

Majluf, P., Matthews, K., Pauly, D., Skerritt, D.J., Palomares, M.L.D., 2024. A review of the global use of fishmeal and fish oil and the fish In: fish Out metric. Sci. Adv. 10 (42). doi:10.1126/sciadv.adn5650.

Maulu, S., Hasimuna, O.J., Haambiya, L.H., Monde, C., Musuka, C.G., Makorwa, T.H., Munganga, B.P., Phiri, K.J., Nsekanabo, J.D.M., 2021. Climate change effects on aquaculture production: sustainability implications, mitigation, and adaptations. Front. Sustain. Food Syst. 5. doi:10.3389/fsufs.2021.609097.

Mavraganis, T., Constantina, C., Kolygas, M., Vidalis, K., Nathanailides, C., 2020. Environmental issues of aquaculture development. Egypt. J. Aquat. Biol. Fish. 24 (2), 441–450. doi:10.21608/EJABF.2020.85857.

Medhe, S.V., Kamble, M.T., Kumar, A., Daunde, V.Y., Kettawan, A., Nirmal, N., Pirarat, N., 2025. Fish protein hydrolysate as a food and feed ingredient. In: Nirmal, N., Santivarangkna, C., Bekhit, A.E.A., Barba, F.J. (Eds.), *Fish Protein Hydrolysates*. Academic Press, pp. 249–276. doi:10.1016/B978-0-443-21654-1.00010-0.

Mohammed, E.A.H., Kovács, B., Kuunya, R., Mustafa, E.O.A., Abbo, A.S.H., Pál, K., 2025. Antibiotic resistance in aquaculture: challenges, trends analysis, and alternative approaches. Antibiotics 14 (6). doi:10.3390/antibiotics14060598.

Nederlof, M.A.J., Verdegem, M.C.J., Smaal, A.C., Jansen, H.M., 2022. Nutrient retention efficiencies in integrated multi-trophic aquaculture. Rev. Aquac. 14 (3), 1194–1212. doi:10.1111/raq.12645.

Ramírez-Coronel, F.J., Rodríguez-Elías, O.M., Esquer-Miranda, E., Pérez-Patricio, M., Pérez-Báez, A.J., Hinojosa-Palafox, E.A., 2024. Non-invasive fish biometrics for enhancing precision and understanding of aquaculture farming through statistical morphology analysis and machine learning. Animals 14 (13). doi:10.3390/ani14131850.

Shete, R.P., Bongale, A.M., Dharrao, D., 2024. IoT-enabled effective real-time water quality monitoring method for aquaculture. MethodsX 13. doi:10.1016/j.mex.2024.102906.

Stentiford, G.D., Bremner, J., Bass, D., 2025. A blue transformation for global One Health. Front. Aquac. 4. doi:10.3389/faquc.2025.1627309.

Tamilarasu, R., Vignesh, P., Paplin Prince, B., Ezhilarasi, V., 2024. Aquaculture park: a revolutionary model for sustainable aquaculture. J. Aquac. Trop. 39 (1–4), 95–101. doi:10.32381/jat.2024.39.1-4.11.

Troell, M., Costa-Pierce, B., Stead, S., Cottrell, R.S., Brugere, C., Farmery, A.K., Little, D.C., Strand, Å., Pullin, R., Soto, D., Beveridge, M., Salie, K., Dresdner, J., Moraes-Valenti, P., Blanchard, J., James, P., Yossa, R., Allison, E., Devaney, C., Barg, U., 2023. Perspectives on aquaculture's contribution to the sustainable development goals for improved human and planetary health. J. World Aquac. Soc. 54 (2), 251–342. doi:10.1111/jwas.12946.

Yadav, N.K., Patel, A.B., Singh, S.K., Mehta, N.K., Anand, V., Lal, J., Dekari, D., Devi, N.C., 2024. Climate change effects on aquaculture production and its sustainable management through climate-resilient adaptation strategies: a review. Environ. Sci. Pollut. Res. 31 (22), 31731–31751. doi:10.1007/s11356-024-33397-5.

Yadav, S.R., Chavan, B.R., Chadha, N.K., Naik, S.D., Krishnani, K.K., Sawant, P.B., 2022. Algal-bacterial intervention as a management tool for next-generation aquaculture sustainability. J. Environ. Biol. 43 (4), 485–497. doi:10.22438/jeb/43/4/MRN-2001.

Wasave, S.S., Chavan, B.R., Naik, S.D., Wasave, S.M., Pawase, A.S., Tibile, R.M., Ghode, G.S., Meshram, S.J., Shivalkar, V.S., 2020. Role of microbes in bifloc systems: a review. J. Exp. Zool. India 23, 903–906.

Yadav, S.R., Chavan, B.R., Chadha, N.K., Naik, S.D., Krishnani, K.K., Sawant, P.B., 2021. Effect of stocking density on growth performance, plankton abundance, body composition and haematological parameters of Etroplus suratensis (Bloch, 1790). Aquac. Res. 52 (10), 4752–4766. doi:10.1111/are.15309.

Yang, H., Feng, Q., Xia, S., Wu, Z., Zhang, Y., 2025. AI-driven aquaculture: a review of technological innovations and their sustainable impacts. Artif. Intell. Agric. 15 (3), 508–525. doi:10.1016/j.aiia.2025.01.012.

Zhang, L., Li, B., Sun, X., Hong, Q., Duan, Q., 2023. Intelligent fish feeding based on machine vision: a review. Biosystems Eng. 231, 133–164. doi:10.1016/j.biosystemseng.2023.05.010.

CHAPTER

2

Aquaculture wastewater treatment technologies: Updates on recent system advancements

Kalapala Satvik[a], *Dharmendra Kumar Meena*[b] *and Aruna Kamble*[c]

[a]Kerala University of Fisheries and Ocean Studies, Kochi, Ernakulam, Kerala, India [b]ICAR-Central Inland Fisheries Research Institute, Barrackpore, Kolkata, West Bengal, India [c]College of Fisheries, Rangeilunda, Odisha University of Agriculture & Technology (OUAT), Bhubaneswar, Odisha, India

2.1 Introduction

The global growth of aquaculture has changed food production significantly. This sector now provides more than half of the fish that people eat (FAO, 2020). However, this rapid increase has put more stress on both freshwater and coastal ecosystems due to the release of nutrient-rich and chemically diverse wastewater. In particular, recirculating aquaculture systems (RASs), marine cage culture, and high-density pond production produce large amounts of effluent. If not treated, this waste can seriously harm nearby aquatic environments (Crab et al., 2007; Praveen et al., 2021). The ecological effects of this waste go beyond local pollution. They also add to global issues such as climate change, loss of biodiversity, and the rise of antimicrobial resistance (Cabello, 2006; Chowdhury et al., 2023).

Aquaculture wastewater is a complex mix that contains suspended solids, dissolved organic matter, nitrogen, phosphorus, microbial pathogens, and emerging pollutants such as antibiotics, heavy metals, and microplastics (Muir, 2005; Wahid et al., 2020). Each of these components presents unique risks. Excess nitrogen and phosphorus can cause eutrophication, leading to algal blooms and low-oxygen levels. Residual antibiotics promote resistant microbial strains, which can harm both human and animal health (Shah et al., 2014). Heavy metals and microplastics worsen ecotoxicology issues by building up in aquatic food webs and lowering water quality for those downstream.

In the past, managing aquaculture wastewater relied on simple physical methods such as sedimentation and filtration. While these methods reduce suspended solids, they cannot

Zero-Waste Technologies in Aquaculture: Global Perspectives and Methods toward a Circular Economy
DOI: https://doi.org/10.1016/B978-0-443-34689-7.00009-4

effectively remove dissolved nutrients, pathogens, and trace contaminants. As a result, these techniques are increasingly being supplemented or replaced by better systems, such as biological reactors, constructed wetlands, microalgae cultivation platforms, and membrane technologies (Hai and Yamamoto, 2018; Telli and Pulatsu, 2021). These technologies are not only more effective at removing pollutants but also allow for resource recovery, including the production of protein-rich microbial biomass, biofertilizers, and renewable bioenergy.

Among the new solutions, several approaches stand out. Biofloc technology (BFT) can turn nitrogen waste into microbial biomass that serves as added feed, reducing feed costs and boosting overall productivity (Avnimelech, 2009; Crab et al., 2012). Likewise, microalgal cultivation systems that use strains such as *Chlorella vulgaris* and *Scenedesmus obliquus* can effectively take in nitrogen and phosphorus while producing valuable algal biomass for aquafeed, fertilizers, and even biofuel (Rawat et al., 2011; Abreu et al., 2012). Aquaponic systems, which combine aquaculture with hydroponic crop farming, illustrate circular economy principles. They create nutrient recycling loops that reduce wastewater discharge while also producing vegetables and fish (Rakocy et al., 2006; Love et al., 2015).

Hybrid and integrated systems are becoming more popular, especially those that combine membrane bioreactors (MBRs) with biological oxidation or electrocoagulation (EC)-enhanced sedimentation. These systems can effectively manage the complex pollution found in aquaculture wastewater. They also provide high removal efficiency and steady water quality for reuse (Kundu et al., 2020; Zhang et al., 2023). However, the broader use of these technologies is still limited by economic and operational issues. High capital and energy costs, the need for skilled labor, and changes in wastewater characteristics make implementation difficult, especially for small operators in developing areas. In addition, new pollutants such as pharmaceuticals, hormones, and microplastics challenge traditional treatment methods. This calls for new approaches that balance efficiency with care for the environment (Boyd et al., 2020; Zhao et al., 2025).

This chapter aims to give a clear and critical overview of recent advancements in aquaculture wastewater treatment. Specifically, it (1) classifies and describes major treatment technologies, (2) reviews innovative systems such as BFT, aquaponics, and advanced membrane applications, (3) explores hybrid and integrated models, and (4) highlights the challenges and future directions needed for sustainable aquaculture development. By bringing together findings from recent studies and real-world examples, this work offers important insights for researchers, practitioners, and policymakers focused on balancing aquaculture productivity with environmental protection.

2.2 Overview of aquaculture wastewater characteristics

The composition of aquaculture wastewater is highly variable and depends on multiple factors such as the cultured species, feed type, system design, and water exchange rate. In intensive systems such as RAS, marine cages, and high-density pond cultures, effluents typically carry elevated concentrations of suspended solids, organic matter, dissolved nutrients, pathogens, trace chemicals, and emerging pollutants, including microplastics and pharmaceuticals (Muir, 2005; Crab et al., 2007). The interactions among these components not only influence water quality within aquaculture facilities but also determine the ecological impact of discharges on receiving environments. Understanding the physical, chemical, and biological characteristics of aquaculture wastewater is therefore critical for the design and selection of appropriate treatment technologies.

2.2.1 Suspended solids

Suspended solids are among the most visible and disruptive components of aquaculture waste. They mainly come from uneaten feed, fish waste, mucosal shedding, and biofilm shedding from tank or pond surfaces. These particles increase turbidity, cut down water clarity, and settle into sediments. There microbial breakdown creates low-oxygen conditions that harm bottom-dwelling organisms (Liu et al., 2016). Typical concentrations range from 45 to 214 mg/L in standard aquaculture settings, while more intensive systems can go beyond 300 mg/L (Crab et al., 2007). The amount of suspended solids is closely linked to feeding methods. When feeding is excessive or poorly managed, it leads to more particles. Besides ecological impacts, high levels of suspended solids also lower photosynthesis in ponds and disrupt water flow, making wastewater management even more challenging.

2.2.2 Organic matter

In addition to particulate solids, aquaculture effluents contain large amounts of dissolved and particulate organic matter. Much of this comes from uneaten feed and metabolic waste. Proteins and lipids are the main organic components. Their microbial breakdown increases biochemical oxygen demand (BOD) and chemical oxygen demand (COD). These measures are commonly used to assess wastewater strength. In aquaculture systems, BOD values usually range from 5 to 40 mg/L. COD can exceed 100 mg/L, especially in nutrient-intensive operations such as shrimp farming (Wahid et al., 2020). High BOD and COD levels lead to oxygen depletion in receiving waters. This situation contributes to hypoxia and fish kills. In addition, the decomposition of organic matter can produce greenhouse gases such as methane and nitrous oxide. This links aquaculture effluent management to larger climate change issues (Schrader et al., 2011).

2.2.3 Nutrients: Nitrogen and phosphorus

Nutrient enrichment from nitrogen and phosphorus is a key characteristic of aquaculture wastewater. Nitrogen primarily comes from ammonia (NH_3/NH_4^+) released during protein metabolism and excretion. When levels exceed 0.1 mg/L, unionized ammonia becomes toxic to fish, especially in alkaline pH conditions (Ip et al., 2001). Other nitrogen compounds, such as nitrite (NO_2^-) and nitrate (NO_3^-), also build up and can harm aquatic health over time. Total nitrogen levels in aquaculture wastewater usually range from 10 to 100 mg/L (Crab et al., 2007). Phosphorus mostly appears as orthophosphate (PO_4^{3-}) from feed additives and fecal excretion. Normal concentrations range from 1 to 15 mg/L (Mmojieje et al., 2022). When released untreated, nitrogen and phosphorus significantly contribute to eutrophication. This process leads to harmful algal blooms and a drop in oxygen levels.

2.2.4 Biological contaminants

Aquaculture wastewater contains biological contaminants, including bacteria, viruses, and parasites. Pathogenic bacteria, such as *Vibrio* spp. and *Aeromonas* spp., are common in effluents, along with viral agents such as koi herpesvirus (KHV) (Bondad-Reantaso et al., 2005; FAO, 2020). These pathogens threaten not only cultured fish stocks but also wild aquatic populations and, in some cases, human health. Discharging effluent can spread diseases across production facilities and into natural ecosystems.

2.2.5 Emerging pollutants

In recent years, emerging pollutants have become important issues in aquaculture effluents. These include leftover antibiotics such as oxytetracycline, which are often used preventively in aquaculture. They also include hormones, heavy metals such as copper and zinc from antifouling products and feed additives, and microplastics from broken nets and equipment. Even at low levels (µg/L), their combined ecological effects can be serious. These effects include disrupting hormones, building up in organisms, and increasing the spread of antimicrobial resistance genes (Cabello, 2006; Shah et al., 2014). These pollutants are especially difficult to treat because conventional biological systems are not designed to remove them.

2.2.6 Gaseous metabolites

Anaerobic decomposition of organic matter in sediments or poorly aerated systems can produce toxic gases such as hydrogen sulfide (H_2S), methane (CH_4), and nitrous oxide (N_2O). H_2S is harmful to fish at very low levels. Methane and nitrous oxide also contribute to climate change as strong greenhouse gases (Kampbell et al., 2009). The presence of these gases is closely related to inadequate solid management and poor aeration practices. This emphasizes the need for effective system design in wastewater control.

2.2.7 General water quality indicators

Several general parameters give a clear view of aquaculture effluent quality. The pH of aquaculture wastewater usually falls between 6.5 and 8.5, depending on the feed composition and the water's ability to buffer changes. Dissolved oxygen (DO) levels should ideally stay above 5 mg/L, but they often drop below this level in crowded or poorly managed systems, particularly where there is a lot of organic matter. Turbidity frequently rises, which reduces light penetration and affects photosynthesis. In addition, temperature changes can worsen metabolic stress in both the cultured organisms and microbial treatment communities (Wahid et al., 2020).

2.2.8 Environmental impacts

The combined impact of untreated aquaculture wastewater is significant. Nutrient enrichment leads to eutrophication. Meanwhile, the breakdown of organic matter causes oxygen depletion and hypoxic conditions. Suspended solids and organic deposits change benthic habitats, lowering biodiversity and ecological resilience. Most concerning is how effluents contribute to antimicrobial resistance, as leftover antibiotics favor resistant strains that can survive in both aquatic and land ecosystems (Cabello, 2006). These risks highlight the urgent need for improved wastewater treatment technologies that can tackle both traditional pollutants and new contaminants.

2.3 Classification of aquaculture wastewater treatment technologies

Aquaculture wastewater treatment systems aim to reduce the environmental impact of effluent discharges by removing suspended solids, organic matter, nutrients, pathogens, and trace contaminants. These systems fall into four main categories: Physical, chemical, biological, and hybrid or integrated methods. Each category relies on different principles to remove pollutants, and their effectiveness depends on the wastewater's composition and the operational goals of the aquaculture facility. In practice, treatment technologies are often used one after another or in combination. The choice of system is influenced by specific factors such as production scale, cost, regulatory requirements, and how the treated effluent will be reused or disposed of (Cripps and Bergheim, 2000; Telli and Pulatsu, 2021).

2.3.1 Physical treatment methods

Physical treatment technologies mainly focus on removing suspended solids and coarse particles. These make up the largest portion of pollutants in aquaculture. The simplest and most common method is mechanical screening. This process uses sieves or drum filters with pore sizes from 40 to 500 μm to capture uneaten feed and feces before they dissolve or decompose. Another popular option is sedimentation tanks and clarifiers, which rely on gravity to separate solids from the water. In more complex setups, dissolved air flotation (DAF) attaches fine particles to microbubbles. These bubbles then rise to the surface, where the particles can be skimmed off. Filtration methods using sand, gravel, or membrane media help retain finer particles. Adsorption onto activated carbon can also remove leftover organics and odors.

While physical methods are relatively simple, inexpensive, and effective for pretreatment, they have limitations. They cannot effectively remove dissolved nutrients, pathogens, or trace pollutants. For this reason, physical treatments typically serve as the first stage in a multi-step process. They provide effluent quality that is suitable for later biological, chemical, or hybrid technologies (Cripps and Bergheim, 2000).

2.3.2 Chemical treatment methods

Chemical treatment technologies target dissolved pollutants and microbial contaminants that physical methods cannot manage well. Coagulation and flocculation are established approaches. Coagulants such as aluminum or iron salts, or synthetic polymers, destabilize colloidal particles and improve their aggregation into larger flocs, which can then be removed through sedimentation or filtration. This method is particularly effective for phosphorus removal, achieving efficiencies of more than 90% under optimized conditions (Zhang et al., 2023).

In addition to coagulation, chemical disinfection, and oxidation methods are widely used. Ozone, chlorine, and UV-C radiation can degrade organic pollutants and kill pathogens. However, these treatments need careful management to prevent harmful byproducts, such as chloramines or excess oxidants. Recently, electrochemical processes, such as EC and electrooxidation (EO), have gained attention. EC uses electric current to release ions from electrodes, promoting the destabilization of suspended solids and phosphorus removal. EO generates hydroxyl radicals and chlorine species that can degrade pharmaceuticals, hormones, and other micropollutants (Kundu et al., 2020; Parsa et al., 2021).

Despite their effectiveness, chemical treatments often have high operational costs, complex dosing requirements, and produce secondary residues such as chemical sludge, which must be managed responsibly. Because of this, chemical treatments are generally used as polishing steps or in hybrid systems rather than as standalone solutions.

2.3.3 Biological treatment methods

Biological treatment methods use the metabolic activity of microorganisms and plants to break down organic matter and take in nutrients. Conventional biological reactors, such as activated sludge systems and moving bed biofilm reactors (MBBRs), are commonly used for aerobic degradation of organic matter and nitrification of ammonia. In MBBRs, biofilm grows on suspended plastic carriers. This setup provides a large surface area for microbial communities to effectively process dissolved wastes.

Constructed wetlands are another important nature-based solution. They involve macrophytes, soils, and microbial groups working together to remove nitrogen, phosphorus, organic matter, and pathogens. These systems not only treat wastewater but also offer ecological benefits, such as creating habitats and enhancing landscapes (Vymazal, 2011). Likewise, microalgal cultivation systems are increasingly used for nutrient recovery. Algae take in nitrogen and phosphorus and convert them into biomass. This biomass can then be used for feed, fertilizer, or biofuel applications (Rawat et al., 2011; Abreu et al., 2012).

More integrated approaches include aquaponics and integrated multitrophic aquaculture (IMTA). Aquaponic systems combine fish farming with hydroponic plant growing. This creates closed nutrient loops that improve water quality while generating crop yields (Rakocy et al., 2006). In contrast, IMTA cocultures organisms such as fish or shrimp with extractive species such as seaweeds and bivalves. These organisms use waste nutrients and organic particles, supporting aquaculture within circular economy principles (Troell et al., 2009).

Anaerobic digestion systems also play important roles in managing aquaculture wastewater. Upflow anaerobic sludge blanket (UASB) digesters and other anaerobic bioreactors turn organic matter into biogas, mainly methane. This can be captured as renewable energy (Pang et al., 2021). While these systems are sensitive to environmental changes, anaerobic treatments provide pollution control and possibilities for energy recovery.

2.3.4 Hybrid and integrated technologies

As aquaculture effluents are chemically and biologically complex, single treatment methods are rarely enough. Hybrid and integrated technologies are increasingly used to achieve higher removal efficiencies and better operational resilience. MBRs pair biological treatment with ultrafiltration (UF) or microfiltration membranes. These systems can remove more than 95% of BOD, suspended solids, ammonia, and pathogens (Qin et al., 2023). Electrobiosystems combine EC with biological filtration. This approach allows for the simultaneous removal of nutrients, pathogens, and micropollutants.

Nature-based hybrid systems are also emerging. Multistage constructed wetlands, for example, combine sedimentation ponds, vertical-flow wetlands, and horizontal reed beds. They achieve significant nutrient removal while providing landscape and biodiversity benefits (Zhou et al., 2021). Similarly, algae–bacteria hybrid systems take advantage of mutualistic interactions. In these systems, algae produce oxygen through photosynthesis, and bacteria break down organic matter. This leads to better pollutant removal and biomass recovery (García et al., 2020).

The flexibility of hybrid systems makes them ideal for various operational contexts. They can be used in small-scale farms in low-resource regions or in high-value RASs in developed countries.

2.3.5 Technology selection considerations

Selecting the right wastewater treatment technology depends on the type of effluent, the operational goals, and the socioeconomic situation of the aquaculture system. For instance, sedimentation and DAF work well for high suspended solids. In contrast, biofilters, MBBRs, and algae-based systems are more effective for removing nitrogen. You can remove phosphorus using coagulation or constructed wetlands. Reducing pathogens requires methods such as UV disinfection, ozonation, or MBRs. In low-resource settings, nature-based systems such as wetlands, algae ponds, and BFT provide cost-effective and sustainable options. On the other hand, high-tech systems such as MBRs and electrochemical treatments fit better where water reuse and strict environmental compliance are essential.

In practice, no single method works for every situation. System designers need to combine multiple technologies in sequence or together to ensure reliable treatment that meets both environmental rules and production targets.

2.4 Microalgal cultivation in aquaculture wastewater treatment

Microalgae have become one of the most promising biological agents for treating aquaculture wastewater. They can take in nutrients, produce oxygen, and create biomass that can be turned into valuable products. Unlike traditional bacterial systems, which mainly break down organic matter, microalgae use dissolved nitrogen and phosphorus as nutrients for growth. They effectively convert pollutants into cellular biomass through photosynthesis (Rawat et al., 2011). This dual role of removing nutrients while producing useful biomass makes microalgal systems particularly appealing for integration into sustainable aquaculture practices. Recently, both open pond and closed photobioreactor (PBR) systems have been tested and used with varying success. These systems show high efficiency in nutrient uptake and can be applied in a range of aquaculture environments (Abreu et al., 2012; Li et al., 2019).

2.4.1 Mechanisms of nutrient removal

The ability of microalgae to remove pollutants relies on their process of photosynthesis and how they take in nutrients. While growing, microalgae absorb ammonium (NH_4^+), nitrate (NO_3^-), and phosphate (PO_4^{3-}). They incorporate these nutrients into proteins, nucleic acids, and other cell parts. This reduces the amount of dissolved nutrients in the water and stops their release into surrounding areas, where they could cause eutrophication (Craggs et al., 2012). Photosynthesis also produces oxygen, which improves water quality for farmed fish and boosts aerobic microbial activity in mixed systems. Furthermore, microalgae compete with harmful bacteria for nutrients and space, helping to prevent diseases in aquaculture. Some species can also remove carbon dioxide and toxic ammonia from the water, stabilizing pH and enhancing overall water quality.

2.4.2 Removal efficiencies and system performance

Experimental and field-scale studies have shown that microalgae can achieve very high nutrient removal rates, often getting close to completely eliminating dissolved nitrogen and phosphorus under ideal conditions. For example, *C. vulgaris* and *S. obliquus* have been documented to remove nearly 100% of both nitrogen and phosphorus in controlled wastewater treatment systems. They also produce biomass yields of up to 880 mg/L/day (Li et al., 2019; Abreu et al., 2012). Other species, such as *Tetraselmis suecica* and *Nannochloropsis oculata*, have slightly lower efficiencies, but they are still significant because of their high lipid content and potential for biofuel production (Mata et al., 2010; Abdel-Raouf et al., 2012). The effectiveness of nutrient removal relies on several factors, including species choice, light availability, temperature, and the nutrient makeup of the wastewater. Closed PBRs usually provide better control and higher productivity than open systems, but they also come with higher initial and operational costs.

2.4.3 Resource recovery and circular benefits

A key benefit of microalgal systems is their fit with circular economy principles. They treat wastewater while generating biomass that can be turned into various products. Algal biomass is rich in proteins, carotenoids, and polyunsaturated fatty acids. This makes it a good supplement for aquafeeds or livestock feed (Becker, 2007). In agriculture, dried algae are increasingly used as biofertilizers to improve soil fertility and crop productivity (Markou and Georgakakis, 2011). In addition, lipid-rich strains such as *Nannochloropsis* are appealing for biodiesel production. Algal polysaccharides are being studied as feedstocks for bioplastics (Mata et al., 2010). These uses show how microalgal cultivation can change aquaculture wastewater from a problem into a resource that supports food, energy, and material production.

2.4.4 Environmental and operational benefits

Microalgal wastewater treatment systems provide various ecological and operational benefits compared to traditional biological or chemical treatments. They use photosynthesis, which helps reduce greenhouse gas emissions by capturing carbon dioxide. This process also lowers the energy needed for aeration, a common requirement in bacterial systems such as activated sludge or BFT (Pittman et al., 2011). The continuous generation of oxygen in algal cultures boosts DO levels. This improvement lowers stress on fish and stabilizes cocultured microbial communities.

Moreover, algal systems decrease the need for chemical additives, such as flocculants or synthetic fertilizers. This results in lower operational costs and fewer chemical residues. They can easily scale and adapt, making them suitable for both small, low-resource settings and large industrial operations. In addition, microalgae can work well with other aquaculture practices, such as aquaponics, constructed wetlands, or biofloc systems. This integration further improves nutrient cycling and treatment efficiency.

2.4.5 System designs

Several system designs have been developed to grow microalgae in aquaculture wastewater. Each has its own strengths and weaknesses. High-rate algal ponds (HRAPs) are among the most common and affordable designs. They consist of shallow raceway ponds with paddlewheel circulation. These ponds work well in warm climates with plenty of sunlight, but they are more susceptible to contamination and changes in the environment. In contrast, PBRs are closed systems made of transparent tubes or panels. They offer better control over environmental conditions and protection from pathogens. Although they are more expensive, PBRs usually achieve higher productivity and better pathogen control. Algal turf scrubbers are another promising design. In this setup, microalgae form biofilms on sloped surfaces exposed to wastewater flow. This design allows for efficient nutrient removal and makes it easier to harvest biomass (Craggs et al., 2012).

2.4.6 Limitations and challenges

Despite their benefits, microalgal systems have several limitations that need to be resolved for wider use. Light penetration is a common issue in dense cultures, where self-shading reduces photosynthesis. This can be improved with thin-layer cultivation systems, regular harvesting, or engineered light delivery technologies. Seasonal temperature changes also affect growth rates, which means careful species selection or controlled environments, such as greenhouses, may be necessary. Another major problem is biomass harvesting, which is still expensive and energy-consuming. Methods such as flocculation, filtration, and centrifugation are commonly used, but they can significantly raise operational costs. Nutrient imbalances, such as not enough carbon compared to nitrogen, can limit growth and lower biomass yield, so strategies for supplementation are needed to keep productivity high (Chowdhury et al., 2023).

2.4.7 Integration potential

The integration of microalgal cultivation into larger aquaculture systems is gaining recognition as a way to improve sustainability. In aquaponics, for instance, algae can support hydroponic plants by helping with nutrient removal and adding oxygen. In biofloc systems, algae provide oxygen and stabilize microbial communities, which enhances both water quality and the nutritional value of bioflocs. Hybrid approaches that combine microalgal ponds with constructed wetlands or MBRs are also becoming popular; they allow for multiple treatment stages that can tackle a broader range of pollutants (García et al., 2020). These integrated systems mark a move toward versatile wastewater treatment setups that support food production, resource recovery, and environmental protection at the same time.

2.5 Aquaponics: Linking fish and plant production

Aquaponics has emerged as a nature-based, closed-loop solution for treating aquaculture wastewater and producing food sustainably. It combines aquaculture, which is the farming of aquatic species such as fish or shrimp, with hydroponics, the soil-free cultivation of plants. In this integrated model, the waste produced by fish provides nutrients for plants, while plants and their associated microbes act as filters that improve water quality for the fish. By recycling nutrients within one system, aquaponics reduces water discharge, lessens the need for synthetic fertilizers, and allows for the simultaneous production of fish and vegetables (Rakocy et al., 2006; Goddek et al., 2015). Its ability to improve resource use and food security has made aquaponics an increasingly important area of study and practice, especially in areas facing water shortages and urban growth.

2.5.1 Principles and operational mechanisms

The functioning of aquaponics relies on a series of connected biological and chemical processes that maintain nutrient cycling. Fish release ammonia mainly through their gills and feces. This ammonia, if not treated, can build up to toxic levels. In aquaponic systems, biofilters with nitrifying bacteria convert ammonia first into nitrite (NO_2^-) and then into nitrate (NO_3^-), which plants can easily absorb (Rakocy et al., 2006). Besides nitrogen, other important nutrients, such as phosphorus, potassium, and micronutrients, become available to plants through the breakdown of fish waste. The plants then take these nutrients out of the water, preventing their buildup and turning them into edible biomass. The cleaned water is circulated back into the fish tanks, closing the loop and reducing the need for fresh water. This interaction between fish, microbes, and plants supports the ecological balance of aquaponics, making it an effective way to reuse wastewater.

2.5.2 Major system components

A typical aquaponic system includes several key components that work together. Fish tanks hold the aquatic species and provide the main source of nutrient-rich waste. Biofilters, often filled with materials such as gravel or plastic beads, offer surface area for nitrifying bacteria to grow and turn ammonia into nitrate. Hydroponic grow beds support vegetables or herbs that absorb dissolved nutrients while also cleaning the water. Other parts include solids removal units, such as clarifiers or swirl filters, which capture waste and leftover feed. Sump tanks help maintain water levels and redistribute flow throughout the system. Together, these components ensure stable water quality and ongoing nutrient cycling, both of which are vital for the health of the fish and plants (Somerville et al., 2014).

2.5.3 Advantages of aquaponics

Aquaponics provides various ecological, economic, and social advantages over traditional aquaculture and hydroponic systems used separately. Its water efficiency stands out, with studies indicating that aquaponics can cut freshwater use by up to 90% compared with conventional agriculture (Love et al., 2015). Removing the need for synthetic fertilizers, it reduces nutrient runoff into nearby ecosystems and lowers costs for farmers. The combined production of fish and vegetables boosts profitability while diversifying products, helping to ensure local food security. From an ecological viewpoint, the plants serve as natural filters, removing harmful substances and stabilizing water quality, which improves fish health and growth. In addition, aquaponics systems are flexible and can be set up in rural, suburban, or urban areas, including greenhouses, rooftops, and converted industrial sites, making them suitable for climate-smart urban farming (Endut et al., 2010).

2.5.4 Innovative applications and developments

Recent innovations have further improved the efficiency and flexibility of aquaponic systems. One important advancement is the use of mineralization chambers. These chambers help break down solid fish waste into useful micronutrients for plants. This process enhances nutrient retention and cuts down on waste (Goddek and Körner, 2019). Another significant development is the use of decoupled aquaponic systems. In these systems, the fish and plant components operate in separate but connected loops. This setup allows for independent adjustments of water temperature, pH, and nutrient levels for both fish and plants. It is especially helpful when growing species with different needs. The addition of smart monitoring technologies, such as Internet of Things (IoT) sensors, has also changed how these systems are managed. Tracking pH, DO, electrical conductivity, and nutrient concentrations in real time allows for precise adjustments and better control. This reduces the amount of labor needed and improves system resilience (Goddek et al., 2015).

2.5.5 Crop and fish suitability

The choice of species in aquaponics is essential for achieving high productivity and balanced nutrient cycling. Tilapia (*Oreochromis niloticus*) is one of the most popular fish species because it is hardy, eats a variety of foods, and can tolerate changes in water quality. Catfish (*Clarias gariepinus*) are also frequently used since they can handle higher levels of suspended solids. In cooler systems, carp species are often chosen, while more valuable species, such as barramundi, are used in setups where water quality can be closely managed (Rakocy et al., 2006). On the plant side, leafy greens such as lettuce, spinach, and kale work well due to their low nutrient needs and fast growth. Other crops, such as tomatoes, cucumbers, peppers, and strawberries, have also been successfully grown in aquaponic systems, as long as nutrient needs are met and system adjustments are made (Somerville et al., 2014).

2.5.6 Challenges and considerations

Despite its promise, aquaponics has several challenges that limit its adoption. Nutrient imbalances can arise because fish waste often lacks enough micronutrients, such as iron, calcium, or magnesium, for optimal plant growth. This usually means adding mineral supplements, which raises costs and makes system management more complicated (Endut et al., 2010). If solids build up and are not effectively removed or broken down, they can harm water quality and clog hydroponic systems. Temperature control is another issue since fish and plants may prefer different temperatures. This situation may require separate zones or systems that are not connected. Lastly, aquaponics systems are more complicated to design and run than standalone aquaculture or hydroponics. They need careful balancing of biological loads and regular monitoring to stay stable (Goddek et al., 2015).

2.5.7 Future outlook

The future of aquaponics is in its ongoing growth and integration into larger food production systems. Modular system designs, along with automation and smart controls, are allowing for commercial-scale operations that can supply urban markets. Aquaponics is increasingly combined with technologies such as biofloc systems and microalgal cultivation. This creates highly integrated platforms for recycling nutrients and reusing water. In addition, aquaponics fits well with global sustainability goals, including efficient use of resources, climate resilience, and principles of a circular economy. This positions it as a promising model for future food production (Love et al., 2015; Walsh and Cowx, 2020). As the demand for sustainable protein and vegetable sources rises, aquaponics is likely to become more important, especially in areas dealing with water shortages and rapid urban growth.

2.6 Biofloc technology: A sustainable approach in intensive aquaculture

BFT has become one of the most innovative and sustainable ways to manage wastewater in intensive aquaculture systems. Unlike traditional systems that need regular water changes to keep water quality high, BFT works by using microbial waste assimilation. By adjusting the carbon-to-nitrogen (C:N) ratio, harmful nitrogen compounds, such as ammonia, nitrite, and nitrate, are turned into microbial biomass, commonly called "bioflocs." These bioflocs not only improve water quality but also act as a protein-rich supplement for farmed species. This helps lower feed costs and increases production efficiency (Avnimelech, 2009; Crab et al., 2012). Because it recycles nutrients and provides feed, BFT has attracted attention as a method for both protecting the environment and improving economics in aquaculture.

2.6.1 Operational principles of biofloc technology

The effectiveness of BFT relies on keeping a high C:N ratio, typically above 15:1, by adding external carbon sources such as molasses, tapioca starch, or other carbohydrate-rich materials. This change in nutrient balance encourages the growth of heterotrophic bacteria, which take in dissolved nitrogen and turn it into microbial biomass. Continuous aeration and active mixing are vital for BFT operation. They ensure that oxygen is available and that bioflocs remain suspended in the water. By stopping the settling of solids, these processes improve waste assimilation and help distribute microbial activity evenly. Regular removal of excess solids is also necessary to avoid sludge buildup and oxygen depletion (De Schryver et al., 2008).

2.6.2 Benefits of biofloc systems

The environmental and economic benefits of BFT are significant. By reducing water exchange needs by up to 90%, BFT greatly lowers the amount of waste released into the environment. This helps save freshwater resources and cuts down on nutrient overload in nearby ecosystems (Kuhn et al., 2010). Through microbial recycling, nitrogenous waste turns into bioflocs that cultured species, such as tilapia, catfish, or shrimp, can eat. This process effectively lowers feed conversion ratios and reduces reliance on expensive fishmeal-based diets. Bioflocs usually contain 30%–50% protein, along with lipids, essential amino acids, vitamins, and beneficial microorganisms that improve gut health and boost immunity in cultured species (Ekasari et al., 2016; Kumar et al., 2022). In addition, the dense microbial communities in BFT systems create a competitive environment that hinders the growth of harmful bacteria, helping to prevent disease (Bossier and Ekasari, 2017).

2.6.3 Suitable species and productivity

BFT works best with species that feed on suspended particles, such as filter feeders and bottom grazers. *Litopenaeus vannamei* shrimp have shown particularly high productivity in BFT systems. They achieve yields of 20–25 tons per hectare per year, with feed conversion ratios dropping to about 1.2–1.4 (Crab et al., 2012). Tilapia (*O. niloticus*) is another commonly farmed species in biofloc-based systems. These fish can get up to 30% of their dietary protein directly from eating biofloc (Azim and Little, 2008). Recent trials with African catfish (*C. gariepinus*) have shown better growth and survival rates in BFT systems compared with traditional aquaculture. This highlights how the technology can work well for different species (Timmons et al., 2021).

2.6.4 Limitations and challenges

Despite its benefits, BFT faces several challenges that need to be addressed for wider use. The main issue is the high energy demand for continuous aeration. This can make up to 60% of total energy use in biofloc systems (Kuhn et al., 2010). Dependence on energy-heavy infrastructure can make BFT costly in areas with limited or expensive electricity. It also requires skilled management. Keeping the right C:N ratio, DO levels, and solids concentrations needs constant monitoring and intervention. If not properly managed, excess sludge can build up. This can cause oxygen depletion, lead to harmful metabolites, and worsen water quality. Furthermore, species that cannot tolerate high suspended solid levels may not be suitable for BFT, which restricts its use in certain aquaculture systems (Avnimelech, 2009).

2.6.5 Recent advances in biofloc technology

Recent technological developments have aimed to overcome these limitations and improve the efficiency of BFT. The use of IoT monitoring systems has allowed for real-time control of key parameters such as DO, temperature, and the C:N ratio. This reduces labor needs and human error (Zhou et al., 2023). Targeted probiotics have also been introduced to improve nitrogen absorption, stabilize microbes, and enhance the immune health of cultured species (Bossier and Ekasari, 2017). Hybrid systems that combine BFT with recirculating aquaculture systems (BFT-RAS) show promise in further improving water quality and lowering the environmental impact of intensive aquaculture. In addition, solar-powered aerators and energy-saving blowers are being tested to decrease the high energy demand of BFT, especially in remote or off-grid areas (Taw, 2021).

2.6.6 Integration potential

The integration of BFT with other aquaculture wastewater treatment and recycling technologies offers exciting opportunities for future development. In aquaponics, for example, biofloc water can serve as a nutrient-rich input for plant cultivation. This further improves nutrient recovery and reduces waste discharge. In integrated IMTA, biofloc-enriched effluents can be redirected into tanks for secondary species. Here, organisms such as shellfish or seaweeds can take up leftover nutrients. Algae-based systems also work well with BFT. Algal photosynthesis adds oxygen and stabilizes microbial communities. This creates a more balanced and resilient treatment platform (Crab et al., 2012; Ekasari et al., 2016). Such integrations show the growing recognition of BFT as a flexible technology that not only treats wastewater but also supports circular economy principles in aquaculture.

2.7 Advanced membrane and physicochemical methods in aquaculture wastewater treatment

As aquaculture systems grow more intensive and water reuse becomes more common, treatment methods are essential for maintaining effluent quality, biosecurity, and environmental standards. Membrane-based and physical–chemical technologies offer precise treatment, effectively removing suspended solids, dissolved salts, organic pollutants, pathogens, and new contaminants. Unlike traditional biological systems that depend on microbial metabolism, these technologies use physical separation and electrochemical processes. This allows them to perform consistently, even when the effluent composition changes. While they often come with higher costs and energy needs, their ability to generate water suitable for reuse makes them essential in valuable aquaculture areas, such as RASs, hatcheries, and integrated aquaponic farms (Kim et al., 2020; Qin et al., 2023).

2.7.1 Membrane-based filtration technologies

2.7.1.1 *Ultrafiltration and reverse osmosis*

Membrane filtration technologies have received a lot of attention in aquaculture wastewater treatment because of their reliability and high ability to remove pollutants. UF uses pressure-driven membranes with pore sizes ranging from 0.01 to 0.1 μm. This allows for the removal of suspended solids, colloids, and most bacteria. Reverse osmosis (RO), on the other hand, uses semipermeable membranes under high pressure to remove dissolved ions, heavy metals, and trace organic compounds. RO can achieve more than 98% removal of salts and nutrients (Kim et al., 2020). These systems are especially useful for polishing effluents before discharge or reuse in aquaponic systems. They are also important for desalination in brackish or saline aquaculture facilities. Improvements in membrane material science have greatly boosted the energy efficiency of RO systems. This makes them more practical for aquaculture applications where conserving freshwater is essential.

2.7.1.2 *Membrane bioreactors*

MBRs are some of the most versatile technologies for treating aquaculture wastewater. MBRs combine biological treatment, usually through activated sludge processes, with membrane filtration using UF or microfiltration membranes. This combination allows for the removal of organic matter, ammonia, suspended solids, and pathogens at the same time, resulting in high-quality effluent that is suitable for reuse (Qin et al., 2023). Removal rates over 99% have been noted for BOD, ammonia, and total suspended solids. The compact design of MBRs makes them ideal for facilities with limited space. They can also handle varying influent loads, which adds resilience to operations. However, the high initial costs, problems with membrane fouling, and the need for skilled maintenance limit their use.

2.7.2 Electrochemical treatment methods

Electrochemical processes are being used more frequently in aquaculture wastewater treatment because they effectively remove nutrients, heavy metals, and new contaminants. EC depends on the in situ release of metal ions, usually aluminum or iron, from sacrificial electrodes when an electric current passes through. These ions destabilize colloids and help them clump together. This makes EC useful for removing phosphorus, reducing turbidity, and precipitating heavy metals (Kundu et al., 2020).

EO uses anodic reactions to create hydroxyl radicals and active chlorine species. These substances can oxidize organic pollutants, pharmaceutical leftovers, and hormones. EO is particularly important for addressing antibiotic residues and concerns about antimicrobial resistance in aquaculture wastewater (Parsa et al., 2021).

Although electrochemical treatments offer high versatility and effectiveness, they use a lot of energy and need regular electrode replacement, which increases operational costs. However, their accuracy in targeting micropollutants makes them an important addition to traditional biological or physical systems.

2.7.3 Hydrodynamic cavitation

Hydrodynamic cavitation has recently emerged as a promising method for treating wastewater in aquaculture. The process forces water through narrow channels, creating low-pressure areas where vapor bubbles form and then collapse. The collapse of these bubbles generates localized temperatures more than 5000 K and high shear forces, producing reactive oxygen species that can break down organic compounds and kill microorganisms (Gogate, 2011). Pilot studies on shrimp farm waste show reductions of 70%–90% in COD and ammonia levels after cavitation treatment (Zhang et al., 2022). Besides removing pollutants, cavitation also provides disinfection benefits and requires relatively few chemicals. However, the need for high-pressure systems and specialized equipment limits its widespread use.

2.7.4 Advantages and limitations

Each advanced technology has its own advantages and challenges that need to be tackled for sustainable use. UF and RO systems offer high removal efficiency, including the rejection of pathogens and ions, but they face issues with membrane fouling and energy demands. MBRs have a compact design and produce high-quality effluent, but they need skilled maintenance and can be expensive to install. Electrochemical methods effectively remove nutrients and micropollutants, but they use a lot of energy and are limited by electrode wear. Hydrodynamic cavitation works well for breaking down organics and disinfecting with little chemical input, but it requires a large investment in high-pressure systems (Kim et al., 2020; Parsa et al., 2021; Qin et al., 2023).

2.7.5 Integration and hybridization

Given their limitations, advanced membrane and electrochemical technologies are increasingly being integrated into hybrid treatment systems. For instance, MBRs often combine with algae-based polishing systems to improve nutrient removal and support biomass recovery. Electrochemical units can work with biological reactors to ensure the simultaneous removal of traditional pollutants and trace contaminants. In some situations, RO systems connect with anaerobic digestion or solar-assisted units to recover energy and lower operational costs. These integrations not only boost overall treatment efficiency but also fit with circular economy models by enabling resource recovery and water reuse (González-Fernández et al., 2020).

2.7.6 Outlook and emerging trends

Emerging innovations are expected to improve the performance and sustainability of wastewater treatment in aquaculture. Researchers are developing nanomaterial-enhanced membranes that have better permeability and resist fouling. These membranes offer greater durability and require less maintenance. Forward osmosis (FO) is being considered as a low-pressure alternative to RO. It shows promise in reducing energy use while still achieving similar levels of salt and nutrient rejection. Self-cleaning membranes that use nanobubbles or vibration to prevent fouling are also under investigation. In addition, the use of digital monitoring tools, such as IoT-enabled sensors, allows for real-time tracking of transmembrane pressure, fouling rates, and conductivity. This technology provides predictive control and cuts down on downtime (Wang et al., 2022; Zhou et al., 2023).

In the long term, refining advanced membranes and physicochemical methods, along with improving energy efficiency and system automation, will be crucial for achieving the sustainability goals of intensive aquaculture. These technologies are likely to be key in high-value, biosecure production systems, where reusing effluent and strict environmental compliance are essential.

2.8 System integration and hybrid approaches in aquaculture wastewater treatment

Aquaculture wastewater is very diverse. It contains a range of pollutants, including suspended solids, organic matter, nutrients, pathogens, and new contaminants such as antibiotics and microplastics. No single treatment method can effectively manage this complexity in every situation. Because of this, there has been an increase in the use of system integration and hybrid approaches. These methods combine different physical, chemical, biological, and ecological processes into multistage or multifunctional systems. These systems aim to take advantage of the strengths of individual technologies while addressing their weaknesses. This leads to better treatment efficiency, stronger operational resilience, and greater sustainability (González-Fernández et al., 2020; Chaudhary et al., 2021).

2.8.1 Definitions

Integrated systems usually use different treatment processes in sequence. Each stage targets a specific class of pollutants. For example, sedimentation may come first, followed by biological filtration and then constructed wetlands for nutrient polishing. Hybrid systems combine two or more processes into one operational unit or loop. This creates interactions that improve performance. MBRs pair biological oxidation with membrane separation. They are a prime example of hybrid systems, allowing for the simultaneous removal of organic matter, nutrients, and pathogens (Qin et al., 2023).

2.8.2 Key hybrid and integrated system designs

One of the most widely used designs is the membrane-based hybrid system, where MBRs are paired with additional polishing units, such as algae cultivation or advanced oxidation processes. These systems consistently achieve more than 99% removal of suspended solids, ammonia, phosphorus, and pathogens, producing water suitable for reuse in aquaponics or irrigation (Wang et al., 2022). FO membranes combined with nanobubble technology are also being explored for aquaculture effluents. The collapse of nanobubbles provides a self-cleaning mechanism that reduces fouling and maintains long-term membrane performance (Qin et al., 2023).

Another promising approach is the development of multistage constructed wetlands. These systems typically integrate sedimentation ponds, vertical-flow wetlands, aeration chambers, and horizontal-flow reed beds into a sequential treatment setup. Studies have shown that such designs can remove more than 60% of nitrogen, phosphorus, and COD from aquaculture effluents. They also support biodiversity and landscape multifunctionality (Zhou et al., 2021). Because they are low-cost and easy to maintain, multistage wetlands are particularly suitable for rural and peri-urban aquaculture operations where land is not a limiting factor.

IMTA is another notable system that uses ecological engineering principles. IMTA combines feed species such as fish or shrimp with extractive species, including seaweeds, shellfish, or filter-feeding organisms. The waste nutrients from the fed species become inputs for the extractive species. This process reduces effluent discharge while diversifying farm outputs. This approach improves nutrient cycling and provides extra economic resilience by producing multiple high-value products (Troell et al., 2009; Chopin et al., 2013).

Algal–biofilm hybrid reactors are also gaining popularity. These systems combine HRAPs with attached biofilms to improve nutrient uptake and make biomass harvesting easier. By integrating suspended algal growth with biofilm-based processes, these reactors achieve high nutrient removal and simplify downstream processing of biomass for conversion into feed or biofertilizers (García et al., 2020).

Finally, anaerobic–aerobic treatment chains offer a robust hybrid approach. High-strength organic loads are first processed in anaerobic digesters, such as UASB reactors. The anaerobic stage generates biogas, contributing to energy recovery. Subsequent aerobic polishing, such as wetlands or trickling filters, removes leftover nutrients and pathogens (Pang et al., 2021). This cascade approach combines pollution control with renewable energy production, fitting well with circular economy strategies.

2.8.3 Emerging innovations

The integration of digital monitoring and artificial intelligence (AI) into hybrid systems is a new area of exploration. IoT-enabled sensors now allow for real-time monitoring of key parameters such as pH, DO, turbidity, and nutrient levels. With AI-driven models, these tools support predictive system control, provide early warnings of potential failures, and optimize aeration or dosing processes. This significantly improves efficiency and reduces costs (Zhou et al., 2023).

Renewable energy is also making progress in hybrid wastewater treatment. Solar-powered pumps and aerators, along with biogas recovery from anaerobic digesters, are being used to improve the energy self-sufficiency of aquaculture farms. In coastal regions, systems that use wind or hybrid energy are being explored to further cut down on fossil fuel use (CORDIS, 2022).

Urban and modular applications highlight another key innovation. Rooftop aquaponics systems that include compact RAS–MBR units are being tested for small-scale food production in cities. Containerized treatment units are being developed for remote aquaculture sites or as emergency solutions during fish kills or disease outbreaks. These modular systems offer scalability and flexibility, making them suitable for various environmental and socioeconomic situations (Walsh and Cowx, 2020).

2.8.4 Summary of benefits

Overall, hybrid and integrated approaches provide several advantages over single-process technologies. By combining different treatment methods, they achieve higher rates of pollutant removal, offer better resilience against varying effluent loads, and support resource recovery based on circular economy principles. Multistage wetlands promote biodiversity and ecological functions. IMTA generates multiple revenue streams. Algal biofilm systems make biomass harvesting easier, and anaerobic-aerobic cascades recover renewable energy. These varied benefits make hybrid and integrated systems essential for the sustainable growth of aquaculture globally.

2.9 Challenges and future perspectives in aquaculture wastewater treatment

Despite significant progress in developing and implementing treatment systems, aquaculture wastewater management still faces many technical, economic, and environmental challenges. These obstacles not only limit the scalability of new solutions but also affect the long-term sustainability of aquaculture production. It is essential to understand these challenges to create future strategies that support global goals for food security, environmental protection, and climate resilience.

2.9.1 Current challenges

One of the biggest challenges is the high operational and capital costs tied to advanced systems. Technologies such as MBRs, RASs, and PBRs need a significant investment in infrastructure and skilled workers. For smallholder farmers and producers in developing countries, these costs can be too high, making low-cost, nature-based options more attractive even if they are less effective at removing pollutants (FAO, 2020; Praveen et al., 2021).

Energy consumption is another issue related to cost. Continuous aeration, which is crucial in BFT and MBRs, can make up to 60% of total energy use. This becomes a sustainability problem when powered by fossil fuels (Kuhn et al., 2010; Taw, 2021). Without renewable energy, these systems may end up increasing greenhouse gas emissions while addressing water quality issues.

The need for skilled labor and effective management is another limitation. Hybrid and integrated treatment systems, especially those that depend on real-time monitoring and IoT-enabled automation, require technical knowledge that might not be easily found in low-resource environments. Mistakes in system calibration or poor maintenance can significantly reduce treatment efficiency and even threaten fish health (Chowdhury et al., 2023).

The changing characteristics of aquaculture effluents introduce more complexity. Variations in feed type, stocking density, water temperature, and salinity affect nutrient load and microbial community stability. This makes it tough to create one-size-fits-all treatment systems (Zheng et al., 2025). In addition, the increasing presence of new contaminants, such as antibiotics, hormones, and microplastics, is a new challenge in wastewater management. Traditional biological systems often struggle to tackle these pollutants, which need advanced oxidation, electrochemical treatments, or adsorption technologies (Nguyễn et al., 2024).

Managing sludge and byproducts continues to be a significant issue. Systems such as BFT, constructed wetlands, and anaerobic digesters create residual biomass and sludge that must be either disposed of or reused. Without sound management strategies, sludge can turn into an environmental problem instead of a resource (Alumichem, 2024). Lastly, land and space limitations restrict the use of nature-based solutions, such as wetlands or algae ponds, in urban or densely populated coastal areas, where aquaculture is often growing the fastest (Zhou et al., 2021).

2.9.2 Future perspectives

To address these challenges, several promising directions are emerging. First, integrated, modular, and scalable systems are expected to lead future aquaculture wastewater management. These systems combine multiple treatment stages, such as sedimentation followed by MBRs and algal polishing, in compact units that can be adjusted based on production needs. This flexibility makes them suitable for both smallholder operations and large industrial aquaculture parks (Zheng et al., 2025).

The use of circular economy approaches will also be essential for the future of aquaculture wastewater treatment. Waste-to-resource strategies, such as recovering nutrients as struvite, using biofertilizers, generating biogas, or using microbial and algal biomass as aquafeeds, can transform waste into valuable resources (WAS, 2023). By incorporating resource recovery, wastewater treatment systems can improve profitability while lessening environmental damage.

Another key trend is the adoption of smart monitoring and automation. IoT sensors and AI models are increasingly able to predict failures, optimize aeration, and adjust nutrient dosing. These tools not only cut labor costs but also improve treatment efficiency and fish welfare by maintaining stable water quality (Zhou et al., 2023).

Incorporating renewable energy is also crucial. Solar panels, wind turbines, and biogas digesters can power aeration, pumping, and monitoring systems, lowering both operational costs and carbon emissions. In coastal areas, hybrid renewable systems can provide reliable energy for off-grid aquaculture facilities (CORDIS, 2022).

At the cutting edge of research, biotechnological innovations are creating new possibilities. Designer microbial groups, genetically altered algae, and microbial fuel cells are being studied to improve pollutant removal while also generating energy (Zheng et al., 2025). Although still experimental, these methods show potential for linking wastewater treatment to broader bioeconomy frameworks.

Finally, deploying decentralized treatment solutions will be vital for expanding access to sustainable wastewater management. Mobile or containerized units can support remote aquaculture farms, respond to emergencies during disease outbreaks, or fit urban setups that combine fish production with wastewater recycling (Walsh and Cowx, 2020). Policy frameworks and financial incentives will also play a key role. Supportive regulations and targeted subsidies can speed up the adoption of low-discharge, resource-recovery systems (Praveen et al., 2021).

2.9.3 Concluding remarks

Aquaculture is at a crucial point in its global development. It needs to increase production for food security while also protecting aquatic ecosystems and public health. Wastewater treatment is key to this shift. Aquaculture effluents are now seen as environmental risks and also as sources of valuable resources such as nutrients, microbial biomass, and organic matter. The development of treatment technologies, from traditional sedimentation ponds to MBRs, biofloc systems, aquaponics, and hybrid ecological-engineered platforms, shows the industry's growing commitment to sustainability. However, although many of these innovations have shown promise in controlled research and pilot studies, real-world application still encounters challenges related to cost, energy demands, management complexity, and scalability across different geographic and socioeconomic contexts.

Looking ahead, the future of managing aquaculture wastewater will involve combining multiple approaches. Modular and hybrid systems that integrate physical, chemical, and biological treatments will likely take the lead. These systems can flexibly handle changing effluent loads and adjust to specific farm conditions. Strategies from the circular economy, such as recovering nutrients as biofertilizers, using microbial biomass as feed, and generating energy from anaerobic digestion, will turn aquaculture waste into new value chains. Digital technologies such as IoT-enabled monitoring, AI, and predictive control will improve system reliability and reduce labor needs, making advanced systems easier for small- and medium-scale producers to use. Integrating renewable energy solutions will also be crucial. These can lower the high energy costs of aeration and pumping while reducing carbon footprints, aligning aquaculture with broader goals for addressing climate change.

At the same time, future research must keep tackling ongoing challenges. Emerging contaminants, such as antibiotics, hormones, and microplastics, create complex treatment needs that traditional systems cannot fully address. Innovations in biotechnology, such as engineered microbial groups, nanomaterial-enhanced membranes, and algal-bacterial communities, are expected to be vital in dealing with these pollutants and improving system resilience. Policy frameworks and financial incentives will also be important for speeding up adoption, especially in developing countries where aquaculture is growing rapidly and where environmental protection often struggles due to limited resources.

In conclusion, managing aquaculture wastewater sustainably requires a shift in perspective. We need to see effluents not just as waste but as opportunities for resource recovery and the development of a circular bioeconomy. By integrating ecological principles, engineering innovations, digital technologies, and supportive policies, aquaculture can move from being an extractive industry to a regenerative one. By adopting these strategies, the sector can secure its environmental and economic future while also contributing significantly to global goals around food security, water sustainability, and climate resilience.

References

Abdel-Raouf, N., Al-Homaidan, A.A., Ibraheem, I.B.M., 2012. Microalgae and wastewater treatment. Saudi J. Biol. Sci. 19 (3), 257–275.

Abreu, A.P., Fernandes, B., Vicente, A.A., Teixeira, J., Dragone, G., 2012. Effect of light and temperature on *Scenedesmus obliquus* biomass productivity and nutrient removal in wastewater. Bioresour. Technol. 123, 488–494.

Alumichem, 2024. 6 challenges faced by wastewater treatment plants and their solutions. Technical Report. https://alumichem.com/6-challenges-faced-by-wastewater-treatment-plants-and-their-solutions.

Avnimelech, Y., 2009. Biofloc Technology: A Practical Guide Book. World Aquaculture Society, Baton Rouge. Louisiana, USA.

Azim, M.E., Little, D.C., 2008. The biofloc technology (BFT) in indoor tanks: water quality, biofloc composition, and growth and welfare of Nile tilapia (*Oreochromis niloticus*). Aquaculture 283 (1–4), 29–35.

Becker, E.W., 2007. Micro-algae as a source of protein. Biotechnol. Adv. 25 (2), 207–210.

Bondad-Reantaso, M.G., Subasinghe, R.P., Arthur, J.R., Ogawa, K., Chinabut, S., Adlard, R., Tan, Z., Shariff, M., 2005. Disease and health management in Asian aquaculture. Vet. Parasitol. 132 (3–4), 249–272. https://doi.org/10.1016/j.vetpar.2005.07.005.

Bossier, P., Ekasari, J., 2017. Biofloc technology: a practical review for aquaculture. Aquacult. Int. 25, 139–146.

Boyd, C.E., McNevin, A.A., Davis, R.P., Godumala, R., 2020. Perspectives on the responsible use of aquaculture drugs. Rev. Aquac. 12 (2), 664–672.

Cabello, F.C., 2006. Heavy use of prophylactic antibiotics in aquaculture: a growing problem for human and animal health and for the environment. Environ. Microbiol. 8 (7), 1137–1144. https://doi.org/10.1111/j.1462-2920.2006.01054.x.

Chaudhary, R., Pathak, V.V., Ojha, S., 2021. Hybrid and integrated wastewater treatment technologies: a review. Environ. Technol. Innov. 23, 101757.

Chopin, T., Cooper, J.A., Reid, G.K., Cross, S.F., Moore, C., 2013. Open-water integrated multi-trophic aquaculture: environmental biomitigation and economic diversification of fed aquaculture by extractive aquaculture. Rev. Aquac. 4 (4), 209–220. doi:10.1111/j.1753-5131.2012.01074.x.

Chowdhury, M.A., et al., 2023. Challenges and potential solutions of microalgae-based aquaculture effluent treatment. Front. Microbiol. 14, 10277499.

CORDIS, 2022. Integrated aquaculture based on sustainable water management. European Commission Project 689427 https://cordis.europa.eu/project/id/689427

Crab, R., Avnimelech, Y., Defoirdt, T., Bossier, P., Verstraete, W., 2007. Nitrogen removal techniques in aquaculture for sustainable production. Aquaculture 270 (1–4), 1–14.

Crab, R., Defoirdt, T., Bossier, P., Verstraete, W., 2012. Biofloc technology in aquaculture: beneficial effects and future challenges. Aquaculture 356–357, 351–356. doi:10.1016/j.aquaculture.2012.04.046.

Craggs, R.J., Heubeck, S., Lundquist, T.J., Benemann, J.R., 2012. Algal biofilm wastewater treatment for nutrient removal and biomass production. Water Sci. Technol. 66 (2), 303–309.

Cripps, S.J., Bergheim, A, 2000. Solids management and removal for intensive land-based aquaculture production systems. Aquacultural Engineering 22 (1–2), 33–56. doi:10.1016/s0144-8609(00)00031-5.

Ekasari, J., Crab, R., Verstraete, W., Bossier, P., 2016. Biofloc technology in low salinity shrimp culture: water quality, biofloc composition, and growth performance. Aquaculture 452, 273–279.

Endut, A., Jusoh, A., Ali, N., Wan Nik, W.B., Hassan, A., 2010. Nutrient removal from aquaculture wastewater by vegetable production in aquaponics system. Desalin. Water Treat. 24 (1–3), 246–252.

FAO, 2020. The State of World Fisheries and Aquaculture 2020. FAO, Rome.

García, D., et al., 2020. Algal–biofilm hybrid systems for wastewater valorization. Bioresour. Technol. 314, 123737.

Goddek, S., Delaide, B., Mankasingh, U., Ragnarsdottir, K.V., Jijakli, H., Thorarinsdottir, R., 2015. Challenges and technological innovations in aquaponics. Sustainability 7 (4), 4199–4224. https://doi.org/10.3390/su7044199.

Goddek, S., Körner, O., 2019. A fully integrated model of nutrient dynamics in aquaponics: a case study for phosphorus. Aquacult. Environ. Interact. 11, 61–76.

Gogate, P.R., 2011. Hydrodynamic cavitation for food and environmental applications. Curr. Sci. 100 (5), 690–701.

González-Fernández, C., Sialve, B., Bernet, N., 2020. Membrane-based hybrid biorefineries for aquaculture wastewater valorization. Renew. Sustain. Energy Rev. 115, 109386.

Hai, F.I., Yamamoto, K., 2018. Membrane Biological Reactors: Theory, Modeling, Design, Management and Applications to Wastewater Reuse, 2nd edn IWA Publishing, London.

Ip, Y.K., Chew, S.F., Randall, D.J., 2001. Ammonia toxicity, tolerance, and excretion. Mar. Pollut. Bull. 42 (12), 1023–1030.

Kampbell, D.H., Wilson, J.T., Vandegrift, S.A., 2009. Toxic gas emissions from aquaculture sediments. Aquacult. Eng. 41 (3), 121–129.

Kim, H., et al., 2020. Water reclamation using membrane processes in aquaculture systems. Water Res. 185, 116243.

Kumar, S., et al., 2022. Nutritional characterization of biofloc in tilapia aquaculture. Aquac. Rep. 25, 100884.

Kundu, N., et al., 2020. Electrochemical technologies for aquaculture wastewater treatment: recent advances and future perspectives. J. Environ. Manage. 256, 109959.

Kuhn, D.D., Boardman, G.D., Lawrence, A.L., Marsh, L., Flick, G.J., 2010. Microbial floc meal as a replacement ingredient for fishmeal and soybean protein in shrimp feed. Aquaculture 296 (1–2), 51–57.

Li, T., Zheng, Y., Yu, L., Chen, S., 2019. Enhancing nutrient removal and biomass production by *Chlorella vulgaris* in wastewater treatment. Bioresour. Technol. 275, 34–41.

Liu, Y., et al., 2016. The fate of suspended solids in recirculating aquaculture systems. Aquacult. Eng. 72–73, 30–38.

Love, D.C., Fry, J.P., Genello, L., Hill, E.S., Frederick, J.A., Li, X., Semmens, K., 2015. An international survey of aquaponics practitioners. PLoS One 10 (7), e0133670.

Markou, G., Georgakakis, D., 2011. Cultivation of filamentous cyanobacteria (*Spirulina sp.*) in agro-industrial wastes and wastewaters. Bioresour. Technol. 102 (17), 10093–10100.

Mata, T.M., Martins, A.A., Caetano, N.S., 2010. Microalgae for biodiesel production and other applications: a review. Renew. Sustain. Energy Rev. 14 (1), 217–232.

Mmojieje, C., et al., 2022. Nutrient dynamics in aquaculture systems: implications for treatment strategies. Environ. Sci. Pollut. Res. 29 (4), 5501–5515.

Muir, J.F., 2005. Managing to harvest? Perspectives on the potential of aquaculture. Philos. Trans. R. Soc. B Biol. Sci. 360 (1453), 191–218.

Pang, Y., et al., 2021. Anaerobic–aerobic hybrid systems for aquaculture waste. Process Biochem. 103, 23–34.

Parsa, J.B., et al., 2021. Application of electrocoagulation in aquaculture wastewater treatment: review. Environ. Sci. Pollut. Res. 28 (2), 12345–12359.

Pittman, J.K., Dean, A.P., Osundeko, O., 2011. The potential of sustainable algal biofuel production using wastewater resources. Bioresour. Technol. 102 (1), 17–25.

Praveen, N., et al., 2021. Aquaculture wastewater treatment technologies and their sustainability. Sci. Total Environ. 770, 145379.

Qin, L., et al., 2023. Advances in membrane bioreactor systems for aquaculture wastewater. Water Res. 229, 119416.

Rakocy, J.E., Masser, M.P., Losordo, T.M., 2006. Recirculating Aquaculture Tank Production Systems: Aquaponics—Integrating Fish and Plant Culture, 454. Southern Regional Aquaculture Center (SRAC) Publication No, Stoneville, MS.

Rawat, I., Kumar, R., Mutanda, T., Bux, F., 2011. Dual role of microalgae: phycoremediation of domestic wastewater and biomass production for sustainable biofuels. Appl. Energy 88 (10), 3411–3424.

Schrader, K.K., Green, B.W., Perschbacher, P.W., 2011. Microbial metabolism of aquaculture waste: oxygen demand and greenhouse gas emissions. Aquacult. Eng. 45 (2), 87–92.

Shah, S.Q.A., et al., 2014. Antibiotic resistance in aquaculture: mechanisms and solutions. Environ. Pollut. 191, 27–35.

Somerville, C., Cohen, M., Pantanella, E., Stankus, A., Lovatelli, A., 2014. Small-Scale Aquaponic Food Production: integrated Fish and Plant Farming, 589. Food and Agriculture Organization (FAO), Rome FAO Fisheries and Aquaculture Technical Papers.

Taw, N., 2021. Sustainable aquaculture energy management: aeration challenges and solutions. Aquac. Eng. 94, 102145.

Telli, K., Pulatsu, S., 2021. Innovations in aquaculture wastewater treatment: a global review. Aquac. Res. 52 (4), 1553–1571.

Timmons, M.B., Ebeling, J.M., Wheaton, F.W., Summerfelt, S.T., Vinci, B.J., 2021. Recirculating Aquaculture Systems, 3rd ed. 401. NRAC Publication.

Troell, M., Halling, C., Neori, A., Chopin, T., Buschmann, A.H., Kautsky, N., Yarish, C., 2009. Ecological engineering in aquaculture: potential for integrated multi-trophic aquaculture (IMTA) in marine offshore systems. Aquaculture 297 (1–4), 1–9.

Vymazal, J., 2011. Constructed wetlands for wastewater treatment: five decades of experience. Environ. Sci. Technol. 45 (1), 61–69.

Wahid, F., et al., 2020. Characterization and management of aquaculture wastewater: a review. Environ. Rev. 28 (1), 1–12.
Walsh, M., Cowx, I., 2020. Urban wastewater reuse in aquaculture: opportunities and risks. Aquacult. Env. Interact. 12, 145–158.
Wang, Z., et al., 2022. Nanobubble-assisted forward osmosis for aquaculture effluent treatment. Chem. Eng. J. 433, 134519.
Nguyễn, T., Tran-Nguyen, P.L., Vo, T., 2024. Advances in aeration and wastewater treatment in shrimp farming: Emerging trends, current challenges, and future perspectives. AQUA—Water Infrastructure, Ecosystems and Society 73, 902–916. https://doi.org/10.2166/aqua.2024.328.
WAS (World Aquaculture Society), 2023. Circular economy approaches in aquaculture wastewater management. [Technical Report]. World Aquaculture Society. Baton Rouge, LA
Zhang, D., et al., 2023. Hybrid membrane technologies for advanced aquaculture effluent treatment: progress and perspectives. Sep. Purif. Technol. 318, 123982.
Zhang, Y., et al., 2022. Hydrodynamic cavitation for aquaculture wastewater treatment. Environ. Technol. Innov. 27, 102475.
Zhao, X., et al., 2025. Addressing emerging contaminants in aquaculture wastewater: towards next-generation treatment strategies. J. Cleaner Prod. 410, 137523.
Zheng, Y., Khan, A.Z., Li, Z., Huang, G., Liu, L., Jiang, F., Hu, Z., 2025. Advances in microalgal-bacterial granular sludge systems for sustainable aquaculture wastewater treatment: Mechanisms, optimization, and future perspectives. SSRN (Elsevier), Rochester, NY. https://doi.org/10.2139/ssrn.5141541.
Zhou, H., et al., 2021. Multi-stage wetland designs for aquaculture effluent treatment. Ecol. Eng. 162, 106168.
Zhou, H., et al., 2023. Smart hybrid aquaculture water treatment with AI-enabled sensors. Environ. Syst. Res. 12, 66.

CHAPTER

3

Recirculation and recycling: Innovations in hybrid recirculating aquaculture systems

Lokesh Pawar[a,*], *Kaliyaperumal Sriranjani*[b] *and Sruti T.N.*[c]

[a]Faculty of Sciences, UGent Campus Sterre, University of Ghent, Krijgslaan, Gent, Belgium [b]Institute of Fisheries Post Graduate Studies, Vaniyanchavadi, Tamil Nadu Dr. J. Jayalalithaa Fisheries University, Tamil Nadu, India [c]Dr. M.G.R. Fisheries College and Research Institute, Thalainayeru, Tamil Nadu, India

*Corresponding author: pawarlokesh2405@gmail.com

3.1 Introduction

Aquaculture is undergoing a transformation toward more intensive yet sustainable methods to meet the rising demand for seafood and fish. Recirculating aquaculture systems (RASs), which rear aquatic organisms at high density in controlled, filtration-recycled water loops, have emerged as a key approach for this intensification. In recent years, hybrid RAS designs have gained attention, combining conventional RAS with complementary technologies (e.g., biofloc reactors, integrated multitrophic aquaculture, and constructed wetlands) to enhance performance and net output from the entire production cycle. These integrated systems promise improved nutrient recycling, water efficiency, and environmental control, while also aiming to overcome the limitations of standalone RAS or traditional open aquaculture. The rise of hybrid RAS in the past decade reflects a pursuit of sustainable intensification, meeting global food security needs through closed-loop, ecologically and economically balanced aquaculture. This chapter reviews the state-of-the-art in hybrid RAS, focusing on system design innovations, environmental impacts, and economic feasibility, based on current literature, industry reports, and the authors' personal experience and exposure (Table 3.1).

Zero-Waste Technologies in Aquaculture: Global Perspectives and Methods toward a Circular Economy
DOI: https://doi.org/10.1016/B978-0-443-34689-7.00007-0

TABLE 3.1 Hybrid RAS integration strategies: Key features, benefits, and challenges (Correia et al., 2020; Li et al., 2025; Betanzo-Torres et al., 2025).

Integration strategy	Key features and components	Environmental benefits	Economic considerations
RAS + Biofloc (BFT-RAS)	Conventional RAS with microbial floc cultivated in tanks (high C: N). Bioflocs act as additional biofilters and natural feed	**Nutrient recycling:** Heterotrophic microbes assimilate ammonia, reducing toxic nitrogen and producing protein-rich flocs for fish **Water savings**: >90% water reuse achievable, minimal exchange needed. **Waste reduction:** 40%–50% lower N and P emissions vs. traditional systems; solids partly consumed by microbes	**Feed cost reduction:** FCR improved by ~20%, lowering feed inputs (biofloc supplies up to 30% of fish nitrogen needs), boosting profitability if managed well **High operational skill:** Requires careful control of C/N ratio, oxygen, and solids; risk of system crashes if imbalance occurs **Energy intensive:** Strong aeration needed for floc; increases energy cost (aeration ~40% of OPEX)
RAS + IMTA (multitrophic)	RAS linked with other trophic levels (e.g., algae, bivalves, and detritivores) in either the same loop or side stream reactors	**Nutrient uptake:** Algae and filter feeders extract excess nutrients (nitrogen, phosphorus) and organic matter, turning waste into biomass **Ecosystem mimicry:** Creates a balanced mini ecosystem, improving water quality and stability; effluent nutrient discharge greatly reduced **Biodiversity and habitat**: Provides habitat for additional species, potentially enhancing system resilience	**Product diversification:** Generates secondary products (seaweed, shellfish) for extra income, enhancing ROI and spreading risk **Complex management:** Must meet husbandry needs of multiple species; increases labor and management complexity **Infrastructure cost:** Additional tanks or raceways needed for other species; moderate capital cost increase, but some components (e.g., algal tanks) can be low cost
RAS + Constructed Wetland	RAS coupled with a biofilter wetland or plant hydroponic system for wastewater polishing. Water flows through plant root zones before recirculation	**Natural filtration:** Wetland plants and microbes remove ~90%–97% of nutrients (ammonia, nitrate) and sediments, enabling near-zero nutrient discharge. **Water reuse:** Allows high-rate water recycling by biologically regenerating water quality; aligns with circular water economy **Low energy:** Relies on passive ecological treatment, reducing the need for energy-intensive chemical filtration	**Low operating cost:** Plants generally require little upkeep; wetlands can be cost-effective once installed. **Land requirement:** Wetlands need significant space and proper design; scaling to large RAS may be limited by footprint availability **Seasonal performance:** Efficacy may vary with climate (plant growth slows in cold seasons), which could necessitate backup treatment or larger area

(continued on next page)

TABLE 3.1 Hybrid RAS integration strategies: Key features, benefits, and challenges (Correia et al., 2020; Li et al., 2025; Betanzo-Torres et al., 2025)—cont'd

Integration strategy	Key features and components	Environmental benefits	Economic considerations
RAS + Aquaponics (fish–plant)	A specialized IMTA where fish RAS is directly coupled with hydroponic plant cultivation (e.g., leafy greens, herbs). Nutrient-rich fish water irrigates plants, which strip nitrates and then return to clean water	**Closed nutrient loop:** "Fish-vegetable" synergy converts fish waste to fertilizer for plants, achieving nearly complete nutrient recycling **Water efficiency:** Same water produces two crops; <5% of water added periodically, making it one of the most water-efficient food production methods. **Urban farming potential:** Can be set up in greenhouses near cities, reducing food miles and enabling local sustainable produce	**Dual revenue:** Sale of both fish and high-value vegetables improves overall farm income and can shorten payback period **Balance and risk:** Nutrient supply and pH must balance fish and plant needs; system upset (disease, failures) can affect both crops **Capital and skill:** Requires greenhouse infrastructure and knowledge of aquaculture and horticulture; higher initial cost, but with integrated outputs

BFT, biofloc technology; *IMTA*, integrated multitrophic aquaculture; *RAS*, recirculating aquaculture system; *ROI*, return on investment.

3.2 System design and integration strategies

3.2.1 Key components of recirculating aquaculture systems

A modern RAS consists of tanks for fish culture coupled with treatment units that continuously purify and recirculate water. Core technologies include mechanical filters to remove solids, biofilters housing nitrifying bacteria to convert toxic ammonia to nitrate, oxygenation/aeration systems, and ultraviolet or ozone units for disinfection (Martins et al., 2010; Badiola et al., 2018). These components maintain water quality within narrow parameters, enabling high stocking densities. Control systems (increasingly IoT-driven) monitor parameters such as dissolved oxygen, pH, ammonia, and temperature in real time to ensure stability. Recent innovations focus on automation and precision control; for example, online sensors with machine learning can dynamically adjust aeration or carbon addition to prevent ammonia or nitrite spikes in dense cultures, and these technological refinements are critical as systems scale up (Badiola et al., 2018; Kumar et al., 2023). A well-designed RAS creates a closed-loop habitat; however, designers are now integrating additional biological and ecological units into RAS to form "hybrid" systems that further enhance nutrient utilization and system resilience. One prominent hybrid model combines traditional flow-through aquaculture with recirculation (Martins et al., 2010; Assiddiqi et al., 2025) (Fig. 3.1).

Fig. 3.1 Schematic representation of the recirculating aquaculture system (RAS) loop, illustrating the sequential components involved in water treatment and recirculation, including the culture tank, solid removal system, biological filtration, degassing, reservoir and pumping unit, UV and ozone treatment, and oxygenation.

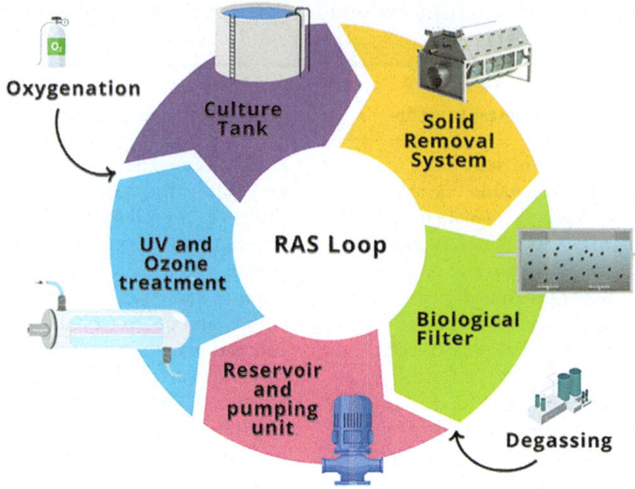

FIGURE 3.1 Schematic representation of the recirculating aquaculture system (RAS) loop, illustrating the sequential components involved in water treatment and recirculation, including the culture tank, solid removal system, biological filtration, degassing, reservoir and pumping unit, UV and ozone treatment, and oxygenation.

3.2.2 Hybrid flow-through systems (Partial reuse)

In a hybrid flow-through system (HFTS), fresh water (often from a river, lake, or sea) is continuously added like a flow-through farm, but a substantial portion (up to ~65%) of the water is recirculated through RAS components instead of being discharged. Essentially, the system reuses part of the water via filtration while still flushing some of it to the environment. This partial reuse approach, originally developed as "partial water reuse" in the 1980s, has recently been rebranded as HFTS in the salmon industry (Summerfelt et al., 2004; Dalsgaard et al., 2013; Badiola et al., 2018; Kassem et al., 2021). The advantage is a balance between water and energy use: by allowing some flow-through, pumping requirements and energy costs can be lower than in a full closed RAS, while water use and effluent discharge are still greatly reduced versus 100% flow-through. HFTS also tends to be simpler in design (since biofilters handle only a fraction of flow) and can dilute potential buildups of contaminants by blending with new water (Dalsgaard et al., 2013; Badiola et al., 2018). Notably, HFTS has gained investor attention for salmon grow-out on land. For instance, recent projects in Norway and Iceland have used HFTS to farm Atlantic salmon with good results. The RAS component provides water quality control and pathogen reduction, yielding biosecurity benefits, while the flow-through aspect helps manage temperature and chemistry naturally (Lund et al., 2021; Cermaq, 2022). Moreover, effective disease control in hybrid RAS depends on integrated water quality and biosecurity measures. Disinfection methods such as ozone, UV, and low-dose hydrogen peroxide are used alongside oxidation–reduction potential (ORP) monitoring to protect fish and biofilters. Preventive tools, such as vaccines, probiotics, and immunostimulants, are replacing antibiotics, enhancing sustainability. These strategies work together to lower the risk of disease while keeping the system running smoothly (Holan et al., 2020).

Initial data indicate that HFTS salmon farms are producing fish that are healthy and of high quality. Salmon Evolution in Norway reports superior growth and lower mortality rates in its hybrid system, even if energy use is higher (Salmon Evolution, 2023). Indeed, HFTS can require

1–8 kWh of energy per kilogram of fish, reflecting the pumping of large volumes of water, but operators find the trade-off acceptable because of robust fish performance and reduced disease issues (Badiola et al., 2018; Lund et al., 2021). In sum, hybrid flow-through designs offer an intermediate solution between fully closed RAS and open flow-through, leveraging the strengths of each. These systems are viewed to scale up land-based aquaculture with potentially lower operational risk and cost, albeit at the expense of using more water than pure RAS.

3.2.3 Coupling with biofloc technology

One prominent hybridization is the biofloc technology (BFT)-RAS hybrid, which unites conventional RAS filtration with biofloc technology. In biofloc systems, high C:N ratios and vigorous aeration promote dense microbial communities (flocs) that assimilate waste nutrients and serve as natural feed. Combining BFT with RAS leverages the strengths of both: The closed, controlled environment of RAS provides optimal conditions for floc microbes, while the bioflocs, in turn, improve water quality and nutrition. Studies report that in a BFT–RAS hybrid, the microbial flocs are rich in protein and essential nutrients, partially replacing formulated feed for the fish/shrimp (Avnimelech, 2015; da Silva et al., 2021; Li et al., 2025).

This approach can lower the feed conversion ratio (FCR) by ~20%–30% compared with standard RAS or pond culture, as observed in trials with white shrimp (*Litopenaeus vannamei*). Flocs were able to supply around 30% of the nitrogen requirements of shrimp, and probiotic microbes in the floc even enhanced digestive enzyme activity (Avnimelech, 2015; Hendearti et al., 2022). Because of this, cultured species in BFT–RAS systems often grow and survive better. For instance, Indian prawn (*Penaeus indicus*) raised in a hybrid biofloc–RAS showed significantly higher growth performance and immune indicators (e.g., elevated total serum protein and polyphenol oxidase activity) compared with clear-water RAS, alongside lower ammonia and nitrite levels (Li et al., 2025). These outcomes highlight the synergistic design principle: Nutrient recycling within the system (via microbial loops) purifies water and boosts productivity. However, BFT–RAS integration introduces new technical considerations and operational complexity. Managing the balance between heterotrophic biofloc bacteria and autotrophic nitrifiers is crucial; fluctuations in carbon input or dissolved oxygen can upset this balance, demanding careful monitoring (Emerenciano et al., 2017; Badiola et al., 2018). The accumulation of suspended solids (floc biomass) can also stress mechanical filters or cause biofilter clogging if not controlled.

Innovations such as dual-stage filtration (settlers + biofilters) and the use of moving bed biofilm reactors (MBBR) alongside biofloc have been explored to increase stability. Indeed, coupling biofloc with biofilm reactors has shown complementary removal of ammonia; heterotrophic flocs assimilate ammonia into biomass, while biofilm nitrifiers oxidize ammonia to nitrate, together achieving more complete nitrogen removal (Crab et al., 2009; Li et al., 2025). This dual approach improved water quality and shrimp growth in high-density trials. In summary, the BFT–RAS hybrid design represents an innovative integration of microbial technology into RAS, yielding a more self-contained and feed-efficient system. Key design challenges remain in scaling up large-scale BFT–RAS farms; they must prevent oxygen crashes and biofilter fouling and may require modular tank designs or decoupled loops so that floc culture can be managed separately from fish tanks (Emerenciano et al., 2017;

Wasielesky et al., 2020). Nonetheless, pilot studies demonstrate that with proper control, BFT–RAS systems can sustain intensive production with minimal water exchange and improved feed utilization, a significant step toward sustainable aquaculture intensification (Martins et al., 2010; Li et al., 2025).

3.2.4 Integration with integrated multitrophic aquaculture

Another hybridization strategy is integrating RAS with multitrophic aquaculture principles. Integrated multitrophic aquaculture (IMTA) involves rearing species from different trophic levels in tandem so that waste from one is a resource for another. In an RAS context, this approach often means adding units for extractive species (such as algae, bivalves, or detritivores) to capture nutrients from fish effluent before water is recirculated. For example, a land-based IMTA–RAS might circulate water from a fish tank to tanks containing shellfish or seaweed that filter out organic particles and absorb dissolved nutrients, then return the cleansed water to the fish. Such arrangements "close the loop" biologically; fish feed and excreta become inputs for algae and filter feeders, which in turn can be harvested as additional products. IMTA thus transforms waste into cobenefits, greatly improving nutrient cycling efficiency (Li et al., 2025). One laboratory RAS–IMTA setup cultivated European seabass in one tank, sea urchins in the next, and Ulva seaweed in another, all connected in series with recirculating seawater (Correia et al., 2020). The sea urchins consumed particulate wastes (feces, uneaten feed) from the fish, and the seaweed absorbed inorganic nitrogen, resulting in a system where very little nutrient left as effluent. Generally, IMTA additions have been shown to reduce nutrient discharge and improve water quality by treating RAS outflow as a feed source rather than a pollutant (Li et al., 2025). This not only benefits the environment but also diversifies production; farmers obtain secondary crops (e.g., seaweed biomass and shellfish) that provide extra income and spread economic risk (Correia et al., 2020).

Moreover, the presence of extractive species can enhance public perception of intensive aquaculture. By demonstrating that fish waste is recycled into valuable products, IMTA–RAS is considered a more circular and eco-friendly approach, potentially reducing local opposition to fish farms. However, designing an effective RAS–IMTA system requires carefully matching species and managing their interactions. Each added organism has its own husbandry needs (e.g., light and carbon dioxide for algae, flow rates for filter feeders). Integration strategies often decouple the sub-systems to some extent, for instance, using a side stream of effluent to feed an algae raceway or shellfish basin, then returning treated water. This allows control of conditions in each unit. Research in recent years (e.g., trials combining shrimp RAS with oyster filtration or fish RAS with macroalgae bioreactors) shows promising nutrient removal rates and growth of extractive species (Correia et al., 2020). One study added *Crassostrea* oysters to a shrimp RAS and observed that while total suspended solids were not greatly lowered, the oysters selectively grazed certain microbes, indicating potential for refining microbial composition in the system (Emerenciano et al., 2023). Scaling IMTA–RAS beyond the pilot scale remains an active area of experimentation. Key design questions include how large the extractive components need to be relative to fish biomass, how to automate solid waste delivery to detritivores, and ensuring the biosecurity of multi-species systems. Nonetheless, IMTA integration is a compelling strategy to enhance RAS sustainability by biologically engineering a mini ecosystem where wastes are minimized and outputs maximized.

3.2.5 Integrating constructed wetlands for water treatment

A hybrid RAS design can also incorporate constructed wetlands (CWs) or plant-based treatment units to naturally process wastewater. In this approach, water from the RAS is routed through a wetland cell or plant hydroponic bed, where aquatic plants and their microbial root biofilms remove nutrients and organic matter via phytoremediation. The treated water is then recirculated back to the culture tanks, drastically reducing discharge. CWs offer a low-energy, ecologically inspired filtration stage that complements the high-tech mechanical filters of RAS. Recent literature indicates growing interest in RAS–CW integration as part of a circular aquaculture model (Betanzo-Torres et al., 2025). A systematic review (2003–2023) identified numerous plant species, e.g., reeds (*Phragmites australis*), cattails (*Typha latifolia*), water hyacinth (*Eichhornia crassipes*), effective at treating aquaculture effluents (Chen et al., 2024). In experimental RAS-coupled wetlands, these plants achieved substantial nutrient reductions. For instance, water hyacinth floating in a treatment basin removed ~90% of ammonia and nitrate from fishpond effluent, and the full wetland system achieved a 97% reduction in inorganic nitrogen overall. Similarly, pilot-scale wetlands for intensive shrimp RAS have shown >90% removal of total suspended solids and nitrite and ~75% of nitrate, especially when a sedimentation prefilter is used (Betanzo-Torres et al., 2025). These results demonstrate that a properly engineered wetland unit can polish RAS water to near-background nutrient levels, enabling safe reuse.

The design of RAS–CW hybrids typically involves a reservoir to collect RAS effluent, a pumping system to feed the wetland (often a gravel bed or floating plant system), and a holding time sufficient for plant uptake and sedimentation. Because wetland processes are slower, larger footprint areas and longer hydraulic retention (days) are needed compared to conventional filters. This can conflict with the intensive space efficiency that makes RAS attractive, and the greater need for effective and efficient water recycling. Indeed, surveys show that while many lab studies prove CW efficacy, only ~40% of reviewed cases actually integrated wetlands into a recirculating loop for water reuse, and full-scale commercial implementations are still limited (Betanzo-Torres et al., 2025). Challenges include the land area required, seasonal variation in plant performance, and ensuring the treated water meets the biosecurity and quality needs of sensitive aquaculture species. One promising application is using wetland plants to not only treat water but also yield useful biomass (e.g., constructing wetlands that grow salicornia or duckweed, which can be harvested as a crop or even feed ingredient). Overall, CW integration is a nascent yet innovative design strategy to improve RAS sustainability. It aligns with circular economy principles by treating waste nutrients as inputs to plant production (Betanzo-Torres et al., 2025). Future designs are investigating modular "wetland filters" that can be scaled and attached to RAS facilities and exploring a wider diversity of plant species for different climates. If technical hurdles of scale and consistency can be overcome, RAS wetland hybrids could allow zero-discharge aquaculture with nature-based water recycling.

3.2.6 Zero-water discharge systems

Zero-water discharge (ZWD) provides environmental benefits, but its low carrying capacity limits them. In contrast, RASs support high-density culture but are resource-intensive. Hybrid

ZWD–RAS systems integrate microbial nutrient recycling and advanced water treatment to improve water quality, stabilize rearing conditions, and enhance growth and survival. Pilot studies indicate that such systems can increase productivity with minimal water exchange, supporting sustainable intensification of aquaculture(Assiddiqi et al., 2025).

3.2.7 Scalability and technological innovations

As hybrid RAS designs mature, a major focus is on scalability, translating successful pilots into industrial-scale operations. Large commercial RAS (especially for species such as salmon, trout, or shrimp) are already being built globally, but integrating multi-component systems at scale poses engineering and management challenges. One issue is that added complexity (e.g., biofloc reactors, plant beds, and multispecies units) can complicate control and increase the risk of failures if not properly decoupled. Modular design is a key innovation to address this; instead of one giant recirculation loop, modern approaches use multiple linked modules that can be individually controlled or isolated. For example, decoupled aquaponics systems have separate loops for fish and plants with intermediate nutrient transfers, improving stability. Similarly, a scaled-up hybrid RAS might partition the system, e.g., parallel biofilter/floc units for different tank groups, to localize any water quality issues and facilitate maintenance. Researchers emphasize enhancing stability through modular filtration and precise nutrient regulation, calling for flexible system architectures as scale increases (Betanzo-Torres et al., 2025).

Innovations in automation and monitoring also play a pivotal role. The use of distributed IoT sensors and AI analytics in RAS is on the rise (Kumar et al., 2023). At a large scale, real-time data on water quality in each component (fish tanks, biofloc reactor, wetland sump, etc.) can be fed into predictive models that adjust operations, for example, turning on backup aeration or adjusting feed rates. Researchers have tested machine learning algorithms to improve carbon dosing in BFT–RAS and find early signs of parameter drift that human operators might miss (Kumar et al., 2023). These smart controls are essential for managing the tighter tolerances required in intensive, integrated systems. On the engineering front, new low-energy equipment is being developed to reduce the heavy power requirements of RAS. High-efficiency pumps, automated feeding systems, and improved oxygenation (such as microbubble diffusers) can cut energy per unit biomass. Some large RAS farms are integrating renewable energy (e.g., solar panels and waste heat recovery) to lower operating costs and carbon emissions (Badiola et al., 2018).

A notable advancement is the concept of zero-liquid discharge RAS, where all water is treated and recirculated indefinitely, with only periodic removal of concentrated sludge. Prototypes by technology providers have combined ultrafiltration, reverse osmosis, and evaporative techniques to achieve near-zero effluent release, albeit with high capital cost (Badiola et al., 2018). Hybrid systems (such as RAS + wetlands or RAS + algal bioreactors) can aid this goal by breaking down and uptaking waste so thoroughly that make-up water needs are minimal. Indeed, the hybrid BFT–RAS discussed earlier achieved >90% water reuse rates, pointing to the feasibility of minimal discharge systems (Li et al., 2025). In summary, scaling hybrid RAS to commercial levels is underway through technological innovation (automation and efficient equipment) and systemic innovation (modularity and decoupling). While current integrated designs are often at pilot scale, progress in the last 5–10 years suggests that industrial-scale hybrid RAS can become a reality. Continued interdisciplinary research and engineering refinement are needed to ensure that as systems grow, they remain stable, cost-effective, and manageable in practice (Betanzo-Torres et al., 2025).

3.3 Environmental impact and sustainability

3.3.1 Nutrient cycling and water use efficiency

A primary motivation for hybrid RAS is to improve nutrient cycling and minimize resource use, thereby reducing aquaculture's environmental footprint. Traditional flow-through or pond aquaculture discharges nutrient-rich water (containing nitrogen, phosphorus, and organic matter) directly into the environment, risking eutrophication of natural waters. RAS, by contrast, recirculates the water and concentrates wastes for treatment, massively cutting direct nutrient discharge. By integrating additional treatment or utilization steps (e.g., biofloc and IMTA), hybrid RAS can achieve even tighter closed-loop nutrient cycling. In practical terms, this means more of the feed nutrients are converted to harvestable products (fish or secondary species) and less to pollution. For example, the BFT–RAS hybrid model synchronizes microbial nutrient uptake with fish production: Heterotrophic bacteria immobilize inorganic nitrogen into biomass that fish can later consume. This yields remarkable efficiency gains; one study reports that the coupled BFT–RAS reduced nitrogen and phosphorus emissions by 40%–50% compared with an equivalent standalone RAS or pond system (Emerenciano et al., 2017). Similarly, integrating plants or algae in RAS (aquaponics or wetland systems) closes the nutrient loop by taking up residual nitrates and phosphates for plant growth. In one case, an aquaponic RAS loop (tilapia with hydroponic vegetables) was able to recirculate water for months with minimal buildup of nitrate, as the plants removed on average 90% of the nitrate generated (Goddek et al., 2016). Water use efficiency in these systems is accordingly very high.

Conventional pond culture often exchanges large volumes of water to maintain quality, whereas a well-run RAS might replace only 5%–10% of total water per day (mainly to remove nitrate or maintain buffering capacity). Hybrid approaches drive water reuse even further. The incorporation of biofloc or advanced filters has enabled >90% water savings relative to open systems (Emerenciano et al., 2017). Case studies show hybrid RAS models can operate with only negligible water addition to compensate for evaporation and sludge removal. For instance, a biofloc-enhanced RAS unit reported a water-saving rate over 90% while sustaining a high stocking density of shrimp (Li et al., 2025). Such a drastic reduction in water demand is crucial for sustainability, especially in water-scarce regions. It implies that large quantities of fish can be produced with only a fraction of the freshwater footprint of traditional aquaculture.

Effective nutrient recycling also means that effluent discharge is minimized, protecting surrounding ecosystems. Many hybrid RAS aim for "zero exchange," where practically no liquid effluent is released aside from deliberate sludge concentrate removals. The nutrient cycling benefits extend beyond the farm, as captured nutrients (in sludge or plant biomass) can potentially be repurposed as fertilizers for agriculture, closing loops between aquaculture and farming. Overall, by treating waste as a resource, hybrid RAS drastically improves the nutrient conversion efficiency of aquaculture. Each unit of feed leads to more biomass and less pollution. This not only mitigates local environmental impacts (preventing algal blooms, oxygen depletion in waterways) but also enhances the global sustainability profile of aquaculture. As one review succinctly noted, hybrid models "optimize nutrient cycling and feed efficiency, reduce feed conversion, and achieve more than 90% water savings" relative to traditional systems (Li et al., 2025). These are key metrics that make recirculating systems an ecologically attractive route for expanding fish production without proportionally increasing the burden on land and water resources.

3.4 Waste management and emissions control

Intensive aquaculture inevitably generates waste streams, chiefly in the form of solid organic waste (feces, uneaten feed) and dissolved metabolites (ammonia, CO_2). RAS are designed to intercept and treat these wastes internally. Mechanical filters and settling units capture solids as sludge, whereas biofilters convert dissolved nitrogen wastes to less harmful forms. A well-managed RAS thus prevents most waste from entering the environment untreated. Hybrid RAS further enhances waste management by actively converting or removing these wastes through natural processes. For example, in BFT-based systems, the solid waste is largely consumed by microbes within the tank, reducing the volume that must be removed as sludge (Avnimelech, 2015; Emerenciano et al., 2017). The remaining sludge from RAS (a concentrated suspension of organic matter and bacterial biomass) can be thickened and reused; current research is exploring its use as a crop fertilizer or as an input for bioenergy (e.g., anaerobic digestion to produce biogas) (Badiola et al., 2018). However, it should be noted that in practice, many commercial RAS still dispose of sludge via conventional waste channels (land application or landfill), incurring costs and environmental considerations. Hybrid approaches, such as wetlands or IMTA, can alleviate these issues by either mineralizing the sludge on-site (wetland microbes breaking it down) or consuming it (detritivores feeding on it). For instance, the inclusion of detritivore species (such as sea cucumbers or certain polychaetes) in some RAS has been used to directly ingest settleable solids, effectively turning sludge into animal biomass (Zamora and Jeffs, 2012). Another emission of concern is nutrient-laden water. Even RAS with high recirculation may need to purge some water periodically to prevent accumulation of nitrate or salts. The goal of hybrid systems is to treat this water such that it can be reused or safely discharged. As discussed, CWs can strip out nutrients to very low levels (up to 97% nitrogen removal), and biofloc systems assimilate nutrients into biomass, meaning that any overflow water has substantially less pollutant load than typical aquaculture effluent. A study on a hybrid tilapia system with floating macrophyte treatment reported more than 90% removal of nitrite and solid particles, demonstrating how well-designed natural filters can polish RAS effluents (Betanzo-Torres et al., 2025). Thus emissions to the environment, whether solid or liquid, are greatly reduced in magnitude and impact.

In addition, RAS virtually eliminates the problem of chemical and therapeutic discharge that is common in open net-pen farms; any antibiotics or treatments (ideally minimized through good biosecurity) are confined to the system and can be treated, e.g., ozone breaks down many compounds before water release (Martins et al., 2010; Badiola et al., 2018). One environmental benefit sometimes overlooked is that RAS prevents fish escapes and the associated ecological risks, such as genetic dilution of wild stocks or invasive species introductions. By physically isolating cultures, hybrid RAS contribute to biodiversity protection, which is an important sustainability facet, though not typically quantified in nutrient terms (Tlusty, 2002). In summary, hybrid RAS designs excel in waste capture and conversion. Sludge outputs are concentrated, and potentially useful dissolved nutrients are recaptured into secondary biomass or inert forms, and virtually no untreated emissions leave the farm. This not only ensures compliance with stringent environmental regulations but also actively turns what was waste into value. The net effect is a dramatic lowering of the ecosystem impacts per unit of fish produced. Aquaculture can thus grow in scale without a proportional increase in eutrophication, habitat degradation, or chemical pollution, marking a significant advancement over earlier aquaculture systems.

3.5 Energy use and carbon footprint

A critical environmental consideration for RAS-based systems is their energy intensity, which in turn influences the carbon footprint of the produced seafood. Operating pumps, blowers, heaters/chillers, and treatment units continuously can make RAS far more energy-demanding per kilogram of fish than extensive pond or cage culture. This higher energy use can lead to higher greenhouse gas (GHG) emissions, unless renewable energy or efficiency measures are used to lower them. Recent analyses highlight this challenge: For example, a life-cycle assessment found that producing Atlantic salmon in a recirculating system in the United States had an on-farm carbon footprint of about 7.0 t CO_2/ton of fish, roughly double the ~3.4 t CO_2/ton for salmon grown in open net pens in Norway. The difference was mainly because of electricity and fuel demands of RAS water treatment and climate control (Ayer and Tyedmers, 2009; Badiola et al., 2018). However, the same study noted an intriguing twist: when considering the full supply chain to a North American consumer, locally grown RAS salmon could avoid the high emissions from transoceanic air freight of fresh Norwegian salmon. After accounting for transport, the delivered carbon footprint of RAS salmon was ~7.4 t CO_2/ton of fish versus ~15.2 t CO_2/ton for imported net-pen salmon (Ayer and Tyedmers, 2009). This indicates that RAS can leverage locality to offset some production-phase emissions. Nonetheless, reducing the energy use of RAS remains paramount for sustainability. Optimizing energy use in hybrid RAS requires careful design, planning, and control. Studies have tested solar, wind, and biogas integration, such as photovoltaic systems in salmon farms or solar-powered heating in aquaponics, to boost sustainability. Innovations such as RES-driven oxygen generation show promise, though most findings are site-specific and lack broadly applicable optimization models (Badiola et al., 2018; Liu et al., 2022; Zhang et al., 2024).

The energy intensity of RAS-based systems is an important environmental factor that affects the carbon footprint of the seafood they produce. Operating pumps, blowers, heaters/chillers, and treatment units continuously can make RAS far more energy-demanding per kilogram of fish than extensive pond or cage culture. This higher energy use can lead to higher GHG emissions, unless renewable energy or efficiency measures are used to lower them. Recent analyses highlight this challenge: For example, a life-cycle assessment found that producing Atlantic salmon in a recirculating system in the United States had an on-farm carbon footprint of about 7.0 t CO_2/ton of fish, roughly double the ~3.4 t CO_2/ton for salmon grown in open net pens in Norway. The difference was mainly because of the electricity and fuel demands of RAS water treatment and climate control (Ayer and Tyedmers, 2009; Badiola et al., 2018). However, the same study noted an interesting fact that when considering the full supply chain to a North American consumer, locally grown RAS salmon could avoid the high emissions from transoceanic air freight of fresh Norwegian salmon. After accounting for transport, the delivered carbon footprint of RAS salmon was ~7.4 t CO_2/ton of fish versus ~15.2 t CO_2/ton for imported net-pen salmon (Ayer and Tyedmers, 2009). This indicates that RAS can leverage locality to offset some production-phase emissions. Nonetheless, reducing the energy use of RAS remains paramount for sustainability. Optimizing energy use in hybrid RAS requires careful design, planning, and control. Studies have tested solar, wind, and biogas integration, such as photovoltaic systems in salmon farms or solar-powered heating in aquaponics to boost sustainability. Innovations such as RES-driven oxygen generation show promise, though most findings are site-specific and lack broadly applicable optimization models (Badiola et al., 2018; Liu et al., 2022).

In summary, the environmental profile of hybrid RAS is a mix of outstanding achievements in water and nutrient stewardship, combined with the challenge of high energy use. They virtually eliminate many local environmental impacts of aquaculture (no effluent, escapes, and habitat damage), but to truly be sustainable, the energy and climate footprint must be managed. Encouragingly, modern designs show pathways to do so, through efficiency innovations, renewable energy integration, and smart system choices. With these measures, hybrid RAS can deliver fish with a far lower total environmental impact (from eutrophication to land use to carbon emissions) than both traditional aquaculture and many forms of livestock production (Li et al., 2025). The continued improvement of energy efficiency will ensure that the impressive gains in water and nutrient performance are matched by progress in climate sustainability.

3.6 Economic feasibility and market perspectives

3.6.1 Capital and operating costs

Economic viability is a key factor in the adoption of hybrid RAS. These systems are capital-intensive and require substantial upfront investment in tanks, life-support equipment, sensors, and often climate-controlled facilities. Compared with traditional pond farms or sea cages, a commercial RAS may cost several times more per unit of production capacity in initial construction. For example, small-scale farmers often find the entry barrier prohibitive; one review noted that high initial equipment investment and specialized operation costs limit RAS uptake by small operators (Li et al., 2025). The integration of additional components (biofloc reactors, wetland units, etc.) can further add to complexity and capital cost, though some components (such as wetlands) are relatively low-tech.

Operationally, RAS also incurs higher ongoing expenses for energy, skilled labor, and equipment maintenance. A recent case study from Hungary on carp production in RAS vividly illustrates the cost structure: the unit production cost was €5.47 per kg, with energy (47%) and feed (24%) identified as the largest cost drivers (Mihályr-Karnai et al., 2025). This cost slightly exceeded the market price of carp, resulting in a net loss of ~€0.41 per kg. The study underlined that while RAS achieves high productivity, economic viability is challenged by high energy and feed costs and the depreciation of expensive infrastructure. Many RAS enterprises face similar hurdles; profitability can be elusive unless costs are controlled or compensated by higher fish prices. Hybrid RAS might impose additional costs (e.g., a supplemental carbon source for biofloc and additional tanks for IMTA species), but they can also yield cost savings in some areas through improved feed efficiency, higher biomass yields, or marketable coproducts.

BFT–RAS systems have demonstrated reductions in feed costs by generating natural feed on-site. In a shrimp biofloc–RAS trial, feed input was reduced by 30%–35% without loss of production because of biofloc consumption by *L. vannamei* (Wasielesky et al., 2006). This translated into improved feed conversion and could boost overall profitability despite higher capital outlay. Similarly, integrated systems that produce secondary crops (e.g., shellfish, seaweed, or plants) create additional revenue streams. These byproducts improve return on investment (ROI) by utilizing waste nutrients to grow salable products. For instance, a salmon RAS that incorporates a sidestream *Ulva* seaweed culture can sell the biomass as a nutraceutical or feed additive to offset operational costs. Another economic benefit of hybrids is potential savings on waste disposal and water. If a wetland or biofloc unit significantly reduces water exchange, the farm saves on water purchase or wastewater treatment fees. Large RAS projects have reported substantial costs for effluent permits and sludge disposal; integrating internal treatment could lower those regulatory and handling costs.

Despite these advantages, the financial risk of RAS remains nontrivial. Engle (2020) analyzed commercial RAS farms and found that many were profitable only when very high production densities and biological performance were maintained. Break-even often requires optimal survival, rapid growth, and minimal downtime. This fragility is why early RAS ventures frequently failed; small deviations in performance or system failure often tipped the balance into loss. To mitigate this, strategies now include scaling up (to capture economies of scale), targeting high-value species, and improving reliability. Proven profitable use cases tend to focus on niches: Salmon smolt production in RAS is commercially successful because of high survival rates, consistent year-round output, and downstream integration with sea cages (Badiola et al., 2018). In the United States, some long-running RAS operations have succeeded by farming specialty fish for live markets or overwintering pond fingerlings to extend growing seasons.

These cases show that with the right market and operational strategy, RAS can be economically viable. Hybrid methods could lead to new ways to make money by increasing total output (through multiple product streams) and lowering input costs (through better FCRs and water reuse). However, they also require multidomain expertise (e.g., aquaculture, horticulture, and microbial management), which can increase labor or management complexity and cost. A holistic economic evaluation must therefore account for both the additional revenues and the cross-sector operational burdens. In essence, hybrid RAS economics require high upfront investment and running costs, balanced against improved resource efficiencies and diversified outputs.

As technology advances and costs decrease (e.g., for sensors, automation, and renewable integration), the economic equation is expected to improve. Furthermore, environmental regulations and sustainability incentives such as grants for water-saving technology or price premiums for eco-labeled seafood can shift the cost–benefit balance in favor of RAS (Badiola et al., 2018). Many governments and investors recognize the long-term need for sustainable aquaculture, and increasing support for RAS innovation and infrastructure has helped alleviate capital barriers recently (Engle, 2020).

In conclusion, while economic challenges for hybrid RAS are real, they are being addressed through scale, innovation, and integrated value generation. Careful planning, especially optimizing energy use, nutrient recovery, and labor, remains critical. One recent review concluded that achieving profitability in RAS "requires making more efficient use of energy, water, capital, and labor inputs"—in essence, doing more with less, which is exactly the design principle behind most hybrid RAS systems (Martins et al., 2010). As technologies mature and scale economies are realized, the cost per unit of fish is likely to decline, inching these systems closer to broad commercial viability.

3.7 Profitability and market adoption

Profitability of hybrid RAS varies widely by region, species, and market conditions. In high-income markets, such as North America, Europe, and East Asia, consumers are increasingly willing to pay a premium for locally produced, environmentally sustainable seafood, which can help RAS farms secure higher prices (Bergman et al., 2020). For instance, land-based RAS salmon is often positioned as a sustainable alternative to ocean-raised salmon, potentially commanding premium retail pricing.

Market acceptance of RAS products is generally positive when quality is well managed. Fillets from RAS-grown fish can equal or even surpass conventional products in taste and texture, provided the fish are purged correctly and systems are managed to control water quality (Davidson et al., 2016). However, challenges remain in consumer perception. Some buyers are skeptical of "tank-grown" fish, associating them with unnatural conditions or inferior taste. A recurring technical issue is the presence of off-flavor compounds, especially geosmin and 2-methylisoborneol (MIB), produced by microbial activity in recirculating systems. These can impart musty or earthy flavors if not effectively removed before harvest. Past failures to address this issue have led to consumer rejection and contributed to what some researchers have termed a "sociocultural discredit" of RAS products, damaging their image in competitive seafood markets (Bergman et al., 2020).

Modern RAS operations devote significant effort to depuration, purging fish in clean, off-flavor-free water before harvest, and employ additional treatment techniques such as ozonation, activated carbon filtration, and improved biofilter management to ensure sensory quality (Azaria et al., 2018a). As a result, many RAS-raised fish now reach the market with excellent flavor and texture profiles. Ongoing research continues to develop strategies to mitigate off-flavor compounds such as geosmin and MIB, further strengthening consumer confidence in these products.

Importantly, consumer studies indicate that when people are informed about the sustainable and local attributes of RAS fish, their acceptance tends to improve, especially when flavor expectations are met (Bergman et al., 2020). Another crucial aspect of market adoption is the reception from retailers and supply chains. Major supermarket chains and seafood distributors increasingly value the year-round production stability that indoor RAS offers, as it is independent of seasonal fluctuations and environmental uncertainty. This consistent supply, coupled with strong traceability and biosecurity assurances such as lower pathogen risk and no exposure to ocean lice, has led to growing partnerships between RAS producers and retailers (Kok et al., 2020). For example, several RAS salmon ventures have secured long-term retail contracts based on their sustainability narrative, including reduced antibiotic use, water recycling, and animal welfare benefits.

Nonetheless, for hybrid RAS ventures to fully penetrate markets, they must also demonstrate economic sustainability. Long-term competitiveness will hinge not only on product quality and environmental credentials but also on cost control, efficient operations, and consumer trust.

The ROI in successful hybrid RAS cases can be attractive but often materializes only after several years of operation once biofilters mature, operational protocols are optimized, and market channels are secured (Martins et al., 2010). High-value species such as eel, sturgeon (for caviar), barramundi, or shrimp in premium markets tend to offer quicker ROI because of their higher unit prices and stronger niche appeal. In contrast, producing low-value whitefish species, such as carp, catfish, or tilapia, in RAS systems remains challenging unless production costs are significantly reduced or the product is differentiated (e.g., local, organic, and antibiotic-free labeling) (Badiola et al., 2018).

Regionally, adoption dynamics vary. In water-scarce or environmentally sensitive areas such as parts of the Middle East or Southern Europe, RAS is emerging as the only scalable aquaculture solution, enabling production without placing stress on coastal zones or limited freshwater supplies (Mihályr-Karnai et al., 2025). In these contexts, public support and infrastructure investment have helped normalize RAS-sourced fish in domestic markets. In contrast, in regions where pond-raised fish dominate and prices are low, RAS must target quality-driven niches to justify premium pricing.

A wider view looks at the social and environmental benefits of hybrid RAS in addition to its direct profitability. When factoring in externalities such as nutrient discharge, land use, or water withdrawals, studies often demonstrate that RAS delivers greater net positive societal value even if private margins are modest (Bergleiter et al., 2017). These findings have implications for policy and investment: Eco-labeling, carbon credit systems, or sustainable seafood certifications could strengthen RAS market competitiveness. Certifications such as the Aquaculture Stewardship Council (ASC) and Best Aquaculture Practices (BAP) already include RAS facilities and reassure consumers about the sustainability of land-based products (ASC, 2024; GSA, 2024). Importantly, hybrid RAS also aligns with environmental, social, and governance (ESG) investment frameworks, which increasingly emphasize circularity, carbon efficiency, and resource stewardship.

Recently, substantial venture capital has flowed into large-scale RAS projects, driven by the promise of sustainable, local protein production. This investment surge is part of the broader "blue economy" wave, which seeks to align economic growth with ocean stewardship (SeafoodSource, 2023). While early concerns lingered over the unproven economic viability of RAS, successful operations are gradually building investor confidence and market credibility. The path to profitability for hybrid RAS is steadily improving, with key success factors including selection of high-value or differentiated species, continuous technical optimization to reduce operational costs, and effective use of sustainability narratives to secure premium pricing and improved market access (Mihályr-Karnai et al., 2025). Consumer acceptance has also become more favorable, particularly when product quality in terms of taste and freshness meets expectations. Challenges such as off-flavor compounds are being actively managed through improved purging protocols and system design (Azaria et al., 2018). As wild fisheries plateau and environmental regulations constrain conventional aquaculture expansion, hybrid RAS are well-positioned to capture a larger share of the global seafood market. While profitability remains variable, it is expected to stabilize as technology matures, costs decline, and environmental benefits become more monetarily valued by markets and regulators alike.

3.8 Conclusion

Hybrid RASs are a promising way to combine technology and ecology in fish farming. By integrating biofloc microbial loops, multitrophic species, and natural treatment processes into closed RAS, these systems achieve levels of resource efficiency and environmental control unparalleled in traditional aquaculture. The literature of the past decade documents substantial advancements: nutrient recycling is maximized, water use is cut by an order of magnitude, and waste emissions are turned into inputs or valuable byproducts. At the same time, the challenges are clearer than ever, chiefly the high energy demand and capital cost that make economics the limiting factor for widespread adoption. Yet, ongoing innovations in system design (modularity, decoupling, and smart automation) and improvements in feed and energy efficiency are steadily addressing these bottlenecks. Ongoing innovations in system design (modularity, decoupling, and smart automation) and improvements in feed and energy efficiency are steadily addressing these bottlenecks. Already, certain hybrid RAS models are operating profitably for specific species and markets, and they offer a template for sustainable intensification of aquaculture. In an era where sustainability is paramount, the ability of hybrid RAS to deliver fish with minimal water, land, and environmental impact is a significant advantage. Continued interdisciplinary research spanning engineering, biology, and economics will be crucial to refining these systems. If progress continues at the current pace, hybrid RAS could evolve from niche innovations into a mainstream aquaculture model, supporting global seafood supply with improved ecological and economic outcomes. The rise of hybrid RAS thus signals a shift toward a more resilient and sustainable aquaculture industry, one that harmonizes technological precision with ecological principles for the benefit of producers, consumers, and the environment alike.

Abbreviation

AI	Artificial intelligence
ASC	Aquaculture Stewardship Council
BAP	Best aquaculture practices
BFT	Biofloc technology
CW	Constructed wetlands
ESG	Environmental, social, and governance
FCR	Feed conversion ratio
GHG	Greenhouse gas
HFTS	Hybrid flow-through system
IMTA	Integrated multitrophic aquaculture
IoT	Internet of Things

MBBR	Moving bed biofilm reactor
ORP	Oxidation–reduction potential
RAS	Recirculating aquaculture system
RES	Renewable energy sources
ROI	Return on investment
ZWD	Zero-water discharge

References

Aquaculture Stewardship Council (ASC). (2024). Certification for land-based and recirculating aquaculture systems (RAS). https://www.asc-aqua.org .

Assiddiqi, T.D., Setiawan, B.I., Kurniawan, A., 2025. Hybrid Recirculating Aquaculture Systems with Ultrafiltration and Diffused Aeration: Experimental Verification and Predictive Nitrogen Modeling Using Monod-Based Simulation. SSRN Electron. J. doi:10.2139/ssrn.5262873. https://www.ssrn.com/index.cfm/en/.

Assiddiqi, M.A., Effendi, H., Kusnadi, E., Rachmansyah, A., 2025. Hybrid zero-water discharge and recirculating aquaculture systems: a sustainable solution for intensive aquaculture. Aquac. Res. 56 (2), 789–804.

Avnimelech, Y., 2015. Biofloc Technology: A Practical Guidebook, 3rd ed. The World Aquaculture Society, Baton Rouge, LA. https://www.cabidigitallibrary.org/doi/full/10.5555/20113266301.

Ayer, N.W., Tyedmers, P.H., 2009. Assessing alternative aquaculture technologies: life cycle assessment of salmonid culture systems in Canada. J. Cleaner Prod. 17 (3), 362–373. doi:10.1016/j.jclepro.2008.08.002.

Azaria, A., Mozes, N., Avnimelech, Y., Tirosh, O., 2018. Geosmin and 2-methylisoborneol (MIB) off-flavours in recirculating aquaculture systems (RAS): a review. Trends Food Sci. Technol. 79, 33–43.

Badiola, M., Basurko, O.C., Piedrahita, R., Hundley, P., Mendiola, D., 2018. Energy use in recirculating aquaculture Systems (RAS): a review. Aquacult. Eng. 81, 57–70. doi:10.1016/j.aquaeng.2018.03.003.

Bergleiter, S., Heiland, A., Lorkowski, I., 2017. Life cycle assessment of closed-containment aquaculture systems for rainbow trout and African catfish in Germany. Aquaculture 477, 1–12.

Bergman, K., van Veen, M., Henriksson, P.J.G., 2020. Consumer perception of land-based salmon in Europe: a study of acceptance and product quality expectations. Trends Food Sci. Technol. 101, 296–303.

Betanzo-Torres, E.A., Ballut-Dajud, G., Aguilar-Cortés, G., Delfín-Portela, E., Herazo, L.C.S., 2025. Plants used in constructed wetlands for aquaculture: a systematic review. Sustainability 17 (14), 6298. doi:10.3390/su17146298.

Betanzo-Torres, E.A., Ballut-Dajud, G., Aguilar-Cortés, G., Delfín-Portela, E., Sandoval-Herazo, L.C., 2025. Performance of constructed wetlands integrated with recirculating aquaculture systems: a circular water reuse model. Ecol. Eng. 192, 106944.

Cermaq, 2022. Land-Based Salmon Farming Project Update: Hybrid Flow-Through System. Sustainability Report 2022. Cermaq Group AS., Oslo. https://www.cermaq.com/assets/Global/PDFs-sustainability/Rev-Cermaq-Sustainability-Report-2022-GRI.pdf.

Chen, Y., Lee, Y.J., Wang, H., 2024. Phytoremediation potential of aquatic plants for aquaculture wastewater: a systematic review (2003–2023). Environ. Sci. Pollut. Res. 31 (2), 1993–2011. https://www.tandfonline.com/doi/abs/10.1080/10643389.2022.2164154.

Correia, M., Azevedo, I.C., Peres, H., Magalhães, R., Oliva-Teles, A., Almeida, C.M.R., Guimarães, L., 2020. Integrated multi-trophic aquaculture: a laboratory and hands-on experimental activity to promote environmental sustainability awareness and value of aquaculture products. Front. Mar. Sci. 7. doi:10.3389/fmars.2020.00156.

Crab, R., Defoirdt, T., Bossier, P., Verstraete, W., 2009. Biofloc technology in aquaculture: beneficial effects and future challenges. Aquaculture 289 (1–2), 1–8.

Da Silva, J.M.V., Emerenciano, M.G.C., Gálvez, A.O., Wasielesky, W., 2021. Biofloc technology and its application in intensive shrimp farming: a review. Rev. Aquac. 13 (2), 836–861.

Dalsgaard, J., Lund, I., Thorarinsdottir, R., Drengstig, A., Arvonen, K., Pedersen, P.B., 2013. Farming different species in RAS in Nordic countries: current status and future perspectives. Aquacult. Eng. 53, 2–13. doi:10.1016/j.aquaeng.2012.11.008.

Davidson, J., Good, C., Barrows, F.T., Welsh, C., Kenney, P.B., Summerfelt, S.T., 2016. Comparing the effects of two commercial diets on water quality, growth, nutrient retention, and waste production of Atlantic salmon (Salmo salar) in recirculating aquaculture systems. Aquacult. Eng. 74, 38–51.

Emerenciano, M., Gaxiola, G., Cuzon, G., 2017. Biofloc technology (BFT): a review for aquaculture application and animal food industry. In: Torrissen, G., Hansen, S. (Eds.), Advances in Aquaculture Nutrition. Academic Press, pp. 56–73.

Emerenciano, M., Gomes, G.R., Wasielesky, W., Gaxiola, G., 2023. Bivalve-based microbial refinement in shrimp-RAS: exploring selective grazing potential of Crassostrea spp. in floc-dominated water. Aquacult. Rep. 30, 101589.

Engle, C.R., 2020. Economics of aquaculture systems: recirculating aquaculture systems. Aquac. Econ. Manag. 24 (4), 447–465. doi:10.1080/13657305.2020.1775785.

Global Seafood Alliance (GSA). (2024). Best Aquaculture Practices (BAP) Certification Program. https://www.globalseafood.org.

Goddek, S., Espinal, C.A., Delaide, B., Jijakli, M.H., Schmautz, Z., Wuertz, S., Keesman, K.J., 2016. Navigating towards decoupled aquaponic systems: a system dynamics design approach. Water 8 (7). doi:10.3390/W8070303.

Hendearti, N., Aslamyah, S., Sari, C.K., 2022. Growth performance and immune response of white shrimp (Litopenaeus vannamei) in a biofloc-based RAS with reduced feeding frequency. Aquacult. Rep. 23, 101073.

Holan, A.B., Good, C., Powell, M.D., 2020. Health management in recirculating aquaculture systems (RAS). In: Lucas, A.D. (Ed.), Aquaculture Health Management. Academic Press, London, pp. 281–318.

Kassem, T., Shahrour, I., El Khatabi, J., Raslan, A., 2021. Smart and sustainable aquaculture farms. Sustainability 13 (19). doi:10.3390/su131910685.

Kok, B., Van Duren, I., Vellinga, T.V., de Boer, I.J.M., 2020. Strengthening the sustainability performance of aquaculture supply chains: the role of retailers and consumers. Aquacult. Rep. 17, 100319.

Kumar, R., Pradhan, C., Mishra, S., Ghosh, K., Singh, R.K., 2023. Smart aquaculture systems: recent trends, challenges, and future prospects. Aquac. Fish. 8 (3), 242–255.

Li, C., Ge, Z., Dai, L., Chen, Y., 2025. Integrated application of biofloc technology in aquaculture: a review. Water 17 (14), 2107. doi:10.3390/w17142107.

Liu, Y., Chen, D., Lin, H., Hu, Q., 2022. Optimization of renewable energy integration in recirculating aquaculture systems: a review of technologies and future directions. Renew. Sustain. Energy Rev. 158, 112128.

Lund, I., Dalsgaard, J., Jokumsen, A., 2021. Developments and performance of partial reuse and hybrid flow-through systems for Atlantic salmon farming. Aquac. Int. 29, 1101–1116.

Martins, C.I.M., Eding, E.H., Verdegem, M.C.J., Heinsbroek, L.T.N., Schneider, O., Blancheton, J.P., d'Orbcastel, E.R., Verreth, J.A.J., 2010. New developments in recirculating aquaculture systems in Europe: a perspective on environmental sustainability. Aquacult. Eng. 43 (3), 83–93. doi:10.1016/j.aquaeng.2010.09.002.

Mihályr-Karnai, L., Fehér, M., Bársony, P., Szűcs, I., Mihály, T., Fróna, D., Szőllősi, L., 2025. Sustainability in intensive aquaculture—profitability of common carp (cyprinus carpio) production in recirculating aquaculture systems based on a hungarian case study. Animals 15 (7), 1055. https://www.mdpi.com/2076-2615/15/7/1055.

Salmon Evolution, 2023. Operational Update Q4 2023: Hybrid Flow-Through Performance and Outlook. Salmon Evolution ASA, Bud, Norway. https://salmonevolution.no/investor/reports-presentations/.

SeafoodSource. (2023). North American RAS firms are tapping into the "blue economy" investment wave. https://www.seafoodsource.com/news/aquaculture/north-american-ras-firms-are-tapping-into-the-blue-economy-investment-wave.

Summerfelt, S.T., Davidson, J., Waldrop, T., Tsukuda, S., Bebak, J., 2004. A partial-reuse system for coldwater aquaculture. Aquacult. Eng. 31 (3–4), 157–181. https://www.sciencedirect.com/science/article/pii/S0144860904000196.

Tlusty, M.F., 2002. The benefits and risks of aquacultural production for the aquarium trade. Aquaculture 205 (3–4), 203–219. https://www.sciencedirect.com/science/article/abs/pii/S0044848601006834.

Wasielesky, W., Atwood, H., Stokes, A., Browdy, C.L., 2006. Effect of natural production in a zero exchange suspended microbial floc based super-intensive culture system for white shrimp Litopenaeus vannamei. Aquaculture 258 (1–4), 396–403. doi:10.1016/j.aquaculture.2006.04.030.

Wasielesky, W., Ballester, E.L.C., Poersch, L.H., 2020. Development and application of biofloc technology for the nursery of marine shrimp. Aquacult. Rep. 17, 100384.

Zamora, L.N., Jeffs, A.G., 2012. Feeding ecology of the deposit-feeding sea cucumber Australostichopus mollis (Echinodermata: Holothuroidea) in shallow seagrass and sandflat habitats. J. Mar. Biol. Assoc. U. K. 92 (3), 563–575. https://www.sciencedirect.com/science/article/abs/pii/S0044848611009045.

Zhang, R., Chen, T., Wang, Y., Short, M., 2024. An optimisation approach for the design and operation of recirculating aquaculture systems integrated with sustainable hybrid energy systems. J. Cleaner Prod. 477, 143860. doi:10.1016/j.jclepro.2024.143860.

CHAPTER

4

Trapping nutrients through constructed wetlands: An ecosystem engineered approach to mitigate nutrient pollution

Pronob Das[a], Basanta Kumar Das[b], Dibakar Gogoi[a], Niti Sharma[a], Shyamal Chandra Sukla Das[a] and Sullip Kumar Majhi[a]

[a]ICAR-Central Inland Fisheries Research Institute, Regional Centre, Guwahati, Assam, India [b]ICAR-Central Inland Fisheries Research Institute, Barrackpore, Kolkata, West Bengal, India

4.1 Introduction

Nutrient pollution, primarily originating from agricultural runoff, industrial discharge, and urban wastewater, has emerged as a critical environmental challenge with far-reaching ecological, economic, and public health implications. This form of pollution results from excessive inputs of nitrogen (N) and phosphorus (P) into aquatic environments, surpassing the natural assimilative capacity of these ecosystems. As a consequence, nutrient overenrichment leads to a cascade of detrimental effects, most notably eutrophication, which fosters the rapid proliferation of harmful algal blooms (HABs). These blooms can produce toxins that threaten aquatic life, reduce oxygen levels in the water, and create hypoxic or anoxic conditions that can lead to massive fish kills and biodiversity losses (Das et al., 2025a; Smith and Schindler, 2009; Das et al., 2025b). In addition, the decline in water quality from excessive nutrient loading disrupts the structure and function of aquatic food webs, altering species composition and ecosystem dynamics. Beyond ecological consequences, nutrient pollution also poses significant risks to human health, as contaminated water supplies can contain harmful toxins and pathogens, leading to outbreaks of waterborne diseases and increased costs for water

Zero-Waste Technologies in Aquaculture: Global Perspectives and Methods toward a Circular Economy
DOI: https://doi.org/10.1016/B978-0-443-34689-7.00011-2

treatment (Carpenter et al., 1998). Furthermore, the economic burden of degraded water resources extends to industries such as fisheries, tourism, and recreation, which suffer from declining water quality and habitat loss. Given the escalating prevalence of nutrient pollution worldwide, effective and sustainable mitigation strategies are urgently required to restore and protect the health of affected water bodies.

Among the various approaches to nutrient pollution remediation, constructed wetlands (CWs) have gained considerable attention as an environmentally sustainable and cost-effective solution for nutrient removal and water quality improvement. CWs are engineered ecosystems designed to replicate the natural functions of wetlands, utilizing a combination of vegetation, microbial activity, and substrate interactions to filter, assimilate, and transform pollutants. These systems operate as biofilters, capturing and processing excess nutrients, suspended sediments, and other contaminants before they enter downstream water bodies (Vymazal, 2011). By leveraging the natural processes of nutrient uptake, microbial denitrification, and sediment retention, CWs provide a multifunctional approach to water purification. In addition to nutrient removal, they offer a range of ecological and socioeconomic benefits, including habitat creation for wildlife, carbon sequestration, flood control, and aesthetic value. Because of their adaptability, CWs can be implemented in diverse settings, including agricultural landscapes, urban stormwater systems, and wastewater treatment facilities, making them a viable strategy for mitigating nutrient pollution in both rural and urban environments (Kadlec and Wallace, 2009). As nutrient pollution continues to threaten water resources globally, expanding the use of CWs represents a promising step toward more sustainable water management practices.

4.2 Sources and impacts of nutrient pollution

The primary sources of nutrient pollution include agricultural runoff, industrial effluents, and municipal wastewater discharges. Agricultural practices, particularly the excessive application of fertilizers and improper manure management, contribute significant amounts of nitrogen and phosphorus to surface and groundwater systems (Sharpley et al., 2013). Industrial discharges containing phosphorus-laden detergents, chemicals, and organic waste further exacerbate the problem (Jarvie et al., 2015). Similarly, urban wastewater from domestic sources, including sewage and stormwater runoff, introduces substantial nutrient loads into receiving water bodies, promoting eutrophication and algal proliferation (Conley et al., 2009).

Eutrophication, driven by elevated nutrient levels, manifests through various ecological and economic repercussions. The rapid growth of phytoplankton and cyanobacteria during algal blooms reduces light penetration, inhibiting photosynthesis in submerged aquatic vegetation (Paerl and Otten, 2013). Consequently, oxygen depletion occurs as microbial decomposition of algal biomass increases, leading to hypoxic or anoxic conditions that threaten fish and invertebrate populations (Diaz and Rosenberg, 2008). Moreover, the production of harmful algal toxins poses risks to aquatic life, livestock, and human populations dependent on contaminated water sources (Brooks et al., 2016). Addressing these adverse effects requires innovative strategies to manage and reduce nutrient inputs effectively.

4.3 Mechanisms of nutrient trapping in constructed wetlands

CWs are engineered ecosystems designed to replicate the natural functions of wetlands, particularly their ability to filter, retain, and transform pollutants through physical, chemical, and biological mechanisms. These systems are increasingly utilized for wastewater treatment, agricultural runoff management, and stormwater filtration because of their cost-effectiveness, ecological benefits, and sustainable approach to water quality improvement. By leveraging the natural filtration and nutrient cycling processes found in wetlands, CWs play a crucial role in reducing the concentrations of excess nitrogen (N) and phosphorus (P)—two key nutrients responsible for eutrophication and HABs in aquatic ecosystems (Kadlec and Wallace, 2009). Eutrophication, the excessive enrichment of water bodies with nutrients, leads to oxygen depletion, loss of biodiversity, and deteriorating water quality, making nutrient removal an essential component of wetland-based water treatment strategies (Vymazal, 2011).

The effectiveness of CWs in nutrient removal depends on a combination of sedimentation, adsorption, and precipitation, microbial transformations, plant uptake, and organic matter accumulation—processes that collectively contribute to nutrient retention and removal within the wetland system. Sedimentation allows particulate-bound nutrients to settle out of the water column, while adsorption and precipitation facilitate phosphorus retention by binding it to minerals in the wetland substrate. Microbial transformations, including nitrification, denitrification, and anaerobic ammonium oxidation (anammox), are key processes responsible for converting nitrogen compounds into less harmful forms or releasing them into the atmosphere. In addition, wetland vegetation plays an integral role in nutrient uptake, absorbing nitrogen and phosphorus for growth while also enhancing microbial activity in the rhizosphere. Lastly, organic matter accumulation, including the formation of peat and microbial biomass, contributes to the long-term sequestration of nutrients, preventing their reintroduction into the water system (Kadlec and Wallace, 2009; Vymazal, 2011). These mechanisms collectively make CWs an effective and environmentally friendly solution for mitigating nutrient pollution and enhancing water quality in both natural and built environments.

4.3.1 Sedimentation

Sedimentation is one of the most fundamental physical processes in CWs that facilitates the removal of particulate-bound nutrients from the water column. As water enters a CW, its velocity significantly decreases due to friction with vegetation, substrate, and extended hydraulic retention time (HRT), allowing heavier particles to settle out of suspension (Kadlec and Wallace, 2009). These particles, which include organic matter, suspended solids, and soil particles, often serve as carriers for nitrogen (N) and phosphorus (P) compounds. The gradual accumulation of these sediments in the wetland substrate contributes to the long-term storage of nutrients, effectively reducing their mobility and preventing their reintroduction into downstream aquatic ecosystems. The efficiency of sedimentation depends on several factors, including water depth, vegetation density, retention time, and particle size—with larger and denser particles settling more quickly than finer materials (Vymazal, 2011).

Over time, sediment layers in CWs stabilize and compact, further immobilizing nutrients within the wetland system. However, under certain environmental conditions, these deposited sediments can become a secondary source of nutrient pollution if they are disturbed or resuspended. Factors such as strong hydrodynamic forces (e.g., high flow rates and flooding), bioturbation by benthic organisms (e.g., burrowing invertebrates and fish activity), and extreme weather events (e.g., storms and heavy rainfall) can lead to the rerelease of phosphorus and nitrogen back into the water column (Smith and Schindler, 2009). This process, known as internal nutrient loading, can diminish the effectiveness of CWs in nutrient removal and contribute to eutrophication if not properly managed.

To maximize sedimentation efficiency, CWs must be strategically designed and maintained to optimize water residence time, enhance sediment retention, and minimize resuspension risks. Design considerations include the incorporation of shallow zones, vegetation buffers, and sedimentation basins to facilitate particle settling (Kadlec and Wallace, 2009). Regular sediment monitoring and management, such as periodic dredging or controlled flushing, may also be required to prevent excessive nutrient buildup and maintain wetland functionality over time. By effectively integrating sedimentation into CW design, these systems can provide sustained nutrient removal benefits, reducing the risk of water quality deterioration in receiving aquatic environments.

4.3.2 Adsorption and precipitation

Adsorption and precipitation are two key chemical processes that regulate phosphorus (P) retention in CWs, ensuring long-term nutrient removal and water quality improvement. Phosphorus removal in CWs primarily depends on interactions between dissolved phosphorus and the wetland substrate, where chemical binding and mineral formation occur. Adsorption is the process by which phosphorus attaches to soil particles and mineral surfaces, while precipitation involves the transformation of dissolved phosphorus into insoluble mineral compounds that settle within the wetland sediments (Reddy and D'Angelo, 1997). These mechanisms significantly reduce phosphorus bioavailability, preventing its reintroduction into the water column and minimizing the risk of eutrophication (Vymazal, 2011).

Wetland substrates rich in iron (Fe), aluminum (Al), and calcium (Ca) play a crucial role in phosphorus adsorption, as these elements form strong chemical bonds with phosphate ions (PO_4^{3-}). Iron and aluminum oxides and hydroxides, commonly found in wetland soils, provide positively charged adsorption sites that attract and immobilize phosphate under acidic to neutral pH conditions. In contrast, calcium-based phosphorus retention is more effective in alkaline environments, where calcium ions (Ca^{2+}) react with phosphate to form calcium phosphate [$Ca_3(PO_4)_2$], an insoluble precipitate that remains trapped in the substrate (Kadlec and Wallace, 2009). The efficiency of these adsorption processes depends on several factors, including soil pH, redox potential, temperature, and the concentration of available binding sites. For instance, under anaerobic (low oxygen) conditions, iron-bound phosphorus may become unstable, leading to phosphorus release from sediments, a phenomenon known as internal phosphorus loading (Smith and Schindler, 2009).

In addition to adsorption, precipitation reactions further enhance phosphorus retention in CWs. This process occurs when dissolved phosphorus reacts with metal ions (Fe^{3+}, Al^{3+}, and Ca^{2+}) to form stable, insoluble mineral precipitates such as iron phosphate ($FePO_4$), aluminum phosphate ($AlPO_4$), and calcium phosphate [$Ca_3(PO_4)_2$] (Vymazal, 2011). These mineral compounds settle into wetland sediments, providing a long-term phosphorus sink that prevents remobilization under stable environmental conditions. The rate and efficiency of phosphorus precipitation are influenced by wetland hydrology, redox conditions, and the chemical composition of the substrate, making it essential to optimize wetland design to enhance phosphorus retention capacity (Reddy and D'Angelo, 1997).

However, the long-term efficiency of adsorption and precipitation in CWs may decline over time as substrate binding sites become saturated and phosphorus accumulation reaches equilibrium. If phosphorus retention capacity is exceeded, the wetland substrate may no longer effectively capture additional phosphorus, leading to potential phosphorus leaching back into the water. To prevent saturation and maintain wetland performance, periodic substrate replenishment, amendment with phosphorus-binding materials (e.g., iron-rich sands, limestone, or alum-based compounds), or sediment dredging may be necessary (Kadlec and Wallace, 2009). Proper monitoring and management strategies are essential to ensure sustained phosphorus removal efficiency in CWs, particularly in systems receiving high phosphorus loads from agricultural or wastewater sources.

By integrating adsorption and precipitation into CW design, these systems can effectively reduce phosphorus concentrations, mitigate nutrient pollution, and support long-term water quality improvements. Understanding the chemical interactions governing phosphorus retention is crucial for optimizing CW performance and ensuring their continued success as a nature-based solution for wastewater treatment and environmental conservation.

4.3.3 Microbial processes

Microbial communities within CWs play an essential role in nitrogen transformation and removal, primarily through biological processes such as nitrification, denitrification, and anaerobic ammonium oxidation (anammox). These microbial-driven pathways are crucial for removing excess nitrogen from wastewater, agricultural runoff, and stormwater, preventing eutrophication and water quality degradation in downstream ecosystems. The efficiency of these processes is influenced by oxygen availability, organic carbon levels, temperature, HRT, and wetland design (Vymazal, 2011). By optimizing these factors, CWs can enhance nitrogen cycling and achieve significant nutrient load reductions.

4.3.3.1 *Nitrification*

Nitrification is a two-step aerobic process in which ammonium (NH_4^+) is oxidized to nitrate (NO_3^-) by specialized nitrifying bacteria. This process occurs in the oxygen-rich (aerobic) zones of CWs, particularly in the rhizosphere of wetland plants, where oxygen is actively transported through roots. The first step of nitrification is mediated by ammonia-oxidizing bacteria (AOB) such as *Nitrosomonas*, which convert ammonium (NH_4^+) into nitrite (NO_2^-). In the second step, nitrite-oxidizing bacteria (NOB) such as *Nitrobacter* further oxidize nitrite into nitrate (NO_3^-) (Vymazal, 2011). The overall reactions are as follows:

(1) Ammonium oxidation: $NH_4^+ + 1.5\ O_2 \rightarrow NO_2^- + H_2O + 2H^+$
(2) Nitrite oxidation: $NO_2^- + 0.5\ O_2 \rightarrow NO_3^-$

Nitrification plays a critical role in nitrogen cycling by converting less mobile ammonium into highly soluble nitrate, which can either be assimilated by plants or further removed via denitrification in anaerobic conditions (Kadlec and Wallace, 2009). However, nitrification is highly dependent on oxygen availability, making it more effective in free water surface

(FWS) wetlands or subsurface flow (SSF) wetlands with oxygen diffusion through plant roots (Reddy and D'Angelo, 1997).

4.3.3.2 *Denitrification*

Denitrification is the primary nitrogen removal pathway in CWs, occurring under oxygen-depleted (anoxic) conditions where nitrate (NO_3^-) serves as an electron acceptor instead of oxygen. This process is carried out by denitrifying bacteria, including *Pseudomonas*, *Paracoccus*, and *Bacillus*, which reduce nitrate to nitrogen gas (N_2), thereby permanently removing nitrogen from the wetland system (Kadlec and Wallace, 2009). The sequential reactions are:

$$NO_3^- \rightarrow NO_2^- \rightarrow NO \rightarrow N_2O \rightarrow N_2 \text{ (gas)}$$

The release of nitrogen gas (N_2) into the atmosphere is a desirable outcome because it prevents nitrogen accumulation in water bodies. However, the efficiency of denitrification depends on several factors, including the availability of organic carbon (as an electron donor), nitrate concentration, and HRT (Vymazal, 2011). A lack of organic carbon can limit denitrification rates, which is why carbon amendments (e.g., adding wood chips, organic matter, or external carbon sources such as acetate) are sometimes used to enhance microbial activity in CWs (Reddy and D'Angelo, 1997).

4.3.3.3 *Anammox: The direct conversion of ammonium and nitrite to nitrogen gas*

Anaerobic ammonium oxidation (anammox) is an alternative nitrogen removal pathway in CWs, in which ammonium (NH_4^+) reacts with nitrite (NO_2^-) to form nitrogen gas (N_2) without passing through the nitrate (NO_3^-) stage. This process occurs under strictly anaerobic (oxygen-free) conditions and is mediated by anammox bacteria, such as *Brocadia* and *Kuenenia* (Vymazal, 2011). The overall reaction is:

$$NH_4^+ + NO_2^- \rightarrow N_2 + 2H_2O$$

Anammox is particularly effective in low-oxygen environments and can significantly improve nitrogen removal efficiency when combined with partial nitrification. However, anammox bacteria grow slowly, and their presence in CWs is often limited compared to traditional denitrifying bacteria. Despite this, anammox provides a cost-effective nitrogen removal pathway that requires less organic carbon than denitrification, making it beneficial in CWs treating low-carbon wastewater (Kadlec and Wallace, 2009).

4.3.4 Plant uptake

Wetland vegetation plays a fundamental role in nutrient assimilation within CWs by directly absorbing nitrogen (N) and phosphorus (P) from the water column and incorporating these nutrients into plant tissues. This biological mechanism reduces nutrient concentrations in wastewater, stormwater, and agricultural runoff, thereby mitigating eutrophication risks and improving overall water quality (Brix, 1997). The effectiveness of plant uptake as a nutrient removal strategy depends on plant species selection, growth rates, seasonal variation, and biomass management. In addition to nutrient uptake, wetland plants contribute to other ecosystem services, including habitat provision, sediment stabilization, microbial activity enhancement, and aesthetic improvement.

4.3.4.1 *Mechanisms of nutrient uptake in wetland plants*

4.3.4.1.1 Nitrogen uptake and storage

Nitrogen removal in CWs occurs through plant uptake, microbial transformation, and gaseous emissions. Plants primarily absorb nitrogen in inorganic forms, such as nitrate (NO_3^-) and ammonium (NH_4^+), through their root systems. Once taken up, nitrogen is translocated from the roots to aerial tissues, including stems, leaves, and flowers, where it plays a crucial role in protein synthesis and plant growth. During the growing season, plants exhibit high nitrogen demand, leading to its accumulation in biomass as a form of seasonal storage. If the plants remain unharvested, nitrogen can be temporarily sequestered within the wetland system and subsequently recycled through litter decomposition and leaching, contributing to the internal nitrogen cycle. This dynamic process influences nitrogen availability and retention within the ecosystem, ultimately impacting the efficiency of nitrogen removal in CWs (Reddy and D'Angelo, 1997; Vymazal, 2011).

4.3.4.1.2 Phosphorus uptake and retention

Phosphorus removal via plant uptake is slower than nitrogen removal because phosphorus is less mobile and often exists in organic or particulate-bound forms. Wetland plants primarily take up phosphorus as orthophosphate (PO_4^{3-}) through root absorption, incorporating it into cell structures, nucleic acids, and energy molecules (ATP). However, because phosphorus is not lost to the atmosphere (unlike nitrogen, which can be converted to nitrogen gas), long-term phosphorus removal relies on plant harvesting to prevent nutrient recycling (Vymazal, 2011).

4.3.4.2 *Key wetland plant species for nutrient uptake*

Emergent macrophytes play a crucial role in nutrient uptake within CWs because of their extensive root systems, which enhance nutrient absorption, stabilize sediments, and support microbial activity. Among the commonly utilized species, *Typha* spp. (cattail) is highly efficient in removing nitrogen and phosphorus, making it a widely used plant in CWs (Brix, 1997). *Phragmites australis* (common reed) is another key species known for its tolerance to fluctuating water levels and its ability to facilitate oxygen diffusion into sediments. *Scirpus* spp. (bulrush) contributes to microbial activity in root zones and aids in sediment trapping, further improving water quality. In addition, *Juncus* spp. (rushes) and *Carex* spp. (sedges) are effective in nutrient retention and provide essential habitat for wetland fauna. The selection of plant species for CWs should be based on site-specific factors, including climate, water depth, HRT, and pollutant load, to ensure optimal performance in nutrient removal and ecosystem sustainability.

4.3.4.3 *The role of plant harvesting in nutrient removal*

A major challenge in nutrient removal via plant uptake is that nutrients can be reintroduced into the wetland if plant material decomposes within the system. Decomposition releases stored nitrogen and phosphorus back into the water, potentially undermining the benefits of nutrient sequestration (Kadlec and Wallace, 2009). To prevent this, harvesting above-ground biomass at the end of the growing season is an effective strategy for permanently removing accumulated nutrients from the system (Reddy and D'Angelo, 1997). Research has shown that annual harvesting can enhance long-term nutrient removal efficiency and extend wetland lifespan by reducing phosphorus saturation in sediments (Vymazal, 2011).

4.3.4.4 *Additional benefits of wetland vegetation*

Beyond nutrient uptake, wetland plants provide a range of essential ecosystem services that contribute to both environmental sustainability and human well-being. One of their primary functions is supporting habitat and biodiversity, as wetlands with diverse vegetation serve as breeding grounds for birds, fish, and invertebrates, thereby fostering rich ecological communities (Mitsch and Gosselink, 2015). In addition, wetland plants play a crucial role in sediment stabilization, with their root systems binding sediments to prevent erosion and reduce the resuspension of nutrients, ultimately improving water quality (Kadlec and Wallace, 2009). Some species also facilitate oxygen transfer to sediments, enhancing aerobic microbial processes such as nitrification, which is vital for nitrogen transformation and overall nutrient cycling in wetlands (Brix, 1997). Beyond ecological benefits, wetlands offer significant aesthetic and recreational value, contributing to landscape beauty and promoting ecotourism opportunities, which can support local economies and conservation efforts (Verhoeven et al., 2006). These combined functions highlight the multifaceted importance of wetland vegetation in maintaining ecosystem health and resilience.

4.3.5 Organic matter accumulation

CWs function as long-term nutrient sinks by facilitating the accumulation of organic matter in sediments. This process plays a crucial role in nutrient retention, especially for carbon (C), nitrogen (N), and phosphorus (P), as organic material derived from plant litter, microbial biomass, and detritus settles within the wetland substrate. Over time, these organic-rich deposits contribute to peat formation, carbon storage, and long-term nutrient sequestration, making CWs effective for both water quality improvement and climate change mitigation (Kadlec and Wallace, 2009). However, excessive organic matter accumulation can lead to anoxic conditions, altering microbial processes and potentially reducing treatment efficiency. Therefore, proper wetland management, including sediment removal and hydrological adjustments, is necessary to sustain long-term nutrient retention.

4.3.5.1 *Mechanisms of organic matter accumulation in constructed wetlands*

4.3.5.1.1 Deposition of plant and microbial biomass

In CWs, organic matter accumulates primarily through plant litterfall, root turnover, and microbial biomass production. Emergent and submerged macrophytes such as *Typha* (cattail), *Phragmites* (common reed), and *Scirpus* (bulrush) contribute significant amounts of organic material as they grow, die, and decompose (Vymazal, 2011). As plant tissues break down, organic particles settle in the substrate, forming organic-rich sediments that act as a nutrient sink. Microbial communities, particularly heterotrophic bacteria and fungi, further contribute to organic matter accumulation by cycling nutrients through their metabolic activities. Decomposition of organic material is slower under anaerobic conditions, leading to progressive accumulation of organic deposits (Reddy and D'Angelo, 1997).

4.3.5.1.2 Peat formation and long-term carbon sequestration

In wetlands with slow decomposition rates, such as those with low oxygen levels, acidic pH, or waterlogged conditions, organic matter can persist for centuries in the form of peat deposits. Peat accumulation occurs when organic material is only partially decomposed because of oxygen limitations, resulting in the buildup of carbon-rich sediments (Kadlec and Wallace, 2009). This process is particularly significant in carbon-rich wetlands, where peat formation contributes to long-term carbon sequestration and climate regulation. In addition to storing carbon, peat deposits trap nitrogen and phosphorus, effectively preventing these nutrients from being released back into the water column.

4.3.5.2 *Effects of organic matter accumulation on wetland function*

Organic matter accumulation plays a crucial role in shaping wetland functionality by influencing nutrient retention, microbial activity, and water treatment efficiency. While moderate levels of organic buildup enhance nutrient cycling and provide habitat for various microorganisms, excessive accumulation can introduce several challenges that may compromise the wetland's ability to perform essential ecological functions effectively (Vymazal, 2011).

One significant impact of excessive organic matter accumulation is the reduction in HRT. As organic sediments build up over time, they can decrease the wetland's overall water storage capacity, limiting the time available for natural filtration and treatment processes to occur. This reduction in retention time can lead to less effective pollutant removal, undermining the wetland's role in water quality improvement, particularly in CWs designed for wastewater treatment (Vymazal, 2011).

Another critical consequence is the development of anoxic conditions, where oxygen availability in water and sediment layers becomes severely limited. Thick organic deposits create oxygen-depleted zones that disrupt microbial processes, leading to increased methane (CH_4) production. Methane is a potent greenhouse gas (GHG) with a much higher global warming potential than carbon dioxide, making its release from wetlands a significant concern in the context of climate change (Reddy and D'Angelo, 1997). The accumulation of organic matter, therefore has implications not only for wetland health but also for GHG emissions and global carbon cycling.

Furthermore, the composition and activity of microbial communities within the wetland can be altered due to high organic content. In oxygen-limited environments, microbial processes shift, favoring anaerobic pathways such as fermentation and sulfate reduction over essential nitrogen removal processes, such as nitrification and denitrification. This shift can reduce the wetland's ability to remove nitrogen efficiently, potentially leading to elevated nitrogen levels in downstream water bodies, which can contribute to eutrophication and other water quality issues (Reddy and D'Angelo, 1997). While organic matter accumulation is a natural and beneficial component of wetland ecosystems, excessive buildup can lead to reduced water treatment efficiency, increased GHG emissions, and altered microbial processes. Effective wetland management strategies should aim to balance organic matter inputs and decomposition rates to maintain optimal wetland function while minimizing potential negative impacts (Reddy and D'Angelo, 1997; Vymazal, 2011).

4.3.5.3 *Management strategies for sustaining organic matter accumulation*

To ensure that CWs maintain optimal nutrient retention while preventing functional decline, several management strategies must be implemented. These strategies focus on controlling organic matter accumulation to sustain wetland efficiency, enhance water treatment performance, and mitigate negative environmental impacts.

4.3.5.3.1 *Sediment and organic matter removal*

Over time, organic matter and sediments accumulate in CWs, reducing hydraulic capacity and altering wetland function. Periodic sediment excavation is a crucial strategy for restoring water storage capacity and maintaining effective wastewater treatment. By removing excess organic material, wetlands can continue to provide efficient nutrient retention without compromising their long-term sustainability. In mature wetlands, where nutrient trapping efficiency has declined because of sediment saturation, more intensive restoration techniques such as dredging or wetland rejuvenation may be necessary. These methods help reset nutrient retention capacity and prolong the functional lifespan of the wetland system (Kadlec and Wallace, 2009).

4.3.5.3.2 *Hydrological management*

Proper hydrological management plays a key role in regulating organic matter dynamics within wetlands. Adjusting water levels and flow rates can optimize oxygen availability, which is essential for microbial processes that break down organic matter. Maintaining adequate water movement helps prevent excessive organic accumulation and reduces the formation of anaerobic conditions that contribute to methane production. In addition, implementing intermittent drying cycles has been found to enhance organic decomposition rates by promoting aerobic microbial activity. Controlled drying phases can help manage peat buildup, ensuring sustained nutrient sequestration while preventing long-term saturation of sediments (Vymazal, 2011).

4.3.5.3.3 *Vegetation and biomass harvesting*

Managing plant biomass is another essential approach to controlling organic matter accumulation in CWs. The removal of above-ground plant material before seasonal die-off can significantly reduce the accumulation of decomposing organic matter in sediments. This practice prevents the release of stored nutrients back into the wetland system, thereby improving long-term nutrient retention and water quality. In addition, selecting high-biomass wetland species that are periodically harvested can enhance nutrient uptake while minimizing organic matter buildup. These species can be strategically integrated into CWs to maximize efficiency, as their regular harvesting helps remove nutrients from the system and limits excessive organic deposition (Reddy and D'Angelo, 1997).

4.4 Constructed wetlands as a solution

CWs offer a nature-based solution for nutrient removal by leveraging biological, chemical, and physical processes to filter and assimilate pollutants. These engineered systems are classified into surface flow (SF) wetlands, SSF wetlands, and hybrid wetlands, each with distinct design features and treatment efficiencies (Wu et al., 2014, 2015). SF wetlands resemble natural marshes, where water flows over a vegetated substrate, facilitating nutrient uptake by macrophytes and microbial communities. In contrast, SSF wetlands direct water through porous media, promoting anaerobic microbial processes such as denitrification, which enhances nitrogen removal (Brix, 1997). Hybrid wetlands integrate both SF and SSF systems to optimize treatment performance across various contaminant loads (Akratos and Tsihrintzis, 2007).

The mechanisms underlying nutrient removal in CWs encompass multiple biogeochemical pathways. Nitrogen is primarily removed through microbial-mediated processes such as nitrification and denitrification. Nitrification, an aerobic process facilitated by AOB, converts ammonium (NH_4^+) to nitrate (NO_3^-), which subsequently undergoes denitrification under anoxic conditions to produce inert nitrogen gas (N_2) (Seitzinger et al., 2006). Phosphorus removal, on the other hand, occurs through adsorption onto substrate materials, precipitation with metal cations, and uptake by wetland vegetation (Reddy et al., 1999). The efficiency of these processes depends on factors such as HRT, substrate composition, plant species selection, and seasonal variations (Zhao et al., 2010a, 2010b, 2010c, 2010d, 2010e, 2013).

4.5 Design considerations for constructed wetlands

The successful implementation of CWs for nutrient removal requires careful consideration of design parameters that influence treatment efficacy. Several key factors, including wetland size, HRT, hydrology, vegetation selection, substrate composition, and climatic conditions, must be optimized to enhance nutrient removal efficiency and ensure the long-term sustainability of the system (Kadlec and Wallace, 2009). Each of these factors plays a vital role in regulating physical, chemical, and biological processes, which govern the removal of nitrogen (N), phosphorus (P), organic matter, and other contaminants from wastewater, stormwater, and agricultural runoff.

4.5.1 Wetland size and hydraulic retention time

The size of a CW is a fundamental design consideration, as it determines treatment capacity, pollutant loading rate, and retention efficiency. Larger wetlands generally provide greater surface area for sedimentation, microbial activity, and plant uptake, improving overall treatment performance (Kadlec and Wallace, 2009). However, the efficiency of nutrient removal is not solely dependent on wetland size but also on HRT, which refers to the duration that water remains within the wetland before exiting the system.

A longer HRT enhances microbial processes, allowing sufficient time for nitrification-denitrification, phosphorus adsorption, and organic matter decomposition (Verhoeven and Meuleman, 1999). Studies have shown that HRTs of 5–10 days are generally effective for nitrogen and phosphorus removal, although higher pollutant loads may require extended retention to achieve optimal treatment levels (Wu et al., 2014, 2015). Proper flow regulation and water level management are essential to prevent short-circuiting, where water bypasses treatment zones and reduces removal efficiency.

4.5.2 Hydrology: Water depth and flow dynamics

The hydrological characteristics of a CW, including water depth, flow patterns, and residence time, significantly influence nutrient removal processes. Wetlands can be designed with SF or SSF configurations, each offering distinct advantages:

- **Surface flow-constructed wetlands (SF-CWs):** These systems maintain a shallow water column (~10–50 cm), allowing for direct plant interaction with the water, atmospheric oxygen exchange, and enhanced sedimentation (Kadlec and Wallace, 2009). However, SF wetlands may have higher evaporative losses and are more susceptible to seasonal temperature variations.
- **Subsurface flow-constructed wetlands (SSF-CWs):** These wetlands direct water through gravel, sand, or porous substrates, minimizing surface exposure while promoting microbial-driven nutrient removal under controlled redox conditions (Wu et al., 2014, 2015). SSF wetlands are more efficient in cold climates, as the subsurface environment provides thermal insulation, maintaining microbial activity during winter months (Zhao et al., 2010a, 2010b, 2010c, 2010d, 2010e, 2013).

A hybrid CW system, combining both SF and SSF designs, can optimize aerobic and anaerobic processes, maximizing nitrogen removal efficiency. Furthermore, water depth adjustments can help regulate oxygen availability, supporting processes such as nitrification, denitrification, and phosphorus precipitation.

4.5.3 Vegetation selection and role in nutrient removal

The selection of wetland vegetation is crucial for nutrient uptake, root oxygenation, and microbial interactions within the rhizosphere. Emergent macrophytes, such as *P. australis* (common reed), *Typha latifolia* (cattail), and *Scirpus* spp. (bulrush), are commonly used in CWs due to their high-biomass productivity, adaptability to waterlogged conditions, and extensive root networks (Brix, 1997). These plants enhance nutrient removal through multiple pathways:

- Direct uptake of nitrogen and phosphorus, storing them in plant tissues.
- Oxygen transport to the root zone, creating microaerophilic conditions that support nitrifying bacteria and enhance organic matter decomposition.
- Rhizodeposition, where organic compounds secreted by roots stimulate microbial activity, facilitating denitrification and phosphorus binding.

To maintain optimal treatment performance, periodic harvesting of above-ground biomass is recommended to prevent nutrient recycling and improve system efficiency (Reddy and D'Angelo, 1997). The effectiveness of plant species varies depending on climate, water chemistry, and wetland design, requiring careful selection based on local environmental conditions.

4.5.4 Substrate selection and phosphorus retention

The composition of the substrate in CWs plays a crucial role in phosphorus retention by influencing adsorption capacity, microbial colonization, and hydraulic conductivity. Various substrate materials are commonly used to optimize nutrient removal efficiency. Gravel and sand serve as a stable medium for plant anchoring and microbial attachment, thereby enhancing biological nutrient removal through biofilm formation and microbial activity (Molle et al., 2005a, 2005b, 2006). In addition, calcium-rich materials such as limestone and crushed shells contribute to phosphorus precipitation by forming insoluble calcium phosphate compounds, which are resistant to leaching and help reduce phosphorus concentrations in wetland effluent. Similarly, iron- and aluminum-rich substrates facilitate phosphorus adsorption through the formation of Fe–P and Al–P complexes, effectively reducing phosphorus availability in the water column and improving overall water quality (Vymazal, 2011). The selection of an appropriate substrate is essential for maximizing phosphorus retention and ensuring long-term wetland performance.

Substrate longevity is an important consideration, as binding sites for phosphorus can become saturated over time, reducing treatment efficiency. Periodic substrate replacement or rejuvenation through lime or iron amendments can help sustain long-term phosphorus removal capacity.

4.5.5 Influence of climate and seasonal variability

Climatic conditions, particularly temperature fluctuations, seasonal plant growth cycles, and hydrological variations, play a crucial role in determining the overall performance and efficiency of CWs. These factors influence key biogeochemical processes, microbial activity, and nutrient removal efficiency, making it essential to incorporate adaptive management strategies to optimize CW function under varying climatic conditions.

Temperature is one of the most significant climatic factors affecting CW performance. Microbial activity, which is responsible for essential processes such as nitrification, denitrification, and phosphorus precipitation, is highly temperature-dependent. Higher temperatures enhance microbial metabolism, increasing the rates of nitrogen and phosphorus removal. Conversely, colder temperatures slow down these biological and chemical processes, reducing treatment efficiency and potentially leading to nutrient accumulation or inadequate wastewater treatment (Zhao et al., 2010a, 2010b, 2010c, 2010d, 2010e, 2013). The impact of temperature is particularly pronounced in regions with strong seasonal variations, where winter conditions can significantly reduce the effectiveness of CWs.

In addition to temperature, seasonal plant growth cycles also influence wetland performance. Vegetation plays a vital role in nutrient uptake, oxygen transfer to root zones, and organic matter stabilization. During the growing season, plants actively absorb nutrients, enhancing the removal of nitrogen and phosphorus from wastewater. However, in colder months, plant growth slows, leading to reduced nutrient uptake and an increased likelihood of organic matter accumulation. To counteract these seasonal shifts, CWs can be designed with a mix of plant species that have different growth cycles, ensuring continuous nutrient uptake throughout the year. A diverse plant community also enhances habitat resilience and supports microbial diversity, further stabilizing wetland function (Zhao et al., 2010a, 2010b, 2010c, 2010d, 2010e, 2013).

Hydrological variations, including changes in precipitation and water flow rates, further influence CW performance. Seasonal fluctuations in water availability can impact HRT, affecting the contact time between pollutants and treatment surfaces. Excessive rainfall or flooding may lead to short-circuiting, where water bypasses treatment zones, reducing pollutant removal efficiency. On the other hand, prolonged dry periods can limit microbial activity and nutrient processing. To mitigate these effects, CWs can incorporate adaptive hydrological management strategies, such as adjusting hydraulic loading rates seasonally. By modifying water flow rates in response to seasonal changes, wetland managers can maintain optimal conditions for microbial and plant activity, ensuring consistent pollutant removal throughout the year (Zhao et al., 2010a, 2010b, 2010c, 2010d, 2010e, 2013).

4.5.5.1 *Management strategies to mitigate seasonal impacts*

To enhance CW resilience against climate and seasonal variability, several key strategies can be implemented:

Designing deeper subsurface flow wetlands to insulate microbial communities from temperature extremes, maintaining stable biological activity even during colder months.
Incorporating diverse plant species with varying seasonal growth patterns to ensure continuous nutrient uptake and system stability throughout the year.
Adjusting hydraulic loading rates seasonally to compensate for fluctuations in microbial activity and plant productivity, ensuring optimal retention times and treatment efficiency.

4.5.6 Adaptive management strategies

To ensure the long-term treatment efficiency of CWs, adaptive management strategies should be implemented to maintain system functionality and nutrient removal performance. Regular performance monitoring is essential to assess nutrient removal rates and overall system health, allowing for timely interventions when treatment efficiency declines (Kadlec and Wallace, 2009). Periodic plant harvesting and biomass removal help prevent nutrient recycling by extracting accumulated nitrogen and phosphorus from the system, thereby enhancing long-term nutrient retention (Brix, 1997). In addition, sediment and substrate maintenance is necessary to sustain hydraulic function and phosphorus adsorption capacity, preventing clogging and ensuring optimal water flow through the wetland (Vymazal, 2011). Flow regulation adjustments are also crucial, as they help optimize aerobic and anaerobic conditions, which are vital for microbial nutrient transformations such as nitrification, denitrification, and phosphorus sequestration (Verhoeven et al., 2006). By implementing these adaptive management strategies, CWs can continue to provide effective wastewater treatment and ecosystem services over extended operational periods.

4.6 Effectiveness and challenges of constructed wetlands

CWs have gained recognition as an effective and sustainable method for nutrient removal from wastewater, stormwater, and agricultural runoff. By utilizing physical, chemical, and biological mechanisms, CWs significantly reduce nitrogen (N) and phosphorus (P) concentrations, mitigating the risk of eutrophication in receiving water bodies. Research indicates that well-designed CWs can achieve nitrogen removal efficiencies ranging from 50% to 90% and phosphorus removal efficiencies between 40% and 80%, depending on system configuration, HRT, and operational conditions (Kadlec and Wallace, 2009). Despite their proven efficacy,

CWs also face design, operational, and environmental challenges that may limit their long-term performance and widespread adoption. Addressing these challenges requires ongoing research, adaptive management strategies, and integration with complementary treatment technologies.

4.6.1 Effectiveness of constructed wetlands in nutrient removal

4.6.1.1 Nitrogen removal

Nitrogen removal in CWs is achieved through a combination of sedimentation, microbial transformation, plant uptake, and organic matter accumulation. The primary microbial pathways responsible for nitrogen reduction include nitrification, denitrification, and anaerobic ammonium oxidation (anammox) (Vymazal, 2011). In aerobic zones, ammonium (NH_4^+) is converted into nitrate (NO_3^-) through nitrification, while in anaerobic zones, denitrifying bacteria reduce nitrate to gaseous nitrogen (N_2), which is released into the atmosphere. Studies have shown that SSF-CWs tend to achieve higher nitrogen removal efficiencies than SF systems because of better oxygen distribution and enhanced microbial interactions (Wu et al., 2014, 2015).

The efficiency of nitrogen removal in CWs is governed by several interrelated factors that influence microbial activity, biochemical processes, and overall system performance. Key parameters such as HRT, oxygen availability, and organic carbon supply play crucial roles in determining the extent to which nitrogen is transformed and removed from wastewater. Understanding these factors allows for the optimization of CW design and management to enhance nitrogen removal efficiency.

4.6.1.1.1 Factors influencing the effectiveness of nitrogen removal

4.6.1.1.1.1 Hydraulic retention time

One of the primary factors affecting nitrogen removal is HRT, which refers to the duration wastewater remains within the wetland system. Longer retention times provide more opportunities for microbial communities to process nitrogen compounds through nitrification and denitrification, leading to improved removal efficiency. Extended HRT allows for the gradual transformation of ammonium (NH_4^+) into nitrate (NO_3^-) via nitrification, followed by the reduction of nitrate to nitrogen gas (N_2) in anoxic conditions through denitrification. However, excessively long retention times can lead to organic matter buildup, anoxic conditions, and potential release of unwanted byproducts such as methane (CH_4) (Verhoeven and Meuleman, 1999).

4.6.1.1.1.2 Oxygen availability

It directly influences the balance between nitrification and denitrification. Nitrification, the process of converting ammonium into nitrate, is an aerobic reaction that requires sufficient oxygen supply. In contrast, denitrification—the conversion of nitrate into nitrogen gas—occurs under anoxic conditions, where oxygen is absent or limited. Maintaining a strategic balance between aerobic and anoxic zones within CWs is essential to optimize both processes. SSF wetlands, for example, can incorporate design elements such as aeration zones or alternating flow conditions to enhance oxygen diffusion and support effective nitrogen cycling. The availability of oxygen thus dictates the efficiency of nitrogen transformations within the system (Kadlec and Wallace, 2009).

4.6.1.1.1.3 Organic carbon supply

Denitrification is a microbial process that relies on an adequate supply of organic carbon as an electron donor to facilitate the conversion of nitrate into nitrogen gas. In cases where organic carbon availability is insufficient, denitrification efficiency can be significantly reduced, leading to incomplete nitrogen removal. Wetlands receiving wastewater with low organic carbon content may require additional carbon inputs, such as plant-derived organic matter or external carbon amendments, to sustain microbial activity. Natural sources of carbon, such as wetland vegetation, can also contribute to maintaining optimal conditions for denitrification while supporting overall ecosystem function (Kadlec and Wallace, 2009).

4.6.1.2 Phosphorus removal

Phosphorus removal in CWs primarily occurs through sedimentation, adsorption, and precipitation. Phosphorus binds to iron (Fe), aluminum (Al), and calcium (Ca) minerals in the wetland substrate, forming stable precipitates that are less bioavailable. However, substrate selection plays a crucial role in maintaining phosphorus removal efficiency. Materials such as limestone, crushed shells, and iron-rich soils enhance phosphorus binding, preventing its release back into the water column (Reddy et al., 1999; Vymazal, 2011).

Phosphorus (P) removal in CWs is governed by a combination of physical, chemical, and biological processes. The efficiency of phosphorus retention and long-term sequestration depends on several key factors, including substrate composition and adsorption capacity, wetland design, and plant uptake with biomass harvesting. Understanding these factors is crucial for optimizing wetland performance and preventing phosphorus release into downstream water bodies, which can contribute to eutrophication and water quality degradation.

One of the most important determinants of phosphorus removal efficiency is substrate composition and adsorption capacity. Phosphorus is primarily removed from wastewater through adsorption onto wetland substrates, including soil, gravel, and reactive materials such as limestone or iron-rich media. The effectiveness of this process depends on the availability of binding sites within the substrate, which gradually become saturated over time. Once the adsorption capacity is reached, phosphorus removal efficiency declines, potentially leading to increased phosphorus release back into the water column. To enhance long-term phosphorus retention, substrates with high adsorption potential, such as materials rich in aluminum (Al), calcium (Ca), or iron (Fe), can be incorporated into CW designs (Kadlec and Wallace, 2009).

Wetland design also plays a critical role in phosphorus removal by influencing the interaction between phosphorus and reactive substrates. Shallow water depths promote greater sedimentation and increase the contact time between phosphorus and adsorption surfaces, enhancing phosphorus retention. In addition, extended HRT allows for prolonged interactions with reactive materials, improving overall phosphorus removal efficiency. SSF wetlands, which promote greater filtration through substrates, often demonstrate enhanced phosphorus removal compared with SF wetlands, where phosphorus may remain in suspension for longer periods (Vymazal, 2011). Proper design considerations, including flow path configuration and sedimentation areas, can help optimize phosphorus sequestration within the wetland system.

The role of vegetation in phosphorus uptake and retention is another key factor influencing removal efficiency. Wetland plants absorb phosphorus during growth, temporarily storing it in their tissues. However, unlike nitrogen, phosphorus does not readily volatilize and remains within the system unless it is physically removed. As plants undergo senescence and decay, stored phosphorus can be rereleased into the wetland, potentially reducing long-term removal effectiveness. To prevent phosphorus release from decaying biomass, harvesting of wetland vegetation is necessary. By periodically removing plant biomass, stored phosphorus is effectively extracted from the system, reducing the risk of remobilization into the water column (Reddy and D'Angelo, 1997). Selecting high-biomass plant species with efficient phosphorus uptake can further enhance this process and contribute to improved long-term phosphorus management.

4.6.1.3 Field applications and benefits

CWs have been widely implemented in various environmental and industrial applications due to their ability to effectively treat wastewater while providing ecological benefits. These natural treatment systems offer a cost-effective, low-energy alternative to conventional wastewater treatment technologies and are particularly valuable in addressing nutrient pollution, contaminant removal, and ecosystem restoration. Key applications of CWs include their use in agricultural drainage systems, urban stormwater management, and industrial wastewater treatment.

One of the most important applications of CWs is in agricultural drainage systems, where they play a crucial role in reducing fertilizer runoff and preventing nutrient pollution in rivers, lakes, and coastal waters. Agricultural runoff is often rich in nitrogen and phosphorus from fertilizers, which can contribute to eutrophication and HABs. CWs act as natural filters by trapping sediments, facilitating nutrient uptake by vegetation, and promoting microbial processes such as nitrification and denitrification. By intercepting and treating drainage water before it enters surface water bodies, CWs help protect water quality while also enhancing soil moisture retention and agricultural sustainability (Kadlec and Wallace, 2009).

CWs are also widely used in urban stormwater management, where they help mitigate the impacts of stormwater runoff from roads, rooftops, and urban landscapes. During heavy rainfall, stormwater carries a variety of contaminants, including heavy metals, oils, sediments, and nutrients, which can degrade aquatic ecosystems. CWs provide a natural solution by slowing down runoff, allowing pollutants to settle, and promoting biological uptake and microbial degradation of contaminants. By integrating CWs into urban green infrastructure, cities can improve water quality, reduce flood risks, and enhance urban biodiversity (Vymazal, 2011).

Another significant application of CWs is in industrial wastewater treatment, particularly for industries that produce nutrient-rich effluents, such as food processing, livestock operations, and mining activities. These industries generate large volumes of wastewater containing organic matter, nitrogen, phosphorus, and heavy metals, which require effective treatment before discharge. CWs provide an efficient and sustainable treatment method by utilizing natural processes to break down organic pollutants, remove excess nutrients, and immobilize heavy metals. Industries adopting CW technology benefit from reduced energy and maintenance costs compared to traditional treatment plants while also contributing to environmental sustainability (Kadlec and Wallace, 2009).

4.6.2 Challenges of constructed wetlands

Despite their advantages, CWs face several challenges that may affect their long-term efficiency and practicality in large-scale applications.

4.6.2.1 *Land requirement*

One of the primary limitations of CWs is their space requirement. Compared to traditional wastewater treatment plants, CWs require larger land areas to achieve similar nutrient removal efficiencies (Verhoeven and Meuleman, 1999). This limitation is particularly challenging in urban areas, where land availability is restricted and expensive. To address space constraints in CWs, several strategies can be implemented to optimize treatment efficiency within a limited footprint. One effective approach is the use of hybrid wetland systems, which integrate CWs with other treatment technologies such as biofilters or activated sludge systems. This combination enhances pollutant removal while reducing land requirements (Kadlec and Wallace, 2009). Another strategy involves the adoption of vertical-flow wetlands, which require less horizontal space compared with traditional horizontal-flow systems while maintaining high treatment efficiency by promoting oxygen transfer and microbial activity (Vymazal, 2011). In addition, optimizing hydraulic loading rates allows for the maximization of nutrient removal per unit area, ensuring that the available space is utilized effectively without compromising treatment performance (Tanner, 2001). Implementing these strategies can help overcome spatial limitations while maintaining the effectiveness of wetland-based wastewater treatment systems.

4.6.2.2 *Declining phosphorus retention capacity*

One of the primary challenges associated with phosphorus (P) removal in CWs is the gradual decline in phosphorus adsorption capacity over time. Phosphorus is primarily removed through adsorption onto wetland substrates, including soil, gravel, and reactive media. However, as these materials reach saturation, their ability to retain phosphorus diminishes, potentially leading to reduced removal efficiency and even phosphorus release back into the water column under changing environmental conditions (Reddy et al., 1999). Without proper management and periodic interventions, CWs may become less effective at controlling phosphorus levels in wastewater.

4.6.2.2.1 Factors contributing to phosphorus saturation

The decline in phosphorus retention capacity is mainly because of the saturation of binding sites within the wetland substrate. This occurs when the available adsorption sites on materials such as iron (Fe), calcium (Ca), and aluminum (Al) compounds become fully occupied by phosphate ions over extended periods of operation. Once saturation is reached, any additional phosphorus introduced into the system may remain in the water column, increasing the risk of downstream eutrophication. In addition, certain environmental conditions—such as changes in pH, redox potential, or temperature fluctuations—can further influence phosphorus binding efficiency and, in some cases, trigger the desorption and release of previously retained phosphorus (Vymazal, 2011).

4.6.2.2.2 Management strategies to sustain phosphorus removal

To maintain long-term phosphorus removal efficiency in CWs, several strategies can be implemented to counteract the effects of substrate saturation:

- **Periodic replacement of substrate materials**: Replacing or supplementing phosphorus-saturated substrates with fresh materials enriched in calcium, iron, or aluminum can restore phosphorus adsorption capacity. Common reactive media used for phosphorus binding include limestone, crushed shells, and iron-rich sands. By periodically adding or replacing these materials, CWs can sustain their long-term effectiveness in phosphorus removal (Reddy et al., 1999).
- **Utilizing reactive filter materials**: Incorporating specialized phosphorus-binding materials into CW designs can enhance phosphorus retention. Lightweight expanded clay aggregates (LECA), slag, and engineered reactive media have been shown to improve phosphorus adsorption due to their high porosity and surface reactivity. These materials can be integrated into wetland substrates or used in designated filter zones to extend phosphorus removal capacity (Kadlec and Wallace, 2009).
- **Regular harvesting of wetland plants**: Wetland vegetation plays an important role in phosphorus uptake, temporarily storing it in plant tissues. However, if plants are allowed to decay within the system, stored phosphorus may be released back into the water. Regular harvesting of biomass, particularly high-phosphorus-accumulating species, can effectively remove phosphorus from the system and prevent rerelease, thereby contributing to sustained phosphorus management (Vymazal, 2011).

4.6.2.3 Impact of extreme weather events

Climate variability and extreme weather events, including heavy rainfall, prolonged droughts, and temperature fluctuations, can significantly influence the hydrology and treatment efficiency of CWs. These environmental factors affect key processes such as nutrient retention, microbial activity, and HRT, ultimately determining the long-term performance of CW systems. As climate change intensifies, it is essential to understand these impacts and implement adaptive wetland management strategies to enhance system resilience.

4.6.2.3.1 Effects of extreme weather on wetland function

- **Heavy rainfall and hydraulic short-circuiting**: Intense rainfall events can lead to hydraulic short-circuiting, where large volumes of water pass through the wetland too quickly, reducing HRT and lowering nutrient removal efficiency. Under these conditions, phosphorus and nitrogen may be washed out before adequate microbial or plant-based treatment can occur, leading to potential downstream pollution. In addition, excessive surface runoff can erode wetland substrates and disturb established vegetation (Kadlec and Wallace, 2009).
- **Prolonged droughts and reduced microbial activity**: Extended dry periods can desiccate wetland substrates, leading to a decline in microbial populations responsible for essential nutrient transformation processes. Denitrification and phosphorus adsorption are particularly sensitive to moisture levels, and drying can result in the oxidation of previously stored nutrients, increasing the risk of nutrient release once rehydration occurs. In addition,

droughts can alter plant community composition, favoring species less effective in nutrient uptake and retention (Vymazal, 2011).

- **Cold temperatures and decreased nitrogen removal efficiency**: Low temperatures slow down microbial metabolism, significantly affecting nitrification and denitrification processes, which are crucial for nitrogen removal. At temperatures below 10°C, nitrifying bacteria exhibit reduced efficiency, leading to increased ammonia and nitrate concentrations in wetland effluents (Zhao et al., 2010a, 2010b, 2010c, 2010d, 2010e, 2013). Cold stress can also impact plant uptake rates and decomposition dynamics, further affecting overall treatment performance.

4.6.2.3.2 Adaptation strategies for climate resilience

To mitigate the impacts of extreme weather and enhance the resilience of CWs, several design and operational modifications can be implemented:

- **Adjustable water level controls**: Incorporating water level management structures, such as adjustable weirs or automated flow regulators, can help manage seasonal fluctuations in water availability. These controls allow wetlands to retain stormwater during heavy rainfall while maintaining optimal water levels during drought conditions.
- **Hybrid wetland systems**: Combining SSF and SF wetlands improves system resilience by balancing water retention, microbial activity, and plant-based treatment. SSF wetlands offer protection against cold temperature impacts by insulating microbial communities, while SF wetlands provide flexibility in handling variable hydrological conditions (Kadlec and Wallace, 2009).
- **Diverse vegetation selection**: Planting a mix of seasonally adaptive wetland species enhances year-round treatment efficiency. Deep-rooted species help stabilize substrates against erosion, while cold-tolerant plants ensure nutrient uptake continues during winter months. In addition, selecting flood- and drought-resistant vegetation can improve wetland functionality under extreme weather conditions (Vymazal, 2011).

4.6.2.4 Potential for greenhouse gas emissions

While CWs offer significant environmental benefits, including nutrient removal and water quality improvement, they also have the potential to contribute to GHG emissions. The primary GHGs emitted from CWs are methane (CH_4) and nitrous oxide (N_2O), both of which have significantly higher global warming potentials than carbon dioxide (CO_2). Methane is 25 times more potent than CO_2, while nitrous oxide has a global warming potential approximately 298 times greater over a 100-year period (Wu et al., 2014, 2015). Understanding the mechanisms behind these emissions and implementing appropriate mitigation strategies is essential for maximizing the sustainability of CW systems.

4.6.2.4.1 Sources of greenhouse gas emissions in constructed wetlands

- **Methane (CH_4) production in anaerobic zones**: CH_4 emissions primarily occur in the oxygen-depleted (anaerobic) zones of CWs, where microbial decomposition of organic matter leads to methanogenesis. This process is especially prevalent in SF wetlands with deep organic-rich sediments, where conditions favor methanogenic bacteria over methane-oxidizing microbes. Factors such as excessive organic loading, stagnant water conditions, and high temperatures further promote methane generation (Vymazal, 2011).

- **Nitrous oxide (N_2O) from incomplete denitrification**: N_2O emissions result from the incomplete microbial reduction of nitrate (NO_3^-) to nitrogen gas (N_2) during denitrification. When oxygen levels fluctuate between aerobic and anaerobic conditions, denitrifying bacteria may release N_2O as an intermediate byproduct instead of fully converting nitrogen compounds into harmless dinitrogen gas. High N_2O emissions are commonly associated with low carbon-to-nitrogen (C:N) ratios, fluctuating water levels, and excessive nitrogen inputs (Wu et al., 2014, 2015).

4.6.2.4.2 Strategies to minimize greenhouse gas emissions

To reduce the carbon footprint of CWs, management strategies should focus on optimizing hydrology, plant selection, and nutrient inputs to create conditions that minimize CH_4 and N_2O production while maintaining treatment efficiency:

- **Optimizing aeration and flow patterns**: Enhancing oxygen availability in CWs can suppress methanogenesis while promoting complete denitrification, thereby reducing CH_4 and N_2O emissions. Strategies such as shallow water depths, intermittent drainage, and aerated flow-through designs can improve oxygen diffusion, shifting microbial processes toward CO_2-dominated respiration instead of methane production (Kadlec and Wallace, 2009).
- **Selecting Plant Species That Enhance Oxygen Diffusion**: Certain wetland plant species, such as *P. australis* (common reed), *Typha* spp. (cattails), and *Scirpus* spp. (bulrushes), can transport atmospheric oxygen to their root zones, enhancing aerobic microbial activity in sediments. These plants help create microoxic zones that reduce CH_4 emissions while supporting efficient nitrogen removal through complete denitrification pathways (Vymazal, 2011).
- **Balancing carbon and nitrogen inputs**: Excessive organic carbon loading, such as from agricultural runoff or high-strength wastewater, can promote methane-producing conditions. Managing C:N ratios by controlling nutrient inputs and integrating carbon sources that favor complete denitrification (e.g., wood chips in SSF wetlands) can help reduce N_2O emissions while maintaining nitrogen removal efficiency (Wu et al., 2014, 2015).

4.6.3 Addressing challenges: Future research and innovations

CWs have demonstrated their effectiveness in wastewater treatment, offering sustainable and cost-efficient alternatives to conventional treatment systems. However, several challenges remain, including declining phosphorus retention, GHG emissions, and performance variability because of climatic factors. Ongoing research and technological advancements are essential to address these issues and further enhance CW efficiency. Emerging innovations focus on improving nutrient removal, optimizing hydrological processes, and integrating smart monitoring systems.

4.6.3.1 Key innovations in constructed wetland technology

Engineered substrates for enhanced phosphorus adsorption: Traditional wetland substrates lose their phosphorus adsorption capacity over time, leading to saturation and potential phosphorus release. Engineered materials such as LECA, iron- and aluminum-enriched substrates, and biochar-based media offer improved phosphorus binding and longevity. These materials enhance adsorption efficiency while maintaining hydraulic conductivity for optimal wetland function (Kadlec and Wallace, 2009).

Integration of CWs with advanced treatment technologies: Hybrid CW systems combine natural treatment processes with engineered enhancements to improve pollutant removal efficiency. Examples include:

Biochar filters: Improve phosphorus and heavy metal retention while providing additional carbon sources for microbial denitrification.

Electrocoagulation: Uses an electric field to precipitate phosphorus and organic matter, enhancing removal efficiency in high-load wastewater (Wu et al., 2014, 2015).

Membrane bioreactors (MBRs): Combining CWs with MBRs improves water quality by increasing microbial degradation of contaminants (Vymazal, 2011).

Microbial inoculants for enhanced nutrient removal: Microbial communities play a critical role in nitrogen and phosphorus transformation within CWs. Research has focused on introducing specialized microbial inoculants, such as:

Denitrifying bacteria (e.g., *Paracoccus denitrificans*) to enhance nitrogen removal.

Phosphorus-accumulating organisms (PAOs) to improve phosphorus uptake and storage.

Methanotrophic bacteria reduce methane emissions by oxidizing CH_4 before it reaches the atmosphere (Wu et al., 2014, 2015).

Remote monitoring and AI-based optimization: Advances in sensor technology and artificial intelligence (AI) enable real-time monitoring of CW performance, allowing adaptive management strategies to be implemented. Smart monitoring systems utilize:

- Internet of Things (IoT) sensors for tracking water quality parameters (e.g., dissolved oxygen, pH, and nutrient concentrations).
- Machine learning algorithms to predict and optimize CW function under variable conditions.
- Automated flow control mechanisms to dynamically adjust water levels and HRTs based on real-time environmental data.

4.7 Conclusion

CWs represent a highly effective, environmentally sustainable, and cost-efficient approach to mitigating nutrient pollution in aquatic ecosystems. Through a combination of physical, chemical, and biological processes, CWs efficiently remove excess nitrogen and phosphorus from wastewater, stormwater, and agricultural runoff. These systems leverage natural filtration mechanisms such as sedimentation, adsorption, microbial transformations, and plant uptake to improve water quality while simultaneously supporting biodiversity, carbon sequestration, and habitat restoration. One of the most significant advantages of CWs is their ability to replicate and enhance natural wetland functions, making them a viable solution for both rural and urban settings. In agricultural landscapes, CWs play a critical role in intercepting and treating nutrient-rich runoff, preventing excess fertilizers from reaching rivers and lakes. Similarly, in urban environments, CWs help manage stormwater runoff, reducing the burden on conventional wastewater treatment facilities while mitigating flood risks. In addition, CWs are widely employed in industrial wastewater treatment, offering an energy-efficient alternative to mechanical and chemical-based treatment systems.

Despite their effectiveness, CWs face several challenges that must be addressed to maximize their long-term sustainability. One of the primary concerns is land availability, as CWs require significant space compared to traditional wastewater treatment plants. This limitation can be mitigated by adopting space-efficient designs such as hybrid wetlands or vertical-flow systems, which optimize land use while maintaining high treatment efficiency. Another challenge is the gradual decline in phosphorus retention capacity because of substrate saturation. Periodic substrate replacement, the incorporation of high-capacity phosphorus-binding materials, and regular plant harvesting can help sustain phosphorus removal efficiency over time. Climate variability also poses a challenge to CW performance, as extreme weather events such as heavy rainfall and prolonged droughts can impact HRTs and microbial activity. Implementing adaptive water level management strategies and selecting resilient plant species can enhance the robustness of CWs under changing climatic conditions. In addition, GHG emissions from CWs, particularly methane (CH_4) and nitrous oxide (N_2O), must be carefully managed. Optimizing aeration, selecting plant species that facilitate oxygen diffusion, and maintaining a balanced carbon-to-nitrogen ratio can significantly reduce emissions while ensuring efficient nutrient removal.

CWs provide a promising and sustainable solution to one of the most pressing environmental issues—nutrient pollution. By leveraging natural processes to remove pollutants, these systems not only safeguard water quality but also promote ecological health and climate resilience. Addressing the challenges associated with CW implementation through innovative design, adaptive management, and scientific advancements will further cement their role as a cornerstone of sustainable water management practices worldwide. CWs represent a cost-effective, environmentally sustainable solution for nutrient pollution mitigation. By leveraging natural biogeochemical processes, CWs improve water quality while supporting biodiversity and ecosystem resilience. Continued research and technological advancements will further optimize their effectiveness in sustainable water management.

References

Akratos, C.S., Tsihrintzis, V.A., 2007. Effect of temperature, HRT, vegetation and porous media on removal efficiency of pilot-scale horizontal subsurface flow constructed wetlands. Ecol. Eng. 29 (2), 173–191. doi:10.1016/j.ecoleng.2006.06.013.

Brix, H., 1997. Do macrophytes play a role in constructed treatment wetlands? Water Sci. Technol. 35 (5), 11–17. doi:10.2166/wst.1997.0154.

Brooks, B.W., Lazorchak, J.M., Howard, M.D.A., Johnson, M.V.V., Morton, S.L., Perkins, D.A.K., Reavie, E.D., Scott, G.I., Smith, S.A., Steevens, J.A., 2016. Are harmful algal blooms becoming the greatest inland water quality threat to public health and aquatic ecosystems? Environ. Toxicol. Chem. 35 (1), 6–13. doi:10.1002/etc.3220.

Carpenter, S.R., Caraco, N.F., Correll, D.L., Howarth, R.W., Sharpley, A.N., Smith, V.H., 1998. Nonpoint pollution of surface waters with phosphorus and nitrogen. Ecol. Appl. 8 (3), 559–568. doi:10.1890/1051-0761(1998)008[0559:nposww]2.0.co;2.

Conley, D.J., Paerl, H.W., Howarth, R.W., Boesch, D.F., Seitzinger, S.P., Havens, K.E., Lancelot, C., Likens, G.E., 2009. Ecology - Controlling eutrophication: nitrogen and phosphorus. Science 323 (5917), 1014–1015. doi:10.1126/science.1167755.

Das, P., Das, B. K., Meena, D. K., Singh, S. K., Hoque, M., Majhi, S. K., 2025a. Harmful algal biomass vis-à-vis aquatic ecosystem health. Food Sci. Rep. 6 (12), 56–63. https://foodandscientificreports.com/details/december-2025.html.

Das, P., Hoque, M., Saikia, B., Borah, S., Majhi, S. K., 2025b. Toxic Trophic Transfers: Biomagnification and Bioaccumulation in Aquatic Ecosystems. Food Sci. Rep. 6(7), 82-92.https://foodandscientificreports.com/details/july-2025.html.

Diaz, R.J., Rosenberg, R., 2008. Spreading dead zones and consequences for marine ecosystems. Science 321 (5891), 926–929. doi:10.1126/science.1156401.

Jarvie, H.P., Sharpley, A.N., Withers, P.J., Scott, J.T., Haggard, B.E., Neal, C., 2015. Phosphorus mitigation to control river eutrophication: murky waters, inconvenient truths, and "postnormal" science. J. Environ. Qual. 44 (2), 1–27.

Kadlec, R.H., Wallace, S.D., 2009. Treatment Wetlands, second ed. CRC Press, Boca Raton, FL. https://doi.org/10.1201/9781420012514.

Mitsch, W.J., Gosselink, J.G., 2015. Wetlands, fifth ed. Wiley, Hoboken. https://www.wiley.com/en-gb/Wetlands%2C%2B5th%2BEdition-p-9781119019787.

Molle, P., Liénard, A., Grasmick, A., Iwema, A., 2006. Effect of reeds and feeding operations on hydraulic behaviour of vertical flow constructed wetlands under hydraulic overloads. Water Res. 40 (3), 606–612. doi:10.1016/j.watres.2005.11.026.

Molle, P., Liénard, A., Boutin, C., Merlin, G., Iwema, A., 2005a. How to treat raw sewage with constructed wetlands: an overview of the French systems. Water Sci. Technol. 51 (9), 11–21. doi:10.2166/wst.2005.0277.

Molle, P., Liénard, A., Grasmick, A., Iwema, A., 2005b. Effect of reeds and gravel size on hydraulic and treatment behaviour of vertical flow constructed wetlands. Water Sci. Technol. 51 (9), 75–82.

Paerl, H.W., Otten, T.G., 2013. Harmful cyanobacterial blooms: causes, consequences, and controls. Microb. Ecol. 65 (4), 995–1010. doi:10.1007/s00248-012-0159-y.

Reddy, K.R., Kadlec, R.H., Flaig, E., Gale, P.M., 1999. Phosphorus retention in streams and wetlands: a review. Crit. Rev. Environ. Sci. Technol. 29 (1), 83–146. doi:10.1080/10643389991259182.

Reddy, K.R., D'Angelo, E.M., 1997. Biogeochemical indicators to evaluate pollutant removal efficiency in constructed wetlands. Water Sci. Technol. 35 (5), 1–10. doi:10.2166/wst.1997.0152.

Seitzinger, S., Harrison, J.A., Böhlke, J.K., Bouwman, A.F., Lowrance, R., Peterson, B., Tobias, C., Van Drecht, G., 2006. Denitrification across landscapes and waterscapes: a synthesis. Ecol. Appl. 16 (6), 2064–2090. doi:10.1890/1051-0761(2006)016[2064:dalawa]2.0.co;2.

Sharpley, A., Jarvie, H.P., Buda, A., May, L., Spears, B., Kleinman, P., 2013. Phosphorus legacy: overcoming the effects of past management practices to mitigate future water quality impairment. J. Environ. Qual. 42 (5), 1308–1326. doi:10.2134/jeq2013.03.0098.

Smith, V.H., Schindler, D.W., 2009. Eutrophication science: where do we go from here? Trends Ecol. Evol. 24 (4), 201–207. doi:10.1016/j.tree.2008.11.009.

Tanner, C.C., 2001. Plants as ecosystem engineers in subsurface-flow treatment wetlands. Water Sci. Technol. 44 (11–12), 9–17. doi:10.2166/wst.2001.0804.

Verhoeven, J.T.A., Arheimer, B., Yin, C., Hefting, M.M., 2006. Regional and global concerns over wetlands and water quality. Trends Ecol. Evol. 21 (2), 96–103. doi:10.1016/j.tree.2005.11.015.

Verhoeven, J.T.A., Meuleman, A.F.M., 1999. Wetlands for wastewater treatment: opportunities and limitations. Ecol. Eng. 12 (1–2), 5–12. doi:10.1016/S0925-8574(98)00050-0.

Vymazal, J., 2011. Constructed wetlands for wastewater treatment: five decades of experience. Environ. Sci. Technol. 45 (1), 61–69. doi:10.1021/es101403q.

Wu, S., Wallace, S., Brix, H., Kuschk, P., Kirui, W.K., Masi, F., Dong, R., 2015. Treatment of industrial effluents in constructed wetlands: challenges, operational strategies and overall performance. Environ. Pollut. 201, 107–120.

Wu, S., Wallace, S., Brix, H., Kuschk, P., Kirui, W.K., Masi, F., Dong, R., 2014. Treatment of industrial effluents in constructed wetlands. Ecol. Eng. 70, 244–254.

Zhao, Y., Collum, S., Phelan, M., Goodbody, T., Doherty, L., Hu, Y., 2013. Preliminary investigation of constructed wetland incorporating microbial fuel cell: batch and continuous flow trials. Chem. Eng. J. 229, 364–370. doi:10.1016/j.cej.2013.06.023.

Zhao, Y., Collum, S., Phelan, M., Goodbody, T., 2010a. Assessing the influence of design and environmental parameters on the performance of constructed wetlands. Water Res. 44 (12), 3763–3770.
Zhao, Y., Liu, S., Zhang, W., 2010b. Effects of temperature on nitrogen removal efficiency and bacterial community composition in subsurface flow constructed wetlands. Ecol. Eng. 36 (12), 1807–1814.
Zhao, Y., Liu, S., Liu, Y., Guo, L., Chen, F., 2010c. Effects of temperature on nitrogen removal efficiency and microbiological characteristics in vertical flow constructed wetlands. Ecol. Eng. 36 (12), 1789–1794.
Zhao, Y., Liu, S., Quan, X., Zhang, Y., 2010d. Effect of plant harvest on nitrogen removal in constructed wetlands treating polluted river water. J. Environ. Sci. 22 (10), 1604–1610.
Zhao, Y., Xia, S., Yang, L., Samson, R., 2010e. Role of constructed wetlands in alleviating climate change through carbon sequestration: a review. Int. J. Environ. Res. Public Health 7 (7), 1–15.

CHAPTER

5

Aquaponics: A sustainable system for producing plants and aquatic organisms

Enric Gisbert[a], *Cristóbal Aguilera*[a] *and Baldassare Fronte*[b]

[a] Aquaculture Program, Institute of Agrifood Research and Technology (IRTA), La Ràpita, Spain [b] Department of Veterinary Sciences, University of Pisa, Pisa, Italy

5.1 Introduction

Sustainability has become a guiding principle across economic sectors worldwide. While most industries, including aquaculture, commit to sustainable practices, implementation remains challenging. Developing sustainable strategies for blue food economies is crucial for transitioning toward more responsible and comprehensive food production systems with positive impacts (Partelow et al., 2023). The United Nations ("Food Systems Summit 2021") emphasized that diversity in food production systems enhances resilience against socioeconomic disruptions such as COVID-19 and extreme climate events, creating opportunities for developing food systems that are healthy, resilient, circular, and sustainable. Under this frame, aquaponics stands out as an alternative sustainable technology that integrates fish farming with hydroponics, offering benefits including improved sustainability, reduced resource utilization, and minimized environmental footprint compared to traditional aquaculture and agricultural methods (Baganz et al., 2021).

Based on the above-mentioned principles, aquaponics may be defined as an integrated food production system that combines aquaculture (raising aquatic animals such as fish, snails, or worms) with hydroponics (cultivating plants in water without soil) in a symbiotic environment (Baganz et al., 2021). In an aquaponic system, water from the aquaculture section cycles through the system, with waste from the aquatic animals being broken down by beneficial bacteria into nutrients that plants can use. The plants then filter and clean the water, which is recirculated back to the aquatic animals.

Zero-Waste Technologies in Aquaculture: Global Perspectives and Methods toward a Circular Economy
DOI: https://doi.org/10.1016/B978-0-443-34689-7.00025-2

This technology offers several important connections to sustainability and food security challenges. For instance, the resource use efficiency of aquaponics systems is remarkable, using up to 90% less water than conventional agriculture. Simultaneously, it produces both protein (fish) and vegetables in the same system, maximizing output from limited resources (Rakocy et al., 2016). Besides, these systems can be established in urban environments, on rooftops, or in areas with poor soil quality, bringing food production closer to consumers and utilizing spaces not suitable for traditional farming or other uses.

In controlled environments, aquaponics can operate continuously regardless of seasonal changes or climate conditions, providing consistent food availability year-round. Being less dependent on external inputs, such as soil quality, rainfall patterns, and large land areas, these systems are more resilient to environmental challenges and climate change impacts. Furthermore, local production reduces the carbon footprint and costs associated with long-distance food transportation while improving access to fresh produce in food deserts (Somerville et al., 2014).

As aquaponics systems simultaneously produce protein-rich fish and nutrient-dense vegetables, they may potentially address multiple nutritional needs from a single production unit. Aquaponics typically uses fewer pesticides and chemical fertilizers, producing food with fewer chemical residues. Communities can establish localized food production systems that reduce dependence on distant supply chains and price fluctuations in global markets, with that enhancing food sovereignty. These systems create avenues for small-scale entrepreneurship and employment in food production, particularly in urban or challenging environments. In addition, aquaponics serves as a practical teaching tool for sustainable food systems, potentially encouraging wider adoption of efficient growing methods. These attributes make aquaponics particularly valuable for addressing food security challenges in regions facing water and food scarcity, limited arable land, urbanization pressures, or climate instability (Mchunu et al., 2018; Baganz et al., 2021).

Overall, aquaponics clearly aligns with several Sustainable Development Goals (SDGs) of the 2030 Agenda for Sustainable Development, adopted by all United Nations Member States in 2015, which provides a shared blueprint for peace and prosperity for people and the planet. In particular, among the 17 SDGs, aquaponics, through this dual benefit to aquatic and terrestrial ecosystems, represents a holistic solution aligning with multiple UN sustainability objectives. For instance, aquaponics directly supports the SDG 2 (zero hunger) by providing a sustainable food production method that can increase food security and nutrition. The system's ability to produce both protein (bluefood) and vegetables efficiently in limited spaces addresses the goal's targets of ending hunger, achieving food security, improving nutrition, and promoting sustainable agriculture.

Beyond the SDG 2, aquaponics contributes to several other goals, such as the SDG 3 (Good health and well-being). Thus, aquaponics supports a healthy, balanced, and diversified diet based on aquatic and terrestrial foods since fish contain nutritional properties (i.e., essential fatty acids, proteins, vitamins, and minerals) and plant products associated with aquaponic systems promote a diet rich in vitamins, minerals, and fiber. Aquaponics also aligns with the SDG 6 (Clean water and sanitation) since aquaponics uses up to 90% less water than conventional agriculture through its recirculating design, supporting sustainable water management. Furthermore, urban aquaponics systems can transform cities by enabling local food production, reducing transportation emissions, and utilizing underused spaces, which aligns with the SDG 11 (Sustainable cities and communities). The circular nature of aquaponics, where fish waste becomes plant nutrients, and water is recycled, exemplifies sustainable production patterns proposed by the SDG 12 (Responsible consumption and production). Similarly, by reducing water usage, minimizing land requirements, and potentially decreasing food transportation distances, aquaponics can help mitigate climate change impacts, which addresses the SDG 13 (Climate action). Furthermore, aquaponics significantly contributes to both the SDG 14 (Life below water) and the SDG 15 (Life on land) through its innovative approach. This integrated system presents viable alternatives to harvesting wild fish populations while simultaneously expanding the diversity of aquaculture production methods. In addition, aquaponics minimizes the demand for traditional agricultural land, promotes more sustainable soil management practices, and substantially decreases dependence on chemical fertilizers in food production. As highlighted, the versatility of aquaponics in addressing multiple sustainability challenges makes it an important technology for achieving the integrated vision of the SDGs, particularly in regions facing resource constraints or climate vulnerabilities.

Having established the significant benefits of aquaponic systems for enhancing food security and their alignment with the United Nations' SDGs, this chapter provides a comprehensive examination of aquaponics from its historical origins to contemporary applications. The discussion traces the evolution of aquaponic technology through time, exploring how traditional practices have been transformed into modern integrated food production systems. The chapter also explores system configurations and operational dynamics, offering insights from biological, chemical, and engineering perspectives. This multidisciplinary approach highlights the complex interactions between fish, plants, and microorganisms that form the foundation of successful aquaponic ecosystems. By examining the nitrogen cycle, water chemistry parameters, and essential design considerations, readers will gain a thorough understanding of the scientific principles underpinning aquaponic productivity.

The chapter concludes by identifying emerging trends and promising research directions in aquaponic technology, with particular emphasis on technological innovations and social sustainability factors that will shape the future of this field. By addressing both technical advancements and community integration aspects, this overview provides a holistic examination of aquaponics as a continuously evolving solution for sustainable food production in the 21st century.

5.2 The inside story of aquaponics: A condensed history

Although aquaponics may be perceived as a recent sustainable and innovative technology, its origins trace back to ancient agricultural practices that combined aquaculture with plant cultivation. These integrated systems emerged independently across several civilizations along the history, demonstrating humanity's innate understanding of ecological relationships.

5.2.1 The ancient beginnings

The Neolithic marked the origin of agriculture and sedentism and is associated with the emergence of new techniques for making stone tools and the origin of writing. Its beginning varies temporally between 8,000 BCE and 4,500 BCE, and this explosion of human technology occurred in the Fertile Crescent area, encompassing the Mediterranean Levant, Mesopotamia, and Egypt. If livestock domestication and sedentism are two of the main characteristics of the Neolithic, with evidence found in China, it is highly probable that similar associations with fishing, shellfish harvesting, and integration with agriculture were also developed in other areas during this period. It is only a matter of time before archaeological evidence is uncovered.

With the incorporation of what is known as the Neolithic package, which included certain tools and agricultural knowledge, humanity significantly increased its productive capacity. This change was necessary to adapt to sedentism and population growth. These new productive skills evolved, and it did not take long to discover the advantageous symbiotic effects of combining the farming of plants and animals. Later, fish and crustaceans were incorporated, creating a model of polyculture that is considered as the precursor to what we now know as aquaponics. This involves using fish fecal wastes to fertilize plants. This method existed for thousands of years in ancient civilizations in both Asia and South America. The production of surpluses, resulting from intervention in the natural cycles of plants and animals to enhance their use, led to a process of specialization. It is difficult to believe that some of these progresses were not involved in managing aquatic resources. This is the moment when the earliest references to aquaculture appear, dating back approximately 4,000 years in China and 3,500 years in Mesopotamia and other Eurasian and American cultures.

5.2.2 Aquaponics in the millenary China

In China, approximately 4,000 years ago, farmers recognized the symbiotic relationships between plants and animals. They positioned duck cages or livestock pens above fishponds, allowing animal waste to fertilize the water and feed fish. During the Zhou Dynasty (1,046–256 BCE), this nutrient-rich water irrigated plants and rice fields, significantly improving their yields. This led to the common practice of cultivating fish in rice fields, significantly improving yields because of water enrichment from residues from carp, catfish, small crustaceans, and mollusks. This primary method expanded into various types of cultivation and is one of the oldest and most effective examples of what is now known as integrated agroaquaculture systems (Edwards, 2025). These systems share resources to produce multiple species simultaneously through the synergy generated by integrating two farming systems. This strategy increased production, diversified crops, efficiently used water and nutrients, and promoted multitrophic crops with socioeconomic benefits.

Fan Li's treatise "The Classic of Fish Culture," written in 475 BCE, represents the first documented guide on aquatic animal husbandry. He advocated for an integrated multitrophic aquaculture (IMTA) model that combined carp cultivation with mulberry production for silkworms, creating a circular system where goat droppings fertilized ponds and pond water irrigated mulberries.

5.2.3 The pre-Columbian American systems

Chinampas—floating gardens built on shallow lakebeds—are one of the most ingenious agricultural systems in human history. The technique dates back roughly 2,000 years, long before the Aztec Empire existed, but it was the Aztecs who perfected and expanded it on a massive scale around Tenochtitlán (present-day Mexico City). Each chinampa was constructed as a raft-like platform of woven logs and reeds, layered with nutrient-rich mud and organic matter dredged from the lake bottom, then anchored in place by willow trees planted along its edges. The genius of the system lay in its closed-loop design. Canal sediment and organic waste enriched the soil, while the surrounding water provided constant natural irrigation, eliminating the need for artificial watering. Farmers also raised fish and waterfowl in the canals, which supplied food and, through their waste, further fertilized the plots—an early form of integrated aquaculture and agriculture. This efficiency made chinampas extraordinarily productive: a single hectare could yield several harvests a year of maize, beans, squash, tomatoes, chili peppers, and flowers. By the 16th century, the chinampas surrounding the lakes of Xochimilco and Chalco were feeding an estimated 200,000 people, making them a cornerstone of the Aztec food supply. Beyond their agricultural output, the canals fostered a rich ecosystem, sustaining fish, birds, and other wildlife alongside the crops—a rare example of farming and biodiversity reinforcing one another rather than competing.

5.2.4 The Egyptian innovations

In the Nile Valley approximately 2,500 years ago, Egyptians combined rice production with tilapia farming on lands enriched by seasonal floods. Evidence from artwork in the tomb-chapel of Nebamun (ca. 1,400–1,350 BCE) depicts ponds containing ducks, lotus flowers, and tilapias, surrounded by various plants. These illustrations suggest that integrated fish and plant systems existed in Egypt over 3,500 years ago.

5.2.5 The Mesopotamian connections

Although there are no definitive references confirming the existence of the Hanging Gardens of Babylon (6th century BCE) and their historical reality remains debated, some descriptions cited by Stephanie Dalley in The Mystery of the Hanging Garden of Babylon refer to the gardens of Nineveh as including "fish ponds." While the interpretation of these accounts is not without uncertainty, they suggest an early association between cultivated landscapes and aquatic organisms. In this sense, the concept bears some resemblance to modern aquaponic principles, particularly given references to the presence of fish and other aquatic life within the gardens.

5.2.6 The European developments

Although Europe did not have traditional aquaponic systems such as chinampas or the Asian rice-fish systems, similar practices integrating agriculture and aquaculture did exist. It is well-known that the Romans had a particular fondness for oysters, often associated with certain halophytic plants they passionately consumed, complemented by sauces, as evidenced by abundant archaeological data. The combination of these spaces suggests they may have been used jointly, leveraging the benefits of both types of cultivation systems. In some regions, such as medieval monasteries, monks raised fish in ponds and used the aquatic waste to fertilize nearby crops. While this was not aquaponics in the modern sense, it laid the groundwork for understanding the relationship between aquatic and terrestrial systems, eventually evolving into more complex systems.

During the Agricultural Revolution (17th–18th centuries), European agriculture experienced significant innovations, focusing more on crop rotation and chemical fertilization. Contrary to Asian countries, integrated fish and plant systems did not develop widely due to the abundant availability of fertile lands and reliance on natural fertilizers, such as manure.

5.2.7 Modern aquaponics

The modern practice of aquaponics, as we know it today with water recirculation systems, originated in the 1970s. It was developed at the University of the Virgin Islands (USA) by Professor James Rakocy, known as the UVI System, and was the first to integrate aquaculture and hydroponic systems, based on the breeding of tilapia and the cultivation of vegetables such as lettuce and aromatic plants such as basil. However, it seems that this idea had already been circulating in the minds of several colleagues for some time, inspired by existing commercial systems.

Regarding the origin of the term and the activity, there is some controversy with The New Alchemy Institute (Massachusetts, USA), a research center founded in 1969 that designed "The Ark" in 1976, a residential-experimental complex in Prince Edward Island (Canada), as an example of a self-sufficient lifestyle model. Motivated by the quest to reduce overdependency on finite natural resources such as land and water, these scientists developed the modern and efficient integration of the aquaculture and hydroponics system. The Ark incorporated sustainable architecture technologies developed by the New Alchemy Institute. The end goal of the project was to test whether a sustainable building could be built in a harsh northern climate. Futhermore, a secondary objective of the Canadian government was to test green technologies, including a unit to farm fish and plants.

European interest in aquaponics grew in the late 20th century, with pioneering research at Germany's University of Rostock and Scandinavian institutions developing cold-climate models. Environmental concerns about aquifer pollution and resource overexploitation sparked renewed interest in aquaponics as a sustainable solution. In the early 21st century, European aquaponics have incorporated advanced technologies such as smart sensors, automated nutrient control systems, and recirculating aquaculture systems (RASs). Urban aquaponics farms have emerged in major cities such as Barcelona, Berlin, London, and Amsterdam, producing fresh food for local markets.

The historical development of aquaponics demonstrates humanity's long-standing recognition of the potential for symbiotic relationships between aquatic and terrestrial production systems. From ancient Chinese rice-fish cultivation to Aztec chinampas and modern high-tech systems, aquaponics continues to evolve as a solution for sustainable food production that addresses contemporary challenges of resource scarcity and environmental protection.

5.3 Aquaponics designs and configurations

5.3.1 System components

Aquaculture can be conducted using a variety of production systems (Somerville et al., 2014; Junge et al., 2020). Among them, aquaponics is considered one of the most efficient and environmentally sustainable systems. As already mentioned, it consists of a combination of fish and plant culture, within a recirculating loop system (Somerville et al., 2014; Oladimeji et al., 2020; Okomoda et al., 2023). Its design may vary depending on specific characteristics, such as scale (e.g., large or small), production (plant and fish species), hosting infrastructure (e.g., open field, greenhouse, or other urban facilities), and climatic region, among others. Over time, several authors have designed aquaponic systems (Goddek et al., 2019a), highlighting also the key components and role for proper system functionality. According to the description provided by Palm et al. (2018), an aquaponic system essentially consists of the following components: (1) Fish tanks; (2) solid removal unit; (3) biofilter; (4) hydroponic unit; (5) pump unit; and (6) blowers or liquid oxygen for water oxygenation. More components may also be considered for achieving easier system management and reliability. In fact, a well-designed aquaponic system should be considered as a complex "organism" where different "organs" (components) work together to maintain stable and balanced conditions to provide a proper environment to fish, plants, and bacteria (Kazosi et al., 2021). Therefore, an accurate design of the whole aquaponic system is critical to achieving long-term success and functional reliability.

The main components of an aquaponic system and some relevant tips are schematically given as follows:

(1) *Fish tank*
- **(a) Function**: Hosting the reared fish.
- **(b) Tank capacity**: The size of the tank depends on the fish biomass to be reared, and normally it is interdependent on the plant growing area. It may vary from less than 1 to 30 m^3 or even more.
- **(c) Materials**: Must be safe for both fish and plants. Despite being normally more expensive, thermoplastics such as polyethylene (PE), polypropylene (PP), and polyvinyl chloride (PVC) are the best options since they are normally resistant to ultraviolet light (mainly PE), less porous (PP is more porous than PE), and thermally stable (PVC becomes brittle below 0°C). In general, PE is the material to select for greenhouses or outdoors and long-lasting installations, thanks to its high resistance to harsh weather conditions. Galvanized materials should always be avoided because of possible zinc toxicity. Other less efficient alternative materials can be concrete and fiberglass. Digging basins in the ground and using plastic sheeting for impermeability may be possible for very poor aquaponic designs.

(2) *Water pump and circulation system*
- **(a) Function**: Circulate water between different system sections (e.g., from sump to fish tank and plant growing beds).
- **(b) Pump type**: Submersible or external pumps are suggested for small-scale or large-scale systems, respectively. They must be as energetically efficient as possible, since they are the main energy-demanding component of an aquaponic system, when a heat pump or chiller is not implemented.

(3) *Water filtration system*
- **(a) Function**: Removing suspended solids and converting ammonia into.
- **(b)** Subcomponents:
  - **(i) Mechanical filters** are required for removing water-suspended solids (e.g., feed particles, fish and plant debris, and fish feces). They may be sand, socks, or drum filters, which are generally used for other aquaculture applications.

(ii) **Biofilter** hosts media where nitrifying bacteria grow and convert ammonia into nitrites and nitrates. The media consist of porous material (e.g., bioballs, chips, and sponges). Coconut fibers or other similar organic materials can sometimes be used (e.g., small-scale and hobby systems).

(4) *Degassing and aeration system*

(a) Function: Removing gases such as carbon dioxide from the water and maintaining the DO water level above the required threshold (e.g., above 5 mg/ L).

(b) Subcomponents:

(i) Air blowers can ensure the removal of gases from water and the maintenance of dissolved oxygen (DO) levels. Just in some cases, it may be necessary to install an additional specific degassing unit.

(ii) Oxygen diffusers, such as air stones optimize oxygen water dispersion. In this case, an oxygen storage system must be considered. Microperforated tubes like those used in hydroponics may also be implemented in case a high oxygen-rich environment is required.

(5) *Plant section*

(a) Function: Hosting growing plants.

(b) Subcomponents:

(i) Components of this section may be highly variable depending on the adopted system based on the hydroponics technology [i.e., media-based grow beds (MGBs), deep water culture (DWC), nutrient film technique (NFT), among others] and the plant species (e.g., leafy green of fruity plants). Basically, tanks where plants can be hosted are necessary for both MGB and DWC systems, while for NFT or similar, a pipeline network is required.

(ii) Inert substrates are necessary for MGB-like systems. For the DWC system, floating panels (rafts) are used to support cultivated plants, together with plant pots and inert media; these latter are also necessary for the NFT system.

(iii) Substrate material can consist of sand, expanded clay, gravel, perlite, vermiculite, or other synthetic substrates, such as bottle caps or bioballs.

(6) *Monitoring and control system*

(a) Function: Detecting and adjusting crucial parameters such as pH, temperature, oxygen levels, salinity, and water flow.

(b) Subcomponents:

(i) **Sensors and probes:** pH, temperature, ORP (oxidation-reduction potential), conductivity and/or salinity, DO sensors.

(ii) **Programmable logic controller (PLC)** that automatically controls and regulates the functionality of system components such as water pumps, aeration system, lights, heaters/chillers, nutrients, and buffer distributors.

(c) Water quality control: In the case of a large-scale system, it is highly suggested its implementation. On the contrary, in small-scale systems, water quality monitoring can be led using manual probes (temperature, pH, DO, and conductivity).

(7) *Pipes and pipelines*

(a) Function: Transport water between various system components (fish tank, grow beds, and biofilter).

(b) Material: PVC or PE (durable and safe for water).

(c) Flow dynamics: Pipes should be sized according to water flow rate and arranged to allow optimal flow between the various compartments of the system.

(d) Drainage and water return systems:
  (i) Allow water to return to the fish tank after passing through the growing bed.
  (ii) A well-designed drainage system prevents standing water and allows efficient and clean return to the tank.
  (iii) Channels and grates can be used to control flow and prevent blockages.

(8) *Lighting systems*
  (a) Function: Provide artificial light for plants if the system is located indoors or with insufficient natural light.
  (b) Features:
    (i) LED lights specifically designed for plant growth are very energy efficient.
    (ii) Lights' intensity and spectrum should be adjustable to efficiently simulate the natural cycle of light and darkness to stimulate plant growth.

(9) Composting containers
  (a) Function: In some systems, organic waste (e.g., feed not consumed by fish) can be treated organically to improve soil quality or supplement the aquaponic system.
  (b) Composting and decomposition of waste that may be used for enriching the substrate with nutrients.

In more detail, within the aquaponics system, each component plays a specific role, and its characteristics may affect the system's efficiency and performance. Therefore some aspects should be carefully considered when designing and building an aquaponic system. To this purpose, further are described functions and technical aspects of the main components of an aquaponic system. For those components which will not be discussed, it is suggested to consider other bibliographic references (Rakocy et al., 2004; Somerville et al., 2014; Gosh and Chowdhury, 2019; Junge et al., 2020).

Although any shape of tank can be used for holding fish, round tanks with flat or slightly conic bottoms are recommended due to their circular hydraulic pattern that implies self-cleaning properties (Somerville et al., 2014). In particular, in a round tank two different water flows can be described: (1) Primary rotating flow starts from the water-inlet and clockwise/anticlockwise stream around the tank, determining a uniform distribution of inflow water and fish feed, as well as a slow settlement of suspended solids (uneaten feed and feces); (2) the primary rotating flow creates secondary radial flow that convey settled suspended particles to the center of the tank, where a discharging water outlet is present. Hence, together, primary and secondary water flows generate a consistent current that makes round tanks self-cleaning. Alternatively, square flow channels can be used for raising fish. Despite their lower self-cleaning efficiency, square or rectangular tanks allow a more efficient use of the available surface (Junge et al., 2020). Besides tank shape, the materials are also relevant. Because of zinc toxicity, galvanized materials must be avoided, as well as some types of plastic that can be harmful to fish. Although more expensive, thermoplastics such as PE, PP, or PVC are the best option since they present UV resistance (particularly black PE), reduced porosity (PP is more porous than PE), and thermal stability (PVC can glaze at temperatures below 0°C). In conclusion, PE may represent the best option for long-lasting installation, because of its resistance to harsh weather conditions (Junge et al., 2020).

The water **pumping and circulation systems** usually pump water from the lowest to the highest sections of the aquaponic unit. Then, from this latter section, water can flow gravitationally through the different sections of the system. More frequently, water flows from the fish tank to the filtration unit, the plant beds (hydroponic section), and finally back to the fish tank in coupled aquaponic systeems (Rakocy et al., 2004; Krastanova et al., 2022). However, in some other cases, water can also flow from both fish and plant sections to the filter unit (sump), where it is fully mixed with its high and low load of nitrogen (notably nitrates), respectively (Rossi et al., 2021). If well designed, one or a few pumps may be enough to run an aquaponic system. However, since pumps consume most of the energy required to run an aquaponic system, they must be energetically efficient and accurately chosen. Flow rate, lift height, power consumption, noisiness, and reliability are some important aspects to be considered. The use of a simple calculator for pressure losses in the pipeline is highly suggested to decide the characteristics of the pumps to be installed (http://www.pressure-drop.com/Online-Calculator/). Moreover, submersible pumps are ideal for small and medium-sized systems, while external pumps are more efficient and suggested for large-scale systems (Junge et al., 2020).

Maintaining healthy water is an essential aspect for fish and plant well-being, as well as for proper biofilter and whole system efficiency. The **water filtration system** should physically remove suspended solid particles and biologically reduce solute wastes (toxic to fish) into plant nutrients, which are less toxic to fish. The solid waste removal is essential since wastes increase the ammonia water content, reduce DO, and increase the risk of disease outbreaks (Somerville et al., 2014; Junge et al., 2020). To this purpose, radial settling tanks are normally suggested for small-scale systems, because of their low efficiency. Sock filters are more efficient, even though they need frequent sock replacement and imply a higher operational cost. Sand and drum filtration technologies offer enhanced automation capabilities; however, despite their efficacy in suspended solids removal, these systems necessitate periodic backwashing operations that result in water discharge, increasing system's operational costs. This water loss represents a significant operational constraint that diminishes the high water-use efficiency typically characteristic of aquaponic production systems. To minimize the volume of wastewater, sludge can be treated (mineralization) and reused within the system (Rafiee and Saad, 2005). In this way, even a low-tech filtration system can achieve a significant reduction of the final wastewater volume. Finally, a foam fractionator, the so-called skimmer, can remove compounds such as proteins, fatty acids, bacteria, and metals. However, they are mainly used in RAS or marine water, as their efficiency in freshwater is very low (Junge et al., 2020).

The most used biofilter is the so-called Moving Bed Biofilter Reactor (MBBR). It consists of a tank filled with porous media (1-2 cm plastic structures) that provides physical surfaces where bacteria can adhere and grow. The filter media are kept in constant movement by aeration wihin the MBBR, for instance, by installing air plates at the bottom of the biofilter tank. The constant movement of the media determines a self-cleaning effect on the filter media and prevents extensive bacterial growth. Aeration is also required to properly accomplish the nitrification process. The biofilter can be simple or more complex according to the application of small-scale or large-scale systems, respectively. The size of the biofilter depends on several factors, such as temperature, DO, water flow rate, water salinity, stocking density, feed rate, and feed protein content. A special mention is due to the specific surface area (SSA) of the filter medium, which severely affects the entity of the nitrification process (smaller pores increase the bacteria population, However, in turn, the capacity of converting ammonia and nitrites into nitrates) and the possibility of entrapping small waste particles that by clogging creates anoxic zones where denitrification process may take place (Somerville et al., 2014). Ultimately, to detect the proper functionality of the biofilter, it is possible to measure ammonia, nitrites, and nitrates water content (Krastanova et al., 2022) by simply using commercial kits.

An **aeration system** consisting of blowers and pipelines can help to dissolve oxygen of the air in the water at low pressure. In addition, an oxygen distribution system can be used, but only in case of a real need due to its higher operational costs. Maintaining a high level of DO in the water is extremely important for both fish and plant. If the content of DO is low, the plant nutrient uptake decreases, which negatively affects plants' growth (Goddek et al., 2019a; Krastanova et al., 2022). The basic oxygenation technologies are the U-pipe, oxygenation cone, and low head oxygenator, with the first two being more efficient than the third, respectively (Junge et al., 2020).

Regarding the **plants growing beds,** the most common systems are MGBs, DWC or RAFT, and NFT. These main systems may also be modified according to specific needs. For instance, in case of lack of space, Vertical Towers similar to NFT can be implemented for exploiting the available volumes in addition to the surface, even though their yield is generally lower than the other plant growing systems. MGB consists of a tank for hydroponic culture that may be filled with a variety of inert substrates such as expanded clay, perlite, pumice, gravel, and many others (Somerville et al., 2014; Gosh and Chowdhury, 2019). Gravel-based substrates and lightweight expanded clay aggregate (LECA) are media commonly utilized for large-scale aquaponics. An important characteristic of inert substrates is the so-called SSA, which in these cases is between 250 and 400 m^2/m^3, respectively (Alam et al., 2020). These SSA values positively affect water retention capacity and production rate in comparison to other types of media characterized by lower SSA (Oladimeji et al., 2020). In this regard, it must be considered that as the SSA values of an inert media increase, the risk of creating anoxic clogged area in the media bed increases (Somerville et al., 2014). In this case, an effective solid-removal system is highly recommended. The inert substrates just provide physical support to the roots, enabling the plants to maintain their verticality, and do not provide any nutrients to the plants. However, the inert media hosts bacteria that integrate the biofilter activity (Gosh and Chowdhury, 2019).

Regarding DWC, which consists of a large water tank, perforated floating rafts, and plant pots containing some root-supporting substrate like those used for MGB. The pots are set in the raft holes, and plants may extend their roots directly into the water below for nutrient absorption. Similarly to DWC, in NFT, the plant pots are set in the holes of a perforated narrow channel where the plant's roots are partially immersed in a thin layer of running water (Gosh and Chowdhury, 2019; Krastanova et al., 2022). Finally, a survey carried out some years ago suggested that the majority of the aquaponic practitioners used a combination of two or more of the above-mentioned techniques (Love et al., 2014). Thus, aquaponics systems can be configured in various ways based on their hydrological and functional configuration, and a comparative study that provides some key parameters is available (Gosh and Chowdhury, 2019). After all, the technique and the type of beds to be used depend on the high or low assimilation of nutrients by plants. In this perspective, growing several crop species at different growth stages can optimize the production process and income (Krastanova et al., 2022).

For large-scale aquaponic systems, it is highly recommended the implementation of Internet of Things (IoT) and smart systems for **real-time monitoring and physical system control**. Applications that use digital twinning are already available or under development. They are tools that model the relationship between crop size and fresh weight and then estimate crop size as it grows. The various parameters of the aquaponics system and the sensors used in the IoT and digital twin framework system have already been described, as well as the controlling strategies, various smart technologies, and implementation procedures (Samikannu et al., 2023).

5.3.2 Small and large aquaponics systems

Aquaponic systems have been developed as innovative food production systems to be implemented for both small (e.g., backyard) and large production (e.g., "industrial" greenhouses). They are frequently classified as small- and large-scale systems (Somerville et al., 2014), which refer to the size and capacity of aquaponic systems, both in terms of quantitative production and the dimensions of the occupied surface dedicated to food production. Surface dedicated to food production may range from a few square meters up to several hundreds, and production may largely vary accordingly. Generally, large-scale aquaponic systems, where fish and plants are produced at high densities, have the goal of continuous and constant industrial or semiindustrial production scale. These systems are designed for commercial mass production, aiming to meet larger market demands, and consequently, they require a high initial investment. They can be installed in greenhouses or industrial and/or urban facilities, and are designed to optimize productivity, efficiency, and profitability. In large-scale aquaponic systems, automated systems are often employed to control environmental parameters (i.e., temperature, pH, water conductivity, and DO) and nutrient management; therefore are equipped with sensors or probes and related management software. Large-scale aquaponic systems increase cost efficiency and can include integrated production and direct sales to consumers or retailers. Finally, production systems whose plant section is based on MGB are not considered for large-scale production, as the production intensification strategy cannot be applied for this technology. This method is mainly suitable for small-scale production systems such as those meant for hobby and educational projects.

On the other hand, small-scale systems are usually intended for families or small community production and are then limited and targeted to meet local food supply needs. Often, small-scale aquaponic systems are characterized by a lesser complexity in terms of the degree of technology and automation than large-scale systems. Small-scale systems can be semiautomated or even operated manually, equipped with a few sensors and simple automatic control devices. In these systems, management often depends on direct user intervention. These systems do not allow to achieve economies of scale but are more flexible and suitable for decentralized production settings (Paxton, 1994). The characteristics outlined in Section 5.3.1 indicate that small-scale aquaponic installations frequently serve purposes beyond commercial production, functioning as recreational (hobby) and pedagogical resources. Community-based aquaponic systems integrated within educational institutions operate as dynamic research environments where practical application meets theoretical learning

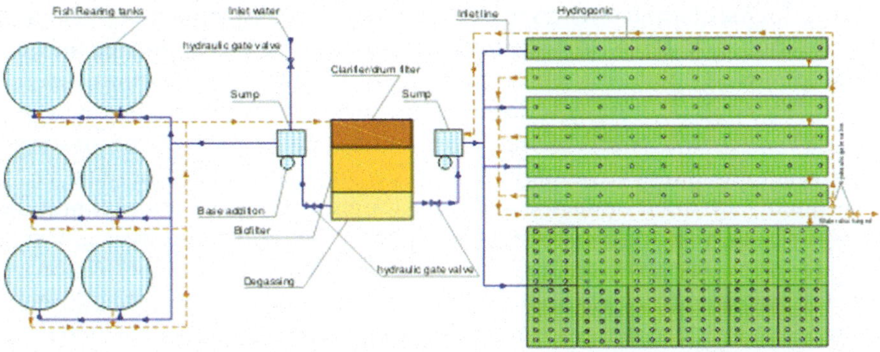

FIGURE 5.1 Synoptic representation of a coupled aquaponic system with its components.

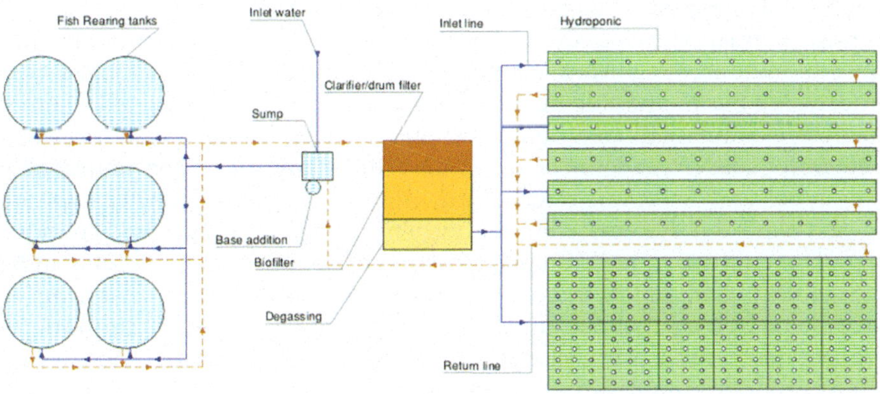

FIGURE 5.2 Synoptic representation of a decoupled aquaponic system with its components.

(Gosh and Chowdhury, 2019). Within this implementation framework, the societal advantages conferred by aquaponics systems surpass considerations of economic viability.

Regarding the environmental impact of the two aquaponic settings (large- and small-scale systems), large-scale systems may generate a more significant local environmental impact, but they may be characterized by a greater efficiency in terms of use of resources (e.g., water and energy). On the contrary, small-scale systems are less impactful, but less efficient from a mass production perspective and use of resources (e.g., unit of energy per unit of product).

In conclusion, both models have great potential in different contexts. Large-scale systems are ideal for commercial production and industrial use, while small-scale systems are more suitable for local, educational, niche, or residential production purposes.

5.3.3 Coupled and decoupled aquaponics systems

Aquaponic systems can be categorized into two distinct classifications based on hydraulic circuit design: Coupled (Fig. 5.1) and decoupled (Fig. 5.2) configurations. This taxonomic division reflects the degree of interconnection between the system's primary water compartments. In coupled aquaponic systems, the aquaculture component (dedicated to fish production) and the hydroponic component (dedicated to plant cultivation) maintain direct hydraulic connectivity, facilitating continuous water exchange between these functional units. This type of system is very efficient in terms of water use, but may require more monitoring and management, as the balance between fish, plants, and bacteria must be maintained very precisely, as all compartments are interconnected and have their own specific optimal requirements. In decoupled aquaponic systems, fish farming and plant cultivation are physically separated. In particular, water from fish tanks is pumped or flows by gravity into a biofiltration tank, where fish wastes and nutrients from uneaten pellets are converted into useful compounds for the plants. When needed, the hydroponic unit where plants are grown is connected, and nutrient-rich water flows by means of a separate irrigation system into the unit dedicated to plant production. In this case, the two components (fish and plants) are more independent, which allows for greater control over each part of the system, especially about nutrient management and environmental parameters (Baganz et al., 2021).

Contemporary aquaponics research still requires more precise taxonomic classification and standardized terminology regarding coupled and decoupled aquaponic subsystem configurations. Addressing this methodological gap, Baganz et al. (2021) conducted a comprehensive systems analysis examining diverse aquaponic architectures to establish more precise descriptive parameters for subsystem connectivity. This analysis introduced the quantifiable concepts of "coupling type" and "coupling degree" into the aquaponics lexicon. The research further delineated and classified systems wherein nutrient-rich water is supplied to independently functioning plant cultivation units on an as-needed basis. Despite operational independence of the horticultural component, these systems are still appropriately classified as coupled aquaponic systems, given that the hydroponic subsystem remains entirely dependent on nutrients generated within the aquaculture subsystem. In addition, the coupling degree framework incorporates multiple parameters to quantitatively assess connectivity efficiency between internal resource streams, particularly regarding water and nutrient transfer dynamics (Baganz et al., 2021).

In relation to the coupled and decoupled system efficiency, studies have demonstrated some nutrient deficiencies and imbalances in coupled systems. In the study carried out by Aslanidou et al. (2023), a negative impact on nutrient assimilation was observed, particularly in fruiting vegetables. However, the average nutrient use efficiency (NUE) for nitrogen, phosphorus, potassium, and calcium was 42% higher in the coupled system than in a hydroponic and decoupled aquaponic systems (Aslanidou et al., 2023). From the perspective of improving the aquaponic system and nutrient reuse efficiency, the opportunity of reusing sludge organic matter has also been considered in a decoupled aquaponic system with an additional sludge treatment loop and distillation loop (Goddek et al., 2019a). Unfortunately, adverse effects of the treatment were observed on symbiotic relationships. Moreover, economic feasibility was also weakened and challenged (Goddek et al., 2019a). As a consequence, a novel dual-loop aquaponic system, coupling the main and side streams, has been proposed by Zi et al. (2024) for the reuse of both dissolved and solid nutrients. Nutrient conversion and retention in the dual-loop aquaponic system proposed improved efficiency through an optimized symbiotic relationship in the main and side streams. The phototrophic bioconversion of sludge produced nutrients and minerals, thereby enhancing the nutrient recovery efficiency of carbon, nitrogen, and phosphorus (Zi et al., 2024).

Summarizing, the operational viability of innovative zero-discharge aquaponic systems has been empirically validated, with documented implementation cases providing substantive evidence (Zhu et al., 2022, 2024). The integration of anaerobic digestion processes with conventional aquaponic configurations demonstrates significant potential for enhanced system efficiency parameters, aligning such integrated systems with fundamental circular economy principles and resource optimization frameworks.

5.4 Water chemistry, nutrient management, and microbiome

Successful aquaponic operations require attentive management of multiple factors that influence system efficiency and economic viability. Fish and plant species must be carefully matched based on their compatibility with the system's environmental conditions, including water temperature, pH range, DO levels, and plants' nutritional requirements, among others (Kushwaha et al., 2025). Maintaining optimal water conditions is essential and demands consistent monitoring and regulation of several parameters such as acidity/alkalinity balance, thermal conditions, nitrogen compound concentrations (ammonia, nitrite, and nitrate), oxygen saturation, and nutrient profiles. Effective solids management practices must be implemented to prevent deterioration of water quality that may affect fish and plant performances (Goddek et al., 2019b). Water circulation rates require optimization to ensure adequate oxygenation while minimizing evaporative losses, and feed distribution must be calibrated according to fish developmental stages. Regular assessment of water quality indicators with appropriate adjustments represents a practical strategy for enhancing overall system performance.

5.4.1 Water quality parameters

Water temperature is a fundamental regulatory factor in aquaponic environments, simultaneously affecting the growth of fish, plants, and microbial communities (Rakocy et al., 2006). The optimal temperature range for aquaponic operations is 18–30°C, varying by species. For instance, for warm-water species such as tilapia and catfish, the optimal temperature is 22–30°C, whereas for cold-water species such as rainbow trout, temperatures are recommended between 10 and 18°C. Regarding plants, most of them grow well between 18 and 26°C, whereas temperatures below this range result in reduced growth rates. Temperature also directly influences DO concentration, a critical parameter for nitrification processes. The communities of nitrifying microorganisms demonstrate optimal metabolic activity within the 25–30°C range, with ammonia-oxidizing bacteria typically exhibiting faster growth rates than nitrite-oxidizing bacteria under these conditions (Zou et al., 2016; Wongkiew et al., 2017). Thermal variations can substantially alter nitrification efficiency and unbalance the transformation of ammonia to nitrates . Research has documented that rapid temperature decreases from 20 to 10°C can diminish nitrification performance by approximately 58% (Monsees et al., 2017). Such thermal shifts modify microbial community structure and function, affecting nitrogen transformation rates within the system (Yildiz et al., 2017; Palm et al., 2018). Understanding these temperature-dependent microbial dynamics is essential for maintaining stable nitrogen cycling in aquaponic systems across seasonal variations and diverse climate conditions.

In addition to temperature, pH is another critical parameter. The optimal pH range of 6.8–7.0 for aquaponics systems represents a carefully balanced compromise that requires a deeper understanding of the biological and chemical interactions within these integrated ecosystems. This may be defined as a three-way compromise, as aquaponics needs to adjust the pH values considering the optimal levels for fish, plants, and bacteria. In particular, fish health is directly impacted by pH levels. At pH values below 6.5, fish experience physiological stress as blood acidosis begins to occur, affecting oxygen transport and metabolic functions. Studies by Monsees et al. (2017) demonstrated that tilapia , one of the most common aquaponics species, showed decreased growth rates and increased cortisol levels (stress biomarker) when maintained at pH levels below 6.5 for extended periods. At extremely low pH levels (<5.5), fish mortality increases dramatically due to gill damage and internal acid-base imbalances.

Plants generally prefer slightly acidic conditions (pH 5.5–6.5) to fish because this pH range optimizes nutrient availability. At higher pH levels (>7.0), several critical micronutrients, including iron, manganese, copper, zinc, and boron, become less soluble in water and therefore, less available for plant uptake. Iron deficiency, manifested as interveinal chlorosis in newer leaves, is particularly common in aquaponic systems operating above pH values of 7.0 (Kasozi et al., 2019). Phosphorus availability also decreases significantly at higher pH values due to precipitation with calcium ions (da Silva Cerozi et al., 2016).

The optimal pH values for the nitrification process for converting toxic ammonia from fish waste into less harmful nitrates are between 7.0 and 8.0. The research conducted by Zou et al. (2016) demonstrated that nitrification rates decrease significantly at pH values below 7.0, with a dramatic reduction below pH 6.5. At pH 6.0, nitrification efficiency may drop by 50% compared with optimal conditions. This occurs because the enzymes involved in ammonia oxidation function optimally in slightly alkaline conditions. Wongkiew et al. (2017) further characterized how pH affects the ratio of unionized ammonia (NH_3) to ammonium ion (NH_4^+), with lower pH shifting the equilibrium toward the less toxic ammonium form but simultaneously reducing bacterial conversion efficiency.

Careful monitoring of all three system components allows for informed adjustments. If fish show signs of stress while plants exhibit nutrient deficiencies, the pH may need to be readjusted to balance the competing needs. However, pH management based on system maturity is also advisable, as Palm et al. (2018) recommended. For instance, in new systems, maintaining pH closer to 7.2–7.5 encourages robust bacterial colonization for promoting nitrification, whereas as systems mature, allowing pH to gradually decline toward 6.8–7.0 optimizes plant nutrient availability while maintaining adequate biofiltration.

Oxygen dissolved in water (DO) is among the most critical parameters for water quality in aquaculture systems. Insufficient DO concentrations can rapidly cause fish stress, disrupt nitrification processes in biofilters, and potentially lead to substantial fish mortality. The oxygen demands of an aquaculture system typically depend on several factors, including the density of fish populations, feeding rates, water temperature, and the particular species' ability to tolerate low-oxygen conditions (Yildiz et al., 2017).

The optimal DO levels for aquaponics systems should be maintained above 5 mg/L, with ideal levels between 6 and 8 mg/L to ensure proper functioning of all system components. For instance, the minimum DO requirements for fish are 5 mg/L, even though some species, such as the Nile tilapia or African catfish, can tolerate lower DO levels (<4 mg/L). In contrast, cold-water species, such as rainbow trout, require higher levels (>7 mg/L). Regarding plants, their roots are the main organ of the plant involved in nutrient absorption, and consequently, a minimum DO of 3 mg/L is required to prevent root hypoxia and guarantee nutrient uptake; however, higher DO levels of 5 mg/L are recommended for optimal nutrient uptake and root health. Considering the microbial communities involved in nitrification, the minimum DO recommended for guaranteeing their basic function is 4 mg/L, whereas the ideal DO for optimal nitrification rates should be higher than 6 mg/L (Yildiz et al., 2017; Palm et al., 2018).

The chemical buffering dynamics of water needs also to be considered in aquaponic units, since the biological processes in aquaponics systems tend to acidify the water over time, similarly to RAS units. In particular, nitrification itself is an acid-producing process, generating H^+ ions as ammonia is converted to nitrate. In addition, fish respiration produces carbon dioxide, which dissolves in water to form carbonic acid. Similarly, the absorption of nutrients by plant roots may also acidify water since during nutrient absorption for preserving the root's ionic equilibrium (Maucieri et al., 2019). Thus, water acidification is generally counterbalanced by the addition of different carbonate/bicarbonate sources. In this sense, Rakocy et al. (2006) emphasized that maintaining adequate carbonate hardness of at least 60–80 mg/L $CaCO_3$ provides crucial pH buffering capacity, preventing rapid pH fluctuations that can stress all biological system components.

The management of dissolved gases in aquaponics extends beyond just providing oxygen to fish, plants, and bacteria. Other dissolved gases in the water can impact fish health if not properly regulated. Elevated levels of dissolved carbon dioxide in water can prevent CO_2 from diffusing out of fish's bloodstream (respiratory acidosis), affecting their physiology and health condition. Species-specific tolerance exists toward CO_2; for instance, Nile tilapia can tolerate CO_2 levels up to 60 mg/L, while more sensitive cold-water species, such as rainbow trout, show distress at levels above 10–15 mg/L. Chronic exposure to elevated CO_2 has been linked to reduced growth rates and increased susceptibility to disease (Yildiz et al., 2017). While plants uptake CO_2 for photosynthesis primarily from the air, not from dissolved CO_2 in water, roots are affected by the water chemistry changes resulting from dissolved CO_2. The associated pH fluctuations can alter nutrient availability, particularly for iron, calcium, and phosphorus, affecting plant nutrient uptake efficiency and consequently, their growth and quality (Delaide et al., 2016). As plants and nitrifying bacteria reduce pH through their metabolic processes, controlled dissolved CO_2 removal through degassing can help maintain pH in optimal ranges without requiring chemical inputs (Goddek et al., 2019b).

Solid waste in aquaponic systems consists primarily of fecal matter and uneaten feed, containing significant quantities of organic nitrogen, phosphorus, and carbon compounds. The management of these solids profoundly impacts nutrient availability and system stability (Monsees et al., 2017). Conventional aquaponic wisdom often emphasizes rapid removal of solids to maintain water clarity and prevent oxygen depletion. However, this approach can lead to significant nutrient losses, particularly of phosphorus and micronutrients that are predominantly bound in solid form (Goddek et al., 2019b). Controlled mineralization offers a solution to this dilemma. By subjecting collected solids to microbial decomposition under managed conditions, organic nutrients can be converted to dissolved inorganic forms available to plants (Delaide et al., 2018). This process occurs through the action of diverse microbial communities, including heterotrophic bacteria, fungi, and protozoa. Aerobic mineralization maintains a higher redox potential, preventing the production of potentially toxic reduced compounds, such as hydrogen sulfide, but requires energy input for aeration (Monsees et al., 2017). Anaerobic mineralization requires less energy but produces intermediate compounds that may require further treatment before introduction to plant systems (Goddek et al., 2019a).

The optimal suspended solids concentration depends on system design. Zou et al. (2016) recommended maintaining total suspended solids (TSS) below 25 mg/L for most aquaponics systems, while Goddek et al. (2019b) suggested that decoupled aquaponics systems can operate effectively with TSS levels up to 100 mg/L within the fish component. Furthermore, Palm et al. (2018) emphasized that finding the balance between sufficient solids removal and nutrient conservation is crucial for optimizing aquaponics productivity. This is of relevance since the level of TSS may significantly impact fish and plant root health. For instance, suspended solids can accumulate on root surfaces, creating a barrier that reduces oxygen and nutrient absorption (Rakocy et al., 2006). This physical blockage interferes with the plant's ability to take up water and essential nutrients. Particulate matter also provides substrate for biofilm growth, which can smother roots if excessive (Goddek et al., 2015). In addition, high suspended solids levels can decrease DO in the water through increased biological oxygen demand, compromising root respiration (Somerville et al., 2014), whereas they can also serve as reservoirs for plant pathogens (Zou et al., 2016). In this sense, Rakocy et al. (2006) found that excessive suspended solids may foster undesirable heterotrophic bacteria that compete with nitrifying bacteria for oxygen and space, potentially compromising the nitrogen cycle. At the fish level, high levels of TSS may result in gill damage that may significantly impair respiratory function and osmoregulation, potentially leading to increased susceptibility to disease and reduced growth performance (Yildiz et al., 2017). For these reasons, effective solids management is critical in aquaponics. Mechanical filtration through settling tanks, screens, and filters should remove particulates before water reaches plant roots (Wongkiew et al., 2017). Despite the potential negative effects of suspended solids on fish and plants, some particulate organic matter can slowly release micronutrients as it decomposes because of mineralization, providing supplemental nutrition to plants (Palm et al., 2018), whereas suspended particles may also harbor beneficial microorganisms that colonize the rhizosphere and may enhance plant growth (Schmautz et al., 2017).

5.4.2 Nutrient cycling in aquaponics

Hydroponics allows for precise management of nutrients, with solutions specifically formulated and regulated to meet the exact requirements of different plant species throughout their growth cycles. In contrast, aquaponics systems rely predominantly on nutrients derived from fish waste and unconsumed feed pellets, integrating aquaculture and hydroponics into a symbiotic relationship that maximizes resource efficiency (Trang et al., 2017)Yang and Kim, 2020). At the heart of this integration lies the nutrient cycle, which is a complex and delicate balance of biological and chemical processes that transform fish waste into plant nutrients. This dependence on fish-derived nutrients makes monitoring and controlling nutrient concentrations in water for plant uptake considerably more challenging than in hydroponics (Yep et al., 2018). This section summarizes the intricate mechanisms of nutrient cycling in aquaponic units, examining the transformation pathways, the roles of microorganisms, the influence of system design on nutrient availability, and the challenges and opportunities in optimizing the nutrient use efficiency.

5.4.3 The nitrogen cycle

The nitrogen cycle forms the cornerstone of all aquaponic systems and is arguably the most critical nutrient transformation process (Zou et al., 2016). In aquaculture components, fish excrete ammonia (NH_4^+) as their primary nitrogenous waste product, primarily through gill diffusion and urine (Timmons and Ebeling, 2013). Left unmanaged, ammonia would quickly accumulate to toxic levels for fish. In essence, the fish tank and biofilters in aquaponic systems resemble those in RAS units in terms of total ammonia nitrogen (TAN) and nitrite oxidation, which suggests that the same nitrifying processes take place in both systems (Wongkiew et al., 2017). However, in aquaponic systems, ammonia is transformed into nitrite and then into nitrate, which becomes a valuable resource through a two-stage biological process known as nitrification. The microbial oxidation of ammonia (NH_4^+) via nitrite (NO_2^-) to nitrate (NO_3^-) in the presence of oxygen, termed nitrification, is a central process of the nitrogen cycle that is chemically represented by the following stoichiometric equations:

$$NH_4^+ + O_2 \rightarrow NO_2^- + 3H^+ + 2e^- \text{ (ammonia oxidation)}$$

$$NO_2^- + H_2O \rightarrow NO_3^- + 2H^+ + 2e^- \text{ (nitrite oxidation)}$$

Two phylogenetically distinct groups of bacteria are collectively involved in the nitrification process. These two groups of bacteria are generally described as chemosynthetic autotrophic bacteria because they derive their energy from inorganic compounds, whereas heterotrophic bacteria derive energy from organic compounds (Moschos et al., 2022). In the first step of nitrification, ammonia-oxidizing bacteria obtain their energy by catabolizing unionized ammonia to nitrite (NO_2^-) and include bacteria of the genera *Nitrosomonas, Nitrosococcus, Nitrosospira, Nitrosolobus*, and *Nitrosovibrio*, whereas nitrite-oxidizing bacteria oxidize nitrite to nitrate (NO_3^-), and include bacteria of the genera *Nitrobacter, Nitrococcus*, and *Nitrospina* (Wongkiew et al., 2017; Preena et al., 2021). However, recent research has expanded our understanding of this process with the discovery of complete ammonia oxidizers (comammox) bacteria that can perform both steps, converting ammonia directly to nitrate (Daims et al., 2015). These bacteria, particularly *Nitrospira* species, have been identified in some aquaponic biofilters, potentially offering more efficient nitrogen conversion pathways (Bartelme et al., 2019). The major factors affecting the rate of nitrification are the ammonia and nitrite levels, the carbon/nitrogen ratio, the DO, pH, temperature, alkalinity values, and water turbulence levels (Chen et al., 2006). For a comprehensive revision on the effects of water quality factors on nitrification kinetics in aquaponics, readers are encouraged to consult the work of Wongkiew et al. (2017), among others.

The nitrate (NO_3^-) produced through nitrification serves as the primary nitrogen source for plants in the hydroponic compartment of the aquaponic system. Unlike ammonia and nitrite, nitrate is relatively nontoxic to fish and can accumulate to much higher concentrations without causing harm to fish. Plants absorb nitrate through their roots via both active and passive transport mechanisms, incorporating the absorbed nitrogen into amino acids, proteins, and other essential cellular components that sustain their growth and development (Bindraban et al., 2015).

5.4.4 Phosphorus dynamics

In aquaponic systems, phosphorus enters primarily through fish feed and is excreted in both solid and dissolved forms. Approximately 30%–40% of phosphorus in feed is retained by fish, whereas the rest is excreted into the water (Schneider et al., 2005), while phosphorous excretion in open-flow aquaculture systems is considered a negative factor because of its contribution to water eutrophication, this represents an opportunity in aquaponic systems, since this nutrient is essential for photosynthesis, energy transfer, and growth in plants. According to Cerozi and Fitzsimmons (2017), phosphorus assimilation by fish and plants may account for 72% of the total phosphorus input, demonstrating the high phosphorus utilization efficiency of aquaponic systems. Nevertheless, the levels of phosphorus in water need to be taken cautiously, since phosphorus deficiency causes stunted plant growth, whereas phosphorus excess may lead to antagonistic interactions with micronutrients, especially zinc. Regardless of the importance of this nutrient in aquaponic systems, phosphorus dynamic modeling in aquaponic systems has been poorly investigated (Cerozi and Fitzsimmons, 2016; Zou et al., 2016; Cerozi and Fitzsimmons, 2017).

The bioavailability of phosphorus in aquaponic systems depends significantly on water pH levels. At pH values above 7.5, phosphorus tends to precipitate with calcium and magnesium ions, forming insoluble compounds that are unavailable to plants (Yogev et al., 2016). This precipitation can also occur on surfaces within the system, including pipes and media in grow beds, creating a phosphorus sink. In this sense, Zou et al. (2016) demonstrated that maintaining slightly acidic conditions (pH 6.0–6.5) in the hydroponic component significantly increased phosphorus availability to plants. Thus the availability of phosphorus tends to decrease with increasing pH values as phosphorus binds to several cations present, leaving fewer free phosphate ions available for uptake by plants (Cerozi and Fitzsimmons, 2016). However, lowering the pH for enhancing the availability of phosphorus presents a challenge in system management, as nitrifying bacteria prefer slightly alkaline conditions (pH 7.0–8.0) for optimal performance, creating competing requirements within a single system (Wongkiew et al., 2017). Based on these results, Cerozi and Fitzsimmons (2016) recommend that pH in aquaponics systems is maintained at a range of 5.5–7.2 for optimal availability and uptake by plants.

Mobilization of precipitated phosphorus can occur through bacterial action, particularly phosphate-solubilizing bacteria such as *Bacillus* sp. and *Pseudomonas* sp. These bacteria produce organic acids that can dissolve calcium phosphate and other insoluble forms, releasing bioavailable orthophosphate that may be absorbed by plants, promoting their growth (Raghu and MacRae, 1966; Rodrıǵuez and Fraga, 1999).

5.4.5 Potassium, calcium, and magnesium cycling

While nitrogen and phosphorus have received considerable attention in aquaponic research, other macronutrients are equally essential for plant growth and development. However, these nutrients may often exist in suboptimal concentrations in typical aquaponic systems.

Plants require potassium for many vital functions, such as activating enzymes, regulating water balance, and supporting photosynthesis. When plants have sufficient potassium, they can perform these functions more effectively, which promotes improved growth and development. In addition, potassium helps move nutrients throughout the plant, enhancing the absorption of other important elements needed for plant health (Johnson et al., 2022). When plants lack sufficient potassium, the initial symptoms appear as yellowish spots along the edges of older leaves that rapidly develop into necrotic (dead) tissue. Plants suffering from potassium deficiency show increased vulnerability to environmental stresses, including sudden temperature decreases and water shortages, and become more susceptible to fungal infection growth (Maucieri et al., 2019). In aquaponics systems, potassium is frequently a limiting nutrient in aquaponics as fish feed generally contains insufficient quantities relative to plant requirements (Duarte and Cerozi, 2024). Unlike nitrogen and phosphorus, potassium is not a structural component of fish tissues and is primarily used for osmotic regulation. Consequently, only small amounts are retained by fish, with most being directly excreted through the gills and urine (Seawright et al., 1998). Even with this direct excretion, potassium levels are typically insufficient for optimal plant growth, often necessitating supplementation (Nicoletto et al., 2018). Research shows that adding potassium to aquaponic systems—whether directly to the water, through foliar sprays, or by enriching fish feed—has been linked to increased plant biomass production. However, the optimal amount of potassium supplementation can differ based on the specific plant species being cultivated and the chemical composition of the water in the system (Duarte and Cerozi, 2024).

Calcium (Ca) and magnesium (Mg) are generally present in adequate quantities in aquaponic systems due to their abundance in most water sources. However, their bioavailability is strongly influenced by pH and interactions with other ions (Delaide et al., 2016). For instance, at a pH above 8.3, Ca^{2+} and Mg^{2+} ions easily precipitate as carbonates by reacting with $CO_3{}^{2-}$ similarly, sulfate forms strong complexes with Ca^{2+} and Mg^{2+}.

Calcium plays several critical roles in plant physiology, such as contributing to cell wall structure, regulating membrane permeability, supporting cell division and elongation, and enhancing plant resistance against pathogens. Unlike some nutrients, calcium cannot be easily redistributed within the plant as it has limited mobility, which means deficiency symptoms first appear in newly formed tissues. When plants lack sufficient calcium, they typically exhibit stunted overall growth, deformed margins on younger leaves, light green or chlorotic (yellowed) new tissues, and underdeveloped root systems with few fine roots (Maucieri et al., 2019). Although calcium is generally not a limiting nutrient in aquaponics systems, calcium may form insoluble precipitates with phosphates at high pH levels, reducing the availability of both nutrients (Goddek et al., 2016).

Magnesium is essential for chlorophyll molecule formation in plants. Its absorption becomes restricted at pH levels under 5.5, and it competes with potassium and calcium for uptake (Akter et al., 2024). When magnesium is deficient, plants display distinctive symptoms, including yellowing between leaf veins and internal chlorosis, specifically in the older, basal leaves (Maucieri et al., 2019). This pattern occurs because magnesium is highly mobile within plants—when supplies are limited, plants break down chlorophyll in older leaves and redistribute the magnesium to newer growth.

5.4.6 Micronutrient dynamics

Iron (Fe), manganese (Mn), zinc (Zn), copper (Cu), boron (B), and molybdenum (Mo) constitute the essential micronutrients required for plant growth. Their availability in aquaponic systems is influenced by complex interactions involving pH, redox conditions, chelation, and microbial activity (Bittsánszky et al., 2016). Most of the available knowledge on micronutrient requirements and dynamics in aquaponics emerged from hydroponic systems, where the application of water-based nutrient solutions allows a proper understanding of plants' nutrient requirements and how different nutrients affect each other's availability, uptake, and utilization by plants.

Iron is among the most important micronutrients because it is key in many biological processes, such as photosynthesis. Iron deficiency is particularly common in aquaponic systems due to the tendency of iron to oxidize from the plant-available ferrous (Fe^{2+}) state to the less available ferric (Fe^{3+}) form at the neutral to slightly alkaline pH typical of most systems (Nicoletto et al., 2018). In addition, the levels of iron in aquafeeds tend to be lower, considering the prooxidant effect of this micronutrient. The deficiency symptoms are characterized by interveinal chlorosis from the young leaves toward the older basal ones, and by reduced root system growth (Maucieri et al., 2019). Supplementation with chelated forms of iron, such as Fe–DTPA or Fe–EDDHA, is often necessary to prevent deficiency symptoms in plants (Bugbee, 2004).

Zinc and copper are present in fish feed and may accumulate in systems over time. While essential in small quantities, these elements can become phytotoxic at high concentrations (Goddek et al., 2015). However, zinc deficiency is uncommon but can be identified by chlorotic (yellowing) spots between leaf veins, significantly shortened internodes, downward curling of leaves (epinasty), and stunted overall plant growth. Copper plays an essential role in plant respiratory and photosynthetic processes. Its absorption is hindered when water pH exceeds 6.5, while pH levels below 5.5 may cause toxic effects. Conversely, molybdenum availability increases at higher pH levels, contrasting with most other micronutrients (Bugbee, 2004). High amounts of ammonia and phosphorus can interact with copper, reducing its availability to plants. When copper is present in excessive quantities, it interferes with the plant's ability to absorb iron, manganese, and molybdenum. Plants suffering from copper deficiency typically display interveinal chlorosis (yellowing between leaf veins), which progressively worsens until the leaf tissues collapse, giving them a dried-out or desiccated appearance. Molybdenum is essential for protein synthesis and nitrogen metabolism in plants. Unlike other micronutrients, it is more readily available to plants at neutral pH values. When plants suffer from molybdenum deficiency, symptoms typically begin with chlorosis (yellowing) and necrosis (tissue death) along the main ribs of older leaves. In addition, younger leaves often appear deformed (Maucieri et al., 2019).

5.4.7 Bacterial communities

The relevance of the microbiome in aquaponic systems cannot be overstated, as it is a cornerstone in optimizing system performance and promoting sustainability. Understanding microbial dynamics enables aquaponics practitioners to optimize nutrient cycling processes and minimize harmful byproduct formation. By developing a deeper comprehension of how microorganisms break down nutrients and how various operational factors affect their efficiency, aquaponic system managers can strategically adjust these parameters. For instance, modifying flow rates or extending retention time in biofilters and sludge digesters can significantly increase performance. Similarly, careful pH monitoring creates ideal conditions for microbial communities to grow out. Thus, depending on the environmental parameters, such as nutrient and oxygen availability, hydrodynamics, and water temperature, the composition and function of the bacterial community change in each compartment of the aquaponic unit. Because of the design, function, and configuration of the aquaponic systems, each of the individual compartments presents different environmental conditions (Section 5.3.1. System components). These conditions will shape the microbial processes occurring in the compartments and affect the quality of the water passing through them. Thus, finding the balance between the requirements for fish, plants, and microorganisms is important for a successful aquaponic production system.

Microorganisms have an essential function in enhancing plant nutrient availability by extracting both macro- and micronutrients from uneaten feed and waste materials. The composition and activity of microbial communities in aquaponic systems are significantly affected by the levels of carbon and nitrogen present in the water. While nitrifying bacteria have received primary attention in aquaponic literature because of their important role in nitrification, heterotrophic bacteria play equally crucial roles in nutrient cycling. These organisms decompose solid waste (feces, uneaten feed) and dissolved organic matter, releasing nutrients in forms available to plants (Schmautz et al., 2017). The ratio of carbon-to-nitrogen (C/N) significantly influences heterotrophic activity. With high organic carbon availability, heterotrophs can immobilize inorganic nitrogen, temporarily reducing its availability to plants (Wongkiew et al., 2017). Conversely, in carbon-limited conditions, heterotrophs mineralize organic nitrogen, releasing ammonia that enters the nitrification pathway (Zou et al., 2016). Several studies have pointed out that autotrophic nitrifying microorganisms typically exist in smaller populations compared with heterotrophic microbes, primarily because of their relatively slow maximum growth rates. The presence of organic carbon can substantially decrease TAN removal. Both nitrifying and heterotrophic bacteria require oxygen for their metabolic processes, creating competition for this resource in aquaponic environments. While heterotrophic bacteria play a vital role in the overall efficiency of aquaponic systems, elevated total carbon concentrations can cause these organisms to impede the biofiltration process (Kasozi et al., 2021). Research has demonstrated that when the carbon-to-nitrogen (C/N) ratio reaches a value of 6, the efficacy of ammonia elimination can decrease up to 90%. The increased heterotrophic assimilation process tends to inhibit nitrification, resulting in reduced nitrate formation. Within biofilters, higher C/N ratios tend to favor heterotrophic microbes over the autotrophic microbes responsible for nitrification. Heterotrophic bacteria also contribute to phosphorus cycling by releasing phosphatase enzymes that hydrolyze organic phosphorus compounds into orthophosphate (Cerozi and Fitzsimmons, 2017). In addition, heterotrophs produce siderophores—organic compounds that chelate iron and other metals, potentially increasing their bioavailability to plants (Bartelme et al., 2018). Maintaining appropriate organic carbon levels and DO concentrations is critical for supporting a balanced microbial ecosystem. Studies have shown that increasing C/N ratios can significantly alter microbial community structures, with nitrogen fixation rates potentially decreasing by half when influent C/N ratios increase from 2 to 5. Table 5.1 summarizes the most important prokaryotic genus involved in different microbiological processes as reviewed by Kasozi et al. (2021).

TABLE 5.1 Primary microbiological activities and key prokaryotic organisms within aquaponic systems.

Process	Microorganism at genus level
Nitrification	
Ammonia oxidation	*Nitrosomonas, Nitrosococcus, Nitrosospira,Nitrosolobus, Nitrosovibrio*
Ammonia oxidation by archaea	*Nitrosopumilus, Nitrososphaera*
Nitrite oxidation	*Nitrobacter, Nitrospira, Nitrococcus, Nitrospina*
Complete ammonia oxidation	*Nitrospira*
Denitrification	*Dokdonella, Thermomonas*
Anaerobic ammonium oxidation (Anammox)	*Brocadia*
Mineralization	*Pseudomonas, Flavobacterium, Arcobacter, Sphingobacterium*
Sulfate reduction	*Fusibacter, Bacteroides, Desulfovibrio, Dethiosulfovibrio*
Organic phosphorus mineralization	*Modestobacter*
Iron cycling	*Acidibacter*
Nitrogen fixation	*Pontibacter, Pseudonocardia*

Several authors have postulated that it is feasible to intentionally manipulate the bacterial communities in aquaponics systems, enhancing system performance and resilience (Bartelme et al., 2018). In this context, plant growth-promoting microorganisms play a pivotal role in sustainable agricultural systems by enhancing crop productivity since these beneficial microbes contribute to phosphorus uptake by plant roots primarily through the mineralization of organic phosphorus compounds and the solubilization of insoluble phosphate forms. The presence of naturally occurring rhizospheric phosphorus-solubilizing microorganisms has been documented since the early 20th century, with several representatives of the genera *Bacillus, Pseudomonas*, and *Enterobacter* spp. Among these, species of the genus *Bacillus* have received considerable attention in the scientific literature because of their well-characterized mechanisms of action and their potential for application as biofertilizers in diverse agroecosystems (Ruzzi and Aroca, 2015), including aquaponics (Cerozi and Fitzsimmons, 2016).

Techniques for microbiome engineering include selective inoculation with beneficial microorganisms, substrate selection to favor desired metabolic pathways, and operating condition adjustments to shape community composition (Wongkiew et al., 2017). For example, Joyce et al. (2019) demonstrated that inoculation with selected plant growth-promoting rhizobacteria enhanced nitrogen use efficiency and plant growth rates in experimental aquaponic systems. In this sense, Cerozi and Fitzsimmons (2016) demonstrate that the addition of a commercial mixture of *Bacillus* spp. (Sanolife PRO-W; 5.0×10^{10} CFU/g, INVE, Belgium) enhanced lettuce growth, increased the accumulation of phosphorus in plant tissues, and increased the chlorophyll content in their leaves. The former authors also postulated that the tested inoculum of *Bacillus* spp. affected the dynamics of ammonia, nitrite, and nitrate in the system, as well as increased the levels of dissolved orthophosphate in the water.

5.5 Plant and fish species selection

Multiple investigators have shown that organisms cultivated within aquaponic systems must exhibit specific degrees of biological and ethological compatibility (Somerville et al., 2014; Junge et al., 2020). They all should live in a similar confined environment, sharing water and water physico-chemical traits such as temperature, DO, pH, salinity, in addition to ammonia, nitrites, and nitrates levels, among many others. This is particularly true when dealing with a coupled aquaponic system (Baganz et al., 2021). Besides that, it is crucial to consider additional aspects such as market conditions (e.g., consumers' preferences and means of production availability), national or regional regulation (e.g., for introducing new species or related to food safety), facility traits (e.g., open door, greenhouse, and urban buildings), and, last but not least, the system design itself. Hence, when choosing animal and plant species to be reared within an aquaponic system, an accurate analysis of these many aspects is then necessary (Knaus et al., 2015). To achieve the best commercial setup and take the best decisions possible, all these aspects should be analyzed using a holistic approach and making the best effort to reduce as much as possible the subjectivity of the choices. To this purpose, a preliminary comprehensive analysis of the available literature and data collection should be carried out. Then, carrying out a simple Strengths, Weaknesses, Opportunities, and Threats analysis (SWOT) may be helpful to better define the whole framework (Mancuso et al., 2016; Gasco et al., 2020). However, to holistically assess the complex multidimensional framework, a multicriteria decision-aid (MCDA) or multicriteria decision-making (MCDM) might be also considered. Both approaches provide a systematic methodology to integrate heterogeneous and uncertain information with cost–benefit information and stakeholders' views, with that creating an understandable framework and ranking of project alternatives (Huang et al., 2011).

There are several available tools that can be used to this purpose. For instance, DEX is a qualitative decision support methodology (Bohanec et al., 2013), implemented in the DEXi software (https://kt.ijs.si/MarkoBohanec/dexi.html) that helps to divide the decision-making problem into less complex subproblems, hierarchically organized (DEXi is distributed as "freeware"). A comprehensive application of this approach has been described by Rossi et al. (2021) for marine organism selection and Puccinelli et al. (2022) for crop species selection.

5.5.1 Optimal plant species for aquaponics systems

Aquaponic cultivation success depends primarily on selecting plant species that demonstrate high adaptability to the unique conditions present in these integrated systems (Knaus and Palm, 2017). Research indicates that plant selection must be calibrated to specific system parameters, particularly regarding coupling degree (Baganz et al., 2021). Decoupled systems facilitate cultivation of nutrient-intensive crops through targeted supplementation protocols, while coupled systems typically demonstrate optimal performance with crops requiring lower nutrient concentrations (Goddek et al., 2016; Palm et al., 2018). Water temperature also significantly influences species selection, with tropical and subtropical plants (optimal range 22–28 °C) demonstrating compatibility with warm-water aquaculture species, while cool-season crops may require temperature management strategies when integrated with warm-water fish production (Rakocy et al., 2006; Love et al., 2015). Different studies have consistently demonstrated that system maturity correlates positively with the performance of nutrient-demanding crops, as microbial communities become more thoroughly established and nutrient cycling efficiencies increase (Tyson et al., 2011; Wongkiew et al., 2017). The following species have demonstrated particular suitability based on empirical research.

5.5.1.1 *Leafy vegetables*

Leafy greens consistently demonstrate superior performance in aquaponic environments due to their relatively modest nutritional requirements, rapid growth cycles, and good adaptability to aquaponic conditions. The most common leafy vegetables in aquaponics configurations are:

Lettuce (*Lactuca sativa*) has multiple varieties (i.e., romaine, iceberg, red leaf, green leaf, Parris Island, and crispy, among others) that demonstrate exceptional adaptability with harvest periods of 24–32 days and optimal growth at nitrate–nitrogen concentrations of 21–42 mg/L (Rakocy et al., 2006; Pantanella et al., 2015).

Basil (*Ocimum basilicum*) exhibits excellent productivity with a higher market value than lettuce, while requiring similar nutritional parameters. Studies report yields of 2.0–2.5 kg/m^2 per harvest cycle under optimal conditions (Rakocy et al., 2004; Somerville et al., 2014).

Spinach (*Spinacia oleracea*) has been successfully cultivated in systems maintaining water temperatures below 18°C, demonstrating particular suitability for cool-season production (Lennard and Leonard, 2006; Endut et al., 2010).

5.5.1.2 *Fruiting vegetables*

Despite higher nutritional requirements, several fruiting species demonstrate commercial viability in well-established aquaponic systems, even though their production and management are more complex compared to leafy vegetables:

Tomato (*Solanum lycopersicum*) requires elevated potassium and calcium concentrations, frequently necessitating supplementation in coupled aquaponics systems. Yields of 2–4 kg/plant have been documented in optimized aquaponic environments (Graber and Junge, 2009; Roosta and Hamidpour, 2011).

Cucumber (*Cucumis sativus*) is particularly suitable for decoupled aquaponic systems where nutritional parameters can be more precisely regulated. Research indicates productivity comparable to conventional hydroponic cultivation when appropriate nutrient supplementation is provided (Tyson et al., 2008; Suhl et al., 2016).

Bell pepper (*Capsicum annuum*) may be successfully cultivated in systems with sufficient phosphorus and potassium supplementation (Rakocy et al., 2004; Pinho et al., 2017).

5.5.1.3 *Herbs and other crops*

Mint (*Mentha* spp.) requires careful management, but providing multiple harvests from single plantings, even though the use of chemical nutrient solutions is encouraged to support high vegetative growth (Buzby and Lin, 2014; Saha et al., 2016).

Watercress (*Nasturtium officinale*) demonstrates exceptional adaptability to aquaponic environments with documented medicinal properties enhancing market value (Endut et al., 2011; Palm et al., 2019).

Pak choi (*Brassica rapa subsp. chinensis*) shows rapid growth cycles and excellent nutrient utilization efficiency (Liang and Chien, 2013; Shete et al., 2016), whereas optimal vegetative growth has been reported when using a substrate-based closed capillary water distribution system compared with floating rafts or NFT systems (Sundar and Chen, 2020).

5.5.2 Optimal fish species for aquaponics systems

The selection of appropriate fish species represents a critical determinant of aquaponic system performance and economic viability. Research indicates that species selection should be calibrated to specific system parameters, including:

Stocking density: Species-appropriate stocking densities range from 15 to 20 kg/m^3 for tilapia to 40–50 kg/m^3 for barramundi in well-managed systems with appropriate biofiltration systems guaranteeing proper nitrification (Rakocy et al., 2006; Palm et al., 2018).
Temperature: Heating and/or cooling infrastructure determines viable species selection, with energy costs frequently representing a decisive factor in commercial applications (Love et al., 2015; Goddek et al., 2016).
Water quality parameters: Species-specific tolerance ranges for DO, pH, total dissolved ammonia, and nitrite concentrations must align with system design and production capacity (Rakocy et al., 2006; Somerville et al., 2014).
Market considerations: Local market preferences, regulatory frameworks, and price significantly influence species selection for commercial applications (Kyaw and Ng, 2017; König et al., 2018).

Based on the former principles, the following species have demonstrated particular suitability based on extensive empirical research:

5.5.2.1 *Tilapia* (**Oreochromis** *spp.)*

Tilapia species, particularly Nile tilapia (*O. niloticus*), is the preferred choice for aquaponics systems due to several advantageous characteristics like its environmental tolerance since this species exhibits exceptional resilience to fluctuating water quality parameters, including temperature (24–30°C optimal range), DO (>4 mg/L), and pH (6.5–7.5) (El-Sayed, 2006; Rakocy et al., 2006). Furthermore, tilapias are characterized by their good adaptability to intensive culture conditions and high growth rates, achieving marketable size (350–500 g) within 6–8 months under optimal conditions (Rakocy et al., 2004; Pantanella et al., 2015). Regardless of its wide use in aquaponic systems, there are still gaps in terms of tilapia management in aquaponic systems, such as determining the optimal dietary crude protein levels in aquafeeds for GIFT tilapia or the optimal feeding frequency for supporting the highest growth of tilapia and plants in aquaponic systems (Hussain and Brown, 2024).

5.5.2.2 *Common carp* (**Cyprinus carpio***)*

Common carp represents another extensively utilized species in aquaponic production, particularly in temperate regions. This species tolerates a wide range of water temperatures (4–34°C), even though in aquaponic systems, they display better results when reared at moderate temperature ranges (22–28°C), facilitating production in transitional climate regions (Masser et al., 1999; Danaher et al., 2013). This omnivorous species can reach the harvest size

(500–600 g) in less than a year. This species benefits from established consumer recognition and acceptance in multiple regional markets (Love et al., 2015; Palm et al., 2018). Common carp exhibits behavioral characteristics compatible with various aquaponic configurations, including DWC and media-based systems (König et al., 2016).

5.5.2.3 *Channel catfish* (Ictalurus punctatus)

Channel catfish demonstrates particular suitability for aquaponic production in North American contexts mainly because of its environmental hardiness since this species of catfish tolerates high stocking density, and suboptimal water quality parameters, including temporarily reduced DO levels (>3 mg/L) and wider pH ranges (6.0–8.5) than many alternative species (Wurts and Durborow, 1992; Palm et al., 2018).

5.5.2.4 *Rainbow trout* (Oncorhynchus mykiss)

Rainbow trout represents the primary cold-water species utilized in aquaponic production. This salmonid species requires water temperatures between 10 and 18°C, facilitating aquaponic production in cool-climate regions where tilapia cultivation would necessitate prohibitive heating costs (Adler et al., 2000; Buzby and Lin, 2014). Trout is characterized by its highly valued flesh with favorable organoleptic properties and nutritional profiles, enhancing market acceptance (Davidson et al., 2009; Trang and Brix, 2014). The use of trout in aquaponics generates effluent with elevated phosphorus and micronutrient concentrations in comparison to warm-water species, potentially reducing supplementation requirements for fruiting crops (Goddek et al., 2015; Suhl et al., 2016).

5.5.2.5 *Barramundi* (Lates calcarifer)

Barramundi demonstrates increasing adoption in commercial aquaponic applications, particularly in Australia and Southeast Asia, because of its exceptional growth rates, frequently exceeding those of tilapia or other farmed species in aquaponics (Partridge et al., 2008; Love et al., 2015). Furthermore, this species has a premium market pricing, enhancing economic viability despite higher infrastructure and operational requirements (Kyaw and Ng, 2017; Palm et al., 2018). Because of its euryhaline characteristics, this species represents a versatile option for brackish or saltwater aquaponics where appropriate salt-tolerant plant species are integrated (Pantanella et al., 2015; Wongkiew et al., 2017).

5.5.2.6 *Emerging species*

Jade perch (*Scortum barcoo*) exhibits favorable growth characteristics in aquaponic systems. Several authors have postulated that integrating aquaponics with Jade perch farming could provide valuable diversification opportunities and boost economic resilience for businesses in the aquaculture sector in areas with stringent nutrient release regulations in Australia, since this combination can enhance nutrient management strategies while addressing the increasing market demand for locally grown native fish (Emerenciano et al., 2025).

Gray mullet (*Mugil cephalus*) has been identified as a promising candidate for integration into aquaponic systems (Rossi et al., 2021). This species is particularly valuable because of its tolerance for varying salinity levels and omnivorous feeding habits, which result in lower protein requirements compared with carnivorous fish species (Gisbert et al., 2024). The versatility of gray mullet makes it an excellent option for addressing nutritional needs in developing regions, either in saltwater or freshwater aquaponic systems, while also being highly regarded in more affluent markets where both its meat and specialty products, such as salted roe (bottarga).

Freshwater prawn (*Macrobrachium rosenbergii*) has shown compatibility with DWC configurations as well as in polyculture systems with tilapia (Sace and Fitzsimmons, 2013; Alberts-Hubatsch et al., 2017).

Murray cod (*Maccullochella peelii*) is a native Australian species with premium market acceptance (firm and white flesh) and moderate temperature requirements (16–24°C) that adapts well to aquaponics (Lennard, 2021). This species may reach up to 2–3 kg in 12–18 months under optimal conditions, and it tolerates higher stocking densities than many other species when properly managed (Trang and Brix, 2014; Love et al., 2015)

5.6 Aquaponics, sustainability, and multifunctionality

In the last decades, a surge in publications focused on the term "sustainability" has been observed, and despite that, "sustainability" remains an open concept with several possible and interchangeable interpretations (Purvis et al., 2019). However, the most known and recognized concept is the one that describes sustainability as three intersecting circles, each one representing society, environment, and economy. Moreover, when dealing with sustainability, it is not always approached holistically, considering equally the three mentioned pillars. The social pillar is frequently underconsidered or neglected, probably because it does not play a crucial role from a technical point of view. Instead, this aspect deserves consideration when it comes to aquaponics. Indeed, by its nature, aquaponics lends itself very well to promoting social inclusion and welfare, since it can be easily implemented in several social contexts, without losing its role in securing food for people (Milliken and Stander, 2019).

5.6.1 Urban aquaponics: Sustainable food production in metropolitan environments

Urban aquaponics represents a promising solution for sustainable food production within metropolitan areas, addressing multiple urban challenges simultaneously. According to Dos Santos (2016), these systems can transform underutilized urban spaces such as rooftops, abandoned buildings, and vacant lots into productive food hubs while providing significant environmental benefits. Proksch et al. (2019) documented numerous case studies demonstrating commercial viability in cities worldwide, with operations ranging from community-scale to industrial implementations. The integration of aquaponics within urban environments creates multiple sustainability benefits beyond food production. For instance, Gould and Caplow (2012) demonstrated how these systems can reduce stormwater runoff and mitigate urban heat island effects when implemented on rooftops. Furthermore, Palm et al. (2018) highlighted how urban aquaponics significantly shortens supply chains, reducing transportation emissions while providing fresher food to urban consumers. Some examples are provided about the potential incorporation of aquaponic systems in urban environments:

Home, according to Duarte et al. (2015), "aquaponics is a climate-smart agriculture system that allows people to bring nature into the comfort of their homes." Tiny aquaponic systems can be implemented even in small apartments to serve as interior decoration (e.g., aquarium with ornamental fish), with that contributing to improve people's feelings and welfare. Definitely, it may also provide aromatic herbs to be used when home-cooking (e.g., parsley, basil, thyme, and mint).

Home backyards, small-scale systems can be installed in home backyards for obbistic purpose, which can be beneficial, among others, for reducing human stress levels related to the hectic and modern urban lifestyle (Beery et al., 2023). Also in this case, the system can provide food for personal consumption (either fish or plants, such as lettuce and other vegetables).

Apartment buildings, small-scale aquaponic systems can be implemented in common areas such as gardens and rooftop. Their management can be shared by condominiums as a best fit to them, for instance, by assigning retirees to take care of daily tasks and granting supervision to an internal and/or external expert. Then, the products (fish and vegetables) can be sold or even shared according to the time spent by each condominium in this activity.

Urban communities, similarly, people living within a specific urban district, can take advantage of unused old buildings to establish an aquaponic production. It might be run according to different associative forms, such as social association or cooperative. In addition to produce food, the core of the activity might be the inclusion of categories of people affected by specific weaknesses.

Young (2019) found that urban aquaponics operations frequently incorporate educational components, connecting urban residents with food production systems and enhancing food literacy. However, as noted by Specht et al. (2019), urban practitioners face unique challenges, including higher land costs, complex regulatory frameworks, and the need for specialized business models that often incorporate direct-to-consumer sales, premium pricing strategies, and diversified revenue streams, including agritourism and educational programming.

Schools, small- or large-scale aquaponic systems can also be implemented in schools for educational purposes. Aquaponics can be considered a living lab, where students can approach basic subjects such as mathematics, physics, chemistry, biochemistry, biology, and microbiology, or even applied sciences such as aquaculture, crop science, food science, and, why not, economy and social sciences (Goddek et al., 2019). The system can be easily run by assessing the students to take care of ordinary tasks, being supervised by specialized technicians or professors. Moreover, the products obtained by this activity can be used by the school canteen for preparing the daily meals for students and other employees.

5.6.2 Aquaponics as a tool for social development

Beyond its environmental advantages and flexibility to meet different climatic scenarios, aquaponics presents a wide array of social benefits that align with global development goals. One of aquaponics' most significant contributions is toward food sovereignty and security. It enables local, continuous production of both animal and plant protein, thereby reducing dependence on external food sources and increasing access to nutritious foods in underserved communities (Love et al., 2014; Somerville et al., 2014). This is particularly valuable in food desert areas or regions where traditional agriculture is not feasible because of land degradation, water scarcity, or urbanization.

Aquaponics also acts as a catalyst for education and capacity building. Community-based systems implemented in schools or training centers serve as living laboratories, enhancing awareness of sustainability while teaching skills in biology, ecology, engineering, and agriculture. Such educational initiatives can empower youth and foster interest in science, technology, engineering, or mathematics, as well as in sustainable practices and entrepreneurial ventures.

From an economic perspective, aquaponics offers a platform for inclusive entrepreneurship and job creation. Small-scale and modular systems are relatively accessible and can be adapted for use in rural cooperatives, backyard aquaculture models, or urban enterprises, offering income-generating opportunities for marginalized groups such as women and youth (Mchunu et al., 2018). The growing demand for local, sustainably produced food further enhances its commercial viability, especially when integrated into local food systems and markets.

Last but not least, aquaponics systems may also be implemented in refugee camps where people frequently live in extreme poverty and food shortage conditions (Milliken and Stander, 2019). Under the supervision of NGO experts, refugees can be educated and involved in the management of an aquaponic system, with that improving their psychological conditions and community resilience.

Moreover, aquaponics enhances community resilience and social cohesion. Its collaborative nature encourages shared management and decision-making, which can strengthen social ties, foster a sense of ownership, and build collective environmental stewardship (Saxena et al., 2022). In postcrisis scenarios or economically vulnerable settings, aquaponics can also contribute to recovery and regeneration efforts by promoting self-reliance and reducing resource dependency.

In conclusion, although aquaponics has been suggested as a solution to "food deserts" in urban settings, food sovereignty and food system resilience (Rizal et al., 2018), it can also promote social sustainability by reconnecting urban life with nature (Beery et al., 2023). This approach also gives the opportunity of involving several categories of citizens, such as retirees, students, and communities, including categories affected by specific weaknesses, with this improving resilience of urban communities.

5.7 Concluding remarks, future trends, and research directions

5.7.1 Aquaponics shows promising future perspectives in several key areas of food security

(1) *Sustainable food production*: Recent research indicates aquaponics systems are gaining traction as sustainable alternatives to conventional horticulture. They offer significant water conservation benefits, using up to 90% less water than traditional farming while producing both fish protein and plant crops in integrated systems (Junge et al., 2017; Goddek et al., 2019a). Furthermore, Palm et al. (2018) demonstrated that well-designed aquaponics systems can achieve near-closed nutrient cycles, significantly reducing fertilizer requirements compared to conventional hydroponics.

(2) *Climate resilience*: As a controlled environment food production method, aquaponics offers protection from extreme weather events and climate variability, especially considering the flexibility of this technology in terms of fish and plant selection to different climatic conditions (Lennard and Goddek, 2019). Forchino et al. (2017) conducted life cycle assessments showing that aquaponics can significantly reduce environmental impacts compared to conventional agriculture. Furthermore, Delaide et al. (2017) demonstrated that decoupled aquaponics systems provide greater flexibility to optimize growing conditions under changing climate scenarios.

(3) *Technological advancements*: The integration of IoT monitoring systems, AI-powered optimization algorithms, and automated control systems is transforming aquaponics (Kyaw and Ng, 2017; Monsees et al., 2017). In this sense, Espinal and Matulić (2019) reported how these technologies enable precise control of water quality parameters, feeding schedules, and environmental conditions, improving both efficiency and yields. Pattillo (2017) showed that automated monitoring reduced labor costs by up to 30% while improving production consistency.

(4) *Urban horticulture solutions*: Studies highlight aquaponics as particularly well-suited for urban food production, with commercial systems operating in repurposed warehouses, shipping containers, and rooftops in cities worldwide (Dos Santos, 2016; Goodman and Minner, 2019). Proksch et al. (2019) analyzed multiple case studies of urban aquaponics facilities, demonstrating their potential to address food deserts while reducing transportation emissions associated with traditional supply chains.

(5) *Economic viability*: While early research indicated challenges with economic sustainability (Engle, 2015; Tokunaga et al., 2015), newer business models incorporating value-added products, direct-to-consumer sales, and diversified revenue streams show improved economic outcomes (Love et al., 2015; König et al., 2018). Greenfeld et al. (2019) identified that successful operations often combined production with tourism, education, or community engagement to create multiple revenue streams.

5.7.2 Bridging the gap: Critical focus areas for advancing aquaponics technology

To fully actualize the considerable potential of aquaponics technology within global food production systems, concentrated research and development efforts must be strategically deployed across several critical domains. Addressing these key areas will be instrumental in overcoming current limitations, enhancing system efficiency, and facilitating broader commercial adoption. The scientific community, industry stakeholders, and policy makers must collaborate to advance methodological standardization, technological innovation, and regulatory frameworks that can support aquaponics' transition from a promising concept to mainstream agricultural practice. The following priority areas require particular attention and resource allocation to catalyze the transformation of aquaponics from an emerging technology to a significant contributor to sustainable food production landscapes.

(1) *Real-time monitoring of nutrient concentrations:* Monitoring nutrient dynamics remains challenging in commercial-scale aquaponic systems. Most operations rely on periodic laboratory analysis or proxy measurements, limiting the potential for precise nutrient management (Goddek et al., 2015). In this sense, current research clearly indicates that nutrient dynamics in aquaponics extend far beyond the basic nitrogen cycle traditionally emphasized in system design. Phosphorus availability, micronutrient transformations, solid waste mineralization, and the roles of diverse microbial communities all significantly influence system productivity and sustainability. Future advances in aquaponic nutrient management will

likely emerge from interdisciplinary approaches combining microbiology, plant physiology, aquaculture science, and engineering. By developing a more sophisticated understanding and control of nutrient cycling processes, the next generation of aquaponic systems may achieve unprecedented levels of efficiency, productivity, and sustainability.

(2) *Aquaponics tailor-made aquafeeds*: Aquaponics research needs to focus on the development of tailor-made species-specific aquafeeds that fulfill the nutrient requirements of fish and plants in the system. As the primary nutrient input, aquafeed composition fundamentally shapes nutrient cycling in aquaponic systems. Conventional fish feeds are formulated primarily for fish growth rather than balanced plant nutrition, contributing to some nutrient deficiencies observed in plants (Goddek et al., 2015). Development of specialized "aquaponic feeds" with modified nutrient profiles could better align fish waste outputs with plant requirements. For instance, Delaide et al. (2017) showed that increasing the potassium content in fish feed improved plant growth without negatively affecting fish performance. Similarly, manipulating dietary phosphorus forms may enhance its eventual availability to plants (Cerozi and Fitzsimmons, 2017).

(3) *Fish and plant selection, toward diversification*: In aquaponic systems, certain fish species have demonstrated superior adaptability, such as Nile tilapia emerging as the predominant choice, followed by common carp and African catfish. These species' success can be attributed to their remarkable tolerance for elevated nitrate concentrations and reduced oxygen levels in the aquatic environment. Despite the theoretical advantages of polyculture approaches—potentially yielding more balanced nutrient profiles for plant cultivation—comprehensive research investigating multiple-species fish integration within aquaponic systems remains notably absent from the literature (Yep and Zheng, 2019). Future investigations exploring polyculture dynamics could significantly advance system optimization and nutrient utilization efficiency. Within the plant component, leafy greens, particularly lettuce varieties, mint, and basil, have exhibited exceptional performance characteristics in aquaponic cultivation. Their widespread adoption stems from relatively modest nutritional requirements, consistent market demand, and favorable economic returns relative to system inputs. Conversely, flowering species remain underrepresented in aquaponics research literature. Systematic evaluation of flowering crop production within aquaponic frameworks represents a promising avenue for expanding this technology's application scope, potentially facilitating its integration into commercial floriculture operations and diversifying production capabilities.

(4) *Pest and disease management and control:* One of the bottlenecks hindering the development of aquaponics is effective pest and disease management. Plant-damaging organisms, including aphids, spider mites, whiteflies, and fungus gnats, alongside bacterial, fungal, and viral pathogens, have been documented causing substantial yield reductions. These challenges necessitate increased pest management investments, directly affecting operational economics (Folorunso et al., 2021). The integrated nature of aquaponics systems presents constraints for conventional pest control approaches. System design variations—whether coupled, decoupled, or incorporating mineralization units—further complicate management strategies, as interventions targeting plant pests risk harming fish populations or beneficial microorganisms. Pesticides, repellents, and even biological control agents must be evaluated for potential cross-system effects before implementation. In this sense, there exists a pressing need to develop specialized pest and disease management protocols specifically tailored for diverse aquaponics configurations. Such protocols must balance efficacy against plant pests while maintaining system integrity and biosecurity across all functional components.

(5) *Decision support systems:* Emerging sensor technologies, including ion-selective electrodes, optical methods, and biosensors, offer promise for continuous, multiparameter monitoring (Yep and Zheng, 2019). When coupled with machine learning algorithms for data interpretation and predictive modeling for the implementation of decision support systems, these technologies could enable dynamic system management responsive to changing conditions and crop requirements (Junge et al., 2017).

References

Adler, P.R., Harper, J.K., Wade, E.M., Takeda, F., Summerfelt, S.T., 2000. Economic analysis of an aquaponic system for the integrated production of rainbow trout and plants. Int. J. Recirc. Aquac. 1 (1), 15–34.

Akter, U., Hashem, S., Farhad, F.B., Rana, K.S., Haque, M.M., Salam, M.A., 2024. Effects of Epsom (magnesium sulfate) salt on growth performance of water spinach (*Ipomoea aquatica*) in nutrient film technique and media-based aquaponics systems. Aquac. Int. 32 (1), 691–707.

Alam, M.N.H.Z., Othman, N.S.I.A., Samsudin, S.A., Johari, A., Hassim, M.H., Kamaruddin, M.J., 2020. Carbonized rice husk and cocopeat as alternative media bed for aquaponic system. Sains Malays. 49, 483–492.

Alberts-Hubatsch, H., Ende, S.S.W., Schuhn, A., Slater, M.J., Slater, M., 2017. Integration of Macrobrachium rosenbergii into tropical freshwater aquaponic systems. Aquac. Int. 25 (1), 1–15.

Aslanidou, M., Elvanidi, A., Mourantian, A., Levizou, E., Mente, E., Katsoulas, N., 2023. Nutrients use efficiency in coupled and decoupled aquaponic systems. Horticulturae 9 (10), 1077.

Baganz, G.F.M., Junge, R., Portella, M.C., Goddek, S., Keesman, K.J., Baganz, D., Staaks, G., Shaw, C., Lohrberg, F., Kloas, W., 2021. The aquaponic principle—it is all about coupling. Rev. Aquac. 13 (3), 1661–1678.

Bartelme, R.P., Oyserman, B.O., Blom, J.E., Sepulveda-Villet, O.J., Newton, R.J., 2018. Stripping away the soil: plant growth promoting microbiology opportunities in aquaponics. Front. Microbiol. 9, 8.

Bartelme, R.P., Smith, M.C., Sepulveda-Villet, O.J., Newton, R.J., 2019. Component microenvironments and system biogeography structure microorganism distributions in recirculating aquaculture and aquaponic systems. Msphere 4, 10–1128.

Beery, T., Stahl Olafsson, A., Gentin, S., Maurer, M., Stålhammar, S., Albert, C., Bieling, C., Buijs, A., Fagerholm, N., Garcia-Martin, M., Plieninger, T., Raymond, C.M., 2023. Disconnection from nature: expanding our understanding of human–nature relations. People Nat. 5 (2), 470–488.

Bindraban, P.S., Dimkpa, C., Nagarajan, L., Roy, A., Rabbinge, R., 2015. Revisiting fertilisers and fertilisation strategies for improved nutrient uptake by plants. Biol. Fertil. Soils 51 (8), 897–911.

Bittsánszky, A., Uzinger, N., Gyulai, G., Mathis, A., Junge, R., Villarroel, M., Kotzen, B., Kőmíves, T., 2016. Nutrient supply of plants in aquaponic systems. Ecocycles 2 (2), 17–20.

Bohanec, M., Žnidaršič, M., Rajkovič, V., Bratko, I., Zupan, B., 2013. DEX methodology: three decades of qualitative multi-attribute modeling. Informatica 37 (1), 49–54.

Bugbee, B., 2004. Nutrient management in recirculating hydroponic culture. Acta Hortic. 648, 99–112.

Buzby, K.M., Lin, L.S., 2014. Scaling aquaponic systems: balancing plant uptake with fish output. Aquacult. Eng. 63, 39–44.

Cerozi, B.S, Fitzsimmons, K., 2016. The effect of pH on phosphorus availability and speciation in an aquaponics nutrient solution. Bioresour. Technol. 219, 778–781.

Cerozi, B.S., Fitzsimmons, K., 2017. Phosphorus dynamics modeling and mass balance in an aquaponics system. Agric. Syst. 153, 94–100.

Chen, S., Ling, J., Blancheton, J.P., 2006. Nitrification kinetics of biofilm as affected by water quality factors. Aquacult. Eng. 34 (3), 179–197.

da Silva Cerozi, B., Fitzsimmons, K., 2016. The effect of pH on phosphorus availability and speciation in an aquaponics nutrient solution. Bioresourc. Technol. 219, 778–781.

Daims, H., Lebedeva, E.V., Pjevac, P., Han, P., Herbold, C., Albertsen, M., Jehmlich, N., Palatinszky, M., Vierheilig, J., Bulaev, A., Kirkegaard, R.H., von Bergen, M., Rattei, T., Bendinger, B., Nielsen, P.H., Wagner, M., 2015. Complete nitrification by *Nitrospira* bacteria. Nature 528 (7583), 504–509.

Danaher, J.J., Shultz, R.C., Rakocy, J.E., Bailey, D.S., 2013. Alternative solids removal for warm water recirculating raft aquaponic systems. J. World Aquac. Soc. 44 (3), 374–383.

Davidson, J., Good, C., Welsh, C., Summerfelt, S.T., 2009. Comparing the effects of high vs. low nitrate on the health, performance and welfare of juvenile rainbow trout *Oncorhynchus mykiss* within water recirculating aquaculture systems. Aquacult. Eng. 59, 30–40.

Delaide, B., Goddek, S., Gott, J., Soyeurt, H., Jijakli, M.H., 2016. Lettuce (*Lactuca sativa* L. var. Sucrine) growth performance in complemented aquaponic solution outperforms hydroponics. Water 8, 467.

Delaide, B., Goddek, S., Gott, J., Soyeurt, H., Jijakli, M.H., 2017. Lettuce (*Lactuca sativa* L. var. Sucrine) growth performance in complemented aquaponic solution outperforms hydroponics. Water 9 (6), 429.

Delaide, B., Goddek, S., Keesman, K., Jijakli, H., 2018. A methodology to quantify aerobic and anaerobic sludge digestion performances for nutrient recycling in aquaponics. Biotechnol. Agron. Soc. Environ. 22, 106–112.

Dos Santos, M.J.P.L., 2016. Smart cities and urban areas—aquaponics as innovative urban agriculture. Urban For. Urban Green. 20, 402–406.

Duarte, A.J., Malheiro, B., Ribeiro, C., Silva, M.F., Ferreira, P., Guedes, P., 2015. Developing an aquaponics system to learn sustainability and social compromise skills. J. Technol. Sci. Educ. 5, 235–253.

Duarte, S.F.P., Cerozi, B.S., 2024. Enhancing plant growth in aquaponic systems via potassium manipulation in fish feeds: a pilot study of tailored feeds bridging nutritional gaps in aquaponics. Agric. Syst. 218, 104001.

Edwards, P., 2025. Co-evolution of fish and rice farming by Austroasiatic communities in the Neolithic Era and early documentation of aquaculture in China. Front. Aquac. 3, 1437833.

El-Sayed, A.F.M., 2006. Tilapia culture. CABI Publishing, Wallingford, UK, p. 277.

Emerenciano, M.G.C., Slinger, J., Koster, G., Aland, J., Lima, P.C., Arsic, M., O'Sullivan, C., 2025. Mineral supplementation in Jade perch (*Scortum barcoo*) aquaponics with lettuce: a comparison with hydroponics and RAS. Animals 15 (3), 317.

Endut, A., Jusoh, A., Ali, N., Wan Nik, W.B., Hassan, A., 2010. A study on the optimal hydraulic loading rate and plant ratios in recirculation aquaponic system. Bioresour. Technol. 101 (5), 1511–1517.

Endut, A., Jusoh, A., Ali, N., Wan Nik, W.B., 2011. Nutrient removal from aquaculture wastewater by vegetable production in aquaponic recirculation system. Desalin. Water Treat. 32 (1–3), 422–430.

Engle, C.R., 2015. Economics of Aquaponics. Southern Regional Aquaculture Center, Stoneville, MS SRAC Publication No. 5006.

Espinal, C.A., Matulić, D., 2019. Recirculating aquaculture technologies. Aquaponics Food Prod. Syst. 1, 35–76.

Folorunso, E.A., Roy, K., Gebauer, R., Bohatá, A., Mraz, J., 2021. Integrated pest and disease management in aquaponics: a metadata-based review. Rev. Aquac. 13 (2), 971–995.

Forchino, A.A., Lourguioui, H., Brigolin, D., Pastres, R., 2017. Aquaponics and sustainability: the comparison of two different aquaponic techniques using the life cycle assessment (LCA). Aquacult. Eng. 77, 80–88.

Gasco, L., Acuti, G., Bani, P., Dalle Zotte, A., Danieli, P.P., De Angelis, A., Fortina, R., Marino, R., Parisi, G., Piccolo, G., Pinotti, L., Prandini, A., Schiavone, A., Terova, G., Tulli, F., Roncarati, A., 2020. Insect and fish by-products as sustainable alternatives to conventional animal proteins in animal nutrition. Ital. J. Anim. Sci. 19 (1), 360–372.

Gisbert, E., Molas, S., Hernández, E., Carbó, R., Ruiz, A., 2024. Production of flathead grey mullet (*Mugil cephalus*) and lettuce (*Lactuca sativa*) in a coupled aquaponic system under suboptimal water temperatures. Fishes 9 (6), 189.

Goddek, S., Delaide, B., Mankasingh, U., Ragnarsdottir, K.V., Jijakli, H., Thorarinsdottir, R., 2015. Challenges of sustainable and commercial aquaponics. Sustainability 7 (4), 4199–4224.

Goddek, S., Espinal, C.A., Delaide, B., Jijakli, M.H., Schmautz, Z., Wuertz, S., Keesman, K.J., 2016. Navigating towards decoupled aquaponic systems: a system dynamics design approach. Water 8 (7), 303.

Goddek, S., Joyce, A., Kotzen, B., Dos-Santos, M., 2019a. Aquaponics and global food challenges. In: Goddek, S., Joyce, A., Kotzen, B., Burnell, G.M. (Eds.), Aquaponics Food Production Systems. Springer-Verlag, Cham., pp. 3–18.

Goddek, S., Joyce, A., Wuertz, S., Körner, O., Bläser, I., Reuter, M., Keesman, K.J., 2019b. Aquaponics and global food challenges. In: Goddek, S., Joyce, A., Kotzen, B., Burnell, G.M. (Eds.), Aquaponics Food Production Systems. Springer-Verlag, Cham, pp. 201–230.

Goodman, W., Minner, J., 2019. Will the urban agricultural revolution be vertical and soilless? A case study of controlled environment agriculture in New York City. Land Use Policy 83, 160–173.

Gosh, K., Chowdhury, S., 2019. Review of aquaponics system: searching for a technically feasible and economically profitable aquaponics system. J. Agric. Environ. Consum. Sci. 19, 5–13.

Gould, D., Caplow, T., 2012. Building-integrated agriculture: a new approach to food production. In: Zeman, F. (Ed.), Metropolitan Sustainability: Understanding and Improving the Urban Environment. Woodhead Publishing, Cambridge, pp. 147–170.

Graber, A., Junge, R., 2009. Aquaponic systems: nutrient recycling from fish wastewater by vegetable production. Desalination 246 (1–3), 147–156.

Greenfeld, A., Becker, N., McIlwain, J., Fotedar, R., Bornman, J.F., 2019. Economically viable aquaponics? Identifying the gap between potential and current uncertainties. Rev. Aquac. 11 (3), 848–862.

Huang, I.B., Keisler, J., Linkov, I., 2011. Multi-criteria decision analysis in environmental sciences: ten years of applications and trends. Sci. Total Environ. 409 (19), 3578–3594.

Hussain, A.S., Brown, P.B., 2024. A literature review of tilapia/lettuce aquaponics—production status, varieties, and research gaps. Aquac. Res. 2024 (1), 2642434.

Johnson, R., Vishwakarma, K., Hossen, M.S., Kumar, V., Shackira, A.M., Puthur, J.T., Abdi, G., Sarraf, M., Hasanuzzaman, M., 2022. Potassium in plants: growth regulation, signaling, and environmental stress tolerance. Plant Physiol. Biochem. 172, 56–69.

Joyce, A., Timmons, M.B., Goddek, S., Pentz, T., 2019. Bacterial relationships in aquaponics: new research directions. In: Goddek, S., Joyce, A., Kotzen, B., Burnell, G.M. (Eds.), Aquaponics Food Production Systems. Springer-Verlag, Cham, pp. 145–161.

Junge, R., König, B., Villarroel, M., Komives, T., Jijakli, M.H., 2017. Strategic points in aquaponics. Water 9 (3), 182.

Junge, R., Antenen, N., Villarroel, M., Griessler Bulc, T., Ovca, A., Milliken, S., 2020. Aquaponics Textbook for Higher Education. Zenodo, Geneva, Switzerland. https://aquateach.files.wordpress.com/2020/07/aquateach_o4_textbook_en-2.pdf.

Kasozi, N., Abraham, B., Kaiser, H., Wilhelmi, B., 2021. The complex microbiome in aquaponics: significance of the bacterial ecosystem. Ann. Microbiol. 71 (1), 1–13.

Kasozi, N., Tandlich, R., Fick, M., Kaiser, H., Wilhelmi, B., 2019. Iron supplementation and management in aquaponic systems: A review. Aquac. Rep. 15, 100221.

Knaus, U., Palm, H.W., 2017. Effects of the fish species choice on vegetables in aquaponics under spring-summer conditions in northern Germany (Mecklenburg Western Pomerania). Aquaculture 473, 62–73.

Knaus, U., Segade, Á.P., Palm, H.W., 2015. Significant factors affecting the economic sustainability of closed aquaponic systems. Part II: Fish and plant growth. AACL Bioflux 8 (5), 763–771.

König, B., Junge, R., Bittsanszky, A., Villarroel, M., Komives, T., 2016. On the sustainability of aquaponics. Ecocycles 2 (1), 26–32.

König, B., Janker, J., Reinhardt, T., Villarroel, M., Junge, R., 2018. Analysis of aquaponics as an emerging technological innovation system. J. Cleaner Prod. 180, 232–243.

Krastanova, M., Sirakov, I., Ivanova-Kirilova, S., Yarkov, D., Orozova, P., 2022. Aquaponic systems: biological and technological parameters. Biotechnol. Biotechnol. Equip. 36 (1), 305–316.

Kushwaha, J., Priyadarsini, M., Rani, J., Pandey, K.P., Dhoble, A.S., 2025. Aquaponic trends, configurations, operational parameters, and microbial dynamics: a concise review. Environ. Dev. Sustain. 27 (1), 213–246.

Kyaw, T.Y., Ng, A.K., 2017. Smart aquaponics system for urban farming. Energy Procedia 143, 342–347.

Lennard, W.A., 2021. A comparison of buffering species and regimes applied within a research-scale, recirculating aquaponics system. Aquac. Fish. 6 (5), 495–505.

Lennard, W., Goddek, S., 2019. Aquaponics: the basics. In: *Aquaponics food production systems: Combined aquaculture and hydroponic production technologies for the future*. Springer International Publishing, Cham, Switzerland, pp. 113–143.

Lennard, W.A., Leonard, B.V., 2006. A comparison of three different hydroponic sub-systems (gravel bed, floating and nutrient film technique) in an aquaponic test system. Aquac. Int. 14 (6), 539–550.

Liang, J.Y., Chien, Y.H., 2013. Effects of feeding frequency and photoperiod on water quality and crop production in a tilapia–water spinach raft aquaponics system. Int. Biodeterior. Biodegrad. 85, 693–700.

Love, D.C., Fry, J.P., Genello, L., Hill, E.S., Frederick, J.A., Li, X., Semmens, K., 2014. An international survey of aquaponics practitioners. PLoS One 9 (7), e102662.

Love, D.C., Fry, J.P., Li, X., Hill, E.S., Genello, L., Semmens, K., Thompson, R.E., 2015. Commercial aquaponics production and profitability: findings from an international survey. Aquaculture 435, 67–74.

Mancuso, T., Baldi, L., Gasco, L., 2016. An empirical study on consumer acceptance of farmed fish fed on insect meals: the Italian case. Aquac. Int. 24, 1489–1507.

Masser, M.P., Rakocy, J.E., Losordo, T.M., 1999. Recirculating Aquaculture Tank Production Systems: Management of Recirculating Systems. Southern Regional Aquaculture Center, Stoneville, MS, pp. 1–12 SRAC Publication No. 452.

Maucieri, C., Nicoletto, C., Van Os, E., Anseeuw, D., Van Havermaet, R., Junge, R., 2019. Hydroponic technologies. In: Goddek, S., Joyce, A., Kotzen, B., Burnell, G.M. (Eds.), Aquaponics Food Production Systems. Springer-Verlag, Cham, pp. 978–983.

Mchunu, N., Lagerwall, G., Senzanje, A., 2018. Aquaponics in South Africa: results from a survey of provincial agricultural departments. Aquac. Rep. 7, 12–19.

Milliken, S., Stander, H., 2019. Aquaponics and social enterprise. In: Goddek, S., Joyce, A., Kotzen, B., Burnell, G.M. (Eds.), Aquaponics Food Production Systems. Springer-Verlag, Cham, pp. 607–619.

Monsees, H., Kloas, W., Wuertz, S., 2017. Decoupled systems on trial: eliminating bottlenecks to improve aquaponic processes. PLoS One 12 (9), e0183056.

Moschos, S., Kormas, K.A., Karayanni, H., 2022. Prokaryotic diversity in marine and freshwater recirculating aquaculture systems. Rev. Aquac. 14 (4), 1861–1886.

Nicoletto, C., Maucieri, C., Mathis, A., Schmautz, Z., Komives, T., Sambo, P., Junge, R., 2018. Extension of aquaponic water use for NFT baby-leaf production: mizuna and rocket salad. Agronomy 8 (5), 75.

Okomoda, V.T., Oladimeji, S.A., Solomon, S.G., Olufeagba, S.O., Ogah, S.I., Ikhwanuddin, M., 2023. Aquaponics production system: a review of historical perspective, opportunities, and challenges of its adoption. Food Sci. Nutr. 11 (3), 1157–1165.

Oladimeji, S.A., Okomoda, V.T., Olufeagba, S.O., Solomon, S.G., Abol-Munafi, A.B., Alabi, K.I., Ikhwanuddin, M., Martins, C.O., Umaru, J.A., Hassan, A., 2020. Aquaponics production of catfish and pumpkin: comparison with conventional production systems. Food Sci. Nutr. 8 (5), 2307–2315.

Palm, H.W., Knaus, U., Appelbaum, S., Goddek, S., Strauch, S.M., Vermeulen, T., Jijakli, M.H., Kotzen, B., 2018. Towards commercial aquaponics: a review of systems, designs, scales and nomenclature. Aquac. Int. 26 (3), 813–842.

Palm, H.W., Knaus, U., Wasenitz, B., Bischoff, A.A., Strauch, S.M., 2019. Proportional up scaling of African catfish (*Clarias gariepinus* Burchell, 1822) commercial recirculating aquaculture systems disproportionally affects nutrient dynamics. Aquaculture 491, 155–168.

Pantanella, E., Cardarelli, M., Colla, G., Rea, E., Marcucci, A., 2015. Aquaponics vs. hydroponics: production and quality of lettuce crop. Acta Hortic. 1062, 43–50.

Partelow, S., Asif, F., Béné, C., Bush, S., Manlosa, A.O., Nagel, B., Schlüter, A., Chadag, V.M., Choudhury, A., Cole, S.M., Cottrell, R.S., Gelcich, S., Gentry, R., Gephart, J.A., Glaser, M., Johnson, T.R., Jonell, M., Krause, G., Kunzmann, A., Kühnhold, H., Little, D.C., Marschke, M.J., Mizuta, D.D., Paramita, A.O., Pin, N., Salayo, N.D., Stentiford, G.D., Stoll, J., Troell, M., Turchini, G.M., 2023. Aquaculture governance: five engagement arenas for sustainability transformation. Curr. Opin. Environ. Sustain. 65, 101379.

Partridge, G.J., Lymbery, A.J., George, R.J., 2008. Finfish mariculture in inland Australia: a review of potential water sources, species, and production systems. J. World Aquac. Soc. 39 (3), 291–310.

Pattillo, D.A., 2017. An Overview of Aquaponic Systems: Hydroponic Components. North Central Regional Aquaculture Center, Ames, IA., p. 19 NCRAC Technical Bulletins, No. 123.

Preena, P.G., Rejish Kumar, V.J., Singh, I.S.B., 2021. Nitrification and denitrification in recirculating aquaculture systems: the processes and players. Rev. Aquac. 13 (4), 2053–2075.

Pinho, S.M., Molinari, D., de Mello, G.L., Fitzsimmons, K.M., Emerenciano, M.G.C., 2017. Effluent from a biofloc technology (BFT) tilapia culture on the aquaponics production of different lettuce varieties. Ecol. Eng. 103, 146–153.

Proksch, G., Ianchenko, A., Kotzen, B., 2019. Aquaponics in the built environment. In: Goddek, S., Joyce, A., Kotzen, B., Burnell, G.M. (Eds.), Aquaponics Food Production Systems. Springer-Verlag, Cham, pp. 523–558.

Puccinelli, M., Fierro-Sañudo, J.F., Bibbiani, C., Fronte, B., Maibam, C., Dubois, T., Pardossi, A., Incrocci, L., Rossi, L., 2022. Multi-criteria DEXi analysis for the selection of crop species for saltwater aquaponics. Horticulturae 8 (8), 703.

Purvis, B., Mao, Y., Robinson, D., 2019. Three pillars of sustainability: in search of conceptual origins. Sustain. Sci. 14, 681–695.

Rafiee, G., Saad, C.R., 2005. Nutrient cycle and sludge production during different stages of red tilapia (*Oreochromis* sp.) growth in a recirculating aquaculture system. Aquaculture 244 (1–4), 109–118.

Raghu, K., MacRae, I.C., 1966. Occurrence of Phosphate-dissolving micro-organisms in the rhizosphere of rice plants and in submerged soils. J. Appl. Bacteriol. 29 (3), 582–586.

Rakocy, J.E., Bailey, D.S., Shultz, R.C., Thoman, E.S., 2004. Update on tilapia and vegetable production in the UVI aquaponic system. In: New Dimensions on Farmed Tilapia: Proceedings of the Sixth International Symposium on Tilapia in Aquaculture. Philippine International Convention Center, Manila, Philippines, pp. 676–690.

Rakocy, J.E., Masser, M.P., Losordo, T.M., 2006. Recirculating aquaculture tank production systems: aquaponics—integrating fish and plant culture. SRAC Publication 454, 1–16.

Rakocy, J.E., Masser, M.P., Losordo, T.M., 2016. Recirculating Aquaculture Tank Production Systems: Aquaponics—Integrating Fish and Plant Culture. Southern Regional Aquaculture Center Publication, Stoneville, MS, p. 454.

Rizal, A., Dhahiyat, Y., Zahidah, Andriani, Y., Handaka, A.A., Sahidin, A., 2018. The economic and social benefits of an aquaponic system for the integrated production of fish and water plants. IOP Conf. Ser.: Earth Environ. Sci. 137, 012098.

Rodrıǵuez, H., Fraga, R., 1999. Phosphate solubilizing bacteria and their role in plant growth promotion. Biotechnol. Adv. 17 (4–5), 319–339.

Roosta, H.R., Hamidpour, M., 2011. Effects of foliar application of some macro- and micro-nutrients on tomato plants in aquaponic and hydroponic systems. Sci. Hortic. 129 (3), 396–402.

Rossi, L., Bibbiani, C., Fierro-Sañudo, J.F., Maibam, C., Incrocci, L., Pardossi, A., Fronte, B., 2021. Selection of marine fish for integrated multi-trophic aquaponic production in the Mediterranean area using DEXi multi-criteria analysis. Aquaculture 535, 736402.

Ruzzi, M., Aroca, R., 2015. Plant growth-promoting rhizobacteria act as biostimulants in horticulture. Sci. Hortic. 196, 124–134.

Sace, C.F., Fitzsimmons, K.M., 2013. Vegetable production in a recirculating aquaponic system using Nile tilapia (*Oreochromis niloticus*) with and without freshwater prawn (*Macrobrachium rosenbergii*). Acad. J. Agric. Res. 1 (12), 236–250.

Saha, S., Monroe, A., Day, M.R., 2016. Growth, yield, plant quality and nutrition of basil (*Ocimum basilicum* L.) under soilless agricultural systems. Ann. Agric. Sci. 61 (2), 181–186.

Samikannu, R., Koshariya, A. K., Poornima, E., Ramesh, S., Kumar, A., Boopathi, S., 2023. Sustainable Development in Modern Aquaponics Cultivation Systems Using IoT Technologies. In: *Human Agro-Energy Optimization for Business and Industry*. IGI Global Scientific Publishing, pp. 105–127.

Saxena, G., Pullanikkatil, D., Thobela, M., 2022. Aquaponics as an alternative livelihood option to enhance community resilience in rural Southern Africa. Sustainability 14 (5), 2754.

Schneider, O., Sereti, V., Eding, E.H., Verreth, J.A.J., 2005. Analysis of nutrient flows in integrated intensive aquaculture systems. Aquacult. Eng. 32 (3–4), 379–401.

Seawright, D.E., Stickney, R.R., Walker, R.B., 1998. Nutrient dynamics in integrated aquaculture–hydroponics systems. Aquaculture 160 (3–4), 215–237.

Shete, A.P., Verma, A.K., Chadha, N.K., Prakash, C., Peter, R.M., Ahmad, I., Nuwansi, K.K.T., 2016. Optimization of hydraulic loading rate in aquaponic system with Common carp (*Cyprinus carpio*) and Mint (*Mentha arvensis*). Aquacult. Eng. 72, 53–57.

Schmautz, Z., Graber, A., Jaenicke, S., Goesmann, A., Junge, R., Smits, T.H., 2017. Microbial diversity in different compartments of an aquaponics system. Arch. Microbiol. 199 (4), 613–620.

Somerville, C., Cohen, M., Pantanella, E., Stankus, A., Lovatelli, A., 2014. Small-Scale Aquaponic Food Production: Integrated Fish and Plant Farming. Food and Agriculture Organization of the United Nations, Rome, p. 288 AO Fisheries and Aquaculture Technical Paper No. 589.

Specht, K., Siebert, R., Hartmann, I., Freisinger, U.B., Sawicka, M., Werner, A., Thomaier, S., Henckel, D., Walk, H., Dierich, A., 2019. Urban agriculture of the future: An overview of sustainability aspects of food production in and on buildings. Agric. Hum. Values 31 (1), 33–51.

Suhl, J., Dannehl, D., Kloas, W., Baganz, D., Jobs, S., Scheibe, G., Schmidt, U., 2016. Advanced aquaponics: Evaluation of intensive tomato production in aquaponics vs. conventional hydroponics. Agric. Water Manage. 178, 335–344.

Sundar, L.S., Chen, G.S., 2020. Study on the growth performance of lettuce (*Lactuca sativa*) and pak choi (*Brassica chinensis*) in different aquaponic growing systems. Horticulturae 6 (4), 69.

Timmons, M.B., Ebeling, J.M., 2013. Recirculating Aquaculture, third ed. Ithaca Publishing Company LLC, Ithaca, NY.

Tokunaga, K., Tamaru, C., Ako, H., Leung, P., 2015. Economics of small-scale commercial aquaponics in Hawaii. J. World Aquac. Soc. 46 (1), 20–32.

Trang, N.T.D., Brix, H., 2014. Use of planted biofilters in integrated recirculating aquaculture-hydroponics systems in the Mekong Delta, Vietnam. Aquac. Res. 45 (3), 460–469.

Trang, N.T.D., Schierup, H.H., Brix, H., 2017. Leaf vegetables for use in integrated hydroponics and aquaculture systems: effects of root flooding on growth, mineral composition and nutrient uptake. Afr. J. Biotechnol. 9 (27), 4186–4196.

Tyson, R.V., Simonne, E.H., Treadwell, D.D., White, J.M., Simonne, A., 2008. Reconciling pH for ammonia biofiltration and cucumber yield in a recirculating aquaponic system with perlite biofilters. HortScience 43 (3), 719–724.

Tyson, R.V., Treadwell, D.D., Simonne, E.H., 2011. Opportunities and challenges to sustainability in aquaponic systems. HortTechnology 21 (1), 6–13.

Wongkiew, S., Hu, Z., Chandran, K., Lee, J.W., Khanal, S.K., 2017. Nitrogen transformations in aquaponic systems: A review. Aquacult. Eng. 76, 9–19.

Wurts, W.A., Durborow, R.M., 1992. In: Interactions of pH, Carbon Dioxide, Alkalinity and Hardness in Fish Ponds, 464. Southern Regional Aquaculture Center, Stoneville, MS, pp. 1–4 SRAC Publication No 464.

Yang, T., Kim, H.J., 2020. Characterizing nutrient composition and concentration in tomato-, basil-, and lettuce-based aquaponic and hydroponic systems. Water 12, 1259.

Yep, B., Gale, N.V., Zheng, Y., 2018. Aquaponic systems: Nutrient recycling from fish waste to plant growth. In: Galanakis, C.M. (Ed.), Sustainable Food Systems from Agriculture to Industry. Academic Press, London, pp. 343–388.

Yep, B., Zheng, Y., 2019. Aquaponic trends and challenges–A review. J. Clean. Prod. 228, 1586–1599.

Yildiz, H.Y., Robaina, L., Pirhonen, J., Mente, E., Domínguez, D., Parisi, G., 2017. Fish welfare in aquaponic systems: its relation to water quality with an emphasis on feed and faeces—a review. Water 9 (1), 13.

Yogev, U., Barnes, A., Gross, A., 2016. Nutrients and energy balance analysis for a conceptual model of a three loops off grid, aquaponics. Water 8 (12), 589.

Young, I., 2019. Farm urban and urban aquaponics: changing perceptions in classrooms and communities. In: Food and Sustainability in the Twenty-First Century: Cross-Disciplinary Perspectives, 9. Berghahn Books, New York and Oxford, p. 103.

Zhu, Z., Yogev, U., Goddek, S., Yang, F., Keesman, K.J., Gross, A., 2022. Carbon dynamics and energy recovery in a novel near-zero waste aquaponics system with onsite anaerobic treatment. Sci. Total Environ. 833, 155245.

Zhu, Z., Yogev, U., Keesman, K.J., Gross, A., 2024. Promoting circular economy: comparison of novel coupled aquaponics with anaerobic digestion and conventional aquaponic systems on nutrient dynamics and sustainability. Resour. Conserv. Recycl. 208, 107716.

Zi, Y., Zhu, D.Z.K., He, X., Chen, A., Li, C., 2024. Enhancing nutrient recovery efficiency through a novel dual-loop aquaponic system coupling mainstream and side stream: a review. Chem. Eng. Trans. 113, 619–624.

Zou, Y., Hu, Z., Zhang, J., Xie, H., Guimbaud, C., Fang, Y., 2016. Effects of pH on nitrogen transformations in media-based aquaponics. Bioresour. Technol. 210, 81–87.

Paxton, A., 1994. The Food Miles Report: the Dangers of Long-distance Food Transport, vol. 1994. SAFE Alliance, London, UK. Available at. http://www.sustai nweb.org/publications/?id¼191. (Accessed 4 December 2025).

CHAPTER

6

Effluent-derived nutrient uptake through algaquaculture concept

Maibam Malemngamba Meitei[a], *Angom Baleshwor Singh*[b], *Maitarambum Wartung Monsang*[c], *Potluri Sai Kishore*[a], *Karthireddy Syamala*[d], *Muralidhar P Ande*[d] and *Sukham Munilkumar*[a]

[a]ICAR-Central Institute of Fisheries Education, Mumbai, Maharashtra, India [b]ICAR-Krishi Vigyan Kendra, Chandpur, Assam, India [c]College of Fisheries, Central Agricultural University (Imphal), Agartala, Tripura, India [d]ICAR-Central Institute of Fisheries Education, Kakinada Centre, Kakinada, Andhra Pradesh, India

6.1 Introduction

Aquaculture has become one of the world's fastest-growing food-producing areas. It is vital to meet the growing demand for seafood while also fostering nutritional stability and economic growth. 59 percent of the world's fisheries and aquaculture production, or about 130.9 million tons, came from aquaculture in 2022 (FAO, 2024). Aquaculture's explosive expansion is indicative of its ability to relieve the strain on capture fisheries, which have been in decline as a result of habitat damage and overfishing (FAO, 2024). Notwithstanding its advantages, aquaculture presents serious environmental problems with increasing efforts to enhance productivity. Water contamination, invasive species introduction, disease transmission to wild populations, and habitat loss are all consequences of unsustainable activities. Aquatic ecosystems are disturbed when aquaculture farms overload their waters with nutrients from the metabolic waste and uneaten feed, causing eutrophication. The environmental impact of the wastewater outflow depends on the kind of aquaculture system, the species being raised, and the management techniques used (Boyd and McNevin, 2015). Proper environmental monitoring, farm location, and waste reduction are crucial for sustainable aquaculture (Carballeira Braña et al., 2021). While aquaculture was initially not viewed as environmentally

Zero-Waste Technologies in Aquaculture: Global Perspectives and Methods toward a Circular Economy
DOI: https://doi.org/10.1016/B978-0-443-34689-7.00003-3

critical, it has become a central concern across various sectors, necessitating improved legislation and regulation (Martinez-Porchas and Martinez-Cordova, 2012; Arshad et al., 2024).

During harvesting or water exchange, pond systems frequently release untreated water that is rich in nutrients and organic matter, whereas cage culture in open waters directly enriches natural water bodies with nutrients. Aquaculture effluents pose significant environmental challenges due to their high nutrient content. These effluents can lead to eutrophication, chemical toxicity, and food insecurity. Various treatment technologies, such as constructed wetlands, biofloc systems, recirculating aquaculture systems (RASs), and integrated multitrophic systems, have been developed to mitigate environmental impacts and recover valuable resources from effluents. Moreover, nutrient recycling is a crucial ecosystem process involving the movement and reuse of nutrients through biological, chemical, and geological mechanisms (Borah and Sekhar Bora, 2020). By incorporating nutrient recycling into aquaculture, for instance, using methods such as aquaponics or integrated multitrophic aquaculture (IMTA), waste from one species can be used as a source of nutrients for another, lowering effluent outflow and increasing production (Chopin et al., 2012). In the context of the circular economy, nutrient recovery technologies are rapidly advancing to address resource depletion and environmental concerns. Further, there is potential for recovering ingredients from effluents for use in food additives, nutraceuticals, and animal feed, contributing to a circular bioeconomy (Venugopal and Sasidharan, 2021).

Microalgae play a promising role for efficient nutrient utilization, reducing the waste either in the aquaculture system (Addy et al., 2017; Kumari et al., 2023) or the other industrial wastewater system (Kong et al., 2010; Li et al., 2011; Wu et al., 2012). Algae use nutrients like phosphorus and nitrogen from the wastewater as a resource for growth, and achieve 100% nutrient removal depending upon the species. The nutrient remove converted into algal biomass as used for various bioeconomy purposes. They can be brown algae (Phaeophyceae), green algae (Chlorophyceae), red algae (Rhodophyceae), blue-green algae (Cyanophyceae), and diatoms (Bacillariophyceae). However, commonly used algae species include *Chlorella sp.*, *Tetraselmis* sp., *Pavlova* sp., *Isochrysis* sp., *Chaetoceros* sp., *Phaeodactylum* sp., *Skeletonema* sp., *Thalassiosira* sp., and *Nannochloropsis* sp. (Shah et al., 2018). Microalgae are rich in protein, carbohydrates, lipids, vitamins, minerals, and other bioactive compounds (Apandi et al., 2019) that have antioxidants, immunostimulants, and growth promoters that can be used as supplements for humans as well as animals (Hamed et al., 2015; Monteiro et al., 2020).

Algaquaculture, the integration of algae cultivation with aquaculture, is emerging as a promising approach for sustainable aquaculture development. In this system, algae are cultured along with the aquaculture to utilize the waste generated by the fed species, and in turn, algae contribute to water purification and offer important biomass for a variety of applications, such as feed, biofuel, and medicines. By incorporating waste nutrients into their biomass, algae can mitigate this problem and lower the nitrogen load in effluents (Azim and Little, 2008). This system offers superior nutrient recycling capabilities compared to traditional aquaponics, as algae efficiently utilize waste-derived nutrients such as nitrate, nitrite, and total ammonia nitrogen (Meitei et al., 2022). This system aligns with the concept of circular economies in aquaculture, promoting more sustainable practices that reduce waste and save resources (Campanati et al., 2022).

In this chapter, the concept of algaquaculture, its components, and how it can be practiced are discussed. The nutrient dynamics and management in this system are also discussed.

Lastly, the various benefits of the algae biomass harvested from the system have also been provided.

6.2 Effluent composition in aquaculture

Intensive aquaculture systems produce various effluents that, when discharged into natural water bodies, can degrade water quality and disturb ecosystem dynamics. Effluents from aquaculture systems can contain a wide range of pollutants, including organic matter, nitrogen compounds, phosphates, suspended solids, pathogens, antibiotics, and other chemical residues. The characteristics and concentrations of these effluents vary depending on factors, including the type of culture system, the species cultured, the quality and type of feed, system design, and management practices (Boyd et al., 2020).

(1) A potential build-up of nitrite and other forms of nitrogenous wastes because of the lack of an efficient water treatment system may deteriorate the effluent quality.
(2) Intensive culture systems, larger fish, and higher stocking densities generate more waste and, thus higher organic and nutrient loads in the effluent.
(3) Poor feed quality, overfeeding, and inefficient feeding practices lead to higher levels of uneaten feed and feces, increasing the organic and nutrient content of the effluent.

6.2.1 Composition and key nutrients of effluents

Aquaculture effluents typically contain a complex mixture of solid and dissolved waste products. These components can be broadly categorized into organic and inorganic wastes, with organic matter being the major contributor to the environmental load from aquaculture systems.

6.2.1.1 Organic solids

Organic solids from aquaculture effluent mainly result from uneaten feed, fish feces, and other biological wastes. The quality and quantity of organic waste are also related to the type of feed used and the feeding efficiency of the cultured species. Feed waste accounts for a significant portion of the organic load, especially in systems where feeding practices are not optimized (Bureau and Hua, 2010). While decomposing in the watercourse environment, organic solids in effluents increase the biological oxygen demand (BOD) of water, meaning more oxygen is required by microorganisms to decompose the organic material. High BOD levels can lead to hypoxic or anoxic conditions, negatively impacting both cultured and wild aquatic organisms (Boyd, 2017). While decomposing, organic solids also release byproduct toxins such as ammonia, nitrite, and methane into the culture system (Hua and Bureau, 2006).

Feces is another significant contributor to organic waste. The digestibility of the feed dictates the quantity and nature of the feces. For instance, diets rich in indigestible fiber will increase fecal output, contributing to the effluent's organic load. Organic solids particle size is crucial since larger-sized particles settle faster in the water column than smaller particles that stay suspended in the water column and are more challenging to remove using filtration processes (Badiola et al., 2012). Effluent management practices such as settling ponds, biofilters, and reducing feed waste are essential to minimize the discharge of organic matter.

6.2.1.2 Nutrients

Nitrogen and phosphorus are two important elements that are found in aquaculture effluents, primarily from uneaten feed, fish excreta, and feces. Nitrogen occurs in effluents as ammonia (NH_3), nitrite (NO_2^-), nitrate (NO_3^-), and organic nitrogen compounds, while phosphorus is mostly an orthophosphate (PO_4^{3-}) and particulate phosphorus (Boyd et al., 2007).

(1) Ammonia: Ammonia is the most important nitrogenous waste in aquaculture, being directly produced by fish through gills and as a part of the decomposition of unused feed and feces. Usually, ammonia can be converted into nitrate through nitrification; however, in systems with poor water circulation, ammonia may accumulate to harmful levels (Martins et al., 2010). High concentrations of ammonia in effluent can be harmful to fish and other aquatic species, leading to growth problems, reproductive disorders, and overall health deterioration (Randall and Tsui, 2002).

(2) Nitrate and nitrite: Nitrate and nitrite are produced during the nitrification process. Nitrifying bacteria convert ammonia to nitrite, and nitrite is further oxidized to nitrate. Nitrate is less toxic than nitrite and ammonia but can accumulate if not removed (Schneider et al., 2005).

(3) Phosphorus: Unlike nitrogen, phosphorus does not volatilize, so it tends to accumulate in sediments and can be released under certain environmental conditions, exacerbating water quality issues (Boyd and Tucker, 2012). High phosphorus concentrations in effluents lead to the eutrophication of the receiving water bodies. High algal blooms lead to oxygen depletion in the same body of water. Dissolved phosphorus and particulate forms are found in aquaculture effluents, but dissolved forms can easily be utilized for algae growth (Hua and Bureau, 2006).

6.2.1.3 Dissolved salinity and soluble salts

Dissolved solids in aquaculture effluents are salts, minerals, and organic molecules that dissolve in water during aquaculture operations. The salinity of the water used in aquaculture systems influences the concentration of dissolved solids. For example, marine aquaculture systems have higher concentrations of dissolved salts compared with freshwater systems (Piedrahita, 2003). Effluent salinity can indeed impact the receiving aquatic environment, especially when saline effluents are discharged into freshwater systems. Elevated salinity increases the osmoregulatory capacity, which can cause physiological stress or changes in community structure (Sahu, 2021).

6.2.1.4 Chemicals and pharmaceuticals

Aquaculture operations rely heavily on chemicals and pharmaceuticals, including antibiotics, pesticides, disinfectants, and antifoulants, to manage diseases, parasites, and biofouling and to maintain fish health and water quality in the culture system. These residues are consequently released into the environment through effluents. This may potentially harm nontarget species and contribute to antimicrobial resistance (Rico et al., 2012). Chemicals such as chlorine, formalin, and copper-based antifoulants are applied to prevent biofouling on cages and nets, which bioaccumulate in sediments, causing toxic effects to benthic organisms. Antibiotics are now banned, but if applied to treat bacterial infections in fish persist in the environment and disrupt the microbial communities and selection of antibiotic-resistant bacteria (Defoirdt et al., 2011).

6.2.2 Impact of effluent discharge on the environment

Uncontrolled and untreated aquaculture effluents in natural water may lead to several detrimental environmental effects. Suspended solids in these effluents can increase turbidity, sedimentation, reduce light penetration, and smother benthic habitats when discharged into natural water bodies (Crab et al., 2007). In addition, the high concentration of nutrients, organic matter, and chemicals contribute to eutrophication, oxygen depletion, and habitat degradation. The major environmental impacts arising from the aquaculture effluent discharge are mentioned in the following subsections:

6.2.2.1 Eutrophication and algal blooms

Excess nutrient inputs, such as nitrogen and phosphorus, from aquaculture effluents, can trigger the growth of algae and aquatic plants in receiving water bodies, leading to eutrophication (Crab et al., 2007). Algal blooms limit light penetration and deplete dissolved oxygen levels, creating hypoxic conditions that are detrimental to fish and other aquatic organisms (Boyd et al., 2007).

6.2.2.2 Oxygen depletion

The microbial decomposition of organic matter in aquaculture effluents consumes dissolved oxygen in the water. High organic loads can lead to oxygen depletion, especially in the bottom of the culture system, where organic solids settle and create anaerobic conditions (Martins et al., 2010). Oxygen depletion can cause the death of fish, invertebrates, and other aquatic organisms, thus disrupting the balance of aquatic ecosystems.

6.2.2.3 Chemical contamination and antimicrobial resistance

The chemicals and pharmaceuticals that are released in the effluents of aquaculture pose a risk to nontarget species and the broader aquatic environment. For instance, residual antibiotics have been shown to disrupt microbial communities and cause the proliferation of antibiotic-resistant bacteria, which is a growing global concern (Rico et al., 2012).

6.2.2.4 Pathogens and disease agents

Effluents from aquaculture systems may also contain pathogens, including bacteria, viruses, and parasites, which can be released into the environment, potentially spreading diseases to wild fish populations and other aquatic organisms (Rodgers and Furones, 2009). This is particularly problematic in open systems such as cage culture, where effluents are discharged directly into natural water bodies. Pathogens such as *Vibrio* and *Aeromonas* species have been detected in aquaculture effluents, raising concerns about their potential to cause disease outbreaks in wild fish populations (Miranda et al., 2018). Parasites such as fish lice are a major problem in aquaculture, and the effluents from a contaminated system can facilitate the spread of these parasites to wild fish, leading to increased mortality in wild populations (Costello, 2009). To mitigate the spread of pathogens, biosecurity measures, such as water treatment, vaccination, and the use of closed or semiclosed systems, such as RAS, are essential.

Effective effluent management strategies were required in aquaculture to improve water quality and minimize these environmental impacts. Advanced aquaculture techniques, such as RAS, biofloc, and IMTA systems, aim to reduce the nutrient load and organic matter in their effluents. IMTA systems integrate several species at different trophic levels, and waste generated by one species can be a nutrient source for another. Seaweed and filter-feeding bivalves can take up excess nutrients and organic matter from finfish culture, thereby reducing the amount of nutrients in the effluent (Chopin, 2014). RASs are designed to minimize water usage and discharge of effluent through recirculation of water through biological and mechanical filtration systems. These systems are effective in removing organic matter, nitrogen, and phosphorus from the water, thus minimizing the environmental impact of effluent discharge (Badiola et al., 2012). The development of low-pollution diets with lower nitrogen and phosphorus content, combined with better precision feeding practices, can significantly reduce the nutrient load in aquaculture effluents (Hua and Bureau, 2006). Automated feeding systems and real-time water quality monitoring can optimize feeding rates to minimize feed losses. Aquaculture management practices, such as water exchange rates, aeration, and the use of biofilters, have a significant impact on effluent composition. High water exchange rates dilute the concentration of waste products in the effluent but increase the total volume of water discharged (Badiola et al., 2012). Proper aeration and the use of biofilters can enhance the removal of ammonia and organic solids, reducing the nutrient load in the effluent (Martins et al., 2010).

6.3 Concept of algaquaculture

Fish and shrimp culture can be integrated with algae cultivation to enhance nutrient recycling and improve water quality. In this integrated system, algae act as natural biofilters by assimilating excess dissolved nutrients, particularly nitrogen and phosphorus, originating from uneaten feed, faecal matter, and metabolic waste. This nutrient uptake reduces the discharge of aquaculture effluents into the environment while converting waste nutrients into valuable algal biomass that can be utilized for food, feed, biofuels, pharmaceuticals, and other bio-based products (Farghali et al., 2023; Neori et al., 2004).

Algaquaculture is an innovative concept integrating algae culture with aquaculture (Yang, 2020). In this system, the microalgae are used instead of the leafy vegetables of the traditional aquaponics system due to slow nutrient uptake, consumption of more energy for aeration, high feed input, and limited market for the leafy vegetables (Yang, 2020). On the other hand, microalgae cultivation in aquaculture systems can remediate wastewater, release oxygen, and yield valuable biomass while promoting fish growth, stimulating immune systems, and enhancing coloration (Meitei et al., 2022). Thus algaquaculture serves dual purposes of nutrient assimilation and production of fish and algal biomass, potentially offsetting capital costs through multiple revenue streams.

Algaquaculture consists of an aquaculture unit, an algae culture unit, and a harvesting unit. The commercially important finfish and shellfish species are cultured in the aquaculture unit with a nutrient-rich diet. The system is rich in nutrients such as ammonia, nitrite, nitrate, and phosphate because of the uneaten feed from the nutrient-rich diet and the metabolic waste. This nutrient-rich waste water is then circulated to the microalgae culture unit. Here, the nutrients, particularly ammonia, were absorbed by the microalgae and transformed into protein, polyunsaturated fatty acids (PUFAs), pigments, and other mineral-rich value-added products (Yang et al., 2020). In addition, the microalgae consume the dissolved carbon dioxide contained in the wastewater from the aquaculture for photosynthesis and release oxygen. This oxygenated and nutrient-free water from the algae culture unit is then recirculated back to the aquaculture unit.

The microalgae culture can be conducted in various ways, such as (1) photobioreactors (PBRs) are the reactors in which microalgae are grown in photobiological reactions (Chini Zittelli et al., 2013). These can be classified into tubular, vertical column, and flat-plate types that can be practiced indoors or outdoors. These reactors reduce contamination risks and better control over conditions, making them ideal for microalgae cultivation (Patil et al., 2005; Bacellar Mendes and Vermelho, 2013), (2) a vertical column PBR consists of a transparent vertical tube (glass or acrylic) allowing light penetration, with a gas sparger at the base to facilitate gas transfer and culture mixing without physical agitation, minimizing cell damage. Based on liquid flow patterns, it can be classified as a bubble column that use upward gas bubbling for mixing or airlift reactor that have two interconnected zones—a gas riser for upward flow and a downcomer for downward circulation (Hamed, 2016), and (3) the flat-panel photobioreactor (FP-PBR) consists of translucent, thin rectangular boxes made of transparent materials like glass or optical light film, vertically oriented, open at one end, illuminated on both sides, and aerated for mixing (Hamed, 2016). FP-PBRs are suitable for outdoor cultivation and easy to clean, but face challenges in scaling up and temperature control. A recent design of a vertical FP-PBR has been proposed that features a transparent plastic bag mounted on a rigid frame, significantly reducing equipment costs (Sierra et al., 2008). Another method of algae that can be cultured in the algaquaculture system is the algal–bioflim system, where algae are attached to a material surface and harvested by scraping. It can be either a stationary biofilm system in which the algal biomass is cultured in a horizontal or vertical film, and the aquaculture wastewater is moved over the surface of the algal film, or a rotating algal biofilm in which the algal biofilm rotates between the air and the liquid phase (Gross et al., 2015).

When the algal biomass increases their growth because of the nutrients from the aquaculture, they need to be harvested so that the culture system can be maintained at an optimum level. According to Shah et al. (2014), the harvesting methods can be various processes such as flocculation, floatation, sedimentation, centrifugation, and filtration (Fig. 6.1).

The system deviates from the traditional aquaponic system. In the traditional aquaponic system, leafy vegetables were used, while the microalgae were used in the algaquaculture system. This is due to the more efficient nutrient recovery of microalgae than that of leafy vegetables. Moreover, the leafy vegetables used in aquaponics have limited market and demand, while the microalgae have huge demand in the world market, fetching a high value. Further, the problems of oxygen depletion in the aquaponics system can be fixed in the algaquaculture with efficient and continuous carbon fixation, producing more oxygen (Yang et al., 2020).

6.4 Nutrient dynamics in algaquaculture systems

Algae have high nutrient removal capabilities because of their ability to assimilate nitrogen, phosphorus, and other pollutants from water as part of their growth process (Ramesh et al., 2023). This nutrient uptake contributes to pollution control and generates valuable algal

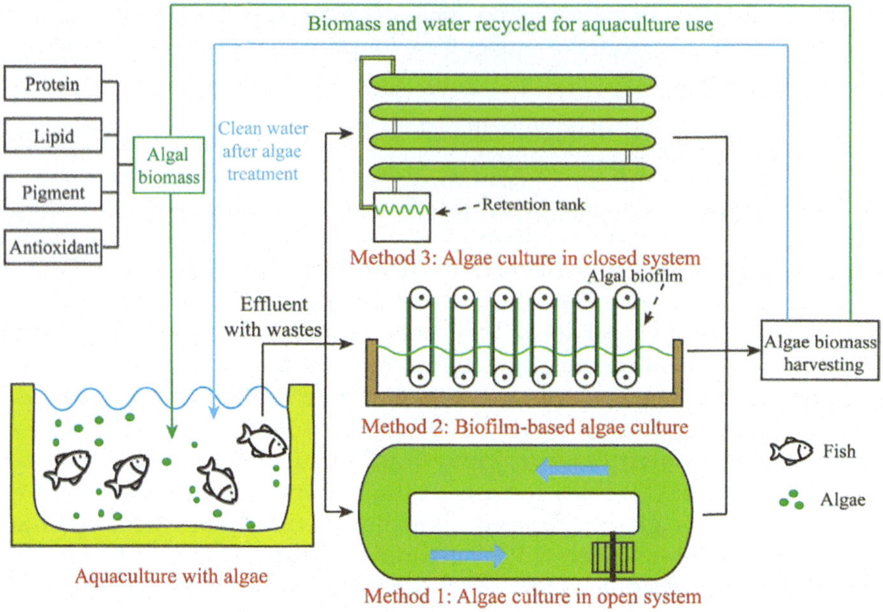

FIGURE 6.1 Concept of algaquaculture (Yang et al., 2020).

biomass that can be repurposed for feed or biofuel. Nutrient dynamics in algaquaculture systems are complex, involving multiple biochemical processes that govern the transformation, uptake, and release of essential nutrients such as nitrogen, phosphorus, and carbon (Viswanaathan et al., 2022). The efficiency of removal varies significantly among algal species, cultivation systems, and environmental conditions. The synergy between algae, bacteria, and aquatic organisms establishes a balanced system that supports optimal nutrient utilization and improves aquaculture productivity.

6.4.1 Algal nutrient uptake mechanisms in aquaculture

Algae absorb nutrients from water through their cell walls, a process regulated by various biochemical pathways. Nitrogen is generally taken up as nitrate (NO_3^-) and ammonium (NH_4^+), while phosphorus is assimilated in the form of phosphate ions (PO_4^{3-}). Other essential nutrients, including potassium, calcium, and trace metals, are absorbed as part of metabolic processes (Fox and Zimba, 2018). Nutrient uptake rates are influenced by factors such as light intensity, temperature, pH, and the specific algal species used.

6.4.2 Nutrient removal efficiency of different algal species

The primary nutrient inputs in algaquaculture systems include nitrogen, phosphorous and carbon compounds from feed inputs, excretions by cultured organisms, and other external nutrient sources (Al-Hafedh et al., 2015). Fish and other aquatic organisms excrete ammonia as

TABLE 6.1 Recent studies on the nutrient removal of algae in different aquaculture systems.

Algae species	Name culture system	Target nutrients	Nutrient uptake/removal efficiency	References
Spirulina platensis	Closed photobioreactor	N, CO_2	N removal: 76%; CO_2 reduction: 60%	Kumari et al. (2023)
Dunaliella salina	Saline aquaculture	N, P	N: 69%; P: 58%	Mirzaei et al. (2024)
Chlorella vulgaris	Closed photobioreactor	N, P	N: 90%; P: 78%	Singh et al. (2017)
Scenedesmus obliquus	Open raceway pond	N	N: 82%	Husin and Basri (2022)
Ulva lactuca	Integrated multitrophic system	N, ammonia	Ammonia: 95%	Al-Hafedh et al. (2015)
Nannochloropsis sp.	Marine aquaculture	NH_4^+	NH_4^+: 85%	Soo et al. (2017)
Gracilaria tikvahiae	Integrated multitrophic system	N, P	N: 73%; P: 68%	Samocha et al. (2015)
Oscillatoria sp.	Open pondwater	NH_4^+, NO_3^-	NH_4^+: 88%; NO_3^-: 72%	Attasat et al. (2013)
Porphyra dioica	IMTA system	NO_3^-	NO_3^-: 78%	Day (2008)

a metabolic waste product, which is the main nitrogen source in these systems (Bucking, 2017). Algae, in turn, utilize these nutrients for growth and reproduction, assimilating inorganic nitrogen and phosphorus. Algae play a central role in nutrient cycling by absorbing these nutrients, converting potential pollutants into biomass that can serve various uses, including as fish feed or biofuel feedstock (Yaakob et al., 2021). This uptake helps maintain water quality and provides a natural nutrient buffer for the aquaculture system. Recent research shows that microalgae, such as *Chlorella* and *Scenedesmus*, can effectively reduce nitrogen and phosphorus concentrations in aquaculture wastewater. For instance, *Chlorella vulgaris* has shown nitrogen removal rates up to 90% in high nutrient environments and also effectively reduces phosphorus (Singh et al., 2017). Macroalgal species such as *Ulva lactuca* are effective in marine aquaculture systems for their nutrient reduction and biomass production capacity (Samocha et al., 2015). They can remove up to 95% of ammonia from wastewater (Al-Hafedh et al., 2015). This reduction is critical for marine species that are sensitive to ammonia. Table 6.1 illustrates the recent studies on the nutrient removal of algae in different aquaculture systems:

Algae can uptake ammonia directly, converting it into proteins and other cellular components necessary for growth (Verdegem, 2013). This process helps reduce toxic levels of ammonia in the water, benefiting both the algae and cultured organisms. Ammonia that algae do not absorb undergoes nitrification, a two-step process facilitated by nitrifying bacteria, transforming ammonia into nitrite and subsequently into nitrate. This symbiotic relationship between algae and bacteria promotes a balanced system, with bacteria transforming nutrients into forms more readily accessible for algae and reducing potential toxicants, such as ammonia (Viswanaathan et al., 2022). While nitrite is also toxic to aquatic organisms, nitrate is less harmful and can serve as an additional nutrient source for algae. This process enables efficient nitrogen cycling, where algae serve as both nutrient sinks and sources within the system, supporting stable nitrogen levels and reducing the need for frequent water exchanges (Wang et al., 2022). In some algaquaculture systems, anaerobic conditions allow denitrification, where nitrate is reduced to nitrogen gas and released into the atmosphere. This process helps prevent nitrogen accumulation and maintains long-term balance in closed systems, although it may require specific conditions to function effectively (Pinho et al., 2022).

Algae take up inorganic phosphate, the form of phosphorus most available in aquatic systems, and store it as polyphosphate or integrate it into cellular structures. This uptake reduces phosphorus levels in the water, mitigating the risk of algal blooms that can degrade water quality and lead to harmful conditions for fish (Wang et al., 2022). Harvesting algal biomass regularly can remove significant amounts of phosphorus from the system, providing a sustainable means of nutrient extraction. This biomass can serve as a source of phosphorus-rich fertilizer or be processed for other agricultural applications, creating a closed-loop approach that minimizes waste (Yaakob et al., 2021).

Carbon dynamics in algaquaculture systems primarily involve carbon dioxide (CO_2) exchange, dissolved organic carbon (DOC), and particulate organic carbon (POC). Algae play a crucial role in capturing CO_2 through photosynthesis, facilitating carbon sequestration and oxygen production, directly benefiting aquatic organisms (Viswanaathan et al., 2022). Algae absorb CO_2, converting it into organic carbon via photosynthesis. This process generates oxygen as a byproduct, which helps maintain aerobic conditions essential for fish and other aerobic organisms in the system. In addition, the photosynthetic carbon fixation by algae enhances overall system productivity and offsets CO_2 produced through respiration (Gao and Beardall, 2022). DOC and POC accumulate from uneaten feed, feces, and dead algae, contributing to the organic load in aquaculture systems. These organic carbon sources decompose and release nutrients that algae can reuse, thereby contributing to nutrient cycling and reducing organic pollution in the water (Verdegem, 2013). Heterotrophic bacteria break down organic matter, including dead algae and fecal material, releasing nutrients back into the system. Decomposition contributes to essential nutrient availability and prevents the build-up of organic sludge (Pinho et al., 2022). This microbial activity complements algal nutrient uptake, enhancing overall nutrient cycling efficiency and stability (Wang et al., 2022).

6.4.3 Factors affecting nutrient removal efficiency in algal systems

(1) Environmental conditions: Light intensity, temperature, and pH play vital roles in nutrient uptake efficiency. Algae require species-specific adequate light conditions and optimal temperatures to maximize nutrient assimilation rates (Borowitzka, 2016).

(2) Algal species: Different species exhibit varying preferences and tolerances for nitrogen and phosphorus forms (NO_3^- vs. NH_4^+). The selection of algae based on target nutrients and wastewater composition is crucial for effective bioremediation (Husin and Basri, 2022).

(3) Culture system type: Closed photobioreactors are ideal for microalgae due to controlled environmental conditions, while raceway ponds and IMTA systems support macroalgae growth in open systems (Kumari et al., 2023).

(4) Wastewater composition: Nutrient levels in aquaculture wastewater fluctuate based on feed type, fish density, and waste management practices. Algae are typically more effective at higher nutrient concentrations, where they can sustain rapid growth and high removal rates (Soo et al., 2017).

6.4.4 Algal biomass growth in response to nutrient availability

Algal growth is closely tied to the availability of N and P, which are essential macronutrients for protein synthesis, nucleic acids, and other cellular structures. Nutrient concentrations that exceed algal uptake capacity may cause growth inhibition because of toxic effects, while nutrient limitations can lead to reduced biomass productivity (Chen et al., 2017). Therefore managing nutrient levels in aquaculture wastewater is crucial to optimize algal growth.

Nitrogen is one of the primary nutrients that influences algal biomass growth. Ammonium (NH_4^+) is often the preferred nitrogen source for many algal species because it can be directly incorporated into amino acids without requiring additional reduction steps (Salbitani and Carfagna et al., 2021). However, high concentrations of NH_4^+ can inhibit growth due to toxicity, so it is essential to maintain balanced concentrations (Nie et al., 2020). Research indicates that nitrate (NO_3^-) is generally less toxic than ammonium and can also support high algal growth rates (Cai et al., 2013). In addition, nitrogen availability influences algal lipid content, with nitrogen-limited conditions often promoting lipid accumulation as a stress response (Chen et al., 2017). Nitrogen deficiency can cause algal cells to enter a stationary growth phase, where biomass production ceases, and cells start accumulating storage compounds such as lipids and carbohydrates. This shift in metabolism is especially beneficial for applications where high lipid content is desired, such as biofuel production. Conversely, high nitrogen availability promotes protein synthesis and rapid cell division, which supports the growth of dense algal cultures suitable for nutrient removal (Salbitani and Carfagna, 2021).

Like nitrogen, phosphorus concentration significantly influences algal growth rates and biomass yield. However, algae require lower amounts of phosphorus than nitrogen, making it a limiting nutrient in many aquatic systems (Barnharst et al., 2018). Optimal phosphorus concentrations for algal growth vary among species, but excessive phosphorus may lead to the precipitation of phosphorus salts, thereby reducing bioavailability. Inadequate phosphorus levels trigger a reduction in cell division rates and may lead to the accumulation of polyphosphate granules within cells, affecting cell function and biomass productivity (Slocombe et al., 2020). A balanced N:P ratio is essential for maximizing growth, as nitrogen and phosphorus often have synergistic effects on algal biomass yield.

The nitrogen-to-phosphorus (N:P) ratio is a critical factor in determining algae's growth performance and nutrient uptake efficiency. The optimal N:P ratio varies across algal species but generally falls within the range of 10:1–20:1 in aquaculture wastewater settings (Cai et al., 2013). When the N:P ratio is too high, phosphorus may become limiting, slowing down growth rates. Conversely, low N:P ratios result in nitrogen limitation, causing algae to shift toward slower growth and higher lipid accumulation (Salbitani and Carfagna, 2021). This balance enables algae to utilize available nitrogen and phosphorus efficiently, thereby promoting robust growth and maximizing nutrient uptake rates.

6.5 Benefits of effluent-derived nutrient uptake by algae

As above-mentioned, algae are the potential biofilter for the aquaculture wastewater. In this process, they absorbed the waste nutrients and converted them into useful biomass, accelerating additional income vis-à-vis reducing environmental stress (Li et al., 2019). This strategy supports sustainable aquaculture methods and is consistent with the ideas of the circular economy (Neori et al., 2004; Chopin et al., 2012).

6.5.1 Nutrients recovery and reduction of eutrophication

Nitrogen and phosphorus play a major role in the metabolic processes of algae. Algae convert the inorganic nitrogen or phosphorus into organic nutrients and help in making proteins, enzymes, chlorophylls, and energy, while phosphorus plays a vital role in building nucleic acids, lipids, and proteins (Gualtieri and Barsanti, 2022; Nguyen et al., 2022). It was also evident that around 70%–80% of the absorbed nitrogen is constituted in the protein of algae (Su, 2021). Algae prefer to consume different kinds of inorganic nitrogenous sources, such as ammonium (NH_4^+), nitrate (NO_3^-), nitrite (NO_2^-), in which NH_4^+ is the most preferred among them (Perez-Garcia et al., 2011; Cai et al., 2013), while the inorganic phosphorus source is dihydrogen phosphate ($H_2PO_4^-$) and hydrogen phosphate (HPO_4^-). It has been reported that algae preferentially consume NO_3^- only when NH_4^+ is limited; however, NO_3^- is more thermodynamically stable in oxidized aquatic environments and thus tends to predominate (Cai et al., 2013). Algae are capable of tolerating high nutrient loads, including ammonia concentrations as high as 300–500 mg L^{-1} (Lu et al., 2018) and NH_4^+ of 20–661 mg L^{-1} depending on the microalgal species (Collos and Harrison, 2014). The microalgae-based photobioreactor effectively remove 99% of nitrogen and phosphate in the Mar Menor lagoon (Gil-Izquierdo et al., 2021). Various microalgae species, including *C. vulgaris* and *Spirulina maxima*, have been effective in removing total phosphorus and ammonia, with removal rates reaching up to 95.8% and 94.6%, respectively (Chan et al., 2014). In addition, Godos et al. (2010) also reported that algae such as *Euglena viridis, Chlorella sorokiniana*, and *Scenedesmus obliquus* can reduce 20%–65% of the phosphorus in the diluted seawater. Further, the microalgae consortium showed better nutrient removal performance than the unialgal cultures (Sriram and Seenivasan, 2012). However, the nutrient uptake by the microalgae depends on the N/P ratio because a higher or lower N/P ratio may affect the growth and survivability of the microalgae and thus the optimum N/P ratio for microalgal productivity is given as 5– 30 (Choi and Lee, 2015). Many freshwater and marine algal species have a high capacity for nutrient assimilation, gaining more biomass, which is shown in Table 6.2. The prolonged containment of these nutrients, nitrogen and phosphorus, leads to eutrophication that promotes the growth of harmful microalgae (Zuo et al., 2024). Thus, the assimilation of excess nutrients helps mitigate the adverse effects of nutrient overloading, reducing the potential for algal blooms and eutrophication in aquatic ecosystems.

In addition, microalgae can remove heavy metals from the effluent waters, ranging from 32% to 100% removal efficiency through biosorption and bioaccumulation (Meitei et al., 2022). There is also evidence on the reduction of many other contaminants, such as pesticides, polycyclic aromatic hydrocarbons, phenolics, and organic solvents, through microalgae treatments (Muñoz and Guieysse, 2006; Singhal et al., 2021). Microalgae such as *C. vulgaris*

TABLE 6.2 Nutrient removal capacity of various microalgal species.

Types of microalgae	Source of wastewater	Nutrient removal	Productivity	References
Chlorophyta species	Aquaculture wastewater	NH_4^+: 97.9%–98.9% NO_2^-: 94.3%–99.8% NO_3: 85.7%–97.1% PO_4^+: 90.2%–98.9%	225–260 mg L^{-1} depending on the species	Liu et al. (2019)
Chlorella pyrenoidosa	Soybean processing wastewater	NH_4^+–N:89.1% TN: 88.8%, TP:70.3%	0.64 g L^{-1} d^{-1}	Hongyang et al. (2011)
Chlorella sp.	Municipal wastewater	Ammonia: 93.9%, TN: 89.1%, TP: 80.9%	0.92 g L^{-1} d^{-1}	Li et al. (2011)
Chlamydomonas sp.	Industrial wastewaters	NH_4^+–N: 100% (38.4 mg/L) NO_3^-–N: 100% (3.1 mg/L) PO_4^{3-}–P: 33% (44.7 mg/L)	—	Wu et al. (2012)
Chlamydomonas reinhardtii	Metropolitan wastewater	N: 55.8 mg/L/day P: 17.4 mg/L/day	2.0 g L^{-1} day^{-1}	Kong et al. (2010)
Galdieria sulphuraria	Landfill leachate	NH_4^+–N: 22.78 mg $L^{-1}d^{-1}$ (99.9%) PO_4–P: 2.91 mg $L^{-1}d^{-1}$ (34%)	—	Pan et al. (2021)

and *Scenedesmus quadricauda* have demonstrated significant wastewater treatment potential, removing up to 80.64% COD and 70.91% BOD, and 70.97% COD and 89.21% BOD, respectively (Kshirsagar, 2013), whereas *C. vulgaris* alone has also been reported to remove 88.0% COD and 89.6% BOD (Azeez, 2010).

6.5.2 Biofuel production

The algal biomass harvested, which was cultured in various systems including waste water system are rich in valuable byproducts such as lipids (oils), carbohydrates, proteins, and various feedstocks that can be converted into biofuels and other useful materials (Hu et al., 2008), contributing to a circular economy approach. Algae can harness around 8%–10% of the solar energy with a productivity of 77 g dcw m^{-2} d^{-1} (Melis, 2009). In addition, this microalgal-based biofuels production approach is an eco-friendly, nontoxic, strong potential for fixing carbon dioxide and low sulfur emission (Khan et al., 2018). These microalgae have a lipid content of 2%–75% depending on the species, which makes them a potential source of good biofuels (Meitei et al., 2022). Some species with the lipid content that has the potential for biofuel production are given in Table 6.3. It was also reported that up to 65% of microalgal biomass can be converted into biogas, whereas only about 40% can be converted into bioethanol and 30% into biodiesel when each process is specifically targeted. Furthermore, some algae species can produce hydrogen gas through photosynthesis, which can be used to produce electricity (Melis and Happe, 2001). The oil generated may cost $8 per kg of biomass, considering the cost of production of dry microalgae with 25% lipid in $2 per kg (Hingsamer et al., 2019).

TABLE 6.3 Lipid content of various algal biomass.

Algal species	Biomass productivity	Lipid productivity	Lipid content	References
Scenedesmus obliquus	292.50 mg L^{-1} d^{-1}	78.73 mg L^{-1} d^{-1}	38.9%	Ho et al. (2010)
Chlorella vulgaris	234.1–532.2 mg L^{-1} d^{-1}	183.7 mg L^{-1} d^{-1}	19.8%–34.8%	Deng et al. (2018)
Parachlorella kessleri	0.82 g L^{-1} d^{-1}	0.59 g L^{-1} d^{-1}	66%	Takeshita et al. (2018)
Tetraselmis suecica	65.7 mg L^{-1} d^{-1}	—	75.8	Andreotti et al. (2019)
Chlorella sorokiniana	498.14 mg $L^{-1}d^{-1}$	150.19 mg $L^{-1}d^{-1}$	26%–39.1%	Guldhe et al. (2017)
Nannochloropsis sp.	0.17–0.25g L^{-1} d^{-1}	37.6–61 g L^{-1} d^{-1}	21.6%–30.9%	Rodolfi et al. (2009)
Botryococcus terribilis	0.11–0.24 g L^{-1} d^{-1}	33.3–73.4 mg L^{-1} d^{-1}	29.7%–31.2%	Nascimento et al. (2015)
Botryococcus braunii	0.12–0.24 g L^{-1} d^{-1}	36.6–74.7 mg L^{-1} d^{-1}	30–32	Nascimento et al. (2015)

6.5.3 Microalgae as supplements and its bioactive compounds

Microalgae, as versatile photosynthetic microorganisms, are gaining global recognition for their nutritional richness and health-promoting bioactive compounds. These bioactive substances include proteins, polysaccharides, lipids, PUFAs, pigments, phycobiliproteins, vitamins, and enzymes (Khan et al., 2018). They also provide vital micronutrients such as vitamins A, B-complex (B1, B2, B3, B5, B7, and B9), C, and E, along with key minerals such as iodine, iron, magnesium, potassium, and calcium (Sathasivam et al., 2019). This has led to the commercialization of various microalgal genera in formats such as tablets, powders, and beverages. Prominent genera include Chlorodendrophyceae (*Tetraselmis chui, Tetraselmis striata* CTP4); Chlorophyceae (*Haematococcus lacustris*, formerly *Haematococcus pluvialis*); Coccolithophyceae (*Tisochrysis lutea*); Bacillariophyceae (*Phaeodactylum tricornutum, Skeletonema* sp.); Eustigmatophyceae (*Nannochloropsis* sp.); Porphyridiophyceae (*Porphyridium* sp.); and Cyanophyceae (*Spirulina*). Some species, including *Arthrospira platensis (Spirulina), C. vulgaris, Dunaliella salina, Euglena gracilis,* and *H. pluvialis*, have attained "Generally Recognized as Safe" (GRAS) status by regulatory authorities like the FDA (Mavrommatis et al., 2023). They are a rich source of bioactive compounds such as protein, amino acids, fatty acids, vitamins, polyphenols, and antioxidants.

6.5.3.1 *Proteins, peptides, and amino acids*

Microalgal proteins play a vital role in cell growth, enzyme activity, hormonal balance, and immune responses. Some species store soluble proteins in their cytoplasm and chloroplasts. Examples such as *A. platensis, Chlamydomonas reinhardtii,* and *Neochloris cohaerens* have phycobilisomes attached to thylakoid membranes, whereas species such as *C. vulgaris* and *Botryococcus braunii* are rich in essential amino acids, especially arginine and cysteine, which can make up to 44.7% of their amino acid profile (Dolganyuk et al., 2020). *Spirulina platensis* and *C. vulgaris* can contain protein levels as high as 60%–65% and 51%–58% of their dry weight, respectively (Matos, 2017). These proteins are complemented by other nitrogen-rich components, such as nucleic acids and cell wall materials (Tibbetts et al., 2015). Although there are slight deficiencies in sulfur-containing amino acids such as methionine and cysteine, the overall amino acid profile of microalgae generally meets FAO standards. Their environmentally friendly production enhances their attractiveness as alternative protein sources (Koyande et al., 2019).

6.5.3.2 *Lipids and polyunsaturated fatty acids*

Lipids in microalgae encompass fatty acids, sterols, waxes, and pigments that function in energy storage and membrane structure. Lipid content can range from 20% to 50% of the dry biomass, often rising under nutrient stress or during stationary growth phases (Borowitzka, 2013; Barkia et al., 2019). Microalgae are especially valuable for their production of essential PUFAs, such as omega-3 (arachidonic acid, ALA; eicosapentaenoic acid, EPA; and eicosapentaenoic acid, DHA) and omega-6 (linoleic acid, LA; γ-linolenic acid, GLA; arachidonic acid, ARA) fatty acids. Specific species show selective accumulation: *Crypthecodinium cohnii* primarily contains DHA, while *Nannochloropsis* sp. and *Phaeodactylum* sp. are EPA-dominant (Matos, 2017). *C. vulgaris* and *Spirulina* also demonstrate high PUFA yields, with *Auxenochlorella protothecoides* reaching up to 70% lipid content (Griffiths and Harrison, 2009).

6.5.3.3 *Carbohydrates and polysaccharides*

Microalgal carbohydrates serve as vital energy reserves and structural materials. They occur as starch or as cell wall components such as cellulose and sulfated polysaccharides (Zhang and Liu, 2021). The sugar composition varies by species, with significant proportions of glucose and mannose found in *Chlorococcum* sp. and *Dunaliella salina* (Chen et al., 2013). Cyanobacteria, red algae, and green algae synthesize glycogen-like, floridean, and amylopectin-like carbohydrates, respectively (Ciferri, 1983). Bioactive polysaccharides, such as 1,3-β-glucan, found in *Chlorella* and *Euglena*, act as immunostimulants and cholesterol regulators (Koller et al., 2014). Sulfated variants show promise in treating microbial infections and skin conditions (Dineshbabu et al., 2019.

6.5.3.4 *Vitamins*

Microalgae are dense sources of both water- and fat-soluble vitamins. They uniquely supply bioavailable vitamin B12, which is typically absent in plant-based diets (Croft et al., 2005). Species such as *Dunaliella tertiolecta, Tetraselmis suecica, Arthrospira* sp., and *Chlorella* sp. are rich in vitamins B1–B12, A, C, D, and E. *D. salina*, for instance, accumulates high levels of β-carotene, while Chlorella species have notable vitamin C and B12 content (Watanabe et al., 2002; Del Mondo et al., 2020).

6.5.3.5 *Pigments*

Pigments such as chlorophylls, carotenoids, and phycobiliproteins play key roles in photosynthesis and antioxidative defense. Green microalgae contain chlorophylls, while red and cyanobacteria produce phycobiliproteins. Carotenoids such as astaxanthin and β-carotene, produced by *Haematococcus* sp. and *Dunaliella* sp., are in high commercial demand due to their potent antioxidant properties (Barros de Medeiros et al., 2022). *Chlorella* species contain chlorophyll levels up to 7% of biomass (Khanra et al., 2018), and *Spirulina* is a major source of the health-promoting pigment phycocyanin (Ji et al., 2020).

6.5.3.6 Polyphenols and antioxidants

Microalgae also synthesize secondary metabolites, such as polyphenols, including flavonoids and phenolic acids, which contribute to antioxidant potential. The phenolic content of a strain often correlates directly with its bioactivity and oxidative stress protection capabilities (Coulombier et al., 2021).

6.5.4 Health benefits of microalgal bioactive compounds

Microalgae-derived compounds such as omega-3 fatty acids, carotenoids, polysaccharides, and vitamins play significant roles in supporting human and animal health (Kaur et al., 2023). For humans, omega-3s, especially EPA and DHA, help reduce inflammation, enhance cardiovascular function, and support neural development (Eladl et al., 2024). Carotenoids such as astaxanthin and β-carotene counteract oxidative stress, thereby lowering risks of chronic diseases, including cancer and Alzheimer's (Ayub et al., 2025).

Sulfated polysaccharides and extracellular polymeric substances enhance immune defense by stimulating natural killer cells and blocking viral entry (Ayub et al., 2025). Clinical studies with *Chlorella* supplementation (3–10 g/day) report improvements in lipid metabolism, blood pressure, and antioxidant capacity.

In livestock, microalgal supplementation improves feed efficiency, growth rate, and immune responses. High-protein microalgal species have shown positive effects on body weight gain and fatty acid composition in ruminants, poultry, and swine (Mavrommatis et al., 2023). Polysaccharides from *Spirulina* and *Chlorella* also enhance the expression of immune markers like lysozyme and nitric oxide.

Microalgae-derived dietary fibers and polyphenols contribute to gut health by promoting beneficial microbes and increasing short-chain fatty acid production, thus reinforcing gut integrity and nutrient absorption (Krohn et al., 2022).

In aquaculture, microalgae such as *Spirulina, H. pluvialis*, and *Nannochloropsis gaditana* are utilized as feed additives to enhance fish growth, immune response, and resistance to stressors. Their incorporation into diets results in higher survival rates, better FCR, and improved fillet nutritional quality (Chénais, 2021; Trevi et al., 2023).

6.5.5 Microalgae as sustainable feed and food supplements

6.5.5.1 Aquaculture applications

In aquaculture, microalgae such as *S. platensis* and *Schizochytrium* have proven effective in improving growth rates, feed conversion ratios, and immune resilience across multiple fish species. A meta-analysis on 26 finfish species found that even 0.025%*Spirulina* supplementation significantly improved growth outcomes, while *Schizochytrium* helped retain omega-3 levels in fish tissue at 10% inclusion (Trevi et al., 2023). In tilapia, Spirulina boosted weight gain and nutrient utilization, and combinations of *A. platensis* and *N. gaditana* with phytase improved gut health in seabass (Peralta-Sánchez et al., 2024). Thus algae are a good source of growth promoters as well as immunostimulants.

6.5.5.2 Human nutrition

For human consumption, *C. vulgaris* and *S. platensis* are highly regarded due to their complete protein content, B-complex vitamins, antioxidants, and fatty acids (Caporgno and Mathys, 2018). However, cell wall resistance to digestion necessitates processing methods such as enzymatic hydrolysis or ultrasonication to enhance nutrient bioavailability (Ali et al., 2025). Enrichment of microalgal biomass with EPA and docosahexaenoic acid (DHA) is being scaled through hatchery cultivation systems, offering a viable alternative to marine-derived omega-3 supplements.

6.5.6 Carbon sequestration and reduction of carbon footprint

As we all know, CO_2 is one of the major greenhouse gases, and it is critical to take urgent action to sequester it. Microalgae have the potential for biofuel production as discussed in the earlier section. In the production process, carbon dioxide plays a major role in the growth and increase in biomass of the microalgae. For instance, the diatom can tolerate 20%–25% of the carbon dioxide, leading to total lipid content increased from 8.91% to 10.57 % in cell dry mass (Artamonova et al., 2017). Similarly, Cuellar-Bermudez et al. (2015) also observed that 3% CO_2, *Synechocystis* increased the lipid content of 14% of dry weight basis. Thus microalgae can assimilate CO_2 through photosynthesis for metabolic process vis-à-vis reducing atmospheric carbon dioxide. It has been reported that 1 kg of algal biomass can fix 1.83 kg of CO_2 (Gendy and El-Temtamy, 2013). Besides, the CO_2 emissions from the traditional fuel production can be replaced by microalgae-based biofuel production with less CO_2 emissions, thereby mitigating CO_2 emissions. Microalgae exhibit a CO_2-fixation efficiency that is approximately 10–50 times higher than that of terrestrial plants (Wang et al., 2008). The sequestration involves three processes: (1) CO_2 uptake, (2) fixation through photosynthesis, and (3) carbon storage in the form of biomass (Tarafdar et al., 2023). Microalgae species such as *Chlorella* sp. have carbon fixation efficiency up to 90% (Tarafdar et al., 2023). Thus microalgae can act as a carbon sink and reduce carbon footprint.

6.5.7 Biostimulants

Biostimulants are compounds or molecules that regulate and enhance the physiological processes of plants (Ronga et al., 2019). The biostimulant effects can be obtained through phytohormones, exopolysaccharide (EPSs) release, and antifungal activity, as well as protein and amino acid activity. Phytohormones such as cytokinins, gibberellins, indole acetic acid, salicylic acid, abscisic acid, and jasmonic acid are present in the microalgae such as *Spirulina, Chlorella,* and *Scenedesmus* (Bayona-Morcillo et al., 2020; Navarro-López et al., 2020). That plays a major role in seed germination, bud development, leaf blade development, and stem elongation (Tarakhovskaya et al., 2007). This phytohormone can be delivered to the plant either through the incubation of seeds and roots in a solution containing microalgae or their extracts or by applying their extracts to leaves or seeds (Alvarez et al., 2021).

Microalgae are also a source of EPS, composed of neutral sugars, uronic acids, and sulfate, which are also known for their biostimulating effect on plant growth (EL Arroussi et al., 2018). The EPS of microalgae can enhance the enzymatic activity in plant leaves, such as nitrate reductase and NAD-glutamate dehydrogenase (Rachidi et al., 2020, while the EPS of *P. triocnutum, Desmodesmus* sp., and *Porphyridium* sp. improved the phenylalanine ammonia lyase, chitinase, β-1,3-glucanase, and peroxidase activities (Rachidi et al., 2021). It was evident that the topical application of microalgae containing EPS promoted the growth of tomato plants under salt stress, leading to increased shoot height, dry weight, chlorophyll, and protein content, while reducing stress indicators such as proline and phenolic compounds

(EL Arroussi et al., 2018). Moreover, Ranglová et al. (2021) also showed that *C. vulgaris* has antifungal and antibacterial properties, resulting in the inhibition of *Clavibacter, Fusarium, Rhizoctonia*, and *Phytopthora* in domestic wastewater.

Microalgae also exhibit protein hydrolysate, which is one of the important ingredients of plant biostimulants that enhance the growth and development of plants (Ronga et al., 2019). They also contain proteins and amino acid, which provide nitrogen and help in other metabolic activities (Braun and Colla, 2023).

6.6 Economic viability for microalgae production

The world commercial microalgae industry is dominated by only a few microalgae, such as *Chlorella, Dunaliella, Haematococcus, Isochrysis* (Spolaore et al., 2006), *Nannochloropsis, Porphyridium*, and *Spirulina*, in which Spirulina and Chlorella are the leading microalgae used in the production of functional foods and dietary supplements, whereas species such as *D. salina* and *H. pluvialis* are primarily cultivated for their valuable antioxidant compounds (González-Pérez et al., 2022). In the past decades, 75% of the yearly microalgal biomass production has been used in the health food production, including powders, tablets, capsules, or pastilles (Pulz and Gross, 2004). The value of *Chlorella* and *Spirulina* ranges from \$40 to \$50 per kg in the global market (Spolaore et al., 2006), which value increased from \$300 to \$3000 per kg when pure-fine chemicals such as β-carotene are obtained (Ben-Amotz, 2004).

The cost of production of microalgae depends on the production system, inputs such as light, labor, and scale of farming. For example, the tubular photobioreactor costs 290–329 € kg^{-1} microalgae production while the bubble column reactor costs €573–587 kg^{-1} (Oostlander et al., 2020). However, the production can be minimized by scaling up the production area, for instance, the production from 25 m^2 to 1,500 m^2 slashes costs by 55%–65%, reaching €23.47/kg in the greenhouse TPBR system. Further, Vázquez-Romero et al. (2022) proved that increasing the production area from 1 ha to 10 ha may reduce the production cost by 17.99%. Integrating microalgae cultivation with wastewater treatment can further lower costs to \$0.39–\$0.92/kg (Chaudry, 2021). Thus the cost of microalgae can be minimized by scaling up the microalgae production and by integrating in the wastewater, where microalgae can utilize the readily available nutrients at no additional cost.

6.7 Challenges and future prospects

Algaquaculture is a potential technology that reutilizes the wastewater and enhances the biomass, thereby promoting the circular economy. However, there are some challenges and limitations. The bacterial contamination and unfavorable effluent conditions need to be addressed for the successful implementation of microalgae-based pollution control in aquaculture (Li et al., 2021). Besides, closed PBRs have significantly higher capital and operating costs (Clippinger and Davis, 2019). More than 3%–15% of the total cost of production is contributed by the harvesting in commercial algal photobioreactor plants (Fasaei et al., 2018). So, currently, low-cost, large-scale microalgae cultivation is limited due to high operational costs, expensive infrastructure and maintenance requirements, and difficulties in scaling up the production process (Richardson et al., 2012). Moreover, difficulties in algal biomass separation, prolonged retention times, and low organic matter removal efficiency (Tawfik et al., 2022).

Furthermore, the growth of algae in aquaculture wastewater introduces unique challenges that are absent in synthetic or controlled nutrient media. Factors such as fluctuating pH, dissolved oxygen levels, and the presence of other contaminants can impact algal growth (Nie et al., 2020). In high-density aquaculture operations, dissolved oxygen levels can drop significantly, affecting both algae and fish, necessitating aeration to sustain algal cultures.

To address these challenges, pretreatment methods such as pH adjustment, dilution, or controlled nutrient addition can help create a more favorable growth environment for algae. In addition, certain algal strains have demonstrated resilience to fluctuating conditions, making them suitable candidates for cultivation in aquaculture wastewater (Rani et al., 2021). The use of mixed algal cultures, composed of species with complementary nutrient uptake profiles, can enhance biomass yield and nutrient removal efficiency, allowing for more effective wastewater remediation.

Looking ahead, the future of aquaculture lies in eco-friendly and integrated approaches that combine algae in aquaculture for aquaculture waste treatment for resource-efficient systems that minimize environmental impacts while maximizing productivity. Algaquaculture, the integration of algae in aquaculture, remains a concept till today, though it is a promising technology toward sustainability and the circular economy. The work done by Addy et al. (2017), in which algae were incorporated in the aquaponics system, is the only work related to the concept. Advancements in biotechnology, genetic engineering, and system automation will further enhance the efficiency of nutrient removal and biomass production. Moreover, strong policy support, economic incentives, and interdisciplinary research are essential to accelerating the adoption of algaquaculture on a larger scale. By leveraging algae as a key component in sustainable aquaculture, the industry can move toward a more environmentally responsible, economically viable, and resource-efficient future, ensuring long-term food security and ecological balance. By adopting these innovative strategies, the aquaculture industry can contribute to a more sustainable and responsible global food system. While algaquaculture shows great potential for sustainable aquaculture, further research is needed to develop commercial-scale technologies and address challenges such as food safety and consumer (Chinnasamy et al., 2014; Campanati et al., 2022).

6.8 Conclusion

The integration of algae into aquaculture systems, algaquaculture, offers a sustainable solution for managing effluent discharge while enhancing resource efficiency. Algae play a vital role in nutrient uptake, particularly in removing excess nitrogen and phosphorus from aquaculture wastewater, thereby mitigating the risks of eutrophication and water pollution. Their natural ability to absorb and recycle nutrients not only improves water quality but also supports the overall health and productivity of aquaculture systems. In addition, the harvested algal biomass has significant economic potential, with applications in biofuels, pharmaceuticals, animal feed, and bioplastics, promoting a circular economy in aquaculture. Further, microalgae represent a paradigm shift in carbon footprint reduction, transforming CO_2 from a pollutant into a resource, enabling the trading of carbon credits in markets.

References

Addy, M.M., Kabir, F., Zhang, R., Lu, Q., Deng, X., Current, D., Griffith, R., Ma, Y., Zhou, W., Chen, P., Ruan, R., 2017. Co-cultivation of microalgae in aquaponic systems. Bioresour. Technol. 245, 27–34. doi:10.1016/j.biortech.2017.08.151.

Al-Hafedh, Y.S., Alam, A., Buschmann, A.H., 2015. Bioremediation potential, growth and biomass yield of the green seaweed, Ulva lactuca in an integrated marine aquaculture system at the Red Sea coast of Saudi Arabia at different stocking densities and effluent flow rates. Rev. Aquac. 7 (3), 161–171. doi:10.1111/raq.12060.

Ali, S.S., Al-Tohamy, R., Al-Zahrani, M., Schagerl, M., Kornaros, M., Sun, J., 2025. Advancements and challenges in microalgal protein production: a sustainable alternative to conventional protein sources. Microb. Cell Fact. 24 (1). doi:10.1186/s12934-025-02685-1.

Alvarez, A.L., Weyers, S.L., Goemann, H.M., Peyton, B.M., Gardner, R.D., 2021. Microalgae, soil and plants: a critical review of microalgae as renewable resources for agriculture. Algal Res. 54. doi:10.1016/j.algal.2021.102200.

Andreotti, V., Solimeno, A., Chindris, A., Marazzi, F., García, J., 2019. Growth of tetraselmis suecica and dunaliella tertiolecta in aquaculture wastewater: numerical simulation with the BIO_ALGAE model. Water Air Soil Pollut. 230 (3). doi:10.1007/s11270-019-4122-0.

Apandi, N.M., Mohamed, R.M.S.R., Al-Gheethi, A., Kassim, A.H.M., 2019. Microalgal biomass production through phycoremediation of fresh market wastewater and potential applications as aquaculture feeds. Environ. Sci. Pollut. Res. 26 (4), 3226–3242. doi:10.1007/s11356-018-3937-3.

Arshad, S., Arshad, S., Afzal, S., Tasleem, F., 2024. Environmental impact and sustainable practices in aquaculture: a comprehensive review. Haya: Saudi J. Life Sci. 9 (11), 447–454. doi:10.36348/sjls.2024.v09i11.005.

Artamonova, E.Y., Vasskog, T., Eilertsen, H.C., 2017. Lipid content and fatty acid composition of Porosira glacialis and Attheya longicornis in response to carbon dioxide (CO_2) aeration. PLoS One 12 (5), e0177703. doi:10.1371/journal.pone.0177703.

Attasat, S., Wanichpongpan, P., Ruenglertpanyakul, W., 2013. Cultivation of microalgae (Oscillatoria okeni and Chlorella vulgaris) using tilapia-pond effluent and a comparison of their biomass removal efficiency. Water Sci. Technol. 67 (2), 271–277. doi:10.2166/wst.2012.505.

Ayub, A., Rahayu, F., Khamidah, A., Antarlina, S.S., Iswari, K., Supriyadi, K., Mufidah, E., Singh, A., Chopra, C., Wani, A.K., 2025. Harnessing microalgae as a bioresource for nutraceuticals: advancing bioactive compound exploration and shaping the future of health and functional food innovation. Discov. Appl. Sci. 7 (5). doi:10.1007/s42452-025-06916-3.

Azeez, R.A., 2010. A study on the effect of temperature on the treatment of industrial wastewater using Chlorella vulgaris alga. Eng. Technol. J. 28 (4), 785–792.

Azim, M.E., Little, D.C., 2008. The biofloc technology (BFT) in indoor tanks: water quality, biofloc composition, and growth and welfare of Nile tilapia (Oreochromis niloticus). Aquaculture 283 (1–4), 29–35. doi:10.1016/j.aquaculture.2008.06.036.

Bacellar Mendes, L.B., Vermelho, A.B., 2013. Allelopathy as a potential strategy to improve microalgae cultivation. Biotechnol. Biofuels 6 (1). doi:10.1186/1754-6834-6-152.

Badiola, M., Mendiola, D., Bostock, J., 2012. Recirculating Aquaculture systems (RAS) analysis: main issues on management and future challenges. Aquacult. Eng. 51, 26–35. doi:10.1016/j.aquaeng.2012.07.004.

Barkia, I., Saari, N., Manning, S.R., 2019. Microalgae for high-value products towards human health and nutrition. Mar. Drugs 17 (5). doi:10.3390/md17050304.

Barnharst, T., Rajendran, A., Hu, B., 2018. Bioremediation of synthetic intensive aquaculture wastewater by a novel feed-grade composite biofilm. Int. Biodeterior. Biodegrad. 126, 131–142. doi:10.1016/j.ibiod.2017.10.007.

Barros de Medeiros, V.P., da Costa, W.K.A., da Silva, R.T., Pimentel, T.C., Magnani, M., 2022. Microalgae as source of functional ingredients in new-generation foods: challenges, technological effects, biological activity, and regulatory issues. Crit. Rev. Food Sci. Nutr. 62 (18), 4929–4950. doi:10.1080/10408398.2021.1879729.

Bayona-Morcillo, P.J., Plaza, B.M., Gómez-Serrano, C., Rojas, E., Jiménez-Becker, S., 2020. Effect of the foliar application of cyanobacterial hydrolysate (Arthrospira platensis) on the growth of Petunia x hybrida under salinity conditions. J. Appl. Phycol. 32 (6), 4003–4011. doi:10.1007/s10811-020-02192-3.

Ben-Amotz, A., 2004. Industrial production of microalgal cell-mass and secondary products—major industrial species—Dunaliella. In: Richmond, A (Ed.), Handbook of microalgal culture: biotechnology and applied phycology. Blackwell Publishing Ltd, pp. 273–280. https://doi.org/10.1002/9780470995280.ch13.

Borah, D, Sekhar Bora, S., 2020. In-situ nutrient recycling and management. Res. Biotica 2 (3), 91–96. doi:10.54083/resbio/2.3.2020.91-96.

Borowitzka, M.A., 2013. High-value products from microalgae-their development and commercialisation. J. Appl. Phycol. 25 (3), 743–756. doi:10.1007/s10811-013-9983-9.

Borowitzka, M.A., 2016. Algal physiology and large-scale outdoor cultures of microalgae. In: The physiology of microalgae. Developments in applied phycology, 6. Springer International Publishing, Cham, Switzerland, pp. 601–652. https://doi.org/10.1007/978-3-319-24945-2_23.

Boyd, C., McNevin, A., 2015. Aquaculture, resource use and the environment. John Wiley & Sons, Chichester, UK, pp. 61–80.

Boyd, C.E., 2017. General relationship between water quality and aquaculture performance in ponds. In: Fish diseases: prevention and control strategies. Elsevier Inc., United States, pp. 147–166. doi:10.1016/B978-0-12-804564-0.00006-5.

Boyd, C.E., D'Abramo, L.R., Glencross, B.D., Huyben, D.C., Juarez, L.M., Lockwood, G.S., McNevin, A.A., Tacon, A.G.J., Teletchea, F., Tomasso, J.R., Tucker, C.S., Valenti, W.C., 2020. Achieving sustainable aquaculture: historical and current perspectives and future needs and challenges. J. World Aquac. Soc. 51 (3), 578–633. doi:10.1111/jwas.12714

Boyd, C.E., Tucker, C., McNevin, A., Bostick, K., Clay, J., 2007. Indicators of resource use efficiency and environmental performance in fish and crustacean aquaculture. Rev. Fish. Sci. 15 (4), 327–360. doi:10.1080/10641260701624177.

Boyd, C.E., Tucker, C.S., 2012. In: Pond aquaculture water quality management, 1461554071. Springer Science & Business Media, New York, p. 9781461554073. doi:10.1007/978-1-4615-5407-3.

Braun, J.C.A., Colla, L.M., 2023. Use of microalgae for the development of biofertilizers and biostimulants. BioEnergy Res. 16 (1), 289–310. doi:10.1007/s12155-022-10456-8.

Bucking, C., 2017. A broader look at ammonia production, excretion, and transport in fish: a review of impacts of feeding and the environment. J. Comp. Physiol. B 187 (1), 1–18. doi:10.1007/s00360-016-1026-9.

Bureau, D.P., Hua, K., 2010. Towards effective nutritional management of waste outputs in aquaculture, with particular reference to salmonid aquaculture operations. Aquac. Res. 41 (5), 777–792. doi:10.1111/j.1365-2109.2009.02431.x.

Cai, T., Park, S.Y., Li, Y., 2013. Nutrient recovery from wastewater streams by microalgae: status and prospects. Renew. Sustain. Energy Rev. 19, 360–369. doi:10.1016/j.rser.2012.11.030.

Campanati, C., Willer, D., Schubert, J., Aldridge, D.C., 2022. Sustainable intensification of aquaculture through nutrient recycling and circular economies: more fish, less waste, blue growth. Rev. Fish. Sci. Aquac. 30 (2), 143–169. doi:10.1080/23308249.2021.1897520.

Caporgno, M.P., Mathys, A., 2018. Trends in microalgae incorporation into innovative food products with potential health benefits. Front. Nutr. 5. doi:10.3389/fnut.2018.00058.

Carballeira Braña, C.B., Cerbule, K., Senff, P., Stolz, I.K., 2021. Towards environmental sustainability in marine finfish aquaculture. Front. Mar. Sci. 8. doi:10.3389/fmars.2021.666662.

Chan, A., Salsali, H., McBean, E., 2014. Nutrient removal (nitrogen and phosphorous) in secondary effluent from a wastewater treatment plant by microalgae. Can. J. Civ. Eng. 41 (2), 118–124. doi:10.1139/cjce-2013-0405.

Chaudry, S., 2021. Integrating microalgae cultivation with wastewater treatment: a peek into economics. Appl. Biochem. Biotechnol. 193 (10), 3395–3406. doi:10.1007/s12010-021-03612-x.

Chen, B., Wan, C., Mehmood, M.A., Chang, J.S., Bai, F., Zhao, X., 2017. Manipulating environmental stresses and stress tolerance of microalgae for enhanced production of lipids and value-added products–a review. Bioresour. Technol. 244, 1198–1206. doi:10.1016/j.biortech.2017.05.170.

Chen, C.Y., Zhao, X.Q., Yen, H.W., Ho, S.H., Cheng, C.L., Lee, D.J., Bai, F.W., Chang, J.S., 2013. Microalgae-based carbohydrates for biofuel production. Biochem. Eng. J. 78, 1–10. doi:10.1016/j.bej.2013.03.006.

Chénais, B., 2021. Algae and microalgae and their bioactive molecules for human health. Molecules 26 (4), 1185. doi:10.3390/molecules26041185.

Chini Zittelli, G., Rodolfi, L., Bassi, N., Biondi, N., Tredici, M.R., 2013. Photobioreactors for microalgal biofuel production. In: Algae for biofuels and energy. Springer Netherlands, Dordrecht, pp. 115–131. doi:10.1007/978-94-007-5479-9_7.

Chinnasamy, S., Sood, A., Renuka, N., Prasanna, R., Ratha, S.K., Bhaskar, S., Rengasamy, R., Lewis, D.M., 2014. Ecobiological aspects of algae cultivation in wastewaters for recycling of nutrients and biofuel applications. Biofuels 5 (2), 141–158. doi:10.4155/bfs.13.78.

Choi, H.J., Lee, S.M., 2015. Effect of the N/P ratio on biomass productivity and nutrient removal from municipal wastewater. Bioprocess. Biosyst. Eng. 38 (4), 761–766. doi:10.1007/s00449-014-1317-z.

Chopin, T., 2014. Integrated multi-trophic aquaculture (IMTA): an aquanomic approach to farming the sea and how it might work for Brazil. In Proceedings of the Animal production in view of the climate and technological changes: 51st Annual Meeting of the Brazilian Society of Animal Science, Sergipe, Brazil, July 29–August 1, 2014. https://www.cabidigitallibrary.org/doi/full/10.5555/20203170790.

Chopin, T., Cooper, J.A., Reid, G., Cross, S., Moore, C., 2012. Open-water integrated multi-trophic aquaculture: environmental biomitigation and economic diversification of fed aquaculture by extractive aquaculture. Rev. Aquac. 4 (4), 209–220. doi:10.1111/j.1753-5131.2012.01074.x.

Ciferri, O., 1983. Spirulina, the edible microorganism. Microbiol. Rev. 47 (4), 551–578. doi:10.1128/mr.47.4.551-578.1983.

Clippinger, J.N., Davis, R.E., 2019. Techno-economic analysis for the production of algal biomass via closed photobioreactors: future cost potential evaluated across a range of cultivation system designs (No. NREL/TP-5100-72716). National Renewable Energy Lab.(NREL), Golden, CO (United States), p. 42.

Collos, Y., Harrison, P.J., 2014. Acclimation and toxicity of high ammonium concentrations to unicellular algae. Mar. Pollut. Bull. 80 (1–2), 8–23. doi:10.1016/j.marpolbul.2014.01.006.

Costello, M.J., 2009. How sea lice from salmon farms may cause wild salmonid declines in Europe and North America and be a threat to fishes elsewhere. Proc. R. Soc. B Biol. Sci. 276 (1672), 3385–3394. doi:10.1098/rspb.2009.0771.

Coulombier, N.oé, Jauffrais, T., Lebouvier, N., 2021. Antioxidant compounds from microalgae: a review. Mar. Drugs 19 (10), 549. doi:10.3390/md19100549.

Crab, R., Avnimelech, Y., Defoirdt, T., Bossier, P., Verstraete, W., 2007. Nitrogen removal techniques in aquaculture for a sustainable production. Aquaculture 270 (1–4), 1–14. doi:10.1016/j.aquaculture.2007.05.006.

Croft, M.T., Lawrence, A.D., Raux-Deery, E., Warren, M.J., Smith, A.G., 2005. Algae acquire vitamin B12 through a symbiotic relationship with bacteria. Nature 438 (7064), 90–93. doi:10.1038/nature04056.

Cuellar-Bermudez, S.P., Romero-Ogawa, M.A., Vannela, R., Lai, Y.J.S., Rittmann, B.E., Parra-Saldivar, R., 2015. Effects of light intensity and carbon dioxide on lipids and fatty acids produced by Synechocystis sp. PCC6803 during continuous flow. Algal Res. 12, 10–16. doi:10.1016/j.algal.2015.07.018.

Day, J.P., 2008. The development of a modular integrated recirculating aquaculture system using Porphyra (nori) for the bioremediation of marine finfish effluent. University of New Hampshire, Durham, NH, USA.

Defoirdt, T., Sorgeloos, P., Bossier, P., 2011. Alternatives to antibiotics for the control of bacterial disease in aquaculture. Curr. Opin. Microbiol. 14 (3), 251–258. doi:10.1016/j.mib.2011.03.004.

Del Mondo, A., Smerilli, A., Sané, E., Sansone, C., Brunet, C., 2020. Challenging microalgal vitamins for human health. Microb. Cell Fact. 19 (1). doi:10.1186/s12934-020-01459-1.

Deng, X.Y., Gao, K., Addy, M., Li, D., Zhang, R.C., Lu, Q., Ma, Y.W., Cheng, Y.L., Chen, P., Liu, Y.H., Ruan, R., 2018. Cultivation of Chlorella vulgaris on anaerobically digested swine manure with daily recycling of the post-harvest culture broth. Bioresour. Technol. 247, 716–723. doi:10.1016/j.biortech.2017.09.171.

Dineshbabu, G., Goswami, G., Kumar, R., Sinha, A., Das, D., 2019. Microalgae–nutritious, sustainable aqua-and animal feed source. J. Funct. Foods 62, 103545.

Dolganyuk, V., Belova, D., Babich, O., Prosekov, A., Ivanova, S., Katserov, D., Patyukov, N., Sukhikh, S., 2020. Microalgae: a promising source of valuable bioproducts. Biomolecules 10 (8), 1153. doi:10.3390/biom10081153.

Eladl, S.N., Elnabawy, A.M., Eltanahy, E.G., 2024. Recent biotechnological applications of value-added bioactive compounds from microalgae and seaweeds. Bot. Stud. 65 (1). doi:10.1186/s40529-024-00434-y.

EL Arroussi, H., Benhima, R., Elbaouchi, A., Sijilmassi, B., EL Mernissi, N., Aafsar, A., Meftah-Kadmiri, I., Bendaou, N., Smouni, A., 2018. Dunaliella salina exopolysaccharides: a promising biostimulant for salt stress tolerance in tomato (Solanum lycopersicum). J. Appl. Phycol. 30 (5), 2929–2941. doi:10.1007/s10811-017-1382-1.

FAO, 2024. The state of world fisheries and aquaculture: blue transformation in action. Food and Agriculture Organization (FAO) 264. doi:10.4060/cd0683en.

Farghali, M., Mohamed, I. M., Osman, A. I., Rooney, D. W., 2023. Seaweed for climate mitigation, wastewater treatment, bioenergy, bioplastic, biochar, food, pharmaceuticals, and cosmetics: a review. Environmental Chemistry Letters 21 (1), 97–152. https://doi.org/10.1007/s10311-022-01520-y.

Fasaei, F., Bitter, J.H., Slegers, P.M., van Boxtel, A.J.B., 2018. Techno-economic evaluation of microalgae harvesting and dewatering systems. Algal Res. 31, 347–362. doi:10.1016/j.algal.2017.11.038.

Fox, J.M., Zimba, P.V., 2018. Minerals and trace elements in microalgae in health and disease prevention. Microalgae in health and disease prevention 177–193. doi:10.1016/B978-0-12-811405-6.00008-6.

Gao, K., Beardall, J., 2022. Using macroalgae to address UN sustainable development goals through CO_2 remediation and improvement of the aquaculture environment. Appl. Phycol. 3 (1), 360–367. doi:10.1080/26388081.2022.2025617.

Gendy, T.S., El-Temtamy, S.A., 2013. Commercialization potential aspects of microalgae for biofuel production: an overview. Egypt. J. Pet. 22 (1), 43–51. doi:10.1016/j.ejpe.2012.07.001.

Gil-Izquierdo, A., Pedreño, M.A., Montoro-García, S., Tárraga-Martínez, M., Iglesias, P., Ferreres, F., Barceló, D., Núñez-Delicado, E., Gabaldón, J.A., 2021. A sustainable approach by using microalgae to minimize the eutrophication process of Mar Menor lagoon. Sci. Total Environ. 758, 143613. doi:10.1016/j.scitotenv.2020.143613.

Godos, I.de., Vargas, V.A., Blanco, S., González, M.C.G., Soto, R., García-Encina, P.A., Becares, E., Muñoz, R., 2010. A comparative evaluation of microalgae for the degradation of piggery wastewater under photosynthetic oxygenation. Bioresour. Technol. 101 (14), 5150–5158. doi:10.1016/j.biortech.2010.02.010.

González-Pérez, B.K., Rivas-Castillo, A.M., Valdez-Calderón, A., Gayosso-Morales, M.A., 2022. Microalgae as biostimulants: a new approach in agriculture. World J. Microbiol. Biotechnol. 38 (1). doi:10.1007/s11274-021-03192-2.

Griffiths, M.J., Harrison, S.T.L., 2009. Lipid productivity as a key characteristic for choosing algal species for biodiesel production. J. Appl. Phycol. 21 (5), 493–507. doi:10.1007/s10811-008-9392-7.

Gross, M., Jarboe, D., Wen, Z., 2015. Biofilm-based algal cultivation systems. Appl. Microbiol. Biotechnol. 99 (14), 5781–5789. doi:10.1007/s00253-015-6736-5.

Gualtieri, P., Barsanti, L., 2022. Algae: Anatomy, Biochemistry, and Biotechnology, 3rd ed CRC Press, Boca Raton, FL doi:10.1201/9781003187707.

Guldhe, A., Ansari, F.A., Singh, P., Bux, F., 2017. Heterotrophic cultivation of microalgae using aquaculture wastewater: a biorefinery concept for biomass production and nutrient remediation. Ecol. Eng. 99, 47–53. doi:10.1016/j.ecoleng.2016.11.013.

Hamed, I., Özogul, F., Özogul, Y., Regenstein, J.M., 2015. Marine bioactive compounds and their health benefits: a review. Compr. Rev. Food Sci. Food Saf. 14 (4), 446–465. doi:10.1111/1541-4337.12136.

Hamed, I., 2016. The evolution and versatility of microalgal biotechnology: a review. Compr. Rev. Food Sci. Food Saf. 15 (6), 1104–1123. doi:10.1111/1541-4337.12227.

Hingsamer, M., Jungmeier, G., 2019. Biorefineries. In: The role of bioenergy in the bioeconomy. Academic Press, London, UK, pp. 179–222.

Ho, S.H., Chen, W.M., Chang, J.S., 2010. Scenedesmus obliquus CNW-N as a potential candidate for CO_2 mitigation and biodiesel production. Bioresour. Technol. 101 (22), 8725–8730. doi:10.1016/j.biortech.2010.06.112.

Hongyang, S.u, Yalei, Z., Chunmin, Z., Xuefei, Z., Jinpeng, L.i, 2011. Cultivation of Chlorella pyrenoidosa in soybean processing wastewater. Bioresour. Technol. 102 (21), 9884–9890. doi:10.1016/j.biortech.2011.08.016.

Hu, Q., Sommerfeld, M., Jarvis, E., Ghirardi, M., Posewitz, M., Seibert, M., Darzins, A.l, 2008. Microalgal triacylglycerols as feedstocks for biofuel production: perspectives and advances. Plant J. 54 (4), 621–639. doi:10.1111/j.1365-313x.2008.03492.x.

Hua, K., Bureau, D.P., 2006. Modelling digestible phosphorus content of salmonid fish feeds. Aquaculture 254 (1–4), 455–465. doi:10.1016/j.aquaculture.2005.10.019.

Husin, N.A.S., Basri, M.H.H., 2022. Review on factors affecting the effectiveness in removing pollutants by microalgae Scenedesmus obliquus in wastewater treatment. In: AIP Conference Proceedings, 2532. American Institute of Physics Inc doi:10.1063/5.0110998 15517616.

Ji, H., Liu, G., Han, J., Zhu, F., Dong, X., Li, B., 2020. C-phycocyanin inhibits epithelial-to-mesenchymal transition in Caski cells. Cancer Cell Int. 20 (1). doi:10.1186/s12935-020-01384-8.

Kaur, M., Bhatia, S., Gupta, U., Decker, E., Tak, Y., Bali, M., Kumar Gupta, V., Ahmad Dar, R., Bala, S., 2023. Microalgal bioactive metabolites as promising implements in nutraceuticals and pharmaceuticals: inspiring therapy for health benefits. Phytochem. Rev. 22 (4), 903–933. doi:10.1007/s11101-022-09848-7.

Khan, M.I., Shin, J.H., Kim, J.D., 2018. The promising future of microalgae: current status, challenges, and optimization of a sustainable and renewable industry for biofuels, feed, and other products. Microb. Cell Fact. 17 (1). doi:10.1186/s12934-018-0879-x.

Khanra, S., Mondal, M., Halder, G., Tiwari, O.N., Gayen, K., Bhowmick, T.K., 2018. Downstream processing of microalgae for pigments, protein and carbohydrate in industrial application: a review. Food Bioprod. Process. 110, 60–84. doi:10.1016/j.fbp.2018.02.002.

Koller, M., Muhr, A., Braunegg, G., 2014. Microalgae as versatile cellular factories for valued products. Algal Res. 6, 52–63. doi:10.1016/j.algal.2014.09.002.

Kong, Q.X., Li, L., Martinez, B., Chen, P., Ruan, R., 2010. Culture of microalgae chlamydomonas reinhardtii in wastewater for biomass feedstock production. Appl. Biochem. Biotechnol. 160 (1), 9–18. doi:10.1007/s12010-009-8670-4.

Koyande, A.K., Chew, K.W., Rambabu, K., Tao, Y., Chu, D.T., Show, P.L., 2019. Microalgae: a potential alternative to health supplementation for humans. Food. Sci. Hum. Wellness 8 (1), 16–24. doi:10.1016/j.fshw.2019.03.001.

Krohn, I., Menanteau-Ledouble, S., Hageskal, G., Astafyeva, Y., Jouannais, P., Nielsen, J.L., Pizzol, M., Wentzel, A., Streit, W.R., 2022. Health benefits of microalgae and their microbiomes. Microb. Biotechnol. 15 (7), 1966–1983. doi:10.1111/1751-7915.14082.

Kshirsagar, A.D., 2013. Bioremediation of wastewater by using microalgae: an experimental study. Int. J. Life Sci. Biotechnol. Pharma Res. 2 (3), 339–346.

Kumari, P., Shukla, S.P., Rathi Bhuvaneswari, G., Kumar, S., Xavier, M., Kumar, M., 2023. High value pigment production and carbon sequestration through wastewater grown Spirulina (Arthrospira) platensis: a green technology for wastewater utilization. Waste Manag. Bull. 1 (3), 1–10. doi:10.1016/j.wmb.2023.06.003.

Li, H., Chen, S., Liao, K., Lu, Q., Zhou, W., 2021. Microalgae biotechnology as a promising pathway to ecofriendly aquaculture: a state-of-the-art review. Journal of Chemical Technology & Biotechnology 96 (4), 837–852. https://doi.org/10.1002/jctb.6624.

Li, K., Liu, Q., Fang, F., Luo, R., Lu, Q., Zhou, W., Huo, S., Cheng, P., Liu, J., Addy, M., Chen, P., Chen, D., Ruan, R., 2019. Microalgae-based wastewater treatment for nutrients recovery: a review. Bioresour. Technol. 291, 121934. doi:10.1016/j.biortech.2019.121934.

Li, Y., Chen, Y.F., Chen, P., Min, M., Zhou, W., Martinez, B., Zhu, J., Ruan, R., 2011. Characterization of a microalga Chlorella sp. well adapted to highly concentrated municipal wastewater for nutrient removal and biodiesel production. Bioresour. Technol. 102 (8), 5138–5144. doi:10.1016/j.biortech.2011.01.091.

Liu, Y., Lv, J., Feng, J., Liu, Q.i, Nan, F., Xie, S., 2019. Treatment of real aquaculture wastewater from a fishery utilizing phytoremediation with microalgae. J. Chem. Technol. Biotechnol. 94 (3), 900–910. doi:10.1002/jctb.5837.

Lu, Q., Chen, P., Addy, M., Zhang, R., Deng, X., Ma, Y., Cheng, Y., Hussain, F., Chen, C., Liu, Y., Ruan, R., 2018. Carbon-dependent alleviation of ammonia toxicity for algae cultivation and associated mechanisms exploration. Bioresour. Technol. 249, 99–107. doi:10.1016/j.biortech.2017.09.175.

Martinez-Porchas, M., Martinez-Cordova, L.R., 2012. World aquaculture: environmental impacts and troubleshooting alternatives. Sci. World J. 2012. doi:10.1100/2012/389623.

Martins, C.I.M., Eding, E.H., Verdegem, M.C.J., Heinsbroek, L.T.N., Schneider, O., Blancheton, J.P., d'Orbcastel, E.R., Verreth, J.A.J., 2010. New developments in recirculating aquaculture systems in Europe: a perspective on environmental sustainability. Aquacult. Eng. 43 (3), 83–93. doi:10.1016/j.aquaeng.2010.09.002.

Matos, Â.P., 2017. The impact of microalgae in food science and technology. J. Am. Oil Chem. Soc. 94 (11), 1333–1350. doi:10.1007/s11746-017-3050-7.

Mavrommatis, A., Tsiplakou, E., Zerva, A., Pantiora, P.D., Georgakis, N.D., Tsintzou, G.P., Madesis, P., Labrou, N.E., 2023. Microalgae as a sustainable source of antioxidants in animal nutrition, health and livestock development. Antioxidants 12 (10). doi:10.3390/antiox12101882.

Meitei, M.M., Singh, S.K., Mangang, Y.A., Meena, D.K., Debbarma, R., Biswas, P., Waikhom, G., Patel, A.B., Ngasotter, S., Newmei, T., Meena, K., 2022. Effective valorization of precision output of algaquaculture towards eco-sustainability and bioeconomy concomitant with biotechnological advances: an innovative concept. Clean. Waste Syst. 3. doi:10.1016/j.clwas.2022.100026.

Melis, A., 2009. Solar energy conversion efficiencies in photosynthesis: minimizing the chlorophyll antennae to maximize efficiency. Plant Sci. 177 (4), 272–280. doi:10.1016/j.plantsci.2009.06.005.

Melis, A., Happe, T., 2001. Hydrogen production. Green algae as a source of energy. Plant Physiol. 127 (3), 740–748. doi:10.1104/pp.010498.

Miranda, C.D., Godoy, F.A., Lee, M.R., 2018. Current status of the use of antibiotics and the antimicrobial resistance in the chilean salmon farms. Front. Microbiol. 9. doi:10.3389/fmicb.2018.01284.

Mirzaei, M., Jazini, M., Aminiershad, G., Refardt, D., 2024. Biodesalination of saline aquaculture wastewater with simultaneous nutrient removal and biomass production using the microalgae Arthrospira and Dunaliella in a circular economy approach. Desalination 581, 117564. doi:10.1016/j.desal.2024.117564.

Monteiro, M., Santos, R.A., Iglesias, P., Couto, A., Serra, C.R., Gouvinhas, I., Barros, A., Oliva-Teles, A., Enes, P., Díaz-Rosales, P., 2020. Effect of extraction method and solvent system on the phenolic content and antioxidant activity of selected macro- and microalgae extracts. J. Appl. Phycol. 32 (1), 349–362. doi:10.1007/s10811-019-01927-1.

Muñoz, R., Benoit, G., 2006. Algal–bacterial processes for the treatment of hazardous contaminants: a review. Water Res. 40 (15), 2799–2815. doi:10.1016/j.watres.2006.06.011.

Nascimento, I.A., Cabanelas, I.T.D., Santos, J.N.d., Nascimento, M.A., Sousa, L., Sansone, G., 2015. Biodiesel yields and fuel quality as criteria for algal-feedstock selection: effects of CO_2-supplementation and nutrient levels in cultures. Algal Res. 8, 53–60. doi:10.1016/j.algal.2015.01.001.

Navarro-López, E., Ruíz-Nieto, A., Ferreira, A., Acién, F.G., Gouveia, L., 2020. Biostimulant potential of Scenedesmus obliquus grown in brewery wastewater. Molecules 25 (3), 664. doi:10.3390/molecules25030664.

Neori, A., Chopin, T., Troell, M., Buschmann, A. H., Kraemer, G. P., Halling, C., Shpigel, M, Yarish, C., 2004. Integrated aquaculture: rationale, evolution and state of the art emphasizing seaweed biofiltration in modern mariculture. Aquaculture 231 (1–4), 361–391. https://doi.org/10.1016/j.aquaculture.2003.11.015.

Nguyen, L.N., Aditya, L., Vu, H.P., Johir, A.H., Bennar, L., Ralph, P., Hoang, N.B., Zdarta, J., Nghiem, L.D., 2022. Nutrient removal by algae-based wastewater treatment. Curr. Pollut. Rep. 8 (4), 369–383. doi:10.1007/s40726-022-00230-x.

Nie, X., Mubashar, M., Zhang, S., Qin, Y., Zhang, X., 2020. Current progress, challenges and perspectives in microalgae-based nutrient removal for aquaculture waste: a comprehensive review. J. Cleaner Prod. 277, 124209. doi:10.1016/j.jclepro.2020.124209.

Oostlander, P.C., van Houcke, J., Wijffels, R.H., Barbosa, M.J., 2020. Microalgae production cost in aquaculture hatcheries. Aquaculture 525, 735310. doi:10.1016/j.aquaculture.2020.735310.

Pan, S., Dixon, K.L., Nawaz, T., Rahman, A., Selvaratnam, T., 2021. Evaluation of Galdieria sulphuraria for nitrogen removal and biomass production from raw landfill leachate. Algal Res. 54, 102183. doi:10.1016/j.algal.2021.102183.

Patil, V., Reitan, K.I., Knutsen, G., Mortensen, L.M., Källqvist, T., Olsen, E., Gislerød, H.R., Patil, V., Reitan, K.I., Knutsen, G., Mortensen, L.M., Källqvist, T., Olsen, E., Gislerød, H.R., 2005. Microalgae as source of polyunsaturated fatty acids for aquaculture. Plant Biol 6 (6), 57–65.

Peralta-Sánchez, J.M., Rabelo-Ruiz, M., Martín-Platero, A.M., Vizcaíno, A.J., Flores-Moreno, S., Macías-Vidal, J., Martos-Sitcha, J.A., Alarcón-López, F.J., Baños, A., Valdivia, E., Martínez-Bueno, M., 2024. Microalgae and phytase dietary supplementation improved growth and gut microbiota in juvenile European seabass (Dicentrarchus labrax). BMC Genomics 25 (1). doi:10.1186/s12864-024-10760-x.

Perez-Garcia, O., Escalante, F.M.E., de-Bashan, L.E., Bashan, Y., 2011. Heterotrophic cultures of microalgae: metabolism and potential products. Water Res. 45 (1), 11–36. doi:10.1016/j.watres.2010.08.037.

Piedrahita, R.H., 2003. Reducing the potential environmental impact of tank aquaculture effluents through intensification and recirculation. Aquaculture 226 (1–4), 35–44. http://www.journals.elsevier.com/aquaculture/226.

Pinho, S.M., de Lima, J.P., David, L.H., Emerenciano, M.G.C., Goddek, S., Verdegem, M.C.J., Keesman, K.J., Portella, M.C., 2022. FLOCponics: the integration of biofloc technology with plant production. Rev. Aquac. 14 (2), 647–675. doi:10.1111/raq.12617.

Pulz, O., Gross, W., 2004. Valuable products from biotechnology of microalgae. Appl. Microbiol. Biotechnol. 65 (6), 635–648. doi:10.1007/s00253-004-1647-x.

Rachidi, F., Benhima, R., Kasmi, Y., Sbabou, L., Arroussi, H. E., 2021. Evaluation of microalgae polysaccharides as biostimulants of tomato plant defense using metabolomics and biochemical approaches. Scientific reports 11 (1), 930. https://doi.org/10.1038/s41598-020-78820-2.

Rachidi, F., Benhima, R., Sbabou, L., El Arroussi, H., 2020. Microalgae polysaccharides bio-stimulating effect on tomato plants: growth and metabolic distribution. Biotechnol. Rep. 25, e00426. doi:10.1016/j.btre.2020.e00426.

Ramesh, B., Saravanan, A., Senthil Kumar, P., Yaashikaa, P.R., Thamarai, P., Shaji, A., Rangasamy, G., 2023. A review on algae biosorption for the removal of hazardous pollutants from wastewater: limiting factors, prospects and recommendations. Environ. Pollut. 327, 121572. doi:10.1016/j.envpol.2023.121572.

Randall, D.J., Tsui, T.K.N., 2002. Ammonia toxicity in fish. Mar. Pollut. Bull. 45 (1–12), 17–23. doi:10.1016/s0025-326x(02)00227-8.

Ranglová, K., Lakatos, G.E., Câmara Manoel, J.A., Grivalský, T., Suárez Estrella, F., Acién Fernández, F.G., Molnár, Z., Ördög, V., Masojídek, J., 2021. Growth, biostimulant and biopesticide activity of the MACC-1 Chlorella strain cultivated outdoors in inorganic medium and wastewater. Algal Res. 53. doi:10.1016/j.algal.2020.102136.

Rani, S., Gunjyal, N., Ojha, F.A., Singh, R.P., 2021. Review of challenges for algae-based wastewater treatment: strain selection, wastewater characteristics, abiotic, and biotic factors. J. Hazard. Toxic Radioact. Waste 25 (2). doi:10.1061/(ASCE)HZ.2153-5515.0000578.

Richardson, J.W., Johnson, M.D., Outlaw, J.L., 2012. Economic comparison of open pond raceways to photo bioreactors for profitable production of algae for transportation fuels in the Southwest. Algal Res. 1 (1), 93–100. doi:10.1016/j.algal.2012.04.001.

Rico, A., Satapornvanit, K., Haque, M.M., Min, J., Nguyen, P.T., Telfer, T.C., van den Brink, P.J., 2012. Use of chemicals and biological products in Asian aquaculture and their potential environmental risks: a critical review. Rev. Aquac. 4 (2), 75–93. doi:10.1111/j.1753-5131.2012.01062.x.

Rodgers, C.J., Furones, M.D., 2009. Antimicrobial agents in aquaculture: practice, needs and issues. Options Méditerr. 86, 41–59.

Rodolfi, L., Zittelli, G.C., Bassi, N., Padovani, G., Biondi, N., Bonini, G., Tredici, M.R., 2009. Microalgae for oil: strain selection, induction of lipid synthesis and outdoor mass cultivation in a low-cost photobioreactor. Biotechnol. Bioeng. 102 (1), 100–112. doi:10.1002/bit.22033Italy.

Ronga, D., Biazzi, E., Parati, K., Carminati, D., Carminati, E., Tava, A., 2019. Microalgal biostimulants and biofertilisers in crop productions. Agronomy 9 (4), 192. doi:10.3390/agronomy9040192.

Sahu, P., 2021. A comprehensive review of saline effluent disposal and treatment: conventional practices, emerging technologies, and future potential. J. Water Reuse Desalin. 11 (1), 33–65. doi:10.2166/wrd.2020.065.

Salbitani, G., Carfagna, S., 2021. Ammonium utilization in microalgae: a sustainable method for wastewater treatment. Sustainability 13 (2), 956. doi:10.3390/su13020956.

Samocha, T.M., Fricker, J., Ali, A.M., Shpigel, M., Neori, A., 2015. Growth and nutrient uptake of the macroalga Gracilaria tikvahiae cultured with the shrimp Litopenaeus vannamei in an integrated multi-trophic aquaculture (IMTA) system. Aquaculture 446, 263–271. doi:10.1016/j.aquaculture.2015.05.008.

Sathasivam, R., Radhakrishnan, R., Hashem, A., Abd_Allah, E.F., 2019. Microalgae metabolites: a rich source for food and medicine. Saudi J. Biol. Sci. 26 (4), 709–722. doi:10.1016/j.sjbs.2017.11.003.

Schneider, O., Sereti, V., Eding, E.H., Verreth, J.A.J., 2005. Analysis of nutrient flows in integrated intensive aquaculture systems. Aquacult. Eng. 32 (3–4), 379–401. doi:10.1016/j.aquaeng.2004.09.001.

Shah, J.H., Deokar, A., Patel, K., Panchal, K., Mehta, A.V., 2014. A comprehensive overview on various method of harvesting microalgae according to Indian perspective. In: International Conference on Multidisciplinary Research & Practice, 1. Bhadra, Gujarat, India, pp. 313–317.

Shah, M.R., Lutzu, G.A., Alam, A., Sarker, P., Kabir Chowdhury, M.A., Parsaeimehr, A., Liang, Y., Daroch, M., 2018. Springer Netherlands China Microalgae in aquafeeds for a sustainable aquaculture industry. J. Appl. Phycol. 30, 197–213. doi:10.1007/s10811-017-1234-z.

Sierra, E., Acién, F.G., Fernández, J.M., García, J.L., González, C., Molina, E., 2008. Characterization of a flat plate photobioreactor for the production of microalgae. Chem. Eng. J. 138 (1–3), 136–147. doi:10.1016/j.cej.2007.06.004.

Singh, R., Birru, R., Sibi, G., 2017. Nutrient Removal Efficiencies of Chlorella vulgaris from urban wastewater for reduced eutrophication. J. Environ. Prot. 08 (01), 1–11. doi:10.4236/jep.2017.81001.

Singhal, M., Jadhav, S., Sonone, S.S., Sankhla, M.S., Kumar, R., 2021. Microalgae based sustainable bioremediation of water contaminated by pesticides. Biointerface Res. Appl. Chem. 12 (1), 149–169. doi:10.33263/briac121.149169.

Slocombe, S.P., Zúñiga-Burgos, T., Chu, L., Wood, N.J., Camargo-Valero, M.A., Baker, A., 2020. Fixing the broken phosphorus cycle: wastewater remediation by microalgal polyphosphates. Front. Plant Sci. 11. doi:10.3389/fpls.2020.00982.

Soo, C.L., Chen, C.A., Bojo, O., Hii, Y.S., 2017. Feasibility of marine microalgae immobilization in alginate bead for marine water treatment: bead stability, cell growth, and ammonia removal. Int. J. Polym. Sci. 2017. doi:10.1155/2017/6951212.

Spolaore, P., Joannis-Cassan, C., Duran, E., Isambert, A., 2006. Commercial applications of microalgae. J. Biosci. Bioeng. 101 (2), 87–96. doi:10.1263/jbb.101.87.

Sriram, S., Seenivasan, R., 2012. Microalgae cultivation in wastewater for nutrient removal. J. Algal Biomass Util. 3 (2), 9–13.

Su, Y., 2021. Revisiting carbon, nitrogen, and phosphorus metabolisms in microalgae for wastewater treatment. Sci. Total Environ. 762, 144590. doi:10.1016/j.scitotenv.2020.144590.

Takeshita, T., Ivanov, I.N., Oshima, K., Ishii, K., Kawamoto, H., Ota, S., Yamazaki, T., Hirata, A., Kazama, Y., Abe, T., Hattori, M., Bišová, K., Zachleder, V., Kawano, S., 2018. Comparison of lipid productivity of Parachlorella kessleri heavy-ion beam irradiation mutant PK4 in laboratory and 150-L mass bioreactor, identification and characterization of its genetic variation. Algal Res. 35, 416–426. doi:10.1016/j.algal.2018.09.005.

Tarafdar, A., Sowmya, G., Yogeshwari, K., Rattu, G., Negi, T., Awasthi, M.K, Hoang, A.T, Sindhu, R., Sirohi, R., 2023. Environmental pollution mitigation through utilization of carbon dioxide by microalgae. Environ. Pollut. 328, 121623. doi:10.1016/j.envpol.2023.121623.

Tarakhovskaya, E.R., Maslov, Y.I., Shishova, M.F., 2007. Phytohormones in algae. Russ. J. Plant Physiol. 54 (2), 163–170. doi:10.1134/S1021443707020021.

Tawfik, A., Ismail, S., Elsayed, M., Abdul Qyyum, M., Rehan, M., 2022. Sustainable microalgal biomass valorization to bioenergy: key challenges and future perspectives. Chemosphere 296, 133812. doi:10.1016/j.chemosphere.2022.133812.

Tibbetts, S.M., Milley, J.E., Lall, S.P., 2015. Chemical composition and nutritional properties of freshwater and marine microalgal biomass cultured in photobioreactors. J. Appl. Phycol. 27 (3), 1109–1119. doi:10.1007/s10811-014-0428-x.

Trevi, S., Uren Webster, T., Consuegra, S., Garcia de Leaniz, C., 2023. Benefits of the microalgae Spirulina and Schizochytrium in fish nutrition: a meta-analysis. Sci. Rep. 13 (1). doi:10.1038/s41598-023-29183-x.

Vázquez-Romero, B., Perales, J.A., Pereira, H., Barbosa, M., Ruiz, J., 2022. Techno-economic assessment of microalgae production, harvesting and drying for food, feed, cosmetics, and agriculture. Sci. Total Environ. 837, 155742. doi:10.1016/j.scitotenv.2022.155742.

Venugopal, V., Sasidharan, A., 2021. Seafood industry effluents: environmental hazards, treatment and resource recovery. J. Environ. Chem. Eng. 9 (2), 104758. doi:10.1016/j.jece.2020.104758.

Verdegem, M.C.J., 2013. Nutrient discharge from aquaculture operations in function of system design and production environment. Rev. Aquac. 5 (3), 158–171. doi:10.1111/raq.12011.

Viswanaathan, S., Perumal, P.K, Sundaram, S., 2022. Integrated approach for carbon sequestration and wastewater treatment using algal–bacterial consortia: opportunities and challenges. Sustainability 14 (3), 1075. doi:10.3390/su14031075.

Wang, B., Li, Y., Wu, N., Lan, C.Q., 2008. CO_2 bio-mitigation using microalgae. Appl. Microbiol. Biotechnol. 79 (5), 707–718. doi:10.1007/s00253-008-1518-y.

Wang, C., Jiang, C., Gao, T., Peng, X., Ma, S., Sun, Q., Xia, B., Xie, X., Bai, Z., Xu, S., Zhuang, X., 2022. Improvement of fish production and water quality in a recirculating aquaculture pond enhanced with bacteria-microalgae association. Aquaculture 547, 737420. doi:10.1016/j.aquaculture.2021.737420.

Watanabe, F., Takenaka, S., Kittaka-Katsura, H., EBARA, S., Miyamoto, E., 2002. Characterization and bioavailability of vitamin B12-compounds from edible algae. J. Nutr. Sci. Vitaminol. 48 (5), 325–331. doi:10.3177/jnsv.48.325.

Wu, L.F., Chen, P.C., Huang, A.P., Lee, C.M., 2012. The feasibility of biodiesel production by microalgae using industrial wastewater. Bioresour. Technol. 113, 14–18. doi:10.1016/j.biortech.2011.12.128.

Yaakob, M.A., Mohamed, R.M.S.R, Al-Gheethi, A., Gokare, R.A., Ambati, R.R., 2021. Influence of nitrogen and phosphorus on microalgal growth, biomass, lipid, and fatty acid production: an overview. Cells 10 (2), 393. doi:10.3390/cells10020393.

Yang, L., 2020. "Algaquaculture" integrating algae-culture with aquaculture for sustainable development. J. Cleaner Prod. 244, 118765.

Zhang, S., Liu, Z., 2021. Advances in the biological fixation of carbon dioxide by microalgae. J. Chem. Technol. Biotechnol. 96 (6), 1475–1495. doi:10.1002/jctb.6714.

Zuo, J., Xiao, P., Heino, J., Tan, F., Soininen, J., Chen, H., Yang, J., 2024. Eutrophication increases the similarity of cyanobacterial community features in lakes and reservoirs. Water Res. 250, 120977. doi:10.1016/j.watres.2023.120977.

CHAPTER

7

Integrated technological developments in aquaculture systems

Mohammad Hossein Khanjani[a,*], *Saeed Zahedi*[b] and *Moslem Sharifinia*[c]

[a]Department of Fisheries Sciences and Engineering, Faculty of Natural Resources, University of Jiroft, Jiroft, Iran [b]Department of Fisheries, Faculty of Natural Resources and Environment, Ferdowsi University of Mashhad, Mashhad, Iran [c]Aquaculture Department, Shrimp Research Center, Iranian Fisheries Sciences Research Institute, Agricultural Research, Education and Extension Organization (AREEO), Bushehr, Iran

*Corresponding author: m.h.khanjani@ujiroft.ac.ir, m.h.khanjani@gmail.com

7.1 Introduction

The global community is facing a food crisis driven by a burgeoning population, and it is predicted to exceed 9 billion people by the year 2050. Sectors responsible for food production, such as agriculture (and the subsector of aquaculture), are significantly contributing to the fulfillment of the demand for affordable protein sources. In 2022, the total production from global fisheries and aquaculture was 223.2 million tons, reflecting a 4.4% increase from 2020. This figure includes both 185.4 million tons of aquatic animals and 37.8 million tons of algae (FAO, 2024). Aquaculture is raising aquatic animals and plants in different water bodies, such as fresh, brackish, and saline waters (inland and coastal waters). To encounter the global food crisis, the subsector of aquaculture is moving toward intensification and expansion of farming areas. However, constraints such as less land availability, high environmental impact because of intensification, and high production costs, lead to the adaptation of integrating different cultural practices (Khanjani et al., 2022; Khanjani et al., 2025a).

The higher the production from culture systems due to intensive inputs such as feed, fertilizer, and chemical substances has resulted in greater and more concerning impacts

Zero-Waste Technologies in Aquaculture: Global Perspectives and Methods toward a Circular Economy
DOI: https://doi.org/10.1016/B978-0-443-34689-7.00013-6

(Emerenciano et al., 2025). Hence, different strategies, technologies, and culture systems have been adopted to overcome these challenges, which have a less environmental impact and render higher productivity and profitability. For these purposes, culture systems such as recirculating aquaculture systems (RASs), cages, biofloc technology (BFT), integrated multitrophic aquaculture (IMTA) systems, aquaponics, flocponics, and integrated agriculture aquaculture (IAA) systems have been developed.

First, we have RASs, which are aquatic biota production systems where we have a water loop or recirculation, and different water treatment units are imbedded to keep water quality. In this system, the produced nitrogenous wastes remove in the water by different approaches, such as using nitrogenous bacteria. Therefore we can produce aquatics with little water compared to the traditional systems (ponds and raceways) where water exchange rate is as low as 0.1–3 m^3/kg feed (Moore et al., 2020;Mugwanya et al., 2022). RAS delivers the unique opportunity to conduct intensive aquacultural practices with high stocking densities (biomass per culture area) to attain maximum net production and big profits compared to open aquaculture systems such as ponds (Mugwanya et al., 2022).

Second, we come to BFT, which offers a promising alternative to address the critical aspects of water efficiency, environmental impact, and feed costs. Notably, BFT is recognized for its exceptional water efficiency, achieving remarkable results in three key areas: First, it lowers water consumption throughout production cycles by as much as 90%, significantly decreasing the amount of effluent released into the environment; second, it helps maintain optimal water quality parameters and a balanced microbial profile; and third, it allows for the reuse of mature biofloc water after harvesting, further reducing waste (Figueroa-Espinoza et al., 2022; Khanjani et al., 2024a).

Third, the IMTA system seamlessly combines ecological engineering facilities and ecological engineering systems, featuring an innovative aquaculture technology that concurrently cultivates feed species, inorganic extractive species (such as hydrophytes), and organic extractive species (such as shellfish) (Liu et al., 2021). In contrast to conventional aquaculture systems, IMTAs offer a significantly enhanced culture environment, greater ecological advantages, and diversified economic benefits, with successful implementations reported globally (Zhao et al., 2022a).

Fourth, aquaponics represents an integrated polytrophic system that synergistically blends the principles of aquaculture, including BFT and RAS, with hydroponic cultivation (Pinho et al., 2022a). Nitrogen-rich water (containing compounds such as ammonia and nitrite) serves as nourishment for plant growth. After plants absorb water-soluble nitrogen, the purified water is returned to the tank, contributing to the stable development of aquatic animals. The relationship between aquaculture and hydroponics can address the challenges faced by both systems, making it a hopeful, sustainable method for food production in both the agricultural and aquaculture sectors (Goddek et al., 2015). Finally, the integration of aquaculture with agriculture is a reasonable approach, as it transforms the former into a nonwater-consuming system that complements agriculture, thereby enhancing the benefits of sustainable farming practices (Abdul-Rahman et al., 2011). The multifaceted use of water in IAA systems has been shown to be effective in boosting farm and water productivity, improving the quality of water in fish ponds, and mitigating the environmental impacts associated with nutrient-rich water discharge, water costs, and chemical fertilizer usage in crops. One promising strategy for maximizing "crops per drop" is IAA, also referred to as "more crop per drop" (Ibrahim et al., 2023).

The ecological impact and green credentials of specific aquaculture systems or technologies can be evaluated using a variety of approaches and instruments. Moreover, the environmental impact and sustainability of aquaculture operations are influenced by several key factors (Henares et al., 2020): The specific species of aquatic organism being cultivated; the production system's level of intensity (extensive, semiintensive, or intensive); in intensive system, the density at which the organisms are cultivated; the quality of the feed used; the management practices employed; and the success of wastewater treatment (Laktuka et al., 2023). Achieving sustainable aquaculture intensification demands a multifaceted approach, incorporating a range of innovative technological advancements and improved management strategies. To guarantee the durable viability of aquaculture, it is essential to approve a more comprehensive ecosystem-based approach, while also promoting energy-efficient and resource-conserving practices throughout aquaculture operations. This chapter of the book delves into the concept of integrated aquaculture systems, exploring their potential to drive sustainable growth in the industry.

7.2 Aquaculture systems

Conventional aquaculture practices typically involve the integration of agriculture and aquaculture, where waste products, manure, and crop residues from farming activities are utilized (Edwards, 2015). Similarly, periurban aquaculture systems often rely on municipal wastewater and agricultural byproducts as a source of nutrition. Fertilized ponds, particularly those that use waste from other sources as fertilizer, are appealing because of their minimal reliance on external resources for sustenance, energy, and waste management (Boyd et al., 2020). While increasing the amount of feed in fish ponds may seem like a straightforward way to boost fish yields, it is not always that simple. A substantial amount of feed escapes conversion into consumable fish, instead being discharged into the water as waste, resulting in potentially harmful environmental repercussions (Boyd et al., 2020). For instance, the nutrient-rich waste products, including nitrogen, phosphorus, and other minerals, as well as organic waste such as excrement and uneaten food, can stimulate biological activity in the culture system, leading to a significant decrease in dissolved oxygen levels and putting the cultured fish at risk (Laktuka et al., 2023). The waste generated by aquaculture includes uneaten feed, nitrogenous byproducts, excreta, chemicals, and therapeutics, resulting in two primary types of residues: Liquid and solid (Das et al., 2023). The liquid residues consist of wastewater and discharge water from operations, while the solid residues are comprised of feces, feed, and other solid materials (Das et al., 2023). However, aquaculture ponds present a valuable opportunity to reconcile the need for intensification with sustainable practices, allowing residues to be treated internally without placing additional strain on external ecosystems, provided the right balance is achieved (Boyd et al., 2020).

Enhancing pond productivity can be achieved through the use of fertilizers (organic or chemical) that promote plant growth.

The productivity of fertilized ponds can differ significantly, with yields potentially reaching 4–6 tons per hectare, depending on factors such as the length of the growth period, the type of species being farmed, and the specific fertilization strategy used. In certain stagnant water ponds, yields can reach as high as 8–10 tons per hectare or more when supplemented with feeding and aeration, contingent upon the length of the growing season and the physiological characteristics of the species being cultivated. Ultimately, the maximum yield of plant material in fertilized ponds is constrained by the amount of solar radiation available, which imposes a maximum limit on productivity. To further enhance aquaculture productivity beyond what can be achieved through fertilization alone, essential nutrients must be sourced from outside the water itself and transformed into a nutritious and appealing feed. Commercial aquaculture feeds often consist of a mixture of feed constituents (plant and animal). The sustainability, safety, and environmental impacts of feed ingredient production, as well as food security and import dependency, are becoming progressively imperative considerations.

Advanced filters, recirculation loops, and water treatment technologies can facilitate the reuse of treated water. Conversely, the development of new aquaculture systems is hindered by limited land resources, often because of a lack of access to appropriate water sources or prohibitive land costs.

The amount of nutrients released from ponds into the surrounding environment is directly affected by the water exchange rate. Specifically, ponds with low water exchange rates, where water remains stagnant, are able to treat a substantial amount of waste internally, thereby minimizing the release of nutrients into the external environment. In the following subsections, various technologies such as RAS, BFT, IMTA, aquaponics, flocponics, and IAA will be discussed.

7.3 Recirculating aquaculture systems

In recent years, RASs have gathered significant attention for their ability to intensively produce various aquatic species in controlled environments (Mugwanya et al., 2022). Table 7.1 showcases some of the notable research conducted in RAS systems.

The primary advantage of RAS lies in their water-conserving design, which significantly reduces water usage by reusing culture water fully or partially (Verdegem et al., 2023). These systems range from semiintensive open systems to fully RASs (indoor ones), catering to a wide aquatic species (both freshwater and saline ones). While expensive fish are the primary products of RAS, other rearing species such as crustaceans, mollusks, and aquatic plants can also be cultivated, with the latter often used for hatchery and nursery purposes. The simplest RAS employs sedimentation, mechanical filtration, and aeration (such as airlifts) to eliminate suspended solids from the recirculating water. In contrast, more advanced RAS incorporate biofiltration and cutting-edge treatment technologies, including UV treatment/ozonation, and foam fractionation, to maintain optimal water quality. In contrast to conventional aquaculture systems, RAS provide a more precise control over the rearing environment, thereby enabling efficient waste management and minimizing environmental footprint. By significantly reducing the risk of escapees and virtually eliminating it in fully closed systems, RAS technology optimizes water treatment for intake and outlet, thereby enhancing disease control, minimizing the use of medication, and enabling continuous production throughout the year. Moreover, the most advanced RAS facilities can be strategically located near target markets, resulting in reduced transportation costs and guaranteed product traceability (Timmons et al., 2018). In RAS, aquaculturists often focus on calculating tank volume, but neglect mass balance calculations, which can lead to flooding and pump failure.

TABLE 7.1 Study on recirculating aquaculture systems (RAS).

Cultivated species	Condition	Highlight	References
Paralichthys olivaceus	Trial period: 56 days. Juvenile olive flounder (61.0 g) were initially stocked at 3.29, 4.84, 7.14, and 8.56 kg/m^2	Final stocking density of 20 kg/m^2 may be desirable for juvenile *P. olivaceus* in an RAS for fish weighing ~60–180 g.	Seo et al. (2023)
Oncorhynchus tshawytscha	Three feed ration groups (60%, 80%, and 100% and two temperature groups (13°C and 17°C) were set up to evaluate the effects of water temperature and feed ration on salmon gut microbiota diversity	The salmon gut microbiota was dominated by *Photobacterium piscicola*, especially in individuals with high fecal scores.	Zhao et al. (2021)
Salmo salar	Trial period: 14 days. A total of 450 juvenile *S. salar* (12 g), either previously infected with *Y. ruckeri* or uninfected, were distributed among the nine RAS units using the following infection matrix	Developing new knowledge to improve the RAS biosecurity in the *S. salar* aquaculture industry	Mota et al. (2022)
Solea senegalensis	Fish density of 2.5–5 kg/m^2, total water volume of 370 m^3, and a water recirculation rate of 400%/h	New knowledge on the prokaryotic community composition in different units of RAS, essential for the understanding of the microbial community balance in aquaculture	Almeida et al. (2021)
Dicentrarchus labrax	12 tanks of 350 L, supplied with underground filtered and heated saltwater at 20°C and a salinity of 18 ppm, equivalent to a density of 1.01 g/cm^3	Fish produced in RAS are mainly exposed to MPs through water and feed	Matias et al. (2023)
Clarias gariepinus	Trial period: 60 days. The diameter of the pool (3 m), and the pool height (1.25 m), filled with water up to a height of 0.9 m. The stocking density used was 2000 fish/pond using fingerlings (±2.1 g/fish)	A-RAS technology can maintain water quality to be reused in cultivation and increase yields by about 13%	Ekawati et al. (2021)

(*continued on next page*)

TABLE 7.1 Study on recirculating aquaculture systems (RAS)—cont'd

Cultivated species	Condition	Highlight	References
Scylla paramamosain	The rearing water was circulated four times daily with a flow rate of 0.6 m^3. The circulation volume of the protein separator was set at 0.22 m^3/h	RAS induced improvements in water quality indicators because of alterations in the structure and function of bacterial communities in the rearing water.	Lu et al. (2024)
Colossoma macropomum	Larvae were distributed in 15 tanks (28 L) in a RAS with a mechanical and biological filter. Densities of 10, 30, and 50 larvae/L were tested with five replicates each. Water flow in the tanks was 1.85 L/min, and larviculture was carried out in slightly saline water at 2.01 g of salt/L	Larviculture of C. *macropomum* can be performed in slightly saline water in RAS at a density of up to 50 larvae/L with good results for performance and survival.	Santos et al. (2020)
Oreochromis niloticus	100 swim-up tilapia fry were grown in each 20 L rearing aquarium of the RAS system	Tilapia cultured in RAS had a higher growth rate, lower FCR, and more efficient utilization of protein, fat, and energy from feed for their growth.	Amin et al. (2020)
O. niloticus	The RAS was operated with 7 stocking densities broken down into low (2.3, 3.5, and 4.0 kg/m^3), medium (5.0 and 7.0 kg/m^3), and high (9.0 and 10.0 kg/m^3) stocking densities	Ammonia removal was highest at lower flow rates with percentage removals above 70% and reduced gradually to as low as 9.5% at high flow rates. EC increased with increasing stocking densities and flow rates to values above 190.0 mg/L at 9.0 L/min and 10.0 L/min	Wambua et al. (2021)

Accurate sizing of sumps and pump power is crucial to prevent these issues. Stocking densities in experimental RAS typically range from 10 to 42 kg/m^3. However, more intensive RAS systems can reach densities exceeding 50 kg/m^3, with some even surpassing 300 kg/m^3. Research by Kristiansen and Cripps (1996) demonstrated that physical filters (sand and mechanical filters such as microscreen filters) can effectively remove solids and organic matter. In addition, biological processes, including oxidation of organic matter, nitrification, and denitrification, can be facilitated through the use of submerged biofilters, trickling filters, rotating biological contactors, and fluid bed reactors. The breakdown and movement of organic particles are also influenced by mechanical pressure caused by water flow patterns in the tank system. Therefore the internal design of the tank can play a significant role in controlling hydrodynamic characteristics, such as water velocity, pressure, and turbulence (Zhao et al., 2022b).

In aquaculture tank systems, hydrodynamic conditions are crucial for achieving optimal fish culture, survival rates, growth, feed utilization, and overall fish health. The way water flows through the system has a direct impact on fish behavior and the tank's ability to self-clean (Xue et al., 2022). Moreover, hydrodynamics influence the distribution of essential nutrients, including oxygen, organic matter, and inorganic compounds, which in turn affect the system's self-cleaning efficiency in RAS (Yin et al., 2021). A well-designed hydrodynamic system can also bolster biosecurity in closed culture tank systems, thereby mitigating the risk of disease outbreaks (Xue et al., 2022).

The adoption of RAS for fish culture has several advantages, including reduced water consumption because of partial water reuse after mechanical and biological treatments (Martins et al., 2011). This approach also decreases the amount of nutrients released into the environment. In addition, the equipment, such as tanks and piping systems, can be sustained for a longer period. The RAS system can increase the production of safe seafood by reducing the use of antibiotics and therapeutics. It also helps decrease maintenance costs, including food, hazardous animals, birds, and parasites, thereby enhancing biosecurity (Thomas et al., 2018). Moreover, the system allows for the sterilization of potential bacteria in water intake and the monitoring of climate change impacts on air and water. The aquatic species grow rapidly because of favorable growing conditions. RAS systems can be built near markets or unfertilized land, but the available water supply must be considered in terms of quality and etiquette. Notably, RAS buildings resemble industrial buildings.

7.4 Biofloc technology

The BFT is a water quality enhancement method that leverages the strategic addition of carbon to aquaculture systems (Khanjani et al., 2025b). When carbon and nitrogen are harmoniously balanced within the water, a synergistic process occurs, where ammonium and organic nitrogenous waste are efficiently transformed into a valuable byproduct: bacterial biomass (Khanjani et al., 2024b). The BFT holds promise for boosting productivity in shrimp farming and other aquatic cultivation while mitigating environmental footprints (Khanjani, 2026). By serving as a supplementary nutrition source for the reared organisms and harnessing the water-purifying capabilities of its microbial community (Khanjani et al., 2024c), this system can maintain optimal water conditions. However, a key challenge in BFT-based aquaculture lies in the accumulation of excessive nitrogen, primarily resulting from the excretion of reared organisms, uneaten feed, and organic waste (Timmons and Ebeling, 2010), which can have detrimental effects. The buildup of these compounds can be controlled through two primary pathways: Heterotrophic and chemoautotrophic. In the first one, the addition of an organic carbon source (such as molasses) promotes the growth of heterotrophic bacteria, which help to assimilate ammonia by bacterial biomass and their subsequent growth (Khanjani et al., 2024d). In the second one, nitrifying bacteria oxidize ammonia to nitrate through the nitrification process, which is related to the absorption of inorganic carbon present as alkalinity (Ebeling et al., 2006). In the event of inadequate water circulation, phosphorus, total suspended solids (TSSs), and various other essential nutrients have a tendency to build up within the production system (Khanjani et al., 2024c).

The accumulation of inorganic compounds in aquaculture is linked to the excretory waste of animals and the uneaten remnants of their feed. According to Crab et al. (2007), shrimp are only able to absorb a limited percentage of nutrients, specifically 23%–31% of nitrogen and 10%–13% of phosphorus. The unabsorbed nutrients subsequently accumulate in the water, necessitating their removal to prevent water degradation and toxicity to the aquatic life. Moreover, the high levels of TSSs are a consequence of a heterotrophic system dominated by the rapid growth of bacterial biomass.

Systems dominated by chemoautotrophic microorganisms exhibit increased solids, albeit in smaller quantities, which enhances the nitrification process efficiency (Serra et al., 2021). However, the main task in BFT systems is regulating the escalating solid concentrations over time and managing inorganic nutrients generated during production cycles. These issues can be addressed through mechanical means, such as filters, or biological approaches. Therefore an IMTA system offers a viable alternative to mitigate the unwanted surpluses in BFT systems. By integrating organisms from diverse trophic levels, all sharing the same production system, IMTA systems convert organic and inorganic residues into valuable inputs (feed or fertilizers) for lower-trophic level species within the system (Chopin et al., 2001).

The establishment and maintenance of BFT systems are characterized by high costs and energy demands, necessitating a high level of technical expertise (Khanjani et al., 2024a). In addition, outdoor BFT systems can be set up in tanks or ponds (Khanjani et al., 2024b). However, when implementing outdoor BFT systems, it is crucial to carefully consider the location of the water body, light intensity, and time of year as these factors can significantly impact the system's internal balance. In addition to the above, it is crucial to regularly track water quality indicators, such as dissolved oxygen, pH, and ammonia concentrations, as excessive levels can lead to stress and impeded development in aquatic species (Khanjani et al., 2024c).

The integration of BFT with IMTA can lead to enhanced economic benefits (Laktuka et al., 2023). By implementing IMTA, the production of effluents can be reduced, and productivity increased, thanks to the diverse range of species at different trophic levels. This, in turn, enables more efficient use of the nutrients generated within the BFT system, as excess nutrients are consumed by other species, resulting in optimized nutrient utilization.

7.5 Integrated multitrophic aquaculture system

IMTA is a holistic approach that brings together organisms from diverse trophic levels within a shared production system, thereby optimizing nutrient utilization (de Morais et al., 2023). Table 7.2 highlights various studies conducted within IMTA systems. IMTA is a sustainable strategy aimed at mitigating the ecological footprint of finfish mariculture. Through its innovative approach of transforming waste into high-value resources such as fertilizers and food, IMTA pioneers a closed-loop economy. This innovative method integrates the cultivation of fish or shrimp with inorganic extractive organisms (e.g., seaweed) and organic extractive organisms (e.g., shellfish) to establish a balanced ecosystem and a harmony of species to mitigate environmental impacts. The overarching objective of IMTA is to achieve both ecological sustainability and profitability for each production unit over the long term. Its benefits have sparked widespread interest among researchers and commercial farmers globally. Moreover, the public has responded positively to IMTA as an environmentally friendly approach to reduce ecological impact (Ridler et al., 2007). Initially adopted in offshore marine environments, IMTA

TABLE 7.2 Study on integrated multitrophic aquaculture (IMTA) system.

Cultivated species	Condition	Highlight	References
Shrimp (*P. vannamei*) as the main species, tilapia (*O. niloticus*) as an organic consumer, and the seaweed (*Ulva ohnoi*) at different densities as an inorganic consumer	Shrimp (3.82 g) were stocked in 800 L tanks at a density of 275 shrimp/m^3, fish (14.44 g) in 90 L tanks at a density of 267 tilapia/m^3, and 50 and 100 g of seaweed stocked in 50 L tanks, corresponding to treatments with 1 and 2 g/ L, respectively	The presence of *U. ohnoi* in the IMTA system with *P. vannamei* and *O. niloticus* benefited the performance of all species through nitrogen and phosphorus recovery and an increase in total productivity	de Morais et al. (2023)
Holothuria poli	*H. poli* was cultured directly below a fish cage at 0 m, 10 m, and 25 m distances from the cage and at two reference sites located over 800 m away from the fish farm	*H. poli* in IMTA might have the potential to uptake organic farm waste and increase aquaculture production	Cutajar et al. (2022)
Penaeus vannamei, *Crassostrea gigas*	Trial period: 30 days; SM, OM, and CO. *P. vannamei*: 0.44 ± 0.01 g IW, *C. gigas:* 1.55 ± 0.01 g IW	Shrimp intestine and oyster digestive gland in the co-culture system showed a lower relative abundance of *Vibrionaceae* compared to monoculture treatments. IMTA system imposes digestive microbiome dynamics changes that could improve water management and shrimp productivity	Omont et al. (2020)
Homarus gammarus, *Salmo salar*	Trophic interactions between salmon (90–100 g IW) and lobsters were assessed using δ15N and δ13C stable isotope analysis and fatty acid profiling from fish feed as indicators of nutrient flow	A trophic relationship between salmon and lobsters within this IMTA system, with no apparent advantage or disadvantage to growth	Baltadakis et al. (2020)
Macroalgal, *Ulva rigida* Seabream and seabass	The macroalgae tanks (20,000 L per tank) were set to receive water that is pumped from two fish ponds, mechanically filtered to remove particulate matter (>40 µm), and then continuously distributed across several macroalgae tanks in an open-flow regime	IMTA significantly shaped the structure and composition of the microbial community of the rearing water and cultivated *U. rigida*	Califano et al. (2020)

(*continued on next page*)

TABLE 7.2 Study on integrated multitrophic aquaculture (IMTA) system—cont'd

Cultivated species	Condition	Highlight	References
Pangasianodon hypophthalmus, Hordeum vulgare	The experiment followed with three salinity treatments: T0: Control (freshwater mixed with chemical fertilizers), T1: 5,000 ppm, T2: 10,000 ppm, and T3: 15,000 ppm.	The integration of striped catfish and barley production in water salinities below 15,000 ppm could be a feasible alternative in safeguarding food and feed security in regions affected by soil salinity, soil infertility, and freshwater scarcity	Mugwanya et al. (2023)
Meagre (*Argyrosomus regius*), white seabream (*Diplodus sargus*), and gray mullet (*Mugil cephalus*) are three fish with oysters (*Magallana gigas*) and macroalgae (*Ulva flexuosa*).	Three IMTA treatments with different functional groups were compared in duplicate ponds. The combinations in each treatment were as follow: (T1) fish + oysters + phytoplankton + macroalgae; (T2) fish + oysters + phytoplankton; (T3) fish + phytoplankton + macroalgae	Fish, oyster, and macroalgae produced together in earthen ponds are an improved system compared to the usual semiintensive fish polyculture.	Nahon et al. (2023)
Apostichopus japonicus (as primary species), *Strongylocentrotus intermedius* (as subsidiary species)	Trial period: 34 days. Juvenile *S. intermedius* (0.2 g of IW) and *A. japonicus* (1.1 g of IW) were randomly chosen	IMTA system is a promising candidate to promote the production efficiency of juvenile *A. japonicus* and *S. intermedius*	Hu et al. (2022)
Lemna minor + Wastewater was collected from a fish farm, *Oncorhynchus mykiss* and *Perca fluviatilis*	The experiment lasted 4 days, and started with four culture containers for each plant density. Every 24 h, one culture container per density was used to perform water analysis	The present work enables the calculation of potential nutrient uptake by *L. minor* and lays the foundation for a more scientific approach to the design of duckweed-based aquaculture wastewater treatment systems	Paolacci et al. (2021)
Paracentrotus lividus, Holothuria tubulosa	*P. lividus* and *H. tubulosa* were transferred and maintained separately inside floating baskets (at a density of 20 specimens per basket) in a 600 L closed circulation aquarium equipped with a heat thermostat and biological and mechanical filtration	Less than 24% of the organic matter administered with the food remained as waste in our IMTA system after being ingested by the two trophic levels	Grosso et al. (2021)

now offers vast opportunities for adoption in freshwater environments, combining freshwater culture species in open water resources.

IMTA can be viewed as a progressive form of polyculture, where the farming of economic species (fish or shrimp) is supplemented by the growth of macroalgae, which absorb inorganic nutrients, and deposit feeders, such as shellfish and/or sea cucumbers, which consume organic nutrients (Park et al., 2018). A key distinction between polycultures and IMTAs lies in the fact that polycultures involve different species coexisting in the same water body, but without forming a food web where one species relies on the waste of another (Sanz-Lazaro et al., 2020). One of the primary hurdles in successfully implementing IMTA is establishing a balanced ecosystem among different aquatic species, which demands an in-depth understanding of the unique characteristics of each species, including their position in the food chain, dietary habits, oxygen demands, and other species-specific necessities (Nissar et al., 2023). IMTAs can be established in various settings, including open water bodies (sea, ocean) and inland facilities featuring ponds and tanks (Nissar et al., 2023).

Land-based IMTA systems usually comprise a series of interconnected ponds or tanks, designed to optimize water circulation through either a flow-through or recirculating aquaculture operation (Nissar et al., 2023). Research has consistently shown that IMTAs have a positive impact across the triple bottom line of sustainability, encompassing environmental, economic, and social dimensions (Granada et al., 2016). In addition, IMTA systems operate on the principles of a circular economy, fostering resource efficiency, energy savings, and a decrease in pollution risks (Sanz-Lazaro et al., 2020). The superiority of IMTA systems in terms of resource efficiency is evident in their capacity to generate higher protein yields from a fixed amount of feed, surpassing the productivity of traditional production methods (Boyd et al., 2020). A study by Alexander et al. (2016) on public perception of this system in Europe revealed that while communal information and awareness of this farming method are generally low, respondents expressed increased approval after receiving additional information about IMTA principles. Despite being considered underresearched globally (Kleitou et al., 2018), IMTA systems face challenges such as high initial and operating costs, which need to be addressed in the future (Nissar et al., 2023). Last but not least, the government backing and industry involvement, combined with documented financial gains, could serve as a catalyst for the widespread implementation of IMTA), as suggested by Nissar et al. (2023) and Laktuka et al. (2023).

7.6 Aquaponics

Aquaponics is a revolutionary farming approach that converges aquaculture with hydroponics (the method of growing plants in a soilless environment) (da Silva et al., 2022). This symbiotic system allows fish and plants to thrive together, benefiting from each other's presence. Table 7.3 presents an overview of various studies conducted on aquaponic systems.

Originating from integrated aquaculture systems (Palm et al., 2018), this system has emerged as a sustainable and productive food production system, providing both fish and vegetables while contributing to agricultural innovation (Mchunu et al., 2018). It offers locally cultivated, safe agricultural products, generating additional income for practitioners. Aquaponics serves as a novel model for integrating fish and agricultural crop cultivation, boasting several advantages over conventional systems, including reduced land area requirements and water consumption.

TABLE 7.3 Study on aquaponic aquaculture system.

Cultivated species	Condition	Highlight	References
Spirogyra as a freshwater plant + *Brassica rapa* L. and *Beta vulgaris* L. var. cicla L. as vegetables, *Misgurnus anguillicaudatus* as fish	Trial period: 30 days, and 6–8 true leaves, 17–21 cm in height, and 24.21 ± 2.75 g in plant weight were selected and planted in the hydroponic unit with a planting density of 25 plants/m^3	The aquaponic system with macroalgae has a stronger self-regulation ability. PRO-AP is an effective method in urban food production and is a very promising mode of production in the sustainable development of productive cities	Zhang et al. (2022b)
Lettuce as plant, *Clarias gariepinus* as fish	Trial period: 70 days. The water quality from the aquaponic tank was analyzed daily. The catfish growth was measured weekly, while the lettuce growth and its nutrient contents were analyzed at the end of the experiment	Improvement of water quality parameters and growth rate of catfish and lettuce. NH_3 removal: 66.6% TSS removal: 68%	Su et al. (2020)
Three horticultural crops, *Solanum lycopersicum* L., *Petroselinum crispum* (Mill.) Fuss and *Ocimum basilicum* L., with two ornamental species: *Carassius carassius* L. and *Cyprinus carpio* L. Koi variety	The aquaponics unit consisted of two 3.500-L fish tanks stocked, It was provided with a UV sterilizer, a reverse osmosis system, and a static 2.000-L biofilter filled with bio-media and a bottom-up oxygenation system. The flow rate in the biofilter was 2.5 m^3 h^{-1}, and the retention time was 0.8 h ± 45 min	Aquaponic farming method can be an innovative, rapid, and sustainable way of producing quality food	Braglia et al. (2022)
Marigold (*Tagetes erecta*) as a plant *Pangasius hypophthalmus* as a fish species	Trial period: 90 days. For performance assessment, the system was initially stocked with 54 numbers of fish fry/m^3 (153.7 g/m^3) in the culture tank and 27 marigolds (*Tagetes erecta*) plants/m^2 in hydroponics tank	The TAN reduction in biofilter was found to be 61.97 %.	Mohapatra et al. (2020)
Plant+*Tinca tinca* as a fish species	Trial period: 100 days. The study was started with an initial population of 366 tench fingerlings, counting up a total biomass of about 2100 g	The combination of a joint production of fish and plants in simple aquaponic systems would allow savings in fertilizer and water, within a production framework of circular economy	Fernandez-Cabanas et al. (2022)

(*continued on next page*)

TABLE 7.3 Study on aquaponic aquaculture system—cont'd

Cultivated species	Condition	Highlight	References
Lettuce (*Butter leaf lettuce*) as vegetable, *O. niloticus* as fish species	Trial period: 80 days with the addition of HA at concentrations of 0 ppm, 15 ppm, and 30 ppm	Addition of HA to the aquaponic systems yielded more products and improved water quality and NUE by changing the composition of microbial communities based on the threshold for HA input concentration	Gao et al. (2022)
Lettuce (*Lactuca sativa*) as vegetable, *O. mykiss* as fish species	Trial period: 117 days. A total of 123 rainbow trout (IBW: 142 g) were randomly distributed in six 500-L tanks (3 per stocking density)	*O. mykiss* and lettuce productions were successful in the tested aquaponic system, whereas stocking density did not affect fish growth or flesh quality	Birolo et al. (2020)
O. niloticus and *M. rosenbergii*, are aquatic animals. Watercress (*R. nasturtium-aquaticum*) and curly lettuce (*L. sativa*) as vegetables	Trial period: 45 days. The experiment was carried out in a randomized complete block design with two treatments: ASCW, ASBFT	BFT, in conjunction with aquaponics, facilitated the maintenance of zootechnical performance and contributed to an increase in the fat content of *O. niloticus* fillets and better control of nitrite production	Barbosa et al. (2022)

Commercial large-scale aquaponics farming occurs in controlled settings, including greenhouses and outdoor locations with ideal climate conditions, which require specialized facilities and equipment for both fish farming and crop cultivation. This innovative method integrates expertise from aquaculture, hydroponics, engineering, microbiology, and water quality management (Zhang et al., 2022a). Throughout Europe, various aquaponics systems of different scales and designs have been developed, providing a cutting-edge solution to tackle food security and environmental challenges (Yanes et al., 2020).

Aquaponics is ideally suited for urban settings, areas with limited land availability or environmental pollution, and rural regions, as it enables the cultivation of premium aquatic and agricultural products in a controlled environment. This intensive production technique not only optimizes yields but also helps mitigate the urban heat island effect (Junge, 2017). As a sustainable and eco-friendly aquaculture approach, aquaponics offers remarkable productivity in terms of aquatic life and plant growth, while conserving water resources and minimizing the need for feed in a small space (Taha et al., 2022).

Researchers have demonstrated that aquaponics can achieve a remarkable sixfold increase in productivity compared to traditional outdoor farming and aquaculture practices, all while occupying only one-sixth of the space and water (Okomoda et al., 2023). Furthermore, this sustainable method allows for the production of high-value, contaminant-free food items, such as fruits, flowers, and herbs, in close proximity to large cities and arid regions, without relying on antibiotics and pesticides (Pinho et al., 2017).

Aquaponics promotes the growth of a diverse microbial community, which plays a crucial role in natural productivity, water quality, and nutrient cycling (Meng et al., 2023). To mitigate the issue of excessive nutrients in wastewater, implementing water reuse or nutrient recycling strategies within aquaponics systems is a viable solution. Furthermore, fostering a diverse community of microorganisms can significantly boost the system's efficiency (Emerenciano et al., 2013; da Silva et al., 2022). Establishment of a nitrogen cycle is crucial for the stable operation of aquaponics, wherein ammonia is converted to nitrite by ammonia-oxidizing bacteria and subsequently to nitrate by nitrite-oxidizing bacteria (Gao et al., 2022). The nitrates produced in fish tanks serve as essential nutrients for plant growth, promoting healthy development and circulation.

One of the key benefits of integrated systems is their ability to fulfill the nutritional requirements of plants without resorting to chemical fertilizers. By harnessing the nutrient-rich byproducts of cultured organisms, such as their food waste and excrement, plants can receive a diverse range of essential nutrients (Lee et al., 2019). Moreover, this integrated system can facilitate the coexistence and mutual growth of both the cultured organisms and plants, as demonstrated by Deswati et al. (2023).

7.7 FLOCponics

The combined system of BFT and aquaponics offers a sustainable method of food production, focusing on the recycling of nutrients and conservation of water. Pinho et al. (2022a) coined the term "FLOCponics" to describe this innovative aquaponics system, which replaces traditional RAS with BFT. This integrated system has the potential to increase economic diversification by producing high-value crop products while reducing nitrate and phosphorus accumulation in BFT management (Rakocy, 2012). The continuous absorption of nitrates by plants and the maintenance of phosphate levels within the recommended range for plant growth contribute to the system's low nitrate levels.

Recent studies have demonstrated the effectiveness of this system in upholding low nitrate levels. For example, Saseendran et al. (2021) reported a significant decrease in nitrate levels (2.86 mg/L) in an integrated system using bell pepper and GIFT tilapia (genetically improved farmed tilapia) compared to a standalone aquaponics system (5.26 mg/L). However, phosphate levels remained unaffected.

Pinheiro et al. (2017) also reported low nitrate levels in an integrated system using halophytes and shrimp. Although plant presence did not significantly impact nitrate levels, the levels remained consistently low. Orthophosphate levels showed no significant differences between the integrated and standalone systems.

According to Pinho et al. (2017), this system has been proven to effectively regulate nitrogen-based compounds, such as ammonium nitrogen, nitrite, and nitrate, through plant absorption. Notably, the levels of orthophosphate were found to be within the acceptable range for leafy vegetable production. The integrated system has been shown to maintain a sustainable aquaculture environment, regulating water quality parameters such as dissolved oxygen, temperature, pH, and salinity (Poli et al., 2019). The system also effectively reduces TSS and promotes microbial proliferation through root exudation (the release of nutrient-rich substances from the roots) and biofilm (complex microbial communities adhered to surfaces) formation. Halophyte plants have been found to effectively reduce nitrate levels and TSS while endorsing microbial proliferation.

The appropriate consumption of nitrogen and phosphate compounds by *S. ambigua* has been confirmed, and the system has been shown to have no adverse effects on shrimp performance (Pinheiro et al., 2020).

The integrated system has also been found to provide a suitable aquaculture environment for other species, such as *Oreochromis niloticus* and *Macrobrachium rosenbergii*, without opposing impacts. Curly lettuce and watercress, among other vegetables, have a vital function in controlling the amounts of nitrate produced by chemoautotrophic bacteria, as demonstrated by Barbosa et al. (2022).

Despite the promising results, the research on this integrated system is still in its early stages, with limited available papers and an absence of standardization in system design and experimental methodologies. For the long-term viability and maximal efficiency of this system, it is essential to take into account several crucial factors, including the capacity to withstand salinity, the standard of water quality, and the ability to absorb nutrients.

In the realm of mariculture and hydroponics, the cultivation of plants that can thrive in saline environments, known as halophytes, is crucial. The productivity of aquatic organisms and plants is heavily influenced by the quality of the water, which is in turn affected by the density of organisms in the water and can have a significant impact on their growth rates. Moreover, fluctuations in water quality can trigger oxidative stress, ultimately influencing metabolic processes and growth performance, as demonstrated by Saseendran et al. (2021).

Numerous studies have focused on *O. niloticus* and *Litopenaeus vannamei*, but research on other precious aquatics is partial. Biofloc presence in roots can interfere with plant nutrient absorption, potentially inhibiting plant growth. Decoupled aquaponics systems (DAPS) have been projected as a promising substitute to address these limitations (Goddek et al., 2019).

In summary, combining BFT and aquaponics provides a sustainable solution for food production, yet more research and standardization efforts are required to unlock its full potential and maximize its eco-friendly benefits (Yu et al., 2023).

7.8 Integrated agriculture aquaculture

To cater to the growing population's food demands, two primary food production systems, agriculture and aquaculture, play a critical role. Integrating aquaculture into traditional farming systems has proven to be a viable and eco-friendly option for securing the world's food supply in the years to come, as highlighted by (Pretty, 2008).

The convergence of agriculture and aquaculture systems is denoted as integrated agriculture-aquaculture systems (IAAS), a form of crop diversification that involves the simultaneous cultivation of aquatic and terrestrial foods on a single plot of land. A noteworthy variant of IAAS is pond-based IAAS, where fish are raised as the primary crop for production and income generation. This approach is characterized by its eco-friendly nature, offering a viable strategy for water conservation and nutrient recycling, as substantiated by Phong et al. (2010).

In Kilombero, IAA systems, which combine fish and vegetable farming (including B. Rapa Chinensis), have been shown to increase net production by 3 and 2.5 times, respectively, compared to nonintegrated systems (Limbu et al., 2017). Similarly, in certain areas of Bangladesh, it has been reported that agricultural diversification, including crops, poultry, fisheries, and cattle, can lead to increased income for farmers, particularly when combined with large land holdings and pond areas (Sharmin et al., 2012). Studies have shown that implementing IAA systems that include ponds can significantly enhance farm productivity, with boosts of 11% compared to traditional pond-fish farming or non-IAA systems. Moreover, these systems have been found to increase net income per hectare by a substantial 134% (Limbu et al., 2017; Shoko et al., 2019). The pond-based IAA approach is considered environmentally sustainable because of its ability to recycle water and nutrients, leading to increased productivity and water use efficiency (Phong et al., 2010). Besides, fish waste can increase soil fertility by increasing organic fertilizers and replacing nitrogen and phosphorus elements, which can then be used to enhance crop production (Ahmed et al., 2014). Also, growing vegetables and herbs on pond sediments can help prevent erosion of embankments caused by rainfall.

One of the key benefits of IAAS is the recycling of waste from one system to serve as inputs for another, thereby minimizing pollution (Ahmed et al., 2012). Table 7.4 highlights some of the studies conducted on IAAS. In addition to its primary benefits, IAAS also brings about a range of secondary advantages, including improved environmental sustainability, increased productivity, enhanced profitability, and optimized utilization of essential resources such as water, land, and labor, as found by Limbu et al. (2017). Furthermore, IAAS helps increase food production while mitigating the risks associated with water scarcity.

Compared to other food production systems, IAASs exhibit higher sustainability (Pretty, 2008). While the integration of components in IAAS may vary, they share a common characteristic of being located in developing countries, especially in rural areas. It is observed that the integration of more components in the system leads to higher production, reduced waste generation, and lower risks associated with water scarcity, ultimately enhancing the system's sustainability (Ahmed et al., 2014).

As mentioned before, the primary advantage of adopting IAA lies in the conversion of waste into a valuable resource for cultivating a secondary crop. This approach also offers several other benefits, including the generation of additional food sources from the same land, increased production at little or no extra cost, enhanced employment opportunities, and environmentally sustainable farming practices. Furthermore, IAA reduces the need for external inputs, such as feed and fertilizers, promotes high nutrient circulation, and leads to increased productivity and profitability. The system also supports species diversification and is both economically viable and socially acceptable.

The IAA system has immense potential to boost productivity and yields. Farmers can easily adopt these practices with minimal additional investment, which can lead to remarkable production increases from previously underutilized areas. Recent advancements in IAA, including cage-based systems, are already demonstrating their impact on aquaculture production. It is anticipated that IAA will make a significant contribution to total fish production in the future.

Finally, IAA is essential for addressing the worldwide challenges of hunger, malnutrition, and food insecurity. As emphasized by Ahmad et al. (2014), the synergistic approach of IAA has been promoted to boost food production, reduce environmental waste, and enhance overall food security (Hall et al., 2011). Despite its potential, aquaculture production in Africa remains limited, accounting for only 2.7% of global aquaculture output (FAO, 2020). However, IAA systems offer a sustainable solution, allowing farmers to produce multiple outputs from a single farm while minimizing environmental impact. The benefits of IAA are twofold, including water reuse and improved soil quality. Furthermore, IAA systems generally yield higher production levels compared to monoculture systems, resulting in greater social, economic, and environmental benefits. Therefore IAA is a pioneering approach that offers a promising solution to boosting household incomes and ensuring food security globally. Interestingly, if pond-based IAA were

TABLE 7.4 Study on integrated agriculture aquaculture systems (IAAS).

Cultivated species	Condition	Highlight	References
O. niloticus + Wheat	The experimental ponds were equal in water volume (5250 m^3) and dimensions (42 × 100 m), with the same average water depth of 125 cm	Wheat cultivation in ponds during the winter period, taking into account feeding *O. niloticus* on artificial feed containing 25% crude protein with organic fertilization, leads to an increase in the weights of fish, as well as an increase in the yield of wheat product from fish ponds	El-Bab et al. (2014)
O. niloticus + Lettuce	The fish pond was loaded with *O. niloticus* (100 fish/m^3), which had an initial weight of 50 g per fish. About 15% of the water was replaced daily with new fresh water	Reusing drainage water of fish ponds can enhance lettuce yield where it is rich with soluble nitrogen and increase water and nitrogen productivity, thus reducing mineral fertilizer doses	Attafy and Elsbaay (2018)
O. niloticus+ Sesame crop	Field experiments were carried out under the following variables: Two water resources, two drip irrigation systems, and three nitrogen fertilizer rates (20%, 60%, and 100% N)	The optimum conditions for sesame crop were noticed by using aquaculture drainage with 60% nitrogen fertilizer rate and 40% fertilizer saving compared with traditional methods	Soliman et al. (2020)
Shrimp (*M. rosenbergii*, *P. monodon*, *L. vannamei*) + Rice	There are two main forms of rice-shrimp farming: Alternate culture model (rotational culture) and integrated culture model (simultaneous culture)	Rice-shrimp systems have the potential to improve livelihood, food security, and adaptation of coastal farmers	Dang et al. (2020)
Clarias gariepinus+ Lettuce (*Lactuca sativa*)	Fish were grown at a mean density of ~27.4 kg/m^3. Lettuce was planted on floating rafts (50 mm Styrofoam), with 120 plants in each hydroponics bed	Aquacultural and plant wastes from farms can save water and be used effectively as fertilizers for onsite agricultural use.	Zhu et al. (2023)
Marjoram (*Origanum majorana*)+ Aquaculture effluents	A randomized complete block design was followed with three irrigation treatments, i.e., fertigation with chemical fertilizers (control), irrigation with effluents only (effluents), and mixed treatment with 50% effluent and 50% chemical fertilizers (mixed)	Growing *O. majorana* with a mixed treatment would give the best herbage yields and the highest essential oils with reduced environmental impacts	Kimera et al. (2021)

(*continued on next page*)

TABLE 7.4 Study on integrated agriculture aquaculture systems (IAAS)—cont'd

Cultivated species	Condition	Highlight	References
O. niloticus –Maize (*Zea mays*), Tilapia–Radish	Six round 1m^3 tanks with central drains were stocked with size-sorted (17.9 g) Nile tilapia, fingerlings at 10 fish/tank	Aquaculture effluent can supplant inorganic fertilizers and could actually yield better crop production.	Abdul-Rahman et al. (2011)
M. rosenbergii and rice (*Oryzasativa*)	The factors tested were the density of prawns (2 and 5 prawns/m^2) and the use of commercial diet (without and with feed supply)	The rice–prawn simultaneous culture at a stocking density of 2 prawns/m^2 without commercial diets is economically feasible, which can compensate for the low rice prices	Boock et al. (2016)
Two rice varieties (BRRI-51 and BRRI-52), three different fish species. i.e., *Cyprinus carpio var. communis*, *Barbonymus gonionotus*, and *O. niloticus*, and two combinations of vegetables: (1) Red amaranth (*Amaranthus cruentus*) and Indian spinach (*Spinacia oleracea*), (2) Cucumber (*Cucumis sativus*) and Water spinach (*Ipomoea aquatica*)	Two trials were conducted for 4 months under rainfed conditions. The average area of the rice fields was 0.024 ha	Rice-fish-vegetable integrated culture could address the SDG, and therefore policy implications should be considered for institutional support, technical facilities, and extension services to increase the knowledge of farmers and to uplift the productivity and profitability	Jewel et al. (2023)
Squash and tomato + *O. niloticus*	The fish pond was supported with 25,000 fry mono sex *O. niloticus* fed on fish diets of 30%–25% protein. The second category (two plant crops of squash and tomato) was treated with the water of the fish pond, which contains natural fertilizers	Farming of fish increased effluent nutrient levels, and the effluent was subsequently used to irrigate crops and orchards. Moreover, it could reduce the required amounts of fertilizers in agriculture, and improve the quality and quantity of plant crops	Farrag et al. (2021)

to scale up to 25% of its potential capacity, it would lead to a significant increase of 3.08 million tons in global fish production annually, while also resulting in an additional 95.4 million tons of carbon storage per year, as indicated by Ibrahim et al. (2023).

Globally, further research is essential to address the following key questions: (1) How can integrated aquaculture systems, enhanced by smart technology, help alleviate the consequences of climate change and minimize greenhouse gas emissions? (2) How can integrated aquaculture systems be harnessed to produce organic and organic products, aligning with consumer demands for sustainable and eco-friendly options? (3) What strategies can be employed to optimize land and water usage efficiency in integrated aquaculture systems, thereby minimizing environmental footprints?

In addition to the international points mentioned, first of all, there is a need to use mixed aquaculture systems in arid and semiarid areas that encounter water limitations. Investigations are necessary to breed diverse aquatic organisms and crops that can thrive in environments with higher salt concentrations and poorer water quality, as well as exhibit greater resilience to ecological shifts. In addition, there is a need to move towards the use of hybrid systems to mitigate environmental changes in freshwater.

By leveraging cutting-edge innovations, such as automated monitoring systems and harnessing renewable energy sources, such as solar power, hybrid systems can be optimized for even greater efficiency and sustainability. Moreover, the strategic deployment of hybrid systems in aquaculture, backed by sufficient investment and technological breakthroughs, holds the promise of revolutionizing the global food supply chain.

7.9 Conclusion

Conventional aquaculture practices, which involve high-density farming, have devastating environmental consequences along coastal areas, ultimately leading to poor farm conditions and decreased productivity. Using hybrid systems in aquaculture (RAS, IMTA, BFT, IAA, aquaponics), we can solve some of the problems of traditional aquaculture. For instance, by using the BFT, water quality is ensured, and pathogen infections are prevented. This innovative method not only leads to significant operational cost savings but also promotes the well-being and growth of aquatic life. Nevertheless, BFT is not without its obstacles, and it faces some challenges. For instance, it is susceptible to fluctuations in water quality and seasonal variations, and requires aeration to ensure optimal oxygen levels. To surmount these hurdles, integrating BFT with complementary aquaculture technologies, such as aquaponics and vertical farming, is essential.

Aquaponics is a revolutionary technology that integrates closed-loop aquaculture systems with hydroponics, creating a self-sustaining ecosystem. In contrast, vertical aquaculture capitalizes on the vertical dimensions of high-rise structures to cultivate aquatic species in a controlled environment, maximizing productivity. Notably, both aquaponics and vertical aquaculture are well-suited for urban settings, thanks to their ability to optimize space and alleviate environmental concerns. By combining these technologies with BFT, the resulting system is expected to compensate for BFT's shortcomings, providing groundbreaking solutions to the challenges facing the aquaculture industry.

Combining aquaponics with integrated agriculture in various scenarios has proven to be an effective strategy for optimizing water use, boosting crop yields, and mitigating the risks associated with water scarcity globally and locally. The co-cultivation of fish and plants leads to a significant increase in annual net production compared to monoculture practices. Furthermore, utilizing fish pond effluent to enrich soil with low fertility, particularly in arid regions, enhances soil productivity and facilitates its cultivation. These integrated systems not only enhance the efficiency and sustainability of aquaculture but also have the potential to transform the way we produce and consume aquatic products in urban environments. The development of these innovative aquaculture technologies is expected to play a crucial role in meeting future global food demands. Establishing sustainable and safe aquaculture systems will be vital in supporting the food security of rapidly growing populations. Therefore ongoing research and standardization efforts on integrated systems are essential for the long-term viability of the aquaculture industry (Kristiansen and Cripps, 1996; Chopin et al., 2001; Ebeling et al., 2006; Crab et al., 2007; Ridler et al., 2007; Pretty, 2008; Phong et al., 2010; Timmons and Ebeling, 2010; Abdul-Rahman et al., 2011; Hall et al., 2011; Martins et al., 2011; Ahmed et al., 2012; Sharmin et al., 2012; Rakocy, 2012; Emerenciano et al., 2013; Ahmed et al., 2014; El-Bab et al., 2014; Edwards, 2015; Goddek et al., 2015; Alexander et al., 2016; Boock et al., 2016; Granada et al., 2016; Limbu et al., 2017; Junge et al., 2017; Pinheiro et al., 2017; Pinho et al., 2017; Attafy and Elsbaay, 2018; Kleitou et al., 2018; Mchunu et al., 2018; Park et al., 2018; Thomas et al., 2018; Timmons et al., 2018; Palm et al., 2018; Goddek et al. 2019; Lee et al., 2019; Poli et al., 2019; Shoko et al., 2019; Amin et al., 2020; Baltadakis et al., 2020; Birolo et al., 2020; Boyd et al., 2020; Califano et al., 2020; Dang, 2020; Henares et al., 2020; Mohapatra et al., 2020; Moore et al., 2020; Omont et al., 2020; Pinheiro et al., 2020; Santos et al., 2020; Sanz-Lazaro and Sanchez-Jerez, 2020; Soliman et al., 2020; Su et al., 2020; ; Yanes et al., 2020; Almeida et al., 2021Ekawati et al., 2021; Farrag et al., 2021; Grosso et al., 2021; Kimera et al., 2021; Liu et al., 2021; Paolacci et al., 2021; Saseendran et al., 2021; Serra et al., 2021; Wambua et al., 2021; Yin et al., 2021; Zhao et al., 2021; Barbosa et al., 2022; Braglia et al., 2022; Cutajar et al., 2022; Fernández-Cabanás et al., 2022; Figueroa-Espinoza et al., 2022; Gao et al., 2022; Hu et al., 2022; Khanjani et al., 2022; Mota et al., 2022; Mugwanya et al., 2022; Pinho et al., 2022a; Pinho, et al., 2022b; Taha et al., 2022; Xue et al., 2022; Zhang et al., 2022a; Zhang et al., 2022b; Zhao et al., 2022a; Zhao et al., 2022b; da Silva et al., 2022; Das et al., 2023; de Morais et al., 2023; Deswati et al., 2023; Ibrahim et al., 2023; Jewel et al., 2023; Laktuka et al., 2023; Matias et al., 2023; Meng et al., 2023; Mugwanya et al., 2023; Nahon et al., 2023; Nissar et al., 2023; Okomoda et al., 2023; Seo and Park, 2023; Verdegem et al., 2023; Yu et al., 2023; Zhu et al., 2023; Khanjani et al., 2024a; Khanjani et al., 2024b; Khanjani et al., 2024c; Lu et al., 2024; The State of World Fisheries and Aquaculture: Blue Transformation in Action, 2024; Emerenciano et al., 2025; Khanjani et al., 2025a; Khanjani et al., 2025b).

List of abbreviations

ASBFT	Aquaponic system with biofloc technology
ASCW	Aquaponic system with clear water
BFT	Biofloc technology
CO	Coculture system with oyster *C. gigas*
EC	Electrical conductivity
HA	Humic acid
IBW	Initial body weight
IMTA	Integrated multitrophic aquaculture
IW	Initial weight
MPs	Microplastics
NUE	Nitrogen utilization efficiency
OM	Oyster *C. gigas* in monoculture system
PRO-AP	Photosynthesis reoxygenation aquaponics
RASs	Recirculating aquaculture systems
SDGs	Sustainable Development Goals
SM	*P. vannamei* shrimp in monoculture system
TAN	Total ammoniacal nitrogen

References

Abdul-Rahman, S., Saoud, I.P., Owaied, M.K., Holail, H., Farajalla, N., Haidar, M., Ghanawi, J., 2011. Improving water use efficiency in semi-arid regions through integrated aquaculture/agriculture. J. Appl. Aquac. 23 (3), 212–230. doi:10.1080/10454438.2011.600629.

Ahmed, N., Muir, J.F., Garnett, S.T., 2012. Bangladesh needs a "blue-green revolution" to achieve a green economy. Ambio. 41 (2), 211–215. doi:10.1007/s13280-011-0160-6.

Ahmed, N., Ward, J.D., Saint, C.P., 2014. Can integrated aquaculture-agriculture (IAA) produce "more crop per drop"? Food Secur. 6 (6), 767–779. doi:10.1007/s12571-014-0394-9. http://www.springer.com/life+sci/agriculture/journal/12571.

Alexander, K.A., Freeman, S., Potts, T., 2016. Navigating uncertain waters: European public perceptions of integrated multi trophic aquaculture (IMTA). Environ. Sci. Policy 61, 230–237. doi:10.1016/j.envsci.2016.04.020.

Almeida, D.B., Magalhães, C., Sousa, Z., Borges, M.T., Silva, E., Blanquet, I., Mucha, A.P., 2021. Microbial community dynamics in a hatchery recirculating aquaculture system (RAS) of sole (Solea senegalensis). Aquaculture 539. doi:10.1016/j.aquaculture.2021.736592. http://www.journals.elsevier.com/aquaculture/.

Amin, M., Musdalifah, L., Ali, M., 2020. Growth performances of Nile tilapia, Oreochromis niloticus, reared in recirculating aquaculture and active suspension systems. In: IOP Conference Series: Earth and Environmental Science. Institute of Physics Publishing, Indonesia, p. 441.

Attafy, T., Elsbaay, A., 2018. Integrated management of fish (Nile tilapia) and leaf vegetable crop (Head lettuce) culture under drip irrigation system. Misr J. Agric. Eng. 35 (1), 105–124. doi:10.21608/mjae.2018.96038.

Baltadakis, A., Casserly, J., Falconer, L., Sprague, M., Telfer, T.C, 2020. European lobsters utilise Atlantic salmon wastes in coastal integrated multi-trophic aquaculture systems. Aquacult. Env. Interact. 12, 485–494. doi:10.3354/aei00378.

Barbosa, P.T.L., Povh, J.A., Farias, K.N.N., da Silva, T.V., Teodoro, G.C., Ribeiro, J.S., Stringhetta, G.R., dos Santos Fernandes, C.E., Filho, R.A.C.C., 2022. Nile tilapia production in polyculture with freshwater shrimp using an aquaponic system and biofloc technology. Aquaculture 551. doi:10.1016/j.aquaculture.2022.737916.

Birolo, M., Bordignon, F., Trocino, A., Fasolato, L., Pascual, A.ón, Godoy, S., Nicoletto, C., Maucieri, C., Xiccato, G., 2020. Effects of stocking density on the growth and flesh quality of rainbow trout (Oncorhynchus mykiss) reared in a low-tech aquaponic system. Aquaculture 529, 735653. doi:10.1016/j.aquaculture.2020.735653.

Boock, M.V., Marques, H.L.d.A., Mallasen, M., Barros, H.P., Moraes-Valenti, P., Valenti, W.C., 2016. Effects of prawn stocking density and feeding management on rice-prawn culture. Aquaculture 451, 480–487. doi:10.1016/j.aquaculture.2015.10.009.

Boyd, C.E., D'Abramo, L.R., Glencross, B.D., Huyben, D.C., Juarez, L.M., Lockwood, G.S., McNevin, A.A., Tacon, A.G.J., Teletchea, F., Tomasso, J.R., Tucker, C.S., Valenti, W.C., 2020. Achieving sustainable aquaculture: historical and current perspectives and future needs and challenges. J. World Aquac. Soc. 51 (3), 578–633. doi:10.1111/jwas.12714.

Braglia, R., Costa, P., Di Marco, G., D'Agostino, A., Redi, E.L., Scuderi, F., Gismondi, A., Canini, A., 2022. Phytochemicals and quality level of food plants grown in an aquaponics system. J. Sci. Food Agric. 102 (2), 844–850. doi:10.1002/jsfa.11420.

Califano, G., Kwantes, M., Abreu, M.H., Costa, R., Wichard, T., 2020. Cultivating the macroalgal holobiont: effects of integrated multi-trophic aquaculture on the microbiome of Ulva rigida (Chlorophyta). Front. Mar. Sci. 7. doi:10.3389/fmars.2020.00052.

Chopin, T., Buschmann, A.H., Halling, C., Troell, M., Kautsky, N., Neori, A., Kraemer, G.P., Zertuche-González, J.A., Yarish, C., Neefus, C., 2001. Integrating seaweeds into marine aquaculture systems: a key toward sustainability. J. Phycol. 37 (6), 975–986. doi:10.1046/j.1529-8817.2001.01137.x.

Crab, R., Avnimelech, Y., Defoirdt, T., Bossier, P., Verstraete, W., 2007. Nitrogen removal techniques in aquaculture for a sustainable production. Aquaculture 270 (1–4), 1–14. doi:10.1016/j.aquaculture.2007.05.006.

Cutajar, K., Falconer, L., Massa-Gallucci, A., Cox, R.E., Schenke, L., Bardócz, T., Sharman, A., Deguara, S., Telfer, T.C., 2022. Culturing the sea cucumber Holothuria poli in open-water integrated multi-trophic aquaculture at a coastal Mediterranean fish farm. Aquaculture 550. doi:10.1016/j.aquaculture.2021.737881.

Dang, H.D., 2020. Sustainability of the rice-shrimp farming system in Mekong Delta, Vietnam: a climate adaptive model. J. Econ. Dev. 22 (1), 21–45. doi:10.1108/JED-08-2019-0027.

Das, S.K., Mondal, B., Sarkar, U.K., Das, B.K., Borah, S., 2023. Understanding and approaches towards circular bioeconomy of wastewater reuse in fisheries and aquaculture in India: an overview. Rev. Aquac. 15 (3), 1100–1114. doi:10.1111/raq.12758.

Deswati, O.N.T., Yusuf, Y., Dewi, P., Pardi, H., 2023. Application of FLOCponics to improve water quality (phosphate, sulphate, calcium, potassium). AACL Bioflux 16 (1), 483–495. http://www.bioflux.com.ro/docs/2023.483-495.pdf.

Ebeling, J.M., Timmons, M.B., Bisogni, J.J., 2006. Engineering analysis of the stoichiometry of photoautotrophic, autotrophic, and heterotrophic removal of ammonia-nitrogen in aquaculture systems. Aquaculture 257 (1–4), 346–358. doi:10.1016/j.aquaculture.2006.03.019.

Edwards, P., 2015. Aquaculture environment interactions: Past, present and likely future trends. Aquaculture 447, 2–14. doi:10.1016/j.aquaculture.2015.02.001.

Ekawati, A.W., Ulfa, S.M., Dewi, C.S.U., Amin, A.A., Salamah, L.N., Yanuar, A.T., Kurniawan, A., 2021. Analysis of aquaponic-recirculation aquaculture system (A-ras) application in the catfish (clarias gariepinus) aquaculture in Indonesia. Aquac. Stud. 21 (3), 93–100. doi:10.4194/2618-6381-v21_3_01.

Emerenciano, M., Gaxiola, G., Cuzo, G., 2013. Biofloc technology (BFT): a review for aquaculture application and animal food industry. Biomass now: cultivation and utilization. InTechOpen, Rijeka doi:10.5772/53902.

Emerenciano, M.G.C., Khanjani, M.H., Sharifinia, M., Miranda-Baeza, A., 2025. Could biofloc technology (BFT) pave the way toward a more sustainable aquaculture in line with the circular economy? Aquac. Res. 2025 (1). doi:10.1155/are/1020045.

FAO, 2020. The State of World Fisheries and Aquaculture 2020. Sustainability in action. Rome. https://doi.org/10.4060/ca9229en.

FAO, 2024. Brief to The State of World Fisheries and Aquaculture 2024. Blue Transformation in action, 2024. Rome. https://doi.org/10.4060/cd0690en.

Farrag, M.M.S., Toutou, M.M.M., Sedik, F.S., Mursy, E.E.D.I.A., Osman, A.G.M., 2021. Towards the integrated agri-aquaculture in the desert using groundwater reservoirs for plants and nile tilapia farming "evaluating study in upper Egypt. Egypt. J. Aquat. Biol. Fish. 25 (2), 215–235. doi:10.21608/ejabf.2021.161839.

El-Bab, A.F.F., Mostafa, M.A., Hassan, A.S., 2014. Effect of Exploitation of fish ponds in the cultivation of wheat in the winter season on growth performance and total yield of Oreochromis Niloticus. Egypt. J. Aquat. Biol. Fish. 18 (3), 53–66. doi:10.12816/0011106.

Fernández-Cabanás, V.íM., Delgado, A., Lobillo-Eguíbar, J.R., Pérez-Urrestarazu, L., 2022. Early production of strawberry in aquaponic systems using commercial hydroponic bands. Aquacult. Eng. 97, 102242. doi:10.1016/j.aquaeng.2022.102242.

Figueroa-Espinoza, J., Rivas-Vega, M.E., Mariscal-López, M.Á., Emerenciano, M.G.C., Martínez-Porchas, M., Miranda-Baeza, A., 2022. Reusing water in a biofloc culture system favors the productive performance of the Nile tilapia (Oreochromis niloticus) without affecting the health status. Aquaculture 558. doi:10.1016/j.aquaculture.2022.738363.

Gao, X.T, Zhang, H.G, Xu, Y.F, Ni, Q.i, Zhang, Y.uL., Tan, H.X, Gu, C.C, 2022. Effects of humic acid on the nitrogen utilization efficiency and microbial communities in aquaponic systems. Aquaculture 547, 737475. doi:10.1016/j.aquaculture.2021.737475.

Goddek, S., Joyce, A., Wuertz, S., Körner, O., Bläser, I., Reuter, M., Keesman, K.J., 2019. Decoupled aquaponics systems aquaponics food production systems: combined aquaculture and hydroponic production technologies for the future. Springer International Publishing, Netherlands, pp. 201–230. doi:10.1007/978-3-030-15943-6_8.

Goddek, S., Delaide, B., Mankasingh, U., Ragnarsdottir, K., Jijakli, H., Thorarinsdottir, R., 2015. Challenges of sustainable and commercial aquaponics. Sustainability 7 (4), 4199–4224. doi:10.3390/su7044199.

Granada, L., Sousa, N., Lopes, S., Lemos, M.F.L., 2016. Is integrated multitrophic aquaculture the solution to the sectors' major challenges? – a review. Rev. Aquac. 8 (3), 283–300. doi:10.1111/raq.12093.

Grosso, L., Rakaj, A., Fianchini, A., Morroni, L., Cataudella, S., Scardi, M., 2021. Integrated multi-trophic aquaculture (IMTA) system combining the sea urchin Paracentrotus lividus, as primary species, and the sea cucumber Holothuria tubulosa as extractive species. Aquaculture 534, 736268. doi:10.1016/j.aquaculture.2020.736268.

Hall, S.J., Delaporte, A., Phillips, M.J., Beveridge, M., O'Keefe, M., 2011. Blue frontiers: managing the environmental costs of aquaculture. The WorldFish Center, Penang, Malaysia, p. 92.

Henares, M.N.P., Medeiros, M.V., Camargo, A.F.M., 2020. Overview of strategies that contribute to the environmental sustainability of pond aquaculture: rearing systems, residue treatment, and environmental assessment tools. Rev. Aquac. 12 (1), 453–470. doi:10.1111/raq.12327.

Hu, F., Wang, H., Tian, R., Gao, J., Wu, G., Yin, D., Zhao, C., 2022. A new approach to integrated multi-trophic aquaculture system of the sea cucumber Apostichopus japonicus and the sea urchin Strongylocentrotus intermedius. J. Mar. Sci. Eng. 10 (12), 1875. doi:10.3390/jmse10121875.

Ibrahim, L.A., Abu-Hashim, M., Shaghaleh, H., Elsadek, E., Hamad, A.A.A., Hamoud, Y.A., 2023. A comprehensive review of the multiple uses of water in aquaculture-integrated agriculture based on international and national experiences. Water 15 (2). doi:10.3390/w15020367.

Jewel, M.A.S., Haque, M.A., Ali, S.M.W., Pervin, M.E., Ahmed, M.G.U., Islam, M.S., Hossain, M.B., Albeshr, M.F., Arai, T., 2023. Integration of vegetables and fish with rice in rain-fed farmland: towards sustainable agriculture. Agriculture 13 (4). doi:10.3390/agriculture13040755.

Junge, R., König, B., Villarroel, M., Komives, T., Jijakli, M., 2017. Strategic points in aquaponics. Water 9 (3), 182. doi:10.3390/w9030182.

Khanjani, M.H., Zahedi, S., Sharifinia, M., Hajirezaee, S., Singh, S.K., 2025. Biological removal of nitrogenous waste compounds in the biofloc aquaculture system - a review. Ann. Anim. Sci. 25 (1), 3–21. doi:10.2478/aoas-2024-0060.

Khanjani, M.H., Mozanzadeh, M.T., Sharifinia, M., Emerenciano, M.G.C., 2024. Broodstock and seed production in biofloc technology (BFT): an updated review focused on fish and penaeid shrimp. Aquaculture 579. doi:10.1016/j.aquaculture.2023.740278.

Khanjani, M.H., Sharifinia, M., Akhavan-Bahabadi, M., Emerenciano, M.G.C., 2024. Probiotics and phytobiotics as dietary and water supplements in biofloc aquaculture systems. Aquacult. Nutr. 2024. doi:10.1155/anu/3089887.

Khanjani, M.H., 2026. Evaluating the performance of Pacific white shrimp (*Penaeus vannamei* Boone, 1931) reared under different stocking densities: a study of their biochemical, immune, metabolic and antioxidant responses in a biofloc aquaculture system. Ann. Anim. Sci. 26 (3), 813–823. doi:10.2478/aoas-2025-0087.

Khanjani, M.H., Eslami, J., Emerenciano, M.G.C., 2025. Wheat flour as carbon source on water quality, growth performance, hemolymph biochemical and immune parameters of Pacific white shrimp (Penaeus vannamei) juveniles in biofloc technology (BFT). Aquac. Rep. 40. doi:10.1016/j.aqrep.2025.102623.

Khanjani, M.H., Mohammadi, A., Emerenciano, M.G.C., 2024. Water quality in biofloc technology (BFT): an applied review for an evolving aquaculture. Aquac. Int. 32 (7), 9321–9374. doi:10.1007/s10499-024-01618-w.

Khanjani, M.H., Sharifinia, M., Emerenciano, M.G.C., 2024. Biofloc technology (BFT) in aquaculture: what goes right, what goes wrong? a scientific-based snapshot. Aquacult. Nutr. 2024. doi:10.1155/2024/7496572.

Khanjani, M.H., Zahedi, S., Mohammadi, A., 2022. Integrated multitrophic aquaculture (IMTA) as an environmentally friendly system for sustainable aquaculture: functionality, species, and application of biofloc technology (BFT). Environ. Sci. Pollut. Res. 29 (45), 67513–67531. doi:10.1007/s11356-022-22371-8.

Kimera, F., Sewilam, H., Fouad, W.M., Suloma, A., 2021. Efficient utilization of aquaculture effluents to maximize plant growth, yield, and essential oils composition of Origanum majorana cultivation. Ann. Agric. Sci. 66 (1), 1–7. doi:10.1016/j.aoas.2020.11.002.

Kleitou, P., Kletou, D., David, J., 2018. Is Europe ready for integrated multi-trophic aquaculture? A survey on the perspectives of European farmers and scientists with IMTA experience. Aquaculture 490, 136–148. doi:10.1016/j.aquaculture.2018.02.035.

Kristiansen, R., Cripps, S.J., 1996. Treatment of fish farm wastewater using sand filtration. J. Environ. Qual. 25 (3), 545–551. doi:10.2134/jeq1996.00472425002500030020x.

Laktuka, K., Kalnbalkite, A., Sniega, L., Logins, K., Lauka, D., 2023. Towards the sustainable intensification of aquaculture: exploring possible ways forward. Sustainability 15 (24), 16952. doi:10.3390/su152416952.

Lee, D.H., Kim, J.Y., Lim, S.R., Kim, D.Y., Kim, J.M., Shin, S.J., Kim, J.D., 2019. Effect of dietary monobasic potassium phosphate levels on water quality and the growth of far eastern catfish Silurus asotus and four leafy vegetables in a hybrid biofloc technology aquaponic system. Korean J. Fish. Aquat. Sci. 52, 159–172.

Limbu, S.M., Shoko, A.P., Lamtane, H.A., Kishe-Machumu, M.A., Joram, M.C., Mbonde, A.S., Mgana, H.F., Mgaya, Y.D., 2017. Fish polyculture system integrated with vegetable farming improves yield and economic benefits of small-scale farmers. Aquac. Res. 48 (7), 3631–3644. doi:10.1111/are.13188.

Liu, X.G., Shao, Z., Cheng, G., Lu, S., Gu, Z., Zhu, H., Shen, H., Wang, J., Chen, X., 2021. Ecological engineering in pond aquaculture: a review from the whole-process perspective in China. Rev. Aquac. 13 (2), 1060–1076. doi:10.1111/raq.12512.

Lu, Z., Lin, W., Li, Q., Wu, Q., Ren, Z., Mu, C., Wang, C., Shi, C.e, Ye, Y., 2024. Recirculating aquaculture system as microbial community and water quality management strategy in the larviculture of Scylla paramamosain. Water Res. 252, 121218. doi:10.1016/j.watres.2024.121218.

Martins, C.I.M., Eding, E.H., Verreth, J.A.J., 2011. The effect of recirculating aquaculture systems on the concentrations of heavy metals in culture water and tissues of Nile tilapia Oreochromis niloticus. Food Chem. 126 (3), 1001–1005. doi:10.1016/j.foodchem.2010.11.108.

Matias, R.S., Gomes, S., Barboza, L.G.A., Salazar-Gutierrez, D., Guilhermino, L., Valente, L.M.P., 2023. Microplastics in water, feed and tissues of European seabass reared in a recirculation aquaculture system (RAS). Chemosphere 335. doi:10.1016/j.chemosphere.2023.139055.

Mchunu, N., Lagerwall, G., Senzanje, A., 2018. Aquaponics in South Africa: results of a national survey. Aquac. Rep. 12, 12–19. doi:10.1016/j.aqrep.2018.08.001.

Meng, L.J., Hu, X., Wen, B., Liu, Y.H., Luo, G.Z., Gao, J.Z., Chen, Z.Z., 2023. Microplastics inhibit biofloc formation and alter microbial community composition and nitrogen transformation function in aquaculture. Sci. Total Environ. 866. doi:10.1016/j.scitotenv.2022.161362.

Mohapatra, B.C., Chandan, N.K., Panda, S.K., Majhi, D., Pillai, B.R., 2020. Design and development of a portable and streamlined nutrient film technique (NFT) aquaponic system. Aquacult. Eng. 90. doi:10.1016/j.aquaeng.2020.102100.

Moore, E., Ward, J., Lennard, W., 2020. End-of-pipe horticultural reuse of recirculating aquaculture system effluent: comparing the hydro-economics of two horticulture systems. Water 12 (5), 1409. doi:10.3390/w12051409.

de Morais, A.P.M., Santos, I.L., Carneiro, R.F.S., Routledge, E.A.B., Hayashi, L., de Lorenzo, M.A., Vieira, F.do N, 2023. Integrated multitrophic aquaculture system applied to shrimp, tilapia, and seaweed (Ulva ohnoi) using biofloc technology. Aquaculture 572. doi:10.1016/j.aquaculture.2023.739492.

Mota, V.C., Striberny, A., Verstege, G.C., Difford, G.F., Lazado, C.C., 2022. Evaluation of a recirculating aquaculture system research facility designed to address current knowledge needs in Atlantic Salmon production. Front. Anim. Sci. 3. doi:10.3389/fanim.2022.876504.

Mugwanya, M., Dawood, M.A.O., Kimera, F., Sewilam, H., 2022. A review on recirculating aquaculture system: influence of stocking density on fish and crustacean behavior, growth performance, and immunity. Ann. Anim. Sci. 22 (3), 873–884. doi:10.2478/aoas-2022-0014.

Mugwanya, M., Kimera, F., Madkour, K., Dawood, M.A.O., Sewilam, H., 2023. Infuence of salinity on the biometric traits of striped catfsh (Pangasianodon hypophthalmus) and barley (Hordeum vulgare) cultivated under an integrated aquaculture-agriculture system. BMC Plant Biol. 23, 417.

Nahon, S., de Brito, G.V., Quental-Ferreira, H., Aubin, J., Jaeger, C., Menniti, C., Kerhervé, P., Larroquet, L., Cunha, M.E., 2023. Food web in Mediterranean coastal integrated multi-trophic aquaculture ponds: learnings from fatty acids and stable isotope tracers. Aquaculture 567. doi:10.1016/j.aquaculture.2023.739292.

Nissar, S., Bakhtiyar, Y., Arafat, M.Y., Andrabi, S., Mir, Z.A., Khan, N.A., Langer, S., 2023. The evolution of integrated multi-trophic aquaculture in context of its design and components paving way to valorization via optimization and diversification. Aquaculture 565. doi:10.1016/j.aquaculture.2022.739074.

Okomoda, V.T., Oladimeji, S.A., Solomon, S.G., Olufeagba, S.O., Ogah, S.I., Ikhwanuddin, M., 2023. Aquaponics production system: a review of historical perspective, opportunities, and challenges of its adoption. Food Sci. Nutr. 11 (3), 1157–1165. doi:10.1002/fsn3.3154.

Omont, A., Elizondo-González, R., Quiroz-Guzmán, E., Escobedo-Fregoso, C., Hernández-Herrera, R., Peña-Rodríguez, A., 2020. Digestive microbiota of shrimp Penaeus vannamei and oyster Crassostrea gigas co-cultured in integrated multi-trophic aquaculture system. Aquaculture 521, 735059. doi:10.1016/j.aquaculture.2020.735059.

Palm, H.W., Knaus, U., Appelbaum, S., Goddek, S., Strauch, S.M., Vermeulen, T., Haïssam Jijakli, M., Kotzen, B., 2018. Towards commercial aquaponics: a review of systems, designs, scales and nomenclature. Aquac. Int. 26 (3), 813–842. doi:10.1007/s10499-018-0249-z.

Paolacci, S., Stejskal, V., Jansen, M.A.K., 2021. Estimation of the potential of Lemna minor for effluent remediation in integrated multi-trophic aquaculture using newly developed synthetic aquaculture wastewater. Aquac. Int. 29 (5), 2101–2118. doi:10.1007/s10499-021-00736-z.

Park, M., Shin, S.K., Do, Y.H., Yarish, C., Kim, J.K., 2018. Application of open water integrated multi-trophic aquaculture to intensive monoculture: a review of the current status and challenges in Korea. Aquaculture 497, 174–183. doi:10.1016/j.aquaculture.2018.07.051.

Serra, F.P, Wasielesky, W., Abreu, P.C, 2021. Nitrogen salt fertilization vs. substrate availability: two strategies to improve nitrification during the production of the white shrimp Litopenaeus vannamei. Aquaculture 543, 736997. doi:10.1016/j.aquaculture.2021.736997.

Phong, L.T., van Dam, A.A., Udo, H.M.J., van Mensvoort, M.E.F., Tri, L.Q., Steenstra, F.A., van der Zijpp, A.J., 2010. An agro-ecological evaluation of aquaculture integration into farming systems of the Mekong Delta. Agric. Ecosyst. Environ. 138 (3–4), 232–241. doi:10.1016/j.agee.2010.05.004.

Pinheiro, I., Carneiro, R.F.S., Vieira, F.d.N., Gonzaga, L.V., Fett, R., Costa, A.C.d.O., Magallón-Barajas, F.J., Seiffert, W.Q., 2020. Aquaponic production of Sarcocornia ambigua and Pacific white shrimp in biofloc system at different salinities. Aquaculture 519. doi:10.1016/j.aquaculture.2019.734918.

Pinheiro, I., Arantes, R., do Espírito Santo, C.M., do Nascimento Vieira, F., Lapa, K.R., Gonzaga, L.V., Fett, R., Barcelos-Oliveira, J.L., Seiffert, W.Q., 2017. Production of the halophyte Sarcocornia ambigua and Pacific white shrimp in an aquaponic system with biofloc technology. Ecol. Eng. 100, 261–267. doi:10.1016/j.ecoleng.2016.12.024.

Pinho, S.M., Flores, R.M.V., David, L.H., Emerenciano, M.G.C., Quagrainie, K.K., Portella, M.C., 2022. Economic comparison between conventional aquaponics and FLOCponics systems. Aquaculture 552. doi:10.1016/j.aquaculture.2022.737987.

Pinho, S.M., Molinari, D., de Mello, G.L., Fitzsimmons, K.M., Coelho Emerenciano, M.G., 2017. Effluent from a biofloc technology (BFT) tilapia culture on the aquaponics production of different lettuce varieties. Ecol. Eng. 103, 146–153. doi:10.1016/j.ecoleng.2017.03.009.

Pinho, S.M., de Lima, J.P., David, L.H., Emerenciano, M.G.C., Goddek, S., Verdegem, M.C.J., Keesman, K.J., Portella, M.C., 2022. FLOCponics: the integration of biofloc technology with plant production. Rev. Aquac. 14 (2), 647–675. doi:10.1111/raq.12617.

Poli, M.A., Legarda, E.C., de Lorenzo, M.A., Martins, M.A., do Nascimento Vieira, F., 2019. Pacific white shrimp and Nile tilapia integrated in a biofloc system under different fish-stocking densities. Aquaculture 498, 83–89. doi:10.1016/j.aquaculture.2018.08.045.

Pretty, J., 2008. Agricultural sustainability: Concepts, principles and evidence. Philos. Trans. R. Soc. B Biol. Sci. 363 (1491), 447–465. doi:10.1098/rstb.2007.2163.

Rakocy, J.E., 2012. Aquaponics—Integrating Fish and Plant Culture. In Aquaculture Production Systems. Wiley, Hoboken, NJ, USA, pp. 344–386. doi:10.1002/9781118250105.ch14.

Ridler, N., Wowchuk, M., Robinson, B., Barrington, K., Chopin, T., Robinson, S., Page, F., Reid, G., Szemerda, M., Sewuster, J., Boyne-Travis, S., 2007. Integrated multi-trophic aquaculture (IMTA): a potential strategic choice for farmers. Aquac. Econ. Manag. 11 (1), 99–110. doi:10.1080/13657300701202767.

Santos, F.A.C., Costa, Julio, G.S., Luz, R.K., 2020. Stocking density in Colossoma macropomum larviculture, a freshwater fish, in recirculating aquaculture system. Aquac.Res 1–7. doi:10.1111/are.14976.

Sanz-Lazaro, C., Sanchez-Jerez, P., 2020. Regional integrated multi-trophic aquaculture (RIMTA): spatially separated, ecologically linked. J. Environ. Manage. 271, 110921. doi:10.1016/j.jenvman.2020.110921.

Saseendran, S., Dube, K., Chandrakant, M.H., Babitha Rani, A.M., 2021. Enhanced growth response and stress mitigation of genetically improved farmed Tilapia in a biofloc integrated aquaponic system with bell pepper. Aquaculture 533, 736200. doi:10.1016/j.aquaculture.2020.736200.

Seo, J., Park, J., 2023. Does stocking density affect growth Performance and hematological parameters of juvenile olive Flounder Paralichthys olivaceus in a recirculating aquaculture system? Animals 13 (1), 44. doi:10.3390/ani13010044.

Sharmin, S., Islam, M.S., Hasan, M.K., 2012. Socioeconomic analysis of alternative farming systems in improving livelihood security of small farmers in selected areas of Bangladesh. Agriculturists 10 (1), 51–63. doi:10.3329/agric.v10i1.11065.

Shoko, A.P., Limbu, S.M., Lamtane, H.A., Kishe-Machumu, M.A., Sekadende, B., Ulotu, E.E., Masanja, J.C., Mgaya, Y.D., 2019. The role of fish-poultry integration on fish growth performance, yields and economic benefits among smallholder farmers in sub-Saharan Africa, Tanzania. Afr. J. Aquat. Sci. 44 (1), 15–24. doi:10.2989/16085914.2018.1555512.

da Silva, H.V., Martins, M.A., do Espírito Santo, C.M., do Nascimento Vieira, F., Rezende, P.C., Gonzaga, L.V., Fett, R., Seiffert, W.Q., 2022. Aquaponic production of sea asparagus and Pacific white shrimp using biofloc technology: different irrigation regimes affect plant production of bioactive compounds and antioxidant capacity. Aquac. Res. 53 (3), 1001–1010. doi:10.1111/are.15641.

Soliman, A.I.E., Morad, M.M., Wasfy, K.I., Moursy, M.A.M., 2020. Maximize water productivity using aquaculture water for sesame crop under drip irrigation systems. Zagazig J. Agric. Res. 47 (4), 989–998. doi:10.21608/zjar.2020.110327.

Su, M.H., Azwar, E., Yang, Y.F., Sonne, C., Yek, P.N.Y., Liew, R.K., Cheng, C.K., Show, P.L., Lam, S.S., 2020. Simultaneous removal of toxic ammonia and lettuce cultivation in aquaponic system using microwave pyrolysis biochar. J. Hazard. Mater. 396. doi:10.1016/j.jhazmat.2020.122610.

Taha, M.F., ElMasry, G., Gouda, M., Zhou, L., Liang, N., Abdalla, A., Rousseau, D., Qiu, Z., 2022. Recent advances of smart systems and Internet of Things (IoT) for aquaponics automation: a comprehensive overview. Chemosensors 10 (8). doi:10.3390/chemosensors10080303.

Thomas, W., Tierney, M.S., Ray, A.J., 2018. Comparing biofloc, clear-water and hybrid RAS systems as shrimp nurseries. Global Aquac. Advocate. https://www.aquaculturealliance.org/advocate/comparing-biofloc-clear-water-hybrid-ras-systems-shrimp-nurseries/?headlessPrint=AAAAAPIA9c8r7.

Timmons, M.B., Guerdat, T., Vinci, B.J., 2018. Recirculating aquaculture, 4th ed. Ithaca Publishing Company LLC, Ithaca, NY, p. 775.

Timmons, M.B., Ebeling, J.M., 2010. Recirculating aquaculture systems, 2nd ed. Cayuga Aqua Ventures LLC, Ithaca, NY, p. 939.

Verdegem, M., Buschmann, A.H., Latt, U.W., Dalsgaard, A.J.T., Lovatelli, A., 2023. The contribution of aquaculture systems to global aquaculture production. J. World Aquac. Soc. 54 (2), 206–250. doi:10.1111/jwas.12963.

Wambua, D.M., Home, P.G., Raude, J.M., Ondimu, S., 2021. Environmental and energy requirements for different production biomass of Nile tilapia (Oreochromis niloticus) in recirculating aquaculture systems (RAS) in Kenya. Aquac. Fish. 6 (6), 593–600. doi:10.1016/j.aaf.2020.07.019.

Xue, B., Zhao, Y., Bi, C., Cheng, Y., Ren, X., Liu, Y., 2022. Investigation of flow field and pollutant particle distribution in the aquaculture tank for fish farming based on computational fluid dynamics. Comput. Electron. Agric. 200, 107243. doi:10.1016/j.compag.2022.107243.

Yanes, A.R., Martinez, P., Ahmad, R., 2020. Towards automated aquaponics: a review on monitoring, IoT, and smart systems. J. Cleaner Prod. 263, 121571. doi:10.1016/j.jclepro.2020.121571.

Yin, G., Ong, M.C, Lee, J., Kim, T., 2021. Numerical simulation of oxygen transport in land-based aquaculture tank. Aquaculture 543, 736973. doi:10.1016/j.aquaculture.2021.736973.

Yu, Y.B., Choi, J.H., Lee, J.H., Jo, A.H., Han, S.W., Han, S.H., Choi, H.J., Choi, C.Y., Kang, J.C., Min, E.Y., Kim, J.H., 2023. Biofloc application using aquaponics and vertical aquaculture technology in aquaculture: review. Fishes 8 (11). doi:10.3390/fishes8110543.

Zhang, S., Guo, Y., Li, S., Ke, Z., Zhao, H., Yang, J., Wang, Y., Li, D., Wang, L., Yang, W., Zhang, Z., 2022. Investigation on environment monitoring system for a combination of hydroponics and aquaculture in greenhouse. Inf. Process. Agric. 9 (1), 123–134. doi:10.1016/j.inpa.2021.06.006.

Zhang, Y., Zhang, Y., Li, Z., 2022. A new and improved aquaponics system model for food production patterns for urban architecture. J. Cleaner Prod. 342, 130867. doi:10.1016/j.jclepro.2022.130867.

Zhao, R., Symonds, J.E., Walker, S.P., Steiner, K., Carter, C.G., Bowman, J.P., Nowak, B.F., 2021. Effects of feed ration and temperature on Chinook salmon (Oncorhynchus tshawytscha) microbiota in freshwater recirculating aquaculture systems. Aquaculture 543. doi:10.1016/j.aquaculture.2021.736965.

Zhao, Y.P., Xue, b., Bi, C., Ren, X., Liu, Y., 2022. Influence mechanisms of macro infrastructure on micro-environments in the recirculating aquaculture system and biofloc technology system. Rev. Aquac. 15 (3), 991.

Zhao, Y.X., Liu, X.G., Lu, M., Zhou, R.F., Sun, Z.Y., Xiao, S.W., 2022. Characteristics of bacterial community in Pelteobagrus fulvidraco integrated multi-trophic aquaculture system. Water 14 (20). doi:10.3390/w14203192.

Zhu, Z., Yogev, U., Keesman, K.J., Rachmilevitch, S., Gross, A., 2023. Integrated hydroponics systems with anaerobic supernatant and aquaculture effluent in desert regions: nutrient recovery and benefit analysis. Sci. Total Environ. 904, 166867. doi:10.1016/j.scitotenv.2023.166867.

CHAPTER

8

Global solutions for sustainable aquaculture: Valorizing solid waste through insect meal production

Moslem Sharifinia[a] *and Mohammad Hossein Khanjani*[b]

[a]Aquaculture Department, Shrimp Research Center, Iranian Fisheries Sciences Research Institute, Agricultural Research, Education and Extension Organization (AREEO), Bushehr, Iran [b]Department of Fisheries Sciences and Engineering, Faculty of Natural Resources, University of Jiroft, Jiroft, Iran

8.1 Introduction

As the global population steadily increases, with projections suggesting it could reach nearly 10 billion by 2050, the urgency for sustainable food sources has become more pronounced. This rising population is driving a demand for food that traditional farming methods may struggle to meet. In fact, estimates indicate that global food demand could surge by 35%–56% by 2050, highlighting the critical need for innovative and sustainable agricultural practices (Berners-Lee et al., 2018; Van Dijk et al., 2021). Consequently, there is a pressing need to stabilize food production and enhance food security through sustainable practices, ensuring that resources are preserved for future generations. Aquaculture, which encompasses the farming of fish, shellfish, and other aquatic species, is vital to the global food supply, currently accounting for more than 50% of seafood consumed worldwide (FAO, 2024; Sharifinia, 2025). Fish is recognized as one of the richest sources of protein in our diets, with global fish production having quadrupled in the last 50 years (Hannah et al., 2021). However, the rapid expansion of aquaculture has introduced a range of challenges, including environmental degradation, overfishing of wild stocks, reliance on unsustainable feed sources, and significant organic waste generation (Yeganeh et al., 2020; Sharifinia, 2024). These issues pose significant risks to the sustainability of aquaculture and the ecosystems that support it.

Traditionally, aquaculture has relied on fishmeal and fish oil sourced from wild fish, which places additional strain on marine resources and contributes to the decline of fish populations (Majluf et al., 2024). Furthermore, aquaculture operations produce considerable solid waste,

Zero-Waste Technologies in Aquaculture: Global Perspectives and Methods toward a Circular Economy
DOI: https://doi.org/10.1016/B978-0-443-34689-7.00014-8

including leftover feed, feces, and dead organisms. If not managed properly, this waste can lead to poor water quality, nutrient pollution, and other negative environmental impacts (Dauda et al., 2019). Recently, there has been a growing interest in recycling waste generated from food processing as a substitute for fishmeal in aquaculture (Ma et al., 2019; Nasr et al., 2021). Therefore utilizing protein-rich food waste for aquaculture feeds presents a practical solution for recycling waste within a circular economy.

One innovative strategy to enhance the sustainability of aquaculture is the incorporation of insect meal (Sharifinia et al., 2023a, 2023b), which transforms solid waste into valuable protein (Hameed et al., 2022; Sampathkumar et al., 2023). Insects, such as black soldier flies (BSFs), mealworms (MWs), and crickets, are highly effective at converting organic waste into high-quality protein and fats. Insect farming, which involves feeding insects with waste, is gaining traction as a viable approach (Hameed et al., 2022; Sampathkumar et al., 2023). The bioconversion of insects addresses the drawbacks of directly feeding food waste to fish and enhances the nutritional quality of feeds (Rumpold and Schlüter, 2013; Sánchez et al., 2021). These insects can thrive on various organic materials, including agricultural byproducts, food scraps, and aquaculture waste, making them well-suited for waste recycling (Hameed et al., 2022; Sampathkumar et al., 2023).

Insect meal also provides significant nutritional advantages. It is abundant in essential amino acids, fatty acids, vitamins, and minerals, allowing it to effectively replace traditional feed ingredients such as fishmeal and soybean meal without compromising the health and growth of farmed fish (Ameixa et al., 2020; Sharifinia et al., 2023b; Fantatto et al., 2024). In addition, incorporating insect meal into aquaculture diets can improve the nutritional quality of fish and shellfish, thereby contributing to enhanced food security and public health. Moreover, insect farming is environmentally friendly (Huis and Oonincx, 2017). Unlike traditional livestock, insect cultivation typically requires less space, allowing for higher-density farming in limited areas, which results in lower greenhouse gas emissions (Sharifinia, 2024).

From an economic standpoint, insect meal production offers a promising alternative to conventional feed sources (Sánchez-Muros et al., 2014; Auzins et al., 2024). As the demand for sustainable aquaculture practices grows, the market for insect-based feeds is expanding, creating new opportunities for farmers and entrepreneurs. Insect farming can also create jobs and stimulate local economies, particularly in regions rich in agricultural waste (Chia et al., 2019; Cadinu et al., 2020). Most importantly, insects are omnivorous and can consume a wide range of diets, including industrial food waste (Rumpold and Schlüter, 2013), making them a sustainable option. Despite these benefits, the integration of insect meal into aquaculture is not without challenges. Regulatory frameworks concerning the use of insects in animal feed differ significantly across countries, and ensuring the safety and quality of insect-based products is essential (Van der Fels-Klerx et al., 2018; Okyere, 2023). In addition, consumer acceptance of insect meal as a feed ingredient is a critical factor for its broader adoption (Sogari et al., 2019; Pakseresht et al., 2023). The environmental and economic impacts, along with the future challenges of insect farming and bioconversion, are presented in Fig. 8.1.

This chapter aims to explore the bioconversion potential of various insect species in transforming organic waste into high-quality protein and fats for aquafeed production. It seeks to evaluate how the utilization of insects for waste bioconversion can address current waste management challenges, reduce environmental impact, and contribute to the establishment of a circular economy. To optimize production techniques, the chapter will identify and promote

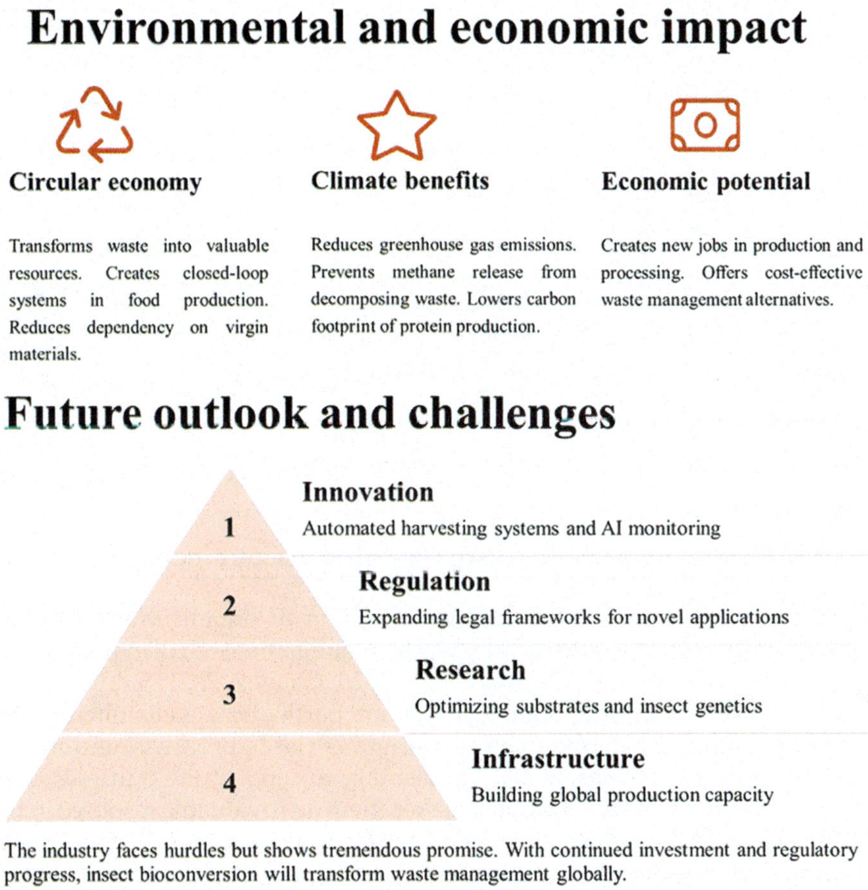

FIGURE 8.1 Environmental and economic impacts, along with the future challenges of insect farming and bioconversion.

innovative methods and technologies that enhance the efficiency and scalability of insect farming systems, including automated farming and advanced bioprocessing techniques. Furthermore, it will examine the nutritional value of insect meals and explore strategies for optimizing their composition, such as enhancing essential fatty acids and managing chitin content to meet the dietary needs of aquatic species. The chapter will also address regulatory frameworks and economic considerations impacting the integration of insect meal into mainstream aquaculture practices, proposing solutions to facilitate acceptance and implementation. By advocating for the adoption of insect-based bioconversion systems, the chapter aims to foster environmental sustainability, resource efficiency, and resilience in food production systems globally. Finally, it will highlight the need for ongoing research to further explore the potential of insect-based feeds, including the evaluation of optimal inclusion levels in aquafeeds and the exploration of various insect parts for diverse applications in aquaculture.

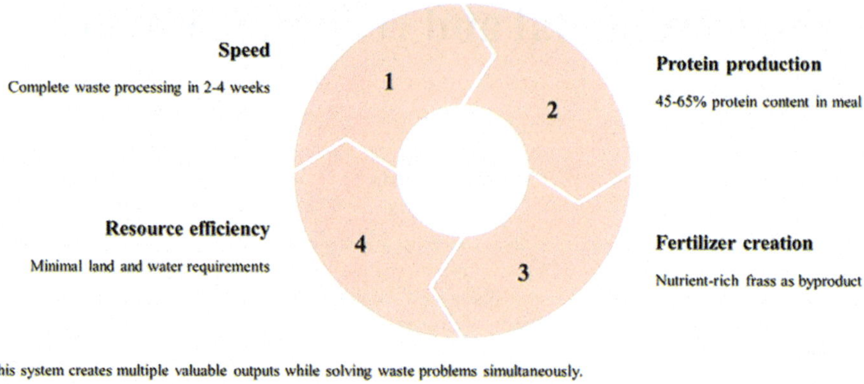

FIGURE 8.2 **Benefits of insect-based waste management.**

8.2 Exploring organic waste types for insect bioconversion

The utilization of insect larvae for the bioconversion of organic waste has emerged as a highly effective strategy for addressing waste management challenges and promoting sustainable food production (Fig. 8.2) (Ravi et al., 2020; Chavez, 2021; Siddiqui et al., 2024b). Insects that thrive on decomposing organic matter are particularly well-suited for biorefinery applications, waste valorization, the treatment of industrial byproducts, and the conversion of agroindustrial residues (Fowles, 2021). Their ability to concentrate nutrients from various waste materials, even at low concentrations, makes them an invaluable resource in this context (Ravi et al., 2020). Notable species such as BSFs (*Hermetia illucens*), MWs (*T. molitor*), and several cricket species are adept at transforming diverse organic waste streams into high-value products, including protein-rich biomass, organic fertilizers, and biofuels. The bioconversion process yields a variety of valuable byproducts, such as proteins, lipids, chitin, and frass, which contribute to the nutritional upcycling of waste biomass (Ravi et al., 2020; Triunfo et al., 2022; Bohm et al., 2023). Identifying the types of organic waste that are most conducive to insect bioconversion is essential for optimizing this process and maximizing its advantages.

Agricultural waste represents one of the most significant sources of organic material suitable for insect bioconversion. This category includes crop residues, such as vegetable and fruit waste, which consists of byproducts from the processing of fruits and vegetables, including peels, cores, and overripe produce. These materials serve as nutrient-dense substrates for insect larvae, promoting their growth and development. Furthermore, byproducts from cereal processing, such as bran and husks, can enhance the nutritional value of the larvae, making them an effective feed source. In addition, residues from legumes, including soybeans and peas, are rich in protein and can further support the growth of insect larvae, highlighting the potential of agricultural waste in this domain. The integration of agricultural waste into insect farming not only reduces environmental pollution but also contributes to the circular economy by closing the nutrient loop (Ravi et al., 2020; Chavez, 2021; Fowles, 2021; Siddiqui et al., 2024b).

Food waste generated by households, restaurants, and food processing industries also presents a significant opportunity for bioconversion. The conversion of food waste by insects is particularly promising, as it may lead to innovative approaches for achieving environmental and economic sustainability within the food industry. This process can help ensure the ongoing availability of safe and nutritious food for all, as edible insects are increasingly recognized as novel, nutritious, functional, and safe food alternatives (Salomone et al., 2017; Ravi et al., 2020). Insects can effectively utilize spoiled or expired food items, such as dairy products, bread, and cooked meals, transforming what would otherwise be waste into a valuable resource (Barry, 2004; Makkar et al., 2014; Abd El-Hack et al., 2020). In addition, food processing waste, including pulp from juice extraction and seeds from fruit processing, provides further nutrient sources for insect larvae (Isibika et al., 2019; Ravi et al., 2020; Siddiqui et al., 2024b). This not only aids in minimizing food waste but also contributes to a more sustainable food system. The incorporation of food waste into insect bioconversion systems can significantly reduce greenhouse gas emissions associated with traditional waste disposal methods, such as landfilling and incineration.

Animal manure is another category of organic waste amenable to insect conversion, offering a rich nutrient source that fosters the growth of insect larvae. Poultry manure, in particular, is advantageous because of its high nitrogen content, which can accelerate the growth rates of insect larvae. Similarly, cattle and swine manure can be utilized, although they may require preprocessing to mitigate pathogens and enhance digestibility for insects (Surendra et al., 2016; Xiao et al., 2018; van Huis, 2019). The incorporation of animal manure into bioconversion processes can significantly improve the efficiency and productivity of insect farming. Research indicates that yellow MWs (*T. molitor*) exhibit suboptimal growth and waste conversion rates when fed a starch-rich diet, suggesting that such a diet may lack essential nutrients (Liu et al., 2020). This highlights the importance of formulating balanced diets for insect larvae to optimize their growth and bioconversion efficiency.

Organic industrial waste from sectors such as breweries, distilleries, and food manufacturing also offers valuable prospects for insect bioconversion (Siddiqui et al., 2022; Biteau et al., 2024; Siddiqui et al., 2024a; Kullan et al., 2025). Byproducts from brewing, such as spent grains, are nutrient-rich and can serve as excellent feed for insects. In addition, waste from fruit and vegetable processing often includes substantial amounts of peels, seeds, and cores, which can be effectively repurposed for insect larvae, further reducing waste and enhancing sustainability. The potential for insects to utilize industrial byproducts not only contributes to waste reduction but also provides a cost-effective feed source for insect farming. Contrary to the common belief that urbanization and food production are incompatible, insect-based bioconversion challenges this notion by facilitating food and feed cultivation in densely populated urban areas. For instance, various insect species, including *H. illucens*, *L. migratoria*, and *T. molitor*, can currently achieve the biodegradation of organic waste (Surendra et al., 2020). This urban farming model can enhance local food security and reduce the carbon footprint associated with transporting food over long distances.

Lastly, yard waste, such as grass clippings, leaves, and garden trimmings, is also suitable for insect bioconversion (Barry, 2004; Siddiqui et al., 2024a). Although these materials are typically high in cellulose, certain insect species can effectively decompose fibrous plant material, especially when combined with other nutrient-rich organic waste. The biodegradation mechanisms in larvae of BSFs (*H. illucens*), yellow MWs (*T. molitor*), and dark MWs (*Tenebrio obscurus*) can be influenced by various factors, including temperature, nutrient composition, and photoperiod (Ribeiro et al., 2018). This integration of diverse organic waste types not only enhances the efficiency of the bioconversion process but also contributes to a more comprehensive approach to waste management. The nutrient content, development, and dietary supplementation of these insects can significantly impact their survival, growth rates, and overall performance, which can also affect fertility, egg yield, and quality in broiler chickens (Yang et al., 2023).

8.3 Insect species for waste bioconversion and aquafeed production

Insects are recognized as the most diverse group of animals on the planet and play a crucial role as a natural food source for various fish species, particularly those that are carnivorous or omnivorous and require diets rich in protein (Nogales-Mérida et al., 2019; Alfiko et al., 2022; Zou et al., 2024). Historically, humans have incorporated insects into their diets for more than two millennia across numerous cultures (Shockley and Dossey, 2014; Lesnik, 2019). However, the integration of insects into aquaculture feeds has gained momentum only in the past decades (Mitra, 2021; Fantatto et al., 2024). This shift has led to significant advancements in the selection of insect species, cultivation techniques, nutritional assessments, and practical field applications. Currently, the production of insect meal is rapidly progressing in regions such as China, Europe, North America, Australia, and Southeast Asia, with at least 16 insect species evaluated as potential alternative protein sources in aquafeeds (Henry et al., 2015; Nogales-Mérida et al., 2019).

Among the various insect species studied for their potential in aquafeed production, eight have emerged as particularly promising: Silkworms (*Bombyx mori*), BSFs (*H. illucens*), houseflies (*Musca domestica*), MWs (*Tenebrio molitor*), lesser MWs (*Alphitobius diaperinus*), house crickets (*Acheta domesticus*), banded crickets (*Gryllodes sigillatus*), and Jamaican field crickets (*Gryllus assimilis*). These species have undergone extensive research regarding their ability to replace traditional fish meal in aquafeeds and have received approval for feed production under European Union regulations (Daniel, 2018). Comprehensive information about the biology and cultivation techniques of these insects is available in various scientific publications (Barroso et al., 2014; Henry et al., 2015; Belghit et al., 2018; Ameixa et al., 2020; Chavez, 2021; Hawkey et al., 2021; Alfiko et al., 2022; Hameed et al., 2022; Auzins et al., 2024).

The recent focus on utilizing insect species for bioconversion has gained traction due to their remarkable ability to transform organic waste into valuable resources. Notably, BSFs, MWs, and several cricket species demonstrate unique advantages in this bioconversion process. As the demand for sustainable food and feed production intensifies—coupled with growing environmental, economic, public health, and food safety concerns—there is an urgent need for innovative, economically viable, and environmentally friendly strategies to enhance food production while effectively managing organic waste (van der Wiel et al., 2019).

The nutritional profiles of the eight insect species utilized in aquafeeds have been thoroughly analyzed, focusing on parameters such as crude protein content, amino acid profiles, fat content, fatty acid compositions, and mineral content (Sánchez-Muros et al., 2014). The findings indicate that crude protein levels in these species range from 42.1% to 63.3%. Although these levels are lower than those found in fish meal, they are comparable to soybean meal (Henry et al., 2015; Allegretti et al., 2017). Notably, the larvae of BSFs and houseflies exhibit higher crude protein levels than the adult pupae of silkworms, locusts, and crickets. The amino acid compositions vary among the eight species; for instance, Orthopterans such as locusts and crickets, as well as MWs, typically contain lower levels of lysine compared to fish meal. In contrast, Dipterans such as BSFs and houseflies, along with silkworms, are richer in lysine. Except for silkworms, the sulfur amino acids found in these insects are generally lower than those in fish meal. Threonine levels are consistent across seven insect species, with silkworms exhibiting higher concentrations. Tryptophan levels are typically lower in six species, with the exception of silkworms and housefly maggot meal. To meet the specific growth requirements of various fish species, supplementation with synthetic amino acids may be necessary. Silkworms, BSFs, and houseflies demonstrate superior amino acid profiles compared with soybean meal, making them more suitable alternatives for replacing fish meal in aquafeeds (Sánchez-Muros et al., 2014; Henry et al., 2015).

The fat content in these insect species is generally lower than that of fish meal, with variations ranging from approximately 8% in mature locusts to about 36% in MW larvae (Surendra et al., 2020). Factors such as developmental stage and diet significantly influence fat accumulation in insects. While fish oil is rich in omega-3 fatty acids, insect meals tend to have higher concentrations of saturated fatty acids. Notably, unsaturated fatty acid concentrations are elevated in house crickets, MWs, and housefly maggot meals, while BSF larvae exhibit lower unsaturated fatty acid content. Insect meals typically contain higher levels of polyunsaturated fatty acids (PUFAs), particularly *n*-6 PUFAs, but lower levels of eicosapentaenoic acid and docosahexaenoic acid compared with fish oil (Gasco et al., 2020). The limited presence of these essential fatty acids in these insect species restricts their effectiveness as oil sources in aquafeeds.

In addition to the well-established species, other insects such as waxworms (*Galleria mellonella*) and various beetle larvae are being explored for their potential in bioconversion. For instance, waxworms have shown the ability to degrade polyethylene, a common plastic waste, highlighting the diverse capabilities of insects in addressing environmental challenges. This unique characteristic not only emphasizes the potential of insects in waste management but also opens new avenues for research into bioremediation strategies aimed at mitigating plastic pollution. The versatility of these insects in converting a wide range of organic materials into valuable products underscores their adaptability in various waste management contexts.

The ash content in these insect species is generally low, with the exception of BSF larvae, which contain over 15% ash. BSF larvae also have relatively high calcium levels, while other species exhibit lower calcium concentrations. When substituting fish meal with insect meals, it is crucial to supplement calcium in aquafeeds. The calcium content of insect larvae meals can be enhanced by fortifying the rearing substrate with calcium. The calcium-to-phosphorus ratios in BSF larvae are higher than in other species, which are lower than the recommended nutritional levels for fish (Sánchez-Muros et al., 2014; Henry et al., 2015; Allegretti et al., 2017). The carbohydrate levels in these eight insect species are generally low, typically below 20%. Insect meals also contain chitin, a polymer of glucosamine found in their exoskeletons. The chitin content varies among species and developmental stages, with reports indicating both high and low values. While most fish species cannot absorb chitin, some possess chitinase activity, enabling them to utilize feeds containing chitin. The addition of small amounts of chitin in aquafeeds has shown benefits, although it is generally considered a limiting factor for the application of insect meals in aquafeeds (Ng et al., 2001; Sánchez-Muros et al., 2014).

The selection of insect species for bioconversion largely depends on the type of organic waste available and the desired end products. Factors such as growth rate, nutrient requirements, and environmental conditions are critical in determining the suitability of specific insect species for particular waste streams. For example, BSF larvae thrive in warm, humid environments, making them particularly well-suited for tropical and subtropical regions, while MWs can be cultivated in cooler climates. The effects of light, CO_2 levels, organic matter, and minerals on an insect's performance, physical defenses, body mass, growth, feeding rate, and efficiency can vary depending on the species, developmental stage, and rearing conditions. Ongoing research into the nutritional profiles and metabolic pathways of these insects is essential for optimizing their performance in bioconversion processes. Studies have demonstrated that variations in the diets of insect larvae can significantly influence their growth rates, nutrient composition, and overall efficiency in converting waste into biomass (Dobermann et al., 2017).

Table 8.1 presents a comprehensive analysis of nine insect species and two reference feed sources (fishmeal and soymeal) utilized in organic waste conversion, highlighting their optimal growing conditions, bioconversion efficiencies, and nutritional content. The data reveal that cricket species (*G. sigillatus* and *G. assimilis*) boast the highest protein content, with values of 63.3% and 59.8%, respectively. In contrast, MW species (*T. molitor* and *A. diaperinus*) exhibit strong bioconversion capabilities, making them effective for waste conversion. The BSF (*H. illucens*) demonstrates remarkable versatility across various substrates, achieving prepupation rates between 45% and 78% on animal manure and bioconversion rates of 4.6%–7.1% on mixed organic waste. Meanwhile, the housefly (*M. domestica*) achieves impressive waste reduction rates of 79.1%–83.6%, thriving under optimal conditions of 25–30°C and a pH range of 5–7. Nutritional content across the species varies, with protein levels ranging from 42.1% to 63.3% and lipid content between 9.9% and 26.0%. Notably, most insect species maintain protein levels that are comparable to or even exceed those of traditional feed sources, such as fishmeal (55.5%) and soymeal (51.8%). These data underscore the potential of these insect species as viable alternatives in sustainable aquaculture and waste management strategies.

8.4 Mealworm (*Tenebrio Molitor*)

MWs (*T. Molitor*) have emerged as a promising alternative for sustainable waste bioconversion and aquafeed production. Recognized by the European Commission as a novel food source, these larvae are among the most widely cultivated insects globally (Sangiorgio et al., 2022). They excel at transforming organic waste, such as grains, fruits, and vegetables, into high-quality protein, making them a viable substitute for traditional aquaculture feed ingredients. Particularly adept at utilizing low-quality feedstocks such as food waste and agricultural residues, MWs contribute to waste reduction while generating nutrient-rich biomass. In addition, they play a critical role in recycling nutrients back into the food chain, supporting the principles of a circular economy. Research indicates that MWs can bioaccumulate specific nutrients and beneficial compounds from their diets, enhancing their nutritional value as a food source.

One of the key advantages of MWs is their ability to thrive on diverse organic substrates, including food scraps and agricultural byproducts. Incorporating food waste as feed can

TABLE 8.1 Nutritional composition and bioconversion efficiency of various insect species (Alfiko et al., 2022; Siddiqui et al., 2024a).

Insect species	Organic waste substrate	Optimal conditions	Bioconversion efficiency	Key nutritional content	Notable outcomes
Hermetia illucens	Animal manure (dairy, swine, poultry)	Temperature 27–28°C, Humidity 70%–80%	45%–78% prepupation rate	Protein: 42.1%, Lipid: 26.0%	Peak growth at days 6–7, weight 150–170 mg
Hermetia illucens	Mixed organic waste (kitchen, agricultural, aquaculture)	Temperature 27.1°C ± 0.2, Humidity 76.4% ± 0.3	Bioconversion: 4.6%–7.1%	Protein: 31.6%–56.5%	Highest growth on mixed substrates
Musca domestica	Food waste with adjuvants	Temperature 25–30°C, pH 5–7	Waste reduction: 79.1%–83.6%	Protein: 50.4%, Lipid: 18.9%	Optimal with wheat bran adjuvant
Tenebrio molitor	Wheat bran, vegetable waste	Temperature 22.5 ± 2.5°C, Humidity 60±4%	Bioconversion: 8.13%	Protein: 52.8%	Highest efficiency on wheat bran
Drosophila suzukii	Coconut fiber + yeast	Standard laboratory conditions	Bioconversion: 3%–6%	NS	Best results with coconut fiber + Brewer's yeast
Acheta domesticus	Brewery waste, mixed substrates	Standard laboratory conditions	Conversion efficiency: 0.66 ± 0.99	NS	Maximum weight on GM+HF mixture
Alphitobius diaperinus	Various organic substrates	Standard laboratory conditions	NS	Protein: 57.3%, Lipid: 17.3%	High protein content
Gryllodes sigillatus	Various organic substrates	Standard laboratory conditions	NS	Protein: 63.3%, Lipid: 13.3%	Highest protein content among insects
Gryllus assimilis	Various organic substrates	Standard laboratory conditions	NS	Protein: 59.8%, Lipid: 11.8%	High protein efficiency
Zophobas morio	Various organic substrates	Standard laboratory conditions	NS	Protein: 55.5%, Lipid: 9.9%	Comparable to fishmeal
Fishmeal	—	—	—	Protein: 55.5%, Lipid: 9.9%	—
Soymeal	—	—	—	Protein: 51.8%, Lipid: 2.0%	—

significantly lower production costs and mitigate environmental impacts (Sangiorgio et al., 2022). This versatility enables efficient nutrient recycling, allowing MWs to convert organic materials into biomass rich in protein, fats, and essential amino acids, making them an excellent feed option for various aquaculture species.

Feeding trials with multiple aquaculture species have confirmed that both fresh and dried MWs serve as acceptable alternative protein sources. For instance, research on African catfish (*Clarias gariepinus*) showed that diets with up to 40% MW replacement resulted in growth and feed utilization efficiency comparable to control diets without MWs (Ng et al., 2001). Similarly, a 90-day trial with common catfish (*Ameiurus melas*) demonstrated that replacing half of the fishmeal with MW maintained growth rates, although fish on fishmeal diets exhibited superior growth (Roncarati et al., 2015).

The bioconversion process involves MWs ingesting organic waste and utilizing their digestive enzymes and gut microbiota to break down complex compounds. Studies have shown that MW larvae fed agroindustrial byproducts, such as rapeseed meal and chicken feed, exhibit improved feed conversion and efficiency, highlighting the effectiveness of these dietary combinations for insect rearing (Bordiean et al., 2022). In addition, the metabolic rates and nutritional profiles of MWs can adapt based on rearing conditions, optimizing growth and feed conversion efficiency (Kröncke and Benning, 2023).

In aquafeed production, MWs offer several advantages over conventional ingredients such as fishmeal and soybean meal. They contain approximately 40%–50% protein by dry weight, along with essential fatty acids and micronutrients. For example, in juvenile yellow catfish (*Petteobagrus fulvidraco*), diets incorporating MW meal as a partial substitute for fishmeal showed no negative effects on growth performance, while including 18% MW meal enhanced immune response and resistance to bacterial infections (Su et al., 2017). Trials with gilthead sea bream (*Sparus aurata*) indicated that a 25% substitution of fishmeal with MW meal did not adversely affect growth, whereas a 50% substitution led to reduced growth rates and lower feed conversion ratios (FCRs) (Piccolo et al., 2014).

Moreover, MW production has a relatively low environmental impact, requiring less land and water compared to traditional livestock farming and generating fewer greenhouse gas emissions. The closed-loop system of utilizing organic waste as feed not only aids in waste reduction but also conserves resources, making MW production an environmentally sustainable alternative. However, despite their versatility and potential as a nutritional substitute, legal and behavioral barriers have hindered the widespread adoption of MWs in aquaculture (Moruzzo et al., 2021).

8.5 Black soldier fly (*Hermetia illucens*)

The BSF (*H. illucens*) is a widely distributed species, particularly in tropical climates, known for its significant contributions to waste management and aquaculture (Mohan et al., 2022; Rehman et al., 2023). Its larvae effectively convert organic waste into high-quality protein and fats. Notably, adult BSF do not feed or bite, posing no health risks to humans (Silva and Hesselberg, 2020). The larvae excel at consuming decomposing organic matter, making them a valuable protein source for various animal species (Bonelli et al., 2019). BSF larvae thrive on a diverse range of organic materials, including food scraps and animal byproducts (Sheppard et al., 2002), highlighting their potential to enhance sustainable waste management practices (Ee et al., 2022).

The life cycle of *H. illucens* is characterized by rapid growth and impressive feed conversion rates, with larvae maturing in approximately 3 weeks (Tomberlin et al., 2002). At the prepupal stage, they instinctively migrate to clean areas, simplifying the harvesting process in farming operations (Sheppard et al., 2002). This behavior, combined with their capacity to consume large volumes of organic waste, positions BSF larvae as effective agents for recycling organic materials, reducing landfill waste, and minimizing the environmental impact of organic waste disposal.

Research indicates that BSF larvae can contain up to 41.76% protein on a dry weight basis, along with beneficial fatty acids, enhancing their nutritional profile for aquaculture feeds (Koutsos et al., 2022). Numerous feeding trials across various aquaculture species have validated the feasibility of integrating BSF larvae into fish diets. For instance, an 8-week study with juvenile yellow catfish (*Pelteobagrus fulvidraco*) demonstrated that substituting 20% of fishmeal with BSF meal did not adversely affect growth metrics, including weight gain and FCRs (Hu et al., 2017). Similarly, a 59-day trial with juvenile Jian carp (*Cyprinus carpio*) found that dietary defatted BSF meal did not negatively impact growth while enhancing antioxidant levels (Li et al., 2017). In rainbow trout (*Oncorhynchus mykiss*), studies have shown that BSF meal can be included in diets up to 40% without compromising growth or survival (Renna et al., 2017).

In addition to their nutritional benefits, BSF larvae production can stimulate local economies by creating jobs in insect biomass production for food and feed, selling byproducts, and operating bioconversion systems. As agricultural land becomes increasingly limited, reducing land use for feed and energy production is critical. Insects such as BSF serve as a low-land-use alternative for biodiesel production (Salomone et al., 2017).

Feeding studies have explored the effects of incorporating BSF meal into diets across various fish species. For example, research indicated that replacing fishmeal with BSF meal in juvenile yellow catfish did not negatively impact growth performance (Hu et al., 2017). In rainbow trout, replacing fishmeal with BSF meal up to 50% resulted in improved growth and FCRs, although some studies noted a decrease in protein utilization efficiency when BSF meal was used throughout the entire production cycle (Stadtlander et al., 2017). In Atlantic salmon, studies demonstrated that replacing fishmeal with BSF meal did not adversely affect fillet quality or growth performance, suggesting that BSF meal can be a viable alternative protein source in aquafeeds (Lock et al., 2014; Bruni et al., 2020).

Despite the promising potential of BSF larvae in aquafeed production, challenges persist. The growth and development of BSFL can be influenced by the type of feed they consume. For instance, larvae fed fermented maize straw showed slower development compared with those fed standard feeds, such as wheat bran, affecting their reproductive capabilities (Koutsos et al., 2022). In addition, a large-scale study on the effects of different manure types (swine, dairy, or chicken) on BSF larvae indicated that those fed dairy manure had longer maturation times and lower survival rates compared with those fed poultry or swine manure (Miranda et al., 2020).

Furthermore, the nutritional composition of BSF meal can be affected by various biological factors, including developmental stage, diet, and processing methods (Rabani et al., 2019). Understanding these factors is essential for optimizing BSF meal use in aquafeeds and ensuring consistent performance across different aquaculture species.

8.6 House fly (*Musca domestica*)

The house fly (*M. domestica*) is a widely distributed insect found alongside humans and animals, ranging from tropical regions to the Arctic (Geden et al., 2021). While often viewed negatively as a pest due to its association with various parasites and diseases (Čičková et al., 2015), *M. domestica* presents significant potential for waste bioconversion and aquafeed production. The larvae, commonly known as maggots, excel at consuming a variety of organic materials, including food waste and animal manure, making them a viable option for sustainable waste management.

As global food waste increases, the application of house fly larvae in bioconversion processes has gained traction, allowing for the production of valuable protein, oil, and organic fertilizers (Cheng et al., 2021). The use of *M. domestica* as a dietary supplement in aquaculture has been particularly studied in species such as tilapia and various catfish, including *C. gariepinus* and *Heterobranchus longifilis*. Numerous feeding trials have shown that including house fly maggots in fish diets can enhance growth performance, although it is recommended that maggot meal should not exceed 30% of the diet to avoid negative impacts on growth (Fasakin et al., 2003; Aniebo et al., 2009; Okore, 2016; Saleh, 2020). For instance, a study indicated that live maggots did not provide sufficient nutrients for optimal growth in *C. gariepinus*, suggesting their role should be limited to supplementation (Emeka and Oscar, 2016).

The rapid life cycle of *M. domestica* allows larvae to mature within approximately a week under ideal conditions, facilitating quick biomass production. The protein content of house fly larvae can reach 40%–50% on a dry weight basis, along with a beneficial amino acid profile and essential fatty acids, making them an attractive protein source for aquaculture. A study assessing the economic viability of maggot production found that aluminum boxes were the most efficient culture enclosures for maximizing maggot yield (Kolawole and Ugwumba, 2018).

Feeding trials across various aquaculture species consistently show that incorporating house fly larvae into fish diets can improve growth rates and feed conversion efficiency. For example, in *Oreochromis niloticus*, fish fed a diet containing a 4:1 blend of wheat bran and live maggots exhibited superior growth compared with those receiving only wheat bran (Ebenso and Udo, 2004). In addition, research indicates that including house fly larvae meal can enhance health outcomes, such as improved immune responses in fish, highlighting the larvae's dual role as a nutritional supplement and a means of promoting fish well-being (Ido et al., 2015; Li et al., 2019; Fawole et al., 2023).

The environmental advantages of utilizing house fly larvae for waste bioconversion are considerable. Studies have shown that house flies can significantly reduce chicken manure volume by up to 59.94% (Hasan and Leong, 2018). By consuming organic waste, house flies help reduce landfill waste and the associated greenhouse gas emissions from organic decomposition. Furthermore, the larvae can be cultivated on various organic substrates, including agricultural byproducts and food waste, fostering a circular economy where waste is converted into valuable resources. This bioconversion process not only aids in sustainable waste management but also enhances food security by providing a reliable protein source for animal feed.

Moreover, the oil extracted from house fly larvae can be transformed into biodiesel, achieving a yield of 87.71% through transesterification with methanol (Niu et al., 2017). Despite the numerous benefits of using house fly larvae in aquafeed, challenges persist. The nutritional quality of the larvae can vary depending on the substrate used for rearing, as different organic materials can influence growth rates and nutrient profiles. For instance, it has been suggested that including 75% of wet maggots in commercial fish feed optimizes utilization by *C. gariepinus* juveniles (Ipinmoroti et al., 2019). Therefore optimizing rearing conditions and substrate selection is crucial to maximize the nutritional value of house fly larvae for aquaculture applications. In addition, addressing concerns related to disease transmission and potential contaminants in organic waste is essential to ensure the safety and efficacy of house fly larvae as a feed ingredient.

8.7 House cricket (*Acheta domesticus*)

The house cricket (*A. domesticus*) is increasingly recognized as a significant candidate for waste bioconversion and aquafeed production due to its rapid growth, high protein content, and adaptability to various organic waste materials. As one of the most extensively farmed insects, crickets demonstrate remarkable efficiency in food utilization, making them a compelling option for sustainable animal nutrition (Pilco-Romero et al., 2023).

House crickets efficiently consume a variety of protein-rich organic materials, including vegetable waste, garden refuse, and animal manure (Oonincx et al., 2015). The nutritional value of crickets is significantly influenced by the substrates used in their rearing; inferior-quality substrates can negatively impact their growth and overall nutritional content. Research indicates that crickets fed low-nutrient diets exhibit reduced protein levels and increased fat concentrations, compromising their viability as a feed source. Therefore selecting high-quality substrates is essential to optimize the nutritional output of cricket meal for aquaculture applications.

In urban entomology, *A. domesticus* is often viewed as a nuisance pest due to its loud chirping and potential for causing damage to materials. However, its potential role in establishing resilient and circular food systems is noteworthy, as it provides a reliable source of protein for both human and animal consumption (Kumar et al., 2023). The incorporation of agroindustrial byproducts as substrates for cricket farming has yielded promising results, with various vegetable byproducts assessed for their effectiveness based on metrics such as feed conversion efficiency, biomass yield, growth rates, and feeding preferences (Kee et al., 2023).

The application of *A. domesticus* meal as a protein source in livestock feeds is well-established, while its integration into fish feed is still developing (Nikoletta, 2019). Several feeding trials have explored the potential of cricket meal in different fish species. For instance, a study on hybrid tilapia (*Oreochromis sp.*) demonstrated that a diet comprising 60%*A. domesticus* and 40% rice bran produced optimal results in terms of survival and growth rates (Lee et al., 2017). Conversely, a trial involving perch (*Perca fluviatilis*) that replaced 25% of fish meal with a mixture of house cricket showed no significant differences in survival rates between control and experimental groups; however, it did observe decreased fish growth and an increased FCR. Notably, the use of insect pellets led to a substantial increase in linoleic acid and total *n*-6 fatty acids in fish fillets, although the overall nutritional quality for human consumption remained largely unchanged (Tilami et al., 2020).

Despite the numerous advantages associated with *A. domesticus*, it is not universally effective for bioconverting all types of organic waste. Research has shown that crickets fed diets containing *Brassica rapa* and okara did not achieve satisfactory growth or survival rates compared with those on more balanced diets. While okara can occasionally serve as a substitute for high-quality animal feed, it is not a consistently viable option for breeding *A. domesticus*. In addition, incorporating carrots into the diet of crickets had no significant impact on their growth, weight gain, or feed conversion efficiency (Veenenbos and Oonincx, 2017).

Crickets are also emerging as significant contributors to the bioconversion landscape, recognized for their high nutritional value, including essential amino acids, vitamins, and minerals. They can efficiently convert organic waste into protein-rich biomass, making them an attractive alternative protein source. By utilizing food waste and agricultural residues, crickets contribute to waste reduction while providing a sustainable feed option for livestock and aquaculture. The cultivation of crickets for bioconversion can be seamlessly integrated into urban farming systems, facilitating localized food production and waste management solutions. In addition, crickets have a lower environmental impact compared with traditional livestock, requiring significantly less land, water, and feed to produce equivalent amounts of protein.

8.8 Insects farming and processing for aquafeed

Insect farming, or entomoculture, is emerging as a sustainable alternative to traditional livestock farming because of its reduced environmental impact and efficient resource utilization. This innovative approach involves the regulated breeding and cultivation of insects for various applications, including food production, animal feed, and waste management (Fig. 8.3). Insects exhibit remarkable efficiency in converting feed into body mass, achieving FCRs significantly lower than those of conventional livestock. For instance, crickets and BSFs convert feed into protein at rates of 1.7:1 and 1.5:1, respectively, compared with a ratio of 8:1 for cattle (Oonincx et al., 2015). This efficiency is largely attributed to their high metabolic rates and ability to thrive on organic waste, making them a compelling option for sustainable food systems.

A critical aspect of insect farming is its potential to transform insects into aquafeed ingredients, significantly influencing their effectiveness in aquatic animals. As the potential of insects to replace fishmeal in animal feeds becomes apparent, concerns arise regarding the impact on biodiversity from natural harvesting. Therefore mass production of insects from agricultural organic residues and biowaste for feed purposes presents a promising and sustainable solution (Varelas, 2019). However, variations during the culturing process in controlled environments can lead to fluctuations in nutritional value (Varelas, 2019).

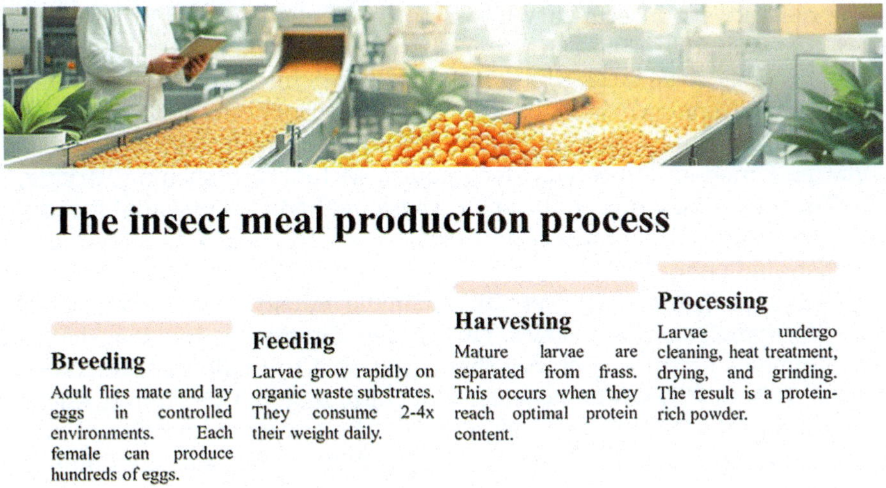

FIGURE 8.3 **Insect farming and meal production processes.**

Cortes Ortiz et al. (2016) highlighted that the artificial diets required by insects vary in form—from liquid to solid—and in nutritional content, feeding adaptations, insect species, and premanufacturing methods. In addition, insects often exhibit low levels of PUFAs, which are essential for human health and crucial for optimal growth and development in children. Implementing PUFA enrichment strategies, such as rearing insects on substrates rich in n-3 PUFAs, is vital (Zarantoniello et al., 2020).

The principle of nutrition upcycling is fundamental to insect farming, as it involves utilizing food waste to extract high-quality digestible micro- and macronutrients that can be reintegrated into the food chain. Insects are recognized for their high nutritional value and are increasingly being adopted in aquaculture as alternatives to traditional protein sources (Hameed et al., 2022; Mannaa et al., 2024). Their capability to process a wide variety of food waste makes them an excellent choice for organic waste management and nutrient recycling, aligning with the global movement toward a circular economy (van Huis, 2022). In the European Union, effective feed utilization can yield 31 g of animal protein per person daily, highlighting the potential of insects in sustainable protein production (Van Hal et al., 2019).

Insect farming is particularly crucial because many insect species are only available in nature during specific seasons. Controlled or indoor farming environments facilitate year-round availability (van Huis, 2020; Alfiko et al., 2022). A vital aspect of insect farming is ensuring the efficient production of eggs in both quantity and quality to maximize the recovery of organic matter, essential for consistent larval production and the maintenance of progenitors (Pastor et al., 2015). To optimize insect growth and development, indoor rearing requires careful management of environmental factors such as temperature, humidity, and light cycles, along with high-quality feed and disease prevention strategies (van Huis, 2020; Alfiko et al., 2022).

Processing methods for insects vary based on their intended use, which can range from whole insect consumption to the extraction of specific biomolecules through various separation techniques (Hajj et al., 2022; Khayrova et al., 2024). Insects harvested from the wild or bred commercially can enhance their nutritional value, sensory qualities, and shelf life through multiple processing methods (Siddiqui et al., 2024a). Manufacturers employ a combination of traditional and modern processing techniques, including blanching, drying, grinding, roasting, and fractionating (Ojha et al., 2020). These methods are essential for improving the nutritional quality and safety of insect products, especially as they become more integrated into human diets and animal feeds.

Blanching serves as an important pretreatment in the industrial production of edible insects, applicable on both small and large scales. This process deactivates enzymes that cause degradation and eliminates harmful microbes that could lead to foodborne illnesses or spoilage (Siddiqui et al., 2024a). However, hot water blanching may not be entirely effective against certain bacterial spores (Alp and Bulantekin, 2021). The nutritional profiles of insects such as *A. domesticus* and *G. assimilis* are rich in protein, essential amino acids, and minerals, making them suitable for various applications (Jucker et al., 2022). The methods used for processing can significantly impact the microbiological and toxicological safety of these insects, which is crucial for consumer health (Mlček et al., 2018).

Drying is one of the most common techniques employed to extend the shelf life of insect products. Various drying methods, such as sun drying, oven drying, and freeze drying, are utilized to reduce moisture levels, thereby inhibiting microbial growth and enzymatic activity (Alp and Bulantekin, 2021). Freeze drying, in particular, has proven effective in preserving proteins and certain vitamins compared to other drying methods (Fröhling et al., 2020). In addition, the drying process can alter the physicochemical properties of insects, affecting their color, size, and nutritional content (Purschke et al., 2018).

Fractionation and extraction techniques are also essential in insect processing, allowing for the recovery of valuable macromolecules such as proteins and lipids. Various extraction methods, including the use of water, organic solvents, and enzymes, can be employed, though many traditional methods rely on chemical solvents that may not be environmentally friendly (Bußler et al., 2016). The rise of "green" extraction methods reflects increasing consumer demand for chemical-free products (Tiwari, 2015). Recently, high hydrostatic pressure processing has been explored to enhance the functional properties of insect proteins, with results varying based on species and treatment conditions (Dion-Poulin et al., 2020).

The global landscape of insect farming is rapidly evolving, with significant investments being made in start-up enterprises focused on insect rearing. By early 2019, at least 42 European companies were actively producing insect meal, generating approximately 6,000 tons annually (Mancuso et al., 2019). Insects are increasingly recognized for their potential in aquaculture, with the European Commission advocating for their inclusion in aquafeeds (Charlton et al., 2015). The industry is projected to grow significantly, with estimates suggesting that insect meal production could reach approximately 0.5 million tons by 2030 (IPIFF).

Despite the promising potential of insect farming, challenges remain, particularly regarding production costs and consumer acceptance. The industry is still largely labor-intensive or nonautomated, especially in regions such as China, the United States, and Europe, where large-scale production of crickets, MWs, and waxworms has been established (van Huis, 2020; Alfiko et al., 2022).

8.9 Challenges in insect-based bioconversion and aquafeed development

Insect farming and bioconversion present a promising pathway for sustainable food production and waste management. However, several challenges must be addressed to fully harness their potential, particularly concerning insect-based bioconversion and aquafeeds. The insect rearing industry is still relatively nascent, with commercial operations lagging behind traditional livestock farming. Numerous international organizations and companies, including AgriProtein, BioflyTech, and Enterra Feed Corporation, are actively engaged in developing industrial-scale production of insect larvae for animal feed (Pastor et al., 2015). Despite these initiatives, current production levels are inadequate to meet the anticipated demand for animal feed and face significant economic hurdles because of high biomass production costs.

A primary challenge lies in the substantial production costs associated with scaling operations. Insect farming is typically labor-intensive and requires considerable investments in infrastructure and technology. The expenses related to feed, breeding, and maintaining optimal environmental conditions can be high, raising concerns about economic viability, especially when competing with traditional protein sources that benefit from established supply chains and economies of scale. In addition, the labor-intensive nature of production, coupled with higher labor costs and land scarcity, is particularly pronounced in urban settings compared to rural areas. To achieve sustainable bioconversion of organic waste, large facilities and automation are essential for cost-effective insect meal production (Cadinu et al., 2020).

Consumer acceptance of insect-based products represents another significant barrier. Many individuals are reluctant to accept insects as a food source or as components of animal feed due to cultural biases and unfamiliarity. Although the concept of edible insects is ancient, there remains a strong aversion to feed containing insect-derived ingredients, particularly in Western nations (Cadinu et al., 2020). To overcome these obstacles, effective marketing strategies, educational initiatives, and awareness campaigns are necessary to emphasize the nutritional advantages and sustainability of insect-based products. Cultivating consumer trust and demand is crucial for the growth of the insect farming sector, as widespread acceptance is vital for its success.

The regulatory framework governing insect farming and insect-based products is still developing, posing challenges in establishing clear and consistent guidelines for production, processing, and marketing. Legal regulations are critical for advancing insect farming and effectively promoting insect-based foods. Insects are often categorized as "impurities" in the Codex Alimentarius, a compilation of international food safety standards (Baiano, 2020). In the United States, the FDA does not classify insects or insect-derived ingredients as "Safe," complicating their application in food and feed. Establishing comprehensive safety standards and guidelines is essential to ensure the safety and quality of insect products intended for human consumption and animal feed. Regulatory challenges can hinder market entry and slow industry growth, making it essential for policymakers to create frameworks that foster innovation while safeguarding public health.

Maintaining nutritional consistency and quality control in insect-based aquafeeds is vital for their acceptance within aquaculture. Variability in the nutritional profiles of various insect species, along with differences in their rearing conditions, can significantly influence feed quality. The quality of insect products can be affected at three critical stages in the production process: The feed provided, the timing of harvest, and the methods used to process larvae into meal. Developing standardized protocols for insect farming and processing is crucial to ensure that aquafeeds fulfill the nutritional requirements of aquatic species and deliver consistent performance. Factors such as insect selection for bioconversion, growth stages, reproduction cycles, breeding conditions, and feeding practices must be optimized for improved bioconversion and insect rearing (Law and Wein, 2018). In addition, operational parameters, including pH, larval density, food-to-larvae ratio, and temperature, should be fine-tuned for optimal bioconversion (Cammack and Tomberlin, 2017).

Scaling up insect farming operations to meet the increasing demand for protein sources presents logistical challenges. While small-scale insect farms have shown success, transitioning to large-scale production requires meticulous planning and investment in automation and technology. Effectively managing large insect populations, maintaining optimal growth conditions, and ensuring biosecurity are critical factors that must be addressed to facilitate large-scale operations. The complexity associated with scaling can deter potential investors and entrepreneurs from entering the market. To achieve production levels comparable to the current volume of fishmeal used in aquafeeds, industries would need access to approximately 100 million metric tons of organic waste substrate (Cadinu et al., 2020).

Although insect farming is often praised for its capacity to convert organic waste into high-quality protein, managing the waste generated during the farming process can pose challenges. It is crucial to ensure that waste is handled sustainably and does not contribute to environmental pollution. Furthermore, sourcing feed for insects, especially if derived from unsustainable practices, can undermine some of the environmental benefits associated with insect farming. Therefore implementing effective waste management strategies is essential for maintaining the sustainability of insect farming operations.

Finally, continuous research and development are necessary to optimize insect farming practices, enhance feed formulations, and improve the nutritional profiles of insect-based aquafeeds. Gaining a deeper understanding of the biology and ecology of various insect species, along with their interactions with different substrates, is vital for maximizing their potential in bioconversion and aquaculture applications. Collaborative efforts among academia, industry, and regulatory bodies can drive innovation and address existing challenges. For example, industries should collaborate with academic or research organizations to develop scalable working models and technologies (Law and Wein, 2018).

8.10 Conclusion

The valorization of solid waste through insect meal production marks a pivotal advancement in sustainable aquaculture and waste management practices. The bioconversion of organic waste by insects, including species such as *T. molitor* (MWs), *H. illucens* (BSFs), *M. domestica* (house flies), and *A. domesticus* (house crickets), presents a comprehensive solution to the challenges of waste disposal while simultaneously generating high-quality protein sources for aquafeed. These insects exhibit remarkable efficiency in converting a wide array of organic substrates—ranging from agricultural and food waste to animal manure—into valuable nutritional products, thereby aligning with the principles of a circular economy. Integrating insect meal into aquaculture not only mitigates waste management costs but also significantly reduces resource consumption compared with traditional protein and fat production methods. This innovative process enhances the value of insect-derived products, contributing to food security and environmental sustainability. Continued research and development are crucial for optimizing production techniques, addressing challenges related to production costs, consumer acceptance, and regulatory frameworks, and ensuring the nutritional consistency of insect meals. Moreover, exploring innovative technologies, such as automated insect farming systems and advanced bioprocessing techniques, will further enhance the scalability and economic viability of insect-based waste conversion systems. As the demand for alternative protein sources continues to rise, it is essential to focus on the specific dietary needs and optimal substrates for insect cultivation to maximize their nutritional profiles and overall effectiveness in aquaculture. With the insect market projected to grow substantially, incorporating insect meal production into aquaculture practices stands as a promising pathway toward fostering environmental sustainability and enhancing global food security. Ultimately, the valorization of solid waste through insect bioconversion not only addresses pressing waste management challenges but also contributes to the establishment of a more resilient and sustainable agricultural landscape.

References

Abd El-Hack, M.E., Shafi, M.E., Alghamdi, W.Y., Abdelnour, S.A., Shehata, A.M., Noreldin, A.E., Ashour, E.A., Swelum, A.A., Al-sagan, A.A., Alkhateeb, M., Taha, A.E., Abdel-moneim, A.M.E., Tufarelli, V., Ragni, M., 2020. Black soldier fly (Hermetia illucens) meal as a promising feed ingredient for poultry: a comprehensive review. Agriculture 10 (8), 1–31. doi:10.3390/agriculture10080339.

Abu Hasan, H., Leong, K.P., 2018. Growth of Musca domestica (Diptera: Muscidae) and Sarcophaga dux (Diptera: Sarcophagidae) larvae in poultry and livestock manures: Implication for animal waste management. J. Asia Pac. Entomol. 21 (3), 880–884. doi:10.1016/j.aspen.2018.07.001.

Alfiko, Y., Xie, D., Astuti, R.T., Wong, J., Wang, L., 2022. Insects as a feed ingredient for fish culture: status and trends. Aquac. Fish. 7 (2), 166–178. doi:10.1016/j.aaf.2021.10.004.

Allegretti, G., Schmidt, V., Talamini, E., 2017. Insects as feed: species selection and their potential use in Brazilian poultry production. World's Poult. Sci. J. 73 (4), 928–937. doi:10.1017/s004393391700054x.

Alp, D., Bulantekin, Ö, 2021. The microbiological quality of various foods dried by applying different drying methods: a review. Eur. Food Res. Technol. 247 (6), 1333–1343. doi:10.1007/s00217-021-03731-z.

Ameixa, O.M.C.C., Duarte, P.M., Rodrigues, D.P., 2020. Insects, food security, and sustainable aquaculture. In: Sustainable food systems from agriculture to industry: improving production and processing. Springer Science and Business Media LLC, Cham, Switzerland, pp. 425–435. doi:10.1007/978-3-319-95675-6_111.

Aniebo, A.O., Erondu, E.S., Owen, O.J., 2009. Replacement of fish meal with maggot meal in African catfish (Clarias gariepinus) diets. Revista Cientifica UDO Agricola 9 (3), 666–671. http://www.bioline.org.br/pdf?cg09081.

Auzins, A., Leimane, I., Reissaar, R., Brobakk, J., Sakelaite, I., Grivins, M., Zihare, L., 2024. Assessing the socio-economic benefits and costs of insect meal as a fishmeal substitute in livestock and aquaculture. Animals 14 (10), 1461. doi:10.3390/ani14101461.

Baiano, A., 2020. Edible insects: an overview on nutritional characteristics, safety, farming, production technologies, regulatory framework, and socio-economic and ethical implications. Trends Food Sci. Technol. 100, 35–50. doi:10.1016/j.tifs.2020.03.040.

Barroso, F.G., de Haro, C., Sánchez-Muros, M.J., Venegas, E., Martínez-Sánchez, A., Pérez-Bañón, C., 2014. The potential of various insect species for use as food for fish. Aquaculture 422–423, 193–201. doi:10.1016/j.aquaculture.2013.12.024.

Barry, T., 2004. Evaluation of the economic, social, and biological feasibility of bioconverting food wastes with the black soldier fly (Hermetia illucens) PhD thesis. University of North Texas, Denton, TX, USA.

Belghit, I., Liland, N.S., Waagbø, R., Biancarosa, I., Pelusio, N., Li, Y., Krogdahl, Å., Lock, E.J., 2018. Potential of insect-based diets for Atlantic salmon (Salmo salar). Aquaculture 491, 72–81. doi:10.1016/j.aquaculture.2018.03.016.

Berners-Lee, M., Kennelly, C., Watson, R., Hewitt, C.N., Kapuscinski, A.R., Locke, K.A., Peters, C.J., 2018. Current global food production is sufficient to meet human nutritional needs in 2050 provided there is radical societal adaptation. Elem. Sci. Anthropocene 6. doi:10.1525/elementa.310.

Biteau, C., Bry-Chevalier, T., Crummett, D., Ryba, R., Jules, M.S., 2024. Is turning food waste into insect feed an uphill climb? a review of persistent challenges. Sustain. Prod. Consum. 49, 492–501. doi:10.1016/j.spc.2024.06.031.

Bohm, K., Hatley, G.A., Robinson, B.H., Gutiérrez-Ginés, M.J., 2023. Analysis of chemical and phytotoxic properties of Frass derived from black soldier fly-based bioconversion of biosolids. Sustainability 15 (15), 11526. doi:10.3390/su151511526.

Bonelli, M., Bruno, D., Caccia, S., Sgambetterra, G., Cappellozza, S., Jucker, C., Tettamanti, G., Casartelli, M., 2019. Structural and functional characterization of Hermetia illucens larval midgut. Front. Physiol. 10. doi:10.3389/fphys.2019.00204.

Bordiean, A., Krzyżaniak, M., Stolarski, M.J., 2022. Bioconversion potential of agro-industrial byproducts by Tenebrio molitor—long-term results. Insects 13 (9). doi:10.3390/insects13090810.

Bruni, L., Belghit, I., Lock, E.J., Secci, G., Taiti, C., Parisi, G., 2020. Total replacement of dietary fish meal with black soldier fly (Hermetia illucens) larvae does not impair physical, chemical or volatile composition of farmed Atlantic salmon (Salmo salar L.). J. Sci. Food Agric. 100 (3), 1038–1047. doi:10.1002/jsfa.10108.

Bußler, S., Rumpold, B.A., Jander, E., Rawel, H.M., Schlüter, O.K., 2016. Recovery and techno-functionality of flours and proteins from two edible insect species: meal worm (Tenebrio molitor) and black soldier fly (Hermetia illucens) larvae. Heliyon 2 (12). doi:10.1016/j.heliyon.2016.e00218.

Cadinu, L.A., Barra, P., Torre, F., Delogu, F., Madau, F.A., 2020. Insect rearing: potential, challenges, and circularity. Sustainability 12 (11), 4567. doi:10.3390/su12114567.

Cammack, J.A., Tomberlin, J.K., 2017. The impact of diet protein and carbohydrate on select life-history traits of the black soldier fly Hermetia illucens (L.), (Diptera: Stratiomyidae). Insects 8 (2). doi:10.3390/insects8020056.

Charlton, A.J., Dickinson, M., Wakefield, M.E., Fitches, E., Kenis, M., Han, R., Zhu, F., Kone, N., Grant, M., Devic, E., Bruggeman, G., Prior, R., Smith, R., 2015. Exploring the chemical safety of fly larvae as a source of protein for animal feed. J. Insects Food Feed 1 (1), 7–16. doi:10.3920/jiff2014.0020.

Chavez, M., 2021. The sustainability of industrial insect mass rearing for food and feed production: zero waste goals through by-product utilization. Curr. Opin. Insect Sci. 48, 44–49. doi:10.1016/j.cois.2021.09.003.

Cheng, Z., Yu, L., Li, H., Xu, X., Yang, Z., 2021. Use of housefly (Musca domestica L.) larvae to bioconversion food waste for animal nutrition and organic fertilizer. Environ. Sci. Pollut. Res. 28 (35), 48921–48928. doi:10.1007/s11356-021-14118-8.

Chia, S.Y., Tanga, C.M., van Loon, J.J., Dicke, M., 2019. Insects for sustainable animal feed: inclusive business models involving smallholder farmers. Curr. Opin. Environ. Sustain. 41, 23–30. doi:10.1016/j.cosust.2019.09.003.

Čičková, H., Newton, G.L., Lacy, R.C., Kozánek, M., 2015. The use of fly larvae for organic waste treatment. Waste Manage. 35, 68–80. doi:10.1016/j.wasman.2014.09.026.

Cortes Ortiz, J.A., Ruiz, A.T., Morales-Ramos, J.A., Thomas, M., Rojas, M.G., Tomberlin, J.K., Yi, L., Han, R., Giroud, L., Jullien, R.L., 2016. Insect mass production technologies insects as sustainable food ingredients: production, processing and food applications. Elsevier, Spain, pp. 153–201. doi:10.1016/B978-0-12-802856-8.00006-5.

Daniel, N., 2018. A review on replacing fish meal in aqua feeds using plant protein sources. Int. J. Fish. Aquat. Stud. 6, 164–179.

da Silva, G.DP., Hesselberg, T., 2020. A review of the use of black soldier fly larvae, Hermetia illucens (Diptera: Stratiomyidae), to compost organic waste in tropical regions. Neotrop. Entomol. 49 (2), 151–162. doi:10.1007/s13744-019-00719-z.

Dauda, A.B., Ajadi, A., Tola-Fabunmi, A.S., Akinwole, A.O., 2019. Waste production in aquaculture: sources, components and managements in different culture systems. Aquac. Fish. 4 (3), 81–88. doi:10.1016/j.aaf.2018.10.002.

Dion-Poulin, A., Laroche, M., Doyen, A., Turgeon, S.L., 2020. Functionality of cricket and mealworm hydrolysates generated after pretreatment of meals with high hydrostatic pressures. Molecules 25 (22), 5366. doi:10.3390/molecules25225366.

Dobermann, D., Swift, J.A., Field, L.M., 2017. Opportunities and hurdles of edible insects for food and feed. Nutr. Bull. 42 (4), 293–308. doi:10.1111/nbu.12291.

Ebenso, I.E, Udo, M.T, 2004. Effect of live maggot on growth of the Nile Perch, *Oreochromis niloticus*(Cichlidae) in South Eastern Nigeria. Glob. J. Agric. Sci. 2 (2). doi:10.4314/gjass.v2i2.2208.

Ee, K.Y., Lam, M.Q., Mah, J.K., Merican, A., 2022. Black soldier fly (Hermetia illucens L.) larvae in degrading agricultural waste as a sustainable protein production: feedstock modification and challenges. Int. J. Trop. Insect Sci. 42 (6), 3847–3854. doi:10.1007/s42690-022-00916-1.

Emeka, A.I., Oscar, E.V., 2016. Comparative study of growth performance, food utilization and survival of the African Catfish Clarias gariepinus (Burchell, 1822) fingerlings fed live maggot (Musca domestica) and coppens commercial feed. Int. J. Sci. Res. Sci. Eng. Technol. 2, 379–386.

Fantatto, R.R., Mota, J., Ligeiro, C., Vieira, I., Guilgur, L.G., Santos, M., Murta, D., 2024. Exploring sustainable alternatives in aquaculture feeding: the role of insects. Aquac. Rep. 37. doi:10.1016/j.aqrep.2024.102228.

FAO, The State of World Fisheries and Aquaculture, 2024. Food and Agriculture Organization of the United Nations. Rome doi:10.4060/cd0683en.

Fasakin, E.A., Balogun, A.M., Ajayi, O.O., 2003. Evaluation of full-fat and defatted maggot meals in the feeding of clariid catfish Clarias gariepinus fingerlings. Aquac. Res. 34 (9), 733–738. doi:10.1046/j.1365-2109.2003.00876.x.

Fawole, F.J., Shamna, N., Memudu, H.A., Abdullahi, N., Hassaan, M.S., Gbadamosi, O.K., 2023. Housefly maggot meal complement soybean meal in a fish-free diet for hybrid catfish (Clarias gariepinus ♀ x Heterobranchus longifilis ♂): effect on growth, body composition, blood biochemistry and antioxidant enzyme activity. Anim. Feed Sci. Technol. 295. doi:10.1016/j.anifeedsci.2022.115543.

Fowles, T.M., 2021. Using insects in conversion of agricultural by-products and applications of their value-added products PhD thesis. University of California, Davis.

Fowles, T.M., Nansen, C., 2019. Insect-based bioconversion: value from food waste. In: Food waste management: solving the wicked problem. Springer International Publishing, Cham, Switzerland, pp. 321–346. doi:10.1007/978-3-030-20561-4_12.

Franco, A., Salvia, R., Scieuzo, C., Schmitt, E., Russo, A., Falabella, P., 2022. Lipids from insects in cosmetics and for personal care products. Insects 13 (1). doi:10.3390/insects13010041.

Fröhling, A., Bußler, S., Durek, J., Schlüter, O.K., 2020. Thermal impact on the culturable microbial diversity along the processing chain of flour from crickets (Acheta domesticus). Front. Microbiol. 11. doi:10.3389/fmicb.2020.00884.

Gasco, L., Acuti, G., Bani, P., Dalle Zotte, A., Danieli, P.P., De Angelis, A., Fortina, R., Marino, R., Parisi, G., Piccolo, G., Pinotti, L., Prandini, A., Schiavone, A., Terova, G., Tulli, F., Roncarati, A., 2020. Insect and fish by-products as sustainable alternatives to conventional animal proteins in animal nutrition. Ital. J. Anim. Sci. 19 (1), 360–372. doi:10.1080/1828051X.2020.1743209.

Geden, C.J., Nayduch, D., Scott, J.G., Burgess, E.R., Gerry, A.C., Kaufman, P.E., Thomson, J., Pickens, V., Machtinger, E.T., 2021. House fly (Diptera: Muscidae): biology, pest status, current management prospects, and research needs. J. Integr. Pest Manag. 12 (1). doi:10.1093/jipm/pmaa021.

Hajj, R.El, Mhemdi, H., Besombes, C., Allaf, K., Lefrançois, V., Vorobiev, E., 2022. Edible insects' transformation for feed and food uses: an overview of current insights and future developments in the field. Processes 10 (5), 970. doi:10.3390/pr10050970.

Hameed, A., Majeed, W., Naveed, M., Ramzan, U., Bordiga, M., Hameed, M., Ur Rehman, S., Rana, N., 2022. Success of aquaculture industry with new insights of using insects as feed: a review. Fishes 7 (6), 395. doi:10.3390/fishes7060395.

Hannah, R., Fiona, S., Max, R., 2021. Fish and overfishing. Our World Data. https://ourworldindata.org/fish-and-overfishing.

Hawkey, K.J., Lopez-Viso, C., Brameld, J.M., Parr, T., Salter, A.M., 2021. Insects: a potential source of protein and other nutrients for feed and food. Ann. Rev. Animal Biosciences 9, 333–354. doi:10.1146/annurev-animal-021419-083930.

Henry, M., Gasco, L., Piccolo, G., Fountoulaki, E., 2015. Review on the use of insects in the diet of farmed fish: Past and future. Anim. Feed Sci. Technol. 203 (1), 1–22. doi:10.1016/j.anifeedsci.2015.03.001.

Hu, J., Wang, G., Huang, Y., Sun, Y., He, F., Zhao, H., Li, N., 2017. Effects of substitution of fish meal with black soldier fly (Hermetia illucens) larvae meal, in yellow catfish (Pelteobagrus fulvidraco) diets. Isr. J. Aquac. - Bamidgeh 69. doi:10.46989/001c.20879.

Huis, A., Oonincx, D.G.A.B., 2017. The environmental sustainability of insects as food and feed. a review. Agron. Sustainable Dev. 37 (5). doi:10.1007/s13593-017-0452-8.

Ido, A., Iwai, T., Ito, K., Ohta, T., Mizushige, T., Kishida, T., Miura, C., Miura, T., 2015. Dietary effects of housefly (Musca domestica), (Diptera: Muscidae) pupae on the growth performance and the resistance against bacterial pathogen in red sea bream (Pagrus major), (Perciformes: Sparidae). Appl. Entomol. Zool. 50 (2), 213–221. doi:10.1007/s13355-015-0325-z.

Ipinmoroti, M.O., Akanmu, O.A., Iyiola, A.O., 2019. Utilisation of house fly maggots (Musca domestica) as replacement for fish meal in the diets ofClarias gariepinus juveniles. J. Insects Food Feed 5 (2), 69–76. doi:10.3920/jiff2017.0057.

Isibika, A., Vinnerås, B., Kibazohi, O., Zurbrügg, C., Lalander, C., 2019. Pre-treatment of banana peel to improve composting by black soldier fly (Hermetia illucens (L.), Diptera: Stratiomyidae) larvae. Waste Manage. 100, 151–160. doi:10.1016/j.wasman.2019.09.017.

Jucker, C., Belluco, S., Oddon, S.B., Ricci, A., Bonizzi, L., Lupi, D., Savoldelli, S., Biasato, I., Caimi, C., Mascaretti, A., Gasco, L., 2022. Impact of some local organic by-products on Acheta domesticus growth and meal production. J. Insects Food Feed 8 (6), 631–640. doi:10.3920/jiff2021.0121.

Kee, P.E., Cheng, Y.S., Chang, J.S., Yim, H.S., Tan, J.C.Y., Lam, S.S., Lan, J.C.W., Ng, H.S., Khoo, K.S., 2023. Insect biorefinery: a circular economy concept for biowaste conversion to value-added products. Environ. Res. 221. doi:10.1016/j.envres.2023.115284.

Khayrova, A., Lopatin, S., Varlamov, V., 2024. A review on characteristics, extraction methods and applications of renewable insect protein. J. Renew. Mater. 12 (5), 923–950. doi:10.32604/jrm.2024.050033.

Kolawole, A.A., Ugwumba, A.AA., 2018. Economic evaluation of different culture enclosures for Musca domestica larval production and their utilization for Clarias gariepinus (Burchell, 1822) fingerlings diets. Not. Sci. Biol. 10 (4), 466–474. doi:10.15835/nsb10410271.

Koutsos, E., Modica, B., Freel, T., 2022. Immunomodulatory potential of black soldier fly larvae: applications beyond nutrition in animal feeding programs. Transl. Anim. Sci. 6 (3). doi:10.1093/tas/txac084.

Kröncke, N., Benning, R., 2023. Influence of dietary protein content on the nutritional composition of mealworm larvae (Tenebrio molitor L.). Insects 14 (3), 261. doi:10.3390/insects14030261.

Kullan, A.R.K., Suresh, A., Choi, H.L., Gabriel Neumann, E., Hassan, F., 2025. Bioconversion of poultry litter into insect meal and organic frasstilizer using black soldier fly larvae as a circular economy model for the poultry industry: a review. Insects 16 (1). doi:10.3390/insects16010012.

Kumar, P., Mehta, N., Abubakar, A.A., Verma, A.K., Kaka, U., Sharma, N., Sazili, A.Q., Pateiro, M., Kumar, M., Lorenzo, J.M., 2023. Potential alternatives of animal proteins for sustainability in the food sector. Food Rev. Int. 39 (8), 5703–5728. doi:10.1080/87559129.2022.2094403.

Law, Y., Wein, L., 2018. Reversing the nutrient drain through urban insect farming—opportunities and challenges. AIMS Bioeng. 5 (4), 226–237. doi:10.3934/bioeng.2018.4.226.

Lee, S.W., Tey, H.C., Wendy, W., Wan Zahari, M., 2017. The effect of house cricket (Acheta domesticus) meal on growth performance of red hybrid tilapia (Oreochromis sp.). Int. J. Aquat. Sci. 8, 78–82.

Lesnik, J.J., 2019. Edible insects and human evolution. University Press of Florida, Gainesville, FL doi:10.5744/florida/9780813056999.001.0001.

Li, S., Ji, H., Zhang, B., Zhou, J., Yu, H., 2017. Defatted black soldier fly (Hermetia illucens) larvae meal in diets for juvenile Jian carp (Cyprinus carpio var. Jian): growth performance, antioxidant enzyme activities, digestive enzyme activities, intestine and hepatopancreas histological structure. Aquaculture 477, 62–70. doi:10.1016/j.aquaculture.2017.04.015.

Li, X., Rahimnejad, S., Wang, L., Lu, K., Song, K., Zhang, C., 2019. Substituting fish meal with housefly (Musca domestica) maggot meal in diets for bullfrog Rana (Lithobates) catesbeiana: effects on growth, digestive enzymes activity, antioxidant capacity and gut health. Aquaculture 499, 295–305. doi:10.1016/j.aquaculture.2018.09.053.

Liu, C., Masri, J., Perez, V., Maya, C., Zhao, J., 2020. Growth performance and nutrient composition of mealworms (Tenebrio Molitor) fed on fresh plant materials-supplemented diets. Foods 9 (2), 151. doi:10.3390/foods9020151.

Lock, E.J., Arsiwalla, T., Waagbø, R., 2014. Insect meal: a promising source of nutrients in the diet of Atlantic salmon (Salmo salar). In: Proceedings of the 1st International Conference "Insects to Feed the World", 14–17 May 2014. Wageningen: Conference Organizers, Wageningen, The Netherlands, p. 67.

Ma, Z., Hassan, M.M., Allais, L., He, T., Leterme, S., Ellis, A., McGraw, B., Qin, J.G., 2019. Comparison of partial replacement of fishmeal with soybean meal and EnzoMeal on growth performance of Asian seabass Lates calcarifer. Comp. Biochem. Physiol. C Toxicol. Pharmacol. 216, 29–37. doi:10.1016/j.cbpc.2018.10.006.

Majluf, P., Matthews, K., Pauly, D., Skerritt, D.J., Palomares, M.L.D., 2024. A review of the global use of fishmeal and fish oil and the fish in:fish out metric. Sci. Adv. 10 (42). doi:10.1126/sciadv.adn5650.

Makkar, H.P.S., Tran, G., Heuzé, V., Ankers, P., 2014. State-of-the-art on use of insects as animal feed. Anim. Feed Sci. Technol. 197, 1–33. doi:10.1016/j.anifeedsci.2014.07.008.

Mancuso, T., Pippinato, L., Gasco, L., 2019. The european insects sector and its role in the provision of green proteins in feed supply. Qual. Access Success 20 (2), 374–381.

Mannaa, M., Mansour, A., Park, I., Lee, D.W., Seo, Y.S., 2024. Insect-based agri-food waste valorization: agricultural applications and roles of insect gut microbiota. Environ. Sci. Ecotechnol. 17. doi:10.1016/j.ese.2023.100287.

Miranda, C.D., Cammack, J.A., Tomberlin, J.K., 2020. Mass production of the black soldier fly, hermetia illucens (L.), (diptera: Stratiomyidae) reared on three manure types. Animals 10 (7), 1–11. doi:10.3390/ani10071243.

Mitra, A., 2021. Thought of alternate aquafeed: conundrum in aquaculture sustainability? Proc. Zool. Soc. 74 (1). doi:10.1007/s12595-020-00352-4.

Mlček, J., Adámková, A., Adámek, M., Borkovcová, M., Bednářová, M., Kouřimská, L., 2018. Selected nutritional values of field cricket (Gryllus assimilis) and its possible use as a human food. Indian J. Tradit. Knowl. 17 (3), 518–524.

Mohan, K., Rajan, D.K., Muralisankar, T., Ganesan, A.R., Sathishkumar, P., Revathi, N., 2022. Use of black soldier fly (Hermetia illucens L.) larvae meal in aquafeeds for a sustainable aquaculture industry: a review of past and future needs. Aquaculture 553. doi:10.1016/j.aquaculture.2022.738095.

Moruzzo, R., Riccioli, F., Diaz, S.E, Secci, C., Poli, G., Mancini, S., 2021. Mealworm (Tenebrio molitor): potential and challenges to promote circular economy. Animals. 11 (9), 2568. doi:10.3390/ani11092568.

Nasr, M.A.F., Reda, R.M., Ismail, T.A., Moustafa, A., 2021. Growth, hemato-biochemical parameters, body composition, and myostatin gene expression of clarias gariepinus fed by replacing fishmeal with plant protein. Animals 11 (3), 1–16. doi:10.3390/ani11030889.

Ng, W.-K., Liew, F.-L., Ang, L.-P., Wong, K.-W., 2001. Potential of mealworm (Tenebrio molitor) as an alternative protein source in practical diets for African catfish, Clarias gariepinus. Aquac. Res. 32 (1), 273–280. doi:10.1046/j. 1355-557x.2001.00024.x.

Nikoletta, H., 2019. Insects as animal feed. Magy. Allatorv. Lapja 141, 117–128.

Niu, Y., Zheng, D., Yao, B., Cai, Z., Zhao, Z., Wu, S., Cong, P., Yang, D., 2017. A novel bioconversion for value-added products from food waste using Musca domestica. Waste Manage. 61, 455–460. doi:10.1016/j.wasman.2016.10.054.

Nogales-Mérida, S., Gobbi, P., Józefiak, D., Mazurkiewicz, J., Dudek, K., Rawski, M., Kierończyk, B., Józefiak, A., 2019. Insect meals in fish nutrition. Rev. Aquac. 11 (4), 1080–1103. doi:10.1111/raq.12281.

Ojha, S., Bußler, S., Schlüter, O.K., 2020. Food waste valorisation and circular economy concepts in insect production and processing. Waste Manage. 118, 600–609. doi:10.1016/j.wasman.2020.09.010.

Okore, Growth and haematological studies of African catfish (Clarias gariepinus) juveniles fed with housefly larva (Musca dometica) as feed supplement. Int. J. Agric. Earth Sci. (2016), 2(3), 21–30.

Okyere, A.A., 2023. Food safety management of insect-based foods food safety management: a practical guide for the food industry. Elsevier, Ghana, pp. 223–233. doi:10.1016/B978-0-12-820013-1.00036-X.

Oonincx, D.G.A.B., Broekhoven, S., Huis, A., Loon, J.J.A., Papadopoulos, N.T, 2015. Feed conversion, survival and development, and composition of four insect species on diets composed of food by-products. PLoS One 10 (12), e0144601. doi:10.1371/journal.pone.0144601.

Pakseresht, A., Vidakovic, A., Frewer, L.J., 2023. Factors affecting consumers' evaluation of food derived from animals fed insect meal: a systematic review. Trends Food Sci. Technol. 138, 310–322. doi:10.1016/j.tifs.2023.05.018.

Pastor, B., Velasquez, Y., Gobbi, P., Rojo, S., 2015. Conversion of organic wastes into fly larval biomass: bottlenecks and challenges. J. Insects Food Feed 1 (3), 179–194. doi:10.3920/jiff2014.0024.

Piccolo, G., Marono, S., Gasco, L., Iannaccone, F., Bovera, F., & Nizza, A. (2014). Use of Tenebrio molitor larvae meal in diets for gilthead sea bream (Sparus aurata) juveniles. Abstract presented at the conference Insects to Feed the World, The Netherlands, p. 76.

Pilco-Romero, G., Chisaguano-Tonato, A.M., Herrera-Fontana, M.E., Chimbo-Gándara, L.F., Sharifi-Rad, M., Giampieri, F., Battino, M., Vernaza, M.G., Álvarez-Suárez, J.M., 2023. House cricket (Acheta domesticus): A review based on its nutritional composition, quality, and potential uses in the food industry. Trends Food Sci. Technol. 142. doi:10.1016/j.tifs.2023.104226.

Purschke, B., Brüggen, H., Scheibelberger, R., Jäger, H., 2018. Effect of pre-treatment and drying method on physico-chemical properties and dry fractionation behaviour of mealworm larvae (Tenebrio molitor L.). Eur. Food Res. Technol. 244 (2), 269–280. doi:10.1007/s00217-017-2953-8.

Rabani, V., Cheatsazan, H., Davani, S., 2019. Proteomics and Lipidomics of black soldier fly (Diptera: Stratiomyidae) and blow fly (Diptera: Calliphoridae) larvae. J. Insect Sci. 19 (3). doi:10.1093/jisesa/iez050.

Ravi, H.K., Degrou, A., Costil, J., Trespeuch, C., Chemat, F., Vian, M.A., 2020. Larvae mediated valorization of industrial, agriculture and food wastes: biorefinery concept through bioconversion, processes, procedures, and products. Processes 8 (7). doi:10.3390/PR8070857.

Rehman, K., Hollah, C., Wiesotzki, K., Rehman, R., Rehman, A.U., Zhang, J., Zheng, L., Nienaber, T., Heinz, V., Aganovic, K., 2023. Black soldier fly, Hermetia illucens as a potential innovative and environmentally friendly tool for organic waste management: a mini-review. Waste Manage. Res. 41 (1), 81–97. doi:10.1177/0734242X221105441.

Renna, M., Schiavone, A., Gai, F., Dabbou, S., Lussiana, C., Malfatto, V., Prearo, M., Capucchio, M.T., Biasato, I., Biasibetti, E., De Marco, M., Brugiapaglia, A., Zoccarato, I., Gasco, L., 2017. Evaluation of the suitability of a partially defatted black soldier fly (Hermetia illucens L.) larvae meal as ingredient for rainbow trout (Oncorhynchus mykiss Walbaum) diets. J. Anim. Sci. Biotechnol. 8 (1). doi:10.1186/s40104-017-0191-3.

Ribeiro, N., Abelho, M., Costa, R., 2018. A review of the scientific literature for optimal conditions for mass rearing Tenebrio molitor (Coleoptera: Tenebrionidae). J. Entomol. Sci. 53 (4), 434–454. doi:10.18474/jes17-67.1.

Roncarati, A., Gasco, L., Parisi, G., Terova, G., 2015. Growth performance of common catfish (Ameiurus melas Raf.) fingerlings fed mealworm (Tenebrio molitor) diet. J. Insects Food Feed 1 (3), 233–240. doi:10.3920/jiff2014.0006.

Rumpold, B.A., Schlüter, O.K., 2013. Potential and challenges of insects as an innovative source for food and feed production. Innov. Food Sci. Emerg. Technol. 17, 1–11. doi:10.1016/j.ifset.2012.11.005.

Saleh, H.H.E., 2020. Effect of feeding on fresh (wet) housefly maggots (Musca domestica) with or without artificial diet on water quality and growth rates of African catfish (Clarias gariepinus Burchell, 1822) fry under laboratory conditions. J. Zool. Res. 2 (2). doi:10.30564/jzr.v2i2.2053.

Salomone, R., Saija, G., Mondello, G., Giannetto, A., Fasulo, S., Savastano, D., 2017. Environmental impact of food waste bioconversion by insects: application of life cycle assessment to process using Hermetia illucens. J. Cleaner Prod. 140, 890–905. doi:10.1016/j.jclepro.2016.06.154.

Sampathkumar, K., Yu, H., Loo, J.S.C., 2023. Valorisation of industrial food waste into sustainable aquaculture feeds. Future Foods 7, 100240. doi:10.1016/j.fufo.2023.100240.

Sánchez, M., Gómez, C., Avendaño, C., Harmsen, I., Ortiz, D., Ceballos, R., Villamizar-Sarmiento, M.íaG., Oyarzun-Ampuero, F., Wacyk, J., Valenzuela, C., 2021. House fly (Musca domestica) larvae meal as an ingredient with high nutritional value: Microencapsulation and improvement of organoleptic characteristics. Food Res. Int. 145, 110423. doi:10.1016/j.foodres.2021.110423.

Sánchez-Muros, M.J., Barroso, F.G., Manzano-Agugliaro, F., 2014. Insect meal as renewable source of food for animal feeding: a review. J. Cleaner Prod. 65, 16–27. doi:10.1016/j.jclepro.2013.11.068.

Sangiorgio, P., Verardi, A., Dimatteo, S., Spagnoletta, A., Moliterni, S., Errico, S., 2022. Valorisation of agri-food waste and mealworms rearing residues for improving the sustainability of Tenebrio molitor industrial production. J. Insects Food Feed 8 (5), 509–524. doi:10.3920/jiff2021.0101.

Sharifinia, M., 2024. Improve aquaculture with insect meal. Science 383 (6685), 838. doi:10.1126/science.ado0380, 838.

Sharifinia, M., 2025. From nutrient bioavailability to disease resistance: The comprehensive benefits of chelated minerals in aquaculture. Fish Shellfish Immunol. 160, 110218. doi:10.1016/j.fsi.2025.110218.

Sharifinia, M., Bahmanbeigloo, Z.A., Keshavarzifard, M., Khanjani, M.H., Daliri, M., Koochaknejad, E., Jasour, M.S., 2023a. Fishmeal replacement by mealworm (Tenebrio molitor) in diet of farmed Pacific white shrimp (Litopenaeus vannamei): effects on growth performance, serum biochemistry, and immune response. Aquat. Living Resour. 36.19. doi:10.1051/alr/2023013.

Sharifinia, M., Bahmanbeigloo, Z.A., Keshavarzifard, M., Khanjani, M.H., Daliri, M., Koochaknejad, E., Jasour, M.S., 2023b. The effects of replacing fishmeal by mealworm (Tenebrio molitor) on digestive enzymes activity and hepatopancreatic biochemical indices of Litopenaeus vannamei. Ann. Anim. Sci. doi:10.2478/aoas-2022-0098.

Sheppard, D.C., Tomberlin, J.K., Joyce, J.A., Kiser, B.C., Sumner, S.M., 2002. Rearing methods for the black soldier fly (diptera: Stratiomyidae). J. Med. Entomol. 39 (4), 695–698. doi:10.1603/0022-2585-39.4.695.

Shockley, M., Dossey, A.T., 2014. Insects for human consumption. In: Mass production of beneficial organisms: invertebrates and entomopathogens. Academic Press (Elsevier), Cambridge, MA, pp. 617–652. doi:10.1016/b978-0-12-391453-8.00018-2.

Siddiqui, S.A., Harahap, I.A., Osei-Owusu, J., Saikia, T., Wu, Y.S., Fernando, I., Perestrelo, R., Câmara, J.S., 2024a. Bioconversion of organic waste by insects – a comprehensive review. Process Saf. Environ. Prot. 187, 1–25. doi:10.1016/j.psep.2024.04.122.

Siddiqui, S.A., Ristow, B., Rahayu, T., Putra, N.S., Widya Yuwono, N., Nisa, K., Mategeko, B., Smetana, S., Saki, M., Nawaz, A., Nagdalian, A., 2022. Black soldier fly larvae (BSFL) and their affinity for organic waste processing. Waste Manage. 140, 1–13. doi:10.1016/j.wasman.2021.12.044.

Siddiqui, S.A., Süfer, Ö., Koç, H.L., Rahayu, T., Castro-Muñoz, R., Fernando, I., 2024b. Enhancing the bioconversion rate and end products of black soldier fly (BSF) treatment–a comprehensive review. Environ. Dev. Sustain. 27, 9673–9741. https://doi.org/10.1007/s10668-023-04306-6.

Sogari, G., Amato, M., Biasato, I., Chiesa, S., Gasco, L., 2019. The potential role of insects as feed: a multi-perspective review. Animals 9 (4), 119. doi:10.3390/ani9040119.

Stadtlander, T., Stamer, A., Buser, A., Wohlfahrt, J., Leiber, F., Sandrock, C., 2017. Hermetia illucens meal as fish meal replacement for rainbow trout on farm. J. Insects Food Feed 3 (3), 165–175. doi:10.3920/jiff2016.0056.

Su, J., Gong, Y., Cao, S., Lu, F., Han, D., Liu, H., Jin, J., Yang, Y., Zhu, X., Xie, S., 2017. Effects of dietary Tenebrio molitor meal on the growth performance, immune response and disease resistance of yellow catfish (Pelteobagrus fulvidraco). Fish Shellfish Immunol. 69, 59–66. doi:10.1016/j.fsi.2017.08.008.

Surendra, K.C., Olivier, R., Tomberlin, J.K., Jha, R., Khanal, S.K., 2016. Bioconversion of organic wastes into biodiesel and animal feed via insect farming. Renew. Energy 98, 197–202. doi:10.1016/j.renene.2016.03.022.

Surendra, K.C., Tomberlin, J.K., van Huis, A., Cammack, J.A., Heckmann, L.H.L., Khanal, S.K., 2020. Rethinking organic wastes bioconversion: evaluating the potential of the black soldier fly (Hermetia illucens (L.)), (Diptera: Stratiomyidae), (BSF). Waste Manage. 117, 58–80. doi:10.1016/j.wasman.2020.07.050.

Tilami, S.K., Turek, J., Červený, D., Lepič, P., Kozák, P., Burkina, V., Sakalli, S., Tomčala, A., Sampels, S., Mráz, J., 2020. Insect meal as a partial replacement for fish meal in a formulated diet for perch perca fluviatilis. Turk. J. Fish. Aquat. Sci. 20 (12), 867–878. doi:10.4194/1303-2712-v20_12_03.

Tiwari, B.K., 2015. Ultrasound: a clean, green extraction technology. TrAC Trends Anal. Chem. 71, 100–109. doi:10.1016/j.trac.2015.04.013.

Tomberlin, J.K., Sheppard, D.C., Joyce, J.A., 2002. Selected life-history traits of black soldier flies (Diptera: Stratiomyidae) reared on three artificial diets. Ann. Entomol. Soc. Am. 95 (3), 379–386. doi:10.1603/0013-8746(2002)095[0379:SLHTOB]2.0.CO;2.

Triunfo, M., Tafi, E., Guarnieri, A., Salvia, R., Scieuzo, C., Hahn, T., Zibek, S., Gagliardini, A., Panariello, L., Coltelli, M.B., De Bonis, A., Falabella, P., 2022. Characterization of chitin and chitosan derived from Hermetia illucens, a further step in a circular economy process. Sci. Rep. 12 (1). doi:10.1038/s41598-022-10423-5.

van der Fels-Klerx, H.J., Camenzuli, L., Belluco, S., Meijer, N., Ricci, A., 2018. Food safety issues related to uses of insects for feeds and foods. Compr. Rev. Food Sci. Food Saf. 17 (5), 1172–1183. doi:10.1111/1541-4337.12385.

van der Wiel, B.Z., Weijma, J., van Middelaar, C.E., Kleinke, M., Buisman, C.J.N., Wichern, F., 2019. Restoring nutrient circularity: a review of nutrient stock and flow analyses of local agro-food-waste systems. Resour. Conserv. Recycl. X 3. doi:10.1016/j.rcrx.2019.100014.

van Dijk, M., Morley, T., Rau, M.L., Saghai, Y., 2021. A meta-analysis of projected global food demand and population at risk of hunger for the period 2010–2050. Nat Food 2 (7), 494–501. doi:10.1038/s43016-021-00322-9.

van Hal, O., de Boer, I.J.M., Muller, A., de Vries, S., Erb, K.-H., Schader, C., Gerrits, W.J.J., van Zanten, H.H.E., 2019. Upcycling food leftovers and grass resources through livestock: Impact of livestock system and productivity. J. Cleaner Prod. 219, 485–496. doi:10.1016/j.jclepro.2019.01.329.

van Huis, A., 2019. Manure and flies: biodegradation and/or bioconversion? J. Insects Food Feed 5 (2), 55–58. doi:10.3920/jiff2019.x002.

van Huis, A., 2020. Insects as food and feed, a new emerging agricultural sector: a review. J. Insects Food Feed 6 (1), 27–44. doi:10.3920/jiff2019.0017.

van Huis, A., 2022. Edible insects: challenges and prospects. Entomol. Res. 52 (4), 161–177. doi:10.1111/1748-5967.12582.

Varelas, V., 2019. Food wastes as a potential new source for edible insect mass production for food and feed: a review. Fermentation 5 (3), 81. doi:10.3390/fermentation5030081.

Veenenbos, M.E., Oonincx, D.G.A.B., 2017. Carrot supplementation does not affect house cricket performance (Acheta domesticus). J. Insects Food Feed 3 (3), 217–222. doi:10.3920/jiff2017.0006.

Xiao, X., Mazza, L., Yu, Y., Cai, M., Zheng, L., Tomberlin, J.K., Yu, J., van Huis, A., Yu, Z., Fasulo, S., Zhang, J., 2018. Efficient co-conversion process of chicken manure into protein feed and organic fertilizer by Hermetia illucens L. (Diptera: Stratiomyidae) larvae and functional bacteria. J. Environ. Manage. 217, 668–676. doi:10.1016/j.jenvman.2018.03.122.

Yang, X.G., Wen, P.P., Yang, Y.F., Jia, P.P., Li, W.G., Pei, D.S., 2023. Plastic biodegradation by in vitro environmental microorganisms and in vivo gut microorganisms of insects. Front. Microbiol. 13. doi:10.3389/fmicb.2022.1001750.

Yeganeh, V., Sharifinia, M., Mobaraki, S., Dashtiannasab, A., Aeinjamshid, K., Borazjani, J.M, Maghsoudloo, T., 2020. Survey of survival rate and histological alterations of gills and hepatopancreas of the Litopenaeus vannamei juveniles caused by exposure of Margalefidinium /Cochlodinium polykrikoides isolated from the Persian Gulf. Harmful Algae 97, 101856. doi:10.1016/j.hal.2020.101856.

Zarantoniello, M., Randazzo, B., Gioacchini, G., Truzzi, C., Giorgini, E., Riolo, P., Gioia, G., Bertolucci, C., Osimani, A., Cardinaletti, G., Lucon-Xiccato, T., Milanović, V., Annibaldi, A., Tulli, F., Notarstefano, V., Ruschioni, S., Clementi, F., Olivotto, I., 2020. Zebrafish (Danio rerio) physiological and behavioural responses to insect-based diets: a multidisciplinary approach. Sci. Rep. 10 (1). doi:10.1038/s41598-020-67740-w.

Zou, X.u, Liu, M., Li, X., Pan, F., Wu, X., Fang, X., Zhou, F., Peng, W., Tian, W., 2024. Applications of insect nutrition resources in animal production. J. Agric. Food Res. 15, 100966. doi:10.1016/j.jafr.2024.100966.

CHAPTER

9

Microbiome-oriented clean water technologies for aquaculture

Prasenjit Mali[a], Dibyendu Kamilya[b] and Md. Idrish Raja Khan[c]

[a]Department of Aquatic Animal Health, Faculty of Fishery Sciences, West Bengal University of Animal and Fishery Sciences, Kolkata, West Bengal, India [b]Agricultural and Food Engineering Department, Indian Institute of Technology Kharagpur, Kharagpur, West Bengal, India [c]Department of Aquatic Animal Health Management, College of Fisheries Science, Chaudhary Charan Singh Haryana Agricultural University, Hisar, Haryana, India

9.1 Introduction

In recent decades, world aquaculture has experienced remarkable growth, largely driven by rising demand for fish and seafood due to an expanding population and shifting dietary preferences (FAO, 2024). This surge in production has been made possible by the intensification of aquaculture practices worldwide. However, such intensification brings its own set of challenges, particularly concerning the decline in water quality associated with these systems. The intensification of aquaculture demands higher inputs, especially feed. The excess feed that is not consumed, along with the fecal matter produced, contributes to a rise in waste generation from the production systems and deteriorates the water quality (Dauda et al., 2019). The primary components of aquaculture wastewater that can cause environmental problems and negatively affect fish health and growth are organic matter, dissolved solids, and nutrients such as nitrogen and phosphorus (Piedrahita, 2003). As such, effective water quality management is essential, not only to safeguard the health of cultured species but also to prevent the complex and harmful ecological changes that can arise in the broader environment.

Traditional physical, chemical, and biological water treatment methods have limitations in their ability to provide long-term, sustainable solutions to water quality issues in aquaculture systems (Tom et al., 2021). In recent years, however, there has been growing interest in microbiome-oriented approaches to water treatment. These innovative technologies leverage the natural microbial communities found in aquatic environments to maintain and restore

Zero-Waste Technologies in Aquaculture: Global Perspectives and Methods toward a Circular Economy
DOI: https://doi.org/10.1016/B978-0-443-34689-7.00016-1

water quality, reduce the need for harmful chemicals, and enhance the well-being and productivity of farmed animals.

The microbiome refers to a complex community of diverse microorganisms, including bacteria, fungi, algae, and viruses, that inhabit an ecosystem. In aquatic systems, the microbiome plays a fundamental role in nutrient cycling, waste degradation, disease prevention, and overall ecosystem health (Loos et al., 2024). These microbial communities are integral to maintaining the stability of aquaculture systems by facilitating key processes such as the nitrogen and phosphorus cycles, biodegradation of wastes, and the suppression of pathogens. This chapter explores the concept of microbiome-oriented clean water technologies for aquaculture and their potential to transform the industry.

9.2 Water probiotics

One of the earliest definitions describes a probiotic as "a live microbial feed supplement which beneficially affects the host animal by improving its intestinal balance" (Fuller, 1989). In another definition, they are "live microorganisms that, when administered in adequate amounts, confer health benefits on the host". Besides these, numerous other definitions have been proposed, and in all the definitions, the probiotic effects are attributed to the active modulation of the gut that confers protection against pathogenic microorganisms, stimulates immunity, and provides nutritional benefits to the host. Eventually, probiotics are contemplated as health-promoting "functional foods" for humans as well as for animals, including aquatic organisms (Kesarcodi-Watson et al., 2008).

Because of the complex relationship between aquatic organisms and their environment, the use of probiotics in aquaculture has expanded beyond health-promoting feed supplements. This broader application includes the use of probiotics as microbial "water additives" to enhance water quality by directly introducing them into the aquaculture environment (Moriarty, 1998). In line with this, Verschuere et al. (2000) proposed an expanded definition of probiotics as "live microbial agents that benefit the host by altering the host-associated or surrounding microbial community, improving feed utilization or its nutritional value, boosting the host's disease resistance, or enhancing the quality of its surrounding environment." Probiotics intended for aquatic usage (often described as "water probiotics"), can modify the ambient microbial community by the competitive exclusion principle (Fachri et al., 2024). One of the obvious benefits of water probiotics is their ability to improve the water quality by degrading the organic wastes present in the aquaculture pond water or sediment (Kamilya and Devi, 2022).

The beneficial effects of water probiotics on improving water quality in aquaculture systems stem from the interaction between the introduced bacteria and the existing microbial community, along with the environmental factors that support these interactions (Kamilya and Devi, 2022). When probiotic cultures are well-adapted to the environmental conditions, a single application can often be enough to achieve successful colonization and long-term presence (Verschuere et al., 2000). However, regular supplementation is recommended to maintain the dominance of these probiotics, especially in environments with stable and well-established microbial communities (Verschuere et al., 2000).

9.2.1 Bioremediation by water probiotics

Generally, Gram-positive bacteria, such as *Bacillus* spp., are more effective at converting organic matter into CO_2 than Gram-negative bacteria. Therefore maintaining a high concentration of Gram-positive bacteria in culture ponds can help reduce the accumulation of dissolved and particulate organic matter during the culture cycle (Verschuere et al., 2000; Balcázar et al., 2006). Many bacterial groups, including *Bacillus*, play key roles in processes such as nitrification, denitrification, nitrogen fixation, and the oxidation and reduction of metals, making them powerful bioremediators of contaminants in aquaculture environments (Priest, 1993; Slepecky and Hemphill, 2006). Several bacterial species have been explored for their ability to improve water quality in aquaculture ponds or tanks. Probiotic bacteria treatments have led to modifications in parameters such as ammonia, nitrite, nitrate, pH, biochemical oxygen demand (BOD), alkalinity, and chemical oxygen demand (COD). For example, the addition of *Bacillus subtilis* in grass carp (*Ctenopharyngodon idellus*) culture tanks resulted in significant reductions in ammonia, nitrite, and total nitrogen levels (Zhang et al., 2013). A microbial consortium of *B. cereus, B. amyloliquefaciens,* and *Pseudomonas stutzeri* effectively reduced ammonia, nitrite, and nitrate levels in tilapia fingerling tanks (John et al., 2020). Similarly, adding *B. subtilis* and *B. licheniformis* to the rearing water of whiteleg shrimp (*Litopenaeus vannamei*) resulted in significant reductions in total ammonia nitrogen and unionized ammonia (Monier et al., 2023). An exogenous bacterial combination of *Rhodopseudomonas palustris* and *B. subtilis* notably reduced the concentrations of COD, total nitrogen, phosphorus, ammonia, nitrite, and nitrate in whiteleg shrimp aquaculture wastewater (Xu et al., 2024). Other studies have also highlighted the bioremediation potential of various probiotic bacteria in improving water quality in aquaculture systems (Dewali et al., 2025; Ghosh, 2025; Soltani et al., 2025).

9.3 Recirculating aquaculture systems

Recirculating aquaculture systems (RASs) have gained popularity in regions facing environmental challenges, stringent wastewater regulations, or limited natural resources. These systems are widely used in hatcheries and for grow-out systems, offering year-round optimal water quality, improved feed efficiency, and high-yield production in controlled environments. RAS also promotes water conservation, reduces waste discharge, and enhances biosecurity (Rurangwa and Verdegem, 2015). To maintain economic viability, RAS operates at high animal densities, generating significant metabolic waste that fosters diverse microbial communities responsible for processing organic and inorganic byproducts (Ende et al., 2024). Chemoautotrophic bacteria are essential for converting ammonia, nitrite, and nitrate (Neissi et al., 2022). However, microbial communities can also host pathogens or produce off-flavors (Podduturi et al., 2020). Microbes enter RAS through water, air, feed, animals, and equipment, with some intentionally introduced to optimize performance (Blancheton et al., 2013). Effective waste management is essential to support sensitive nitrifying bacteria and maintain stable environmental conditions, making microbial dynamics a key focus for sustaining water quality and animal health in RAS (Rurangwa and Verdegem, 2015).

9.3.1 Distribution of microbes in recirculating aquaculture systems

In RAS, microbial communities are distributed across various compartments, with biofilters serving as the primary reservoir of microorganisms (Leonard et al., 2000). Biofilters are categorized as fixed-film, where microbes adhere to surfaces, or single-sludge, where microbes are suspended in water (Suriasni et al., 2023). Organic waste in RAS supports the growth of fungi, bacteria, and algae, which can lead to biofouling and clogging in pipes, tank walls, and air diffusers (Suriasni et al., 2023).

Key microbial groups in RAS biofilters include nitrite-oxidizing bacteria (NOB) such as *Nitrospirae* and ammonia-oxidizing archaea (AOA) such as *Nitrosopumilus*, which thrive under nutrient-limiting conditions, unlike ammonia-oxidizing bacteria (AOB) that require ammonium (Al-Ajeel et al., 2022). Specific nitrifiers, such as *Nitrosopumilus maritimus* and *Nitrosospira*, dominate marine biofilters, while ammonia and nitrite oxidizers vary between marine and freshwater systems (McKnight and Neufeld, 2024). Fungi such as *Aspergillus niger* can assist in nitrogen cycling by reducing nitrates and removing ammonium and nitrite, even at low temperatures (Rurangwa and Verdegem, 2015).

Both autotrophic and heterotrophic microbes contribute to nutrient cycling in RAS. Autotrophs manage carbon and nitrogen transformations, including processes such as anaerobic ammonium oxidation (anammox) and denitrification, while heterotrophs decompose organic matter such as uneaten feed and dead biomass (Bartelme et al., 2017).

Nitrification in RAS relies on AOB and NOB. AOB, such as *Nitrosomonas*, oxidize ammonia to nitrite, while NOB, including *Nitrobacter* and *Nitrospira*, convert nitrite to nitrate. *Nitrospira* is particularly efficient under low oxygen and nitrite conditions compared to *Nitrobacter* (Neissi et al., 2022). In biofilms, AOB dominate oxygen-rich areas, while *Nitrospira* thrives at oxic/anoxic boundaries (Neissi et al., 2022). In oxygen-depleted zones, anaerobic ammonium-oxidizing (anammox) bacteria convert ammonia and nitrite directly into nitrogen gas, reducing the need for traditional denitrification (Tal et al., 2003). These anaerobic ammonium-oxidizing bacteria coexist with AOB and AOA in biofilters and include genera such as *Brocadia*, *Kuenenia*, and *Scalindua*, which are prevalent in marine RAS systems (Sonthiphand et al., 2014).

Heterotrophic bacterial communities in RAS vary by species and system type. In a sea bass RAS, *P. stutzeri*, *Ruegeria* spp., and *Roseobacter* spp. dominated the biofilter, with *Roseobacter* and *Ruegeria* also detected in rearing water. However, potential pathogens such as *Vibrio* spp., *Erwinia* spp., and *Coxiella* spp. were also identified (Michaud et al., 2009). In a common carp RAS, *Alphaproteobacteria* and *Betaproteobacteria* prevailed in the biofilter, while a goldfish RAS exhibited greater microbial diversity, including *Actinobacteria*, *Bacilli*, *Gammaproteobacteria*, and *Nitrospira moscoviensis*. Shared species between these systems included *Hyphomicrobium denitrificans*, *Rhodovulum euryhalinum*, and *N. moscoviensis* (Sugita et al., 2005).

9.3.2 Pathogenic microorganisms management in recirculating aquaculture systems

Pathogens may accumulate in RAS biofilters because of low dilution and high organic loads, increasing infection risks from biofilm detachment (Sharrer et al., 2005). Effective management and control of opportunistic bacteria are essential for biosecurity (Dahle et al., 2023). Key fish pathogens include *Aeromonas salmonicida*, *Vibrio anguillarum*, and *Yersinia ruckeri*, with additional species such as *Bacillus cereus* and *Vibrio* spp. have been reported in RAS

(Drønen et al., 2022). Disinfection methods such as ozone and UV are commonly used but must be carefully managed to avoid disrupting beneficial microbial communities (Blancheton et al., 2013).

(1) Ozonation: Ozone (O_3) is a powerful disinfectant in aquaculture, used to eliminate pathogens, enhance water quality, and oxidize resistant organic molecules (Gorito et al., 2021). It disrupts cell membranes of pathogens, causing leakage of proteins and nucleic acids, while preserving intracellular components. Ozone improves water quality by microflocculating particulates and oxidizing nitrites and refractory organics (Gorito et al., 2021). In brackish and seawater, ozone reacts with bromide ions, forming harmful byproducts such as hypobromous acid and bromate (Reiser et al., 2010). To protect farmed organisms, residual ozone is neutralized using sodium thiosulfate, UV light, or air stripping (Rurangwa and Verdegem, 2015).

(2) Ultraviolet irradiation: UV light, particularly at 254 nm, effectively kills microorganisms by damaging their nucleic acids (Chen et al., 2023). Recommended UV doses range from 100 mWs/cm^2 for bacteria and fungi to 250 mWs/cm^2 for viruses (Runia, 1996). UV disinfection efficiency depends on factors such as irradiation intensity, contact time, and particle size, with better results achieved by reducing particle load (González et al., 2023). A key benefit of UV is its lack of toxic residues, although bacteria can repair UV damage, unlike viruses (Zhu et al., 2023).

(3) Biocontrol: RAS are also used for producing fish and shellfish larvae (Taufik et al., 2024). Because of long water retention and large microbial communities, biofilters in RAS may help stabilize microbial populations, limiting opportunistic pathogens (Attramadal et al., 2012). However, during early larval stages, the microbial community and immune systems are underdeveloped, and feeding on live organisms supports opportunistic bacterial growth (Louvado et al., 2023). The presence of c-Proteobacteria and vibrios in the system can increase mortality in larvae (Manchanayake et al., 2023). Excessive use of antibiotics can lead to resistance, highlighting the importance of using probiotics or alternative methods for disease control. Green water technology using microalgae and microbially matured water are two strategies used to control pathogens in larviculture (Elgendy et al., 2024). Microbial maturation of rearing water can be achieved by filtering water through a maturation unit, promoting recolonization by nonopportunistic species while suppressing pathogenic bacteria. Membrane filtration removes more than 95% of bacteria, allowing slow recolonization by beneficial microbes (Skjermo et al., 1997).

(4) Probiotic treatment: Common probiotic genera in aquaculture include *Lactobacillus, Bacillus, Vibrio, Pseudomonas*, and *Saccharomyces* (Jahangiri and Esteban, 2018). Probiotics can be introduced in RAS through water or food, with fish larvae receiving probiotics via live food (Deng et al., 2022). In addition, RAS may harbor indigenous microbiota that produce inhibitory compounds, potentially controlling pathogen development (Rurangwa and Verdegem, 2015).

9.3.3 Future challenges and scope in recirculating aquaculture systems

RAS faces challenges in managing microbial communities, balancing beneficial bacteria, such as AOB and NOB, with pathogens such as *Vibrio* spp., which threaten fish health and biosecurity. Off-flavors from geosmin and MIB 2-methylisoborneol (MIB), produced by microorganisms, impact product quality and require costly mitigation. High energy consumption from technologies such as ozonation and UV irradiation also poses sustainability concerns. In addition, antibiotic resistance necessitates alternative disease control methods, such as probiotics and green-water technology.

Managing both autotrophic and heterotrophic bacteria is crucial for preserving water quality and ensuring the health of farmed organisms in RAS. Proper management prevents solid accumulation, ammonia/nitrite buildup, and pathogen growth. Heterotrophic bacteria, while containing both harmful and beneficial members, play a key role in system functioning and product quality. Research is needed to understand microbial interactions, particularly among heterotrophic bacteria, to prevent issues such as off-flavors and pathogens.

Future advancements include DNA-based techniques and artificial intelligence for microbial management, biological treatments for off-flavors, and cost-effective advanced oxidation processes. Renewable energy integration, optimized feed formulations, and mixed-species probiotics can enhance sustainability. Scaling RAS for global adoption and addressing regulatory barriers will be crucial for its success in revolutionizing aquaculture and ensuring food security.

9.4 Biofloc technology

Among the microbiome-based emerging technologies, biofloc technology (BFT) has gained prominence for its sustainability benefits, offering a solution to water quality management, waste recycling, and pathogen control (Khanjani et al., 2022). BFT relies on heterotrophic bacteria to metabolize nitrogenous waste, such as ammonia and nitrites, improving water quality and serving as a nutrient-rich food source for farmed species (Ferreira et al., 2015; Wang et al., 2016). By maintaining a carbon-to-nitrogen (C:N) ratio of 10–15:1 through carbon sources such as molasses, BFT fosters microbial communities that convert waste into microbial flocs, enhancing water quality, pathogen control, and immunity in cultured species (Khanjani and Sharifinia, 2020; El-Sayed, 2021).

9.4.1 Microbial communities in biofloc systems

Microorganisms are fundamental to all life forms and ecosystems, playing critical roles in various biological processes (López and Zaballos, 2005). In aquaculture, diverse microbial groups are essential for functions such as contaminant removal, serving as food sources, and recycling organic matter. Understanding the environmental and biological factors that impact microbial functions is crucial for effective aquaculture management (Monroy-Dosta et al., 2015). In biofloc culture, microorganisms are categorized into five overlapping groups: Floc-forming organisms, saprophytes, nitrifying bacteria, algae grazers, and pathogenic bacteria. For BFT systems to function efficiently, these microbial species must remain suspended in the water column, where they interact and perform their roles (Kumar et al., 2021). Floc-forming microorganisms, for example, secrete adhesive extracellular polymeric substances that facilitate floc formation (Manan et al., 2017).

The microbial communities in biofloc systems are primarily composed of heterotrophic bacteria, including *Vibrio, Bacillus,* and *Lactobacillus,* along with algae, fungi, and detritus. Molecular studies using 16S rRNA sequencing have identified dominant bacterial groups such as Proteobacteria, Bacteroidetes, and Cyanobacteria (Tepaamorndech et al., 2020). These microbes enhance host growth and survival while mitigating environmental and pathogenic stressors. Quorum sensing (QS), a bacterial communication mechanism, regulates biofloc formation and promotes microbial aggregation. It also aids in converting harmful compounds, improving protein content, and enhancing host immunity (Fatimah et al., 2019). Understanding QS-regulated microbial phenotypes can further optimize biofloc functionality by enhancing enzyme production and nutrient availability, thereby benefiting aquaculture operations.

(Ju et al., 2008) analyzed microbial biomass in biofloc systems, finding it comprised approximately 24.6% phytoplankton (primarily diatom species such as *Thalassiosira, Chaetoceros,* and *Navicula*), 3% bacteria, a small fraction of protozoans (98% flagellates, 1.5% rotifers, and 0.5% amoebas), 33.2% dead material, and 39.25% ash. While heterotrophic bacteria dominate biofloc communities, other key species include fungi, algae (dinoflagellates and diatoms), flagellates, rotifers, and ciliates (Ju et al., 2008). In a macrocosm–microcosm system associated with tilapia culture, seven genera of ciliates, notably *Vorticella* and *Epistylis,* as well as rotifers from the *Philodina, Lecane,* and *Keratella* genera were identified (Monroy-Dosta et al., 2013). These results emphasize the complexity of microbial communities in biofloc across various farmed species.

BFT systems promote the growth of both autotrophic and heterotrophic microorganisms, with a significant abundance of beneficial heterotrophic species such as *Bacillus, Acinetobacter, Sphingomonas, Pseudomonas, Rhodopseudomonas, Micrococcus, Nitrosomonas, Nitrospira, Nitrobacter, Cellulomonas,* and various yeast species. These microorganisms play a vital role in bioremediation, improving water quality, and enhancing the growth, performance, and health of cultured organisms (Kumar et al., 2021).

Heterotrophic bacteria, particularly species such as *Bacillus,* dominate BFT systems. They decompose waste products, uneaten feed, and dead organic matter, converting ammoniacal nitrogen into innocuous microbial biomass (Kumar et al., 2021). This process is affected by factors such as ammonia levels, the C:N ratio, dissolved oxygen, temperature, alkalinity, salinity, light, and pH. Increasing the C:N ratio above 10 can stimulate heterotrophic bacterial growth, thereby enhancing system efficiency (Ebeling et al., 2006).

Chemoautotrophic bacteria present in the BFT system also contribute to water quality through nitrification, converting ammonia to nitrite and subsequently to nitrate. Genera such as *Nitrosomonas* and *Nitrobacter* are pivotal in this process. Environmental parameters such as ammonia and nitrite concentrations, Dissolve oxygen, temperature, alkalinity, salinity, pH, substrate, and total suspended solids (TSS) significantly impact nitrification efficiency in the biofloc system (Ray, 2012).

Studies have identified diverse bacterial communities in BFT systems. For example, in tilapia cultivation, species such as *Sphingomonas paucimobilis, Pseudomonas luteola, Bacillus* spp., and *Micrococcus* spp. have been observed. In Pacific white shrimp systems, dominant bacteria include *Vibrio* spp., *Photobacterium* spp., and *Marinobacter goseongensis.* The composition of these communities evolves over time, often shifting from pathogenic species in early stages to beneficial heterotrophic bacteria as the system matures (Monroy-Dosta et al., 2013). The abundance and diversity of bacterial populations in BFT systems are influenced by system management practices and environmental conditions. Maintaining optimal water quality parameters and appropriate microbial community structures is crucial for the health and growth of cultured species.

9.4.2 Bioremediation in biofloc technology system

Bioremediation utilizes beneficial microorganisms to transform contaminated water or waste into less toxic forms, offering an eco-friendly solution for waste management in aquaculture (Das et al., 2019; Das et al., 2020). This process can involve indigenous microbes or introduced species, including genetically engineered strains, to enhance its effectiveness (Brim et al., 2003). Bioremediation operates through biological mechanisms that reduce, remove, or convert pollutants, making it a cost-efficient and sustainable approach to contaminant degradation (Gadd, 2000; Jasmin et al., 2020). Its success depends on factors such as the nature of contaminants, hydrogeological conditions, and the ecology of microbial communities (Sarkar et al., 2021). Microbes play a crucial role by using pollutants as energy sources and assimilating nutrients from waste materials (Jones et al., 2018).

Microorganisms involved in bioremediation are categorized based on their carbon utilization: Autotrophs, which synthesize their own food from inorganic substances and include nitrite and AOB, and heterotrophs, which decompose organic material to obtain carbon and act as electron donors in oxidation reactions (Ebeling et al., 2006; Musyoka, 2016). Heterotrophic bacteria are particularly important in breaking down organic waste, converting nitrogenous compounds such as ammonia into harmless microbial agglomerates (Ebeling et al., 2006). Nitrogen, essential for aquatic organisms' physiological processes such as protein and nucleic acid synthesis, must be carefully managed, as excessive accumulation of ammonia (NH_3) and nitrites (NO_2^-) can be toxic to fish (Wei et al., 2016; Emerenciano et al., 2017).

BFT leverages heterotrophic microorganisms to reduce ammonia levels and maintain water quality, making it a valuable tool in aquaculture (Schneider et al., 2005). By fostering the growth of both autotrophic and heterotrophic microbes, including beneficial species such as *Bacillus, Pseudomonas, Nitrosomonas*, and *Nitrobacter*, BFT enhances water quality, growth performance, and the health of cultured aquatic animals (Adel et al., 2017). These microorganisms produce metabolic enzymes that degrade organic matter, converting contaminants into comparatively safer materials (Dash and Das, 2012). For example, heterotrophic bacteria have been found to play a significant role in decomposing organic waste in shrimp (*L. vannamei*) culture tanks, producing flocs that benefit the reared species (Kamilya et al., 2017; Manan et al., 2017) showed ammonia immobilization by heterotrophic bacteria, rather than nitrification, as the dominant mechanism influencing ammonia removal in the biofloc system. Similarly, Hostins et al. (2019) demonstrated that heterotrophic biofloc systems, with or without probiotics, effectively mitigate the impact of acute hepatopancreatic necrosis disease (AHPND), while improving water quality and shrimp survival rates.

The association of bacteria such as *P. stutzeri, Nitrobacter*, and *B. subtilis* with yeast (*Candida tropicalis*) or microalgae (*Schizochytrium* sp.) has also been shown to enhance water quality by reducing total ammonium nitrogen, nitrite, and nitrate concentrations (Gao et al., 2019). Furthermore, Kurniawan et al. (2020) identified diverse biofloc-forming bacteria, including

Proteobacteria, Bacteroidetes, and Actinobacteria, highlighting their potential for bioremediation in aquaculture systems. Recent studies on Nile tilapia (*Oreochromis niloticus*) and Asian stinging catfish (*Heteropneustes fossilis*) have revealed that Proteobacteria and Bacteroidetes dominate the microbial communities in biofloc systems, contributing to maintaining optimal water quality (Mali et al., 2024a,2024b). These findings underscore the importance of microbial diversity in biofloc systems for sustainable aquaculture practices.

9.5 Conclusion

Water probiotics, RAS, and BFT represent sustainable approaches to aquaculture, leveraging microbial communities to maintain water quality, recycle waste, and enhance the health and growth of cultured species. By degrading nitrogenous waste, these technologies reduce water consumption, minimize pollution, and prevent disease outbreaks. Further research into microbial interactions and composition can optimize system efficiency, making the systems a promising solution for global aquaculture and environmental management. The integration of advanced technologies and sustainable practices will ensure the continued success and scalability of these systems, contributing to global food security and environmental conservation.

References

Adel, M., El-Sayed, A.F.M., Yeganeh, S., Dadar, M., Giri, S.S., 2017. Effect of potential probiotic Lactococcus lactis Subsp. lactis on growth performance, intestinal microbiota, digestive enzyme activities, and disease resistance of Litopenaeus vannamei. Probiotics Antimicrob. Proteins 9 (2), 150–156. doi:10.1007/s12602-016-9235-9.

Al-Ajeel, S., Spasov, E., Sauder, L.A., McKnight, M.M., Neufeld, J.D., 2022. Ammonia-oxidizing archaea and complete ammonia-oxidizing Nitrospira in water treatment systems. Water Res. X 15. doi:10.1016/j.wroa.2022.100131.

Attramadal, K.J.K., Øie, G., Størseth, T.R., Alver, M.O., Vadstein, O., Olsen, Y., 2012. The effects of moderate ozonation or high intensity UV-irradiation on the microbial environment in RAS for marine larvae. Aquaculture 330-333, 121–129. doi:10.1016/j.aquaculture.2011.11.042.

Balcázar, J.L., Blas, I., Ruiz-Zarzuela, I., Cunningham, D., Vendrell, D., Múzquiz, J.L., 2006. The role of probiotics in aquaculture. Vet. Microbiol. 114 (3–4), 173–186. doi:10.1016/j.vetmic.2006.01.009.

Bartelme, R.P., McLellan, S.L., Newton, R.J., 2017. Freshwater recirculating aquaculture system operations drive biofilter bacterial community shifts around a stable nitrifying consortium of ammonia-oxidizing archaea and comammox *Nitrospira*. Front. Microbiol. 8, 101.

Blancheton, J.P., Attramadal, K.J.K., Michaud, L., d'Orbcastel, E.R., Vadstein, O., 2013. Insight into bacterial population in aquaculture systems and its implication. Aquacult. Eng. 53, 30–39. doi:10.1016/j.aquaeng.2012.11.009.

Brim, H., Venkateswaran, A., Kostandarithes, H.M., Fredrickson, J.K., Daly, M.J., 2003. Engineering Deinococcus geothermalis for bioremediation of high-temperature radioactive waste environments. Appl. Environ. Microbiol. 69 (8), 4575–4582. doi:10.1128/AEM.69.8.4575-4582.2003.

Chen, H., Moraru, C.I., Protasenko, V.V., 2023. Maximizing the disinfection effectiveness of 254 nm UV-C light with a special design unit: simulation and experimental approaches. Front. Food Sci. Technol. 3, 1223829.

Dahle, S.W., Gaarden, S.I., Buhaug, J.F., Netzer, R., Attramadal, K.J.K., Busche, T., Aas, M., Ribicic, D., Bakke, I., 2023. Long-term microbial community structures and dynamics in a commercial RAS during seven production batches of Atlantic salmon fry (Salmo salar). Aquaculture 565. doi:10.1016/j.aquaculture.2022.739155.

Das, B.K., Behera, B.K., Chakraborty, H.J., Paria, P., Gangopadhyay, A., Rout, A.K., Nayak, K.K., Parida, P.K., Rai, A., 2020. Metagenomic study focusing on antibiotic resistance genes from the sediments of River Yamuna. Gene 758, 144951.

Das, P., Chatterjee, S., Behera, B.K., Dangar, T.K., Das, B.K., Mohapatra, T., 2019. Isolation and characterization of marine bacteria from East Coast of India: functional screening for salt stress tolerance. Heliyon 5 (6). doi:10.1016/j.heliyon.2019.e01869.

Dash, H.R., Das, S., 2012. Bioremediation of mercury and the importance of bacterial mer genes. Int. Biodeterior. Biodegrad. 75, 207–213. doi:10.1016/j.ibiod.2012.07.023.

Dauda, A.B., Ajadi, A., Tola-Fabunmi, A.S., Akinwole, A.O., 2019. Waste production in aquaculture: sources, components and managements in different culture systems. Aquac. Fish. 4 (3), 81–88. doi:10.1016/j.aaf.2018.10.002.

Deng, Y., Verdegem, M.C.J., Eding, E., Kokou, F., 2022. Effect of rearing systems and dietary probiotic supplementation on the growth and gut microbiota of Nile tilapia (Oreochromis niloticus) larvae. Aquaculture 546. doi:10.1016/j.aquaculture.2021.737297.

Dewali, S., Parveen, N., Kathayat, N., Rawat, G., Bora, S., Sharma, N.P., Pathak, V.M., Bisht, S.S., 2025. Biological cleansing: probiotic strategies for tackling contaminants in water and wastewater. Environ. Sci. Eng. 29, 255–279. doi:10.1007/978-3-031-82579-8_11.

Drønen, K., Roalkvam, I., Nilsen, H., Olsen, A.B., Dahle, H., Wergeland, H., 2022. Presence and habitats of bacterial fish pathogen relatives in a marine salmon post-smolt RAS. Aquac. Rep. 26, 101312. doi:10.1016/j.aqrep.2022.101312.

Ebeling, J.M., Timmons, M.B., Bisogni, J., 2006. Engineering analysis of the stoichiometry of photoautotrophic, autotrophic, and heterotrophic removal of ammonia–nitrogen in aquaculture systems. Aquaculture 257 (1–4), 346–358.

Elgendy, M.Y., Ali, S.E., Dayem, A.A., Khalil, R.H., Moustafa, M.M., Abdelsalam, M., 2024. Alternative therapies recently applied in controlling farmed fish diseases: mechanisms, challenges, and prospects. Aquac. Int. 32 (7), 9017–9078. doi:10.1007/s10499-024-01603-3.

El-Sayed, A.F.M., 2021. Use of biofloc technology in shrimp aquaculture: a comprehensive review, with emphasis on the last decade. Rev. Aquac. 13 (1), 676–705. doi:10.1111/raq.12494.

Emerenciano, M.G.C., Martínez-Córdova, L.R., Martínez-Porchas, M., Miranda-Baeza, A., 2017. Biofloc technology (BFT): a tool for water quality management in aquaculture. Biofloc technology–a practical guide book. InTechOpen doi:10.5772/66416.

Ende, S., Henjes, J., Spiller, M., Elshobary, M., Hanelt, D., Abomohra, A.E.-F., 2024. Recent advances in recirculating aquaculture systems and role of microalgae to close system loop. Bioresour. Technol. 407, 131107. doi:10.1016/j.biortech.2024.131107.

Fachri, M., Amoah, K., Huang, Y., Cai, J., Alfatat, A., Ndandala, C.B., Shija, V.M., Jin, X., Bissih, F., Chen, H., 2024. Probiotics and paraprobiotics in aquaculture: a sustainable strategy for enhancing fish growth, health and disease prevention–a review. Frontiers in Marine Science 11, 1499228. doi:10.3389/fmars.2024.1499228.

FAO, 2024. The State of World Fisheries and Aquaculture 2024 – Blue Transformation in action. Food and Agriculture Organization of the United Nations (FAO), Rome.

Fatimah, N., Pande, G.S.J., Natrah, F.M.I., Meritha, W.W., Widanarni, A.S., Ekasari, J., 2019. The role of microbial quorum sensing on the characteristics and functionality of bioflocs in aquaculture systems. Aquaculture 504, 420–426. doi:10.1016/j.aquaculture.2019.02.022.

Ferreira, G.S., Bolívar, N.C., Pereira, S.A., Guertler, C., Vieira, F.d.N., Mouriño, J.L.P., Seiffert, W.Q., 2015. Microbial biofloc as source of probiotic bacteria for the culture of Litopenaeus vannamei. Aquaculture 448, 273–279. doi:10.1016/j.aquaculture.2015.06.006.

Fuller, R., 1989. Probiotics in Man and Animals. Journal of Applied Bacteriology 66 (5), 365–378. doi:10.1111/j.1365-2672.1989.tb05105.x.

Gadd, G.M., 2000. Bioremedial potential of microbial mechanisms of metal mobilization and immobilization. Curr. Opin. Biotechnol. 11 (3), 271–279. doi:10.1016/S0958-1669(00)00095-1.

Gao, F., Liao, S., Liu, S., Bai, H., Wang, A., Ye, J., 2019. The combination use of Candida tropicalis HH8 and Pseudomonas stutzeri LZX301 on nitrogen removal, biofloc formation and microbial communities in aquaculture. Aquaculture 500, 50–56. doi:10.1016/j.aquaculture.2018.09.041.

Ghosh, T., 2025. Recent advances in the probiotic application of the Bacillus as a potential candidate in the sustainable development of aquaculture. Aquaculture 594, 741432. doi:10.1016/j.aquaculture.2024.741432.

González, Y., Gómez, G., Moeller-Chávez, G.E., Vidal, G., 2023. UV disinfection systems for wastewater treatment: emphasis on reactivation of microorganisms. Sustainability 15 (14), 11262. doi:10.3390/su151411262.

Gorito, A.M., Pesqueira, J.F.J.R., Moreira, N.F.F., Ribeiro, A.R., Pereira, M.F.R.P., Nunes, O.C., Almeida, C.M.R., Silva, A.M.T., 2021. Ozone-based water treatment (O_3, O_3/UV, O_3/H_2O_2) for removal of organic micropollutants, bacteria inactivation and regrowth prevention. J. Environ. Chem. Eng. 9 (4), 105315. https://doi.org/10.1016/j.jece.2021.105315.

Hostins, B., Wasielesky, W., Decamp, O., Bossier, P., De Schryver, P., 2019. Managing input C/N ratio to reduce the risk of Acute Hepatopancreatic Necrosis Disease (AHPND) outbreaks in biofloc systems – a laboratory study. Aquaculture 508, 60–65. doi:10.1016/j.aquaculture.2019.04.055.

Jahangiri, L., Esteban, M.Á., 2018. Administration of probiotics in the water in finfish aquaculture systems: a review. Fishes 3 (3). doi:10.3390/fishes3030033.

Jasmin, M.Y., Syukri, F., Kamarudin, M.S., Karim, M., 2020. Potential of bioremediation in treating aquaculture sludge: review article. Aquaculture 519, 734905. doi:10.1016/j.aquaculture.2019.734905.

John, E.M., Krishnapriya, K., Sankar, T.V., 2020. Treatment of ammonia and nitrite in aquaculture wastewater by an assembled bacterial consortium. Aquaculture 526, 735390. doi:10.1016/j.aquaculture.2020.735390.

Jones, D.L., Magthab, E.A., Gleeson, D.B., Hill, P.W., Sánchez-Rodríguez, A.R., Roberts, P., Ge, T., Murphy, D.V., 2018. Microbial competition for nitrogen and carbon is as intense in the subsoil as in the topsoil. Soil Biol. Biochem. 117, 72–82. doi:10.1016/j.soilbio.2017.10.024.

Ju, Z.Y., Forster, I., Conquest, L., Dominy, W., Kuo, W.C., Horgen, F.D., 2008. Determination of microbial community structures of shrimp floc cultures by biomarkers and analysis of floc amino acid profiles. Aquac. Res. 39 (2), 118–133. doi:10.1111/j.1365-2109.2007.01856.x.

Kamilya, D., Debbarma, M., Pal, P., Kheti, B., Sarkar, S., Singh, S.T., 2017. Biofloc technology application in indoor culture of Labeo rohita (Hamilton, 1822) fingerlings: the effects on inorganic nitrogen control, growth and immunity. Chemosphere 182, 8–14. doi:10.1016/j.chemosphere.2017.05.021.

Kamilya, D., Devi, W.M, 2022. Bacillus probiotics and bioremediation: an aquaculture perspective. In: Bacilli in climate resilient agriculture and bioprospecting. Springer Science and Business Media LLC, Cham, pp. 335–347. doi:10.1007/978-3-030-85465-2_15.

Kesarcodi-Watson, A., Kaspar, H., Lategan, M.J., Gibson, L., 2008. Probiotics in aquaculture: The need, principles and mechanisms of action and screening processes. Aquaculture 274 (1), 1–14. doi:10.1016/j.aquaculture.2007.11.019.

Khanjani, M.H., Sharifinia, M., 2020. Biofloc technology as a promising tool to improve aquaculture production. Rev. Aquac. 12 (3), 1836–1850. doi:10.1111/raq.12412.

Khanjani, M.H., Sharifinia, M., Hajirezaee, S., 2022. Recent progress towards the application of biofloc technology for tilapia farming. Aquaculture 552, 738021.

Kumar, V., Roy, S., Behera, B.K., Swain, H.S., Das, B.K., 2021. Biofloc microbiome with bioremediation and health benefits. Front. Microbiol. 12. doi:10.3389/fmicb.2021.741164.

Kurniawan, K., Wheeler, R., Dann, L.M, Mitchell, J.G, 2020. Investigating the effects of urban input on the abundance and diversity of potential bio-floc forming bacteria in the River Murray, South Australia. IOP Conf. Ser.: Earth Environ. Sci. 521 (1), 12015. doi:10.1088/1755-1315/521/1/012015.

Leonard, N., Blancheton, J.P., Guiraud, J.P., 2000. Populations of heterotrophic bacteria in an experimental recirculating aquaculture system. Aquacult. Eng. 22, 109–120.

Loos, D., Filho, A.P.D.C., Dutilh, B.E., Barber, A.E., Panagiotou, G., 2024. A global survey of host, aquatic, and soil microbiomes reveals shared abundance and genomic features between bacterial and fungal generalists. Cell Rep. 43 (4). doi:10.1016/j.celrep.2024.114046.

López, L.A., Zaballos, A., 2005. Diversidad y actividad procariota en ecosistemas marinos. Revista Ecosistemas 14 (2), 30–40.

Louvado, A., Castro, C., Silva, D.A.M., Oliveira, V., Conceição, L.E.C., Cleary, D.F.R., Gomes, N.C.M., 2023. Assessing the effects of rotifer feed enrichments on Turbot (Scophthalmus maximus) larvae and post-larvae gut-associated bacterial communities. Microorganisms 11 (2). doi:10.3390/microorganisms11020520.

Mali, P., Ghosh, S., Dash, G., Chowdhury, S., 2024a. Effects of different carbon-to-nitrogen ratios on Nile tilapia *Oreochromis niloticus* in the biofloc rearing system. Fish. Sci. 90 (2), 239–256.

Mali, P., Ghosh, S., Dash, G., Chowdhury, S., 2024b. Influence of various carbon and nitrogen ratios in a jaggery-based biofloc rearing system on the Asian stinging catfish *Heteropneustes fossilis*. Aquacult. Eng. 107, 102469.

Manan, H., Moh, J.H.Z., Kasan, N.A., Suratman, S., Ikhwanuddin, M., 2017. Identification of biofloc microscopic composition as the natural bioremediation in zero water exchange of Pacific white shrimp, Penaeus vannamei, culture in closed hatchery system. Appl. Water Sci. 7 (5), 2437–2446. doi:10.1007/s13201-016-0421-4.

Manchanayake, T., Salleh, A., Amal, M.N.A., Yasin, I.S.M., Zamri-Saad, M., 2023. Pathology and pathogenesis of Vibrio infection in fish: a review. Aquac. Rep. 28. doi:10.1016/j.aqrep.2022.101459.

McKnight, M.M., Neufeld, J.D., 2024. Comammox Nitrospira among dominant ammonia oxidizers within aquarium biofilter microbial communities. Appl. Environ. Microbiol. 90 (7). doi:10.1128/aem.00104-24.

Michaud, L., Lo Giudice, A., Troussellier, M., Smedile, F., Bruni, V., Blancheton, J.P., 2009. Phylogenetic characterization of the heterotrophic bacterial communities inhabiting a marine recirculating aquaculture system. J. Appl. Microbiol. 107 (6), 1935–1946. doi:10.1111/j.1365-2672.2009.04378.x.

Monier, M.N., Kabary, H., Elfeky, A., Saadony, S., El-Hamed, N.N.B.A., Eissa, M.E.H., Eissa, E.S.H., 2023. The effects of Bacillus species probiotics (Bacillus subtilis and B. licheniformis) on the water quality, immune responses, and resistance of whiteleg shrimp (Litopenaeus vannamei) against Fusarium solani infection. Aquac. Int. 31 (6), 3437–3455. doi:10.1007/s10499-023-01136-1.

Monroy-Dosta, M., Oca, J., Castro-Mejia, D., 2015. Becerril-Cortés, Importance and function of microbial communities in aquaculture systems with no water exchange. Sci. J. Anim. Sci. 4 (9), 103–110.

Monroy-Dosta, M.d.C., de Lara, R.A., Castro-Mejía, J., Castro-Mejía, G., Coelho-Emerenciano, M.G., 2013. Microbiology community composition and abundance associated to biofloc in tilapia aquaculture. Rev. Biol. Mar. Oceanogr. 48 (3), 511–520. doi:10.4067/s0718-19572013000300009.

Moriarty, D.J.W., 1998. Control of luminous *Vibrio* species in penaeid aquaculture ponds. Aquaculture 164 (1–4), 351–358. doi:10.1016/S0044-8486(98)00199-9.

Musyoka, S., 2016. Concept of microbial bioremediation in aquaculture wastes: review. Int. J. Adv. Sci. Technol. Res. 5 (1), 1–10.

Neissi, A., Rafiee, G., Rahimi, S., Farahmand, H., Pandit, S., Mijakovic, I., 2022. Enriched microbial communities for ammonium and nitrite removal from recirculating aquaculture systems. Chemosphere 295, 133811. doi:10.1016/j.chemosphere.2022.133811.

Piedrahita, R.H., 2003. Reducing the potential environmental impact of tank aquaculture effluents through intensification and recirculation. Aquaculture doi:10.1016/S0044-8486(03)00465-4. 00448486 1-4 35 44 Elsevier, United States http://www.journals.elsevier.com/aquaculture/226 .

Podduturi, R., Petersen, M.A., Vestergaard, M., Jørgensen, N.O.G., 2020. Geosmin fluctuations and potential hotspots for elevated levels in recirculated aquaculture system (RAS): a case study from pikeperch (Stizostedion lucioperca) production in Denmark. Aquaculture 514. doi:10.1016/j.aquaculture.2019.734501.

Priest, F.G., 1993. Bacillus subtilis and other Gram-positive bacteria, biochemistry, physiology, and molecular genetics. American Society for Microbiology, Washington, DC, pp. 3–16.

Putu Ayustin Suriasni, F., Faizal, C., Panatarani, W., Hermawan, I., Made, J., 2023. A review of bubble aeration in biofilter to reduce total ammonia nitrogen of recirculating aquaculture system. Water 15 (4), 808. doi:10.3390/w15040808.

Ray, A.J., 2012. Management of Biological and Chemical Constituents for the Advancement of Intensive, Minimal-exchange, Biofloc-based Shrimp (*Litopenaeus vannamei*) Aquaculture (PhD Dissertation). The University of South Mississippi, p. 472.

Reiser, S., Schroeder, J.P., Wuertz, S., Kloas, W., Hanel, R., 2010. Histological and physiological alterations in juvenile turbot (Psetta maxima, L.) exposed to sublethal concentrations of ozone-produced oxidants in ozonated seawater. Aquaculture 307 (1–2), 157–164. doi:10.1016/j.aquaculture.2010.07.007.

Runia, W.van Os, EA (Ed.), 1996. Disinfection of recirculation water from closed production systems 01, 20–24.

Rurangwa, E., Verdegem, M.C.J., 2015. Microorganisms in recirculating aquaculture systems and their management. Rev. Aquac. 7 (2), 117–130. doi:10.1111/raq.12057.

Sarkar, D.J., Sarkar, B.K.D., Sahoo, B.K., Das, A., Nag, S.K., Manna, R.K., Behera, B.K., Samanta, S., 2021. Occurrence, fate, and removal of microplastics as heavy metal vectors in natural wastewater treatment wetland systems. Water Res. 192. https://doi.org/10.1016/j.watres.2021.116853.

Schneider, O., Sereti, V., Eding, E.H., Verreth, J.A.J., 2005. Analysis of nutrient flows in integrated intensive aquaculture systems. Aquacult. Eng. 32 (3-4), 379–401. doi:10.1016/j.aquaeng.2004.09.001.

Sharrer, M.J., Summerfelt, S.T., Bullock, G.L., Gleason, L.E., Taeuber, J., 2005. Inactivation of bacteria using ultraviolet irradiation in a recirculating salmonid culture system. Aquacult. Eng. 33 (2), 135–149. doi:10.1016/j.aquaeng.2004.12.001.

Skjermo, J., Salvesen, I., Øie, G., Olsen, Y., Vadstein, O., 1997. Microbially matured water: a technique for selection of a non-opportunistic bacterial flora in water that may improve performance of marine larvae. Aquac. Int. 5 (1), 13–28. doi:10.1007/bf02764784.

Slepecky, R.A., Hemphill, H.E., 2006. The Genus *Bacillus*—Nonmedical. In: Dworkin, M., Falkow, S., Rosenberg, E.,

Schleifer, KH., Stackebrandt, E. (Eds.), The Prokaryotes. Springer, New York, NY. doi.org/10.1007/0-387-30744-3_16.

Soltani, M., Kuebutornye, F.K.A., Hlordzi, V., Abarike, E.D., Lu, Y., Mraz, J., 2025. Bacillus probiotics as the bioremediatory tools in aquaculture. In: Bacillus probiotics for sustainable aquaculture. CRC Press, Boca Raton, FL,, pp. 128–148.

Sonthiphand, P., Hall, M.W., Neufeld, J.D., 2014. Biogeography of anaerobic ammonia-oxidizing (anammox) bacteria. Front. Microbiol. 5, 399. doi:10.3389/fmicb.2014.00399.

Sugita, H., Nakamura, H., Shimada, T., 2005. Microbial communities associated with filter material in recirculating aquaculture systems of freshwater fish. Aquaculture 243, 403–409.

Tal, Y., Watts, J.E.M., Schreier, S.B., Sowers, K.R., Schreier, H.J., 2003. Characterization of the microbial community and nitrogen transformation processes associated with moving bed bioreactors in a closed recirculated mariculture system. Aquaculture 215 (1–4), 187–202. doi:10.1016/S0044-8486(02)00372-1.

Taufik, M., Tuan Ismail, T.I., Manan, H., Ikhwanuddin, M., Abdul Salam, A.I., Abdul Rahim, A.I., Ishak, A.N., Kamaruzzan, A.S., Draman, A.S., Kasan, N.A., 2024. Synergistic effects of recirculating aquaculture system (RAS) with combination of clear water, probiotic and biofloc technology: a review. Aquac. Fish. 9 (6), 883–892. doi:10.1016/j.aaf.2023.07.006.

Tepaamorndech, S., Nookaew, I., Higdon, S.M., Santiyanont, P., Phromson, M., Chantarasakha, K., Mhuantong, W., Plengvidhya, V., Visessanguan, W., 2020. Metagenomics in bioflocs and their effects on gut microbiome and immune responses in Pacific white shrimp. Fish Shellfish Immunol. 106, 733–741. doi:10.1016/j.fsi.2020.08.042.

Tom, A.P., Jayakumar, J.S., Biju, M., Somarajan, J., Ibrahim, M.A., 2021. Aquaculture wastewater treatment technologies and their sustainability: a review. Energy Nexus 4. doi:10.1016/j.nexus.2021.100022.

Verschuere, L., Rombaut, G., Sorgeloos, P., Verstraete, W., 2000. Probiotic bacteria as biological control agents in aquaculture. Microbiol. Mol. Biol. Rev. 64 (4), 655–671. doi:10.1128/mmbr.64.4.655-671.2000.

Wang, C., Pan, L., Zhang, K., Xu, W., Zhao, D., Mei, L., 2016. Effects of different carbon sources addition on nutrition composition and extracellular enzymes activity of bioflocs, and digestive enzymes activity and growth performance of Litopenaeus vannamei in zero-exchange culture tanks. Aquac. Res. 47 (10), 3307–3318. doi:10.1111/are.12784.

Wei, Y.F., Liao, S.A., Wang, A., 2016. The effect of different carbon sources on the nutritional composition, microbial community, and structure of bioflocs. Aquaculture 465, 88–93.

Xu, J., Nicholaus, R., Wang, Y., Yang, W., Zhu, J., Zheng, Z., 2024. Enhanced aquaculture wastewater treatment using water spinach (Ipomoea aquatica Forsskal) and exogenous compound bacteria. J. Water Process Eng. 63, 105521. doi:10.1016/j.jwpe.2024.105521.

Zhang, X.P., Fu, L.Q., Deng, B., Liang, Q., Zheng, J.J., Sun, J.D., Zhu, H.Y., Peng, L.S., Wang, Y.B., Shen, W.Y., Li, W.F., 2013. *Bacillus subtilis* SC02 supplementation causes alterations of the microbial diversity in grass carp water. World Journal of Microbiology and Biotechnology 29 (9), 1645–1653. doi:10.1007/s11274-013-1327-z.

Zhu, X., Liu, J., Li, L., Zhen, G., Lu, X., Zhang, J., Liu, H., Zhou, Z., Wu, Z., Zhang, X., 2023. Prospects for humic acids treatment and recovery in wastewater: a review. Chemosphere 312 (Pt 2), 137193. doi:10.1016/j.chemosphere.2022.137193.

CHAPTER

10

Bioremediation and reuse in aquaculture: Optimizing resources and reducing impacts

Alfredo Olivera Gálvez[a], *Priscilla Celes Maciel de Lima*[a], *Jessika de Lima Abreu*[a], *Clarissa Vilela Figueiredo da Silva Campos*[b], *Danielle Alves da Silva*[a], *Carlos Yure Barbosa de Oliveira*[c] and *Luis Otavio Brito*[a]

[a] Department of Fisheries and Aquaculture, Federal Rural University of Pernambuco (UFRPE), Recife, Pernambuco, Brazil [b] Academic Unit of Serra Talhada, Federal Rural University of Pernambuco (UFRPE), Serra Talhada, Brazil [c] Department of Botany, Federal University of Santa Catarina (UFSC), Florianópolis, Santa Catarina, Brazil

10.1 Introduction

Aquaculture, a practice dedicated to the cultivation and management of aquatic organisms such as fish, crustaceans, and plants in controlled environments, has become a key alternative to meet the growing global demand for high-quality protein, especially in regions where natural fishery resources are scarce (Troell et al., 2014; Tlusty et al., 2019).

However, this production expansion brings significant environmental challenges. Water pollution arises from various factors, including the use of commercial feed, the accumulation of waste and organic residues in tanks, and disease. As aquaculture grows, improper waste management poses a risk to the balance of aquatic ecosystems, impacting both water quality and the health of cultured organisms (Wang et al., 2012; Dauda et al., 2019).

In light of these issues, bioremediation has emerged as an approach that combines efficiency and sustainability. Based on natural biological processes, this technique leverages living organisms to break down or transform pollutants into harmless compounds. The practice

Zero-Waste Technologies in Aquaculture: Global Perspectives and Methods toward a Circular Economy
DOI: https://doi.org/10.1016/B978-0-443-34689-7.00020-3

aims not only to reduce pollutant load but also to recover and recycle nutrients from waste, fostering a more sustainable production cycle that aligns with circular economy principles (Chary et al., 2023; Perales-Pérez et al., 2023). By incorporating bioremediation into aquaculture management, modern aquaculture can mitigate its environmental impact while contributing to the preservation of the ecosystems in which it operates.

10.1.1 The role of bioremediation in aquaculture

Bioremediation in aquaculture involves the use of living organisms such as bacteria, algae, and aquatic plants to recycle nutrients, treat wastewater, and reduce the toxicity of pollutants. This approach leverages the natural capacity of these organisms to absorb and transform harmful compounds, directly enhancing water quality and the efficiency of aquaculture systems. Acting as biological purifiers, these organisms help stabilize the cultivation environment, promoting a more integrated and sustainable aquaculture (Granada et al., 2016).

Innovative technologies, such as the use of microalgae and biofilters composed of bacteria, have gained prominence for their effectiveness in removing nutrients such as nitrogen and phosphorus, which pose ecological risks in excessive amounts. In water recirculating aquaculture systems (RASs) and integrated multitrophic aquaculture (IMTA), these methods significantly reduce water renewal requirements and optimize resource use (Estim, 2015; Badiola et al., 2018). Beyond environmental benefits, the use of bioflocs, for instance, generates protein-rich biomass that can be reused as feed, creating an efficient production cycle and reducing reliance on external inputs (Sharma et al., 2023). This process not only addresses waste treatment needs but also contributes to the economic viability of aquaculture production.

10.1.2 Resource reuse and optimization in aquaculture

Reusing treated water and recycling nutrients are central aspects of bioremediation applied to sustainable aquaculture. In conventional systems, the intensive use of water to dilute and remove pollutants creates a high demand for this resource, an issue particularly relevant in water-scarce regions. With bioremediation, the need for constant water exchange is drastically reduced, enabling the creation of a closed and efficient system (Marlida, 2020). The integration of microalgae in consortia with bacteria has proven highly effective in nutrient recycling and waste treatment, boosting productivity while minimizing environmental impact (Perales-Pérez et al., 2023).

Solutions such as aquaponics, where plants and fish are cultivated together, demonstrate a practical application of these sustainability principles (Ibrahim et al., 2023). In this model, nutrients from fish waste are used to support plant growth, fostering a closed-loop system that values residual biomass. This approach aligns directly with circular economy principles, where waste reutilization translates into a more complete and environmentally responsible system, supporting sustainable production and natural resource conservation (Wu et al., 2022).

10.2 Chronology of bioremediation

The evolution of bioremediation in aquaculture reflects an increasing pursuit of environmentally sustainable practices in response to the impacts of intensive production. From the initial efforts to control waste to the advanced use of bioengineering, bioremediation in aquaculture has progressed significantly. Key milestones in this trajectory are highlighted in the following sections.

10.2.1 1970–1980: Initial signs of environmental issues

With the rapid expansion of aquaculture in the 1970s and 1980s, particularly in Asia and Europe, large-scale production grew substantially. However, this growth brought a set of challenges: Inadequate infrastructure and management practices led to the accumulation of nutrients, primarily nitrogen and phosphorus, in nearby water bodies. This phenomenon resulted in eutrophication, causing ecological imbalances such as algal blooms and oxygen depletion, which adversely affected aquatic biodiversity. During this period, the main approach to address these issues was frequent water renewal, which required significant water volumes and raised long-term sustainability concerns. This initial phase was critical in fostering environmental awareness and underscoring the need for more effective waste management solutions (Naylor et al., 2000; Temporetti et al., 2000).

10.2.2 1990: Introduction of bioremediation applications

In the 1990s, bioremediation began to emerge as a viable alternative to mitigate aquaculture's environmental impacts. Researchers started exploring the use of microorganisms and aquatic plants to remove excess nutrients and improve water quality. Technologies such as artificial wetlands (Lin et al., 2002) and the first RASs (Schneider et al., 2005) were tested as solutions to transform and recycle waste within the cultivation environment itself. This period also solidified the concept of a circular economy (Verreth et al., 2023), as nutrients removed from the water could be recycled, promoting an integrated and sustainable system that minimized waste and environmental impact.

10.2.3 2000–2010: Consolidation of bioremediation technologies

Between 2000 and 2010, the practice of bioremediation in aquaculture became more widely established. During this period, the use of biofilters and plants for nutrient removal advanced significantly. The integration of aquatic plants, such as water lettuce (*Pistia stratiotes*) (Palomino et al., 2022) and duckweed (*Lemna minor*) (Ceschin et al., 2020), proved effective for removing nitrogen and phosphate compounds, which are responsible for eutrophication. Simultaneously, aquaponic systems—combining fish and plant cultivation in a closed-loop cycle—demonstrated that aquatic waste could be converted into nutrients for plant growth, reducing dependency on fertilizers and water (Estim et al., 2020). This period marked a shift in how aquaculture approached its waste: Instead of discarding it, waste was repurposed, generating added value within the system itself.

10.2.4 2010–2020: Expansion of recirculation and technological integration

Starting in 2010, bioremediation in aquaculture advanced significantly with the evolution of RASs, which provide greater control over water quality and substantial water resource savings. These systems have proven crucial in regions where water availability is limited, such as arid areas or those with strict environmental regulations. The ability to safely reuse water reduced both costs and environmental impacts, making aquaculture economically viable in locations previously considered unsuitable (Badiola et al., 2018; Owatari and Mouriño, 2020). In addition, IMTA improved, supporting the cultivation of organisms such as mollusks and algae, which act as natural purifiers by absorbing residual nutrients and balancing the ecosystem (Nederlof et al., 2021. The implementation of water quality monitoring and control technologies also enabled more efficient use of microorganisms, optimizing the decomposition of organic matter and pollutant control (Gleiser and Moro, 2023). The use of biofloc technology (BFT) also gained popularity during this period. This approach involves the formation of microbial aggregates that treat the water by breaking down nitrogen compounds and transforming waste into protein-rich biomass, which can be used as a supplemental feed for the cultivated organisms. BFT not only enhances sustainability but also improves efficiency by incorporating more nitrogen and phosphorus into animal biomass, reducing the need for frequent water exchanges (Marlida, 2020).

10.2.5 2020–Present: Bioengineering and innovations in bioremediation

In recent years, advances in aquaculture bioremediation have been marked by the use of bioengineering techniques that enhance the ability of microorganisms and plants to break down pollutants and regulate excess nutrients in aquatic environments. One of these innovations involves the development of marine microorganisms capable of producing bioactive compounds that offer sustainable alternatives to antibiotics (Hamza and Zinjarde, 2023). Besides combating pathogens, these biomolecules serve as dietary supplements, improving the overall health of aquaculture species and contributing to water quality and system sustainability (Srirengaraj et al., 2023). In addition, bacteriophage therapy has emerged as an eco-friendly approach to control bacterial pathogens, offering a sustainable solution to the challenges of antibiotic resistance in aquaculture (Çagatay, 2023). The use of bioactive microorganisms thus represents an effective solution to health and environmental quality challenges in cultivation systems.

Another significant advancement is the development of specially designed microbial consortia for bioremediation. By reconfiguring metabolic pathways and optimizing interactions between microbial species, these consortia can effectively degrade pollutants. Engineering these consortia allows for the tuning of microbial interactions to achieve specific functions, such as the decomposition of hazardous pollutants into nontoxic or less harmful byproducts. This approach is gaining traction as a promising alternative, though it still faces challenges related to assembling and maintaining stable, efficient microbial communities (Li et al., 2021; Cao et al., 2022).

In addition, the encapsulation of denitrifying bacteria in RASs is another emerging technique in aquatic bioremediation. These encapsulated bacteria play a crucial role in removing accumulated nitrate, a common byproduct in intensive aquaculture systems. The encapsulation method creates an ideal microenvironment for the bacteria, enabling them to function efficiently and maintain water quality without frequent exchanges. Introducing these capsules helps control nitrogenous compounds, providing a solution for nutrient management in closed systems (Li et al., 2023; Zuo et al., 2023).

In a similar vein, the use of macroalgae in integrated mariculture systems has proven effective in absorbing nitrogen and phosphorus from fish farming. In this context, systems such as IMTA incorporate filter-feeding organisms, such as mussels and macroalgae, which act as natural bioremediators. These organisms not only purify the water column by removing bacteria and other contaminants but also utilize available nutrients for their growth, closing the nutrient cycle and offering additional benefits such as biomass production (Li et al., 2020a; Giangrande et al., 2021).

Lastly, BFT has evolved to incorporate symbiotic systems, also known as "aquamimicry." These systems combine probiotics and prebiotics to replicate natural environmental conditions. This approach not only improves water quality but also fosters ecological stability, promoting the growth of beneficial organisms such as phytoplankton and zooplankton, which serve as supplementary food sources for the cultivated species. This diverse environment creates a healthier ecosystem, enhancing the growth of cultured animals and standing out as a promising innovation for sustainable aquaculture (Khanjani and Sharifinia, 2020; Santhanam et al., 2020).

These recent advances indicate that bioengineering in aquaculture bioremediation is increasingly geared toward integrated and sustainable solutions, where the balance between production and environmental preservation is central.

10.3 Challenges in aquaculture

Modern aquaculture is viewed as a sustainable solution to meet the growing global demand for animal protein. However, the expansion of this sector raises several environmental and operational challenges. Among the primary issues are nutrient pollution, solid waste accumulation, excessive water usage, susceptibility to disease outbreaks, and the release of chemical residues (Campanati et al., 2021).

10.3.1 Nutrient pollution

Nutrient pollution is one of the major challenges in intensive aquaculture systems. The breakdown of uneaten feed and excretions from aquatic organisms releases large amounts of nitrogen and phosphorus into the environment, leading to nutrient buildup in the water. At high concentrations, these nutrients trigger eutrophication, a process where algal blooms rapidly consume oxygen, resulting in fish mortality and the degradation of entire ecosystems. This phenomenon also affects local biodiversity and impairs water quality (Amorim and Moura, 2021; El-Sheekh et al., 2021).

Bioremediation through bioflocs and multitrophic systems emerges as a viable alternative to mitigate this impact. Bioflocs, composed of bacterial and microbial aggregates, enable the absorption and reuse of nutrients, transforming them into protein-rich biomass that can be used to feed the cultivated organisms (Robles-Porchas et al., 2020). Multitrophic systems, integrating species such as mollusks and algae, help absorb nutrients and create a more efficient nutrient recycling cycle (Nederlof et al., 2021). Research indicates that these approaches significantly reduce pollution levels and minimize the need for water replenishment, promoting a more sustainable model that is less reliant on external resources (Mugwanya et al., 2021; Papageorgiou et al., 2023).

10.3.2 Solid and organic waste

The accumulation of solid waste, such as feces and feed remnants, poses another challenge in aquaculture, especially in intensive systems. These wastes tend to settle at the bottom of tanks, directly impacting water quality and reducing oxygen levels, which can facilitate the spread of pathogenic bacteria (Niyi-David et al., 2021). Manual removal of these wastes is costly and inefficient. In this context, biological techniques such as biofiltration and the use of microorganisms to decompose these wastes have proven effective in improving water quality. Besides removing suspended solids, these biological processes generate biomass that can be used as a dietary supplement, promoting a waste valorization cycle. Thus the system becomes more sustainable as wastes are repurposed within the production cycle, reducing waste and minimizing environmental impact (Li et al., 2020b; Chiquito-Contreras et al., 2022).

10.3.3 Inefficient water use

Intensive aquaculture requires large volumes of water to dilute pollutants, which becomes problematic in areas where water resources are limited. In arid regions or areas with restricted water resources, excessive water use presents both economic and environmental obstacles. In conventional systems, frequent water exchanges are necessary to maintain the aquatic environment's quality, but this increases operational costs and the environmental footprint of production (Young-Bin et al., 2023).

RASs and bioremediation offer viable solutions to address this issue, allowing for efficient recycling of water after treatment. These systems utilize biofilters and other purification mechanisms to remove pollutants and organic residues, enabling continuous water reuse. In water-scarce areas, this efficiency is crucial for the economic and environmental viability of aquaculture (Pfeiffer and Ott, 2023; Taufik et al., 2023; Zhou et al., 2023).

10.3.4 Disease outbreaks

High organism density in intensive aquaculture systems increases the risk of disease outbreaks, especially when water quality is compromised by nutrient and organic waste buildup. These conditions create an environment conducive to the proliferation of pathogens, which can quickly spread and cause epidemics among cultured organisms, leading to considerable economic losses (Kousar et al., 2020; Yilmaz et al., 2022).

Bioremediation, by improving water quality and reducing the presence of pathogenic agents, helps prevent these outbreaks. The use of probiotics and the incorporation of bioflocs are effective practices for maintaining a microbiologically healthy environment, fostering a beneficial microbiota that inhibits pathogen growth. In addition, these practices enhance the overall health of fish and increase the resilience of aquaculture systems, contributing to safer and more sustainable production (Nathanailides et al., 2021; Aparna et al., 2023; Kamil et al., 2023; Sohel et al., 2023).

10.3.5 Chemical residues and antibiotics

The use of chemicals and antibiotics was once common in aquaculture for disease and parasite control (Dinesh et al., 2023). However, excessive use of these products can result in the release of toxic residues into the environment, affecting aquatic biodiversity and fostering the development of resistant bacteria. This issue is exacerbated in areas where water is not adequately treated before disposal, posing a risk to the ecosystem (Chen et al., 2020).

Bioremediation, particularly through the use of plants and microorganisms, presents a natural solution for neutralizing these residues. Various species of aquatic plants and algae are capable of absorbing and breaking down toxic compounds, reducing the need for chemicals and improving water quality in a sustainable manner. This practice not only lessens aquaculture's environmental impact but also contributes to a more balanced and safer environment (Kashem et al., 2023; Nasir et al., 2023; Nie et al., 2023).

10.4 Bioremediation techniques

Bioremediation in aquaculture involves techniques that use living organisms, such as microorganisms, plants, and algae, to treat contaminated water, recycle nutrients, and enhance the environmental quality of aquaculture systems. The following are some key approaches, including practical references and recent studies.

10.4.1 Microorganisms and biofilters

Microorganisms play a central role in aquaculture bioremediation. Nitrifying and denitrifying bacteria, for instance, can convert toxic nitrogen compounds, such as ammonia and nitrites, into less harmful nitrogen forms, improving water quality and facilitating recirculation (RAS) in aquaculture (Huang et al., 2020; Shu et al., 2021). Biofilter systems that support these specific bacterial populations ensure continuous removal of nitrogen compounds and minimize the need for water exchanges (Zhou et al., 2022).

Moreover, the use of BFT is an important innovation in aquaculture, where microbial aggregates help metabolize nitrogen compounds, reducing ammonia and nitrite concentrations in the water. This process enhances water quality and decreases the need for frequent water changes. In addition, the protein-rich biomass produced from bioflocs can be used as feed for fish and shrimp, promoting a more efficient nutrient cycle within the system. While BFT does not directly reduce eutrophication in external environments, it minimizes the potential for eutrophication by preventing the discharge of nutrient-rich water, thus contributing to the sustainability of intensive systems (Crab et al., 2007; Emerenciano et al., 2013).

10.4.2 Aquatic plants and phytoremediation

Phytoremediation with aquatic plants is widely used in aquaculture bioremediation (Gao et al., 2023). Species such as water lettuce (*P. stratiotes*) and duckweed (*L. minor*) are recognized for their ability to absorb nutrients, particularly nitrogen and phosphorus, which contribute to pollution in cultivation systems. Studies have shown that *P. stratiotes* can remove pollutants, improving water quality in eutrophic ecosystems and aquaculture environments (Nahar and Hoque, 2021). In the experimental system, *L. minor* achieved a removal rate of 83.33% for N–NO_3^-, highlighting its value in phytoremediation systems and aquaculture (Alam et al., 2021).

10.4.3 Algae and microalgae

Algae and microalgae play a crucial role in bioremediation due to their ability to rapidly absorb dissolved nutrients such as nitrogen and phosphorus, while also providing oxygen and capturing carbon dioxide—key elements for the environmental stability of aquaculture systems

(Wang et al., 2023). Studies on the microalga *Tetraselmis suecica* in recirculating systems have demonstrated its high nutrient removal capacity, making it an excellent candidate for large-scale bioremediation systems. This alga has proven effective in absorbing nitrogen compounds and generating biomass that can be used as a dietary supplement (Andreotti et al., 2020). *Arthrospira platensis* (Spirulina) and other microalgae have been studied as natural biofilters and have shown efficacy in aquaculture systems by contributing protein-rich and antioxidant biomass, adding value to the system (Cardoso et al., 2022). In the context of IMTA, algae such as *Gracilaria* and *Saccharina latissima* assist in absorbing nutrients released by fish, promoting a sustainable environment and reducing the need for external inputs (Rugiu et al., 2020; Sawant and Jaiswar, 2024).

10.4.4 Integrated multitrophic aquaculture systems

IMTA is an approach that combines different species, such as fish, mollusks, and algae, within the same system. The waste produced by one species serves as nutrients for another, creating a nutrient cycle and promoting a balanced and sustainable environment (Knoop et al., 2022). Studies on macroalgae, such as *Gracilaria*, have shown that IMTA can significantly reduce nutrient loads, especially nitrogen and phosphorus, and increase productivity (Pawar et al., 2020; Narvarte et al., 2023).

10.4.5 Synbiotic systems and aquamimicry

Symbiotic technology, also known as aquamimicry, combines the use of probiotics and prebiotics to create an aquatic environment that simulates natural conditions, promoting the growth of phytoplankton and zooplankton. These organisms, including copepods, serve as additional food sources for shrimp and other cultured organisms while helping to control pathogen proliferation, fostering a healthier and more ecologically stable system (Romano, 2017; Khanjani et al., 2022). This technique also reduces the need for supplemental feed and operational costs by supporting a more self-sustaining cultivation environment with less reliance on chemical inputs, such as antibiotics (Panigrahi et al., 2019; Deepak et al., 2020; Nisar et al., 2022). In smaller-scale production units, this approach has shown great efficiency, providing organic and sustainable management for aquaculture, aligned with environmentally responsible and green practices (Kumar et al., 2018; Santhanam et al., 2020).

10.5 Transforming effluents into resources: Bioremediation by aquatic organisms

10.5.1 Microalgae

Aquaculture has experienced significant growth in recent years, and according to FAO data (2024), global aquaculture production reached approximately 130.9 million tons in 2022. This activity requires a high demand for water and generates large quantities of effluents rich in nutrients, containing organic and inorganic compounds such as nitrogen, phosphorus, and carbon, which can lead to the eutrophication of adjacent water bodies (De Lacerda et al., 2006; Cyrino et al., 2010; Crab et al., 2012; Hu et al., 2012). Aquaculture effluents contain high levels of nutrients, residual feed, feces from cultivated animals, and antibiotics

TABLE 10.1 Main groups and species of phytoplankton used in the bioremediation of effluents.

Microalgae species	Effluent source	Bioremediation	References
Scenedesmus obliquus	Wasterwater aquaculture	Reduced nitrate by 77.7%, orthophosphate by 100%, and nitrite 73.83%	Ansari et al. (2017)
Chlorella Sorokiniana		Reduced nitrate by 75.76%, orthophosphate by 100%, and nitrite 81.79%	
Ankistrodesmus falcatus		Reduced nitrate by 80.85%, orthophosphate by 98.52%, and nitrite 99.73%	
Platymonas subcordiformis	Wastewater aquaculture	Reduced nitrogen by 87%–95% and phosphorus by 98%–99%	Guo et al. (2013)
Chlorella vulgaris	Aquaculture wastewater	Reduced nitrogen by 100% and phosphorus 100%	Viegas et al. (2021)
Arthrospira platensis	Aquaculture wastewater from Nile tilapia	Reduced nitrate by 99%, and nitrite 90%	Moraes et al. (2021)
Chlorella vulgaris	Aquaculture wastewater from Nile tilapia	Reduced nitrate by 80.1%, orthophosphate by 48.3%, and nitrite 93.6%	Oliveira et al. (2020)

(Huang et al., 2019; Li et al., 2020; Kurniawan et al., 2021). As a result, many researchers have been working to develop cost-effective and efficient approaches for treating the effluents generated by aquaculture activities, with the aim of reducing their adverse environmental impact and preserving natural water resources (Table 10.1).

In recent years, several studies have been conducted using microalgae in the bioremediation of different types of effluents (Daneshvar et al., 2019; Oliveira et al., 2020; Chai et al., 2021; Moraes et al., 2021; Oliveira et al., 2024). When cultivated in media rich in nitrogen, phosphorus, and organic carbon, microalgae are capable of growing rapidly through heterotrophic or mixotrophic pathways — depending on the availability of light energy (Perez-Garcia et al., 2011). Microalgae are photosynthetic organisms that use sunlight as an energy source to convert carbon dioxide and nutrients into biomass (Richmond, 2004). Currently, they have gained attention for their ability to generate value-added biomass through nutrient recovery in wastewater treatment (Ahmad et al., 2022), thereby improving environmental sustainability.

As bioremediators, microalgae have the ability to reduce the organic load present in effluents, acting in the recovery of phosphorus and nitrogen (Suleiman et al., 2020), absorbing heavy metal ions into their tissues (biosorption), storing them (bioaccumulation) (Shah and Shah, 2020), and transforming them into less toxic compounds (Xia et al., 2019). They also have potential in the removal of pathogens present in liquid effluents (Silva et al., 2019). According to

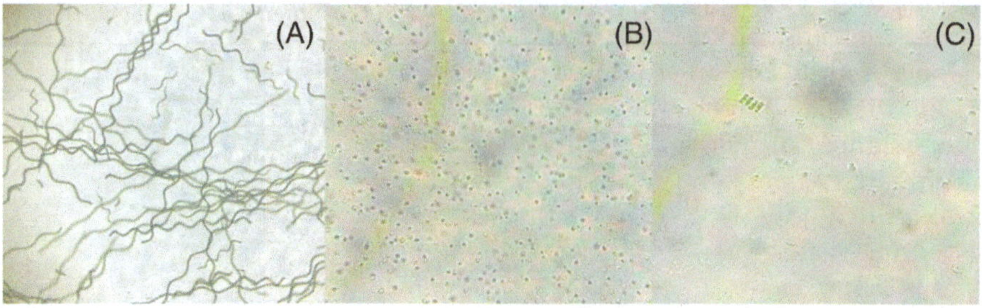

FIGURE 10.1 Illustrative photos (magnification 100x) of the main microalgae used in bioremediation. *Arthrospira platensis* (A), *Chlorella vulgaris* (B), and *Scenedesmus* spp. (C). Photos taken by Abreu, J L., Universidade Federal Rural de Pernambuco, Brazil.

Godos et al. (2011), some species also exhibited coagulating, flocculating, and sedimenting potential, which are interesting characteristics in the biological treatment of effluents. Furthermore, algae have the ability to synthesize both primary and secondary metabolites, and according to Lopes da Silva et al. (2019), they can also produce high-value products using liquid and solid effluents as resources.

One of the advantages of using microalgae in this process is their ability to transform what they absorb into bioactive compounds of high value. These include phycobiliproteins, fatty acids, chlorophylls, carotenoids, and vitamins, which can be widely used in pharmaceutical products, cosmetics, food additives, and ingredients (Raposo et al., 2013). According to Khatoon et al. (2016), microalgae-based products have great commercialization potential, as pure biomass is in high market demand and has significant value. The biomass can be used in several ways: (1) Live food in larviculture, (2) raw material in the pharmaceutical and cosmetic industries, (3) premix for dietary supplementation, and (4) production of organisms focused on improving health.

10.5.1.1 Main species used

A. platensis: This species is capable of removing nitrogen and phosphorus, as well as being effective in the removal of copper and zinc (Cordeiro et al., 2025). It is of great commercial importance due to its ability to synthesize proteins, vitamins, essential amino acids, gamma-linolenic acid (omega-6), and other polyunsaturated fatty acids, minerals, and pigments such as beta-carotene (Vonshak, 2014) (Fig. 10.1A).

Chlorella spp.: This species has a high capacity for nitrogen and phosphorus removal and is widely used in effluent treatment systems due to its rapid growth and high efficiency in nutrient absorption (Ruiz et al., 2011; Toyama et al., 2018; Gao et al., 2019). It is capable of synthesizing chlorophylls, carotenoids, proteins, and essential fatty acids (such as omega-3) (Fig. 10.1B).

Scenedesmus spp.: This species is effective in nutrient removal and adapts well to different environmental conditions, making it an advantageous choice for bioremediation processes (Gao et al., 2019). As bioactive compounds, it produces carotenoids, vitamin B12, and polyunsaturated fatty acids (Fig. 10.1C).

10.5.2 Zooplankton and bioremediation

Among the most commonly used organisms in the process of bioremediation and nutrient cycling, zooplankton has stood out for its ability to act as a natural biofilter in aquatic systems (Rehman et al., 2008; Marinho-Soriano et al., 2011; Kostopoulou et al., 2012; Pau et al., 2013; Gómez et al., 2019; Prado-Cabrero et al., 2022). Organisms such as cladocerans, rotifers, and copepods are widely studied due to their ability to filter large volumes of water, removing suspended particles, organic matter, microorganisms, and even heavy metals (Gómez et al., 2019; Chen et al., 2020; Gorokhova et al., 2020; Pous et al., 2021; Sadaiappan et al., 2021; Prado-Cabrero et al., 2022). By feeding on phytoplankton and suspended particles, zooplankton helps reduce nutrient loads in the water, preventing the excessive growth of algae and bacteria that affect water quality and compromise ecosystem health. Species such as the water flea *Daphnia magna*, for instance, are widely used as bioremediation agents, as these microcrustaceans act as biological filtration agents, removing organic matter, microalgae, and bacteria, improving water transparency and quality in secondary and tertiary effluent treatment systems (Pau et al., 2013; Serra et al., 2014; Pous et al., 2021).

In the specific context of aquaculture, the use of zooplankton in effluent bioremediation has gained prominence. The presence of high nutrient concentrations in the aquatic environment favors the excessive growth of algae, affecting water quality and compromising the health of cultivated organisms, such as fish and crustaceans. In this scenario, introducing zooplankton species into aquaculture systems can serve as an effective strategy for controlling and mitigating these effluents. Organisms such as cladocerans, rotifers, copepods, and branchiopods help filter suspended organic particles (Marinho-Soriano et al., 2011; Ha et al., 2013; Pau et al., 2013; Prado-Cabrero et al., 2022), controlling nutrient excess and promoting ecological balance in the production system. Studies report new strategies for achieving more sustainable aquaculture, such as aquamimicry, which involves inducing the formation of a zooplankton community, especially copepods, which will serve not only as live feed for cultivated organisms but also in nutrient cycling, such as nitrogen and phosphorus, as well as phytoplankton control, preventing algal blooms and restricting the entry of pathogenic bacteria into the cultivation environment (Deepak et al., 2020; Khanjani et al., 2022; Deepti et al., 2024). In addition to improving water quality, using zooplankton in these systems can reduce the need for chemical interventions, such as using products to control algae, making production more sustainable and environmentally friendly. Thus using zooplankton for effluent bioremediation in aquaculture not only contributes to the health of the aquatic environment but also promotes more sustainable and cost-effective practices, aligning with conservation principles and efficiency in aquaculture production.

10.5.2.1 Main species used

The use of zooplankton for the bioremediation of aquaculture effluents has emerged as a promising strategy to mitigate the environmental impacts associated with intensive aquaculture. Several zooplankton species have demonstrated the ability to remove suspended particles, assimilate excess nutrients, and reduce organic matter and other pollutants, thereby improving water quality and promoting ecological balance within aquaculture production systems (Table 10.2). The principal zooplankton groups that have been investigated and applied for aquaculture effluent bioremediation are described below.

Cladocera: This group includes important species such as *Daphnia* sp. and *Moina* sp. The water flea *D. magna* (Fig. 10.2A) is capable of removing suspended particles, such as phytoplankton,

TABLE 10.2 Main groups and species of zooplankton used in the bioremediation of effluents.

Zooplankton group	Zooplankton species	Effluent source	Performance	Bioremediation	References
Cladocera	*Daphnia magna*	Urban wastewater treatment plant	Singular	39% reduction in particles	Pau et al. (2013)
	Daphnia magna	Urban wastewater treatment plant	Singular	Up to 35% of particles and inactivated *E. coli* in 1.2 log units	Serra et al. (2014)
	Daphnia similis	Nile tilapia *Oreochromis niloticus* farming (BFT)	Consortium (microalgae)	Reduced nitrate by 70.37%, orthophosphate by 75.74%, and total suspended solids by 90.74%	Campos et al. (2022)
	Daphnia magna	—	Consortium (microalgae)	Nitrogen reduction reaching removal rates of approximately 35%	Chen et al. (2020)
	Daphnia magna	—	Consortium (macrophyte and microalgae)	redução de aproximadamente 30% do nitrogênio	Chen et al. (2020)
	Daphnia magna	Urban wastewater treatment plant	Consortium (macrophyte, bacteria, and microalgae)	Reduced concentrations of nitrogen, organic carbon, suspended solids, and *E. coli* bacteria	Pous et al. (2021)
	Daphnia galeata	Lake Shirakaba, Japan	Consortium (Rainbow trout)	Biomanipulation for phosphorus reduction and microalgal *blooms*	Ha et al. (2013)
	Moina macrocopa	Nile tilapia *Oreochromis niloticus* farming (BFT)	Singular	Total suspended solids reduction	Shidik et al. (2021)
Rotifera	*Brachionus plicatilis*	Aquaculture	Singular	Phytoplankton and nitrate control	Kostopoulou et al. (2012)
	Philodina roseola	Industrial wastewater (Kasur district, Lahore, Pakistan)	Singular	Reduced 88% Cr^{6+}, 77% Cu^{2+}, and 89% Cd^{2+}	Rehman et al. (2008)
	Notommata copeus	Industrial wastewater (Kasur district, Lahore, Pakistan)	Singular	Reduced 82% Cr^{6+}, 61% Cu^{2+}, and 97% Cd^{2+}	Rehman et al. (2008)

(*continued on next page*)

Copepoda	*Tigriopus californicus*	Green water aquaculture	Consortium (microalgae)	Phytoplankton and nitrate control	Prado-Cabrero et al. (2022)
	Acartia bifilosa, Eurytemora affinis, Pseudocalanus acuspes, and Limnocalanus macrurus	Northern Baltic Sea	Consortium (bacteria)	Mercury cycling	Gorokhova et al. (2020)
	Pleuromamma spp., and *Temora* spp.	—	Consortium (bacteria)	Iron reduction	Sadaiappan et al. (2021)
	Pleuromamma spp.	—	Consortium (bacteria)	Nitrogen fixation	Sadaiappan et al. (2021)
	Temora spp	—	Consortium (bacteria)	Sulfate reduction	Sadaiappan et al. (2021)
Branchiopoda	*Artemia franciscana*	Aquaculture	Singular	Removal of nitrogen compounds, especially nitrite (100%) and nitrate (70%)	Marinho-Soriano et al. (2011)
	Artemia franciscana	Aquaculture	Consortium (Macroalgae)	Reduced 29.8% of ammonia, 100% of nitrite, 72.4% of nitrate, and 44.5% of dissolved inorganic nitrogen	Marinho-Soriano et al. (2011)
Polychaeta	*Perinereis nuntia vallata*	Aquaculture		Assimilated 62.82% of the nitrogen contained in the feces of Japanese flounder, converting 50% of this assimilated nitrogen into body protein	Honda and Kikuchi (2002)
	Abarenicola pusilla	Aquaculture (RAS)	Singular	Reduced 85.37% of total organic matter	Gómez et al. (2019)

BFT, biofloc technology; *RAS*, recirculating aquaculture systems

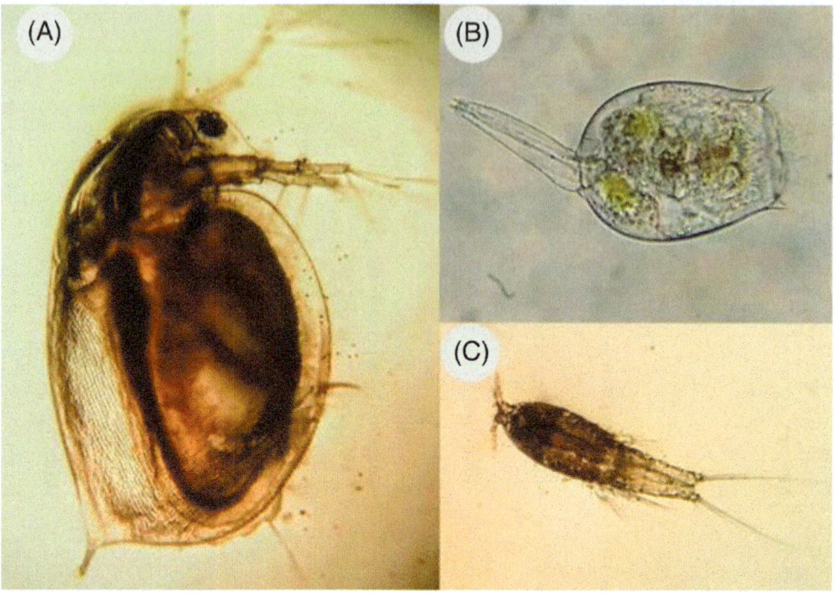

FIGURE 10.2 Illustrative photos (100x magnification) of some zooplankton groups. Cladocera: Represented by *Daphnia magna* (A), Rotifera: Represented by *Brachionus plicatilis* (B), and Copepoda: Represented by *Tisbe holothuriae* (C). Photos taken by Campos, C. V. F. S. and Brandão, B. C. S., Universidade Federal Rural de Pernambuco, Brazil.

organic matter, and bacteria, thereby improving water quality in effluent treatment systems. Pau et al. (2013) found that in secondary effluent treatment, the reduction in particle concentration attributed to daphnia filtration ranged between 2.5% and 39%, corresponding to daphnia concentrations of between 5 and 100 ind L^{-1}. Similarly, Serra et al. (2014) reported that in tertiary effluent treatment, *D. magna* reduced the concentration of particles with diameters below 30 µm by 35%, depending on abiotic parameters such as water temperature and hydraulic retention time (HRT). Furthermore, *D. magna* treatment inactivated *E. coli* by 1.2 log units. Da Silva Campos et al. (2019) cultivated *D. similis* in tilapia biofloc effluent and obtained results in terms of growth and nitrate reduction. In parallel, *Moina macrocopa*, similar to *Daphnia* sp., adapts well to environments with high concentrations of organic matter, feeding on microalgae, decomposing material, and organic debris. *M. macrocopa* was studied by Shidik et al. (2021), who used effluent from tilapia biofloc culture for cladoceran production and obtained biomass production and total suspended solids reduction after 5 days of culture.

Rotifera: Rotifers are commonly used in aquaculture systems, especially for the culture of fish and crustacean larvae (Fig. 10.2B). However, the use of these organisms as bioremediators in aquaculture is still underexplored. Rotifers are excellent candidates for controlling algal blooms in lakes through food chain biomanipulation technology. With this approach, aquatic communities are manipulated to maintain phytoplankton biomass at low levels (Ha et al., 2013), having direct implications for water quality. Rotifers have a short developmental time, a high filtration rate, and can quickly reach high densities, with *Brachionus plicatilis* being notable. This species was found in aquaculture sedimentation basins, contributing to nutrient cycling and phytoplankton control (Kostopoulou et al., 2012). In addition, some rotifer species have shown tolerance to high concentrations of heavy metals, reducing these concentrations in the water, as observed in *Philodina roseola* and *Notommata copeus*. *P. roseola* could decrease 40% Cr^{6+}, 67% Cu^{2+}, and 76% Cd^{2+} from the medium within 2 days. By the end of eight days, this decrease for Cr^{6+}, Cu^{2+}, and Cd^{2+} was 88%, 77%, and 89%, respectively (Rehman et al., 2008). Meanwhile, *N. copeus* proved to be a more effective accumulator of heavy metals, with reductions of 82%, 61%, and 97% for Cr^{6+}, Cu^{2+}, and Cd^{2+}, respectively, after 2 days, and 96%, 81%, and 99%, respectively, after 8 days (Rehman et al., 2008). These species thus demonstrate potential for heavy metal cycling in effluents, as well as in aquaculture effluents.

Copepoda: Copepods are planktonic crustaceans that also play an important role in effluent bioremediation (Fig. 10.2C). They feed on suspended particles, phytoplankton, and bacteria, also acting as filters. Species such as *Tigriopus californicus* are noteworthy for phytoplankton control and nitrate reduction (Prado-Cabrero et al., 2022), while *Acartia bifilosa*, *Eurytemora affinis*, *Pseudocalanus acuspes*, and *Limnocalanus macrurus* are significant in mercury cycling (Gorokhova et al., 2020). In addition, *Pleuromamma* spp. and *Temora* spp. contribute to iron reduction (Sadaiappan et al., 2021).

Branchiopoda: Although widely known as live feed for fish and shrimp larvae, *Artemia spp.* also has potential for use in bioremediation because of their ability to filter organic particles and microorganisms such as microalgae and bacteria (Lenz and Browne 1991; Dhont and Sorgeloos, 2002). *Artemia franciscana*, for instance, has been noted for its efficiency in removing nitrogenous compounds, specifically nitrite (100%) and nitrate (70%) from aquaculture wastewater (Marinho-Soriano et al., 2011).

Polychaeta: Polychaetes possess the ability to assimilate particulate organic matter from intensive aquaculture systems, making them leading candidates for the bioremediation of aquaculture effluents (Robinson et al., 2015; Palmer, 2018). Certain groups, such as Nereididae and Arenicolidae, are notable for their sediment-feeding behavior and have shown efficacy in cycling organic matter derived from aquaculture activities (Guerrero and Cremades, 2012; Pajand et al., 2017). Key species include *Arenicola marina* (Guerrero and Cremades, 2012), *Nereis diversicolor* (Pajand et al., 2017), *Abarenicola pusilla* (Gómez et al., 2019), *Perinereis nuntia vallata* (Honda and Kikuchi, 2002), and *Hediste diversicolor* (Marques et al., 2018). *P. nuntia vallata* assimilated 62.82% of the nitrogen in the feces of Japanese flounder, converting approximately 50% of this assimilated nitrogen into body protein (Honda and Kikuchi, 2002). Similarly, *A. pusilla* (Quatrefages, 1866), when fed with organic sludge from a marine RAS for 45 days, achieved an 85.37% removal rate of total organic matter, equivalent to 35.77 ± 0.05 g m^{-2} day^{-1}, 33.63 ± 0.003 mg day^{-1} of total N, and 236.78 ± 0.003 mg day^{-1} total C when cultivated with 10% sludge and 200 organisms m^{-2} (Gómez et al., 2019).

These zooplankton species are extensively studied for their capacity to enhance water quality in aquaculture systems, while also providing an ecologically sustainable alternative for effluent management. They act as primary consumers, controlling phytoplankton and microalgae populations, as well as filtering agents, removing suspended particles and toxic compounds. The use of zooplankton in bioremediation can reduce the need for chemical treatments, promoting sustainability and the conservation of water resources.

10.5.2.2 Working with consortia

The use of consortia involving zooplankton, algae, fungi, and even multiple zooplankton species in the bioremediation of aquaculture effluents has proven to be an effective and sustainable approach to improving water quality. The combination of different organisms enhances the removal capacity of nutrients, organic particles, and toxic pollutants, creating more balanced and resilient systems. The following is a summary of case studies that illustrate this consortium-based practice.

Daphnia and microalgae: In studies conducted by Campos et al. (2022), the combination of *Daphnia similis* with the microalga *Chlorella vulgaris* demonstrated efficiency in removing up to 70.37% of nitrate, 75.74% of orthophosphate, and 90.74% of total suspended solids from aquaculture effluent from Nile Tilapia farming in low salinity. The microalgae assimilate inorganic nutrients such as orthophosphate and nitrate, while the zooplankton feed on microalgae particles and suspended solids, promoting greater water clarity. In addition, *D. magna* was investigated in a consortium with three microalgae species, *Cyclotella sp.*, *Microcystis aeruginosa*, and *C. vulgaris*, to reduce nitrogen in phosphorus-rich waters (2 mg L^{-1}), achieving approximately 35% removal rates in 16 days (Chen et al., 2020).

Daphnia, aquatic macrophytes, and microalgae: *D. magna*, along with the aquatic macrophyte *Ceratophyllum demersum* and three microalgae species, *Cyclotella sp.*, *M. aeruginosa*, and *C. vulgaris*, enabled a reduction of approximately 30% in nitrogen in waters with a 2 mg L^{-1} total phosphorus concentration over 16 days (Chen et al., 2020). Pous et al. (2021) found that *D. magna*, in cooperation with the floating aquatic macrophyte *Lemna* and bacterial/algal biofilm in tertiary effluent treatment, reduced concentrations of nitrogen, organic carbon, suspended solids, and *E. coli* bacteria.

Daphnia and fish: A study by Ha et al. (2013) demonstrated the biomanipulation potential of a consortium involving rainbow trout (*Oncorhynchus mykiss*) and *Daphnia galeata* to reduce phosphorus and microalgae blooms, increasing water clarity up to 4 meters in Lake Shirakaba, Japan, from 1997 to 2006.

Artemia and macroalgae: Research conducted by Marinho-Soriano et al. (2011) reported the use of a consortium of *A. franciscana* with the macroalga *Gracilaria caudata* for the bioremediation of aquaculture effluents, achieving a 29.8% reduction in ammonia, 100% nitrite removal, 72.4% nitrate reduction, and 44.5% decrease in dissolved inorganic nitrogen.

Copepods and microalgae: The work by Prado-Cabrero et al. (2022) documented the cultivation of the harpacticoid copepod *T. californicus* along with the microalga *Tetraselmis chui* in a bioreactor using effluent from Atlantic salmon juvenile culture over a period of 145 days. In addition to producing *T. californicus* biomass (3.9 kg), nutrient cycling was observed, with nitrate and orthophosphate concentrations reaching zero within the first 30 days.

Copepods and bacteria: Studies have shown that the association of copepod species with specific bacteria can contribute to nutrient cycling (Feng et al., 2023), such as nitrogen fixation (*Pleuromamma spp.*), sulfate reduction (*Temora spp.*), methane oxidation (*Acartia spp.*, *Calanus spp.*, *Centropages sp.*, *Pleuromamma spp.*, and *Temora spp.*), and even metal reduction, such as iron (*Pleuromamma spp.* and *Temora spp.*) (Sadaiappan et al., 2021) and mercury (*A. bifilosa*, *E. affinis*, *P. acuspes*, and *L. macrurus*) (Gorokhova et al., 2020). This demonstrates the potential of these microorganisms, in consortium with bacteria (bacteriobiomes), as a promising biotechnological tool for aquaculture applications, especially in aquatic biogeochemical cycling.

These case studies demonstrate that combining different organisms, such as zooplankton, algae, and bacteria, maximizes the efficiency of effluent bioremediation, offering a promising strategy for application in aquaculture systems. These symbiotic interactions allow for greater control over nutrients and pollutants, providing viable solutions for large-scale effluent management.

10.5.2.3 Future prospects

The use of zooplankton in the bioremediation of aquaculture effluents presents a promising and sustainable strategy for wastewater treatment, offering a natural and efficient alternative for the removal of excess nutrients, organic particles, and even certain pollutants. Key strengths include the ability of some species to filter large volumes of water, control phytoplankton growth, improve water clarity, and reduce nitrogen and phosphorus compounds, thereby promoting ecological balance in aquaculture systems. However, challenges and limitations exist, such as the need to adjust environmental conditions to optimize zooplankton performance, which can be sensitive to changes in temperature, salinity, and oxygen levels. In addition, the effectiveness of each species may vary depending on the effluent type and pollutant composition, necessitating specific studies for each case.

The importance of this approach is evident in the pursuit of more sustainable aquaculture practices, reducing reliance on chemical treatments and high-cost filtration technologies. Zooplankton, in addition to enhancing water quality, can be integrated into live feed production systems, adding value to the bioremediation process. Future research should focus on optimizing consortia with different organisms, such as algae and fungi, and exploring the potential of new zooplankton species for the removal of specific pollutants. It is also important to investigate the scalability of these systems for large aquaculture operations and to develop protocols for their effective management in various farming environments.

10.5.3 Macroalgae and their efficiency in nutrient removal

The global production of macroalgae has shown significant growth in recent years, reflecting the increased demand for these organisms. In 2022, approximately 36.4 million tons of macroalgae were produced, marking a 9.1% increase compared to 2018 production levels (FAO, 2024). This growth is driven by the versatility of macroalgae across various sectors and the expansion of production, particularly in Asia, which accounts for 99.3% of global production. China, in particular, stands out as the largest producer, with 22.4 million tons produced (FAO, 2024). Among the 32 cultivated species, the main ones include: *Laminaria japonica*, *Kappaphycus alvarezii*, *Gracilaria* spp., *Eucheuma* spp., *Porphyra* spp., *Undaria pinnatifida*, and *Caulerpa* spp.

The increase in macroalgae production not only demonstrates their economic value but also reveals an opportunity for incorporating these algae into aquaculture. With unique properties, macroalgae are capable of absorbing nutrients and contaminants, such as heavy metals, present in aquatic effluents, thus helping to mitigate the environmental impacts generated by aquaculture. Consequently, the integration of macroalgae in bioremediation processes or within IMTA emerges as a promising strategy, aligned with the needs for sustainable development and conservation of water resources. Incorporating macroalgae into aquaculture systems not only aids in nutrient recycling and significantly reduces nutrient levels in effluents (bioremediation) but also produces valuable macroalgal biomass. This biomass can be commercially utilized for

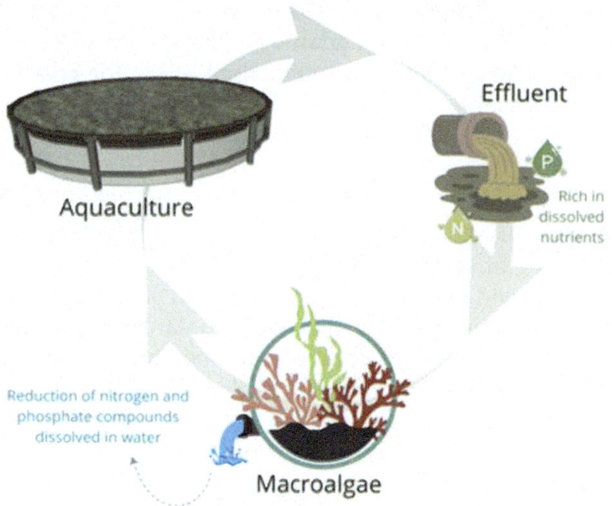

FIGURE 10.3 Bioremediation cycle in aquaculture systems with macroalgae.

producing food for human and animal consumption, cosmetics, pharmaceuticals, fertilizers, and bioactive compounds (Torres et al, 2019; Gelli et al., 2020; Rao and Ravishankar, 2022; Bhushan et al., 2023).

Aquaculture effluent treatment through bioremediation can be structured in distinct stages (primary, secondary, and tertiary treatment) applied independently or in combination, depending on the culture system's conditions and the levels of dissolved solids and nutrients (Rajanandini and Kumar, 2021). In primary treatment, there is mechanical (decantation or filtration) or biological (bivalve mollusks) reduction of solids; in secondary treatment, the degradation of dissolved organic contents occurs by microorganisms, aerobically and/or anaerobically; meanwhile, tertiary treatment involves the biological removal of organic and inorganic compounds, such as nitrogenous and phosphate compounds. Macroalgae are excellent candidates for tertiary effluent treatment, since they are capable of removing organic, inorganic substances, and toxic metals. Thus the integration of algae in cultures can balance the concentration of nutrients, offering an excellent strategy for sustainable systems (Fig. 10.3) (Neori et al., 2004).

Several species of macroalgae from the genera *Caulerpa, Ulva, Gracilaria, Solieria, Acanthophora*, and *Sargassum* have been used to remove nutrients in effluent treatment and improve water quality in crops (bioremediation and multitrophic aquaculture) (Msuya et al., 2008; Chaitanawisuti et al., 2011; Marinho-Soriano et al., 2011; Samocha et al., 2015; Kunawongdet, 2020). However, species selection should be based on their ecophysiological parameters, such as biofiltration capacity, biochemical composition, growth rate, and economic value. These characteristics aid in identifying the adaptive limitations of species in relation to the culture environment, optimizing the use of macroalgae (Kang et al., 2013).

To excel in bioremediation and IMTA systems, algae must demonstrate ease of culture, high absorption rates, and strong efficiency in removing nutrients present in effluents (Chopin et al., 2008; Mawi et al., 2020; Chirapart and Ruangchuay, 2022; Zrimec et al., 2022). As a result, Kang et al. (2013) recently developed a selection index to identify seaweed species most suitable for integrated aquaculture systems based on their economic value, ease of culture, application in aquaculture, and nutrient removal capacity, based on an analysis of existing literature, reference

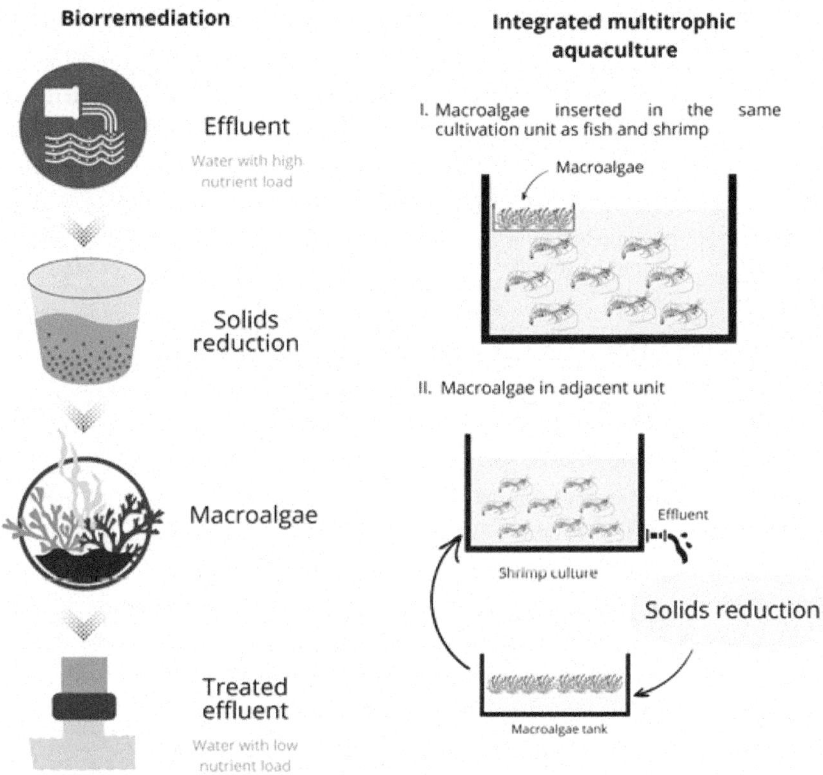

FIGURE 10.4 Applied strategies for using macroalgae to mitigate the environmental impacts of aquaculture.

data, and physiological experiments. In the study, the authors found that the seaweeds with the highest indexes and, consequently, the most suitable for use in aquaculture were: *U. pinnatifida, Porphyra yezoensis, Ulva compressa* e *Gracilaria incurvata* (Kang et al., 2013).

Various strategies can be implemented to utilize macroalgae in mitigating the environmental impacts of aquaculture. Macroalgae culture can occur as a specific stage within the effluent treatment flow, either as an integral part of recirculation systems or through static bioremediation (Jones et al., 2001; Jones et al., 2002; Guttman et al., 2018). In IMTA systems, macroalgae can be cultivated within the same unit as the primary species, directly utilizing available nutrients, or in adjacent units connected through water recirculation, where they receive nutrient-rich effluent (Fig. 10.4) (Barrington et al., 2009; Barrington et al., 2010; Chirapart and Ruangchuay, 2022; Huo et al., 2024).

The efficiency of macroalgae in nutrient removal in aquatic environments has shown promising results, with significant reductions in chemical oxygen demand and concentrations of nitrogen and phosphorus in both effluents and in closed-culture systems as well as natural environments (Abreu et al., 2011), as shown in Table 10.3. However, the nutrient absorption rate may depend on the form of nitrogen (NO_3^-, NO_2^-, or NH_4^+), concentration, relative molar proportions (nitrate/ammonia, nitrogen/phosphorus and carbon/nitrogen) and the species of macroalgae, in addition to factors such as initial concentration of ammonia and phosphate,

TABLE 10.3 Nutrient removal rates of some macroalgae species in aquaculture wastewater and integrated multitrophic aquaculture.

Macroalgae	Use	Removal rate (%)	References
Gracilaria birdiae	Bioremediation of shrimp biofloc wastewater	63.6% TAN 56.7% NO_2-N 22.5% PO_4	Brito et al. (2018)
Gracilaria birdiae	Bioremediation of shrimp wastewater	34% NH_4 100% NO_3 93.5% PO_4-P	Marinho-Soriano et al. (2009)
Gracilaria lichevoides	Bioremediation of shrimp wastewater	88.9%–93.4% NH_4–N 30.95%–55.2% NO_3–N	Zhou et al. (2021)
Gracilaria vermiculophylla	Laboratory test of absorption efficiency of N sources	63%–100% NH_4^+ 19%–53% NO_3^-	Abreu et al. (2011)
Gracilaria verrucosa	Bioremediation in an enclosed sea area (China)	54.1% NH_4–N 49.8% de NO_2–N 75.5% NO_3–N 49% de PO_4–P	Huo et al. (2011)
Gracilaria spp.	Integrated multitropic aquaculture system	22.2% da TAN 45.9% de NO_2–N 33.2% de NO_3-N	Lima et al. (2025)
Gracilaria edulis *Gracilaria changii*	Bioremediation of shrimp wastewater	72.5% NH_4 58.8% NO_3 45.9% PO_4 71% NH_4 56.8% NO_3 43.5% PO_4	Mawi et al. (2020)
Ulva lactuca	Integrated multitropic recirculating aquaculture system	62.5%–100% de TAN 22.5%–33.9% de PO_4–P	Huo et al. (2024)
Ulva lactuca	Bioremediation of fish and shrimp wastewater	33.3%–98.7%NH_4 52.4%–99.1% NO_2 84%–91.7% NO_3 60.9%–84.9% PO_4	Sebök and Hanelt (2023)
Ulva compressa.	Laboratory test absorption of ammonium	> 99% NH_4	Kang et al. (2013)
Ulva pseudorotundata *Ulva rigida*	Bioremediation of fish wastewater	99.3% NH_4 36.9% NO_3 94.7% NH_4 27.9% NO_3	Massocato et al. (2022)

(continued on next page)

TABLE 10.3 Nutrient removal rates of some macroalgae species in aquaculture wastewater and integrated multitrophic aquaculture—cont'd

Macroalgae	Use	Removal rate (%)	References
Porphyra yezoensis	Bioremediation in open sea	50%–94% NH_4–N 42%–91% NO_2–N 21%–38% NO_3–N 42%–67% PO_4–P	He et al. (2008)
Caulerpa macrophysa	Bioremediation of shrimp wastewater	50% NH_4–N 16.6% NO_2–N 38.3% NO_3–N 73.9% PO_4–P	Limhang et al. (2014)

NH_4–N, ammonia nitrogen; *NO_2*, nitrite; *NO_3*, nitrate; *NO_2–N*, nitrite nitrogen; *NO_3–N*, nitrate nitrogen; *PO_4*, orthophosphate; *PO_4–P*, orthophosphate phosphorus; *TAN*, total ammonia nitrogen.

salinity, temperature, turbidity and storage of macroalgae biomass (Chopin et al., 2001; Choi et al., 2010; Abreu et al., 2011; Zrimec et al., 2022).

Macroalgae possess various mechanisms to absorb and utilize nitrogen and phosphorus available in water. In all mechanisms, nitrogen enters the cells either actively or passively, crossing the cell membrane. Once inside the cell, nitrogen compounds are converted into amino acids and incorporated into macromolecules such as proteins, which are essential for growth. However, the development of macroalgae, as well as the absorption rate, can be limited by the concentration and source of nutrients present in the medium, since macroalgae, when cultivated in environments with different nitrogen sources, show a preference for NH_4^+ compared to NO_3^- (Choi et al., 2010; Abreu et al., 2011; Ale et al., 2011). This preference may be due to energy efficiency in utilizing NH_4^+; nitrate must undergo a reduction reaction before assimilation, which requires the synthesis of a greater number of enzymes—nitrate reductase, nicotinamide adenine dinucleotide, nitrite reductase, and ferredoxin (Abreu et al., 2011).

In addition to nutrient removal, macroalgae have also gained attention as an alternative for the removal of heavy metals from water, due to the properties of their cell walls, which readily bind to metals (Klimmek et al., 2001; Mehta and Gaur, 2005; Henriques et al., 2017). When evaluating the removal of metal ions (Cd^{2+}, Ni^{2+}, Pb^{2+} , and Zn^{2+}) by algae, Klimmek et al. (2001) observed a greater removal of Cd and Pb compared to Ni and Zn, while Henriques et al. (2017) found that *Ulva lactuca* was capable of removing more than 95% of Cd, Hg, and Pb from monometallic solutions, highlighting the promising potential of macroalgae for water decontamination treatments.

10.5.4 Use of bivalve mollusks for aquaculture bioremediation

Bivalve mollusks are aquatic invertebrates, represented by species such as oysters, mussels, scallops, and clams, and because of their adaptability and filtering capacity, these organisms are valuable resources for bioremediation in aquaculture systems (Barrington et al., 2009; Newell, 2004; Martinez-Cordova and Martinez-Porchas, 2006; 2010). Bivalves can play a crucial role in mitigating nutrient and organic waste pollution in aquatic ecosystems, contributing to nutrient cycling and improving ecosystem health (Peterson et al., 2010). The growing interest in using bivalves for bioremediation and environmental sustainability in aquaculture has also led to increased research and application of these organisms in IMTA systems (Barrington et al., 2010).

TABLE 10.4 Efficiency of nutrient and solid reduction in aquaculture systems by bivalve species.

Species	Use	Removal rate (%)	References
Anomalocardia flexuosa	Bioremediation of shrimp biofloc wastewater	63.6% TAN 78.4 NO_2–N 69.5% TSS 74.3% VSS 22.5% PO_4	Brito et al. (2018)
Crassostrea rivularis	Bioremediation of shrimp-intensive wastewater	25.2%–35.8% TAN 48.5%–75.1% NO_2–N 71.2%–96.5% NO_3–N 43.6%–49% PO_4–P 50.75%–80.1% TSS 94.1%–98.8% Total coliforms	Lich (2023)
Corbicula fluminea		20.9%–39.3% TAN 57.65%–75.5% NO_2–N 31%–47.8% NO_3–N 44.1%–57.5% PO_4–P 44.8%–56% TSS 80.6%–83.7% Total coliforms	
Saccostrea commercialis	Integrated treatment of shrimp effluent	88% TSS 32.6% total N 37.1% total P 68.4% bactérias 91.5% Clorofila a	Jones et al. (2001)
Saccostrea commercialis	Bioremediation of shrimp wastewater	35.8% total N 44.8% Total P 21.4% TSS 69.9% chlorophylla 61% bacteria	Jones et al. (2002)
Dreissena bugensis	Bioremediation of wastewater	99.8% NH_4–N 73.9 NO_2–N 2.1% PO_4–P	van der Meer et al. (2023)
Mytilus galloprovincialis	Integrated multitropic recirculating aquaculture system	16.1%–18.9% Total solid	Huo et al. (2024)

NO_2, nitrite; *NO_3*, nitrate; *NO_2–N*, nitrite nitrogen; *NO_3–N*, nitrate nitrogen; *PO_4*, orthophosphate; *TAN*, total ammonia nitrogen; *TSS*, total suspended solids; *VSS*, volatile suspended solids.

Studies indicate that bivalves, such as *Mytilus edulis, Perna canaliculus, Crassostrea* sp., among others, have a high growth and survival potential in these culture systems and can contribute significantly to reducing the nutrient load, such as total nitrogen and total phosphorus, the concentration of bacteria and turbidity of aquaculture effluent, reaching a filtration rate of 66.03 mg h^{-1} g^{-1} and a biodeposition rate of 63.6 mg h^{-1} g^{-1}(Jones et al., 2002; Ren et al., 2012; Azevedo et al., 2015; Granada et al., 2016; Ertl et al., 2016; Nascimento et al., 2022) (Table 10.4).

By capturing and removing these residues (organic and inorganic particles), bivalves contribute significantly to improving water quality and promoting a more sustainable environment for aquaculture production. The filtration process of bivalves includes the ingestion of particles suspended in the water column, such as phytoplankton, organic debris, and small inorganic particles (Hopkins et al., 1993; Jones et al., 2001; Jones et al., 2002). Throughout this process, the organisms capture and select nutrients and suspended solids which, under normal conditions, would be accumulating in the culture system (Jmeliova and Sanz, 1969). The particles captured by the bivalves can be rejected (pseudofeces) or filtered through the gills into the labial palps, selected, digested, and the material not used is eliminated as feces (Deslous-Paoli et al., 1992; Smaal et al., 2019). In this way, inorganic particles are generally aggregated and excreted as pseudofeces, whereas organic particles are digested and incorporated into the bivalve tissue, leading to a reduction in excess nutrient and heavy metal loads in the system (Tenore and Dunstan, 1973; Allison et al., 1998; Boening, 1999). During filtration, regardless of the material captured, bivalve mollusks coagulate small particles (seston) into larger, more complex particles through the aggregation of expelled material, aiding in the sedimentation (clarification) of suspended solids (Fig. 10.5) (Kautsky and Evans, 1987; Jones et al., 2001).

In bioremediation, bivalves can be used in closed cultures or in coastal and estuarine areas with high concentrations of nutrients and pollutants, contributing to the improvement of water quality. In closed bioremediation systems, to ensure the success of effluent biofiltration, it is essential to maintain adequate water circulation to evenly distribute food particles, allowing balanced growth of bivalves and efficient removal of aquatic waste (Thielker, 1981). This practice minimizes the accumulation of metabolites and particle sedimentation, factors that could compromise the effectiveness of the bioremediation process, water reuse, and the health of the mollusks, as excess solids may cause gill clogging (Lima et al., 2022).

In IMTA systems, bivalves such as oysters, mussels, and scallops function as biofilters, removing suspended particles, nutrients, and other waste from the water (Oliveira et al., 2014; Royan, 2020; Lima et al., 2021; Lima et al., 2022). In this system, these organisms can be used alone or combined with other species from different trophic levels, such as algae (Samocha et al., 2015; Brito et al., 2016; Lima et al., 2022) and even fish (Holanda et al., 2020), forming an integrated system that not only maximizes resource utilization and reduces environmental impact but also diversifies production and enhances the economic return of enterprises (Soto, 2009; Troell et al., 2009; Barrington et al., 2010; Ahmed and Glaser, 2016; Biswas et al., 2020).

For the ideal conditioning of bivalves in a multitrophic system, it is essential to consider the positioning in water columns, the density of culture, and compatibility with other species to ensure efficient filtration and prevent stress on the organisms (Barrington et al., 2009). However, when mollusks are cultivated in contact with the bottom, solids can accumulate in the gills, jeopardizing their survival (Lima et al., 2022). In addition, the culture structure used can also influence the efficiency of solid and nutrient removal and the performance of the mollusks, since when these animals are stored separately (in adjacent units), a water flow and the presence of organic particles, microalgae and planktonic organisms must be guaranteed so that they continue to feed during culture (Pearse et al., 1987; Ward, 1996; Lima et al., 2022).

In addition to improving water quality, the presence of bivalves in systems can serve as a natural substrate for the development of a microbial consortium (biofilm), which can assist in nitrification processes (Arnold et al., 2009; Zhang et al., 2016), as the bacteria and microalgae present in the biofilm are capable of utilizing these compounds (Moss and Moss, 2004;

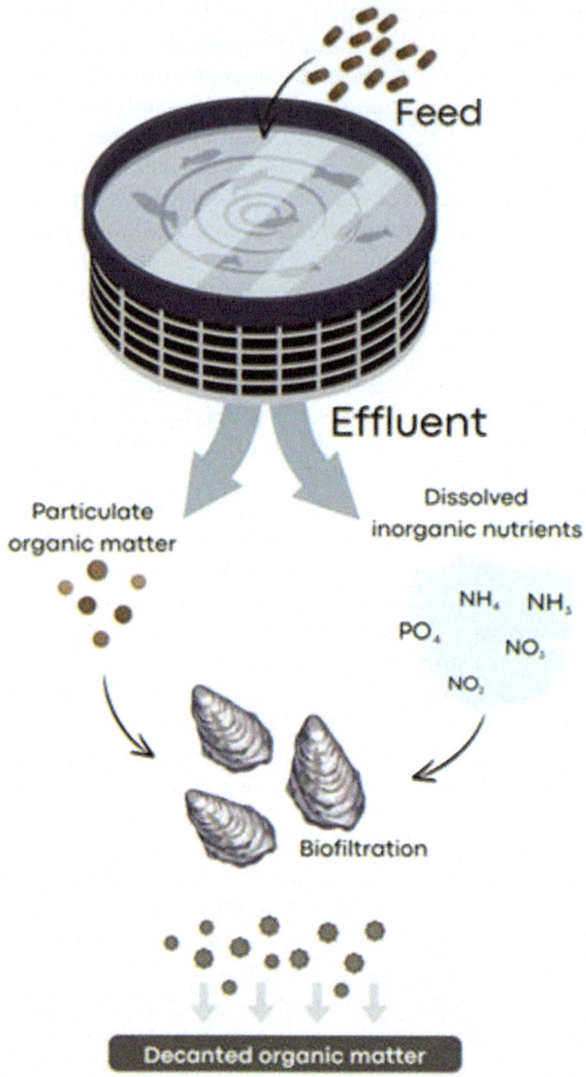

FIGURE 10.5 Bioremediation of aquaculture effluents by bivalve mollusks.

Ballester Wasielesky and Cavalli, 2006; Santos et al., 2020). Furthermore, the biofilm formed on the shells can provide supplemental food for shrimp. The biofilm is composed of a range of microorganisms, including protozoa, algae, bacteria, and fungi (Costerton and Irvin, 1981), and since these components are already part of the natural diet of various shrimp species, the biofilm can result in an increase in the animals' zootechnical performance (Ballester et al., 2007).

Although shellfish integration offers ecological benefits, it can also present risks. One of the challenges in this context is to identify suitable stocking densities to optimize filtration without compromising the main aquaculture production (Cho et al., 2019). In multitrophic systems,

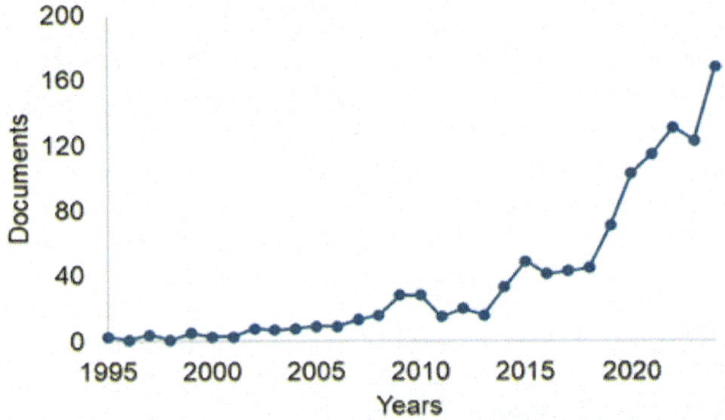

FIGURE 10.6 Evolution in the number of published documents on bioremediation of aquaculture wastewater from 1995 to 2024.

the use of high densities of bivalves can cause an imbalance in quality, increasing dissolved nutrients through excreta and significantly reducing suspended particles, especially in intensive systems where the abrupt reduction in solids has a direct impact on the bacterial community (reduction of 37.32% of *Vibrio* in water) (Lima et al., 2021). In addition, bringing in new organisms without proper inspection can facilitate the entry of pathogens, posing a health risk to cultivated organisms.

10.6 Bibliometric overview

10.6.1 Data collection and analyses

A search in the Elsevier Scopus database (obtained on October 01, 2024) was carried out using [TITLE-ABS-KEY (bioremediation) AND (aquaculture)] as a search query. This search resulted in 1,120 documents after limiting the search timescale to 2024. The results were processed by author keywords with identical meanings and by discarding keywords not related to the bioremediation of aquaculture wastewater. The obtained documents were organized and processed using OpenRefine software, following procedures such as those employed in previous studies published in other contexts (Oliveira et al. 2020). The bibliometric overview was based on both qualitative information and quantitative data. Herein, the elements investigated were: The number of publications per year, the distribution of publications by country/territory, the keywords, and the sources. Moreover, a keyword community analysis was conducted using the VOSviewer software (version 1.6.14).

10.6.2 Overview

The first indexed scientific document on bioremediation of aquaculture effluent was published in 1995, covering an investigation into the potential development of marine biotechnology in aquaculture, industry, and agriculture. It was only 12 years after the publication of this first document (i.e., in 2007) that this research field exceeded 10 publications per year. And it was not until 2020 that more than 100 documents were published in a single year for the first time (Fig. 10.6).

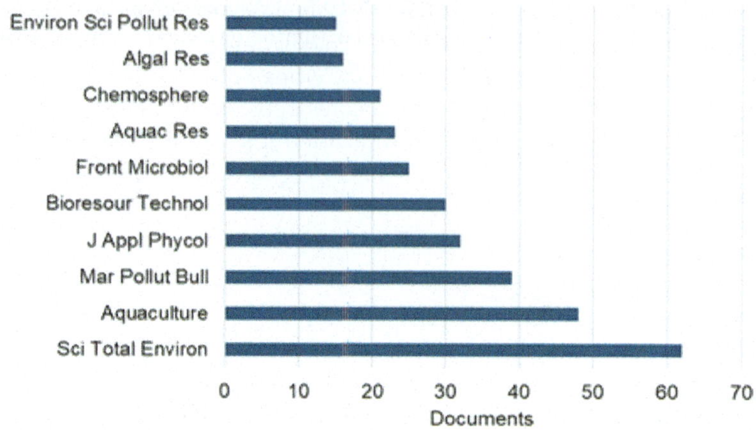

FIGURE 10.7 Top 10 journals with published documents on bioremediation of aquaculture wastewater between 1993 and 2024.

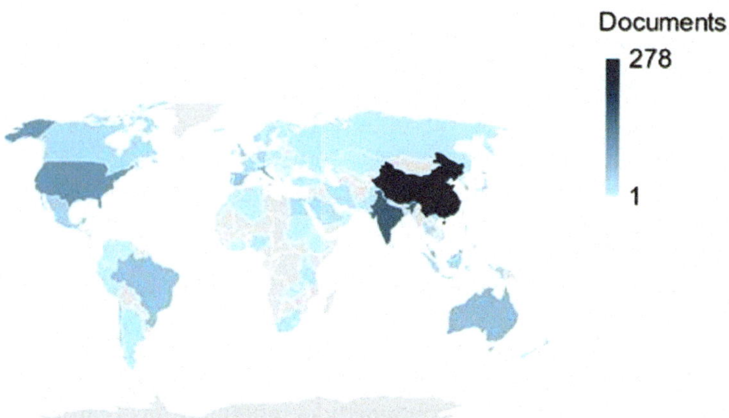

FIGURE 10.8 World map according to the number of published documents on bioremediation of aquaculture wastewater.

The journal with the highest number of published documents on bioremediation of aquaculture effluents is *Science of the Total Environment* (64 published documents), followed by *Aquaculture* (48 documents) and *Marine Pollution Bulletin* (39 documents). Among the top 10 journals in this field, *Bioresource Technology* (with 30 published documents) holds the highest impact factor (9.7) and *CiteScore* (20.8) (Fig. 10.7).

A world map of scientific production indicates that research on the bioremediation of aquaculture wastewater has been conducted across all continents (Fig. 10.8). With the exception of China (top 1; 279 documents) and India (top 2; 166 documents), countries that rank among the top aquaculture producers, such as Indonesia (29 documents), Egypt (21 documents), and Vietnam (12 documents), do not appear among the top 10 countries with the highest number of publications in this field.

In the keyword community analysis, 41 terms were plotted and grouped into five communities. Notably, only the terms "microalgae," "macroalgae," and "bacteria" were plotted as target organisms for bioremediation (Fig. 10.9).

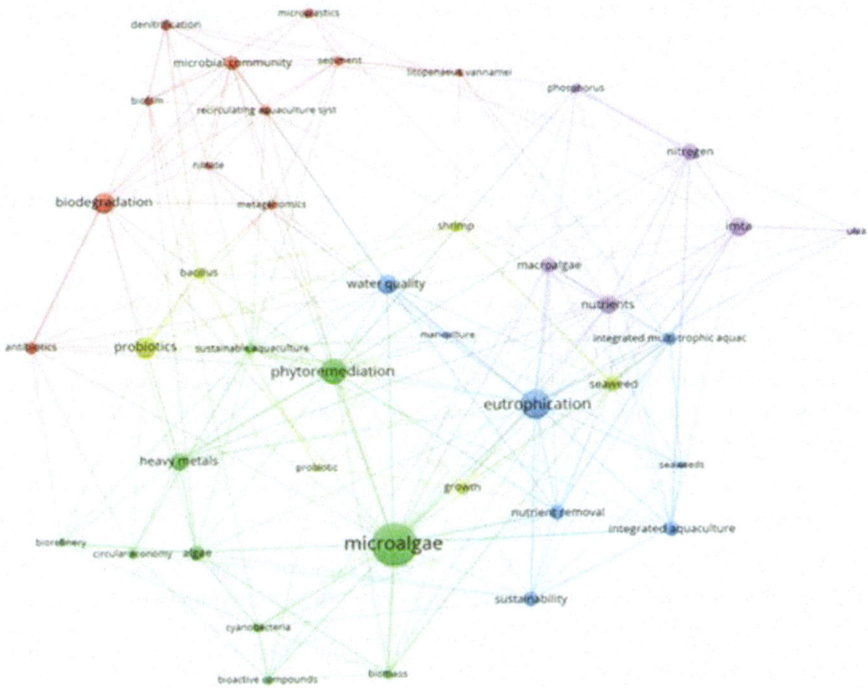

FIGURE 10.9 Keywords community into the bioremediation of aquaculture wastewater topic.

10.7 Conclusion

Bioremediation using aquatic organics is an effective and sustainable strategy for managing aquaculture effluents. By removing excess nutrients and organic matter from the water, macroalgae and bivalve mollusks not only help to mitigate environmental impacts, but also offer an additional source of income through biomass production. Integrated systems, such as multitrophic aquaculture, demonstrate how these organisms can be used to create a system with maximum nutrient utilization, promoting sustainability and the circular economy within aquaculture. In addition, the inclusion of macroalgae and bivalve mollusks in aquaculture crops not only benefits the system by reducing nutrients and pollutants but also generates economic opportunities by transforming aquatic byproducts into commercially valuable species. Given their versatility and efficiency, the use of microalgae, zooplankton, macroalgae, and bivalve mollusks in bioremediation will continue to be a central tool for the development of sustainable aquaculture in the coming decades.

References

Abreu, M.H., Pereira, R., Buschmann, A.H., Sousa-Pinto, I., Yarish, C., 2011. Nitrogen uptake responses of *Gracilaria vermiculophylla* (Ohmi) Papenfuss under combined and single addition of nitrate and ammonium. J. Exp. Mar. Biol. Ecol. 407 (2), 190–199. doi:10.1016/j.jembe.2011.06.034.

Ahmad, A., Hassan, S.W., Banat, F., 2022. An overview of microalgae biomass as a sustainable aquaculture feed ingredient: food security and circular economy. Bioengineered 13 (4), 9521–9547. doi:10.1080/21655979.2022.2061148.

Ahmed, N., Glaser, M., 2016. Can "Integrated Multi-Trophic Aquaculture (IMTA)" adapt to climate change in coastal Bangladesh? Ocean & Coastal Management 132, 120–131.

Alam, R., Khan, S., Basheer, F., Farooqi, I., 2021. Nutrients and organics removal from slaughterhouse wastewater using phytoremediation: a comparative study on different aquatic plant species. IOP Conference Series: Materials Science and Engineering, 1058 doi:10.1088/1757-899X/1058/1/012068.

Ale, M.T., Mikkelsen, J.D., Meyer, A.S., 2011. Differential growth response of *Ulva lactuca* to ammonium and nitrate assimilation. J. Appl. Phycol. 23 (3), 345–351. doi:10.1007/s10811-010-9546-2.

Allison, N., Millward, G.E., Jones, M.B., 1998. Particle processing by Mytilus edulis: effects on bioavailability of metals. J. Exp. Mar. Biol. Ecol. 222 (1–2), 149–162. doi:10.1016/S0022-0981(97)00136-6.

Amorim, C.A, Moura, A.do N., 2021. Ecological impacts of freshwater algal blooms on water quality, plankton biodiversity, structure, and ecosystem functioning. Sci. Total Environ. 758, 143605. doi:10.1016/j.scitotenv.2020.143605.

Andreotti, V., Solimeno, A., Rossi, S., Ficara, E., Marazzi, F., Mezzanotte, V., García, J., 2020. Bioremediation of aquaculture wastewater with the microalgae *Tetraselmis suecica*: semi-continuous experiments, simulation and photo-respirometric tests. Sci. Total Environ. 738, 139859. doi:10.1016/j.scitotenv.2020.139859.

Ansari, F.A., Singh, P., Guldhe, A., Bux, F., 2017. Microalgal cultivation using aquaculture wastewater: integrated biomass generation and nutrient remediation. Algal Res. 21, 169–177. doi:10.1016/j.algal.2016.11.015.

Aparna, Y., Banafsha, S.H, Reddy, M.S, 2023. Efficacy of probiotics and biofloc system in maintenance of water quality and growth performance in shrimp Litopenaeus vannamei. Int. J. Fish. Aquat. Stud. 11 (6), 07–20. doi:10.22271/fish.2023.v11.i6a.2873.

Arnold, S.J., Coman, F.E., Jackson, C.J., Groves, S.A., 2009. High-intensity, zero water-exchange production of juvenile tiger shrimp, Penaeus monodon: an evaluation of artificial substrates and stocking density. Aquaculture 293 (1–2), 42–48. doi:10.1016/j.aquaculture.2009.03.049.

Badiola, M., Basurko, O.C., Piedrahita, R., Hundley, P., Mendiola, D., 2018. Energy use in recirculating aquaculture systems (RAS): a review. Aquacult. Eng. 81, 57–70. doi:10.1016/j.aquaeng.2018.03.003.

Ballester Wasielesky, J.R., Cavalli, W., Abreu, R.O., 2006. Special Topics in Aquatic Biology and Aquaculture 191 202 AQUABIO: São Paulo – Artificial substrates and biofilm utilization in the farming of the pink shrimp *Farfantepenaeus paulensis*.

Ballester, E.L.C., Wasielesky, W., Cavalli, R.O., Abreu, P.C., 2007. Nursery of the pink shrimp Farfantepenaeus paulensis in cages with artificial substrates: biofilm composition and shrimp performance. Aquaculture 269 (1–4), 355–362. doi:10.1016/j.aquaculture.2007.04.003.

Barrington, K., Chopin, T., Robinson, S., 2009. Integrated mariculture: a global review. FAO Fish. Aquac. Tech. Pap. 529, 7–46.

Barrington, K., Ridler, N., Chopin, T., Robinson, S., Robinson, B., 2010. Social aspects of the sustainability on integrated multi trophic aquaculture. Aquac. Int. 18 (2), 201–211.

Bhushan, S., Veeragurunathan, V., Bhagiya, B.K., Krishnan, S.G., Ghosh, A., Mantri, V.A., 2023. Biology, farming and applications of economically important red seaweed *Gracilaria edulis* (S. G. Gmelin) P. C. Silva: a concise review. J. Appl. Phycol. 35 (3), 983–996. doi:10.1007/s10811-023-02955-8.

Biswas, G., Kumar, P., Ghoshal, T.K., Kailasam, M., De, D., Bera, A., Mandal, B., Sukumaran, K., Vijayan, K.K., 2020. Integrated multi-trophic aquaculture (IMTA) outperforms conventional polyculture with respect to environmental remediation, productivity and economic return in brackishwater ponds. Aquaculture 516, 734626. doi:10.1016/j.aquaculture.2019.734626.

Boening, D.W., 1999. An evaluation of bivalves as biomonitors of heavy metals pollution in marine waters. Environ. Monit. Assess. 55 (3), 459–470. doi:10.1023/A:1005995217901.

Brito, L.O., Cardoso Junior, L.d.O., Lavander, H.D., Abreu, J.L.d., Severi, W., Gálvez, A.O., 2018. Bioremediation of shrimp biofloc wastewater using clam, seaweed and fish. Chem. Ecol. 34 (10), 901–913. doi:10.1080/02757540.2018.1520843.

Brito, L.O., Chagas, A.M., da Silva, E.P., Soares, R.B., Severi, W., Gálvez, A.O., 2016. Water quality, Vibrio density and growth of Pacific white shrimp Litopenaeus vannamei (Boone) in an integrated biofloc system with red seaweed Gracilaria birdiae (Greville). Aquac. Res. 47 (3), 940–950. doi:10.1111/are.12552.

Çagatay, İ.T., 2023. Bacteriophage applications in aquaculture. Isr. J. Aquac.-Bamidgeh 75 (2), 1–26. doi:10.46989/001c.78119.

Campanati, C., Willer, D.F., Schubert, J., Aldridge, D., 2021. Sustainable intensification of aquaculture through nutrient recycling and circular economies: more fish, less waste, blue growth. Rev. Fish. Sci. Aquac. 30 (2), 143–169. doi:10.1080/23308249.2021.1897520.

Campos, C.V.F.d.S., Oliveira, C.Y.B., dos Santos, E.P., de Abreu, J.L., Severi, W., da Silva, S.M.B.C., Brito, L.O., Gálvez, A.O., 2022. *Chlorella-Daphnia* consortium as a promising tool for bioremediation of Nile tilapia farming wastewater. Chem. Ecol. 38 (9), 873–895. doi:10.1080/02757540.2022.2120612.

Cao, Z., Yan, W., Ding, M., Yuan, Y., 2022. Construction of microbial consortia for microbial degradation of complex compounds. Front. Bioeng. Biotechnol. 10. doi:10.3389/fbioe.2022.1051233.

Cardoso, L.G., Lemos, P.V.F., de Souza, C.O., Oliveira, M.B.P.P., Chinalia, F.A., 2022. Current advances in phytoremediation and biochemical composition of Arthrospira (Spirulina) grown in aquaculture wastewater. Aquac. Res. 53 (14), 4931–4943. doi:10.1111/are.15996.

Ceschin, S., Crescenzi, M., Iannelli, M.A, 2020. Phytoremediation potential of the duckweeds *Lemna minuta* and *Lemna minor* to remove nutrients from treated waters. Environ. Sci. Pollut. Res. 27 (13), 15806–15814. doi:10.1007/s11356-020-08045-3.

Chai, W.S., Tan, W.G., Halimatul Munawaroh, H.S., Gupta, V.K., Ho, S.H., Show, P.L., 2021. Multifaceted roles of microalgae in the application of wastewater biotreatment: a review. Environ. Pollut. 269. doi:10.1016/j.envpol.2020.116236.

Chaitanawisuti, N., Santhaweesuk, W., Kritsanapuntu, S., 2011. Performance of the seaweeds *Gracilaria salicornia* and *Caulerpa lentillifera* as biofilters in a hatchery scale recirculating aquaculture system for juvenile spotted babylons (Babylonia areolata). Aquac. Int. 19 (6), 1139–1150. doi:10.1007/s10499-011-9429-9.

Chary, K., van Riel, A.J., Muscat, A., Wilfart, A., Harchaoui, S., Verdegem, M., Filgueira, R., Troell, M., Henriksson, P., de Boer, I.J.M., Wiegertjes, G., 2023. Transforming sustainable aquaculture by applying circularity principles. Rev. Aquac. 16 (2), 656–673. doi:10.1111/raq.12860.

Chen, J., Sun, R., Pan, C.-G., Sun, Y., Mai, B., Li, Q.X., 2020. Antibiotics and Food Safety in Aquaculture. J. Agric. Food Chem. 68 (43), 11908–11919. doi:10.1021/acs.jafc.0c03996.

Chen, Y., Yang, P., Fan, M., Lin, Q., Jin, T., Yu, F., Li, Y., Ma, J., 2020. Influence of *Daphnia magna* and *Ceratophyllum demersum* on the control of algae under different phosphorus concentrations. Ann. Limnol. – Int. J. Lim. 56, 12. doi:10.1051/limn/2020006.

Chiquito-Contreras, R.G., Hernandez-Adame, L., Alvarado-Castillo, G., Martínez-Hernández, M.J., Sánchez-Viveros, G., Chiquito-Contreras, C.J., Hernandez-Montiel, L.G., 2022. Aquaculture—production system and waste management for agriculture fertilization—a review. Sustainability 14 (12), 7257. doi:10.3390/su14127257.

Chirapart, A., Ruangchuay, R., 2022. Seaweed in aquaculture: an overview sustainable global resources of seaweeds volume 1: bioresources, cultivation, trade and multifarious applications. Springer International Publishing, Thailand, pp. 381–397. doi:10.1007/978-3-030-91955-9_20.

Cho, Y., Shim, W.J., Jang, M., Han, G.M., Hong, S.H., 2019. Abundance and characteristics of microplastics in market bivalves from South Korea. Environ. Pollut. 245, 1107–1116. doi:10.1016/j.envpol.2018.11.091.

Choi, T.-S., Kang, E.-J., Kim, J.-H., Kim, K.-Y., 2010. Effect of salinity on growth and nutrient uptake of *Ulva pertusa* (Chlorophyta) from an eelgrass bed. Algae 25 (1), 17–26. doi:10.4490/algae.2010.25.1.017.

Chopin, T., Buschmann, A.H., Halling, C., Troell, M., Kautsky, N., Neori, A., Kraemer, G.P., Zertuche-González, J.A., Yarish, C., Neefus, C., 2001. Integrating seaweeds into marine aquaculture systems: a key toward sustainability. J. Phycol. 37 (6), 975–986. doi:10.1046/j.1529-8817.2001.01137.x.

Chopin, T., Robinson, S.M.C., Troell, M., Neori, A., Buschmann, A.H., Fang, J., 2008. Multitrophic integration for sustainable marine aquaculture. Encycl. Ecol. 1–5, 2463–2475. doi:10.1016/B978-008045405-4.00065-3.

Costerton, J.W., Irvin, R.T., 1981. The bacterial glycocalyx in nature and disease. Annu. Rev. Microbiol. 35, 299–324.

Cordeiro, D. L., Bôas, R. N. V., Mendes, M. F., Mendonça, H. V. 2025. Use of the microalga Spirulina platensis in wastewater treatment and biofuel production Revista Em Agronegócio E Meio Ambiente, 18, e13077. https://doi.org/10.17765/2176-9168.2025v18e13077.

Crab, R., Avnimelech, Y., Defoirdt, T., Bossier, P., Verstraete, W., 2007. Nitrogen removal techniques in aquaculture for a sustainable production. Aquaculture 270 (1–4), 1–14. doi:10.1016/j.aquaculture.2007.05.006.

Crab, R., Defoirdt, T., Bossier, P., Verstraete, W., 2012. Biofloc technology in aquaculture: beneficial effects and future challenges. Aquaculture 356–357, 351–356. doi:10.1016/j.aquaculture.2012.04.046.

Cyrino, J.E.P., Bicudo, Á.D.A., Sado, R.Y., Borghesi, R., Dairiki, J.K., 2010. Fish farming and the environment – the use of environmentally friendly feeds in fish farming. Rev. Bras. Zootec. 39 (1), 68–87. doi:10.1590/S1516-35982010001300009.

Daneshvar, E., Zarrinmehr, M.J., Koutra, E., Kornaros, M., Farhadian, O., Bhatnagar, A., 2019. Sequential cultivation of microalgae in raw and recycled dairy wastewater: microalgal growth, wastewater treatment and biochemical composition. Bioresour. Technol. 273, 556–564. https://doi.org/10.1016/j.biortech.2018.11.059.

da Silva Campos, C.V.F., da Silva Farias, R., da Silva, S.M.B.C., Severi, W., Brito, L.O., Gálvez, A.O., 2019. Production of *Daphnia similis* Claus, 1876 using wastewater from tilapia cultivation in a biofloc system. Aquac. Int. 28 (1), 403–419. https://doi.org/10.1007/s10499-019-00470-7.

Dauda, A.B., Ajadi, A., Tola-Fabunmi, A.S., Akinwole, A.O., 2019. Waste production in aquaculture: sources, components and management in different culture systems. Aquac. Fish. 4, 81–88. doi:10.1016/j.aaf2018.10.002.

de Azevedo, R.V, Tonini, W.C.T., dos Santos, M.J.M., Braga, L.G.T, 2015. Biofiltration, growth and body composition of oyster Crassostrea rhizophorae in effluents from shrimp Litopenaeus vannameil. Rev. Ciênc. Agron. 46 (1), 193–203. doi:10.1590/S1806-66902015000100023.

Deepak, A.P., Vasava, R.J., Elchelwar, V.R., Tandel, D.H., Vadher, K.H., Shrivastava, V., Prabhakar, P., 2020. Aquamimicry: new an innovative approach for sustainable development of aquaculture. J. Entomol. Zool. Stud. 8 (2), 1029–1031.

Deepti, M., Meinam, M., Devi, N.C., Singh, S.K., Devi, W.M.A., 2024. Copefloc technology): an innovative approach for sustainable organic farming with special reference to shrimp aquaculture. Blue Biotechnol. 1 (1). https://doi.org/10.1186/s44315-024-00005-7.

De Lacerda, L.D., Vaisman, A.G., Maia, L.P., Silva, C.A.R., Cunha, S.E.M., 2006. Relative importance of nitrogen and phosphorus emissions from shrimp farming and other anthropogenic sources for six estuaries along the NE Brazilian coast. Aquaculture 253 (1–4), 433–446.

Deslous-Paoli, J.M., Lannou, A.M., Geairon, P., Bougrier, S., Raillard, O., Héral, M., 1992. Effect of the feeding behaviour of *Crassostrea gigas* (bivalve molluscs) on biosedimentation of natural particulate matter Hydrobiological 231, 85–91.

Dhont, J., Sorgeloos, P., 2002. Applications of *Artemia*. In: Abatzopoulos, T.J., Beardmore, J.A., Clegg, J.S., Sorgeloos, P. (Eds.), Artemia: basic and applied biology. Springer, Netherlands, pp. 251–271.

Dinesh, R., Anand, C., John, K.R., George, M.R., Bharathi, S., Kumar, J., 2023. An overview of chemicals and drugs in aquaculture disease management. Indian J. Anim. Health 62 (1), 1–20. doi:10.36062/ijah.2022.05222.

El-Sheekh, M., Abdel-Daim, M.M., Okba, M., Gharib, S., Soliman, A.M., El-Kassas, H., 2021. Green technology for bioremediation of the eutrophication phenomenon in aquatic ecosystems: a review. Afr. J. Aquat. Sci. 46 (3), 274–292. doi:10.2989/16085914.2020.1860892.

Emerenciano, M., Gaxiola, G., Cuzon, G., 2013. Biofloc technology (BFT): a review for aquaculture application and animal food industry. Biomass now – cultivation and utilization. InTechOpen, Rijeka, pp. 301–328. doi:10.5772/53902.

Ertl, N.G., O'Connor, W.A., Papanicolaou, A., Wiegand, A.N., Elizur, A., 2016. Transcriptome analysis of the Sydney rock oyster, Saccostrea glomerata: Insights into molluscan immunity. PLoS One 11, e0156649.

Estim, A., 2015. Integrated multitrophic aquaculture. Aquaculture Ecosystems: Adaptability and Sustainability. Wiley, New York, pp. 164–181.

Estim, A., Shaleh, S.R.M., Shapawi, R., Saufie, S., Mustafa, S., 2020. Maximizing efficiency and sustainability of aquatic food production from aquaponics systems - a critical review of challenges and solution options. Turk. J. Fish. Aquat. Sci. 20 (1), 65–72. doi:10.4194/2618-6381-v20_1_08.

FAO, 2024. The State of World Fisheries and Aquaculture 2024: Blue Transformation in Action. Food and Agriculture Organization of the United Nations, Rome.

Feng, J., Mazzei, M., Di Gregorio, S., Niccolini, L., Vitiello, V., Ye, Y., Guo, B., Buttino, I., 2023. Marine copepods as a microbiome hotspot: revealing their interactions and biotechnological applications. Water 15 (24), 4203. https://doi.org/10.3390/w15244203.

Gao, F., Cui, W., Xu, J.P., Li, C., Jin, W.H., Yang, H.L., 2019. Lipid accumulation properties of *Chlorella vulgaris* and *Scenedesmus obliquus* in membrane photobioreactor (MPBR) fed with secondary effluent from municipal wastewater treatment plant. Renew. Energy 136 (6), 671–676. https://doi.org/10.1016/j.renene.2019.01.038.

Gao, X., Bi, Y., Su, L., Lei, Y., Gong, L., Dong, X., Li, X., Yan, Z., 2023. Unveiling the nitrogen and phosphorus removal potential: Comparative analysis of three coastal wetland plant species in lab-scale constructed wetlands. J. Environ. Manage. 351, 119864. doi:10.1016/j.jenvman.2023.119864.

Gelli, V.C., Patino, M.T.O., Rocha, J.V., Barbieri, E., Miranda-Filho, K., Henriques, M.B., 2020. Production of the *Kappaphycus alvarezii* extract as a leaf biofertilizer: technical and economic analysis for the north coast of São Paulo-Brazil. Bol. Inst. Pesca 46, 1–12. doi:10.20950/1678-2305.2020.46.2.568.

Giangrande, A., Gravina, M.F., Rossi, S., Longo, C., Pierri, C., 2021. Aquaculture and Restoration: perspectives from Mediterranean sea experiences. Water 13 (7), 991. doi:10.3390/W13070991.

Gleiser, M., Moro, S., 2023. Implementation of an IoT-based water quality monitoring system for aquaculture. Int. J. Res. Publ. Rev. 4 (5), 1449–1452. doi:10.55248/gengpi.234.5.38043.

Godos, I., Guzman, H.O., Soto, R., García-Encina, P.A., Bécarés, E., Muñoz, R., Vargas, V.A., 2011. Coagulation/flocculation-based removal of algal-bacterial biomass from piggery wastewater treatment. Bioresource Technology 102 (2), 923–927. doi:10.1016/j.biortech.2010.09.036.

Gómez, S., Hurtado, C.F., Orellana, J., 2019. Bioremediation of organic sludge from a marine recirculating aquaculture system using the polychaete *Abarenicola pusilla* (Quatrefages, 1866). Aquaculture 507, 377–384. https://doi.org/10.1016/j.aquaculture.2019.04.033.

Gorokhova, E., Soerensen, A.L., Motwani, N.H., 2020. Mercury-methylating bacteria are associated with copepods: a proof-of-principle survey in the Baltic Sea. PLoS One 15 (3), e0230310. https://doi.org/10.1371/journal.pone.0230310.

Granada, L., Sousa, N., Lopes, S., Lemos, M.F., 2016. Is integrated multitrophic aquaculture the solution to the sectors' major challenges?–a review. Rev. Aquac. 8, 283–300. doi:10.1111/raq.12093.

Guerrero, S., Cremades, J., 2012. Integrated Multi-trophic Aquaculture (IMTA): A sustainable, pioneering alternative for marine cultures in Galicia. In: Guerrero, S., Cremades, J. (Eds.), Regional Government of Galicia (Spain), p. 111. https://hal.archives-ouvertes.fr/hal-00743395/.

Guo, Z., Liu, Y., Guo, H., Yan, S., Mu, J., 2013. Microalgae cultivation using an aquaculture wastewater as growth medium for biomass and biofuel production. Journal of Environmental Sciences 25 (Suppl. 1), S85–S88. doi:10.1016/S1001-0742(14)60632-X.

Guttman, L., Boxman, S.E., Barkan, R., Neori, A., Shpigel, M., 2018. Combinations of *Ulva* and periphyton as biofilters for both ammonia and nitrate in mariculture fishpond effluents. Algal Res. 34, 235–243.

Ha, J.-Y., Saneyoshi, M., Park, H.-D., Toda, H., Kitano, S., Homma, T., Shiina, T., Moriyama, Y., Chang, K.-H., Hanazato, T., 2013. Lake restoration by biomanipulation using piscivore and *Daphnia* stocking: results of the biomanipulation in Japan. Limnology 14, 19–30. https://doi.org/10.1007/s10201-012-0381-9.

Hamza, F., Zinjarde, S., 2023. Use of marine microorganisms in designing anti-infective strategies for sustainable aquaculture production. J. Appl. Microbiol. 134 (7), lxad128. doi:10.1093/jambio/lxad128.

He, P., Xu, S., Zhang, H., Wen, S., Dai, Y., Lin, S., Yarish, C., 2008. Bioremediation efficiency in the removal of dissolved inorganic nutrients by the red seaweed, Porphyra yezoensis, cultivated in the open sea. Water Res. 42 (4–5), 1281–1289.

Henriques, B., Rocha, L.S., Lopes, C.B., Figueira, P., Duarte, A.C., Vale, C., Pardal, M.A., Pereira, E., 2017. A. macroalgae-based biotechnology for water remediation: Simultaneous removal of Cd, Pb and Hg by living *Ulva lactuca*. J. Environ. Manage. 191, 275–289.

Holanda, M., Santana, G., Furtado, P., Rodrigues, R. V., Cerqueira, V. R., Sampaio, L. A., WASIELESKYJR., W., Poersch, L. H. (2020). Evidence of total suspended solids control by Mugil liza reared in an integrated system with pacific white shrimp Litopenaeus vannamei using biofloc technology. *Aquaculture Reports*, *18*, 100479.

Honda, H., Kikuchi, K., 2002. Nitrogen budget of polychaete *Perinereis nuntia vallata* fed on the feces of Japanese flounder. Fish. Sci. 68 (6), 1304–1308. https://doi.org/10.1046/j.1444-2906.2002.00568.x.

Hopkins, J.E., Hamilton, R., Sandifer, P., Browdy, C., Stokes, A., 1993. Effect of water exchange rate on production, water quality, effluent characteristics and nitrogen budgets of intensive shrimp ponds. J. World Aquacult. Soc. 24 (3), 304–320.

Hu, J., Qiao, Y., Zhou, L., Li, S., 2012. Spatiotemporal distributions of nutrients in the downstream from Gezhouba Dam in Yangtze River, China. Environ. Sci. Pollut. Res. 19 (7), 2849–2859.

Huang, F., Pan, L., He, Z., Zhang, M., Zhang, M., 2020. Identification, interactions, nitrogen removal pathways and performances of culturable heterotrophic nitrification-aerobic denitrification bacteria from mariculture water by using cell culture and metagenomics. Sci. Total Environ. 732, 139268. doi:10.1016/j.scitotenv.2020.139268.

Huang, X., Ye, G., Yi, N., Lu, L., Zhang, L., Yang, L., Xiao, L., Liu, J., 2019. Effect of plant physiological characteristics on the removal of conventional and emerging pollutants from aquaculture wastewater by constructed wetlands. Ecol. Eng. 135, 45–53. https://doi.org/10.1016/j.ecoleng.2019.05.017.

Huo, Y., Stuart, K., Rotman, F., Ernst, D., Drawbridge, M., 2024. The culture of fish, mussels, sea cucumbers and macroalgae in a modular integrated multi-tropic recirculating aquaculture system (IMTRAS): performance and waste removal efficiencies. Aquaculture 585, 740720. https://doi.org/10.1016/j.aquaculture.2024.740720.

Huo, Y.Z., Xu, S.N., Wang, Y.Y., Zhang, J.H., Zhang, Y.J., Wu, W.N., Chen, Y.Q., He, P.M., 2011. Bioremediation efficiencies of *Gracilaria verrucosa* cultivated in a enclosed sea area of Hangzhou Bay China. J. Appl. Phycol. 23 (2), 173–182.

Ibrahim, L., Shaghaleh, H., El-Kassar, G., Abu-hashim, M., Elsadek, E., Hamoud, A., Aquaponics, Y., 2023. A sustainable path to food sovereignty and enhanced water use efficiency. Water 15 (24), 4310. doi:10.3390/w15244310.

Jmeliova, N.N., Sanz, J., 1969. Respiration and some feeding characteristics of the oyster *Crassostrea rhizophorae* (Guilding, 1828). Serie oceanología 3, 1–20.

Jones, A.B., Preston, N.P., Dennison, W.C., 2002. The efficiency and condition of oysters and macroalgae used as biological filters of shrimp pond effluent. Aquacult. Res. 33, 1–19.

Jones, A.B., Dennison, W.C., Preston, N.P., 2001. Integrated treatment of shrimp effluent by sedimentation, oyster filtration and macroalgal absorption: a laboratory scale study. Aquaculture 193 (3–4), 155–178.

Kamil, M., Istyadji, M., Khairunnisa, Y., 2023. The effect of use of commercial probiotics in feed on water quality and growth performance of Bawal Fish (*Colossoma macropomum*) bioflok system. Sch. Int. J. Biochem. 6 (6), 79–85. doi:10.36348/sijb.2023.v06i06.001.

Kang, Y.H., Hwang, J.R., Chung, I.K., Park, S.R., 2013. Development of a seaweed species-selection index for successful culture in a seaweed-based integrated aquaculture system. J. Ocean Univ. China 12, 125–133.

Kashem, A., Das, P., Quadir, M.A., Khan, S., Thaher, M., Alghasal, G., Hawari, A., Al-Jabri, H., 2023. Microalgal bioremediation of brackish aquaculture wastewater. Sci. Total Environ. 873, 162384. doi:10.1016/j.scitotenv.2023.162384.

Kautsky, N., Evans, S., 1987. Role of biodeposition by *Mytilus edulis* in the circulation of matter and nutrients in a Baltic coastal system. Mar. Ecol. Prog. Ser. 38, 201–212. doi:10.3354/meps038201.

Khanjani, M.H., Mozanzadeh, M.T., Fóes, G.K., 2022. Aquamimicry system: a suitable strategy for shrimp aquaculture – a review. Ann. Anim. Sci. 22 (4), 1201–1210. doi:10.2478/aoas-2022-0044.

Khanjani, M.H., Sharifinia, M., 2020. Biofloc technology as a promising tool to improve aquaculture production. Rev. Aquac. 12 (3), 1836–1850. doi:10.1111/raq.12412.

Khatoon, H., Banerjee, S., Syahiran, M.S., Noordin, N.B.M., Bolong, A.M.A., Endut, A., 2016. Re-use of aquaculture wastewater in cultivating microalgae as live feed for aquaculture organisms. Desalin. Water Treat. 57, 29295–29302. https://doi.org/10.1080/19443994.2016.1156030.

Klimmek, S., Stan, H., Wilke, A., Bunke, G., Buchholz, R., 2001. Comparative analysis of the biosorption of cadmium, lead, nickel, and zinc by algae. Environ. Sci. Technol. 35, 4283–4288.

Knoop, J., Barrento, S., Lewis, R., Walter, B., Griffin, J., 2022. Incorporating concepts of biodiversity into modern aquaculture: macroalgal species richness enhances bioremediation efficiency in a lumpfish hatchery. Algae 37 (3), 213–226. doi:10.4490/algae.2022.37.5.12.

Kostopoulou, V., Carmona, M.J., Divanach, P., 2012. The rotifer *Brachionus plicatilis*: An emerging bio-tool for numerous applications. J. Biol. Res. 17, 97–112.

Kousar, R., Shafi, N., Andleeb, S., Ali, N., Akhtar, T., Khalid, S., 2020. Assessment and incidence of fish associated bacterial pathogens at hatcheries of Azad Kashmir, Pakistan. Braz. J. Biol. 80 (3), 607–614. doi:10.1590/1519-6984.217435.

Kumar, V., Jat, H.S., Sharma, P.C., Balwinder-Singh, G., Gathala, M.K., Malik, R.K., Kamboj, B.R., Yadav, A.K., Ladha, J.K., Raman, A., Sharma, D.K., McDonald, A., 2018. Can productivity and profitability be enhanced in intensively managed cereal systems while reducing the environmental footprint of production? Assessing sustainable intensification options in the breadbasket of India. Agric. Ecosyst. Environ. 252, 132–147. doi:10.1016/j.agee.2017.10.006.

Kunawongdet, P., 2020. Biological water treatment in recirculating aquaculture system by using *Caulerpa lentillifera*. BUU Sci. J. 509–523.

Kurniawan, S.B., Ahmad, A., Said, N.S.M., Imron, M.F., Abdullah, S.R.S., Othman, A.R., Purwanti, I.F., Hasan, H.A., 2021. Macrophytes as wastewater treatment agents: Nutrient uptake and potential of produced biomass utilization toward circular economy initiatives. Sci. Total Environ. 790, 148–219. https://doi.org/10.1016/j.scitotenv.2021.148219.

Lenz, P.H., Browne, R.A., 1991. Ecology of artemia. In: Browne, R.A., Sorgeloos, P., Trotman, C.N.A. (Eds.), Artemia biology, 237. CRC Press, Boca Raton, Florida, USA, p. 253.

Li, H., Duan, Y., Dong, H., Zhang, J., 2023. Encapsulation of denitrifying bacteria for potential recirculating aquaculture system denitrification. Aquaculture 573, 739584. doi:10.1016/j.aquaculture.2023.739584.

Li, X., Deng, Y., Li, X.-Y., Ma, X., Wang, J., Li, J., 2020a. Integration of marine macroalgae (*Chaetomorpha maxima*) with a moving bed bioreactor for nutrient removal from maricultural wastewater. Archaea 1, 8848120. doi:10.1155/2020/8848120.

Li, X., Wu, S., Dong, Y., Fan, H., Bai, Z., Zhuang, X., 2021. Engineering microbial consortia towards Bioremediation. Water 13 (20), 2928. doi:10.3390/w13202928.

Li, Z.-f., Yu, E., Zhang, K., Gong, W., Xia, Y., Tian, J., Wang, G., Xie, J., 2020b. Water treatment effect, microbial community structure, and metabolic characteristics in a field-scale aquaculture wastewater treatment system. Front. Microbiol. 11, 930. doi:10.3389/fmicb.2020.00930.

Lich, N.Q., 2023. Assessment of the bovalves stoking density for maximizing nutrients removal from shrimp farm effluents. AACL BiofluxI 16, 1648–1657.

Lima, P.C.M., da Silva, A.E.M., da Silva, D.A., de Oliveira, C.Y.B., Severi, W., Brito, L.O., Olivera Gálvez, A., 2022. Use of recirculation and settling chamber in synbiotic multi-trophic culture of *Crassostrea* sp. with *Litopenaeus vannamei*. Aquac. Res. 53 (18), 6626–6640.

Lima, P.C.M., da Silva, A.E.M., Silva, D.A., da Silva, S.M.B.C., Brito, L.O., Gálvez, A.O., 2021. Effect of stocking density of *Crassostrea* sp. in a multitrophic biofloc system with *Litopenaeus vannamei* in nursery. Aquaculture 530, 735913. doi:10.1016/j.aquaculture.2020.735913.

Lima, P.C.M., dos Santos, de Macedo Dantas, Brito, L.O., Gálvez, A.O., 2025. Integrated multi-trophic culture of Penaeus vannamei with *Crassostrea* sp. and *Gracilaria* sp. in a nursery system. Aquacultural Engineering 111, 102595.

Limhang, K., Saeheng, L., Khunpol, K., 2014. Efficiency for treatment of nitrogen compound of the water from rearing pacific white shrimp (*Litopenaeus vannamei*) using sea grapes (*Caulerpa macrophysa*). Proceedings of 52nd Kasetsart University Annual Conference: Fisheries, Agricultural Extension and Home Economics. Kasetsart University, Bangkok, Thailand, pp. 226–233.

Lin, Y.-F., Jing, S., Lee, D.-Y., Wang, T.-W., 2002. Nutrient removal from aquaculture wastewater using a constructed wetlands system. Aquaculture 209, 169–184. doi:10.1016/S0044-8486(01)00801-8.

Lopes da Silva, T., Moniz, P., Silva, C., Reis, A., 2019. The dark side of microalgae biotechnology: A heterotrophic biorefinery platform directed to ω-3 rich lipid production. Microorganisms 2019 (7), 670.

Marinho-Soriano, E., Azevedo, C.A.A., Trigueiro, T.G., Pereira, D.C., Carneiro, M.A.A., Camara, M.R., 2011. Bioremediation of aquaculture wastewater using macroalgae and *Artemia*. Int. Biodeterior. Biodegrad. 65 (1), 253–257. https://doi.org/10.1016/j.ibiod.2010.10.001.

Marinho-Soriano, E., Panucci, R.A., Carneiro, M.A.A., Pereira, D.C., 2009. Evaluation of *Gracilaria caudata* J. Agardh for bioremediation of nutrients from shrimp farming wastewater. Bioresour. Technol. 100 (24), 6192–6198.

Marlida, R., 2020. Biofloc as a Solution to Environmental Challenges for Sustainable Future Aquaculture: A Review. Rawa Sains: Jurnal Sains STIPER Amuntai 10 (1), 38–44. doi:10.36589/rs.v10i1.116.

Marques, B., Lillebo, A.I., Ricardo, F., Nunes, C., Coimbra, M.A., Calado, R., 2018. Adding value to ragworms (*Hediste diversicolor*) through the bioremediation of a super-intensive marine fish farm. Aquacult. Env. Interact. 10, 79–88. https://doi.org/10.3354/aei00255.

Martinez-Cordova, L.R., Martinez-Porchas, M., 2006. Polyculture of Pacific white shrimp, *Litopenaeus vannamei*, giant oyster, *Crassostrea gigas* and black clam, *Chione fluctifraga* in ponds in Sonora, Mexico. Aquaculture 258, 321–326.

Massocato, T.F., Robles-Carnero, V., Moreira, B.R., Castro-Varela, P., Pinheiro-Silva, L., Oliveira, W.D.S., Vaga, J., Avilés, A., Bonomi-Barufi, A., Rörig, L.R., Figueroa, F.L., 2022. Growth, biofiltration and photosynthetic performance of *Ulva* spp. cultivated in fishpond effluents: An outdoor study. Front. Mar. Sci. 9, 981468.

Mawi, S., Krishnan, S., Din, M.F.M., Arumugam, N., Chelliapan, S., 2020. Bioremediation potential of macroalgae *Gracilaria edulis* and *Gracilaria changii* co-cultured with shrimp wastewater in an outdoor water recirculation system. Environ. Technol. Innov. 17, 100571.

Mehta, S.K., Gaur, J.P., 2005. Use of algae for removing heavy metal ions from wastewater: progress and prospects. Crit. Rev. Biotechnol. 25, 113–152.

Moraes, L.B.S., Mota, G.C.P., Farias, R.S., Marinho, Y.F., Brito, L.O., Gálvez, A.O., 2021. Growth and nitrogen uptake by *Arthrospira platensis* cultured in aquaculture wastewater from Nile tilapia reared in biofloc system. Acta Sci. Biol. Sci. 43, e57549.

Moss, K.R.K., Moss, S.M., 2004. Effects of artificial substrate and stocking density on the nursery production of Pacific white shrimp Litopenaeus vannamei. J. World Aquac. Soc. 35 (4), 536–542.

Msuya, F., Neori, A., 2008. Effect of water aeration and nutrient load level on biomass yield, N uptake and protein content of the seaweed *Ulva lactuca* cultured in seawater tanks. J. Appl. Phycol. 20, 1021–1031. doi:10.1007/s10811-007-9300-6.

Mugwanya, M., Dawood, M., Kimera, F., Sewilam, H., 2021. Biofloc systems for sustainable production of economically important aquatic species: a review. Sustainability 13 (13), 7255. doi:10.3390/su13137255.

Nahar, K., Hoque, S., 2021. Phytoremediation to improve eutrophic ecosystem by the floating aquatic macrophyte, water lettuce (*Pistia stratiotes* L.) at lab scale. Egypt. J. Aquat. Res. 47 (2), 231–237. doi:10.1016/j.ejar.2021.05.003.

Narvarte, B.C.V., Hinaloc, L.A.R., Gonzaga, S.M.C., Roleda, M., 2023. Impacts of aquaculture nutrient sources: ammonium uptake of commercially important eucheumatoids depends on phosphate levels. J. Appl. Phycol. 36, 557–565. doi:10.1007/s10811-023-03073-1.

Nascimento, V.S., Lapa, K.R., de Miranda Gomes, C.H.A., Gray, M., da Silva, G., Garbossa, L.H.P., Suplicy, F.M., de Melo, C.M.R., 2022. Filtration and biodeposition rates of Crassostrea oysters for southern Brazilian waters. Reg. Stud. Mar. Sci. 56, 102677.

Nasir, N.M., Jusoh, A., Harun, R., Ibrahim, N.N.L., Rasit, N., Ghani, W., Kurniawan, S., 2023. Nutrient consumption of green microalgae, *Chlorella* sp. during the bioremediation of shrimp aquaculture wastewater. Algal Res. 72, 103110. doi:10.1016/j.algal.2023.103110.

Nathanailides, C., Kolygas, M., Choremi, K., Mavraganis, T., Gouva, E., Vidalis, K., Athanassopoulou, F., 2021. Probiotics have the potential to significantly mitigate the environmental impact of freshwater fish farms. Fishes 6 (4), 76. doi:10.3390/fishes6040076.

Naylor, R., Goldburg, R., Primavera, J., Kautsky, N., Beveridge, M., Clay, J., Folke, C., Lubchenco, J., Mooney, H., Troell, M., 2000. Effect of aquaculture on world fish supplies. Nature 405, 1017–1024. doi:10.1038/35016500.

Nederlof, M.A.J., Verdegem, M., Smaal, A., Jansen, H., 2021. Nutrient retention efficiencies in integrated multi-trophic aquaculture. Rev. Aquac. 14 (3), 1194–1212. doi:10.1111/raq.12645.

Neori, A., Chopin, T., Troell, M., Buschmann, A.H., Kraemer, G.P., Halling, C., Shpigel, M., Yarish, C., 2004. Integrated aquaculture: rationale, evolution and state of the art emphasizing seaweed biofiltration in modern mariculture. Aquaculture 231, 361–391.

Newell, R.I.E., 2004. Ecosystem influences of natural and cultivated populations of suspension-feeding bivalve molluscs: a review. J. Shellfish Res. 23, 51–62.

Nie, Z., Zheng, Z., Zhu, H.-J., Sun, Y., Gao, J., Xu, G., 2023. Effects of submerged macrophytes (*Elodea nuttallii*) on water quality and microbial communities of largemouth bass (*Micropterus salmoides*) ponds. Front. Microbiol. 13. doi:10.3389/fmicb.2022.1050699.

Nisar, U., Peng, D., Mu, Y., Sun, Y., 2022. A solution for sustainable utilization of aquaculture waste: a comprehensive review of biofloc technology and aquamimicry. Front. Nutr. 8, 791738.

Niyi-David, C.C., Wemedo, S., Akani, N.P., Douglas, S., 2021. Culture-based characterization of bacteria associated with fish pond wastewater undergoing treatment using plants. J. Adv. Microbiol. 21 (5), 1–10. doi:10.9734/JAMB/2021/V21I530346.

Oliveira, C.Y.B., Abreu, J.L., Brandão, B.C., Oliveira, D.W.S., de Sena, P.R., da Silva, W.A., Gálvez, A.O., 2024. A holistic approach to producing anti-Vibrio metabolites by an endosymbiotic dinoflagellate using wastewater from shrimp rearing. Microorganisms 12 (8), 1598. https://doi.org/10.3390/microorganisms12081598.

Oliveira, C.Y.B., Abreu, J.L., Oliveira, C.D.L., Lima, P.C., Gálvez, A.O., Dantas, D.M.M., 2020. Growth of *Chlorella vulgaris* using wastewater from Nile tilapia (*Oreochromis niloticus*) farming in a low-salinity biofloc system. Acta Sci. Technol. 42. https://doi.org/10.4025/actascitechnol.v42i1.46232.

Oliveira, C.Y.B., Oliveira, C.D.L., Müller, M.N., Santos, E.P., Dantas, D.M., Gálvez, A.O., 2020. A scientometric overview of global dinoflagellate research. Publications 8 (4), 50. https://doi.org/10.3390/publications8040050.

Oliveira, K.F., Azevedo, R.V., Pereira, M.C., Santos, M.J.M., Carvalho, J.S.O., Braga, L.G.T., 2014. Use of the oyster *Crassostrea rhizophorae* as a biological filter for the treatment of shrimp farming effluents. Semina Ciênc. Agrár. 35 (4), 2789–2798.

Owatari, M.S., Mouriño, J., 2020. Practical and essential information on water reuse systems in experimental aquaculture production: a descriptive review. J. Agric. Food Sci. 3 (1), 186–202. doi:10.25177/jaffs.3.1.ra.10628.

Pajand, Z.O., Soltani, M., Bahmani, M., Kamali, A., 2017. The role of polychaete *Nereis diversicolor* in bioremediation of wastewater and its growth performance and fatty acid composition in an integrated culture system with Huso huso (Linnaeus, 1758). Aquac. Res. 48 (10), 5271–5279. https://doi.org/10.1111/are.13340.

Palmer, P.J., Wang, S., Mann, D., Nash, W.J., 2018. Technical Report. Queensland Government, Brisbane, Australia, p. 44. http://era.daf.qld.gov.au/id/eprint/6150/.

Palomino Seguil, Y., Vilchez Garay, L., Matencios Cortez, C., Cornejo Tueros, J.C., Camargo Hinostroza, S., Canales Guerra, V., 2022. Systematic review of the efficiency of aquatic plants in the wastewater treatment. IOP Conf. Ser.: Earth Environ. Sci. 1009, 012004. doi:10.1088/1755-1315/1009/1/012004.

Panigrahi, A., Otta, S.K., Vasagam, K., K., P., Anand, S., P., S., Biju, I.F., Aravind, R., 2019. Training manual on Biofloc technology for nursery and grow out aquaculture. CIBA TM Series 15, 172.

Papageorgiou, N., Dimitriou, P.D., Chatzivasileiou, D., Tsapakis, M., Karakassis, I., 2023. Can IMTA provide added ecosystem value services in the fish farms of Greece? Front. Mar. Sci. 10, 1083099. doi:10.3389/fmars.2022.1083099.

Pau, C., Serra, T., Colomer, J., Casamitjana, X., Sala, L., Kampf, R., 2013. Filtering capacity of *Daphnia magna* on sludge particles in treated wastewater. Water Res. 47 (1), 181–186. https://doi.org/10.1016/j.watres.2012.09.047.

Pawar, L., Nag, M., Sidiq, M.J., 2020. Integrated multi-trophic aquaculture systems (IMTA): a sustainable approach for better resource utilization. J. Aquac. 28, 19–26. doi:10.61885/joa.v28.2020.255.

Pearse, V., Pearse, J., Buchsbaum, M., Buchsbaum, R., 1987. Living invertebrates. Blackwell Scientific Publications, Palo Alto.

Perales-Pérez, Á., Macías-Sánchez, M., Ruiz, J., Perales, J., Garrido-Pérez, C., 2023. Process for nutrient recycling from intensive aquaculture through microalgae-bacteria consortium. Sci. Total Environ. 901, 165848. doi:10.1016/j.scitotenv.2023.165848.

Perez-Garcia, O., Escalante, F.M.E., de Bashan, L.E., Bashan, Y., 2011. Heterotrophic cultures of microalgae: metabolism and potential products. Water Res. 45 (1), 11–36. http://dx.doi.org/10.1016/j.watres.2010.08.037.

Petersen, C., Costa-Pierce, B.A., Dumbauld, B.R., Friedman, C., Hoffmann, E.E., Kite-Powell, H., Manahan, D.T., O'beirn, F., Paine, R.T., Thompson, P., Whitlach, R., 2010. Ecosystem concepts for sustainable bivalve mariculture. The National Academies Press, Washington, DC.

Pfeiffer, T., Ott, B., 2023. Biofiltration reactivation kinetics of a Cyclobio fluidized sand filter in a warmwater recirculating aquaculture system. North Am. J. Aquacult. 85 (2), 166–173. doi:10.1002/naaq.10282.

Pous, N., Barcelona, A., Sbardella, L., Hidalgo, M., Colomer, J., Serra, T., Salvadó, V., 2021. Zooplankton-based reactors for tertiary wastewater treatment: a pilot-scale case study. J. Environ. Manage. 278, 111538. https://doi.org/10.1016/j.jenvman.2020.111538.

Prado-Cabrero, A., Herena-Garcia, R., Nolan, J.M., 2022. Intensive production of the harpacticoid copepod *Tigriopus californicus* in a zero-effluent 'green water' bioreactor. Sci. Rep. 12 (1), 466. https://doi.org/10.1038/s41598-021-04516-w.

Rajanandini, M., Kumar, S.N., 2021. A biological and technological approach to treat wastewater by using macroalgae and microalgae. Res. J. Chem. Environ. 25 (6), 181–187.

Rao, A.R., Ravishankar, G.A., 2022. Sustainable global resources of seaweeds volume 1: bioresources, cultivation, trade and multifarious applications. Springer Nature, Cham, Switzerland.

Raposo, M.F.D., de Morais, R.M.S.C., de Morais, A.M.M.B., 2013. Health applications of bioactive compounds from marine microalgae. Life Sci. 93 (15), 479–486. https://doi.org/10.1016/j.lfs.2013.08.002.

Rehman, A., Shakoori, F.R., Shakoori, A.R., 2008. Heavy metals resistant rotifers from a chromium contaminated wastewater can help in environmental clean-up. Pak. J. Zool. 40 (5), 309–316.

Ren, J.S., Stenton-Dozey, J., Plew, D.R., Fang, J., Gall, M., 2012. An ecosystem model for optimising production in integrated multitrophic aquaculture systems. Ecol. Modell. 246, 34–46. doi:10.1016/j.ecolmodel.2012.07.020.

Richmond, A., 2004. Handbook of microalgal culture: biotechnology and applied phycology. Blackwell Publishing Company, Iowa, Ames, Iowa, USA.

Robinson, G., Caldwell, G.S., Jones, C.J.W., Slater, M.J., Stead, S.M., 2015. Redox stratification drives enhanced growth in a commercially important deposit-feeding invertebrate: implications for bioremediation and integrated aquaculture. Aquacult. Env. Interact. 8, 1–13. https://doi.org/10.3354/aei00158.

Robles-Porchas, G.R., Gollas-Galván, T., Martínez-Porchas, M., Martínez-Córdova, L.R., Miranda-Baeza, A., Vargas-Albores, F., 2020. The nitrification process for nitrogen removal in biofloc system aquaculture. Rev. Aquac. 12 (4), 2228–2249. doi:10.1111/raq.12431.

Romano, N., 2017. Aquamimicry: a revolutionary concept for shrimp farming. Glob. Aquac. Adv. https://www.aquaculturealliance.org/advocate/aquamimicry-a-revolutionary-concept-for-shrimp-farming.

Royan, M.R., 2020. Integrated multi-trophic aquaculture (IMTA) as a solution for shrimp aquaculture side effects on Northern Coast of Java, Indonesia. Grouper: J. Ilm. Perikan. 11 (1), 31–37.

Rugiu, L., Hargrave, M., Enge, S., Sterner, M., Nylund, G., Pavia, H., 2020. Kelp in IMTAs: small variations in inorganic nitrogen concentrations drive different physiological responses of *Saccharina latissima*. J. Appl. Phycol. 33, 1021–1034. doi:10.1007/s10811-020-02333-8.

Ruiz, J., Álvarez, P., Arbib, Z., Garrido, C., Barragán, J., Perales, J.A., 2011. Effect of nitrogen and phosphorus concentration on their removal kinetic in treated urban wastewater by *Chlorella vulgaris*. Int. J. Phytoremediation 13 (9), 884–896. https://doi.org/10.1080/15226514.2011.573823.

Sadaiappan, B., PrasannaKumar, C., Nambiar, V.U., Subrramanian, M., Gauns, M.U., 2021. Meta-analysis cum machine learning approaches address the structure and biogeochemical potential of marine copepod associated bacteriobiomes. Sci. Rep. 11, 3312. https://doi.org/10.1038/s41598-021-82482-z.

Samocha, T.M., Fricker, J., Ali, A.M., Shpigel, M., Neori, A., 2015. Growth and nutrient uptake of the macroalga *Gracilaria tikvahiae* cultured with the shrimp *Litopenaeus vannamei* in an integrated multi-trophic aquaculture (IMTA) system. Aquaculture 446, 263–271. doi:10.1016/j.aquaculture.2015.05.008.

Santhanam, P., Manickam, N., Perumal, P., 2020. Biofloc-copefloc: a novel technology towards sustained aquaculture. J. Indian Soc. Coast. Agric. Res. 38 (2), 43–50. doi:10.1016/j.aquaeng.2020.123.

Santos, J., Nunes, C.A.R., Tavechio, W.L.G., Ledo, C.A.S., Macedo, C.F., Pereira, C.M., Santos, A.A., Santana, I.C., 2020. Effect of grazing by the marine shrimp (*Litopenaeus vannamei*) on microorganisms associated with artificial substrates. Rev. Sertão Sustent. 2 (1), 29–34.

Sawant, O., Jaiswar, S., 2024. Assessing the nitrate and phosphate uptake kinetics potential and growth performance of *Gracilaria corticata* var. *cylindrica* in shrimp farm water (SFW). Discov. Ocean 1 (7). doi:10.1007/s44289-024-00007-8.

Schneider, O., Sereti, V., Eding, E., Verreth, J., 2005. Analysis of nutrient flows in integrated intensive aquaculture systems. Aquacult. Eng. 32 (3–4), 379–401. doi:10.1016/j.aquaeng.2004.09.001.

Sebök, S., Hanelt, D., 2023. Cultivation of the brackish-water macroalga *Ulva lactuca* in wastewater from land-based fish and shrimp aquacultures in Germany. Aquaculture 571, 739463.

Serra, T., Colomer, J., Pau, C., Marín, M., Sala, L., 2014. Tertiary treatment for wastewater reuse based on the Daphnia magna filtration – comparison with conventional tertiary treatments. Water Sci. Technol. 70 (4), 705–711. https://doi.org/10.2166/wst.2014.284.

Shah, A., Shah, M., 2020. Characterisation and bioremediation of wastewater: A review exploring bioremediation as a sustainable technique for pharmaceutical wastewater. Groundwater for Sustainable Development 11, 100383. doi:10.1016/j.gsd.2020.100383.

Sharma, M., Ravi, O.P.K., Kumari, S., Singh, A.K., 2023. Biofloc technology - aquaculture a way towards sustainable. J. Aqua. 30, 264. doi:10.61885/joa.v30.2022.264.

Shidik, T.S.A.N., Ekasari, J., Jusadi, D., Setiawati, M., 2021. Productivity and quality of *Moina* sp. cultivated with various culture medium. J. Akuakult. Indones. 20 (2), 148–162. https://doi.org/10.19027/jai.20.2.148-162.

Shu, H.-K., Sun, H., Huang, W., Zhao, Y., Ma, Y., Chen, W., Sun, Y., Chen, X., Zhong, P., Yang, H., Wu, X., Huang, M., Liao, S., 2021. Nitrogen removal characteristics and potential application of the heterotrophic nitrifying-aerobic denitrifying bacteria *Pseudomonas mendocina* S16 and *Enterobacter cloacae* DS'5 isolated from aquaculture wastewater ponds. Bioresour. Technol. 345, 126541. doi:10.1016/j.biortech.2021.126541.

Smaal, A.C., Ferreira, J.G., Grant, J., Petersen, J.K., Strand, Ø., 2019. Goods and services of marine bivalves. SpringerOpen, Cham, p. 591. https://doi.org/10.1007/978-3-319-96776-9.

Sohel, A.M., Shahjahan, M., Hossain, M., Sumi, K.R., Hossain, M.S., Kari, Z.A., Tahiluddin, A.B., Téllez-Isaías, G., 2023. Effects of multispecies probiotics on growth, hematology, and gut health of stinging catfish (*Heteropneustes fossilis*) in biofloc system. Water 15 (14), 2519. doi:10.3390/w15142519.

Soto, D., 2009. Integrated mariculture: a global review. Food and Agriculture Organization of the United Nations, Rome FAO Fisheries and Aquaculture Technical Paper No. 529.

Srirengaraj, V., Razafindralambo, H., Rabetafika, H., Nguyen, H.-T., Sun, Y.-Z., 2023. Synbiotic agents and their active components for sustainable aquaculture: concepts, action mechanisms, and applications. Biology 12 (12), 1498. doi:10.3390/biology12121498.

Suleiman, A.K.A., Lourenço, K.S., Clark, C., Luz, R.L., Silva, G.H.R., Vet, L.E.M., Cantarella, H., Fernandes, T.V., Kuramae, E.E., 2020. From toilet to agriculture: Fertilization with microalgal biomass from wastewater impacts the soil and rhizosphere active microbiomes, greenhouse gas emissions and plant growth. Resources, Conservation and Recycling 161, 104924. doi:10.1016/j.resconrec.2020.104924.

Taufik, M., Ismail, T.I., Manan, H., Ikhwanuddin, M., Salam, A.I.A., Rahim, A.I.A., Ishak, A.N., Kamaruzzan, A.S., Draman, A.S., 2023. Synergistic effects of recirculating aquaculture system (RAS) with combination of clear water, probiotic and biofloc technology: a review. Aquac. Fish. 9 (6), 883–892. doi:10.1016/j.aaf.2023.07.006.

Temporetti, P., López, W., Lucas, P., Pedrozo, F., 2000. Limnological effects of intensive aquaculture on lakes and reservoirs and reduction of nutrient input. SIL Proc. 27 (4), 1812–1815. doi:10.1080/03680770.1998.11901554.

Tenore, K.R., Dunstan, W.M., 1973. Comparison of feeding and biodeposition of three bivalves at different food levels. Mar. Biol. 21, 190–195.

Thielker, J.L., 1981. Design and test operation of an intensive controlled-environment oyster production system. J. World Aquac. Soc. 12 (1), 77–93.

Tlusty, M.F., Tyedmers, P., Bailey, M., Ziegler, F., Henriksson, P.J.G., Béné, C., Bush, S., Newton, R., Asche, F., Little, D.C., Troell, M., Jonell, M., 2019. Reframing the sustainable seafood narrative. Glob. Environ. Chang. 59, 101991. doi:10.1016/j.gloenvcha.2019.101991.

Torres, P., Santos, J.P., Chow, F., dos Santos, D.Y., 2019. A comprehensive review of traditional uses, bioactivity potential, and chemical diversity of the genus *Gracilaria* (Gracilariales, Rhodophyta). Algal Res. 37, 288–306.

Toyama, T., Kasuya, M., Hanaoka, T., Kobayashi, N., Tanaka, Y., Inoue, D., Sei, K., Morikawa, M., Mori, K., 2018. Growth promotion of three microalgae, *Chlamydomonas reinhardtii, Chlorella vulgaris and Euglena gracilis*, by in situ indigenous bacteria in wastewater effluent. Biotechnol. Biofuels 11 (1), 176–187. https://doi.org/10.1186/s13068-018-1174-0.

Troell, M., Joyce, A., Chopin, T., Neori, A., Buschmann, A.H., Fang, J.G., 2009. Ecological engineering in aquaculture — Potential for integrated multi-trophic aquaculture (IMTA) in marine offshore systems. Aquaculture 297, 1–9. doi:10.1016/j.aquaculture.2009.09.010.

Troell, M., Naylor, R.L., Metian, M., Beveridge, M., Tyedmers, P.H., Folke, C., Williams, M., 2014. Does aquaculture add resilience to the global food system? Proc. Natl. Acad. Sci. USA 111 (37), 13257–13263. doi:10.1073/pnas.1404067111.

van der Meer, T.V., Davey, C.J., Verdonschot, P.F., Kraak, M.H., 2023. Removal of nutrients from WWTP effluent by an algae-mussel trophic cascade. Ecol. Eng. 190, 106930.

Verreth, J., Roy, K., Turchini, G., 2023. Circular bio-economy in aquaculture. Rev. Aquac. 15 (3), 944–946. doi:10.1111/raq.12812.

Viegas, C., Gouveia, L., Gonçalves, M., 2021. Aquaculture wastewater treatment through microalgal biomass: potential applications on animal feed, agriculture, and energy. J. Environ. Manage. 286. https://doi.org/10.1016/j.jenvman.2021.112187.

Vonshak, A., 2014. Spirulina platensis (Arthrospira): physiology, cell-biology and biotechnology. Taylor & Francis/CRC Press, London. UK, p. 252. https://doi.org/10.1201/9781482272970.

Wang, J., Chen, W., Zhang, S., Li, T., Zhang, L., Xu, J., 2023. Removal of nitrogen and phosphorus using macroalgae in aquaculture systems: a review. Water 15 (3), 396. doi:10.3390/w15030396.

Wang, X., Olsen, L.M., Reitan, K.I., Olsen, Y., 2012. Discharge of nutrient wastes from salmon farms: environmental effects, and potential for integrated multi-trophic aquaculture. Aquacult. Env. Interact. 2 (3), 267–283. doi:10.3354/aei00044.

Ward, J.E., 1996. Biodynamics of suspension-feeding in adult bivalve molluscs: particle capture, processing, and fate. Invertebr. Biol. 115 (3), 218–231. https://doi.org/10.2307/3226932.

Wu, J., Rogers, S., Schaummann, R., Higgins, C., Price, N., 2022. Bioextractive aquaculture as an alternative nutrient management strategy for water resource recovery facilities. Water Res. 212, 118092. doi:10.1016/j.watres.2022.118092.

Xia, S., Song, Z., Jeyakumar, P., Shaheen, S. M., Rinklebe, J., Ok, Y. S., Bolan, N., Wang, H. 2019. A critical review on bioremediation technologies for Cr(VI)-contaminated soils and wastewater. Critical Reviews in Environmental Science and Technology, 49(12), 1027–1078. https://doi.org/10.1080/10643389.2018.1564526.

Yilmaz, S., Ergün, S., Yiğit, M., Yilmaz, E., 2022. An extensive review on the use of feed additives against fish diseases and improvement of health status of fish in Turkish aquaculture sector. Aquac. Stud. 22 (3). doi:10.4194/aquast710.

Yu, Y.-B., Choi, J.-H., Lee, J.-H., Jo, A.-H., Han, S.W, Han, S.-H., Choi, H.J., Choi, C.Y., Kang, J.-C., Min, E., Kim, J.-H., 2023. Biofloc application using aquaponics and vertical aquaculture technology in aquaculture: review. Fishes 8 (11), 543. doi:10.3390/fishes8110543.

Zhang, J., Chen, L., Dong, H., Duan, Y., Li, Z., Wen, G., Chen, J., Feng, Z., Xu, W., Xie, J., 2016. Artificial substrates in zero-water-exchange culture system regulate the rearing performance of Pacific white shrimp *Litopenaeus vannamei* (Boone, 1931) under the winter indoor condition. Aquac. Res. 47 (1), 91–100. doi:10.1111/are.12473.

Zhou, Q., Wu, K., Yao, L., Chen, R., Liu, S., Xing, H., Nie, L., Wu, Z., 2023. Bio-ecological remediation of freshwater aquaculture environments: a systematic review and bibliometric analysis. Water Biol. Secur. 3 (1), 100229. doi:10.1016/j.watbs.2023.100229.

Zhou, T., He, J., Lin, Z., He, L., 2021. Treatment of shrimp effluent by integrated culture of bivalves and macroalgae. Asian Agric. Res. 13 (5), 33–39.

Zhou, Y., Wang, Y.-P., Fu, S., Qiao, W., Zhao, H., Zhu, L., 2022. Enhanced nitrogen removal of aquaculture wastewater in the combined biological aerated filter: the effect of GAC location setting. J. Chem. Technol. Biotechnol. 97 (9), 2519–2527. doi:10.1002/jctb.7112.

Zrimec, M.B., Malta, E., Dunbar, M.B., Cerar, A., Reinhardt, R., Mihelič, R., 2022. Wastewater cultivated macroalgae as a bio-resource in agriculture. In: Rao, A.R., Ravishankar, G.A. (Eds.), Sustainable global resources of seaweeds volume 1: bioresources, cultivation, trade and multifarious applications. Springer Nature, Switzerland, pp. 435–449.

Zuo, X., Xu, W., Wei, S., Jiang, S., Luo, Y., Ling, M., Zhang, K., Gao, Y., Wang, Z., Hu, J., Grossart, H., Luo, Z., 2023. Aerobic denitrifying bacterial-fungal consortium mediating nitrate removal: dynamics, network patterns and interactions. iScience 26 (6), 106824. doi:10.1016/j.isci.2023.106824.

CHAPTER

11

Microbial fermentation and resource recovery from biowaste

Girija Saurabh Behere[a], *Sudhanshu Raman*[a], *Ganesh Kumar*[a], *Tanuj Misra*[a] and *Dharmendra Kumar Meena*[b]

[a]College of Fisheries, Datia, Rani Lakshmi Bai Central Agricultural University, Jhansi, Uttar Pradesh, India [b]ICAR-Central Inland Fisheries Research Institute, Barrackpore, Kolkata, West Bengal, India

11.1 Introduction

11.1.1 Overview of biowaste management in aquaculture

Aquaculture, which includes the farming of fish, shellfish, and other aquatic organisms, is one of the fastest-growing sectors in food production, providing approximately 50% of the world's seafood (FAO, 2020). While aquaculture plays a vital role in global food security, it also generates substantial amounts of organic waste that must be properly managed to minimize its environmental impact. Biowaste in aquaculture refers to the organic materials discarded during production and processing, including uneaten feed, fish excreta, dead fish, and processing byproducts (Mook et al., 2020). These wastes pose significant challenges in terms of disposal, often contaminating surrounding water bodies and contributing to pollution.

Uneaten feed, which can account for up to 20% of total feed input, is one of the primary contributors to aquaculture waste (Tacon and Metian, 2008). This feed, once discarded, can degrade water quality, leading to high nutrient levels that can cause eutrophication—a process where excess nutrients in the water promote the rapid growth of algae, depleting oxygen levels and harming aquatic life (Scarlat et al., 2019). Fish excreta, rich in nitrogen and phosphorus, also contribute to nutrient pollution when not properly managed. In addition, dead fish and other animal remains from harvest or disease outbreaks generate large volumes of organic material, which can lead to environmental contamination if not treated appropriately (Kumar et al., 2021).

Zero-Waste Technologies in Aquaculture: Global Perspectives and Methods toward a Circular Economy
DOI: https://doi.org/10.1016/B978-0-443-34689-7.00018-5

TABLE 11.1 Types of biowaste generated in aquaculture systems and their components.

Type of biowaste produced	Main components	References
Uneaten Feed	Organic matter, nitrogen, phosphorus	Ajani et al. (2011)
Excreta	Ammonia, urea, feces (organic matter, nitrogen, phosphorus)	Costello et al. (2001)
Fish Offal (heads, guts, bones, scales, skins, fins)	Proteins, lipids, collagen, chitin	Mille and Semmens (2002)
Shell Waste (e.g., shrimp shells, mollusk shells)	Chitin, calcium carbonate, proteins	Mohapatra et al. (2017)
Sludge (settled organic particles)	Organic matter, nutrients (nitrogen and phosphorus)	Piedrahita (2003)
Algal Biomass (from cleaning tanks or net pens)	Organic matter, pigments, lipids	Akinwole et al. (2016)
Sediment Accumulation	Organic material, heavy metals, nutrient buildup	Akinwole et al. (2016)

Types of biowaste generated in aquaculture systems and their components are given in Table 11.1. The disposal of these wastes, if not handled efficiently, can lead to several adverse environmental effects, such as water contamination, methane emissions, and landfill overflow, posing threats to both aquatic ecosystems and surrounding communities (FAO, 2020). Therefore developing effective systems for managing aquaculture biowaste is crucial for reducing environmental impacts and improving the overall sustainability of the industry.

11.1.2 Challenges in aquaculture biowaste management

Aquaculture biowaste management presents a unique set of challenges that need to be addressed to reduce negative environmental impacts and promote resource recovery. Improper waste management in aquaculture operations can result in environmental degradation, eutrophication, and methane emissions from landfills, all of which contribute to the global climate crisis (Scarlat et al., 2019). If aquaculture waste is not properly disposed of, the resulting water pollution can lead to habitat destruction, biodiversity loss, and the decline of water quality, affecting both the aquatic organisms being farmed and the surrounding ecosystems (Kumar et al., 2021).

One of the main environmental concerns associated with aquaculture waste is the release of excessive nutrients (primarily nitrogen and phosphorus) into the water. When large quantities of uneaten feed and fish excreta accumulate in water bodies, they lead to eutrophication, causing harmful algal blooms, oxygen depletion, and the eventual death of aquatic life (Mayer et al., 2016). This process not only degrades the quality of water in aquaculture systems but also impacts surrounding freshwater and marine environments, leading to long-term ecological damage.

Another significant challenge is the generation of methane emissions from the decomposition of organic matter, such as dead fish, uneaten feed, and fecal waste, in landfills or open dumps. Methane is a potent greenhouse gas with a much higher global warming potential than carbon dioxide (CO_2), contributing to climate change (Scarlat et al., 2019). The improper disposal of biowaste in landfills results in the anaerobic decomposition of organic material, releasing methane into the atmosphere. This process not only contributes to climate change but also exacerbates the environmental footprint of aquaculture operations, making waste management a critical issue for the industry's sustainability.

Finally, waste management practices in aquaculture vary across regions, and in many parts of the world, effective waste treatment technologies are not yet widely implemented. While some farms rely on effluent treatment systems such as settling ponds or aerated lagoons, these systems often lack the capacity to manage large volumes of waste efficiently. As aquaculture production continues to increase, the need for innovative waste management solutions will become even more urgent.

11.1.3 Sustainability in aquaculture: Role of microbial fermentation

To address these challenges, aquaculture must adopt more sustainable practices that reduce waste, minimize environmental impacts, and promote the efficient use of resources. Microbial fermentation presents a promising solution for transforming aquaculture biowaste into valuable byproducts such as biofuels, bioplastics, and biofertilizers (Sindhu et al., 2020; Srivastava et al., 2020). Through fermentation, organic matter in aquaculture waste is metabolized by microorganisms into products that can be reintegrated into the economy, reducing waste and closing nutrient loops.

Microbial fermentation allows for the conversion of fish excreta, uneaten feed, and fish processing byproducts into biofuels such as bioethanol and biogas, which can be used as renewable energy sources for aquaculture operations (Kumar et al., 2021). In addition, organic acids such as lactic acid, produced during fermentation by bacteria such as *Lactobacillus* spp., can be utilized in the production of biodegradable bioplastics (Kourmentza et al., 2017b). Moreover, microbial fermentation facilitates the recovery of essential nutrients such as phosphorus and nitrogen, which can be transformed into biofertilizers for agricultural use, thus closing the nutrient cycle within the ecosystem (Mayer et al., 2016).

The implementation of microbial fermentation technologies in aquaculture waste management can enhance the circular economy model by ensuring that waste is recycled into valuable resources. By converting organic waste into biobased products, microbial fermentation supports the principles of resource recovery, energy generation, and sustainable product manufacturing. This not only contributes to the sustainability of aquaculture but also reduces the industry's reliance on fossil fuels, petroleum-based plastics, and chemical fertilizers, all of which have significant environmental footprints.

Furthermore, microbial fermentation aligns with global efforts to transition to a more sustainable food system. As aquaculture remains a primary source of protein for millions of people worldwide, ensuring its environmental sustainability is essential. By adopting microbial fermentation and other waste-to-resource technologies, the aquaculture sector can reduce its carbon footprint, minimize pollution, and improve resource use efficiency (Gupta and Patel, 2021; Kumar et al., 2021).

11.1.4 Introduction to circular economy: Role of microbial fermentation

The concept of the circular economy has gained significant attention in recent years as a sustainable alternative to the traditional linear economy, where resources are used, products are consumed, and waste is discarded. The circular economy emphasizes the continuous use of resources through recycling, reusing, and reprocessing (UN, 2015). Microbial fermentation plays a crucial role in this model by converting waste into valuable byproducts, allowing for the recycling of nutrients and the recovery of energy from waste materials.

In aquaculture, microbial fermentation allows the industry to move from a linear waste model (where waste is disposed of) to a circular model (where waste is recycled and converted into useful products). For example, uneaten feed and fish excreta, which are typically considered waste, can be transformed through fermentation into biofuels or bioplastics, thereby reducing the need for external resources and minimizing waste generation (Treu et al., 2021). In addition, phosphorus and nitrogen recovered from wastewater can be used to produce biofertilizers, which can be applied to agriculture, completing a closed-loop system (Mayer et al., 2016).

By integrating microbial fermentation into aquaculture waste management, the industry can adopt a more sustainable and resource-efficient approach. Decentralized fermentation systems, where aquaculture farms can treat their own waste on-site, could be the future of waste management in the sector. These systems reduce transportation costs, lower environmental impacts, and provide local resource recovery options, all while contributing to the goals of the circular economy (Ghimire and Lee, 2021a).

Moreover, as the demand for sustainable and eco-friendly products grows, fermentation-derived products such as bioplastics and biofertilizers are gaining popularity across various industries, including packaging, agriculture, and food production. This shift toward biobased products further supports the transition to a circular economy, where waste becomes a resource for new products.

11.2 Microbial fermentation in aquaculture

Microbial fermentation has emerged as a highly effective method for converting aquaculture waste into valuable byproducts, thereby promoting sustainability and resource recovery in aquaculture systems. This process involves the metabolic conversion of organic compounds by microorganisms, often under anaerobic or limited oxygen conditions. In aquaculture, microbial fermentation plays a pivotal role in managing the large volumes of organic waste produced, such as fish excreta, uneaten feed, and processing residues. Through this process, valuable compounds such as biofuels, bioplastics, organic acids, and biogas can be generated, offering economic and environmental benefits (Zhang et al., 2021).

This section delves into the mechanisms underlying microbial fermentation in aquaculture, with a focus on microbial metabolism, the microorganisms involved, fermentation pathways for biowaste valorization, and innovations in bioreactor designs for efficient aquaculture waste management. Overview of microbial fermentation and resource recovery from aquaculture biowaste is depicted in Fig. 11.1.

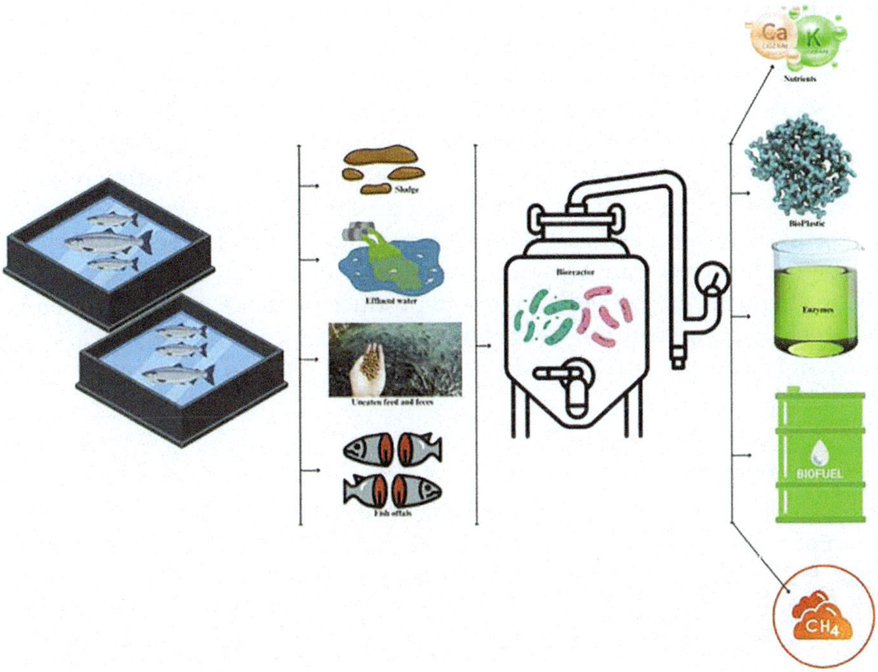

FIGURE 11.1 Overview of microbial fermentation and resource recovery from aquaculture biowaste.

11.2.1 Microbial metabolism in fermentation

Microbial metabolism in fermentation refers to the biochemical processes by which microorganisms convert organic substrates into simpler compounds in the absence or limited presence of oxygen. These processes are fundamental to the conversion of aquaculture waste into valuable products. Several metabolic pathways are involved in microbial fermentation, including glycolysis, acetogenesis, and methanogenesis, each contributing to the breakdown of organic material and the production of energy-rich byproducts.

11.2.1.1 Glycolysis

The first step in microbial fermentation is glycolysis, a central metabolic pathway that breaks down glucose and other simple sugars into pyruvate. Glycolysis occurs in the cytoplasm and is an anaerobic process, meaning it does not require oxygen. This makes glycolysis an ideal pathway for the fermentation of organic substrates in aquaculture waste, as many of these wastes are rich in carbohydrates such as sugars and starches. In the context of aquaculture, glucose from fish feed and other carbohydrate-rich byproducts serve as the primary substrate for glycolysis. The pathway produces two molecules of pyruvate, along with energy intermediates in the form of adenosine triphosphate (ATP) and nicotinamide adenine dinucleotide (NADH), which are essential for microbial growth and maintenance (Yadav et al., 2020). The pyruvate formed in glycolysis can be further metabolized through various pathways, depending on the microbial species and fermentation conditions.

11.2.1.2 *Acetogenesis*

Acetogenesis is a critical metabolic pathway following glycolysis, where pyruvate is converted into acetate through the action of acetogenic bacteria. Acetogenesis is an anaerobic process that plays a central role in the microbial fermentation of organic materials, including aquaculture waste. Acetogenic bacteria such as *Acetobacterium woodii* and *Clostridium* spp. convert pyruvate into acetate via acetyl-CoA intermediates. This pathway is particularly important in anaerobic digesters used for biogas production, as acetate serves as a precursor for methanogenesis (Zhao et al., 2023a). Acetogenic bacteria also contribute to improving biogas production by converting various organic acids and alcohols into acetate. This acetate can then be further utilized by methanogens to produce methane, an energy-rich biofuel. The role of acetogenesis in aquaculture waste fermentation is significant, as it helps in optimizing the efficiency of biogas production and the recovery of renewable energy from aquaculture byproducts.

11.2.1.3 *Methanogenesis*

Methanogenesis is the final step in anaerobic fermentation, where methanogenic archaea convert organic substrates, particularly acetate and carbon dioxide, into methane (CH_4). Methanogenesis is carried out by specialized microorganisms called methanogens, such as *Methanosarcina* and *Methanobacterium*, which thrive in low-oxygen environments. These archaea are capable of breaking down organic materials, such as aquaculture waste, to produce methane as a byproduct.

The process of methanogenesis is integral to the production of biogas, a renewable energy source. In the anaerobic digestion of aquaculture waste, methanogens convert acetate produced by acetogenic bacteria into methane, which can be captured and used as a biofuel. The production of biogas from aquaculture waste offers a sustainable way to reduce waste volume while generating energy (Ghimire and Lee, 2021b).

11.2.2 Types of microorganisms involved

The success of microbial fermentation processes depends largely on the types of microorganisms involved. Different microorganisms specialize in distinct metabolic pathways, each contributing to the breakdown of organic waste and the production of valuable byproducts. The key microorganisms involved in the fermentation of aquaculture waste include yeasts, bacteria, and archaea, each playing a vital role in converting aquaculture byproducts into renewable resources. Microbes used in resource recovery from the aquaculture waste are tabulated in Table 11.2.

11.2.2.1 *Saccharomyces cerevisiae*

Saccharomyces cerevisiae, commonly known as baker's yeast, is one of the most widely used microorganisms in industrial fermentation. This yeast is renowned for its ability to convert sugars, particularly glucose and maltose, into ethanol and carbon dioxide. In the context of aquaculture waste management, *S. cerevisiae* can be employed to ferment the sugars present in uneaten feed and other carbohydrate-rich aquaculture byproducts (Angelidaki et al., 2018).

TABLE 11.2 Microbes used in resource recovery from the aquaculture waste.

Microbes	Raw substratum	Value-added products	References
Lactococcus lactis and *Pediococcus acidilactici*	Fish viscera	Peptones	Vázquez et al. (2020)
Lactobacillus curvatus and *Enterococcus faecium*	Cod and albacore muscle	Preservatives compounds	Gómez-Sala et al. (2016)
Aspergillus oryzae	Fish skin	Antioxidant peptides (preservatives)	Fang et al. (2017)
Enterococcus gallinarum and *Streptococcus* spp.	Head	Oil	Özyurt et al. (2018)
Leuconostoc mesenteroides	Bones	Calcium	Tang et al. (2018)
Bacillus subtilis	Scales and intestine	Bioplastic	Mohapatra et al. (2017)
Lactobacillus pentosus	Catfish and tilapia viscera	Lactic acid	Shi et al. (2018)
Citrobacter freundii and *Comamonas sp.*	Slurry	Production of organic acids	Lee et al. (2019)
Sporolactobacillus inulinus, Lactobacillus mali, Lactobacillus casei, and Clostridium tyrobutyricum)	Uneaten feed and feces	Production of organic acids	Lee et al. (2019)
Acinteobacteria sp., Microlunatus phosphovorus, Arthrobacter sp., Lampropedia spp., *Tetrasphaera* spp., *M. phosphovorus, Pseudomonas putida* and *Burkholderiacepaci*	Sludge	Nutrient (phosphorus)	Luo (2023)

The primary advantage of using *S. cerevisiae* in aquaculture waste fermentation is its high tolerance to ethanol, which allows it to effectively ferment substrates in anaerobic conditions. The ethanol produced can be utilized as a biofuel, replacing fossil fuels and reducing the carbon footprint of aquaculture systems (Zhao et al., 2023a).

11.2.2.2 Lactobacillus *spp.*

Lactobacillus spp., a genus of lactic acid bacteria, plays a crucial role in the production of lactic acid from the fermentation of sugars. Lactic acid is an important organic acid used in food preservation, pharmaceuticals, and the production of biodegradable plastics. In aquaculture waste management, *Lactobacillus* spp. can ferment the sugars present in fish excreta and feed residues into lactic acid, which can be used in various industrial applications (Sindhu et al., 2020). These bacteria are well-suited for fermentation in acidic environments and can efficiently convert a wide range of organic substrates into lactic acid. The ability of *Lactobacillus* spp. to produce lactic acid from aquaculture waste offers an opportunity to generate biodegradable materials such as polylactic acid (PLA), a plastic alternative that can help mitigate the global plastic waste crisis (Kourmentza et al., 2017a).

11.2.2.3 Clostridium *spp*

Clostridium spp. are anaerobic bacteria that are key players in the production of butanol and other solvents. Butanol is a biofuel that can be used as a substitute for gasoline in transportation and energy generation. In aquaculture waste management, *Clostridium* spp. can ferment organic compounds such as carbohydrates, proteins, and lipids from fish excreta or processing residues into butanol and other solvents (Ghimire and Lee, 2021b). *Clostridium* spp. are particularly effective at fermenting complex organic compounds and have been engineered for enhanced tolerance to butanol, a solvent that can be toxic to microorganisms at high concentrations. The fermentation of aquaculture waste by *Clostridium* spp. offers a promising pathway for the production of renewable biofuels and other valuable chemicals.

11.2.3 Fermentation pathways for biowaste valorization

Microbial fermentation is a powerful method for converting aquaculture waste into valuable products such as biofuels, bioplastics, organic acids, and biogas. These fermentation pathways not only provide an alternative source of energy and materials but also contribute to reducing the environmental impact of aquaculture by recycling waste and minimizing pollution.

11.2.3.1 *Biofuels production*

Biofuels, including bioethanol and butanol, are among the most widely produced products through microbial fermentation. Yeasts such as *S. cerevisiae* convert sugars into bioethanol, while bacteria such as *Clostridium* spp. produce butanol. Both bioethanol and butanol are renewable fuels that can be used in transportation, electricity generation, and other industrial processes. In aquaculture systems, biofuels can be produced from organic waste such as fish excreta and uneaten feed. The conversion of these waste materials into biofuels not only provides an alternative energy source but also helps reduce greenhouse gas emissions and the dependency on fossil fuels (Sarkar et al., 2012; Kar and Aich, 2020).

11.2.3.2 *Bioplastics production*

Bioplastics, including polyhydroxyalkanoates (PHA) and PLA, are produced through microbial fermentation of organic substrates. PHA is a biodegradable plastic that can be used in packaging and medical applications, while PLA is widely used for packaging and as a biodegradable alternative to petroleum-based plastics. Bacteria such as *Cupriavidus necator* are known for producing PHA, while PLA is synthesized from sugars through fermentation (Kourmentza et al., 2017a). The use of aquaculture waste as a feedstock for bioplastics production presents an opportunity to address the growing issue of plastic pollution. By recycling aquaculture byproducts into biodegradable plastics, microbial fermentation provides an environmentally friendly alternative to conventional plastics (Liu et al., 2021).

11.2.3.3 *Organic acids production*

Fermentation can also yield organic acids such as lactic acid, acetic acid, and citric acid. These organic acids are used in a wide range of industrial applications, including food preservation, pharmaceuticals, and cosmetics. Lactic acid, for example, can be used in the production of biodegradable plastics, while acetic acid is essential in vinegar production.

The production of organic acids from aquaculture waste through microbial fermentation offers a valuable means of resource recovery. By utilizing waste streams as feedstocks for organic acid production, aquaculture systems can reduce waste while generating valuable industrial products (Sindhu et al., 2020).

11.2.3.4 *Biogas production*

Biogas production is one of the most effective ways to recover energy from aquaculture waste. Through anaerobic digestion, microorganisms break down organic waste in the absence of oxygen to produce biogas, primarily composed of methane. Biogas can be used as a renewable energy source for heating, electricity generation, or transportation fuel.

Anaerobic digestion of aquaculture waste is highly effective in reducing waste volume while generating biogas, which can be captured and used to meet energy demands. The integration of biogas production with aquaculture systems not only reduces environmental pollution but also offers a sustainable source of energy (Ghimire and Lee, 2021a). Zhuang et al. (2021) explored the application of anaerobic digestion in the valorization of aquaculture waste, highlighting its potential to convert organic byproducts into valuable resources such as biogas, thus contributing to more sustainable waste management practices in the aquaculture industry.

11.2.4 Innovations in bioreactor designs for aquaculture waste management

Advancements in bioreactor designs have played a significant role in optimizing microbial fermentation processes for aquaculture waste management. Bioreactors provide controlled environments that enable microorganisms to efficiently convert organic waste into valuable products. Several innovative bioreactor technologies have been developed to improve the efficiency and scalability of microbial fermentation processes in aquaculture.

11.2.4.1 *Anaerobic digesters*

Anaerobic digesters are commonly used for the treatment of organic waste in aquaculture systems. These bioreactors create anaerobic conditions that support the growth of methanogens and other anaerobic microorganisms, which break down organic material into methane and other byproducts. Innovations in anaerobic digestion include the development of high-rate digesters, which can process large volumes of aquaculture waste while maximizing methane production (Ghimire and Lee, 2021b). High-rate digesters can be integrated into aquaculture farms, treating waste in real time while generating renewable energy in the form of biogas. This approach not only reduces waste but also contributes to energy recovery, creating a more sustainable and circular aquaculture system (Zhao et al., 2023a).

11.2.4.2 *Membrane bioreactors*

Membrane bioreactors (MBRs) combine biological treatment with membrane filtration to separate solids from liquids, improving waste treatment efficiency. MBRs are particularly useful for recovering nutrients and organic compounds from aquaculture wastewater while maintaining high water quality. The integration of MBRs with microbial fermentation processes can enhance the recovery of valuable byproducts such as organic acids, biofuels, and bioplastics, while also improving wastewater treatment (Zhao et al., 2023a).

11.2.4.3 *Integrated fermentation systems*

Innovative approaches have also focused on integrating microbial fermentation directly into aquaculture systems. Fermentation ponds and integrated bioreactors allow microbial communities to break down organic matter in situ, converting waste into valuable products while treating aquaculture effluent. These systems are cost-effective and can operate with minimal energy input, making them ideal for sustainable aquaculture waste management (Angelidaki et al., 2018).

Microbial fermentation is a promising strategy for the valorization of aquaculture waste, offering multiple pathways for producing biofuels, bioplastics, organic acids, and biogas. By harnessing the metabolic capabilities of microorganisms such as yeasts, bacteria, and archaea, aquaculture systems can convert organic waste into valuable products, reducing environmental impact and promoting sustainability. Innovations in bioreactor design, such as anaerobic digesters, MBRs, and integrated fermentation systems, have significantly improved the efficiency and scalability of these processes (Wang and Zhan, 2021).

11.3 Resource recovery pathways in aquaculture waste management

Resource recovery pathways, particularly through microbial fermentation, are central to transforming aquaculture waste into valuable products. The application of microbial processes not only helps in managing aquaculture waste but also provides sustainable solutions for producing biofuels, bioplastics, fertilizers, and other high-value products. This section explores key resource recovery pathways, including biofuels production, bioplastics synthesis, nutrient recovery, and the conversion of aquaculture waste into high-value products, highlighting the potential for a circular economy in aquaculture systems. Advanced methods used in the microbial fermentation process are summed up in Table 11.3.

11.3.1 Biofuels production from aquaculture waste

Biofuels, including bioethanol and biogas, are renewable energy sources that can be produced from organic aquaculture waste. The microbial fermentation of organic materials such as fish excreta, uneaten feed, and processing residues allows the conversion of these materials into valuable biofuels, which can replace fossil fuels in energy production. This pathway not only reduces the environmental impact of aquaculture but also contributes to the development of sustainable energy systems.

11.3.1.1 *Bioethanol production from aquaculture waste*

Bioethanol is a widely used biofuel that can be produced from fermentable sugars, including those derived from aquaculture waste. The fermentation of organic substrates, particularly carbohydrates, by *S. cerevisiae* (baker's yeast), results in the production of ethanol and carbon

TABLE 11.3 Advanced methods used in the microbial fermentation process.

Fermentation method	Target compound	Recovery (%)	References
Electro-fermentation	Protein	72.23	Gu et al. (2024)
Solid-state fermentation	Bioplastics and biofuels	80	Sharma et al. (2021)
Submerged culture fermentation	Polysaccharide	79	Sánchez et al. (2015)
In situ product recovery method	Protein	90	Salas-Villalobos et al. (2021)
Continuous stirred tank reactors	Trace metals	60	Sampaio et al. (2009)
Integrated bioprocessing	Biofuel	99.7	Rathore et al. (2018)

dioxide. The process begins with the breakdown of complex organic materials into simpler sugars, which are then fermented into ethanol under anaerobic conditions.

Aquaculture waste, such as fish feed residues, fish excreta, and plant-based byproducts, is often rich in carbohydrates, making it an ideal substrate for ethanol production. Studies have shown that organic materials from aquaculture waste can be effectively fermented to produce bioethanol, providing a renewable fuel source that helps mitigate the environmental impact of aquaculture (Angelidaki et al., 2018). In industrial-scale ethanol production, the use of waste products such as fish processing residues or algal biomass from aquaculture ponds can significantly reduce production costs and enhance sustainability (Ghimire and Lee, 2021a).

11.3.1.2 *Biogas production from aquaculture waste*

Biogas, primarily composed of methane, is another valuable biofuel that can be produced through the anaerobic digestion of organic aquaculture waste. Anaerobic digestion involves the microbial breakdown of organic matter in the absence of oxygen, resulting in the production of biogas. This process is carried out by a consortium of microorganisms, including methanogens, which convert organic substrates into methane and carbon dioxide.

Biogas production from aquaculture waste offers several advantages, including waste reduction and energy recovery. The anaerobic digestion of fish waste, including excreta and uneaten feed, produces biogas that can be used for heating, electricity generation, or as a transportation fuel. Industrial-scale biogas plants, such as those operating in Europe and Asia, have successfully integrated aquaculture waste as a feedstock for biogas production, demonstrating the scalability of this process (Ghimire and Lee, 2021b).

A notable example of biogas production from aquaculture waste comes from pilot-scale plants in Denmark, where fish processing residues are used to generate biogas. These plants contribute to reducing waste volumes while generating renewable energy, demonstrating the feasibility of large-scale biogas production from aquaculture waste (Angelidaki et al., 2018).

11.3.2 Bioplastics from aquaculture byproducts

The conversion of aquaculture byproducts into bioplastics is an emerging area in microbial fermentation, offering sustainable alternatives to petroleum-based plastics. Two prominent bioplastics—PHA and PLA—are produced through microbial fermentation processes and can be synthesized from aquaculture waste. These bioplastics have applications in packaging, agriculture, and the medical industry, among others.

11.3.2.1 *Polyhydroxyalkanoates*

PHA is a biodegradable polymer produced by bacteria through the fermentation of carbon-rich substrates, including organic waste from aquaculture. PHA is considered a promising alternative to conventional plastic due to its biodegradability and renewable nature. Microorganisms such as *C. necator* and *Pseudomonas* spp. are commonly used to produce PHA from a variety of carbon sources, including aquaculture waste. Aquaculture waste, especially lipids and carbohydrates found in fish excreta and other organic byproducts, can serve as feedstocks for PHA production. Studies have shown that bacteria can effectively convert aquaculture byproducts into PHA, thus addressing the environmental concerns associated with plastic pollution. For example, research has demonstrated the successful production of PHA from fish processing residues, highlighting the potential for large-scale production of bioplastics in the aquaculture industry (Kourmentza et al., 2017b).

11.3.2.2 *Polylactic acid*

PLA is another biodegradable polymer that can be synthesized through microbial fermentation. It is produced from lactic acid, which is derived from the fermentation of sugars and carbohydrates. The fermentation of sugars from aquaculture byproducts such as uneaten feed and fish excreta can provide an ideal feedstock for PLA production. *Lactobacillus* spp., a common lactic acid bacterium, is used to produce lactic acid from carbohydrates, which is then polymerized to form PLA. The use of aquaculture waste for PLA production is an innovative approach to reducing the reliance on petroleum-based plastics. PLA is widely used in packaging, food containers, and other applications, offering a biodegradable alternative to conventional plastics. The synthesis of PLA from aquaculture byproducts has been demonstrated in pilot-scale fermentations, further validating the feasibility of this approach in aquaculture waste valorization (Liu et al., 2021).

11.3.3 Nutrient recovery and fertilizer production

One of the most significant challenges in aquaculture is managing the nutrient load in wastewater, particularly nitrogen and phosphorus, which are major contributors to water pollution. Microbial processes can effectively recover these nutrients from aquaculture wastewater and convert them into valuable products such as fertilizers.

11.3.3.1 *Nitrogen and phosphorus recovery*

Aquaculture systems generate large quantities of nitrogen and phosphorus, primarily in the form of ammonia and phosphates, which can lead to eutrophication and degradation of water quality if not properly managed. Microbial processes such as nitrification and denitrification are keys to nitrogen recovery. In nitrification, ammonia-oxidizing bacteria (AOB) convert ammonia into nitrites, which are further converted into nitrates by nitrite-oxidizing bacteria (NOB). Denitrification, on the other hand, involves the reduction of nitrates into nitrogen gas, a process facilitated by denitrifying bacteria. Phosphorus recovery is typically achieved through struvite precipitation, a process in which magnesium ammonium phosphate (MAP) crystals are formed from ammonium and phosphate ions in the presence of magnesium. This process results in the recovery of phosphorus, which can then be used as a slow-release fertilizer. Struvite precipitation has gained attention as a sustainable method for nutrient recovery from aquaculture wastewater, particularly in large-scale aquaculture operations (Mayer et al., 2016).

11.3.3.2 *Fertilizer production from aquaculture waste*

The recovered nutrients—nitrogen and phosphorus—can be used to produce high-quality fertilizers, contributing to the circular economy by reducing the need for synthetic fertilizers. Struvite, for example, is a slow-release fertilizer that provides a controlled supply of nutrients to plants, reducing the risk of nutrient runoff and environmental pollution. The use of aquaculture waste for fertilizer production not only helps mitigate water pollution but also offers a sustainable way to recycle valuable nutrients back into the agricultural cycle (Mayer et al., 2016). Several studies have demonstrated the feasibility of using aquaculture wastewater for nutrient recovery and fertilizer production. For example, pilot-scale systems have been implemented in fish farms in Asia, where struvite precipitation is used to recover phosphorus and nitrogen from aquaculture effluent. The recovered struvite is then used as a slow-release fertilizer for agricultural purposes, demonstrating the potential for integrating nutrient recovery and fertilizer production into aquaculture waste management systems (Mayer et al., 2016).

11.3.4 Waste-to-value products: Contributing to the circular economy

Microbial fermentation plays a crucial role in the concept of the circular economy, where waste products are converted into valuable products, thereby minimizing environmental impact and promoting sustainability. Aquaculture waste, traditionally viewed as an environmental burden, can be transformed into high-value products such as biofertilizers, bioplastics, and biofuels through microbial processes (Jahid et al., 2020).

11.3.4.1 *Biofertilizers*

Biofertilizers are products containing living microorganisms that promote plant growth by supplying essential nutrients. Through microbial fermentation, aquaculture waste can be processed to produce biofertilizers containing beneficial microorganisms such as nitrogen-fixing bacteria, phosphate-solubilizing bacteria, and plant growth-promoting rhizobacteria (PGPR). These biofertilizers can be used to improve soil fertility, reduce the need for chemical fertilizers, and promote sustainable agriculture. Studies have shown that aquaculture waste, particularly fishmeal and fish excreta, can be fermented to produce biofertilizers that enhance soil health and promote plant growth. The use of aquaculture byproducts in biofertilizer production is a prime example of how waste can be converted into a valuable resource, contributing to the circular economy and reducing the environmental impact of both aquaculture and agriculture (Treu et al., 2021).

11.3.4.2 *Bioplastics*

As previously discussed, bioplastics such as PHA and PLA can be produced through microbial fermentation of aquaculture waste, contributing to the reduction of plastic pollution and promoting the use of renewable resources. The production of bioplastics from aquaculture byproducts highlights the potential for waste valorization and underscores the importance of microbial fermentation in the circular economy. Resource recovery from aquaculture waste through microbial fermentation offers a sustainable and economically viable solution for reducing environmental pollution and generating valuable products. Biofuels, bioplastics, fertilizers, and biofertilizers are just a few examples of how aquaculture waste can be transformed into high-value products through microbial processes. As the demand for sustainable solutions in aquaculture continues to grow, the development of efficient microbial fermentation pathways and innovative waste management technologies will play a key role in advancing the circular economy in aquaculture.

11.4 Technological innovations and advancements in microbial fermentation for resource recovery

Technological innovations in microbial fermentation are driving significant advancements in the valorization of aquaculture waste. With the emergence of cutting-edge techniques such as microbial engineering, bioreactor advancements, and the integration of artificial intelligence (AI) and digitalization, the efficiency and effectiveness of microbial processes in converting aquaculture biowaste into valuable resources have reached new heights. These advancements not only enhance the sustainability of aquaculture systems but also contribute to the broader goals of waste-to-resource conversion in circular economies. This section delves into these innovations, highlighting their impact on microbial fermentation processes, resource recovery, and waste management.

11.4.1 Microbial engineering: Enhancing waste-to-resource conversion

The field of microbial engineering has witnessed groundbreaking advancements in recent years, particularly in genetic modification, Clustered Regularly Interspaced Short Palindromic Repeats (CRISPR)/CRISPR-associated protein 9 (Cas9), and synthetic biology. These technologies have significantly improved the efficiency of microbial fermentation, enabling microorganisms to be tailored for specific waste-to-resource conversion processes, thereby enhancing the recovery of valuable products such as biofuels, bioplastics, and fertilizers from aquaculture waste.

11.4.1.1 *Genetic modification and metabolic engineering*

One of the most significant advancements in microbial engineering is the genetic modification of microorganisms to optimize their metabolic pathways for waste valorization. Through genetic engineering, microorganisms can be engineered to enhance their ability to degrade complex organic compounds in aquaculture waste, such as proteins, lipids, and carbohydrates.

This process can be achieved through the insertion, deletion, or modification of specific genes involved in the degradation of target substrates.

For example, the genetic modification of bacteria such as *Escherichia coli* and *S. cerevisiae* has been employed to optimize the production of biofuels and bioplastics from aquaculture byproducts. By introducing genes that enhance the ability to utilize complex substrates, these engineered microorganisms can more efficiently convert aquaculture waste into valuable products such as bioethanol and PHA (Lee et al., 2019).

11.4.1.2 *CRISPR/Cas9 technology for microbial improvement*

The advent of CRISPR/Cas9 technology has revolutionized microbial engineering by allowing precise genome editing in microorganisms. This powerful tool enables the modification of microbial genomes with high accuracy, facilitating the creation of strains that can better tolerate harsh environmental conditions and enhance the yield of desired products. For example, CRISPR/Cas9 has been used to engineer microorganisms capable of producing high yields of biofuels and bioplastics from aquaculture waste, such as fish processing residues or fishmeal. By targeting specific genes involved in key metabolic pathways, CRISPR/Cas9 technology enables the development of strains with enhanced capabilities for converting organic substrates into valuable products. This technology has been applied to optimize the fermentation of aquaculture waste for bioethanol and PHA production, with notable improvements in both yield and process efficiency (Lee et al., 2019; Zhao et al., 2023b).

11.4.1.2.1 Synthetic biology and pathway optimization

In addition to genetic modification, synthetic biology approaches have provided new ways to design and construct synthetic microbial pathways that enhance the conversion of aquaculture waste into high-value products. Synthetic biology involves the design and assembly of novel biological parts, systems, and pathways that do not naturally exist in microorganisms. By constructing custom-designed metabolic pathways, synthetic biology enables microorganisms to perform more efficient conversions of aquaculture waste into biofuels, bioplastics, and other valuable chemicals. For instance, synthetic biology techniques have been employed to optimize the pathways for the production of PLA and PHA from aquaculture byproducts, resulting in higher yields and improved process efficiency. Synthetic biology tools also enable the development of microorganisms that can metabolize a wider range of organic materials, thus expanding the feedstock options for fermentation processes and making them more versatile in the context of waste valorization (Lee et al., 2019; Zhao et al., 2023b).

11.4.2 Bioreactor advancements: Innovations for enhanced fermentation efficiency

The development of advanced bioreactor technologies is another key factor contributing to the efficiency of microbial fermentation processes for resource recovery. Bioreactors are essential for providing controlled environments where microbial fermentation can occur, and recent advancements have led to the creation of more modular, scalable, and automated systems that optimize the fermentation of aquaculture waste.

11.4.2.1 *Modular and scalable bioreactor systems*

One of the most significant innovations in bioreactor technology is the development of modular and scalable systems that allow for greater flexibility in fermentation operations. Modular bioreactors are designed to be easily expanded or reconfigured, making them ideal for both small-scale pilot studies and large-scale industrial applications. This scalability is particularly important for aquaculture waste management, where the volume of waste can vary depending on the size of the aquaculture operation. By using modular bioreactors, aquaculture farms can adjust the fermentation capacity based on the amount of available waste, improving resource efficiency and reducing the overall cost of fermentation processes. This flexibility also facilitates the integration of fermentation systems into existing aquaculture facilities, where bioreactors can be scaled up or down according to operational needs (Treu et al., 2021).

11.4.2.2 *Automated and intelligent bioreactor systems*

Recent advancements in automation and intelligent bioreactors have further enhanced the efficiency of fermentation processes. These bioreactors incorporate sensors, real-time monitoring systems, and automated control mechanisms to maintain optimal fermentation conditions, such as temperature, pH, and nutrient levels. This automation not only improves the precision of fermentation but also reduces the need for manual intervention, thereby increasing the overall efficiency of the process. Moreover, intelligent bioreactors can automatically adjust operational parameters based on real-time data, optimizing microbial growth and product yield. For example, bioreactors can be programmed to adjust nutrient feeding rates or oxygen levels based on the metabolic needs of the engineered microorganisms, ensuring optimal conditions for the conversion of aquaculture waste into biofuels, bioplastics, or fertilizers (Liu et al., 2021).

11.4.2.3 *Integration of advanced bioreactor designs with waste management systems*

Bioreactors designed for microbial fermentation are increasingly being integrated with existing waste management systems in aquaculture facilities. For instance, bioreactors can be coupled with anaerobic digesters to process aquaculture waste and generate both biogas and nutrient-rich byproducts. This integration creates a more streamlined and efficient waste-to-resource conversion process, reducing waste volumes while simultaneously producing renewable energy and high-value products. In addition, innovations such as continuous-flow bioreactors and fed-batch systems have been developed to optimize the fermentation of aquaculture (Treu et al., 2021).

11.4.3 Artificial intelligence and digitalization in fermentation: Optimizing processes with machine learning

The integration of AI and machine learning (ML) into fermentation processes has opened up new possibilities for optimizing microbial activity and improving resource recovery efficiency. AI and ML algorithms can analyze large datasets, predict fermentation outcomes, and fine-tune fermentation conditions to maximize product yields from aquaculture waste.

11.4.3.1 *Predictive modeling and process optimization*

One of the most promising applications of AI and ML in microbial fermentation is the development of predictive modeling tools. These tools use historical data and ML algorithms to predict the outcome of fermentation processes, such as product yield and quality, based on a variety of parameters, such as temperature, pH, nutrient concentrations, and microbial strain used. By incorporating AI into bioprocess design, researchers and engineers can optimize fermentation conditions and minimize the trial-and-error approach traditionally used in fermentation optimization. For instance, AI-based models can predict how specific waste feedstocks, such as fish excreta or algae, will perform in fermentation systems, allowing for the design of more efficient waste-to-resource conversion processes. These models can also identify optimal microbial strains or genetic modifications that improve product yields, such as bioethanol or bioplastics (Mohammad et al., 2021).

11.4.3.2 *Real-time monitoring and control systems*

AI and digitalization also play a critical role in the real-time monitoring and control of fermentation processes. Through the use of Internet of Things (IoT) sensors and advanced data analytics, fermentation systems can be continuously monitored, with real-time feedback used to adjust operational parameters automatically. AI algorithms can process these data to predict future conditions, allowing for proactive adjustments to optimize microbial growth and product formation. For example, AI-powered systems can analyze the growth rates of engineered microorganisms, identify nutrient deficiencies, and recommend adjustments to the feeding schedule or other parameters to maintain optimal fermentation conditions. This level of automation reduces human error, enhances process stability, and increases the overall efficiency of fermentation-based resource recovery from aquaculture waste (Mohammad et al., 2021). Technological advancements in microbial engineering, bioreactor design, and digitalization are revolutionizing the way aquaculture waste is processed and converted into valuable resources. The integration of genetic modification, CRISPR/Cas9, synthetic biology, modular bioreactors, automated fermentation systems, and AI-powered predictive modeling is enhancing the efficiency, scalability, and sustainability of microbial fermentation processes. These innovations not only improve the conversion of aquaculture waste into biofuels, bioplastics, and fertilizers but also contribute to the broader goals of reducing environmental impacts and promoting a circular economy. As these technologies continue to evolve, they hold great promise for advancing resource recovery in aquaculture and other industries reliant on microbial fermentation.

11.5 Challenges and future opportunities in microbial fermentation for resource recovery

Microbial fermentation has the potential to transform aquaculture waste into valuable resources, contributing to a more sustainable and circular economy. However, several challenges impede the widespread adoption of microbial fermentation for aquaculture waste management. These challenges include high operational costs, feedstock variability, regulatory constraints, and issues related to the commercialization of fermentation processes. Despite these barriers, there are numerous opportunities for innovation in the sector, particularly through decentralized systems, integrated biorefinery, and renewable energy integration. This section explores these challenges and discusses the future opportunities in microbial fermentation for aquaculture waste management.

11.5.1 Barriers to widespread adoption of microbial fermentation in aquaculture waste management

Although microbial fermentation has great potential for resource recovery in aquaculture, several barriers hinder its widespread adoption in the industry. These challenges primarily relate to high operational costs, feedstock variability, and regulatory constraints that make it difficult to implement microbial fermentation systems at scale.

11.5.1.1 High operational costs

One of the primary barriers to the adoption of microbial fermentation is the high operational costs associated with the technology. The capital and operational costs of setting up fermentation systems, including the bioreactor infrastructure, equipment for monitoring and control, and energy consumption, can be prohibitively high for small to medium-scale aquaculture operations. The need for continuous optimization of process parameters, such as pH, temperature, and nutrient concentrations, adds complexity and cost to the system (Kumar et al., 2021). In addition, the cost of microbial strains and the required feedstock for fermentation can also contribute to the overall expense. Although aquaculture waste itself is often considered a low-cost feedstock, the variability in waste composition, nutrient content, and availability can lead to unpredictable fermentation performance, further driving up costs (Sindhu et al., 2020).

11.5.1.2 Feedstock variability and availability

Aquaculture waste is often heterogeneous in nature, containing various organic compounds such as proteins, lipids, and carbohydrates, which vary significantly depending on the aquaculture species and feed used. This variability poses a challenge for microbial fermentation processes, as microorganisms may not be able to efficiently degrade all types of organic matter present in the waste. Inconsistent feedstock quality can result in suboptimal microbial growth and reduced resource recovery, leading to lower product yields and reduced process efficiency (Kumar et al., 2021). Moreover, the seasonal fluctuations in aquaculture production can affect the availability of waste for fermentation. In regions with cyclical or seasonal aquaculture activities, there may be periods of low waste availability, making it challenging to maintain continuous fermentation operations. This issue can be mitigated by utilizing alternative or supplementary feedstocks, but the cost of sourcing these materials can further complicate the economic feasibility of microbial fermentation for waste management (Sindhu et al., 2020).

11.5.1.3 Regulatory constraints and compliance

The regulatory environment surrounding aquaculture waste management can also limit the widespread adoption of microbial fermentation. Many countries have stringent regulations regarding waste disposal and the treatment of organic materials, particularly with regard to environmental impact. Regulatory compliance often requires substantial investments in research and development to ensure that microbial fermentation processes are effective and meet environmental standards (Kumar et al., 2021). In some cases, regulatory frameworks may not yet account for the benefits of microbial fermentation for waste valorization, and policymakers may not fully recognize the potential of this technology in the context of sustainable aquaculture practices. This lack of regulatory support can slow the development and implementation of fermentation-based waste management systems (Angelidaki et al., 2018). Furthermore, inconsistent or conflicting regulations across regions can create uncertainties for aquaculture businesses seeking to adopt these technologies.

11.5.2 Economic feasibility and commercialization

The economic feasibility of microbial fermentation for aquaculture waste management is a critical factor that will determine its long-term success and commercial viability. While microbial fermentation offers the potential for significant resource recovery, the commercial implementation of fermentation processes requires careful consideration of infrastructure investments, cost-effectiveness, and market demand for the products generated.

11.5.2.1 Investment in infrastructure

The initial investment in fermentation infrastructure can be substantial. Setting up a microbial fermentation facility, including bioreactors, nutrient supply systems, and monitoring equipment, requires a significant capital investment. In addition, maintenance and operational costs can be high, particularly in large-scale systems where multiple bioreactors and continuous processes are involved (Valenti et al., 2018). In many cases, aquaculture businesses may be hesitant to invest in these technologies without a clear and predictable return on investment. The costs associated with scaling up microbial fermentation systems from pilot-scale experiments to industrial-scale operations can also be a challenge. The transition from laboratory-scale to commercial-scale fermentation often involves issues related to process optimization, feedstock consistency, and microbial strain performance. As a result, large-scale commercialization of microbial fermentation systems for aquaculture waste management can be a slow and costly process (Valenti et al., 2018).

11.5.2.2 Cost-effectiveness of fermentation products

While microbial fermentation can produce valuable products such as biofuels, bioplastics, and fertilizers, the cost-effectiveness of these products depends on market demand, production efficiency, and product pricing. The market for biofuels, for instance, is highly competitive, with fossil fuels and alternative energy sources often being more cost-effective. Similarly, the market for bioplastics, such as PHA and PLA, is still developing, and their production costs need to be reduced to make them commercially viable (Kourmentza et al., 2017a). The economic viability of microbial fermentation will be significantly influenced by the ability to reduce production costs through process optimization and technological advancements. For example, improving the efficiency of microbial strains through genetic engineering and metabolic optimization can increase product yields and lower production costs, making microbial fermentation more economically viable (Treu et al., 2021).

11.5.2.3 *Commercial viability and market demand*

In addition to production costs, the commercial viability of microbial fermentation for aquaculture waste management depends on the demand for the products generated. There must be a consistent market for biofuels, bioplastics, and other fermentation-derived products to ensure the sustainability of fermentation operations. Developing new markets for these products, especially in the context of the growing global demand for sustainable and circular economy solutions, will be a key to the long-term success of microbial fermentation (Ghimire et al., 2021b).

11.5.3 Opportunities for sustainable aquaculture practices

Despite the challenges, there are significant opportunities for integrating microbial fermentation into sustainable aquaculture practices. The growing emphasis on sustainability, circular economies, and waste-to-resource conversion provides a strong foundation for the development and adoption of fermentation-based technologies in aquaculture.

11.5.3.1 *Decentralized fermentation systems*

One promising opportunity for sustainable aquaculture is the development of decentralized microbial fermentation systems. Decentralized systems allow for localized processing of aquaculture waste, reducing the need for centralized waste treatment facilities and transportation costs. These systems can be designed to process aquaculture waste on-site or within a local community, improving waste management efficiency and reducing environmental impacts (Treu et al., 2021). Decentralized systems also offer flexibility, allowing smaller-scale aquaculture operations to integrate fermentation processes without the need for significant capital investments. By using modular bioreactors and automated fermentation systems, these decentralized systems can be adapted to varying scales of operation, enabling diverse aquaculture businesses to benefit from microbial fermentation technology (Ghimire and Lee, 2021a).

11.5.3.2 *Integrated biorefineries*

The concept of integrated biorefineries is another opportunity for enhancing the sustainability of aquaculture practices. In an integrated biorefinery, multiple waste streams from aquaculture are processed using microbial fermentation to produce a range of valuable products, including biofuels, bioplastics, biofertilizers, and other chemicals. This approach not only maximizes resource recovery but also minimizes waste, contributing to the circular economy.

Integrated biorefineries can be designed to optimize the use of both aquaculture waste and other organic materials, such as agricultural byproducts, creating a more efficient and sustainable waste-to-resource conversion system. By combining fermentation with other technologies, such as anaerobic digestion for biogas production, integrated biorefineries offer a comprehensive solution to aquaculture waste management (Treu et al., 2021).

11.5.3.3 *Renewable energy integration*

The integration of renewable energy sources into microbial fermentation systems offers another avenue for enhancing the sustainability of aquaculture practices. For example, biogas produced from anaerobic digestion of aquaculture waste can be used to power fermentation processes, reducing the need for external energy inputs and lowering operational costs. Solar, wind, or other renewable energy sources can also be used to power fermentation systems, further reducing the carbon footprint of microbial fermentation processes (Ghimire and Lee, 2021a).

11.5.4 Regulatory and policy frameworks for sustainable aquaculture

The regulatory and policy environment plays a crucial role in shaping the future of microbial fermentation in aquaculture waste management. Governments and international organizations need to develop and implement policies that support sustainable aquaculture practices and encourage the adoption of innovative technologies, such as microbial fermentation.

11.5.4.1 *Global policies and regulations*

International organizations, such as the United Nations (UN), have recognized the importance of sustainable aquaculture practices in promoting food security and environmental sustainability. In 2015, the UN adopted the 2030 Agenda for Sustainable Development, which includes several goals related to sustainable aquaculture and waste management, such as Goal 14 (Life below water) and Goal 12 (Responsible consumption and production) (UN, 2015). National governments also play a key role in regulating aquaculture waste management and incentivizing the use of microbial fermentation. By providing regulatory frameworks that encourage innovation and support the commercialization of fermentation technologies, governments can help bridge the gap between research and practical implementation in aquaculture industries (Angelidaki et al., 2018).

11.5.4.2 *Promoting sustainable practices*

To accelerate the adoption of microbial fermentation, governments and policymakers need to provide clear and consistent regulations that promote sustainable waste management practices. These regulations should incentivize the development of environmentally friendly technologies, support research and development, and ensure that waste management systems meet environmental standards (Angelidaki et al., 2018). Financial incentives, such as grants, tax breaks, or subsidies, could help reduce the financial barriers to adopting microbial fermentation technologies.

The future of microbial fermentation in aquaculture waste management holds significant promise for contributing to a more sustainable and circular economy. While challenges related to operational costs, feedstock variability, and regulatory constraints persist, innovations in bioreactor technology, microbial engineering, and renewable energy integration offer pathways to overcome these barriers. With the right policy support and a focus on economic feasibility and commercialization, microbial fermentation can play a transformative role in resource recovery, benefiting both the aquaculture industry and the environment.

11.6 Conclusion

Microbial fermentation is emerging as a vital solution for resource recovery in aquaculture waste management, addressing sustainability, waste reduction, and resource optimization. As aquaculture expands to meet global seafood demand, environmental challenges associated with waste management become more significant.

11.6.1 Microbial fermentation's role in the future of aquaculture

Microbial fermentation is key to enhancing aquaculture sustainability through waste valorization and resource recovery. Aquaculture operations produce substantial organic waste, including fish feed residues, fecal matter, and other byproducts. These wastes, if improperly managed, lead to environmental degradation. To address this, microbial fermentation transforms waste into biofuels, bioplastics, and fertilizers, reducing the environmental footprint and reliance on nonrenewable resources such as petroleum-based plastics and chemical fertilizers.

For example, fermentation of aquaculture waste can produce bioethanol and biogas, renewable energy sources for powering aquaculture facilities. Bioplastics, such as PHA and PLA, offer an alternative to conventional plastics harmful to marine environments. Microbial fermentation also produces biofertilizers, recovering nitrogen and phosphorus, thus reducing the need for synthetic fertilizers and supporting circular agriculture (Kumar et al., 2021).

Beyond waste reduction, microbial fermentation minimizes the environmental impact of aquaculture. Traditional waste disposal methods, such as dumping waste into water bodies, contribute to eutrophication, greenhouse gas emissions, and ecosystem contamination. By converting waste into valuable products, fermentation mitigates these negative effects, improving sustainability in the entire food production system (Scarlat et al., 2019). Moreover, fermentation's ability to generate multiple resource streams from aquaculture waste enhances system resilience, making aquaculture operations more adaptable to environmental and economic changes.

11.6.2 Global trends and adoption of microbial fermentation technologies

Globally, microbial fermentation is gaining recognition as a key tool for sustainability in aquaculture. The growing trend toward adopting these technologies is driven by environmental awareness, regulatory pressures, and the need to reduce the industry's carbon footprint. Governments, research institutions, and industries are heavily investing in microbial fermentation for resource recovery from aquaculture waste.

Research in microbial fermentation is intensifying, focusing on improving fermentation processes, microbial strain performance, and bioreactor systems. Advances in synthetic biology and CRISPR/Cas9 technology are enabling the design of microorganisms with tailored metabolic pathways to convert complex aquaculture byproducts into valuable resources (Lee et al., 2019; Zhao et al., 2023a).

Policy support and regulatory incentives are also key drivers. Global initiatives, such as the United Nations' 2030 Agenda for Sustainable Development, promote responsible production and consumption, including in aquaculture (UN, 2015). National regulations are fostering microbial fermentation adoption through tax breaks, grants, and subsidies for businesses using innovative waste-to-resource technologies (Ghimire and Lee, 2021a).

The commercial adoption of microbial fermentation technologies is increasing. Companies in the biotechnology and renewable energy sectors are scaling up fermentation systems to process aquaculture waste efficiently. Integrated biorefineries, processing multiple waste streams to produce a range of products, are becoming popular, offering greater resource recovery and reduced waste (Treu et al., 2021).

Microbial fermentation presents a promising pathway for transforming aquaculture waste management and enhancing industry sustainability. The ability to convert waste into biofuels, bioplastics, and fertilizers supports circular economy goals while reducing environmental impact. As technological advancements, policy support, and industry adoption grow, microbial fermentation will play a crucial role in scaling sustainable aquaculture practices worldwide. The integration of renewable energy, decentralized systems, and biorefinery models will further enhance its potential for global impact.

AI disclosure

During the preparation of this work, the author(s) used ChatGPT to paraphrase already formed content. After using this tool/service, the author(s) reviewed and edited the content as needed and take(s) full responsibility for the content of the publication.

References

Ajani, E.K., Akinwole, A.O., Ayodele, I.A., 2011. Fundamentals of fish farming in Nigeria. Welecrown Ventures, Ibadan, p. 158.

Akinwole, A.O., Dauda, A.B., Ololade, A.O., 2016. Haematological response of Clarias gariepinus juveniles reared in treated wastewater after waste solids removal using alum or Moringa oleifera seed powder. Int. J. Aquac. 6 (11), 1–8.

Angelidaki, I., Alves, M., Lema, J.M., 2018. Biomass conversion technologies for aquaculture waste. Bioresour. Technol. 276, 316–327. doi:10.1016/j.biortech.2018.11.105.

Costello, G., Davies, C., Papoutsoglou, Q., Saroglia, 2001. The control of chemicals used in aquaculture in Europe. J. Appl. Ichthyol. 17 (4), 173–180. doi:10.1046/j.1439-0426.2001.00314.x.

Fang, B., Sun, J., Dong, P., Xue, C., Mao, X., 2017. Conversion of turbot skin wastes into valuable functional substances with an eco-friendly fermentation technology. J. Cleaner Prod. 156, 367–377. doi:10.1016/j.jclepro.2017.04.055.

FAO, 2020. The state of world fisheries and aquaculture: sustainability in action. Food and Agriculture Organization of the United Nations, Rome doi:10.4060/ca9229en.

Ghimire, A., Lee, D.J., 2021a. Aquaculture waste valorization and resource recovery: perspectives on sustainable aquaculture practices. Environ. Sci. Technol. 55 (3), 1448–1460. doi:10.1021/acs.est.0c07472.

Ghimire, S., Lee, J.S., 2021b. Advances in bioconversion of organic waste to biofuels through microbial fermentation. Renew. Sustain. Energy Rev. 145, 111053. doi:10.1016/j.rser.2021.111053.

Gómez-Sala, B., Herranz, C., Díaz-Freitas, B., Hernández, P.E., Sala, A., Cintas, L.M., 2016. Strategies to increase the hygienic and economic value of fresh fish: Biopreservation using lactic acid bacteria of marine origin. Int. J. Food Microbiol. 223, 41–49. doi:10.1016/j.ijfoodmicro.2016.02.005.

Gu, J., Zhou, H., Wang, J., Feng, K., Xie, G., Liu, B., 2024. Defeng Xing, Sequential recovery of protein and ammonium from waste sludge and functional metabolism in a combined process of nutrient recovery electro-fermentation (NREF). Resour. Conserv. Recycl. 203, 107444. doi:10.1016/j.resconrec.2024.107444.

Gupta, R., Patel, S., 2021. Recent advances in microbial fermentation of aquaculture waste. Aquac. Res. 52 (1), 300–315. https://doi.org/10.1111/are.14956.

Jahid, I.K., Hossain, M., 2020. Recent advancements in the microbial transformation of aquaculture waste into value-added products. Waste Biomass Valor. 11 (1), 123–132. doi:10.1007/s12649-019-00772-3.

Kar, D., Aich, S., 2020. Microbial production of biofuels from aquaculture waste. Renew. Sustain. Energy Rev. 128, 109883. doi:10.1016/j.rser.2020.109883.

Kourmentza, C., Plácido, J., Venetsaneas, N., Burniol-Figols, A., Varrone, C., Gavala, H.N., Reis, M.A.M., 2017a. Recent advances and challenges towards sustainable polyhydroxyalkanoate (PHA) production. Bioeng. 4 (2). doi:10.3390/bioengineering4020055.

Kourmentza, C., Stamatelatou, K., Papanikolaou, S., 2017b. Bioplastics production from aquaculture waste via microbial fermentation. Bioresour. Technol. 233, 308–316. doi:10.1016/j.biortech.2017.02.010.

Kumar, M., Singh, L., Sharma, S., 2021. Challenges in microbial fermentation for biowaste valorization. Renew. Sustain. Energy Rev. 133, 110367. doi:10.1016/j.rser.2020.110367.

Lee, J., Kim, I.S., Emmanuel, A., Koh, S.C., 2019. Microbial valorization of solid wastes from a recirculating aquaculture system and the relevant microbial functions. Aquacult. Eng. 87. doi:10.1016/j.aquaeng.2019.102016.

Lee, S.Y., Kim, H.S., Park, J.H., 2019. CRISPR/Cas genome editing for microbial metabolic engineering. Biotechnol. Adv. 37 (1), 26–39. doi:10.1016/j.biotechadv.2018.11.004.

Liu, C., Song, X., Zhu, Q., Zhou, G., 2021. Advances in microbial fermentation for production of bio-based products: advances and future directions. Biotechnol. Progr. 37 (3), e3141. doi:10.1002/btpr.3141.

Luo, G., 2023. Review of waste phosphorus from aquaculture: Source, removal and recovery. Rev. Aquac. 15 (3), 1058–1082. doi:10.1111/raq.12727.

Mayer, B.K., Baker, L.A., Boyer, T.H., Drechsel, P., Gifford, M., Hanjra, M.A., Parameswaran, P., Stoltzfus, J., Westerhoff, P., Rittmann, B.E., 2016. Total value of phosphorus recovery. Environ. Sci. Technol. 50 (13), 6606–6620. doi:10.1021/acs.est.6b01239.

Miller, D., Semmens, K., 2002. Waste management in aquaculture. West Virginia University Extension Service Publication No. AQ02-1, Morgantown, WV, USA, p. 8.

Mohammad, H., Zhang, M., Cheng, Z., 2021. Artificial intelligence and machine learning in microbial fermentation: Applications for process optimization. Ind. Biotechnol. 17 (4), 162–175. doi:10.1089/ind.2020.0043.

Mohapatra, S., Sarkar, B., Samantaray, D.P., Daware, A., Maity, S., Pattnaik, S., Bhattacharjee, S., 2017. Bioconversion of fish solid waste into PHB using Bacillus subtilis based submerged fermentation process. Environ. Technol. 38 (24), 3201–3208. doi:10.1080/09593330.2017.1291759.

Mook, W.T., Chakrabarti, M.H., Aroua, M.K., Khan, G.M.A., Ali, B.S., Islam, M.S., Hassan, A., 2020. Development and applications of bioplastics in aquaculture systems. Bioresour. Technol. Rep. 11, 100468. doi:10.1016/j.biteb.2020.100468.

Özyurt, G., Özkütük, A.S., Uçar, Y., Durmuş, M., Özoğul, Y., 2018. Fatty acid composition and oxidative stability of oils recovered from acid silage and bacterial fermentation of fish (Sea bass – Dicentrarchus labrax) by-products. Int. J. Food Sci. Technol. 53 (5), 1255–1261. doi:10.1111/ijfs.13705.

Piedrahita, R.H., 2003. Reducing the potential environmental impact of tank aquaculture effluents through intensification and recirculation. Aquaculture 226 (1–4), 35–44. http://www.journals.elsevier.com/aquaculture/226.

Rathore, A.S., Kateja, N., Kumar, D., 2018. Process integration and control in continuous bioprocessing. Curr. Opin. Chem. Eng. 22, 18–25. doi:10.1016/j.coche.2018.08.005.

Salas-Villalobos, U.A., Gómez-Acata, R.V., Castillo-Reyna, J., Aguilar, O., 2021. In situ product recovery as a strategy for bioprocess integration and depletion of inhibitory products. J. Chem. Technol. Biotechnol. 96 (10), 2735–2743. doi:10.1002/jctb.6797.

Sampaio, R.M.M., Timmers, R.A., Xu, Y., Keesman, K.J., Lens, P.N.L., 2009. Selective precipitation of Cu from Zn in a pS controlled continuously stirred tank reactor. J. Hazard. Mater. 165 (1–3), 256–265. doi:10.1016/j.jhazmat.2008.09.117.

Sánchez, O.J., Montoya, S., Vargas, L.M., 2015. Polysaccharide production by submerged fermentation. Polysaccharides; Springer, Cham, Switzerland, pp. 451–473.

Sarkar, A., Ghosh, S., Ghosh, S., 2012. Microbial fermentation of aquaculture waste for biofuel production. Renew. Energy 37 (1), 8–13. doi:10.1016/j.renene.2011.04.029.

Scarlat, N., Dallemand, J.F., Fahl, F., 2019. The role of microbial fermentation in reducing environmental impacts in aquaculture. J. Cleaner Prod. 237, 117636 doi.org/10.1016/j.

Sharma, P., Gaur, V.K., Sirohi, R., Varjani, S., Hyoun Kim, S., Wong, J.W.C., 2021. Sustainable processing of food waste for production of bio-based products for circular bioeconomy. Bioresour. Technol. 325. doi:10.1016/j.biortech.2021.124684.

Shi, S., Li, J., Guan, W., Blersch, D., 2018. Nutrient value of fish manure waste on lactic acid fermentation by Lactobacillus pentosus. RSC Adv. 8 (55), 31267–31274. doi:10.1039/c8ra06142d.

Sindhu, R., Binod, P., Pandey, A., 2020. Challenges in biowaste valorization: the role of microbial fermentation in aquaculture waste management. Biotechnol. Adv. 38, 107390. doi:10.1016/j.biotechadv.2020.107390.

Srivastava, V., Singh, A., Kumar, M., 2020. Microbial fermentation for the production of biofuels from aquaculture waste. Environ. Technol. Innov. 18, 100675. doi:10.1016/j.eti.2020.100675.

Tacon, A.G.J., Metian, M., 2008. Global overview on the use of fish meal and fish oil in industrially compounded aquafeeds: trends and future prospects. Aquaculture 285 (1-4), 146–158. doi:10.1016/j.aquaculture.2008.08.015.

Tang, S., Dong, S., Chen, M., Gao, R., Chen, S., Zhao, Y., Liu, Z., Sun, B., 2018. Preparation of a fermentation solution of grass fish bones and its calcium bioavailability in rats. Food Funct. 9 (8), 4135–4142. doi:10.1039/c8fo00674a.

Treu, L., Castaldi, M., Righi, P., 2021. Waste-to-value products: The role of microbial fermentation in aquaculture waste valorization. Waste Manage. 132, 276–286. doi:10.1016/j.wasman.2021.06.014.

UN, 2015. Transforming our world: the 2030 agenda for sustainable development. United Nations, New York. https://sdgs.un.org/2030agenda.

Valenti, W.C., Mazzola, M.S., Viana, L.M., 2018. Economic feasibility of aquaculture waste management strategies. Aquac. Econ. Manag. 22 (4), 409–426. doi:10.1080/13657305.2018.1530726.

Vázquez, J.A., Durán, A.I., Menduíña, A., Nogueira, M., 2020. Biotechnological valorization of food marine wastes: Microbial productions on peptones obtained from aquaculture by-products. Biomolecules 10 (8), 1–18. doi:10.3390/biom10081184.

Wang, J., Zhan, S., 2021. Microbial fermentation-based bioreactors for sustainable aquaculture waste management. Front. Environ. Sci. 8, 608046. doi:10.3389/fenvs.2020.608046.

Yadav, S., Singh, S., Pandey, A., 2020. Microbial metabolism in the conversion of aquaculture waste. Aquac. Res. 51 (2), 433–445. doi:10.1111/are.14168.

Zhang, Y., Zhang, J., Liu, Y., Yang, Y., Li, B., 2021. Sustainable use of aquaculture waste for microbial fermentation: a review. Environ. Technol. Innov. 23, 101626. doi:10.1016/j.eti.2021.101626.

Zhao, H., Liu, Y., Li, Y., 2023a. Microbial fermentation and biogas production from aquaculture waste. Waste Manage. 136, 157–164. doi:10.1016/j.wasman.2021.06.014.

Zhao, M., Xie, W., Zhang, L., 2023b. CRISPR/Cas9-mediated gene editing in microorganisms for improved biofuel production from aquaculture waste. J. Biotechnol. 337, 11–21. doi:10.1016/j.jbiotec.2023.01.003.

CHAPTER

12

Biofloc technology as an advanced zero-waste and water-efficient practice

Mohammad Hossein Khanjani[a], *Saeed Zahedi*[b] *and Moslem Sharifinia*[c]

[a]Department of Fisheries Sciences and Engineering, Faculty of Natural Resources, University of Jiroft, Jiroft, Iran [b]Department of Fisheries, Faculty of Natural Resources and Environment, Ferdowsi University of Mashhad, Mashhad, Iran [c]Aquaculture Department, Shrimp Research Center, Iranian Fisheries Sciences Research Institute, Agricultural Research, Education and Extension Organization (AREEO), Bushehr, Iran

12.1 Introduction

Aquaculture is the fastest-growing sector within the food production industry and plays a crucial role in meeting the global demand for protein (Naylor et al., 2021). By 2022, global fisheries and aquaculture yielded a total production of 223.2 million tons, reflecting a 4.4% increase compared with 2020. This figure comprises 185.4 million tons of aquatic animals and 37.8 million tons of algae (FAO, 2024). The limited availability of land, coupled with a lack of adequate freshwater resources, is anticipated to be one of the most pressing challenges in the years ahead (FAO and UN Water, 2021). The accessibility of freshwater represents a critical challenge for humanity. With the growth of both the economy and population, the demand for water resources intensifies. Concurrently, climate change has led to the occurrence of severe droughts. (Jones, 1999; Emerenciano et al., 2025). If the current level of exploitation persists, the viability of all economic and human survival activities could be dramatically jeopardized (Scheffran and Battaglini, 2011). Freshwater species rely entirely on this vital resource for cultivation. Due to its limited availability, both academic and industrial sectors are compelled to assess and adopt strategies aimed at conserving and recycling water. In this sense, biofloc technology (BFT) is positioned as an alternative to reduce the negative

Zero-Waste Technologies in Aquaculture: Global Perspectives and Methods toward a Circular Economy
DOI: https://doi.org/10.1016/B978-0-443-34689-7.00015-X

consequences of aquaculture activities, as well as limited water use (Khanjani and Alizadeh et al., 2024; Emerenciano et al., 2025).

These systems function with minimal or zero water changes and promote the development of intricate microbial communities (Khanjani et al., 2025a), which protect the appearance of diseases (Khanjani et al., 2023a). BFT systems typically function within greenhouse environments, exhibiting limited interaction with the external environment (Khanjani et al., 2024a). The basis of BFT systems fosters and sustains suitable environments for the growth and multiplication of various microorganisms, including bacteria, fungi, ciliates, protozoa, nematodes, and microalgae, among others (Crab et al., 2012; Khanjani et al., 2023a). Microbial consortia, once formed, develop a complex and interconnected synergistic system that effectively sustains water quality, inhibits the presence of pathogens, enhances immune responses, and plays a significant role in the nutrition of cultured organisms (Liu et al., 2018; Khanjani et al., 2024a).

The generation of waste in intensive aquaculture systems can be defined as the total input of feed, chemicals, pharmaceuticals, and other materials, minus the portion that is assimilated into the biomass. In closed aquaculture systems, there is a growing trend to manage and eliminate waste internally to minimize effluent discharge, in compliance with environmental regulations (Emerenciano et al., 2025).

When comparing waste production relative to live weight, fish generate a similar amount of waste as other animals; however, fish farming results in a significantly higher volume of sludge. The water usage and subsequent waste emissions in aquaculture largely depend on the farming methods utilized (Bergheim, 2007).

In the BFT system, water consumption is very low or near zero exchange (Khanjani et al., 2024a). It helps improve the biosecurity of farms by using less water and requiring less food. Consequently, water consumption is significantly lowered in BFT (Samocha, 2019). For instance, the need for make-up water during prawn *Macrobrachium rosenbergii*, and Nile tilapia (*Oreochromis niloticus*) culture in BFT is only 6.8 and 0.071 m^3 of water, respectively (per kilogram of production) (Perez-Fuentes et al., 2013; Perez-Fuentes et al., 2016). Likewise, a significant decrease in salt make-up water has been reported during the cultivation of shrimp *Litopenaeus vannamei* using BFT (0.098–0.169 m^3 of water per kilogram of production) (Krummenauer et al., 2014). In comparison, conventional freshwater aquaculture typically requires about 16.9 m^3 of water per kilogram of production. Therefore it is important to adopt farming methods that increase production while also using natural resources efficiently, such as BFT (Verdegem and Bosma, 2009). The decrease in water usage with BFT could be more significant if the same water is recycled in multiple culture cycles (Krummenauer et al., 2014).

The development of white shrimp *L. vannamei* was assessed by Krummenauer et al. (2014). Shrimp were raised using various percentages of reused water (between 25% and 100%). It was noted that their survival rate, growth, and feed conversion ratio were superior compared with those observed in clean water. In contrast, a study by Malpartida Pasco et al. (2018) on biofloc culture of Nile tilapia, *O. niloticus*, using 50% biofloc-rich water as inoculum, did not detect adverse effects on the survival and productive performance of the tilapia caused by the water source. In this chapter of the book, the importance of water in aquaculture and BFT as a strategy that aligns with the principles of sustainable aquaculture for optimal water and feed consumption is discussed.

12.2 The importance of water in aquaculture

Water is a very familiar substance to human beings, and is a vital resource that must be protected for both current and future generations. It is not a secret to anyone that both natural ecosystems and human society are highly dependent upon an adequate supply of good-quality water.

In various regions across the globe, there is a scarcity of freshwater, and the quality of both freshwater and marine water is deteriorating in numerous areas because of pollution. All forms of agriculture require water, but it is particularly crucial for aquaculture, as it serves as the medium for the cultivation of aquatic food species.

Because of the water-intensive nature of aquaculture, individuals focused on water conservation have started to raise concerns regarding the potential for excessive water use in aquaculture in this growing industry. Therefore aquaculturists should continually attempt to use water efficiently and to be prepared to defend their water use operations.

The category of water quality required for aquaculture is a very important issue in this industry. Water is a medium in which aquatic organisms are immersed in it. Therefore all of their important biological processes occur in this medium, such as feeding, digestion, excretion, and growth. Aquatic organisms are highly sensitive to any change in water quality, especially when they are cultured in high stocking densities.

Water is the most important resource for aquaculture, and all fish activities, such as breathing, excretion, feeding, osmoregulation, and reproduction, take place in water. The quality of water is determined by factors such as temperature, turbidity, color, carbon dioxide, pH, alkalinity, hardness, nonionized ammonia, nitrite, and nitrate. Poor water quality can lead to low profits, low product quality, and create a risk to human health. Optimal maintenance and management of the physical and chemical factors of water can provide the basis for higher yield with higher quality.

12.3 Biofloc technology

BFT provides a more effective approach for managing water quality, ensuring biosecurity, and feed management compared with conventional pond or cage systems. It can be implemented in freshwater, brackish, and saltwater settings (McCusker et al., 2023).

BFT is a culture system based on microorganisms that functions by adding organic carbon sources into water or enhancing the carbon levels in the feed to raise the carbon-to-nitrogen (C/N) ratio (Khanjani et al., 2024b). BFT systems facilitate nutrient cycling, making them more eco-friendly (Yu et al., 2023). Also, BFT is defined as a suspended biological systems that rely on active phytoplankton, flocculants, protozoa, microalgae, etc., which regulate nonionized ammonia and maintain nutrient levels in water bodies (Hargreaves, 2006). Thus BFT is a novel technology that controls water quality by minimizing water exchange through converting nitrogen wastes such as ammonia and nitrite into less toxic nitrates through microorganisms (Guo et al., 2022). The basis of BFT is the formation of flocculating structures in the water column, which contain miscellaneous microorganisms, such as bacteria, protozoa, algae, and some zooplankton, possibly up to 1000–2000 different species (Avnimelech, 2015).

Bacteria are found in a free state in water. The heterogeneity of water concentration causes nitrogen molecules to diffuse, which hampers the ability of free bacteria to seize them. However, the BFT system effectively addresses this issue (Simon et al., 2002). The inner structure of the bioflocs is permeable. These pores are large enough to allow bacteria to fully absorb the nutrients in the aquaculture water flowing from the pores and gaps. Furthermore, bacteria can utilize bioflocs as safeguards to facilitate their own reproduction. Another significant element affecting the development of bioflocs is the mechanism of flocculation. Bacteria clump together because of extracellular polymeric substances that are made up of various enzymes, polysaccharides, and other organic materials, leading to the formation of floc-like structures (Wang et al., 2022). BFT has a probiotic effect and is therefore able to inhibit pathogenic bacteria (Khanjani et al., 2024a). BFT contains active compounds such as bromophenol, carotenoids, chlorophyll, poly-β-Hydroxybutyrate (PHB), and phytosterol (Ju et al., 2008), some of which have antiseptic properties (Defoirdt et al., 2008). For example, BFT is known to contain polyhydroxyalkanoate (PHA) and PHB that can synthesize PHA particles (Khanjani et al., 2023a).

PHB is a significant polymer synthesized by different microorganisms when faced with physiological stress, serving as an energy reserve. Its proportion can represent 16%–18% of the dry weight, depending on the carbon source used (De Schryver et al., 2009).

PHB can also have a positive impact on the host by promoting the growth of beneficial microbial cells in the intestine and changing the composition of the gut microbiome (Ahmad et al., 2017). The inhibition of pathogenic organisms by BFT occurs through the disruption of quorum sensing. Quorum sensing refers to the process of communication that occurs between bacterial cells. As a result, the presence of infectious bacteria in aquaculture can be managed by disrupting quorum sensing (Crab et al., 2012). BFT is recognized as an important column of the 21st-century water revolution, in which nutrients are continuously recycled and reused. BFT is a technology based on limited water consumption and exchange, regulation of the carbon-to-nitrogen ratio, and intense aeration, which leads to the growth and development of heterotrophic bacteria and various microorganisms in the culture system. Bacteria and microorganisms attached to them, which are called biofloc, play three main roles in the system:

(1) Improving water quality by absorbing inorganic nitrogen compounds.
(2) Producing microbial protein in the same place of cultivation as a supplementary food.
(3) Probiotic properties: Because of the presence of numerous organic compounds in bioflocs, such as polybeta-hydroxybutyrate, bacteria containing peptidoglucan and lipopolysaccharide in their cell walls, carotenoids, chlorophylls, phytosterols, polyphenols, and various vitamins (Khanjani et al., 2022a; Khanjani et al., 2023a; Khanjani et al., 2023b; Khanjani et al., 2024a; Khanjani et al., 2024d; Khanjani et al., 2025a).

The point is that only those species that can tolerate moderate levels of dissolved oxygen in water (3–6 mg/L), suspended solids, moderate water quality, and compressibility are recommended for cultivation in this system (Khanjani et al., 2016; Khanjani et al., 2024a). So far, the use of this system has been reported in aquatic species such as pacific white shrimp (Khanjani et al., 2016), Nile tilapia (Khanjani et al., 2021), common carp (Minabi et al., 2020), catfish (Dauda et al., 2018), banana shrimp (Khanjani and Sharifinia, 2022a), silver carp

(Zhao et al., 2018), crucian carp (Liu et al., 2019), Indian carp such as catla, rohu, and merigall (Deb et al., 2020).

12.4 Water quality in biofloc technology

Microorganisms are used as a tool for water quality management in a biofloc system. There are three different microbial pathways for ammonia conversion, namely, (1) photoautotrophic, which is the process by which algae absorb nitrogenous waste. (2) Chemoautotrophic bacteria, such as nitrifying bacteria, are capable of converting toxic nitrogenous substances into less toxic forms, and (3) the heterotrophic pathway, which is the direct assimilation of ammonia-N to bacterial biomass by heterotrophic bacteria (Avnimelech, 1999; Khanjani et al., 2022a). All three conversion processes exist in a biofloc system, and which one is dominant is related to the management practices of the culture system. Interestingly, we cannot imagine a biofloc system that does not have one of these (Hargreaves, 2006; Ju et al., 2008). Since the growth rate and microbial biomass yield of heterotrophs are ten times greater than those of nitrifying bacteria, they have a far greater capacity to immobilize ammonia than nitrifying bacteria (Khanjani et al., 2024a).

12.5 Aquaculture and waste challenge

In intensive aquaculture systems, waste is the result of all used feed, chemicals, drugs, etc., inputs less the portion that is converted to the biomass (Emerenciano et al., 2025). To lower the effluent discharge governed by environmental regulations, generated wastes in closed systems are increasingly being eliminated within the systems. Fish produce about the same amount of waste as other animals when measured by live weight, but the volume of sludge produced by fish culture is significantly larger. But today, the produced sludge is processed by dewatering, thickening, or stabilizing, and finally, the sludge volume is reduced. Most aquacultural waste can be categorized as solid or dissolved. Solid waste comprises both suspended and settled solids and includes feces, dead or decomposing organic matter, and uneaten feed. Byproducts of metabolism or breakdown are referred to as dissolved waste. Phosphate and ammonia are the two main types of dissolved waste compounds (Dauda et al., 2018). Fecal matter and uneaten feed are the primary sources of waste accumulation in aquaculture systems (Schumann and Brinker, 2020). In fact, fish can assimilate only about ~20%–30% of the ingested nitrogen (N) as biomass, and the rest of the nitrogen is excreted as waste into the water as ammonia (Yogev et al., 2017). As an example of the waste generated, Paez-Osuna et al. (1997) and Casillas-Hernández et al. (2006) state that for the production of every ton of whiteleg shrimp (*P. vannamei*), outputs range from 28.6 to 73 kg of N and 4.6 to 12 kg of P in earthen ponds with water flushing methods.

In a production system, both dissolved and solid wastes are repeatedly transformed by microbial processes. Nitrate accumulation occurs naturally in an aquatic environment via ammonia-oxidizing bacteria, for example, *Nitrosomonas* sp., and nitrite-oxidizing bacteria, for example, *Nitrospirae* sp., in a process known as nitrification. Without treatment, nitrate levels can rise to a point where domesticated animals are poisoned (i.e., for species such as the zebra danio (*Danio rerio*), >200 mg/L) (Learmonth and Carvalho, 2015). In addition, a study by Monsees et al. (2017) has suggested that continued exposure to high levels of nitrate can impair animal growth performance and health. Hence, the culturists should keep the nitrate levels below 20 mg L^{-1}.

Aquaculture is located primarily in rural, inland, or coastal areas with relatively low population densities, where agriculture and fishing have traditionally been the predominant industries. One example is a region that is a coastal part of western and northern Norway, where salmon and trout cage farming account for ca. 75% of P and ca. 50% of N totally discharged from all sources (Bergheim, 2007). Also, Lacerda et al. (2005) found that intensive shrimp culture is a highly significant nutrient source in some estuarine basins of 200–975 km^2 in NE Brazil, where the main origins of N and the second largest of P were shrimp ponds.

In closed pond-based systems, such as producing tilapia in static water, a small portion of the waste produced is released into the surrounding water compared with the burden of cage-based systems or flow-through raceways. In ponds, a large portion of the nutrients supplied by feed and/or fertilizer are effectively trapped as sediments or released as nitrogen into the atmosphere, whereas in open systems, such as floating cages, a large portion of the nutrient waste is continuously removed from the system and released into the environment (Bergheim, 2007).

12.6 Water use

In aquaculture and animal husbandry, there is a difference between on-farm water use for the animals' direct supply and indirect water use required for feed production. Since animals require two to seven kilograms of feed concentrate to gain one kilogram of weight and more than 1 m^3 of water to produce one kilogram of grain used in animal feeds, indirect water use predominates in agriculture. Fish and crustaceans are the most efficient producing animals in terms of feed-associated water use, requiring less than 2 kg of grain concentrate per kilogram produced (Verdegem et al., 2006). For terrestrial animals, on-farm use accounts for only around 1% of the water use associated with feed, whereas direct water use in aquaculture varies greatly and can reach extremely high levels.

Water use in aquaculture is strongly associated with the system applied. It has been determined that the amount of wastewater per ton of fish produced each year is 160 L/min in flow-through systems, 3–4 L/min in ponds, and as low as 0–2 L/min in systems that use water recirculation (True et al., 2004). In other words, the water use vs. fish production varies with a factor of 1:1000 depending on farming system and fish species. As the rate of intensification increases, water use generally decreases. When salmonids are raised on land, one kilogram of fish may use more water than the average household needs for freshwater per person (200–600 L/person/day) in a large flow-through system. However, intensification measures such as adding oxygen and recirculating water can cut the consumption by 90%–95%. Because of various intensification efforts, the amount of water used at Norwegian hatcheries that produce rainbow trout and Atlantic salmon smolt has been gradually decreased. At some hatcheries, the introduction of partial water recirculation and the injection of pure oxygen have reduced the average specific water flow from 0.2 to 0.5 L/kg/min compared with 1–2 L/kg/min 15–20 years ago. The production is now about 180 million smolts annually. Assuming an average biomass of 5000 tons at the 115 hatcheries and a mean smolt size at delivery of 0–1 kg, the total water consumption exceeds the domestic water consumption of the 4.5 million Norwegians (specific water consumption: 200 L/person/day). It has been confirmed that warm water species (such as channel catfish) use less water compared to cold water species (such as trout and salmon). In a review, Chen et al. (2002) referred to make-up water flow rates ranging from flow-through to fully recirculating systems, which range from 6 to 1250 L/min in coldwater systems and 0.2 to 125 L/min in warmwater systems per ton of fish produced annually.

For the production of 1 kg of penaeid shrimp in flow-through systems, approximately 20000 L of water is needed. Unfortunately, a large part of this water is eventually released into receiving water bodies (Timmons and Losordo, 1994). In such systems, water exchange rates are about 10%–15% per day (Hargreaves, 2006). Nowadays, water exchange is known as an important disease-spreading factor. To solve this problem, so-called "zero-exchange" systems have been introduced, which have very low exchange rates (< 5% per day). Introduction of recirculation aquaculture systems (RASs) will further reduce the water use, e.g., as low as 370–480 L/Kg of shrimp wet biomass in "greenwater recirculating raceways" (Moss et al., 2001). In a variety of aquaculture systems, maintenance of water quality depends on an active mass of phytoplankton, free and attached bacteria, live and dead organic matter aggregates, and microbial grazers maintained in suspension. The joint term "photosynthetic suspended-growth" (PSG) system has recently been introduced by Hargreaves (2006), encompassing both conventional and intensive ponds, tanks, and raceways systems, where the water quality is controlled by phytoplankton and bacteria. In these systems, the active participation of bacteria and algae can result in significantly lower water exchange. Production of tilapia mainly takes place in PSG systems, and the reported water use varies from 30 to 40 L in intensive greenhouse tanks to 850 L in combined intensive–extensive ponds per kilogram of fish produced (Bergheim, 2007). These values are very low compared to other systems and cultured species, and it gives hope for a lower water consumption rate for the future.

12.7 Sustainable alternative for water consumption in aquaculture

There are production systems that use less water than traditional systems, such as BFT and RASs, but they are not yet commonly utilized in farms. In particular, BFT can expand aquaculture production without increasing the use of basic natural resources (water and land). BFT has emerged as a promising alternative to promote the sustainability and biosecurity of aquatic animals by using smaller amounts of water, resulting in less damage to the environment.

According to recent research, the conventional systems for shrimp and fish production are not as productive as the production of aquatic organisms in BFT and RAS, where reduced water exchange is needed (Martins et al., 2010; Jatobá et al., 2014; Mesquita et al., 2016; Jatobá et al., 2019).

To achieve sustainable aquaculture, it is necessary to develop aquaculture systems with less water requirement and maintain their quality (Jatobá et al., 2019). In the study by Jatobá et al. (2019), BFT used a lower volume of water, requiring 186X less than in the conventional system (Table 12.1). Also, according to the used production system, for the production of each ton of fish, it is possible that 30g or 2.39 kg of ammonia is formed, demonstrating that the high concentration of nutrients in the water body of the BFT is compensated by the low volume of

TABLE 12.1 Water consumption in different aquaculture systems.

System type	Species	IW (g)	RP (day)	FW (g)	FCR	WC	References
BFT	*O. niloticus*	1.3	42	8.48	1.0	128.21m^3	Jatobá et al. (2019)
CW				9.21	1.39	23,934.42 m^3	
BFT+ Dextrose	*Litopenaeus vannamei*		7			6.49 L	Schveitzer et al. (2015)
BFT+ Molasses						6.89 L	
CW						56.22 L	
CW	*O. niloticus*	56.5	49	137.7	2.07	0% of reused water	Figueroa-Espinoza et al. (2022)
BFT25		55.5		141.9	1.79	25% of reused water	
BFT50		55.1		150.4	1.74	50% of reused water	
BFT75		55.6		151.1	1.59	75% of reused water	
BFT100		56.1		161.2	1.47	100% of reused water	
BFT500	*O. niloticus*	1.17	42	18.99	1.24	52.48 L/kg fish	Lima et al. (2018)
BFT750				17.47	1.30	76.32 L/kg fish	
BFT1000				15.36	1.40	94.77 L/kg fish	
BFT1250				12.40	1.32	101.54 L/kg fish	
Water renewal	*L. vannamei*	0.83	105	12.74	1.32	2.22 m^3/kg shrimp	Zemor et al. (2019)
BFT+ one clarifier				12.14		1.18 m^3/kg shrimp	
BFT+ two clarifiers				11.89		1.20 m^3/kg shrimp	
BFT 4.28 Fish/m^3	*Labeo rohita*	26.56	90	64.16	1.69	No. of WE:3 1050 L (TWR)	Deb et al. (2020)
	Catla catla		19.2	51.70	1.61		
	Cirrihinus mrigala	28.76		59.71	2.23		
BFT 8.57 Fish/m^3	*L. rohita*	20.75		40.75	2.25	No. of WE:3 1050 L (TWR)	
	C. catla	14.6		34.74	1.87		
	C. mrigala	24.26		46.06	2.57		
BFT 12.85 Fish/m^3	*L. rohita*	26.95		50.14	2.12	No. of WE:3 1400 L (TWR)	
	C. catla	13.58		32.03	2.01		
	C. mrigala	29.06		52.73	2.76		

(continued on next page)

TABLE 12.1 Water consumption in different aquaculture systems—cont'd

System type	Species	IW (g)	RP (day)	FW (g)	FCR	WC	References
CW 4.28 Fish/m^3	*L. rohita*	26.93		49.21	2.78	No. of WE:5 1750 L (TWR)	
	C. catla	20.46		36.49	2.92		
	C. mrigala	30.53		52.03	2.56		
CW 8.57 Fish/m^3	*L. rohita*	29.7		52.83	2.99	No. of WE:5 1750 L (TWR)	
	C. catla	16.41		28.51	3.05		
	C. mrigala	29.26		56.36	2.62		
CW 12.85 Fish/m^3	*L. rohita*	27.07		46.87	3.15	No. of WE:5 2100 L (TWR)	
	C. catla	19.8		33.47	3.25		
	C. mrigala	26.93		50.02	2.77		
CW	Gibel Carp (*Carassius auratus* gibelio ♀ × *C. carpio* ♂)	20.65	60	40.56	1.9	31.7 m^3/kg	Cang et al. (2019)
BFT		19.09		44.67	1.6	0.3 m^3/kg	

BFT, biofloc technology; *CW*, conventional water-exchange system; *FCR*, feed conversion ratio; *FW*, final weight; *IW*, initial weight; *RP*, rearing period; *TWR*, total water requirement; *WC*, water consumption.

the used water (Jatobá et al. 2019). Today, it is hopefully possible to achieve yields above 25 kg m^3 of water (Jatobá et al., 2019). As a result, fish culture would become more sustainable by lowering its water demand and effluent volume.

Zapata-Lovera et al. (2017) evaluated water consumption in tilapia culture in a water exchange (1,800 L/kg) and a BFT (90–190 L/kg) and discovered that tilapia can be produced using a biofloc system, particularly in areas with restricted water supplies.

The Nile tilapia raised in BFT use less water and exhibit comparable or superior productive variables to those raised in conventional systems (with continuous renewal). Fish farming is made more sustainable by rearing fish in BFT, which reduces water consumption during fish production and limits the discharge of wastewater into nearby environments (Jatobá et al., 2019).

A study by Schveitzer et al. (2015) determined that the use of biofloc systems without water exchange supplemented with molasses or dextrose as a carbon source results in adequate production indices and water quality during the breeding phase of *L. vannamei*. Since no water is exchanged in these biofloc systems, approximately 12% of the water used in the traditional autotrophic system is required.

In a study by Figueroa-Espinoza et al. (2022) have shown that it is possible to reuse the biofloc water from tilapia cultures up to 100% without adverse effects on production parameters, water quality, and some physiological variables. Overall, the results showed that water reuse in fish BFT systems could be highly effective in promoting fish performance and maintaining safe water quality parameters, even at high levels of reuse. Last but not least, fish health is not compromised by reusing water, suggesting that BFT could use water even more efficiently.

TABLE 12.2 The outcomes of Nile tilapia production among two systems in 35 days of testing (Khanjani and Alizadeh, 2021).

Parameter	BFT	CW
IW (g)	1.79±0.10	1.79±0.10
FW (g)	7.94±0.38	7.46±0.44
WG (g)	6.15±0.38	5.67±0.44
DWG (g/day)	0.175±0.013	0.162±0.014
SGR (%/day)	4.26±0.13	4.08±0.14
B (g)	1033.20±22.43	933.66±25.41
SR (%)	98.82±0.58	96.86±0.34
FCR	1.009±0.09	1.54±0.2
FE (%)	99.1±7.14	64.39±10.47
WC (L/kg)	61.47±1.35	2777.64±75.6
P (kg/m^3)	6.09±0.13	5.49±0.15
FC (g)	1042.5±2.01	1437.83±5.08

B, biomass; *BFT*, biofloc technology; *CW*, conventional water exchange system; *DWG*, daily weight gain; *FC*, feed consumption; *FCR*, feed conversion ratio; *FE*, feed efficiency; *FW*, final weight; *P*, production; *SGR*, specific growth rate; *SR*, survival rate; *WC*, water consumption; *WG*, weight gain.

In general, water exchange in intensive aquaculture systems is a necessary process to maintain water quality. Nowadays, the use of BFT for water exchange in aquaculture is receiving great attention due to the scarcity of freshwater resources. In the study by Deb et al. (2020), the water consumption in the biofloc system was significantly (38%) lower than that of the conventional system, some of the results are mentioned in detail (Table 12.1). In another study, the water exchange requirement in a biofloc system for rearing Indian carp (*Labeo rohita*) was 36% lower than in a conventional system (Mahanand et al., 2013). In a study by Cang et al. (2019), for raising crucian carp (*Carassius auratus gibelio*), in a conventional system, the water exchange rate was 24,012 m^3, and the water consumption was 31.7 kg/m^3, whereas in the biofloc system, the water exchange rate was 0 m^3 and the water consumption was 0.3 kg/m^3. The water consumption in this technology is approximately 6.8 m^3 for the production of each kilogram of giant freshwater shrimp and 0.071 m^3 for the production of each kilogram of Nile tilapia (Perez-Fuentes et al., 2013; Perez-Fuentes et al., 2016).

Table 12.2 presents the production, water, and feed consumption in two conventional and BFT systems in Nile tilapia farming. To illustrate the audience's opinion, it is worth mentioning that in open systems, one ton of rainbow trout is produced per year for 3 L of incoming water per second (which is the same as the production rate of most Iranian water bodies). In cellar operations, the amount is far less than this amount; about 100,000 cubic meters of water are required per year, compared to the biofloc system for tilapia production (70 m^3), which has a saving of 1400%.

In addition to the limited water exchange, biofloc systems offer other advantages such as reducing the environmental impact of aquaculture, reducing feed costs, improving water growth and safety, high biosecurity due to the absence of water from the environment and, as a result, reducing the penetration of pathogenic pathogens into the systems, managing beneficial microbial communities and creating boundaries for pathogens (Moss and Leung, 2007). The use of biofloc as a feed supplement in aquaculture reduces approximately 15% of the total aquaculture costs (Khanjan et al., 2023a). The production of one kilogram of green tiger shrimp (*Penaeus semisulcatus*) using biomass has been associated with a 33% cost reduction (Megahed, 2010) and a 10% cost reduction for tilapia (Schryver and Verstraete, 2009), which has been reported by the type. Nutrition, biomass consumption rate, and price of carbohydrates consumed. The biggest challenge of the biofloc system is the control of suspended solids in this system, because as the amount of suspended solids increases, the turbidity of the water increases and the amount of dissolved oxygen decreases, which reduces growth and increases the length of the cultivation period. and increase the feed conversion ratio during the growing season. Another limitation is the high energy consumption, as many water pumps and electric aerators have to work at the same time to create a current in the pond water and supply the aquatic animals and the system with the oxygen they need. Unfortunately, any technical defect or failure of the ventilation and pumping system can lead to significant losses of aquatic animals in a short period of time. Therefore in addition to its high attractiveness, the biofloc method is also very complex, and those interested in setting up these systems must have sufficient technical knowledge and experience in this field or rely on experienced experts (Ghaedi et al., 2022).

12.8 Biofloc as a complementary feed

These heterotrophic bacteria convert inorganic nitrogen compounds formed from the feces and food residues into microbial protein (Khanjani et al., 2023a), which also provides additional nutrition for the cultured fish (Khanjani and Sharifinia, 2024). The floc that forms in the BFT is nutrient-rich. Aquatic animals can obtain protein in situ from biofloc. It is composed of 6.07% ash, 1.03% lipid, and 41.13% protein (Long et al. 2015), which supplements the feed with extra nutrients. The types of external carbon determine the approximate composition of biofloc. Table 12.3 lists the biofloc's nutritional composition.

The nutrient contained in biofloc represents a significant portion of the nutrient requirements of several aquaculture species (Khanjani et al., 2024a). Biofloc consists of several macronutrients (carbohydrates, proteins, and lipids) and micronutrients (vitamins and minerals), as well as bioactive compounds, antioxidants, enzymes, short-chain fatty acids (SCFAs), etc. (Dauda et al., 2018). Biofloc includes compounds composed of organic materials (60%–70%), heterogeneous combinations of microorganisms (e.g., fungi, algae, bacteria, protozoa, rotifers, and nematodes), and inorganic substances (30%–40%) such as colloids and organic polymers, binary ions, salts, and dead cells (Khanjani et al., 2024a). The growth performance of the culture organisms was enhanced by the application of BFT, which also enhanced the feed conversion ratio and protein retention (Gao et al. 2012; Xu et al. 2012; Khanjani and Sharifinia, 2024).

TABLE 12.3 The nutrient content of harvested biofloc (%DM).

Species	IW (g)	Carbon source used	C:N ratio	CP	CL	Ash	References
		Glucose	15:1	41.2	6.1	15.0	Wei et al. (2016)
		Starch		31.5	8.5	12.4	
		Glycerol		35.5	4.2	15.2	
L. rohita, C. catla, C. mrigala	26.3, 17.3, 28.1	Tapioca flour	10:1	35.14	2.16	15.28	Deb et al. (2020)
O. niloticus	79.28	Molasses	12.5:1	35.90– 36.83	4.14–4.45	11.39–13.61	Gallardo-Collı et al. (2019)
O. niloticus	56.5		12.5:1	24.92–26.30		42.23– 44.15	Figueroa-Espinoza et al. (2022)
Gibelion catla	1.6	Rice bran	10:1	22.66	1.8	11.4	Solanki et al. (2023)
			15:1	24.26	2.1	10.8	
			20:1	26.53	2.2	10.5	
L. vannamei	2.5	Molasses	10:1	23.27	1.09	33.96	Khanjani and Sharifinia (2022a)
			14:1	23.9	1.21	34.54	
			18:1	24.07	1.39	34.66	
			22:1	25.45	1.44	35.15	
Fenneropenaeus merguiensis	4.5 mg	Molasses	15:1	22.83– 30.90	0.8–1.72	32.5– 37.18	Khanjani and Sharifinia (2022b)
O. niloticus	1.3	Molasses	15:1	26.84–31.85	1.68–2.51	30.67–36.44	Khanjani and Sharifinia (2024)

DM: dry matter; *CL*: crude lipid; *CP*, crude protein; *IW*, initial weight.

A number of factors affect the nutritional value and traits of flocs. According to Raza et al (2024), the nutritional composition and features of biofloc include, for instance, the floc density and particle size, the external carbon source that is provided for bacterial growth, environmental factors, and water quality parameters. Particle size, digestibility, and biochemical composition all affect how nutritious bioflocs are for a given organism. Ekasari et al. (2014) reported that flocs with a particle size of >100 and <48 μm are more favorable for shrimp *L. vannamei*, red tilapia *O. niloticus*, and mussel *Perna viridis* because of their higher nutritional value and N recoveries. The flocs with particle size of >100 μm had the highest protein and lipid levels, whereas the flocs <48 μm were rich in essential amino acids. In addition, feed quality and feed particle distribution may affect the performance of animals raised with different feeding behaviors (filter feeders, scavengers, etc.) that utilize the flocs as feed in an in situ biofloc system. Moreover, floc size may relate to the take-up potential of the biofloc by the cultured organism and the digestibility of the flocs (De Schryver et al., 2008).

BFT promotes the growth of cultured organisms by (1) providing supplementary single-cell protein, (2) enhancing the activity of digestive enzymes, and (3) improving the composition of gut microbiota. This approach is more economically advantageous than conventional systems, as it leads to better growth rates and feed conversion ratios (Khanjani and Alizadeh, 2024).

Biofloc is characterized by a high concentration of various probiotic bacteria, bioactive compounds, oxidative enzymes, and certain digestive enzymes, all of which positively influence the digestive processes of aquatic organisms (Ahmad et al., 2019; Khanjani et al., 2025a). The enhancement of digestive enzyme activity facilitates improved nutrient assimilation, thereby promoting the growth of cultured fish (Adineh et al., 2019; Debbarma et al., 2021) and shellfish (Cardona et al., 2015; Khanjani and Sharifinia, 2022a). Fish cultivated in biofloc systems exhibit superior nutrient absorption compared to those raised in nonbiofloc environments. Debbarma et al. (2021) indicated that the enhancement of villi length and width in *O. bimaculatus* contributes to improved nutrient absorption from biofloc culture. In addition, the incorporation of prebiotics such as mannoprotein in situ promotes biofloc, further increasing the internal surface area of the villi.

Biofloc has the potential to act as either a substitute or an additional feed source for various aquaculture species that can directly ingest particulate organic matter, including filter feeders, deposit feeders, detritivores, and certain herbivorous and omnivorous species, owing to its advantageous nutrient composition (Crab et al., 2012). For the species being cultivated to effectively graze on bioflocs, specific morphological adaptations are necessary, along with the ability to digest and assimilate nutrients from microbial aggregates (Ulloa Walker et al., 2020). Tilapia species, in particular, are equipped with micro-gill-rakers that are coated with a viscous mucus, allowing them to filter small particles from the water, ranging from micrometers to millimeters in size (Khanjani et al., 2022a). They have the ability to ingest unattached bacterial cells provided as suspensions in their fry stage. Juveniles of *L. vannamei* are believed to utilize net-like setae found on their third maxillipeds to capture particles as small as 10 μm in diameter, which includes diatoms such as *Thalassiosira* and *Amphiprora*. In addition, other species have been identified with morphological adaptations conducive to particle filtration within the range of 10–1200 μm, including the freshwater prawn *M. rosenbergii*, the silver carp (*Hypophtalmichthys molitrix*), and the flathead gray mullet (*Mugil cephalus*).

The nutritional composition and value of bioflocs represent a variable parameter influenced by various aspects of the culture environment, including temperature, salinity, the species cultivated, and the organic carbon supplied (Raza et al., 2024). In addition, a notable distinction exists between bioflocs derived from green water and those from brown water. Green water bioflocs, which are predominantly composed of specific microalgae, exhibit a more favorable macronutrient profile, particularly in essential fatty acids, compared to bacterial counterparts. These microalgae also contain valuable pigments, such as astaxanthin, which are already incorporated into feed formulations because of their antioxidant properties, and they can synthesize a range of vitamins and immunostimulants (Nagappan et al., 2021).

12.9 Biofloc as probiotics

Bioflocs may also have a probiotic effect and can reduce the number of pathogens in the water (Khanjani et al., 2024a). The parallels between biofloc systems and probiotics arise primarily from their shared characteristic of containing live microorganisms (Padeniya et al., 2022). Probiotics are known to enhance the health of cultured species through mechanisms such as bioremediation, competitive exclusion, and immunomodulation. In addition, bioflocs are composed of various active compounds that exhibit antibacterial properties (Khanjani et al., 2024a). One notable component is poly-β-hydroxybutyrate (PHB), a polymer found in bioflocs that serves a prebiotic function, promoting the proliferation of beneficial microbes within the gastrointestinal tract of organisms. This polymer is typically synthesized by other microorganisms in response to physiological stress (Ahmad et al., 2017). Probiotics employ various mechanisms to inhibit or interfere with quorum sensing, a phenomenon also observed in biofloc systems. In these systems, microorganisms can disrupt the cell-to-cell communication of pathogenic bacteria. Research investigating the survival rates of brine shrimp exposed to *Vibrio harveyi* has provided insights into the effects of biofloc environments. Quorum sensing is a critical factor in regulating the pathogenicity of *V. harveyi*, and findings indicated that Artemia exhibited a marked improvement in survival, achieving complete protection against the pathogen. Common methods of disrupting quorum sensing include the synthesis of antagonistic compounds and the utilization of enzymes that degrade signaling molecules by certain microbes (Defoirdt et al., 2004). Furthermore, recent studies have focused on the health of shrimp cultivated in biofloc systems. For example, Tepaamorndech et al. (2020) investigated the expression of immune-related genes and the composition of the gut microbiome in Pacific whiteleg shrimp. Research findings indicate that biofloc contributes to the preservation of gut microbiota, with *Vibrio* spp. identified as a prominent component of the biofloc. In addition, alterations in the systemic immune response were linked to the practice of biofloc rearing. Notably, Gustilatov et al. (2022) observed that biofloc systems may have the capacity to diminish the levels of *Vibrio parahaemolyticus* and to mitigate the biofilm formation associated with this pathogen. However, there is a scarcity of studies examining the physiological health of shrimp, particularly regarding immune and antioxidant responses in biofloc environments. Jang et al. (2010) reported that *L. vannamei* reared in biofloc systems exhibited a marked increase in the expression of the prophenol oxidase enzyme within hemocytes. Furthermore, another investigation involving the same species revealed an enhancement in superoxide dismutase activity when cultured in biofloc conditions (de Jesús Becerra-Dorame et al., 2014). The significance of these enzymes in innate immunity is attributed to their roles in both cellular and humoral defense mechanisms. Jang et al. (2010) elaborated on this by suggesting that the microbes present in the biofloc can infiltrate the shrimp's body, thereby influencing the immune response. Consequently, biofloc plays a multifaceted and adaptive role within shrimp culture systems, particularly in response to various system inputs. The factors outlined can be strategically adjusted to enhance specific elements of rearing performance and overall health.

12.10 Biofoc technology and sustainable development

Aquaculture growth can be significantly impeded by environmental impacts, alongside the availability of land and water resources. To mitigate the risks associated with eutrophication and other environmental repercussions, aquaculture systems must not only generate income but also deliver broader social benefits. This can be achieved by promoting the cultivation of herbivorous and omnivorous species, as well as by implementing integrated aquaculture practices. In such systems, byproducts or wastes from one species can be repurposed as inputs such as fertilizers, food, and energy for another species (Emerenciano et al., 2025). Practical applications in aquaculture can utilize agricultural byproducts, including cassava meal, molasses, tapioca, wheat bran, rice bran, and brewery residues (Serra et al., 2015; Deng et al., 2018; Khanjani and Sharifinia, 2022a; Li et al., 2023; Khanjani et al., 2023a). Consequently, biofloc, a byproduct of this process, serves as an additional food source for aquatic suspension feeders and deposit feeders, such as herbivorous fish. Research has shown that certain omnivorous aquatic species, including shrimp and tilapia, can be effectively cultivated using BFT (Khanjani et al., 2024a). The implementation of BFT allows aquaculture systems to minimize water exchange, sometimes to the point of requiring none at all, while also decreasing the need for artificial feed inputs through the in situ enhancement of water quality and the provision of organic biofloc as food. Furthermore, the use of wild fish as feed in aquaculture helps maintain ecosystem balance and reduces the risks of eutrophication. In the fish culture component of BFT aquaculture, heterotrophic bacteria are being developed and controlled in flocs. Conventional domestic wastewater treatment systems inform the design of this system, which aims to treat wastewater through the incorporation of carbohydrates (Khanjani et al., 2021a; De Schryver et al., 2008). The introduction of carbohydrate substrates fosters the growth of heterotrophic bacteria, which subsequently absorb nitrogen by synthesizing microbial proteins (Avnimelech, 1999). This heterotrophic biomass serves as a nutritional resource for invertebrates and fish. One of the innovative approaches to enhance sustainability in the aquaculture sector is the implementation of BFT, which achieves several objectives: (1) It enhances nitrogen retention within the harvested biomass; (2) it diminishes the demand for protein in feed; (3) it lowers the levels of potentially harmful total ammonia nitrogen (TAN) and nitrite (NO_2) in the system; and (4) it curtails nitrogen emissions. Given its sustainable characteristics, BFT is poised to play a crucial role in the future of aquaculture. Following a prolonged acclimatization period, heterotrophic bacteria may surpass nitrifiers in competition due to their faster growth rates and greater biomass yield per unit of substrate (Hargreaves, 2006). Consequently, the production of N_2O driven by nitrification will be diminished. Furthermore, heterotrophic bacteria are capable of assimilating inorganic nitrogen that accumulates in aquaculture systems, thereby lowering its concentration and further reducing N_2O emissions (Hu et al., 2012).

12.11 Future perspective with biofloc technology

The planet's freshwater resources are limited and increasingly jeopardized by pollution, alongside a rising demand from agriculture, livestock, industry, and urban areas (Anderson, 2003). In addition, climate change has led to widespread desertification (Huang et al., 2020). It is imperative for aquaculture enterprises to adopt sustainable practices that reduce freshwater consumption while fostering social and economic growth. Research conducted by Figueroa-Espinoza et al. (2022) indicates that the reuse of biofloc water is a viable option, with promising future prospects, although it necessitates pilot-scale testing. Furthermore, water reuse can facilitate the establishment of small-scale farms in semiarid regions or near consumption hubs where water is scarce. Beyond environmental advantages, such practices can stimulate social and economic development; Brunori et al. (2016) highlight that local marketing channels can bolster the economy while reducing environmental harm. The implementation of water reuse strategies can enhance the profitability of aquaculture operations by optimizing the use of natural feed sources such as bioflocs. However, this will require comprehensive redesign and engineering studies for BFT farms, including the development of reservoirs for postharvest water storage and the establishment of reconditioning protocols.

The practical use of BFT, as an environmentally friendly aquaculture system with high production capacity, is one of the solutions to achieve sustainable aquaculture. The authors believe that the development of BFT and farming systems based on it can be an important step in the development of aquaculture, especially in desert and semi-desert areas. With the help of this technology, it is possible to develop aquaculture in the conditions of drought and with the increasing limitations of extracting fresh water resources, and also the neglected potential of using unconventional waters for aquaculture can be activated and put into the production cycle.

Another very important point is that biofloc production, whether separate from aquatic animals or combined with them, does not depend on the import of any items and is completely economical from an economic perspective. In Iran, as an arid and semiarid country, removing the deadlock of tilapia fish production and development, as the first option of fish farming, only in biofloc systems, has the most economic justification where both environmental concerns are minimized because of the closed nature of the system, and saline and low-value desert lands can be used in aquaculture (because of the lack of water and soil connection in this system), and also, there is no pressure on freshwater resources, and maximum efficiency of water use in aquatic production is achieved.

Perhaps the most obvious challenge in promoting BFT is convincing the conservative agricultural community to use brown water for cultivation, where most farmers are old and have low literacy levels and find it difficult to understand the technology. In general, adopting innovation and utilizing new technologies in agriculture is a complex and multi-stage learning process (Salarpour et al., 2022). Promoting and developing BFT-based cultivation systems requires conducting numerous pilot projects and designs at different scales for different species.

In addition, combined aquaculture and biofloc (in situ systems) require high water management and a continuous supply of electricity, which is difficult for ordinary farmers to manage. Perhaps requiring fisheries specialists and graduates to enter such farms by relevant organizations could be a solution and prevent severe losses.

12.12 Conclusion

The water crisis, along with increasing demands on wild fish populations for fish meal and oil, has hindered the advancement of aquaculture. In addition, the rise of diseases and the necessity for isolated and intensive systems in response to disease outbreaks and antibiotic resistance in animal farming highlight the need for improved systems that align with sustainable objectives. The implementation of BFT can facilitate aquaculture practices that are more environmentally sustainable. Our research indicates that this technology offers significant benefits, including reduced water consumption and the recycling of in situ nutrients and organic materials.

BFT effectively addresses the dual objectives of optimizing water usage and managing waste, thereby facilitating enhanced production while preserving the integrity of the ecosystem. In addition, this system meets the in situ nutritional requirements of fish, which subsequently reduces their feeding needs. Crucially, the probiotics present in floc protein act as immune enhancers, improving disease resistance and reducing the reliance on antibiotics in aquaculture practices. The diverse functionalities of this technology can be further refined through targeted research trials and tailored recommendations that consider specific species and environmental contexts. In summary, the biofloc system represents a promising technological advancement for the future, necessitating the development of sound policies and standards to ensure its sustainability over time. By using the biofloc system in arid and semi-arid regions that face limited water resources and species that can tolerate a wide range of environmental conditions, such as tilapia and common carp, an important step can be taken in increasing aquaculture production (Páez-Osuna et al., 1997; Avnimelech, 1999; Jones, 1999; Moss et al., 2001; Chen et al., 2002; Simon et al., 2002; Anderson, 2003; True et al., 2004; Defoirdt et al., 2004; Casillas-Hernández et al., 2006; De Lacerda et al., 2006; Hargreaves, 2006; Verdegem et al., 2006; Bergheim, 2007; Moss and Leung, 2007; De Schryver et al., 2008; Defoirdt et al., 2008; Ju et al., 2008; Schryver and Verstraete, 2009; Verdegem and Bosma, 2009; Jang et al., 2010; Martins et al., 2010; Scheffran and Battaglini, 2011; Crab et al., 2012; Gao et al., 2012; Hu et al., 2012; Xu et al., 2012; Mahanand et al., 2013; Pérez-Fuentes et al., 2013; de Jesús Becerra-Dorame et al., 2014; Ekasari et al., 2014; Jatobá et al., 2014; Krummenauer et al., 2014; Schveitzer et al., 2015; Avnimelech, 2015; Cardona et al., 2015; Learmonth and Carvalho, 2015; Long et al., 2015; Serra et al., 2015; Timmons and Losordo, 1994; Brunori et al., 2016; Khanjani et al., 2016; Mesquita et al., 2016; Perez-Fuentes et al., 2016; Wei et al., 2016; Ahmad et al., 2017; Yogev et al., 2017; Monsees et al., 2017; Cang et al., 2019; Dauda et al., 2018; Deng et al., 2018; Lima et al., 2018; Liu et al., 2018; Malpartida Pasco et al., 2018; Zhao et al., 2018; Nagappan et al., 2021; Zapata et al., 2017; Adineh et al., 2019; Ahmad et al., 2019; Dauda et al., 2019; Gallardo-Collí et al., 2019; Jatobá et al., 2019; Liu et al., 2019; Samocha, 2019; Zemor et al., 2019; Deb et al., 2020; Huang et al., 2020; Minabi et al., 2020; Schumann and Brinker, 2020; Tepaamorndech et al., 2020; Ulloa Walker et al., 2020; FAO and UN Water, 2021; Debbarma et al., 2021; Khanjani and Alizadeh, 2021; Khanjani et al., 2021; Naylor et al., 2021; Ghaedi et al., 2022; Gustilatov et al., 2022; Khanjani and Sharifinia, 2022a; Khanjani and Sharifinia, 2022a; Khanjani and Sharifinia, 2022b; Khanjani et al., 2022a; Khanjani et al., 2022b; Khanjani et al., 2022d; Khanjani et al., 2022c; Padeniya et al., 2022; Salarpour et al., 2022;

Figueroa-Espinoza et al., 2022; Guo et al., 2022; Wang et al., 2022; Li et al., 2023; McCusker et al., 2023; Yu et al., 2023; Raza et al., 2024; Solanki et al., 2023; Khanjani, et al., 2023a; Khanjani, et al., 2023b; Khanjani et al., 2023c; Khanjani and Alizadeh, 2024; Khanjani and Sharifinia, 2024; Khanjani et al., 2024a; Khanjani et al., 2024c; Khanjani et al., 2024d; Raza et al., 2024; FAO, 2024; Khanjani et al., 2025a; Emerenciano et al., 2025; Khanjani et al., 2025b).

List of abbreviations

B	Biomass
BFT	Biofloc technology
BFT500	500 fish/m^3
BFT750	750 fish/m^3
BFT1000	1,000 fish/m^3
BFT1250	1,250 fish/m^3
BWI	Body weight index
CL	Crude lipid
CP	Crude protein
CW	Conventional water exchange system
DM	Dry matter
DWG	Daily weight gain
FC	Feed consumption
FCR	Feed conversion ratio
FE	Feed efficiency
FW	Final weight
IW	Initial weight
P	Production
RP	Rearing period
SGR	Specific growth rate
SR	Survival rate
TWR	Total water requirement
WC	Water consumption
WE	Water exchanges
WG	Weight gain

References

Adineh, H., Naderi, M., Hamidi, M.K, Harsij, M., 2019. Biofloc technology improves growth, innate immune responses, oxidative status, and resistance to acute stress in common carp (Cyprinus carpio) under high stocking density. Fish Shellfish Immunol. 95, 440–448. doi:10.1016/j.fsi.2019.10.057.

Ahmad, I., Babitha Rani, A.M., Verma, A.K., Maqsood, M., 2017. Biofloc technology: an emerging avenue in aquatic animal healthcare and nutrition. Aquac. Int. 25 (3), 1215–1226. doi:10.1007/s10499-016-0108-8.

Ahmad, I., Leya, T., Saharan, N., Asanaru Majeedkutty, B.R., Rathore, G., Gora, A.H., Bhat, I.A., Verma, A.K., 2019. Carbon sources affect water quality and haemato-biochemical responses of Labeo rohita in zero-water exchange biofloc system. Aquac. Res. 50 (10), 2879–2887. doi:10.1111/are.14241.

Anderson, J., 2003. The environmental benefits of water recycling and reuse. Water Supply 3 (4), 1–10. doi:10.2166/ws.2003.0041.

Avnimelech, Y., 2015. Biofloc technology: fifteen years of progress. Hatch. Int. 16, 38–39.

Avnimelech, Y., 1999. Carbon/nitrogen ratio as a control element in aquaculture systems. Aquaculture 176 (3–4), 227–235. doi:10.1016/s0044-8486(99)00085-x.

Bergheim, A., 2007. Global aquaculture water use and pollution. In: Bergheim, A. (Ed.), Aquacultural Engineering and Environment, pp. 1–20.

Brunori, G., Galli, F., Barjolle, D., Van Broekhuizen, R., Colombo, L., Giampietro, M., Kirwan, J., Lang, T., Mathijs, E., Maye, D., De Roest, K., Rougoor, C., Schwarz, J., Schmitt, E., Smith, J., Stojanovic, Z., Tisenkopfs, T., Touzard, J.-M., 2016. Are local food chains more sustainable than global food chains? Considerations for assessment. Sustainability 8 (5), 449. doi:10.3390/su8050449.

Cang, P., Zhang, M., Qiao, G., Sun, Q., Xu, D., Li, Q., Yuan, X., Liu, W., 2019. Analysis of growth, nutrition and economic profitability of Gibel Carp (Carassius auratus gibelio ♀ × C. carpio ♂) cultured in zero-water exchange system. Pak. J. Zool. 51 (2). doi:10.17582/journal.pjz/2019.51.2.619.630.

Cardona, E., Saulnier, D., Lorgeoux, B., Chim, L., Gueguen, Y., 2015. Rearing effect of biofloc on antioxidant and antimicrobial transcriptional response in Litopenaeus stylirostris shrimp facing an experimental sub-lethal hydrogen peroxide stress. Fish Shellfish Immunol. 45 (2), 933–939. doi:10.1016/j.fsi.2015.05.041.

Casillas-Hernández, R., Magallón-Barajas, F., Portillo-Clarck, G., Páez-Osuna, F., 2006. Nutrient mass balances in semi-intensive shrimp ponds from Sonora, Mexico using two feeding strategies: trays and mechanical dispersal. Aquaculture 258 (1–4), 289–298. doi:10.1016/j.aquaculture.2006.03.027.

Chen, S.S., Malone, T.L., 2002. Recirculating systems, effluent, and treatments. In: Aquaculture and the environment in the United States. World Aquaculture Society, Baton Rouge, LA, pp. 119–140.

Crab, R., Defoirdt, T., Bossier, P., Verstraete, W., 2012. Biofloc technology in aquaculture: beneficial effects and future challenges. Aquaculture 356-357, 351–356. doi:10.1016/j.aquaculture.2012.04.046.

Dauda, A.B., Ajadi, A., Tola-Fabunmi, A.S., Akinwole, A.O., 2019. Waste production in aquaculture: sources, components and managements in different culture systems. Aquac. Fish. 4 (3), 81–88. doi:10.1016/j.aaf.2018.10.002.

Dauda, A.B., Natrah, I., Karim, M., Kamarudin, M.S, Bichi, A.H., 2018. African catfish aquaculture in Malaysia and Nigeria: status, trends and prospects. Fish. Aquac. J. 09 (01). doi:10.4172/2150-3508.1000237.

De Lacerda, L.D., Vaisman, A.G., Maia, L.P., Ramos E Silva, C.A., Soares Cunha, E.M., 2006. Relative importance of nitrogen and phosphorus emissions from shrimp farming and other anthropogenic sources for six estuaries along the NE Brazilian coast. Aquaculture 253 (1–4), 433–446. doi:10.1016/j.aquaculture.2005.09.005.

De Schryver, P., Crab, R., Defoirdt, T., Boon, N., Verstraete, W., 2008. The basics of bio-flocs technology: the added value for aquaculture. Aquaculture 277 (3–4), 125–137. doi:10.1016/j.aquaculture.2008.02.019.

De Schryver, P., Verstraete, W., 2009. Nitrogen removal from aquaculture pond water by heterotrophic nitrogen assimilation in lab-scale sequencing batch reactors. Bioresour. Technol. 100 (3), 1162–1167. doi:10.1016/j.biortech.2008.08.043.

Deb, S., Noori, M.T., Rao, P.S., 2020. Application of biofloc technology for Indian major carp culture (polyculture) along with water quality management. Aquacult. Eng. 91. doi:10.1016/j.aquaeng.2020.102106.

Debbarma, R., Biswas, P., Singh, S.K, 2021. An integrated biomarker approach to assess the welfare status of Ompok bimaculatus (Pabda) in biofloc system with altered C/N ratio and subjected to acute ammonia stress. Aquaculture 545, 737184. doi:10.1016/j.aquaculture.2021.737184.

Defoirdt, T., Boon, N., Sorgeloos, P., Verstraete, W., Bossier, P., 2008. Quorum sensing and quorum quenching in Vibrio harveyi: lessons learned from in vivo work. ISME J. 2 (1), 19–26. doi:10.1038/ismej.2007.92.

Defoirdt, T., Boon, N., Bossier, P., Verstraete, W., 2004. Disruption of bacterial quorum sensing: an unexplored strategy to fight infections in aquaculture. Aquaculture 240 (1–4), 69–88. doi:10.1016/j.aquaculture.2004.06.031.

Deng, M., Chen, J., Gou, J., Hou, J., Li, D., He, X., 2018. The effect of different carbon sources on water quality, microbial community and structure of biofloc systems. Aquaculture 482, 103–110. doi:10.1016/j.aquaculture.2017.09.030.

Ekasari, J., Angela, D., Waluyo, S.H., Bachtiar, T., Surawidjaja, E.H., Bossier, P., De Schryver, P., 2014. The size of biofloc determines the nutritional composition and the nitrogen recovery by aquaculture animals. Aquaculture 426-427, 105–111. doi:10.1016/j.aquaculture.2014.01.023.

Emerenciano, M.G.C., Khanjani, M.H., Sharifinia, M., Miranda-Baeza, A., 2025. Could biofloc technology (BFT) pave the way toward a more sustainable aquaculture in line with the circular economy? Aquac. Res. 2025 (1). doi:10.1155/are/1020045.

FAO and UN Water, 2021. Progress on Change in Water-Use Efficiency. Global Status and Acceleration Needs for SDG Indicator 6.4.1. Rome. https://doi.org/10.4060/cb6413en.

Figueroa-Espinoza, J., Rivas-Vega, M.E., Mariscal-López, M.d.l.Á., Emerenciano, M.G.C., Martínez-Porchas, M., Miranda-Baeza, A., 2022. Reusing water in a biofloc culture system favors the productive performance of the Nile tilapia (Oreochromis niloticus) without affecting the health status. Aquaculture 558. doi:10.1016/j.aquaculture.2022.738363.

Gallardo-Collí, A., Pérez-Rostro, C.I., Hernández-Vergara, M.P, 2019. Reuse of water from biofloc technology for intensive culture of Nile tilapia (Oreochromis niloticus): effects on productive performance, organosomatic indices and body composition. Int. Aquat. Res. 11 (1), 43–55. doi:10.1007/s40071-019-0218-9.

Gao, L., Shan, H.W., Zhang, T.W., Bao, W.Y., Ma, S., 2012. Effects of carbohydrate addition on Litopenaeus vannamei intensive culture in a zero-water exchange system. Aquaculture 342–343 (1), 89–96. doi:10.1016/j.aquaculture.2012.02.022.

Ghaedi, A., Sarsangi, H., Akhavan, M., Ali Mahmudi, M., Mohamadi, M., 2022. Biofloc system preparation in simple language. Adv. Aquac. Sci. J. 6 (1), 51–61.

Guo, K., Zhao, Z., Xie, J., Luo, L., Wang, S., Zhang, R., Xu, W., Huang, X., 2022. Combined effects of eco-substrate and carbon addition on water quality, fish performance and nutrient budgets in the pond polyculture system. Fishes 7 (5), 212. doi:10.3390/fishes7050212.

Gustilatov, M., Widanarni, J.E., Pande, G.S.J., 2022. Protective effects of the biofloc system in Pacific white shrimp (Penaeus vannamei) culture against pathogenic Vibrio parahaemolyticus infection. Fish and Shellfish Immunol. 124, 66–73. doi:10.1016/j.fsi.2022.03.037.

Hargreaves, J.A., 2006. Photosynthetic suspended-growth systems in aquaculture. Aquacult. Eng. 34 (3), 344–363. doi:10.1016/j.aquaeng.2005.08.009.

Hu, Z., Lee, J.W., Chandran, K., Kim, S., Khanal, S.K., 2012. Nitrous oxide (N_2O) emission from aquaculture: a review. Environ. Sci. Technol. 46 (12), 6470–6480. doi:10.1021/es300110x.

Huang, J., Zhang, G., Zhang, Y., Guan, X., Wei, Y., Guo, R., 2020. Global desertification vulnerability to climate change and human activities. Land Degrad. Dev. 31 (11), 1380–1391. doi:10.1002/ldr.3556.

Jang, I.K., Pang, Z., Yu, J., Kim, S.K., Seo, H.C., Cho, Y.R., 2010. Selectively enhanced expression of prophenoloxidase activating enzyme 1 (PPAE1) at a bacteria clearance site in the white shrimp, Litopenaeus vannamei. BMC Immunol. 12, 1–11.

Jatobá, A., Da Silva, B.C., Da Silva, J.S., Vieira, F.d.N., Mouriño, J.L.P., Seiffert, W.Q., Toledo, T.M., 2014. Protein levels for Litopenaeus vannamei in semi-intensive and biofloc systems. Aquaculture 432, 365–371. doi:10.1016/j.aquaculture.2014.05.005.

Jatobá, A., Borges, Y.V., Silva, F.A., 2019. Biofloc: sustainable alternative for water use in fish culture. Arq. Bras. Med. Vet. Zootec. 71 (3), 1076–1080. doi:10.1590/1678-4162-10309.

de Jesús Becerra-Dorame, M., Martinez-Cordova, L.R., Martínez-Porchas, M., Hernández-López, J., López-Elías, J.A., Mendoza-Cano, F., 2014. Effect of using autotrophic and heterotrophic microbial-based-systems for the pre-grown of Litopenaeus vannamei, on the production performance and selected haemolymph parameters. Aquac. Res. 45 (5), 944–948. doi:10.1111/are.12033.

Jones, J.A.A., 1999. Climate change and sustainable water resources: placing the threat of global warming in perspective. Hydrol. Sci. J. 44 (4), 541–557. doi:10.1080/02626669909492251.

Ju, Z.Y., Forster, I., Conquest, L., Dominy, W., Kuo, W.C., Horgen, F.D., 2008. Determination of microbial community structures of shrimp floc cultures by biomarkers and analysis of floc amino acid profiles. Aquac. Res. 39 (2), 118–133. doi:10.1111/j.1365-2109.2007.01856.x.

Khanjani, M.H., Mozanzadeh, M.T., Sharifinia, M., Emerenciano, M.G.C., 2023a. Biofloc: a sustainable dietary supplement, nutritional value and functional properties. Aquaculture 562. doi:10.1016/j.aquaculture.2022.738757.

Khanjani, M.H., Zahedi, S., Sharifinia, M., Hajirezaee, S., Singh, S.K., 2025a. Biological removal of nitrogenous waste compounds in the biofloc aquaculture system - a review. Ann. Anim. Sci. 25 (1), 3–21. doi:10.2478/aoas-2024-0060.

Khanjani, M.H., Mozanzadeh, M.T., Sharifinia, M., Emerenciano, M.G.C., 2024a. Broodstock and seed production in biofloc technology (BFT): an updated review focused on fish and penaeid shrimp. Aquaculture 579. doi:10.1016/j.aquaculture.2023.740278.

Khanjani, M.H., Sharifinia, M., Akhavan-Bahabadi, M., Emerenciano, M.G.C., 2024b. Probiotics and phytobiotics as dietary and water supplements in biofloc aquaculture systems. Aquacult. Nutr. 2024. doi:10.1155/anu/3089887.

Khanjani, M.H., Sajjadi, M.M., Alizadeh, M., Sourinejad, I., 2016. Study on nursery growth performance of Pacific white shrimp (Litopenaeus vannamei Boone, 1931) under different feeding levels in zero water exchange system. Iran. J. Fish. Sci. 15 (4), 1465–1484. http://jifro.ir/article-1-2520-en.html.

Khanjani, M.H., da Silva, L.O.B., Fóes, G.K., Vieira, F.d.N., Poli, M.A., Santos, M., Emerenciano, M.G.C., 2023b. Synbiotics and aquamimicry as alternative microbial-based approaches in intensive shrimp farming and biofloc: novel disruptive techniques or complementary management tools? A scientific-based overview. Aquaculture 567. doi:10.1016/j.aquaculture.2023.739273.

Khanjani, M.H., Alizadeh, M., Mohammadi, M., Aliabad, H.S., 2021a. The effect of adding molasses in different times on performance of Nile Tilapia (Oreochromis niloticus) raised in a low-salinity biofloc system. Ann. Anim. Sci. 21 (4), 1435–1454. doi:10.2478/aoas-2021-0011.

Khanjani, M.H., Alizadeh, M., 2021b. Biological and economic performance of Nile tilapia (Oreochromis niloticus) in two conventional and limited water exchange systems. Aquat. Ecol. 11 (3), 12–21.

Khanjani, M.H., Alizadeh, M., 2024. Effects of different salinity levels on performance of Nile tilapia fingerlings in a biofloc culture system. Ann. Anim. Sci. 24 (1), 235–245. doi:10.2478/aoas-2023-0077.

Khanjani, M.H., Eslami, J., Emerenciano, M.G.C., 2025b. Wheat flour as carbon source on water quality, growth performance, hemolymph biochemical and immune parameters of Pacific white shrimp (Penaeus vannamei) juveniles in biofloc technology (BFT). Aquac. Rep. 40. doi:10.1016/j.aqrep.2025.102623.

Khanjani, M.H., Mohammadi, A., Emerenciano, M.G.C., 2022a. Microorganisms in biofloc aquaculture system. Aquac. Rep. 26. doi:10.1016/j.aqrep.2022.101300.

Khanjani, M.H., Mohammadi, A., Emerenciano, M.G.C., 2024c. Water quality in biofloc technology (BFT): an applied review for an evolving aquaculture. Aquac. Int. 32 (7), 9321–9374. doi:10.1007/s10499-024-01618-w.

Khanjani, M.H., Mozanzadeh, M.T., Fóes, G.K., 2022b. Aquamimicry system: a sutiable strategy for shrimp aquaculture - a review. Ann. Anim. Sci. 22 (4), 1201–1210. doi:10.2478/aoas-2022-0044.

Khanjani, M.H., Sharifinia, M., 2022a. Biofloc as a food source for banana shrimp Fenneropenaeus merguiensis postlarvae. North Am. J. Aquacult. 84 (4), 469–479. doi:10.1002/naaq.10261.

Khanjani, M.H., Sharifinia, M., 2022b. Biofloc technology with addition molasses as carbon sources applied to Litopenaeus vannamei juvenile production under the effects of different C/N ratios. Aquac. Int. 30 (1), 383–397. doi:10.1007/s10499-021-00803-5.

Khanjani, M.H., Sharifinia, M., 2024. Feeding Nile tilapia with varying levels of biofloc: effect on growth performance, survival rate, digestive and liver enzyme activities, and mucus immunity. Aquac. Int. 32 (6), 8171–8194. doi:10.1007/s10499-024-01561-w.

Khanjani, M.H., Sharifinia, M., Emerenciano, M.G.C., 2023c. A detailed look at the impacts of biofloc on immunological and hematological parameters and improving resistance to diseases. Fish Shellfish Immunol. 137. doi:10.1016/j.fsi.2023.108796.

Khanjani, M.H., Sharifinia, M., Emerenciano, M.G.C., 2024d. Biofloc technology (BFT) in aquaculture: what goes right, what goes wrong? A scientific-based snapshot. Aquacult. Nutr. 2024. doi:10.1155/2024/7496572.

Khanjani, M.H., Zahedi, S., Mohammadi, A., 2022c. Integrated multitrophic aquaculture (IMTA) as an environmentally friendly system for sustainable aquaculture: functionality, species, and application of biofloc technology (BFT). Environ. Sci. Pollut. Res. 29 (45), 67513–67531. doi:10.1007/s11356-022-22371-8.

Khanjani, M.H, Sharifinia, M., Hajirezaee, S., 2022d. Recent progress towards the application of biofloc technology for tilapia farming. Aquaculture 552, 738021. doi:10.1016/j.aquaculture.2022.738021.

Krummenauer, D., Samocha, T., Poersch, L., Lara, G., Wasielesky, W., 2014. The reuse of water on the culture of pacific white shrimp, Litopenaeus vannamei, in BFT system. J. World Aquac. Soc. 45 (1), 3–14. doi:10.1111/jwas.12093.

Learmonth, C., Carvalho, A.P., 2015. Acute and chronic toxicity of nitrate to early life stages of zebrafish - setting nitrate safety levels for zebrafish rearing. Zebrafish 12 (4), 305–311. doi:10.1089/zeb.2015.1098.

Li, C., Zhang, X., Chen, Y.u, Zhang, S., Dai, L., Zhu, W., Chen, Y., 2023. Optimized utilization of organic carbon in aquaculture biofloc systems: a review. Fishes 8 (9), 465. doi:10.3390/fishes8090465.

Lima, P.C.M., Abreu, J.L., Silva, A.E.M., Severi, W., Galvez, A.O., Brito, L.O., 2018. Nile tilapia fingerling cultivated in a low-salinity biofloc system at different stocking densities. Span. J. Agric. Res. 16 (4). doi:10.5424/sjar/2018164-13222.

Liu, G., Ye, Z., Liu, D., Zhao, J., Sivaramasamy, E., Deng, Y., Zhu, S., 2018. Influence of stocking density on growth, digestive enzyme activities, immune responses, antioxidant of Oreochromis niloticus fingerlings in biofloc systems. Fish Shellfish Immunol. 81, 416–422. doi:10.1016/j.fsi.2018.07.047.

Liu, H., Li, H., Wei, H., Zhu, X., Han, D., Jin, J., Yang, Y., Xie, S., 2019. Biofloc formation improves water quality and fish yield in a freshwater pond aquaculture system. Aquaculture 506, 256–269. doi:10.1016/j.aquaculture.2019.03.031.

Long, L., Yang, J., Li, Y., Guan, C., Wu, F., 2015. Effect of biofloc technology on growth, digestive enzyme activity, hematology, and immune response of genetically improved farmed tilapia (Oreochromis niloticus). Aquaculture 448, 135–141. doi:10.1016/j.aquaculture.2015.05.017.

Mahanand, S.S., Moulick, S., Rao, P.S., 2013. Water quality and growth of rohu, Labeo rohita, in a biofloc system. J. Appl. Aquac. 25 (2), 121–131. doi:10.1080/10454438.2013.788898.

Malpartida Pasco, J.J., Carvalho Filho, J.W., de Espirito Santo, C.M., Vinatea, L., 2018. Production of Nile tilapia Oreochromis niloticus grown in BFT using two aeration systems. Aquac. Res. 49 (1), 222–231. doi:10.1111/are.13451.

Martins, C.I.M., Eding, E.H., Verdegem, M.C.J., Heinsbroek, L.T.N., Schneider, O., Blancheton, J.P., d'Orbcastel, E.R., Verreth, J.A.J., 2010. New developments in recirculating aquaculture systems in Europe: a perspective on environmental sustainability. Aquacult. Eng. 43 (3), 83–93. doi:10.1016/j.aquaeng.2010.09.002.

McCusker, S., Warberg, M.B, Davies, S.J., Valente, C.S, Johnson, M.P., Cooney, R., Wan, A.H.L., 2023. Biofloc technology as part of a sustainable aquaculture system: A review on the status and innovations for its expansion. Aquac. Fish Fish. 3 (4), 331–352. doi:10.1002/aff2.108.

Megahed, M.E., 2010. The effect of microbial biofloc on water quality, survival and growth of the green tiger shrimp (Penaeus semisulcatus) fed with different crude protein levels. J. Arab. Aquac. Soc. 5, 119–142.

Mesquita, R.C.T., Maschio, D., Eloy, L., Godoy, L., Paz, E., Streit, J., Danilo, 2016. Advantages of Nile tilapia (*Oreochromis niloticus*) farming with minimal effluent release. Rev. Bras. Hig. Sanidade Anim. 10 (3), 447–454.

Minabi, K., Sourinejad, I., Alizadeh, M., Ghatrami, E.R., Khanjani, M.H., 2020. Effects of different carbon to nitrogen ratios in the biofloc system on water quality, growth, and body composition of common carp (Cyprinus carpio L.) fingerlings. Aquac. Int. 28 (5), 1883–1898. doi:10.1007/s10499-020-00564-7.

Monsees, H., Klatt, L., Kloas, W., Wuertz, S., 2017. Chronic exposure to nitrate significantly reduces growth and affects the health status of juvenile Nile tilapia (Oreochromis niloticus L.) in recirculating aquaculture systems. Aquac. Res. 48 (7), 3482–3492. doi:10.1111/are.13174.

Moss, S.M., Arce, S.M., Argue, B.J., Otoshi, C.A., Calderon, G.R.O., Tacon, A.G.J., 2001. Greening of the blue revolution: efforts toward environmentally responsible shrimp culture. In: The new wave: proceedings of the special session on sustainable shrimp culture, aquaculture 2001. World Aquaculture Society, Baton Rouge, LA, pp. 1–19.

Moss, S.M., Leung, P., 2007. Comparative cost of shrimp production: earthen ponds versus recirculating aquaculture systems. Shrimp culture: economics, market, and trade 291–300. doi:10.1002/9780470277850.ch19.

Nagappan, S., Das, P., AbdulQuadir, M., Thaher, M., Khan, S., Mahata, C., Al-Jabri, H., Vatland, A.K., Kumar, G., 2021. Potential of microalgae as a sustainable feed ingredient for aquaculture. J. Biotechnol. 341, 1–20. doi:10.1016/j.jbiotec.2021.09.003.

Naylor, R.L., Hardy, R.W., Buschmann, A.H., Bush, S.R., Cao, L., Klinger, D.H., Little, D.C., Lubchenco, J., Shumway, S.E., Troell, M., 2021. A 20-year retrospective review of global aquaculture. Nature 591 (7851), 551–563. doi:10.1038/s41586-021-03308-6.

Padeniya, U., Davis, D.A., Wells, D.E., Bruce, T.J., 2022. Microbial interactions, growth, and health of aquatic species in biofloc systems. Water 14 (24). doi:10.3390/w14244019.

Páez-Osuna, F., Guerrero-Galván, S.R., Ruiz-Fernández, A.C., Espinoza-Angulo, R., 1997. Fluxes and mass balances of nutrients in a semi-intensive shrimp farm in north-western Mexico. Mar. Pollut. Bull. 34 (5), 290–297. doi:10.1016/s0025-326x(96)00133-6.

Pérez-Fuentes, J.A., Pérez-Rostro, C.I., Hernández-Vergara, M.P., 2013. Pond-reared Malaysian prawn Macrobrachium rosenbergii with the biofloc system. Aquaculture 400–401, 105–110. doi:10.1016/j.aquaculture.2013.02.028.

Perez-Fuentes, J.A., Herna´ndez-Vergara, M.P., Perez-Rostro, C.I., 2016. C: N ratios affect nitrogen removal and production of Nile tilapia Oreochromis niloticus raised in a biofloc system under high density cultivation. Aquaculture 425, 247–251.

Raza, B., Zheng, Z., Yang, W., Review on Biofloc System Technology, A., 2024. History, types, and future economical perceptions in aquaculture. Animals 14 (10), 1489. doi:10.3390/ani14101489.

Salarpour, M., Davarpanah, M., Zara, G., 2022. Investigating the effective factors on the acceptance of new irrigation technologies among farmers in Sistan region. J. Water Sustain. Dev. 8 (4), 23–30.

Samocha, T., 2019. Sustainable biofoc systems for marine shrimp. Academic Press, London and San Diego, CA, p. 463.

Scheffran, J., Battaglini, A., 2011. Climate and conflicts: the security risks of global warming. Reg. Environ. Change 11 (S1), 27–39. doi:10.1007/s10113-010-0175-8.

Schumann, M., Brinker, A., 2020. Understanding and managing suspended solids in intensive salmonid aquaculture: a review. Rev. Aquac. 12 (4), 2109–2139. doi:10.1111/raq.12425.

Schveitzer, L.M.A.R., Santo, C.M.E., Candia, E.W.S, Mourino, J.L.P., Legarda, E.C., Seiffert, W.Q., Vieira, F.N., 2015. Intensive hatchery performance of the Pacific white shrimp in biofloc system. Aquacult. Eng. 67, 53–58. doi:10.1016/j.aquaeng.2015.05.007.

Serra, F.P., Gaona, C.A.P., Furtado, P.S., Poersch, L.H., Wasielesky, W., 2015. Use of different carbon sources for the biofloc system adopted during the nursery and grow-out culture of Litopenaeus vannamei. Aquac. Int. 23 (6), 1325–1339. doi:10.1007/s10499-015-9887-6.

Simon, M., Grossart, H.P, Schweitzer, B., Ploug, H., 2002. Microbial ecology of organic aggregates in aquatic ecosystems. Aquat. Microb. Ecol. 28 (2), 175–211. doi:10.3354/ame028175.

Solanki, S., Meshram, S.J., Dhamagaye, H.B., Naik, S.D., Shingare, P.E., Yadav, B.M., Ghafarifarsani, H., 2023. Effect of C/N ratio levels and stocking density of Catla spawn (Gibelion catla) on water quality, growth performance, and biofloc nutritional composition in an indoor biofloc system. Aquac. Res. 2023, 1–11. doi:10.1155/2023/2501653.

Tepaamorndech, S., Nookaew, I., Higdon, S.M., Santiyanont, P., Phromson, M., Chantarasakha, K., Mhuantong, W., Plengvidhya, V., Visessanguan, W., 2020. Metagenomics in bioflocs and their effects on gut microbiome and immune responses in Pacific white shrimp. Fish and Shellfish Immunol. 106, 733–741. doi:10.1016/j.fsi.2020.08.042.

Food and Agriculture Organization of the United Nations (FAO), 2024. The State of World Fisheries and Aquaculture 2024: Blue Transformation in Action. FAO. Rome (2024). https://doi.org/10.4060/cd0683en

Timmons, M.B., Losordo, T.M., 1994. Aquaculture water reuse systems: engineering design and management. Elsevier, Amsterdam, p. 660.

True, B., Johnson, W., Chen, S., 2004. Reducing phosphorus discharge from flow-through aquaculture I: facility and effluent characterization. Aquacult. Eng. 32 (1), 129–144. doi:10.1016/j.aquaeng.2004.08.001.

Ulloa Walker, D.A., Morales Suazo, M.C., Emerenciano, M.G.C., 2020. Biofloc technology: principles focused on potential species and the case study of Chilean river shrimp Cryphiops caementarius. Rev. Aquac. 12 (3), 1759–1782. doi:10.1111/raq.12408.

Verdegem, M.C.J., Bosma, R.H., 2009. Water withdrawal for brackish and inland aquaculture, and options to produce more fish in ponds with present water use. Water Policy 11 (S1), 52–68. doi:10.2166/wp.2009.003.

Verdegem, M.C.J., Bosma, R.H., Verreth, J.A.J., 2006. Reducing water use for animal production through aquaculture. Int. J. Water Resour. Dev. 22 (1), 101–113. doi:10.1080/07900620500405544.

Wang, H., Wu, P., Zheng, D., Deng, L., Wang, W., 2022. N -acyl-homoserine lactone (AHL)-mediated microalgal–bacterial communication driving Chlorella-activated sludge bacterial biofloc formation. Environ. Sci. Technol. 56 (17), 12645–12655. doi:10.1021/acs.est.2c00905.

Wei, Y.F, Liao, S.A., Wang, A., 2016. The effect of different carbon sources on the nutritional composition, microbial community and structure of bioflocs. Aquaculture 465, 88–93. doi:10.1016/j.aquaculture.2016.08.040.

Xu, W.J., Pan, L.Q., Zhao, D.H., Huang, J., 2012. Preliminary investigation into the contribution of bioflocs on protein nutrition of Litopenaeus vannamei fed with different dietary protein levels in zero-water exchange culture tanks. Aquaculture 350–353, 147–153. doi:10.1016/j.aquaculture.2012.04.003.

Yogev, U., Sowers, K.R., Mozes, N., Gross, A., 2017. Nitrogen and carbon balance in a novel near-zero water exchange saline recirculating aquaculture system. Aquaculture 467, 118–126. doi:10.1016/j.aquaculture.2016.04.029.

Yu, Y.B., Lee, K.M., Kim, J.H., Kang, J.C., Kim, J.H., 2023. Comparative analysis of morphological characteristics, hematological parameters, body composition and sensory evaluation in olive flounder, Paralichthys olivaceus raised in biofloc and seawater to evaluate marketability. Aquac. Rep. 30. doi:10.1016/j.aqrep.2023.101616.

Zapata, K.P.L., Brito, L.O., de Lima, P.C.M., Vinatea, L.A.A., Galvez, A.O., Cárdenas, J.M.V., 2017. Culture of Nile tilapia fingerlings in biofloc system under different carbon/nitrogen ratios. Bol. Inst. Pesca 43 (3), 399–407. doi:10.20950/1678-2305.2017v43n3p399.

Zemor, J.C., Wasielesky, W., Fóes, G.K., Poersch, L.H., 2019. The use of clarifiers to remove and control the total suspended solids in large-scale ponds for production of Litopenaeus vannamei in a biofloc system. Aquacult. Eng. 85, 74–79. doi:10.1016/j.aquaeng.2019.03.001.

Zhao, Z., Luo, L., Wang, C., Li, J., Wang, L., Du, X., Xu, Q., 2018. Effects of organic carbon addition on water quality and growth performance of bottom- and filter-feeding carp in a minimum-water-exchange pond polyculture system. Fish. Sci. 84 (4), 681–689. doi:10.1007/s12562-018-1204-7.

CHAPTER

13

Synbiotic-based aquaculture: Insights on water quality management and animal health

Luis Otavio Brito[a], *Otávio Augusto Lacerda Ferreira Pimentel*[a], *Dariano Krummenauer*[b], *Alfredo Olivera Gálvez*[a], *Mohammad Hossein Khanjani*[c] and *Maurício Gustavo Coelho Emerenciano*[d]

[a]Department of Fisheries and Aquaculture, Federal Rural University of Pernambuco (UFRPE), Recife, Pernambuco, Brazil [b]Aquaculture Marine Station, Institute of Oceanography, Federal University of Rio Grande, Rio Grande, Brazil [c]Department of Fisheries Sciences and Engineering, Faculty of Natural Resources, University of Jiroft, Jiroft, Iran [d]Livestock & Aquaculture Program, Aquaculture Systems Team, Commonwealth Scientific and Industrial Research Organization, Bribie Island Research Centre, Woorim, QLD, Australia

13.1 Introduction

The world population has reached approximately 8 billion people, and according to projections, this number could reach 8.5 billion people by the year 2030 and 9 billion by 2037 (UN-United Nations, 2022). This increase in the world population increases the need to guarantee food security for this entire population. Aquaculture plays a key role in this scenario, as fish protein represents 17% of all animal protein consumed in the world (FAO, 2022).

According to FAO (2024), in 2032, the annual per capita consumption of aquatic animal foods will increase from 20.7 kg in 2022 to 21.3 kg, as a result of the high demand for fish. In addition, 90% of all aquatic animal production will be for human consumption. In this sense, aquaculture appears as an important activity for the fish supply. The aquaculture contribution

Zero-Waste Technologies in Aquaculture: Global Perspectives and Methods toward a Circular Economy
DOI: https://doi.org/10.1016/B978-0-443-34689-7.00017-3

to the aquatic animal production in 2024 was 102 million tons, mostly for human use, 38.9 million tons of algae for both food and nonfood uses, and 11,000 tons of shells and pearls for ornamental use, reaching a total of 130.9 million tons in live weight (FAO, 2026). Since 2022, aquaculture production surpassed the production of aquatic organisms through fishing (FAO, 2026). Aquaculture will face many challenges in the coming years to further reduce growth rates. Among these, we can mention: the development of more efficient production systems, the reduction of the impacts caused by diseases, the reduction of production costs, the construction of new products, the expansion of the market, the genetic improvement of species, the use of land with minimal conflict of interests, and climate change (Naylor et al., 2021; Sumalia et al., 2022).

Among the existing models in aquaculture, which minimize the use of water per kilogram of animal produced, use waste, transforming them into microbial and animal biomass, and good economic results. Furthermore, it is important to develop technologies that increase productivity but minimize the use of marine products in aquafeeds, as the demand for fishmeal and fish oil for balanced diets continues to expand and the reduction of fish stocks will result in a lack of raw materials (Naylor et al., 2021; Sumalia et al., 2022). In this context, new microbial biotechnology approaches for waste recovery and use in aquaculture, such as the synbiotic system, emerge as a promising alternative. This chapter will unpack the key premises of this microbial-based system, highlight the differences among other similar culture techniques, and bring insights into key water quality management, fertilization strategies, animal health, and recent R&D.

13.2 Synbiotic system and key management insights

13.2.1 Concept and fundamentals

The synbiotic system (prebiotics + probiotics = synbiotics) is a novel microbial biotechnology approach for waste recovery and use in aquaculture, because it contributes to the water quality control and microbial communities' growth in the aquaculture. This beneficial effect occurs due to the fertilization of the water with vegetable bran processed by fermentation and microbial respiration by probiotic microorganisms (Pimentel et al., 2022; Santos et al., 2022a; Khanjani et al., 2023; Pimentel et al., 2025) (Fig. 13.1). As per (Khanjani et al., 2023), there are other microbial-based systems similar to synbiotics, such as aquamimicry and biofloc. However, key differences include the form and type of carbon sources used to keep the system running. While bioflocs use molasses, aquamimicry, and synbiotics use fermented vegetable bran (Khanjani et al., 2023).

The fertilization in the synbiotic system adopts an organic carbon source to stimulate the growth of the heterotrophic microbial community, which, among its diverse functions, assimilates the toxic nitrogenous compounds produced by the animals and the feed input in the system, improving the water quality. The activity of these bacteria transforms these harmful compounds into microbial protein, providing a natural supplementary source of food for the animals, which has a positive effect on the gut microbial composition and hepatopancreas, improving physiological condition (de Andrade et al., 2021; Hussain et al., 2021; Abdel-Tawwab et al., 2022). In addition to shrimp health and improved water quality,

FIGURE 13.1 Synbiotic system fertilizer using wheat bran in the fermentation phase (A) and microbial respiration phase (B) at Camamum shrimp farming, Bahia, Brazil. *Photos: Luis Otavio Brito.*

FIGURE 13.2 Settable solids from a commercial semiintensive shrimp farming using a synbiotic system. Boca do Rio shrimp farming, Alagoas, Brazil. *Photo: Luis Otavio Brito.*

this strategy promotes a reduction in settleable solids, increases productive performance, and economic results (Romano et al., 2018; Emerenciano et al., 2022; Khanjani et al., 2023) (Fig. 13.2).

The use of fermentation (phase without oxygen) and/or microbial respiration (phase with oxygen) processes by microorganisms (e.g., *Lactobacillus, Pediococcus, Enterococcus, Bacillus, Saccharomyces*, among others) is a way to increase the availability of nutrients in a raw material, especially bran or cereals (Tamang et al., 2016; Dawood and Koshio, 2020; Vassileva et al., 2021). During the fermentation, fiber content decreases, and microbial biomass increases. These microorganisms transform the chemical constituents of the raw materials by breaking complex organic molecules, such as polysaccharides, into simpler molecules, thus increasing the bioavailability and solubility of nutrients, making them more nutritious

(Flesch et al., 2014; Ribeiro et al., 2017). In addition, phytic acid and tannins decrease during fermentation, and there is an increase in the bioavailability of minerals (calcium, phosphorus, zinc, and iron) (Ribeiro et al., 2017). The microorganisms also produce enzymes such as amylases, proteases, phytases, and lipases, improving the quality and quantity of proteins, amino acids, and essential fatty acids (Amadou et al., 2010, Amadou et al., 2011; Tamang et al., 2016). Furthermore, this process generates organic acids such as lactic, acetic, formic, propionic, palmitic, pyruvic, malic, formic, and butyric. These acids reduce the gut pH, reducing the growth of pathogenic microorganisms, restoring the eubiosis (Mukherjee et al., 2016; Wang et al., 2016; Jannathulla et al., 2018; Sharawy et al., 2016; Aliyu-A et al., 2019). However, the generated products and microbial growth can be affected by the source of carbohydrates, oxygen presence, pH values, and colony-forming units (CFU) g^{-1} (Tamanget al.,2016; Charalampopoulos et al., 2002).

The fermented bran addition in the aquaculture ponds fertilization is relatively recent, dating from the mid-1990s in Thailand, when it later came to be called aquamimicry (Romano, 2017). Since then, the popularization of these systems around the world has been accompanied by numerous fertilization protocols using rice, soybean, and wheat bran, generating different effects on microbial flocs, their proximal composition, and fish/shrimp growth performance (Romano, 2017; Romano et al., 2018; de Andrade et al., 2021; Hussain et al., 2021; Liñan-Vidriales et al., 2021; Silva et al., 2021; Santos et al., 2022b).

Although there are few scientific papers about synbiotic addition for aquaculture systems, this has been used in some commercial farming, because of the need for a system that lowers carbon input, thus increasing the possibility of implementation of intensive ponds (Kawahigashi, 2018). Among the characteristics inherent to this system are a more homogeneous distribution between heterotrophic and nitrifying bacteria (allowing the bacterial flocs to grow and residual nitrogen control), stimulation of plankton growth, and reduces organic load (Romano, 2017; Kawahigashi, 2018).

13.2.2 Organic fertilization

For the synbiotic fertilization, some inputs are used, such as: bran, molasses, sugar, dextrose, microorganisms, enzymes, alkalizing (buffer), and water (Romano et al., 2018; Santos et al., 2022a; Pimentel et al., 2024b; Pimentel et al., 2024c). The concentration of the inputs varies according to the stocking density, oxygen concentration in the water, pH, transparency, the color of the water, oxidation/reduction potential (ORP), and *Vibrio* concentration (Table 13.1).

In the synbiotic system fertilizer, the main components are vegetable bran, which acts as a prebiotic, and microorganisms, which act as probiotics. Prebiotics are defined by Collins and Gibson (1999) as nondigestible ingredients that stimulate the growth of beneficial microorganisms in the digestive tract of animals. The polysaccharides used (prebiotics) constitute a group of carbohydrates that are not easily digestible by vibrios, unlike bacteria and/or probiotic microorganisms that are more capable of using this source of carbohydrates, not only inhibiting the growth of pathogens (through the antimicrobial activity of lactic acid), but also lysing toxin molecules (Butt et al., 2021; Goh et al., 2022). The most used prebiotics in the synbiotic system are rice, wheat, and soybean bran.

TABLE 13.1 Inputs used in the synbiotic system fertilization.

Inputs	Source	Role	Concentration
Bran	Rice, wheat, and soybeans	Carbon organic source	0.5–20 g m^{-3}
Sugar	Molasses, sugar, and dextrose	Microbial substrate	0.05–2 g m^{-3}
Probiotics	*Bacillus, Lactobacillus, Pediococcus, Pseudomonas, Bifidobacterium, Saccharomyces, Nitrosomonas,* and *Nitrobacter*	Microorganisms for the bran transformation into more bacterial biomass, organic acids, enzymes, bacteriocins, and other substances	0.02–1.0 g m^{-3}
Enzymes	Protease, amylase, and lipase	Accelerates chemical reactions in the bran	0.5–1.0g kg $bran^{-1}$
Buffer	Sodium bicarbonate, limestone, *Lithothamnium*, and other calcium carbonate equivalents	Avoid reduction and sudden variations in the pH of the fertilizer	0%–10% of the amount bran
Water	Clean water	Solid-state fermentation (SSF) and submerged fermentation (SmF)	2–10 L by kg of bran

Source: Kawahigashi, D., 2018; Romano et al., 2018; Abdel-Tawwab et al., 2022; de Andrade et al., 2021; Santos et al., 2022a, Santos et al., 2022b; Oliveira et al., 2022; Pimentel et al., 2022; Pimentel et al., 2024b; Pimentel et al., 2024c; Coutinho et al., 2024; Nunes et al., 2024; Gonçalves Junior et al., 2025.

To obtain white rice, a series of processes is necessary from brown rice. As rice bran is a byproduct, its use in synbiotic aquaculture can be an interesting approach due to its availability, low cost, and little or no use for human consumption (Ribeiro et al., 2017). According to the International Aquaculture Feed Formulation Database (https://www.iaffd.com/-07.30.2023), the nutritional composition of whole and defatted rice bran varies close to 89% of dry matter, 12.5%–15.5% crude protein, 11.5%–2.4% lipids, 12.7%–10.2% crude fiber, 13.2%–10.3% ash, and 39.7%–50.8% nitrogen-free extract (NFE). However, the proximal composition varies according to the fermentation, the microbial used, pH, etc. (Santos et al., 2022b) evaluated the proximal composition of rice bran with fermentation (anaerobic phase) and microbial respiration (aerobic phase) with *Bacillus* and observed an increase of 28.91% in lipid content with both processing strategies (fermentation plus microbial respiration) and only increase in crude protein (18.27%) (microbial respiration) when compared with rice bran in nature (17.99% lipids and 17.41% crude protein).

Soybeans are one of the most important agricultural activities in the world, and their production is used to obtain oil and bran by the food industry. Soybean meal is obtained from grinding soybeans to extract the oil, which is intended for human consumption, and represents one of the most important ingredients used in animal feed (Halver et al., 2002). As it is a byproduct obtained after the extraction of soybean oil, soybean meal goes through a series of processes that can affect its nutritional quality. Furthermore, soy is a legume plant, which has some antinutritional factors, such as Trypsin inhibitors, hemagglutinins, and others, which must be destroyed by thermal processing before the ingredient is included in feed (do Prado et al., 2022; Elhalis et al., 2024).

According to the International Aquaculture Feed Formulation Database, the nutritional composition of soybean varies close to 88% of dry matter, 39%–68% crude protein, 0.3%–6.0% lipids, 0.1%–6.0% crude fiber, 1.3%–7.0% ash, and 1.6%–35% NFE. However, the proximal composition could increase with the use of fermentation and microbial respiration (El-Dakar et al., 2023) evaluated the proximal composition of fermented soybean meal with *Bacillus subtilis* and observed a increase in crude protein compared to nonfermented soybean meal.

Wheat bran is a byproduct obtained from the grinding of grains to produce wheat flour. This bran is mainly composed of the outer layers of the wheat grain (Wieser et al., 2020). According to the International Aquaculture Feed Formulation Database, the nutritional composition of wheat bran varies close to 87.3% of dry matter, 17.3% crude protein, 3.9% lipids, 10.4% crude fiber, 5.6% ash, and 50.1% NFE. The carbohydrate present in wheat bran is composed of oligosaccharides, while a large part of the fiber is composed of nonstarch polysaccharides, which are complex carbohydrates (Stevenson et al., 2012). The composition of wheat bran suggests that it is processed (e.g., fermentation and/or microbial respiration) prior to its use in aquaculture systems, so that nutrients are made available more simply in the culture medium. Abiri et al. (2022) demonstrated that the fermentation of complex carbohydrates, such as wheat bran, improves the water quality, growth, and health status of common carp (*Cyprinus carpio* L.) with a biofloc system.

Probiotics are live microbial supplements that have a beneficial effect on the host through the modification of its microbial community, which provides higher efficiency in food use, including strengthening the immune system against diseases, and improving water quality (Verschuere et al., 2000; Goh et al., 2022). Among the various probiotic microorganisms that can be used in the synbiotic, we can highlight: (1) Lactic acid bacteria of the genera *Lactobacillus*,

Pediococcus, Bifidobacterium, and *Enterococcus*, among others; (2) Gram-positive spore-forming bacteria of different species such as *B. subtilis, B. licheniformis, B. pumilus, B. megaterium, B. amyloliquefaciens, B. clausii*, among others; (3) Fungi, such as yeasts of the genus *Saccharomyces* (Bourdichon et al., 2012; Tamang et al., 2016; García-Burgos et al., 2020).

Lactic acid bacteria are important for aquaculture because they produce antimicrobial compounds (bacteriocins, hydrogen peroxide, lactic acid, and organic acids) that inhibit the growth of many Gram-negative pathogenic bacteria and stimulate the host's immune system, generating a nonimmune response (Rhee et al., 2011; Bintsis, 2018; Naiel et al., 2021; Wang et al., 2021). Lactic acid bacteria are considered anaerobes tolerant to the absence and presence of oxygen with fermentative metabolism. The aerobic culture of *Lactobacillus plantarum* alters the central metabolism by redirecting pyruvate, generating an accumulation of acetate. However, in an anaerobic condition, lactate is the main product (Ricciardi et al., 2015; Zotta et al., 2017). In an anaerobic environment, the concentration of lactic acid is 2–3 times higher than the concentration in an aerobic condition at optimum pH (between 5 and 6) (Smetanková et al., 2012). Regarding microbial growth, the culture in anaerobic media provides an increase in the growth rate for *Lactobacillus* compared to aerobic conditions (Smetanková et al., 2012).

Bacillus sp. show higher growth and production of lactic acid when cultured in an aerobic environment, even species that can grow by fermentation and anaerobiosis, in the presence of nitrate, nitrite, or pyruvate (Castells et al., 2022). Similar to lactic acid bacteria, *Bacillus* stand out for their characteristics of reducing and/or inhibiting the proliferation of *Vibrio* spp., through the production of natural antibiotics and competitiveness with pathogens for nutrients (Kuebutornye et al., 2019; Knipe et al., 2021; Jiang et al., 2023).

Among yeasts, the genus *Saccharomyces* is widely used in the production of fermentative processes in the food industry (Wang et al., 2017; del Valle et al., 2023). *Saccharomyces* is an ascomycete and grows in both aerobic and anaerobic environments. In an aerobic environment (respiration), sugar (i.e., glucose, fructose, sucrose, or maltose) is oxidized, resulting in carbon dioxide and water. In an anaerobic environment (fermentation), sugar is oxidized, resulting in ethanol and carbon dioxide, and these processes can occur simultaneously (Padilla et al., 2016; Tronchoni et al., 2022). Furthermore, this yeast genus can reduce pathogenic bacteria in the shrimp and fish digestive tract and stimulate their immune system (del Valle et al., 2023).

Enzymes, also known as biocatalysts, are a biological substance that initiates or accelerates the rate of a biochemical reaction, without being consumed in the reaction. Similar to other chemical catalysts, enzymes are also highly effective in increasing the rate of biochemical reactions that otherwise occur very slowly, or in some cases, not at all (Felix et al., 2018; Kuddus, 2018; Liang et al., 2022). A common example is the breakdown of foods, which mainly include proteins, carbohydrates, and fats, into their basic constituents (Ghosh et al., 2006).

Sugar cane molasses and/or sugar are used as a source of carbohydrate in the system, favoring the growth of probiotic microorganisms quickly (Gao et al., 2012; Emerenciano et al., 2013), and may be absent in the preparation of the synbiotic if enzymes are used. In addition to these inputs, clean and/or treated water is used to minimize the risk of undesirable microorganisms entering the culture, as inputs such as molasses stimulate the growth of bacteria, whether beneficial or pathogenic for the system.

In this way, the synbiotic is produced by mixing all inputs and in one or two phases, 12–24 h anaerobic and/or 12–24 h aerobic (Santos et al., 2022a; Khanjani et al., 2023; Pimentel et al., 2024c). During preparation, temperature and pH conditions directly influence microbial growth (Gauvry et al., 2021). Temperature, in turn, directly influences the growth of microorganisms, since all growth processes are dependent on chemical reactions that are affected by temperature. Temperature has a positive or negative influence on microbial activity, and temperatures ranging between 25 and 30°C favor microbial activity in aquaculture systems (Hlordzi et al., 2020). Regarding the ideal pH for the growth of *Lactobacillus*, it is in the range between 4 and 5, while for the growth of yeast, it can vary between 2.4 and 8.6 (Matthews and Webb, 1991; Zhang et al., 2000), while for *Bacillus* it is between 5 and 8 (Asgher et al., 2007; Chen et al., 2015; Sidorova et al., 2020). It is important to highlight that these ranges must be understood in the production of the synbiotic to achieve success in the fertilization process.

It is known, however, that for the synbiotic preparation processes (fermentation and/or microbial respiration) to occur efficiently, it is necessary to know the environmental conditions that favor the metabolism of microorganisms, since variables such as temperature, aeration, pH and fermentation time has a great influence on microbial growth, production, and enzymatic activity (Potumarthi and Jetty, 2007; Shim et al., 2010; Sugiharto and Ranjitkar, 2019; Dawood and Koshio, 2020).

13.3 Impact on animal health and gut microbiome

13.3.1 Microbial diversity and role

The microbes that inhabit the gut play a key role in animal health, driving various metabolic processes in the host, such as vitamin synthesis, physiological, and immunological responses. Therefore the manipulation of this microbial community is of paramount importance in maintaining the health of shrimp farming (Cornejo-Granados et al., 2017; Holt et al., 2021).

The richness and variety of the gut microbiota are influenced by a number of factors, including diet composition (in particular, the sources of protein and fat, as well as probiotics and prebiotics, among other health promoters), water salinity, light intensity, stocking density, water quality, farming system, physiological conditions of the animals, immunity, and the animal's life stage (Romero et al., 2014; El-Saadony et al., 2022). The shrimp gut microbiota has been shown to comprise several distinct groups. Most of the microbiota is composed of Proteobacteria, Bacteroidetes, and Actinobacteria, while a smaller proportion is attributed to Firmicutes (Huang et al., 2016; Vargas-Albores et al., 2017). According to Das et al. (2023), approximately 65% of the shrimp gut microbiota is made up of Bacteroidetes and Proteobacteria, and only 24% of Firmicutes.

Firmicutes is a phylum of bacteria, most of which have a Gram-positive cell wall. The group is typically divided into three classes: Bacilli, facultative or obligate aerobes, Clostridia, anaerobic organisms, and Mollicutes. The Bacilli class is a class that includes Gram-positive bacteria, such as those of the *Bacillus* genus. The organisms included in this class are aerobic and are divided into the orders Bacillales and Lactobacillales (Harwood, C.R., 1989; Zhang and Cai, 2014). These microorganisms are typically present in the probiotics utilized in the preparation of the bran, which is subsequently employed in the synbiotic process (Pimentel et al., 2024b; Pimentel et al., 2024c; Pimentel, et al., 2024d).

The microbiota can be identified using conventional methods, namely the cultivation of samples of gut material in culture media (Table 13.2). Using these culture media, some studies

TABLE 13.2 Culture media used for microbial growth.

Culture media	Acronym	Microbial growth
Eosin-methylene blue	EMB	This agent is selective for Gram-negative bacteria. Bacteria that ferment lactose in the medium form colonies with a distinctive coloration, whereas those that do not ferment lactose appear as colorless colonies
Macconkey agar	MAC	The presence of bacterial colonies that ferment lactose results in a bright pink coloration of the medium, whereas the absence of such colonies gives rise to a light-yellow hue
Mannitol egg yolk agar polymyxin	MYP	Agar has been devised for the isolation of *Bacillus*
Agar Sabouraud dextrose	Agar SD	Culture medium used for the growth of fungi and yeasts
Thiosulfate citrate bile salts sucrose agar	TCBS	Culture medium used for the growth of *Vibrio* spp.
Tryptic soy agar	TSA	Culture medium used for the growth of total heterotrophic bacteria

Source: Silva et al., 2013; de Andrade et al., 2021; Nunes et al., 2024.
Agar SD, agar Sabouraud dextrose; *EMB*, eosin-methylene blue; *MAC*, MacConkey agar; *MYP*, mannitol egg yolk agar polymyxin; *TCBS*, thiosulfate citrate bile salts sucrose agar; *TSA*, tryptic soy agar.

with a synbiotic system for shrimp and fish demonstrate a higher concentration of probiotic microorganisms in relation to pathogenic ones (Table 13.3).

In addition, other techniques such as fluorescent in situ hybridization (FISH), direct nucleic acid stains such as AU and acridine orange can be used to identify and quantify microorganisms, more generally or specifically (Hobbie et al., 1977; Cottrell and Kirchman, 2003). Recently, however, sequencing technology has demonstrated an enormous capacity to sequence millions of DNA fragments simultaneously, allowing for the instant identification of microbial communities. In addition to high precision in the identification of specific bacteria, there is the possibility of discovering previously unidentified species. The use of metagenomics, with a specific focus on the 16S rRNA genes, allows the detection and complete characterization of the entire bacterial community found in the digestive tract of healthy and diseased shrimp (Tepaamorndech et al., 2020).

In the aquamimicry system, when assessing the dissimilarity of the microbial diversity of water, shrimp gut, and sediment based on 16S rRNA sequencing profiles, they observed that the shrimp gut microbiota is closely related to the formation of the environmental microbiota, which is also related to the occurrence of shrimp diseases. The authors also observed that the abundance of some opportunistic pathogens was relatively low (*Aeromonas*, 0.5; 10.7%; *Phascolarctobacterium*, < 0.01%; *Photobacterium*, 0.4%), while *Vibrio* spp. (10.9%) and *Candidatus Bacilloplasma* (10.4%) were present in the shrimp gut (Zeng et al., 2020)

TABLE 13.3 Identification of bacterial cultures in the gut of shrimp and fish reared in the synbiotic system.

Species	Culture medium used (Concentration found in UFC g^{-1})	References
Penaeus vannamei	TSA (4.36×10^6–7.97×10^8) De Man, Rogosa and Sharpe Agar (MRS) (2.81×10^4–8.14×10^6) Sabouraud Agar (AS) (3.39×10^2–1.02×10^3) TCBS (1.38×10^6–1.12×10^7)	Coutinho et al. (2024)
Penaeus vannamei	MYP (70.2×10^5–186.6×10^5) TCBS (42.1×10^4–233.9×10^4)	Silva et al. (2023)
Oreochromis niloticus	TSA (42.1×10^4–233.9×10^4) SBC (1.33×10^6) MAC (4.05×10^5) EMB (1.85×10^5) MYP (5.50×10^5) TCBS (N/D)	Nunes et al. (2024)

EMB, eosin-methylene blue; *MAC*, MacConkey agar; *MYP*, mannitol egg yolk agar polymyxin; *SBC*: standard bacterial count; *TCBS*, thiosulfate citrate bile salts sucrose agar; *TSA*, tryptic soy agar; *N/D*: not detectable.

13.3.2 Eubiosis, dysbiosis, and rebiosis

The gut microbiome of living organisms is a dynamic and interactive system composed of billions of bacteria. The activity of these microorganisms produces signaling molecules and metabolites that are extremely important for the host, with functions ranging from feed digestion to inducing the development and maturation of organs (Iebba et al., 2016). Therefore it is important to maintain the balance of these microorganisms in the digestive system so that their functions are performed correctly. The balance of the gut microbiome is called eubiosis (Rajeev et al., 2021). However, the composition of a host's microbiota can be influenced by several factors, such as genotype, variations in environmental conditions, exposure to pathogens, and diet (Medina-Félix et al., 2023; Mougin and Joyce, 2023).

Subjecting organisms to stressful factors can destabilize the bacterial community, causing loss of diversity in the digestive system and unbalancing the microbiome. This imbalance is known as dysbiosis and can lead to physiological consequences for animals (Vargas-Albores et al., 2021). Changes in the composition of the bacterial community caused by dysbiosis cause the loss of microorganisms beneficial to the host and open space for the establishment of pathobionts or potentially harmful microorganisms (Infante-Villamil et al., 2021). The inclusion of functional additives in the diet or water can prevent dysbiosis or act to reestablish the animals' gut microbiome.

The reestablishment of a healthy microbiome in a dysbiotic host is called rebiosis (Rajeev et al., 2021). Rebiosis in aquatic organisms can be achieved with several tools, such as the use of prebiotics, probiotics, and synbiotics. These strategies promote the recovery of gut functions and strengthen the animals' immune system (Goh et al., 2022). Prebiotics are used to modulate a host's gut microbiota without the need to introduce live microorganisms. When ingested, they naturally stimulate the proliferation of beneficial microorganisms in the host's digestive tract (Goh et al., 2022). Probiotic microorganisms can modulate the microbiota by colonization of the gut mucosa, adherence and competitive exclusion, production of bacteriocins, and enzymes (Vargas-Albores et al., 2021). Synbiotics are a mixture of prebiotics and probiotics that together can result in better health benefits for hosts than prebiotics and probiotics when used separately (Yilmaz et al., 2022). They can improve the growth and resistance of animals to pathogens such as bacteria of the genus *Vibrio* sp. (Yilmaz et al., 2022).

The powerful influence of probiotic microorganisms in the synbiotic can be directly reflected in the maintenance of eubiosis or act on the rebiosis of the animals being cultured. This happens because the bran particles are colonized by probiotic microorganisms during the processing phases and can be consumed when made available in the system, contributing to the colonization of the animals' digestive tract (Khanjani et al., 2023). Liñan-Vidriales et al. (2021) demonstrated that the use of rice bran fermented with *Bacillus* and *Lysinibacillus* reduced the abundance of potential pathogenic species by modulating the gut microbiota of the shrimp *Penaeus vannamei*.

13.4 Recent research on fish, prawn, and shrimp

To date, little scientific information is available for synbiotics and aquamimicry; however, in recent years, there has been an increase in the number of publications on shrimp, prawn, and fish with this topic. The studies have focused on the different applications of fermented products in production units, types of subproducts, different types of bacteria, low salinity, and artificially salinized water. This research has generated information on the planktonic community, bacterial community in the water, and shrimp/fish, water quality focused on nitrogenous control, proximate composition, and growth performance. However, many gaps still require research to expand knowledge about the system. In addition to research papers, there are a few studies on notes and reviews on the subject (Table 13.4 and Table 13.5).

13.5 Future perspectives

The novelty highlighted by the positive effects of using the synbiotic system on the microbial composition of aquaculture systems and animal growth has paved the way for the development of research with this fertilization strategy. However, efforts are needed to understand the effects of synbiotic fertilization alone on several factors that range from intrinsic fertilizer issues, such as the carbon:nitrogen (C:N) ratio, to the economic feasibility of using this system.

In microbial systems, based on fertilization with organic carbon, control of the C:N ratio is a key element in the functioning of the system. In the synbiotic system, there is no control over the C:N ratio as occurs in the biofloc technology (Khanjani et al., 2023). Despite this, vegetable bran has a C:N ratio, and this is complemented with the addition of the simplest organic carbon source in the fertilizer, such as molasses and sugar. This ratio can also be modified by the fermentation and microbial respiration processes that the fertilizer is submitted prior to its application in the system. Therefore synbiotic fertilization maintains a C:N ratio in the system, and this stoichiometry needs to be studied closely to understand the functioning of

TABLE 13.4 Research with fish, prawn, and shrimp reared with synbiotic and/or aquamimicry systems.

References	Water:substrate ratio	pH buffer	Probiotic strains	Cereal substrate	Anaerobic/aerobic	Paper title	Crustacean/fish species
Kawahigashi (2018)	10:1	Sodium bicarbonate or Calcium carbonate	*Probiotic*	Rice bran	24 h aerobic	Synbiotics	*P. vannamei*
Albuquerque (2019)	15:1	Sodium bicarbonate	*B. subtilis*	Rice bran	24 h aerobic	Aquamimicry	*P. vannamei*
Abdel-Tawwab et al. (2022)	2:1	Sodium bicarbonate	*B. subtilis*	Rice bran	24 h aerobic	FRB - fermented rice bran	*P. vannamei*
de Andrade et al. (2021)	10:1	Sodium bicarbonate	*B. subtilis, B. licheniformis,* and *Bacillus* sp.	Rice bran	24 h anaerobic and 24 h aerobic	Biofloc system with rice bran (anaerobic and aerobic) as	*P. vannamei*
Oliveira et al. (2022)	10:1	Sodium bicarbonate	*B. subtilis, B. licheniformis,* and *Bacillus* sp.	Rice bran	24 h anaerobic and 24 h aerobic	Synbiotic	*P. vannamei*
Pimentel et al. (2022)	10:1	Sodium bicarbonate	*Bacillus subtilis, Bacillus licheniformis, Bacillus*	Rice bran	24 h anaerobic and 24 h aerobic	Synbiotic	*P. vannamei*
Catalani et al. (2023)	25–50:1	—	*Bacillus subtilis, B. licheniformis,*	Rice bran	24 h anaerobic	Aquamimicry	*P. vannamei*
Pimentel et al. (2024b)	10:1	Sodium bicarbonate	*Bacillus subtilis* and *Bacillus licheniformis*	Rice bran, wheat bran, and soybean bran	24 h anaerobic and 24 h aerobic	Synbiotic	*P. vannamei*

(continued on next page)

Pimentel et al. (2024)	10:1	Sodium bicarbonate	*Bacillus subtilis and Bacillus licheniformis*	Rice bran	12 and 24 h anaerobic and 12 and 24 h aerobic	Synbiotic	*P. vannamei*
Pimentel et al. (2024d)	10:1	Sodium bicarbonate	*Bacillus subtilis, Bacillus licheniformis, Lactobacillus plantarum*, and *Pediococcus acidilactici*	Rice bran	12h anaerobic and 12h aerobic	Synbiotic	*P. vannamei*
Coutinho et al. (2024)	10:1	—	*Bacillus subtilis, B. licheniformis, B.pumilus, Lactobacillus plantarum. L. acidophilus*, and *Saccharomyces cerevisiae*	Rice bran	24 h anaerobic and 24 h aerobic	Synbiotic	*P. vannamei*
Gonçalves Junior et al. (2025)	10:1	—	*Bacillus subtilis, B. licheniformis,Lactobacillus plantarum*, and *Pediococcus acidilactici*	Rice bran	24 h anaerobic	Synbiotic	*P. vannamei*
Santos et al. (2022b)	10:1	Sodium bicarbonate	*Bacillus subtilis, B. licheniformis, Lactobacillus acidophilus, B. pumilus e Saccharomyces cerevisiae*	Rice bran	12–24 h anaerobic x 12–24 h aerobic	Synbiotic	*M. rosenbergii*
Santos et al. (2022a)	10:1	Sodium bicarbonate	*B. subtilis, B. licheniformis*, and *Bacillus* sp.	Rice bran	24 h anaerobic and 24 h aerobic	Synbiotic	*M. rosenbergii*
Romano et al. (2018)	2:1	—	*B. megaterium and B. licheniformis*	Rice bran	24 aerobic 24 aerobic	FRB	*Clarias gariepinus*
Nunes et al. (2024)	10:1	Sodium bicarbonate	*Bacillus subtilis, Bacillus licheniformis, Bacillus spp.*	Rice bran	24h anaerobic and 24h aerobic	Synbiotic	*Oreochromis niloticus*

TABLE 13.5 Review articles on synbiotic and/or aquamimicry systems.

Title	Journal	References
Back to the basics: Biomimicry in shrimp farming	*International Journal of Current Microbiology and Applied Sciences*	Chakravarty et al. (2018)
Aquamimicry: New an innovative approach for sustainable development of aquaculture	*Journal of Entomology and Zoology Studies*	Deepak et al. (2020)
A solution for sustainable utilization of aquaculture waste: A comprehensive review of biofloc technology and aquamimicry	*Frontiers in Nutrition*	Nisar et al. (2022)
An applied review of advances in production systems, nutrition and breeding	*Animals*	Emerenciano et al. (2022)
Aquamimicry system: A suitable strategy for shrimp aquaculture	*Annals of Animal Science*	Khanjani et al. (2022)
Synbiotics and aquamimicry as alternative microbial-based approaches in intensive shrimp farming and biofloc: Novel disruptive techniques or complementary management tools? A scientific-based overview	*Aquaculture*	Khanjani et al. (2023)
Aquamimicry improves the growth performance and immunity of black tiger shrimp (*Penaeus monodon*) in low saline ponds	*Aquaculture Reports*	Islam et al. (2024)
Aquamimicry (Copefloc technology): An innovative approach for sustainable organic farming with special reference to shrimp aquaculture	*Blue Biotechnology*	Deepti et al. (2024)
The Fertilization of Synbiotic Systems in the Intensive Culture of *Penaeus vannamei*	*Reviews in Aquaculture*	Pimentel et al. (2025)

this system. Some studies reported the theoretical C:N ratio achieved, which varied between 6.8:1 and 10.7:1 (Silva et al., 2021; Pimentel et al., 2024a). However, it is important to analyze the real carbon and nitrogen content to have higher precision in the nutrient supply, which has a direct impact on the functioning of the system. Pimentel et al. (2022) monitored the stoichiometric ratios of water and microbial aggregates formed throughout a *P. vannamei* synbiotic nursery. The authors reported a C:N ratio close to 2:1 in water and 10:1 in microbial aggregates. Therefore studies focused on fertilizer stoichiometry and seeking to address strategies to manipulate the C:N ratio of fertilizer can refine the fertilization process and nitrogenous compounds control in this system. Furthermore, it can determine the dominance of microorganisms.

The microbial composition grown in the water of a synbiotic system is one of the most important factors, considering that this is a microbial-based system. Studying the composition of the microbiome can reveal patterns of bacterial dominance and clarify the unique effects of fertilization on the balance between heterotrophic and chemoautotrophic bacteria in this system. In the same way, it can provide results on the control of bacteria that can be harmful to culture, such as Cyanobacteria and *Vibrio*. Furthermore, this can be reflected in the composition of the animals' gut microbiome, as they directly ingest the microorganisms present in the water. Focused studies on the isolated effect of different fertilization strategies of the synbiotic system, such as the use of different vegetable brans, inoculation of different probiotic microorganisms, and different fertilizer processing phases, on the composition of the shrimp gut microbiome must be developed. This could clarify the best fertilization strategies for fish and shrimp, which have practical implications on feed use, growth, and health status of the animals.

The health of animals produced in aquaculture is seen as a priority, as this determines whether they can be commercialized and consumed. Therefore maintaining a good immune and health status of fish and shrimp is essential for the success of aquaculture. It is a fact that the use of synbiotics in feed improves the health of animals. However, studies need to be carried out testing the effect of the synbiotic system on the resistance of animals to pathogens. Some studies were carried out with challenges against pathogens and analyzing the load of pathogenic bacteria in animals, but they tested the effect of inoculation of phytoplankton, zooplankton, artificial salinization, and environmental variables (da Silva et al., 2021; de Andrade et al., 2021; Silva et al., 2021; Silva et al., 2023; Júnior et al., 2024).

Finally, once a precise fertilization protocol and the best stocking densities for this system have been established, studies need to be carried out on the economic feasibility of using the synbiotic in shrimp and fish aquaculture. It is worth highlighting all the positive impacts of the synbiotic system, but these results need to be reinforced with the production costs in this system. Furthermore, it is important to compare the feasibility of the synbiotic with other culture systems, such as biofloc technology and recirculation systems.

References

Abdel-Tawwab, M., Khalil, R.H., Nour, A.M., Elkhayat, B.K., Khalifa, E., Abdel-Latif, H.M.R., 2022. Effects of Bacillus subtilis-fermented rice bran on water quality, performance, antioxidants/oxidants, and immunity biomarkers of White leg shrimp (Litopenaeus vannamei) reared at different salinities with zero water exchange. J. Appl. Aquac. 34 (2), 332–357. doi:10.1080/10454438.2020.1844110.

Abiri, S.A., Chitsaz, H., Najdegerami, E.H., Akrami, R., Jalali, A.S., 2022. Influence of wheat and rice bran fermentation on water quality, growth performance, and health status of Common carp (Cyprinus carpio L.) juveniles in a biofloc-based system. Aquaculture 555. doi:10.1016/j.aquaculture.2022.738168.

Albuquerque, L.F.G., 2019. Zootechnical and economic performance of Penaeus vannamei cultured using aquamimicry system (in Portuguese). Universidade Federal do Ceará (UFC), Fortaleza, Brazil, p. 77p.

Aliyu-A, A., Aliyu-Paiko, M., Abafi, J., Abdul-Malik, A., Adamu, K.M., King, M.A., 2019. Fermentation of feed ingredients as potential strategy to improve health status and reduce opportunistic pathogens in fish farming. Asian J. Biotechnol. Bioresour. Technol. 1–17. doi:10.9734/ajb2t/2019/v5i230055.

Amadou, I., Amza, T., Foh, M.B.K., Kamara, M.T., Le, G.W., 2010. Influence of Lactobacillus plantarum Lp6 fermentation on the functional properties of soybean protein meal. Emir. J. Food Agric. 22 (6), 456–465. doi:10.9755/ejfa.v22i6.4663.

Amadou, I., Le, G.W., Shi, Y.H., Gbadamosi, O.S., Kamara, M.T., Jin, S., 2011. Optimized Lactobacillus plantarum Lp6 solid-state fermentation and proteolytic hydrolysis improve some nutritional attributes of soybean protein meal. J. Food Biochem. 35 (6), 1686–1694. doi:10.1111/j.1745-4514.2010.00493.x.

Asgher, M., Asad, M.J., Rahman, S.U., Legge, R.L., 2007. A thermostable α-amylase from a moderately thermophilic Bacillus subtilis strain for starch processing. J. Food Eng. 79 (3), 950–955. doi:10.1016/j.jfoodeng.2005.12.053.

Bintsis, T., 2018. Lactic acid bacteria: their applications in foods. J Bacteriol. Mycol. Open Access 6 (2), 89–94. https://doi.org/10.15406/jbmoa.2018.06.00182.

Bourdichon, F., Casaregola, S., Farrokh, C., Frisvad, J.C., Gerds, M.L., Hammes, W.P., Harnett, J., Huys, G., Laulund, S., Ouwehand, A., Powell, I.B., Prajapati, J.B., Seto, Y., Ter Schure, E., Van Boven, A., Vankerckhoven, V., Zgoda, A., Tuijtelaars, S., Hansen, E.B., 2012. Food fermentations: microorganisms with technological beneficial use. Int. J. Food Microbiol 154 (3), 87–97. doi:10.1016/j.ijfoodmicro.2011.12.030.

Butt, U.D., Lin, N., Akhter, N., Siddiqui, T., Li, S., Wu, B., 2021. Overview of the latest developments in the role of probiotics, prebiotics and synbiotics in shrimp aquaculture. Fish Shellfish Immunol. 114, 263–281. doi:10.1016/j.fsi.2021.05.003.

Castells, A., Leon, A., Sosa, D., Cadena, I., Ramirez, D., Serrano, L., Larrea, F., Almeida-Streitwieser, D., Alvarez-Barreto, J.F., 2022. Evaluation of lactic acid production by different Bacillus Subtilis strains isolated from Theobroma Cacao crops in Ecuador. Chem. Eng. Trans 93, 55–60. doi:10.3303/CET2293010.

Catalani, K., Zuñiga, R., de Souza, M.S., Wasielesky, W., Fóes, G., 2023. Comparison between biofloc technology system and aquamimicry in the cultivation of Litopenaeus vannamei in lined ponds in Southern Brazil. Bol. Inst. Pesca 49. doi:10.20950/1678-2305/bip.2023.49.e787.

Chakravarty, S., Kumar, S., Prakash, S., 2018. Back to the basics: biomimicry in shrimp farming. Int. J. Curr. Microbiol. Appl. Sci. 7 (05), 2172–2184. doi:10.20546/ijcmas.2018.705.253.

Charalampopoulos, D., Wang, R., Pandiella, S.S., Webb, C., 2002. Application of cereals and cereal components in functional foods: a review. Int. J. Food Microbiol. 79 (1–2), 131–141. doi:10.1016/s0168-1605(02)00187-3.

Chen, W.C., Juang, R.S., Wei, Y.H., 2015. Applications of a lipopeptide biosurfactant, surfactin, produced by microorganisms. Biochem. Eng. J. 103, 158–169. doi:10.1016/j.bej.2015.07.009.

Collins, M.D, Gibson, G.R, 1999. Probiotics, prebiotics, and synbiotics: approaches for modulating the microbial ecology of the gut. Am. J. Clin. Nutr. 69 (5), 1052S. doi:10.1093/ajcn/69.5.1052s.

Cornejo-Granados, F., Lopez-Zavala, A.A., Gallardo-Becerra, L., Mendoza-Vargas, A., Sánchez, F., Vichido, R., Brieba, L.G., Viana, M.T., Sotelo-Mundo, R.R., Ochoa-Leyva, A., 2017. Microbiome of Pacific Whiteleg shrimp reveals differential bacterial community composition between wild, aquacultured and AHPND/EMS outbreak conditions. Sci. Rep. 7 (1). doi:10.1038/s41598-017-11805-w.

Cottrell, M.T., Kirchman, D.L., 2003. Contribution of major bacterial groups to bacterial biomass production (thymidine and leucine incorporation) in the Delaware estuary. Limnol. Oceanogr. 48 (1 I), 168–178. doi:10.4319/lo.2003.48.1.0168.

Coutinho, A.G., Viana, J.T., Angelim, A.L., Pereira, A.M.L., de Menezes, F.G.R., Gálvez, A.O., Brito, L.O., de Holanda Cavalcante, D., Feijó, R.G., 2024. Evaluation of the stocking density of postlarvae of Penaeus vannamei in synbiotic nursery system. Aquac. Int. 32 (7), 8861–8875. doi:10.1007/s10499-024-01595-0.

Das, S.P., Abidin, Z., Huang, H.T., Lin, Y.R., Huang, C.Y., Wu, Y.S., Hu, Y.F., Nan, F.H., 2023. Deciphering the influence of dietary synbiotics in white shrimp gut and its effects in regulating immune signaling pathways. Front. Mar. Sci. 10. doi:10.3389/fmars.2023.1342708.

da Silva, A.E.M., Brito, L.O., da Silva, D.A., de Lima, P.C.M., da Silva Farias, R., Gálvez, A.O., da Silva, S.M.B.C., 2021. Effect of Brachionus plicatilis and Navicula sp. on Pacific white shrimp growth performance, Vibrio, immunological responses and resistance to white spot virus (WSSV) in nursery biofloc system. Aquaculture 535. doi:10.1016/j.aquaculture.2020.736335.

Dawood, M.A.O., Koshio, S., 2020. Application of fermentation strategy in aquafeed for sustainable aquaculture. Rev. Aquac. 12 (2), 987–1002. doi:10.1111/raq.12368.

de Andrade, R.J.V., dos Santos, E.P., de Almeida Costa, G.K., Campos, C.V.F.d.S., da Silva, S.M.B.C., Gálvez, A.O., Brito, L.O., 2021. Effect of different frequencies of the addition of Brachionus plicatilis on the performance of Litopenaeus vannamei in a nursery biofloc system with rice bran (anaerobic and aerobic) as an organic carbon source. Aquaculture 540. doi:10.1016/j.aquaculture.2021.736669.

Deepak, A.P., Vasava, R.J., Elchelwar, V.R., Tandel, D.H., Vadher, K.H., Shrivastava, V., Prabhakar, P., 2020. Aquamimicry: new an innovative approach for sustainable development of aquaculture. J. Entomol. Zool. Stud. 8 (2), 1029–1031.

Deepti, M., Meinam, M., Devi, N.gC., Singh, S.K, Devi, W.M, 2024. Aquamimicry (Copefloc technology): an innovative approach for sustainable organic farming with special reference to shrimp aquaculture. Blue Biotechnol. 1 (1). doi:10.1186/s44315-024-00005-7.

del Valle, J.C., Bonadero, M.C., Fernández-Gimenez, A.V., 2023. Saccharomyces cerevisiae as probiotic, prebiotic, synbiotic, postbiotics and parabiotics in aquaculture: an overview. Aquaculture 569. doi:10.1016/j.aquaculture.2023.739342.

do Prado, F.G., Pagnoncelli, M.G.B., de Melo Pereira, G.V., Karp, S.G., Soccol, C.R., 2022. Fermented soy products and their potential health benefits: a review. Microorganisms. 10 (8). doi:10.3390/microorganisms10081606.

El-Dakar, A.Y., Elgamal, A.A., Baky Amer, M.A., Mohammed, A.S., Abdel-Aziz, M.F., 2023. Evaluation of fermented soybean meal by Bacillus subtilis as an alternative to fishmeal on the growth, and physiological status of Nile tilapia Oreochromis niloticus fingerlings. Heliyon 9 (9). doi:10.1016/j.heliyon.2023.e19602.

Elhalis, H., Chin, X.H., Chow, Y., 2024. Soybean fermentation: microbial ecology and starter culture technology. Crit. Rev. Food. Sci. Nutr. 64 (21), 7648–7670. doi:10.1080/10408398.2023.2188951.

El-Saadony, M.T., Shehata, A.M., Alagawany, M., Abdel-Moneim, A.M.E., Selim, D.A., Abdo, M., Khafaga, A.F., El-Tarabily, K.A., El-Shall, N.A., Abd El-Hack, M.E., 2022. A review of shrimp aquaculture and factors affecting the gut microbiome. Aquac. Int. 30 (6), 2847–2869. doi:10.1007/s10499-022-00936-1.

Emerenciano, M., Gaxiola, G., Cuzon, G., 2013. Biofloc technology (BFT): a review for aquaculture application and animal food industry. In: Biomass now – cultivation and utilization. InTech, Rijeka, Croatia, pp. 301–328.

Emerenciano, M.G.C., Rombenso, A.N., Vieira, F.D.N., Martins, M.A., Coman, G.J., Truong, H.H., Noble, T.H., Simon, C.J., 2022. Intensification of penaeid shrimp culture: an applied review of advances in production systems, nutrition and breeding. Animals 12 (3). doi:10.3390/ani12030236.

FAO, 2022. The State of World Fisheries and Aquaculture 2022: Towards Blue Transformation. Food and Agriculture Organization of the United Nations., Rome, FAO.

FAO, 2024. The State of World Fisheries and Aquaculture 2024: Blue Transformation in action. Food and Agriculture Organization of the United Nations, Rome.

FAO, 2026. The State of World Fisheries and Aquaculture 2026 – Blue Transformation: turning vision into impact. Rome. https://doi.org/10.4060/cd8357en.

Felix, N., Prabu, E., Kannan, B., Manikandan, K., 2018. An evidential review on potential benefits of enzymes in aqua feed industry. Int. J. Curr. Microbiol. Appl. Sci. 7 (12), 2053–2074. doi:10.20546/ijcmas.2018.712.236.

Flesch, A.G.T., Poziomyck, A.K., Damin, D.C., 2014. The therapeutic use of symbiotics. ABCD Arq. Bras. Cir. Dig. 27 (3), 206–209. https://doi.org/10.1590/S0102-67202014000300012.

Gao, L., Shan, H.W., Zhang, T.W., Bao, W.Y., Ma, S., 2012. Effects of carbohydrate addition on Litopenaeus vannamei intensive culture in a zero-water exchange system. Aquaculture 342–343 (1), 89–96. doi:10.1016/j.aquaculture.2012.02.022.

García-Burgos, M., Moreno-Fernández, J., Alférez, M.J.M., Díaz-Castro, J., López-Aliaga, I., 2020. New perspectives in fermented dairy products and their health relevance. J. Funct. Foods 72, 104059. doi:10.1016/j.jff.2020.104059.

Gauvry, E., Mathot, A.-G., Couvert, O., Leguérinel, I., Coroller, L., 2021. Effects of temperature, pH and water activity on the growth and the sporulation abilities of Bacillus subtilis BSB1. Int. J. Food Microbiol. 337, 108915. doi:10.1016/j.ijfoodmicro.2020.108915.

Ghosh, K., Mukhopadhyay, P.K., 2006. Application of enzymes in aqua feeds. Aqua Feeds Formul. Beyond 3, 7–10.

Goh, J.X.H., Tan, L.T.H., Law, J.W.F., Ser, H.L., Khaw, K.Y., Letchumanan, V., Lee, L.H., Goh, B.H., 2022. Harnessing the potentialities of probiotics, prebiotics, synbiotics, paraprobiotics, and postbiotics for shrimp farming. Rev. Aquac. 14 (3), 1478–1557. doi:10.1111/raq.12659.

Gonçalves Junior, G.F., Wasielesky, W., Cardozo, A., Poersch, L.H.S., Brito, L.O., Krummenauer, D., Fóes, G.K., 2025. Effect of feed restriction for Pacific white shrimp Penaeus vannamei in a semi-intensive synbiotic system: plankton community, growth and economics. Aquaculture 595. doi:10.1016/j.aquaculture.2024.741481.

Halver, J.E., Hardy, R.W., 2002. Fish nutrition. Academic press, San Diego, USA, p. 824.

Harwood, C.R., 1989. Bacillus. In: Biotechnology Handbooks., Vol. 2. Springer, New York, USA, p. 414.

Hlordzi, V., Kuebutornye, F.K.A., Afriyie, G., Abarike, E.D., Lu, Y., Chi, S., Anokyewaa, M.A., 2020. The use of Bacillus species in maintenance of water quality in aquaculture: a review. Aquac. Rep. 18. doi:10.1016/j.aqrep.2020.100503.

Hobbie, J.E., Daley, R.J., Jasper, S., 1977. Use of nuclepore filters for counting bacteria by fluorescence microscopy. Appl. Environ. Microbiol. 33 (5), 1225–1228. doi:10.1128/aem.33.5.1225-1228.1977.

Holt, C.C., Bass, D., Stentiford, G.D., van der Giezen, M., 2021. Understanding the role of the shrimp gut microbiome in health and disease. J. Invertebr. Pathol. 186. doi:10.1016/j.jip.2020.107387.

Huang, Z., Li, X., Wang, L., Shao, Z., 2016. Changes in the intestinal bacterial community during the growth of white shrimp, Litopenaeus vannamei. Aquac. Res. 47 (6), 1737–1746. doi:10.1111/are.12628.

Hussain, A.S., Mohammad, D.A., Sallam, W.S., Shoukry, N.M., Davis, D.A., 2021. Effects of culturing the Pacific white shrimp Penaeus vannamei in "biofloc" vs "synbiotic" systems on the growth and immune system. Aquaculture 542. doi:10.1016/j.aquaculture.2021.736905.

Iebba, V., Totino, V., Gagliardi, A., Santangelo, F., Cacciotti, F., Trancassini, M., Mancini, C., Cicerone, C., Corazziari, E., Pantanella, F., Schippa, S., 2016. Eubiosis and dysbiosis: the two sides of the microbiota. New Microbiol. 39 (1), 1–12. http://www.newmicrobiologica.org/PUB/allegati_pdf/2016/1/1.pdf.

Infante-Villamil, S., Huerlimann, R., Jerry, D.R., 2021. Microbiome diversity and dysbiosis in aquaculture. Rev. Aquac. 13 (2), 1077–1096. doi:10.1111/raq.12513.

Islam, M.d.S., Das, M., Chakroborty, K., Lee, J.M, Islam, M.d.R., Rafiquzzaman, S.M., 2024. Aquamimicry improves the growth performance and immunity of black tiger shrimp (Penaeus monodon) in low saline ponds. Aquac. Rep. 36, 102082. doi:10.1016/j.aqrep.2024.102082.

Jannathulla, R., Dayal, J.S., Ambasankar, K., Muralidhar, M., 2018. Effect of Aspergillus niger fermented soybean meal and sunflower oil cake on growth, carcass composition and haemolymph indices in Penaeus vannamei Boone, 1931. Aquaculture 486, 1–8. doi:10.1016/j.aquaculture.2017.12.005.

Jiang, N., Hong, B., Luo, K., Li, Y., Fu, H., Wang, J., 2023. Isolation of Bacillus subtilis and Bacillus pumilus with anti-Vibrio parahaemolyticus activity and identification of the anti-Vibrio parahaemolyticus substance. Microorganisms 11 (7), 1667. doi:10.3390/microorganisms11071667.

Júnior, ÁC.S, Ferreira, I.FA.A, Costa, G.KA., Michereff, G.S, Silva, L.OB., Santos, F.L, Silva, S.MB.C, 2024. Evaluating the effect of temperature on Penaeus vannamei culture in a heterotrophic system and experimentally infected with the infectious myonecrosis virus. Aquaculture 580, 740348. doi:10.1016/j.aquaculture.2023.740348.

Kawahigashi, D., 2018. Synbiotics as a management tool for improving nursery efficiency. Hatchery Feed, 6 (3)), 36–39.

Khanjani, M.H., da Silva, L.O.B., Fóes, G.K., Vieira, F.d.N., Poli, M.A., Santos, M., Emerenciano, M.G.C., 2023. Synbiotics and aquamimicry as alternative microbial-based approaches in intensive shrimp farming and biofloc: novel disruptive techniques or complementary management tools? A scientific-based overview. Aquaculture 567. doi:10.1016/j.aquaculture.2023.739273.

Khanjani, M.H., Mozanzadeh, M.T., Fóes, G.K., 2022. Aquamimicry system: a suitable strategy for shrimp aquaculture - a review. Ann. Anim. Sci. 22 (4), 1201–1210. doi:10.2478/aoas-2022-0044.

Knipe, H., Temperton, B., Lange, A., Bass, D., Tyler, C.R., 2021. Probiotics and competitive exclusion of pathogens in shrimp aquaculture. Rev. Aquac. 13 (1), 324–352. doi:10.1111/raq.12477.

Kuddus, M., 2018. Introduction to food enzymes enzymes in food biotechnology: production, applications, and future prospects. Elsevier, Saudi Arabia, pp. 1–18. doi:10.1016/B978-0-12-813280-7.00001-3.

Kuebutornye, F.K.A., Abarike, E.D., Lu, Y., 2019. A review on the application of Bacillus as probiotics in aquaculture. Fish Shellfish Immunol. 87, 820–828. doi:10.1016/j.fsi.2019.02.010.

Liang, Q., Yuan, M., Xu, L., Lio, E., Zhang, F., Mou, H., Secundo, F., 2022. Application of enzymes as a feed additive in aquaculture. Mar. Life Sci. Technol. 4 (2), 208–221. doi:10.1007/s42995-022-00128-z.

Liñan-Vidriales, M.A., Peña-Rodríguez, A., Tovar-Ramírez, D., Elizondo-González, R., Barajas-Sandoval, D.R., Ponce-Gracía, E.I., Rodríguez-Jaramillo, C., Balcázar, J.L., Quiroz-Guzmán, E., 2021. Effect of rice bran fermented with Bacillus and Lysinibacillus species on dynamic microbial activity of Pacific white shrimp (Penaeus vannamei). Aquaculture 531. doi:10.1016/j.aquaculture.2020.735958.

Matthews, T.M., Webb, C., 1991. Culture systems. In: Tuite, M.F., Oliver, S.G. (Eds.). In: Saccharomyces. Biotechnology handbooks, 4. Springer, Boston, MA, pp. 163–191.

Medina-Félix, D., Garibay-Valdez, E.ía, Vargas-Albores, F., Martínez-Porchas, M., 2023. Fish disease and intestinal microbiota: a close and indivisible relationship. Rev. Aquac. 15 (2), 820–839. doi:10.1111/raq.12762.

Mougin, J., Joyce, A., 2023. Fish disease prevention via microbial dysbiosis-associated biomarkers in aquaculture. Rev. Aquac. 15 (2), 579–594. doi:10.1111/raq.12745.

Mukherjee, R., Chakraborty, R., Dutta, A., 2016. Role of fermentation in improving nutritional quality of soybean meal — a review. Asian-Australas. J. Anim. Sci. 29 (11), 1523–1529. doi:10.5713/ajas.15.0627.

Naiel, M.A.E., Farag, M.R., Gewida, A.G.A., Elnakeeb, M.A., Amer, M.S., Alagawany, M., 2021. Using lactic acid bacteria as an immunostimulants in cultured shrimp with special reference to Lactobacillus spp. Aquac. Int. 29 (1), 219–231. doi:10.1007/s10499-020-00620-2.

Naylor, R.L., Hardy, R.W., Buschmann, A.H., Bush, S.R., Cao, L., Klinger, D.H., Little, D.C., Lubchenco, J., Shumway, S.E., Troell, M., 2021. A 20-year retrospective review of global aquaculture. Nature 591 (7851), 551–563. doi:10.1038/s41586-021-03308-6.

Nisar, U., Peng, D., Mu, Y., Sun, Y., 2022. A solution for sustainable utilization of aquaculture waste: a comprehensive review of biofloc technology and aquamimicry. Front. Nutr. 8. doi:10.3389/fnut.2021.791738.

Nunes, L.J.L., da Silva Campos, C.V.F., da Silva, S.M.B.C., Gálvez, A.O., Brito, L.O., dos Santos, J.F., 2024. The culture of Nile tilapia (Oreochromis niloticus) juvenile at different culture technologies: autotrophic, bioflocs and synbiotic. Aquaculture 588. doi:10.1016/j.aquaculture.2024.740912.

Oliveira, C.R.d.R., Oliveira, V.Q., Pimentel, O.A.L.F., dos Santos, E.P., de Oliveira Filho, P.R.C., Gálvez, A.O., Brito, L.O., 2022. Growth performance and proximate composition of Penaeus vannamei reared in low-salinity water with different ionic compositions in a synbiotic system. Aquac. Int. 30 (6), 3123–3141. doi:10.1007/s10499-022-00952-1.

Padilla, B., Gil, J.V., Manzanares, P., 2016. Past and future of non-Saccharomyces yeasts: from spoilage microorganisms to biotechnological tools for improving wine aroma complexity. Front Microbiol 7. doi:10.3389/fmicb.2016.00411.

Pimentel, O.A.L.F., de Oliveira, V.Q., do Rêgo Oliveira, C.R., dos Santos, E.P., Severi, W., Pasco, J.M., Emerenciano, M.G.C., Gálvez, A.O., Brito, L.O., 2024a. Nitrogenous compounds and alkalinity patterns in Penaeus vannamei nurseries and pre-grow-out with low salinity water and synbiotic system: a case study. Aquac. Int. 32 (2), 1703–1718. doi:10.1007/s10499-023-01237-x.

Pimentel, O.A.L.F., de Oliveira, V.Q., Oliveira, C.R.d.R., Severi, W., Olivera Gálvez, A., Amado, A.M., Brito, L.O., 2022. Assessment of different ionic adjustment strategies in low-salinity water on the growth of Litopenaeus vannamei and microbial community stoichiometry in a synbiotic nursery system. Aquac. Res. 53 (1), 50–62. doi:10.1111/are.15552.

Pimentel, O.A.L.F., Wasielesky Jr., W., Brito, L.O., Roy, L.A., Romano, N., Krummenauer, D., 2025. The Fertilization of Synbiotic Systems in the. Intensive Culture of Penaeus vannamei. Rev Aquac 17, e13011. https://doi.org/10.1111/raq.13011.

Pimentel, O.A.L.F., Wasielesky, W., da Silva, N.P., do Valle Borges, L., Krummenauer, D., 2024b. Fertilizing synbiotic system with different vegetable brans: effects on nitrification, plankton composition, and growth of Penaeus vannamei in the nursery phase. Aquac. Int. 32 (5), 6407–6429. doi:10.1007/s10499-024-01471-x.

Pimentel, O.A.L.F., Wasielesky, W., Poersch, L.H., Krummenauer, D., 2024c. Effect of different synbiotic fertilizer processing strategies in Penaeus vannamei intensive nurseries. Aquaculture 584. doi:10.1016/j.aquaculture.2024.740667.

Pimentel, O.A.L.F., Wasielesky, W., Sena, R.P.O., Ramiro, B.d.O., Bezerra, A., Krummenauer, D., 2024d. Microbial community composition, nitrification process, and growth of Penaeus vannamei in a synbiotic nursery system inoculated with different probiotic microorganisms. Aquaculture 592. doi:10.1016/j.aquaculture.2024.741254.

Potumarthi, R., Ch, S., Jetty, A., 2007. Alkaline protease production by submerged fermentation in stirred tank reactor using Bacillus licheniformis NCIM-2042: Effect of aeration and agitation regimes. Biochem. Eng. J. 34 (2), 185–192. doi:10.1016/j.bej.2006.12.003.

Rajeev, R., Adithya, K.K., Kiran, G.S., Selvin, J., 2021. Healthy microbiome: a key to successful and sustainable shrimp aquaculture. Rev. Aquac. 13 (1), 238–258. doi:10.1111/raq.12471.

Rhee, S., Lee, J.-E., Lee, C.-H., 2011. Importance of lactic acid bacteria in Asian fermented foods. Microb. Cell Fact. 10 (Suppl 1), S5. doi:10.1186/1475-2859-10-S1-S5.

Ribeiro, A.C., Graça, C.S., Chiattoni, L.M., Massarolo, K.C., Duarte, F.A., Mellado, M.L.S., Soares, L.A.S.S., 2017. Fermentation process in the availability of nutrients in rice bran. Res. Rev.: J. Microbiol. Biotechnol. 6 (2), 45–52.

Ricciardi, A., Castiglione Morelli, M.A., Ianniello, R.G., Parente, E., Zotta, T., 2015. Metabolic profiling and stress response of anaerobic and respiratory cultures of Lactobacillus plantarum C17 grown in a chemically defined medium. Ann. Microbiol. 65 (3), 1639–1648. doi:10.1007/s13213-014-1003-z.

Romano, N., 2017. Aquamimicry: a revolutionary concept for shrimp farming. Global aquaculture advocate, Portsmouth, NH. https://www.globalseafood.org/advocate/aquamimicry-a-revolutionary-concept-for-shrimp-farming/.

Romano, N., Dauda, A.B., Ikhsan, N., Karim, M., Kamarudin, M.S., 2018. Fermenting rice bran as a carbon source for biofloc technology improved the water quality, growth, feeding efficiencies, and biochemical composition of African catfish Clarias gariepinus juveniles. Aquac. Res. 49 (12), 3691–3701. doi:10.1111/are.13837.

Romero, J., Ringø, E., Merrifield, D.L., 2014. The gut microbiota of fish aquaculture nutrition: gut health, probiotics and prebiotics. Wiley, Chile, pp. 75–100. doi:10.1002/9781118897263.ch4.

Santos, R.B., Coelho-Filho, P.A., Assunção, C.S., Santos, T.N., Silva, J.H., Silva, G.C., Brito, L.O., 2022b. The effect of different synbiotic preparation strategies on water fertilization and zootechnical performance of *Macrobrachium rosenbergii* reared in the nursery stage. Aquac. Int. 30, 3159–3178. https://doi.org/10.1007/s10499-022-00955-y.

Santos, R.B, Filho, P.A.C, Gonçalves, A.P, Santos, R.A, Rodrigues, M.L, Correia, E.S, Oliveira, V.Q, Brito, L.O, 2022a. Effects of organic carbon sources on water quality, microbial flocs protein and performance of Macrobrachium rosenbergii post-larvae reared in biofloc and synbiotic systems. Aquac. Res. 53 (2), 388–397. doi:10.1111/are.15580.

Sharawy, Z., Goda, A.M.A.S., Hassaan, M.S., 2016. Partial or total replacement of fish meal by solid state fermented soybean meal with Saccharomyces cerevisiae in diets for Indian prawn shrimp, Fenneropenaeus indicus, postlarvae. Anim. Feed Sci. Technol. 212, 90–99. doi:10.1016/j.anifeedsci.2015.12.009.

Shim, Y.H., Shinde, P.L., Choi, J.Y., Kim, J.S., Seo, D.K., Pak, J.I., Chae, B.J., Kwon, I.K., 2010. Evaluation of multi-microbial probiotics produced by submerged liquid and solid substrate fermentation methods in broilers. Asian-Australas. J. Anim. Sci. 23 (4), 521–529. doi:10.5713/ajas.2010.90446.

Sidorova, T.M., Asaturova, A.M., Homyak, A.I., Zhevnova, N.A., Shternshis, M.V., Tomashevich, N.S., 2020. Optimization of laboratory cultivation conditions for the synthesis of antifungal metabolites by bacillus subtilis strains. Saudi J. Biol. Sci. 27 (7), 1879–1885. doi:10.1016/j.sjbs.2020.05.002.

Silva, D.A., de Lima, P.C.M., da Silva, A.E.M., de Oliveira Filho, P.R.C., da Silva, S.M.B.C., Olivera Gálvez, A., Brito, L.O., 2021. Effects of adding rotifers on the water quality, plankton composition and growth of Pacific white shrimp, Litopenaeus vannamei juvenile, when cultured with biofloc technology. Aquac. Res. 52 (9), 4380–4393. doi:10.1111/are.15276.

Silva, G.C., Limeira, A.C., de Almeida Costa, G.K., da Silva, S.M.B.C., de Oliveira Filho, P.R.C., Brito, L.O., 2023. Effects of different forms of artificially salinized in low-salinity water of Penaeus vannamei in the grow-out phase in a synbiotic system. Aquac. Int. 31 (3), 1303–1324. doi:10.1007/s10499-022-01025-z.

Silva, N., Taniwaki, M.H., Junqueira, V.C.A., Silveira, N.F.A., Okazaki, M.M., Gomes, R.A.R., 2013. Microbiological examination methods of food and water: a laboratory manual. CRC Press/Balkema, Taylor & Francis Group.

Smetanková, J., Hladíková, Z., Valach, F., Zimanová, M., Kohajdová, Z., Greif, G., Greifová, M., 2012. Influence of aerobic and anaerobic conditions on the growth and metabolism of selected strains of Lactobacillus plantarum. Acta Chim. Slovaca. 5 (2), 204–210. doi:10.2478/v10188-012-0031-1.

Stevenson, L., Phillips, F., O'sullivan, K., Walton, J., 2012. Wheat bran: its composition and benefits to health, a European perspective. Int. J. Food Sci. Nutr. 63 (8), 1001–1013. doi:10.3109/09637486.2012.687366.

Sugiharto, S., Ranjitkar, S., 2019. Recent advances in fermented feeds towards improved broiler chicken performance, gastrointestinal tract microecology and immune responses: a review. Anim. Nutrit. 5 (1), 1–10. doi:10.1016/j.aninu.2018.11.001.

Sumaila, U.R., Pierruci, A., Oyinlola, M.A., Cannas, R., Froese, R., Glaser, S., Jacquet, J., Kaiser, B.A., Issifu, I., Micheli, F., Naylor, R., Pauly, D., 2022. Aquaculture over-optimism? Front. Mar. Sci. 9, 984354. https://doi.org/10.3389/fmars.2022.984354.

Tamang, J.P., Watanabe, K., Holzapfel, W.H., 2016. Review: diversity of microorganisms in global fermented foods and beverages. Front. Microbiol. 7. doi:10.3389/fmicb.2016.00377.

Tepaamorndech, S., Nookaew, I., Higdon, S.M., Santiyanont, P., Phromson, M., Chantarasakha, K., Mhuantong, W., Plengvidhya, V., Visessanguan, W., 2020. Metagenomics in bioflocs and their effects on gut microbiome and immune responses in Pacific white shrimp. Fish Shellfish Immunol. 106, 733–741. doi:10.1016/j.fsi.2020.08.042.

Tronchoni, J., Gonzalez, R., Guindal, A.M., Calleja, E., Morales, P., 2022. Exploring the suitability of Saccharomyces cerevisiae strains for winemaking under aerobic conditions. Food Microbiol. 101, 103893. doi:10.1016/j.fm.2021.103893.

UN-United Nations, 2022. Our growing population. https://www.un.org/en/global-issues/population.

Vargas-Albores, F., Porchas-Cornejo, M.A., Martínez-Porchas, M., Villalpando-Canchola, E., Gollas-Galván, T., Martínez-Córdova, L.R., 2017. Bacterial biota of shrimp intestine is significantly modified by the use of a probiotic mixture: a high throughput sequencing approach. Helgol. Mar. Res. 71 (1). doi:10.1186/s10152-017-0485-z.

Vargas-Albores, F., Rafael Martínez-Córdova, L., Hernández-Mendoza, A., Cicala, F., Lago-Lestón, A., Martínez-Porchas, M., 2021. Therapeutic modulation of fish gut microbiota, a feasible strategy for aquaculture? Aquaculture 544, 737050. doi:10.1016/j.aquaculture.2021.737050.

Vassileva, M., Malusà, E., Sas-Paszt, L., Trzcinski, P., Galvez, A., Flor-Peregrin, E., Shilev, S., Canfora, L., Mocali, S., Vassilev, N., 2021. Fermentation strategies to improve soil bio-inoculant production and quality. Microorganisms 9 (6), 1254. doi:10.3390/microorganisms9061254.

Verschuere, L., Rombaut, G., Sorgeloos, P., Verstraete, W., 2000. Probiotic bacteria as biological control agents in aquaculture. Microbiol. Mol. Biol. Rev. 64 (4), 655–671. doi:10.1128/mmbr.64.4.655-671.2000.

Wang, G., Huang, M., Nielsen, J., 2017. Exploring the potential of Saccharomyces cerevisiae for biopharmaceutical protein production. Curr. Opin. Biotechnol. 48, 77–84. doi:10.1016/j.copbio.2017.03.017.

Wang, L., Zhou, H., He, R., Xu, W., Mai, K., He, G., 2016. Effects of soybean meal fermentation by Lactobacillus plantarum P8 on growth, immune responses, and intestinal morphology in juvenile turbot (Scophthalmus maximus L.). Aquaculture 464, 87–94. doi:10.1016/j.aquaculture.2016.06.026.

Wang, Y., Wu, J., Lv, M., Shao, Z., Hungwe, M., Wang, J., Bai, X., Xie, J., Wang, Y., Geng, W., 2021. Metabolism characteristics of lactic acid bacteria and the expanding applications in food industry. Front. Bioeng. Biotechnol. 9. doi:10.3389/fbioe.2021.612285.

Wieser, H., Koehler, P., Scherf, K.A., 2020. Chapter 5 – Wheat-based raw materials. In: Wheat: an exceptional crop—botanical features, chemistry, utilization, nutritional and health aspects. Woodhead Publishing, Duxford, pp. 103–131.

Yilmaz, S., Yilmaz, E., Dawood, M.A.O., Ringø, E., Ahmadifar, E., Abdel-Latif, H.M.R., 2022. Probiotics, prebiotics, and synbiotics used to control vibriosis in fish: a review. Aquaculture 547. doi:10.1016/j.aquaculture.2021.737514.

Zeng, S., Khoruamkid, S., Kongpakdee, W., Wei, D., Yu, L., Wang, H., Deng, Z., Weng, S., Huang, Z., He, J., Satapornvanit, K., 2020. Dissimilarity of microbial diversity of pond water, shrimp intestine and sediment in aquamimicry system. AMB Express 10 (1). doi:10.1186/s13568-020-01119-y.

Zhang, H., Cai, Y., 2014. Lactic acid bacteria: fundamentals and Practice. Springer, Netherlands, p. viii + 535. doi:10.1007/978-94-017-8841-0.

Zhang, J.G., Cai, Y., Kobayashi, R., Kumai, S., 2000. Characteristics of lactic acid bacteria isolated from forage crops and their effects on silage fermentation. J. Sci. Food Agric. 80 (10), 1455–1460. doi:10.1002/1097-0010(200008)80:10⟨1455.

Zotta, T., Parente, E., Ricciardi, A., 2017. Aerobic metabolism in the genus Lactobacillus: impact on stress response and potential applications in the food industry. J. Appl. Microbiol. 122 (4), 857–869. doi:10.1111/jam.13399.

CHAPTER

14

Industrial waste utilization in aquaculture

Sudhanshu Raman[a], Dorothy MS[b,], Ganesh Kumar[a], Girija Saurabh Behere[c], D.K. Meena[a] and Tanuj Misra[a]*

[a]College of Fisheries, Datia, Rani Lakshmi Bai Central Agricultural University, Jhansi, Uttar Pradesh, India [b]National Fisheries Development Board-North Eastern Regional Center, Guwahati, Assam, India [c]ICAR-Central Inland Fisheries Research Institute, Barrackpore, Kolkata, West Bengal, India

***Corresponding author: dorothy.nfdb@gmail.com**

14.1 Introduction

Scientific developments, urbanization, and technological advancement have brought revolutions in industrialization. The rapid growth of industries is coupled with a colossal amount of waste, which often ends up in landfills, incineration mounts, or dumped into open water bodies, thereby severely polluting the environment. However, some of these wastes can be effectively recycled and or reused for other alternate beneficial purposes. Jana (1998) rightly defined waste as "resources out of place." Safe treatment and disposal of industrial byproducts and waste are expensive, whereas untreated dumping is a huge risk to the environment. Efficient utilization of byproducts has a direct impact on the economy and environmental pollution (Irshad and Sharma, 2015). Some of the byproducts from industries, especially the liquid and solid waste, are loaded with beneficial nutrients and microorganisms that have scope for further use in food production after some treatment processes. One of such is in aquaculture industries where the industrial wastewater, semifluids, and solid extracts can be used in fish culture and as fish feed, and or act as important ingredients in aquafeed industries.

Zero-Waste Technologies in Aquaculture: Global Perspectives and Methods toward a Circular Economy
DOI: https://doi.org/10.1016/B978-0-443-34689-7.00012-4

14.2 Industrial waste

Industrial waste is generated from industrial plants during their various processes, such as the extraction of raw materials, processing to intermediate or final products, consumption of final products, and other human activities (Ramachandra, 2006). According to Woodard (2001), industrial waste can be categorized into three groups: (1) Liquid waste, (2) solid waste, and (3) gas/air pollutants.

The comprehensive classification for the aforementioned categories is depicted in figure 2.1. Liquid waste or wastewater are the aqueous discards that result from the use of water in industrial manufacturing processes or the cleaning activities that take place along with that process. Solid waste includes office trash, packaging waste, lunchroom waste, and or manufacturing or processing waste (nonhazardous) that often ends up in trash cans and municipal landfills, whereas the air pollutants are various gases that are released from industries. There are various studies on the utilization of solid and liquid industrial waste in aquaculture, whereas there is no record on the potential use of industrial released gases in aquaculture.

14.3 Application of industrial waste in aquaculture

Fish have been playing a major role in providing nutrient-rich food to a vast population across the globe. According to Allison (2011), fish is the most traded animal protein in the world and is one of the most traded food commodities. With the rapid growth of population, there is also a rapid rise in demand for fish, not only for human consumption alone, but also for various other proposes inclosing animal feed industries. While the capture fisheries have various growth limitations, there is huge pressure on aquaculture to increase fish production to bridge the demand–supply gap. Population growth, urbanization and industrialization have limited various resources for food production while at the same time produce huge wastage on the other hand. Various liquid waste/wastewater and solid waste from industries can be utilized in the aquaculture sector directly or indirectly. Aquaculture using industrial wastewater is proven to promote socioeconomic and environmental benefits while at the same time making a substantial contribution to fish production. Further, the usage of various industrial solid wastes, especially agroindustrial waste, in aquaculture is well documented.

14.3.1 Liquid waste

Industrial wastewater is the aqueous discard that results from the use of water in an industrial manufacturing process or the cleaning activities that take place along with that process (Woodard, 2001). They are often foul-smelling fluids or semifluid with a high load of organic and inorganic solids in dissolved and suspended forms, which are often disposed directly into natural water bodies, posing huge environmental risks (Jana, 1998; Kumar et al., 2014). Plant clean-up operations produce the largest percentage of water waste in any industry. This huge waste water generated from industries can be potentially used for fish farming directly or after undergoing the primary treatment process. The nutrient-loaded wet market wastewater and slaughterhouse wastewaters can be used for culturing microalgae and aquatic macrophytes, which in turn can be used for feeding herbivorous fishes or as ingredients in formulating fish feed (Maizatul et al., 2017). These techniques also serve as a natural water treatment system through phytoremediation, which is cheap, environmentally friendly, and beneficial. Sewage-fed aquaculture is a unique fish farming method that promotes effective utilization of wastewater for growing fish while at the same time acts as a natural water treatment system (Jana, 1998; Kumar et al., 2014).

14.3.1.1 Wastewater fish farming

Wastewater fish farming or sewage-fed aquaculture was started in India in the 1930s in the city of Kolkata, which now contributes to about 8000 tons of fish per annum and is the largest such fish farming system in the world (Jhingran, 1991). Because of the high content of organic and inorganic materials in the wastewater, it is done after a series of natural treatment processes such as sedimentation (to reduce organic loads), oxidation (to reduce obnoxious gases), and dilution with freshwater (1:1–1:4), based on the condition of the water (Jana, 1998). To further reduce the organic loads in fish culture ponds, liming is recommended at 400–500 kg ha^{-1} during the off-season, after drying the ponds and occasionally at 10–15 kg during the rearing period, based on water quality (Jana, 1998; Kumar et al., 2015). Some authors also suggested treating the wastewater for 3–14 days in a macrophyte channel having *Eichhornia, Phragmites, Pistia,* and *Lemna* to remove nutrients, heavy metals, suspended solids, oil, and grease before using it for fish rearing (Jana, 1998).

Not all fish species can be grown in wastewater due to high organic load and ammonia in the water. Therefore selection of fish species is an important step for the success of this technology. Hardy fish species that can tolerate low oxygen levels, a wide range of salinity, and a high level of ammonia are suitable for sewage-fed aquaculture (Kumar et al., 2015). Omnivorous, detritus feeders and bottom-grazing fishes burrow through the sediments and help in releasing obnoxious gases from the pond's bottom (Jana, 1998). Fish species recommended suitable for wastewater aquaculture are: Indian carps, Israeli carp, silver carp, bighead carp, grass carp, common carp, hybrid buffalo, catfish, largemouth bass, tilapia, freshwater prawn, etc. (Ghosh et al., 1985; Sin and Chiu, 1987; Jana, 1998; Kumar et al., 2014; Kumar et al., 2015). Jana (1998) recommended a stocking density of 30,000–50,000 fry or 5000–7000 fingerlings ha^{-1}. Composite fish farming system using different fish species is recommended for effective utilization of different ecological niches of the pond ecosystem (Sin and Chiu, 1987; Jana, 1998; Mishra et al., 2009). Because of rich nutrients in the wastewater, aquatic microorganisms thrive and proliferate in it, which acts as natural food for the rearing fishes and reduces the cost of production by minimizing the cost of fish feed. Darko et al. (2016) reported that there is no evidence of any public health hazard associated with treated wastewater reuse in aquaculture.

The untreated industrial waste water can also be used to grow protein-rich aquatic macrophytes such as azolla (*Azolla pinnata*), duckweeds (*Lemna polyrhiza, L. paucicostata,* and *Spirodela polyrrhiza*), Water hyacinth (*Eichhornia crassipes*), water lettuce (*Pistia stratiotes*), Indian stargrass (*Hydrilla verticillata*), white snow flake (*Nymphoides cristatum*), duck lettuce (*Ottelia alismoides*), bur-reed (*Sparganium spp*), Sedges (*Carex riparia*), and water ferns (*Salvinia cuculata*), which can be further used for feeding fish directly or as ingredients in fish feed formulation after processing (Dorothy et al., 2018). This will also act as natural wastewater treatment through phytoremediation as these aquatic macrophytes absorb nutrient mineral ions, heavy metals, and promote sedimentation of suspended particles, thereby mitigating the pollution level in the wastewater (Srivastava et al., 2008; Lu et al., 2018). Wastewater aquaculture is a proven technology that promotes economic benefits with environmental advantages as the aquatic plants or fish ponds acts are natural water treatment, which reduces the efforts and cost for wastewater treatments while contributing a huge quantity of fish to the fish production basket.

14.3.2 Solid waste

Industrial solid wastes are from varied sources and are produced in huge quantities. Some of the important waste that is often used in aquaculture includes brewer's grain and yeast, poultry and meat wastes from slaughterhouses, fish processing waste, and agricultural waste.

14.3.2.1 Brewer's waste

Beer is the most popular alcoholic beverage consumed worldwide (Bamforth and Cabras, 2016). To meet the global consumer demand, the annual production has reached 1.89 billion hectoliters in 2022 (D'Ascenzo et al., 2024). Barley (*Hordeum vulgare*) is the core ingredient in the production of beer, wherein it undergoes an alcoholic fermentation process using specific yeast species (*Saccharomyces sp.*) that converts simple sugars into ethanol (Karlsen and Skov, 2022; Gokulakrishnan et al., 2023). Brewer's spent grain (BSG) and Brewer's spent yeast (BSY) are the major byproducts generated from beer brewing, which contain relatively high protein and essential amino acids, low market price, and stable annual availability (Gokulakrishnan et al., 2023). The protein content is 49% in BSY and a variable 18%–32% in BSG, with 30% essential amino acids of total protein (Kaur and Saxena, 2004; Zerai et al., 2008; Jayant et al., 2018; Gokulakrishnan et al., 2023). Thus BSG and BSY have been experimented with for use in fish feed industries to replace other expensive protein sources, such as fish meal. The untreated brewer's waste, however, contains high levels of antinutritional factors (ANFs) such as lignin and fire, which are known to reduce digestive performance in fish (Karlsen and Skov, 2022). They suggested fractionation and a conversion-based method of refinement process to remove lignocellulosic material (ANFs) in BSGs before using them in fish feed through chemical extractions (in fractionation) or a combination of chemical, biological, and enzymatic treatments (conversion). San Martin et al. (2020) also recommended an enzymatic hydrolysis and stabilization process to improve the digestibility of brewers' byproducts. The use of brewer's waste has been reportedly successful in replacing fish meal and soybean meal upto 50% in the feed formulation of various marine and freshwater fishes without affecting the growth, nutrient utilization, and feed conversion (Kaur and Saxena, 2004; Zerai et al., 2008; Jayant et al., 2018; San Martin et al., 2020).

14.3.2.2 Slaughterhouse byproducts

Slaughterhouse byproducts, or simply animal byproducts, are defined as every part of a slaughtered animal except the dressed carcass (Irshad and Sharma, 2015). They include the rendered products from butcher's tables and processing units, such as heads, blood, liver, lung, kidney, brains, spleen, tripe, viscera, bones, feet, skin, small meat pieces, and the wastewater discharges (Shapawi et al., 2007; Jayathilakan et al., 2012). The yield of animal byproducts ranges between 50% to 60% of live weight and fetches 10%–15% of the value of live animal (Irshad and Sharma, 2015). These byproducts are rich in nutrients with complex mixtures of fats, proteins, and fibers that can be potentially used in the aquaculture industry (Abdelghany et al., 2005). Direct application of untreated byproducts from slaughterhouses is common in Asian countries, for highly carnivorous fishes such as catfish (*Clarias gariepinus*) and tilapia (*Oreochromis niloticus*), to enhance productivity and profitability (Saeed and Abdel-Mageed et al., 2011). Meat meal is the major secondary byproduct of the slaughtering industry and is an important ingredient in animal and fish feed industries (Irshad and Sharma, 2015).

Poultry byproduct meal (PBM) has been one of the best animal protein alternatives for use in aquafeed industries due to its high nutritional value and amino acid balance (Dawood et al., 2020). It contains 55%–74% protein (dry weight), 10%–19% lipid, and 11%–23% ash and essential amino acids, except lysine and methionine, though the actual composition may, however, vary based on the source of raw materials (Dong et al., 1993). The use of PBM as feed ingredients in fish feed is well documented and successfully used to replace the dietary fish meal protein from 25% to even 100% replacement for carnivorous fishes without adverse effects on the growth of fish (Takagi et al., 2000; Gaylord and Rawles, 2005; Rawles et al., 2011). However, to ensure good growth performance, feed efficiency, and nutrient retention, Dong et al. (1993) and Karapanagiotidis et al. (2019) suggested supplementation of lysine and methionine while using PBM in fish feed. Further, Dawood et al. (2020) suggested that fermentation of FBM better health condition and growth performance of the reared fish.

14.3.2.3 Fish processing waste

The global fish demand has been increasing each year and is expected to continue doing so in the future. Fish production from capture fisheries has various limitations for expansion, and therefore aquaculture has been playing a pivotal role in bridging the fish demand–supply gap. With the advancement of technologies and requirements to produce more fish from minimum resources (land, water, etc.), aquaculture has been undergoing vertical expansion and heavy intensification, which demands more formulated nutrient-rich fish feeds. Fish meal has been the commonly used protein source in aquafeed industries due to high protein content (60%–68%), digestibility, and presence of essential fatty acids, minerals, and trace elements (Mo et al., 2018). The major source of fish meal is trash fish, which comprises low-value small fishes, most of which are actually juveniles of commercially important species, small benthic and mesopelagic fish, crustaceans, and cephalopods (Grainger et al., 2005). Fish meal is expensive and has a high market demand as it is used for various other animal feeds, too. Thus researchers have been exploring ways to replace fish meal in aquafeed with other cheaper, available, and sustainable ingredients. Fish waste from fish processing units can be one of such alternatives.

About 70% of the total global fish production undergoes some kind of processing prior to marketing, leading to the production of about 20 million tons of fish waste (Earth and Environmental, 2003). Fish processing, which includes stunning, grading, slime removal, deheading, washing, scaling, gutting, fin clippings, deboning, filleting, drying, and fermentation, leads to 20%–80% of the fish mass as waste (depending on the level of processing and type of fish) during the process (Ghaly et al., 2013). Fish waste includes bones, intestines, heads, tails, and all other fish tissue that is not suitable for consumption (Mo et al., 2018). This fish waste is equivalent to about 25% of the world's total marine fish production (Racioppo et al., 2021). They are often burned, landfilled, or dumped in the sea. Fish waste is rich in crude protein (58%), fats (19%), and minerals, which makes it a suitable source of fishmeal for fish and other animal feeds (Ghaly et al., 2013). Further, the proteins derived from fish are nutritionally superior to those of plant sources and have a better balance of the dietary essential amino acids compared to animal protein sources (Friedman, 1996).

Fish waste can be processed by drying, powdering, fermentation, and through biotechnological and biopreservation techniques for the extraction of fish oil, fish meal, protein hydrolysates, and fish silage (Afreen and Ucak, 2020). These can be used in the formulation of fish feed to reduce the dependence on fishmeal or indirectly for culturing protein-rich insects, such as black soldier fly, that can be, in turn, used in fish feed industries (Racioppo et al., 2021). The use of these byproducts in fish feed has reportedly enhanced antioxidant activity and immune function in various fish species, including salmon, seabass, carps, tilapia, sturgeon, turbot, catfish, and shrimps (Swanepoel and Goosen, 2018).

14.3.2.4 *Agroindustrial waste*

The rapid growth of the global population, coupled with urbanization, has significantly escalated the demand for processed foods. By 2050, the world population is projected to reach 9.7 billion, with approximately 68% of people living in urban areas (United Nations, 2019). This shift has led to notable changes in dietary preferences, with a growing reliance on processed and convenience foods to suit the fast-paced urban lifestyles where time for food preparation is limited (FAO, 2020). To cater to this burgeoning demand, the food processing industry has expanded substantially, leading to increased production of industrial waste. It is estimated that nearly 40% of global food production is wasted, with a considerable amount originating from agroindustrial processes (IFT, 2021).

Agricultural waste refers to residues generated from the cultivation, harvesting, and processing of crops (Obi et al., 2016). They represent a substantial yet underutilized resource with vast potential for economic value creation and environmental sustainability (Patel et al., 1972). They are often burned, decomposed, or discarded. The untreated disposals decompose and produce foul odors, and also pose severe environmental issues. At the same time, these huge quantities of agricultural wastes present a tremendous opportunity for recycling/reuse for other industries, such as the aquafeed industries, because of their nutritional and bioactive properties that could enhance the sustainability and efficiency of aquaculture operations (Pardo et al., 2016; Santos et al., 2021). Most agricultural wastes have high levels of fiber and carbohydrates. They are loaded with vitamins, minerals, phenolic acids, flavonoids, alkaloids, and other phytochemicals that can improve fish immunity and reduce oxidative stress (Sepehrfar et al., 2023). Extracts from food processing waste can be included in aquafeeds to provide natural antioxidants and antimicrobial agents, improving fish growth performance, immunity, and disease resistance (Omojowo et al., 2012; Outama et al., 2022). These natural extracts can replace synthetic additives, which are often costly and may have adverse environmental effects, making aquaculture practices more green, sustainable, and economically viable (Dawood et al., 2022).

Based on the types of residues, agricultural waste or agroindustry waste (AIW) can be grouped into two major categories: (1) Farm residues and (2) food processing residue.

14.3.2.4.1 Agricultural farm residues

Agricultural farm residues include all those direct byproducts derived from the farm itself during the process of cultivation and harvesting of crops, but before undergoing any processes. This includes stem, leaves, seed and seedpod, stalks, and roots.

Rice is a staple food for a large population across the globe and therefore cultivated widely in different agroclimatic regions. Rice cultivation leads to the production of huge amounts of farm waste, such as straw and husk. These fresh stems and green paddy leaves can be chopped and fed directly to herbivorous fish such as grass carp (*Ctenopharyngodon idella*). In composite fish farming, these semidigested fecal matters of grass carp also served indirectly as feed to other omnivorous and detritus feeder fishes such as common carp (Dorothy et al., 2018). Leaves of sweet potato, cassava, mulberry, peanut, beans, banana, and bamboo leaves are also often used to feed herbivorous fish (Dongmeza et al., 2009). Citrus seeds contain significant amounts of lipids, which can be a beneficial source of essential fatty acids in fish diets (Morais et al., 2017). Cottonseed powder can replace up to 40% of fish meal used in crucian carp diets without impacting the growth of fish (Cai et al., 2011). Use of mango seed kernels in fish diets has also been experimented (Patel et al., 1972). Common carp fed with diets having wheat leaves showed positive growth performance, hematobiochemical responses, and health indices (Ahmed, 2024).

14.3.2.4.2 Food processing residue

Food processing residues are those byproducts derived from the processing units, such as husk, peels, pulps, juice extracts, and cakes. Food processing residues such as peels, pulps, and juice extracts contain carotenoids, chlorophylls, and essential oils that are excellent sources of bioactive compounds with strong antioxidant properties (Okino Delgado and Fleuri et al., 2016; Calinoiu et al., 2023).

Okara, a byproduct of soybean milk extract, contains about 20.9%–39.1% crude protein, while citrus byproducts contain 5.8%–12.87% crude protein (Rahman et al., 2021). Bioactive compounds extracted from citrus peels possess antimicrobial properties that inhibit the growth of harmful bacteria, fungi, and viruses, and can potentially protect fish against pathogens (Kesbiç et al., 2022). Incorporating these peel powders in tilapia diets up to 10%–20% inclusion level demonstrated significantly improved growth, immune markers, and antioxidant status, along with positive changes in the intestinal structure of tilapia (Hosseini et al., 2020; Mohamed et al., 2021; Tawfik et al., 2022; Reda et al., 2024). Essential oils extracted from citrus lemon and mandarin peel have been shown to enhance immune responses and disease resistance in Mozambique tilapia when challenged with the pathogen *Edwardsiella tarda* (Baba et al., 2016). Potato, cassava, and mango peels have been experimented with for potential benefits in fish (Omojowo et al., 2012; Outama et al., 2022). Banana peels contain significant amounts of lipids, which can serve as alternative sources of essential fatty acids in fish diets (Morais et al., 2017). Inclusion of banana peel flour up to 7.5% in common carp (*Cyprinus carpio*) diets enhanced growth rates, improved feed conversion ratios, and maintained fish health without impacting survival rates, making it an economical additive for carp feed (Melanie et al., 2024). Shellfish such as shrimp are more selective in their dietary preferences and have higher protein requirements. Agricultural wastes like citrus peels and okara can be used in shrimp diets when supplemented with additional protein sources to ensure balanced nutrition (Lee et al., 2024). Soybean byproducts and papaya processing waste bioprocessed with yeast are a viable protein source in shrimp diets (Kang et al., 2010).

Rice bran is the predominant feed ingredient often used not only for fish but also for other animal feed, and is considered one of the most economical ingredients in feed industries (Ranjan et al., 2022). Bran comprises approximately 5%–8% of the total grain (Pillaiyar and Mohandoss, 1981). Deoiled rice bran is a low-energy, high-nutrient byproduct containing 14%

crude protein, 0.59% ether extract, 14.9% crude fiber, 11.2% total ash, and 59.1% nitrogen-free extract (Ranjan et al., 2017; Kumar et al., 2018). Small-scale farmers often prepared low-cost farm-made fish using rice bran, broken rice, oil-cakes (from mustard, soybean, sesame, groundnut, etc.) for cost-effectiveness (Mazid et al., 1997). Molasses, a byproduct from sugarcane, is often used in aquaculture as a rich carbon source to promote the growth of heterotrophic microorganisms (biofloc), buffer pond water pH, and optimize water quality (Alfiansah et al., 2022).

14.3.2.4.3 Drawback in the inclusion of agroindustrial waste in fish feed

The use of AIW in aquafeed requires appropriate processing techniques to ensure nutritional quality, feed safety, and stability, as studies indicated the presence of ANFs in plant-derived products (Raman et al., 2024). Common ANFs found in industrial wastes include tannins, saponins, phytates, and alkaloids (Mirzaei-Aghsaghali et al., 2008; Kokou and Fountoulaki, 2018). The presence of these compounds can limit the use of certain wastes in aquafeeds, as they may interfere with nutrient absorption, reduce palatability, or cause physiological stress to the fish (Mirzaei-Aghsaghali et al., 2008). Researchers suggested different kinds of processing procedures that include drying and powdering, soaking and fermenting, or chemical treatments to reduce the ANFs from those plant extracts (Dorothy et al., 2018).

14.3.3 Industrial waste as sustainable fertilizers in aquaculture

Fertilization in aquaculture systems is crucial for promoting the growth of natural food sources, such as phytoplankton and zooplankton, which serve as primary food for fish. Industrial wastes, including press mud, spent wash, and BSG, are rich in essential nutrients such as nitrogen and phosphorus, making them viable fertilizers in fish ponds (Saha et al., 2005; Keshavanath et al., 2006; Surnar et al., 2019). Pressmud is particularly high in organic matter, nitrogen, and phosphorus, essential for phytoplankton growth (Keshavanath et al., 2006). By adding these organic wastes, farmers can foster the natural productivity of ponds, increasing fish yield and reducing reliance on costly commercial feeds. Utilizing these wastes as fertilizers not only reduces feed costs but also recycles nutrients that would otherwise contribute to environmental pollution (Schildbach et al., 1992). Methods such as windrow composting and aerated static piles enhance nutrient content, mitigate pathogens, and prepare the waste for safe and effective use in aquaculture. These techniques are essential for transforming raw agroindustrial byproducts, such as pressmud, into stable, nutrient-rich compost. When added to ponds, composted waste releases essential nutrients, supporting plankton populations that serve as food for fish, thereby increasing fish production and maintaining water quality (Surnar et al., 2019).

14.4 Economic and environmental benefits of using industrial waste in aquaculture

The use of industrial waste in aquaculture offers several economic and environmental benefits. Industrial waste occurs in huge volumes and masses. Proper treatment and disposal of this waste is expensive, time-consuming, and requires a lot of space. Repurposing the waste into useful products in aquaculture industries not only utilizes the benefits of the huge byproducts but also reduces the burden of disposal.

14.4.1 Cost reduction

Feed costs account for a significant portion, up to 60% of the total production costs in aquaculture (Raman et al., 2024). By incorporating industrial waste, which is often available at low or no cost, aquafeed manufacturers can significantly reduce the cost of feed production, which can lead to lower production costs and improved profitability for fish farmers (Dorothy et al., 2018).

14.4.2 Waste minimization and circular economy

The utilization of industrial waste in aquaculture contributes to waste minimization and supports the principles of a circular economy. By converting waste materials into valuable feed ingredients, the aquaculture industry can reduce its environmental footprint, decrease waste disposal costs, and promote sustainable resource use (Kapoor et al., 2020; Hamam et al., 2021; Pereira et al., 2022).

14.4.3 Natural pond fertilizer

Industrial waste fertilizers offer a low-cost solution to complete the biogeochemical cycle of essential nutrients in aquaculture, both for organic and conventional farming. This approach provides an effective means to reduce the environmental and economic burden associated with waste disposal. With the increasing demand for fertilizers in aquaculture due to intensified systems and rising production pressure, these wastes present a viable and eco-friendly alternative to conventional mineral fertilizers (Joshi, 1999). There is significant potential for research and development to enhance the application of these waste products as organic fertilizers in aquaculture, addressing sustainability, productivity goals, and pollution (Schildbach et al., 1992; Joshi, 1999; Surnar et al., 2019; Keshavanath et al., 2006).

14.4.4 Environmental friendly

Industrial waste, if not properly managed, can lead to significant environmental pollution. The incorporation of these wastes into aquafeeds reduces the need for landfill disposal, incineration, or other waste management practices that can have negative environmental impacts. This also reduces greenhouse gas emissions and conserves land and water resources (Pardo et al., 2016; Santos et al., 2021).

14.5 Conclusion

Integration of industrial waste into aquaculture offers a sustainable and economically viable approach to address environmental and industry challenges. By repurposing nutrient-rich byproducts from various industries, aquaculture can reduce feed costs, enhance fish production, and minimize the environmental impact of waste disposal. The use of liquid and solid waste in aquaculture systems not only promotes a circular economy but also supports environmental conservation through waste reduction and resource recycling. Furthermore, the development of innovative processing techniques ensures that industrial byproducts can be safely and effectively utilized, fostering economic growth and environmental sustainability. This approach underscores the importance of continued research and technological advancements to optimize the use of industrial waste in aquaculture, contributing to a greener and more efficient fish production system.

References

Abdelghany, A.E., Ahmad, M.H., Sayed, S.H., Ibrahim, H.I., Abdel-Fatah, M.E., 2005. Replacement of fish meal with poultry by-product meal in diets for mono-sex Nile tilapia, Oreochromis niloticus. Egypt. J. Nutrit. Feeds 8 (1), 1049–1063.

Afreen, M., Ucak, I., 2020. Fish processing wastes used as feed ingredient for animal feed and aquaculture feed. J. Surv. Fish. Sci. 6 (2), 55–64. doi:10.18331/sfs2020.6.2.7.

Ahmed, V.M., 2024. Effects of adding germinated wheat green part and seed to the diet on common carp (Cyprinus carpio L.): growth performance, somatic indices, and hemato-biochemical responses. Kirkuk Univ. J. Agric. Sci. 15 (2), 57–65.

Alfiansah, Y.R., Harder, J., Slater, M.J., Gärdes, A., 2022. Addition of molasses ameliorates water and bio-floc quality in shrimp pond water. Trop. Life Sci. Res. 33 (1), 121–141. doi:10.21315/tlsr2022.33.1.8.

Allison, E.H., 2011. Aquaculture, Fisheries, Poverty and Food Security. In: Working Paper 2011–65. The WorldFish Center, Penang, Malaysia, p. 60.

AMEC Earth & Environmental Limited. (2003). Management of wastes from Atlantic seafood processing operations (Final report submitted to the National Programme of Action (NPA) Atlantic Regional Team, Environment Canada). Dartmouth, Nova Scotia, Canada.

Baba, E., Acar, Ü., Öntaş, C., Kesbiç, O.S., Yılmaz, S., 2016. Evaluation of Citrus limon peels essential oil on growth performance, immune response of Mozambique tilapia Oreochromis mossambicus challenged with Edwardsiella tarda. Aquaculture 465, 13–18. doi:10.1016/j.aquaculture.2016.08.023.

Bamforth, C.W., Cabras, I., 2016. Interesting times: changes for brewing. In: Brewing, beer and pubs: a global perspective. Palgrave Macmillan, London, pp. 15–33. doi:10.1057/9781137466181_2.

Cai, C., Li, E., Ye, Y., Krogdahl, A., Jiang, G., Wang, Y., Chen, L., 2011. Effect of dietary graded levels of cottonseed meal and gossypol on growth performance, body composition and health aspects of allogynogenetic silver crucian carp, Carassius auratus gibelio♀ × Cyprinus carpio♂. Aquacult. Nutr. 17 (4), 353–360. doi:10.1111/j.1365-2095.2010.00801.x.

Calinoiu, L.F., Mitrea, L., Teleky, B.E., Szabo, K., Martău, A.G., Nemes, S.A., Plamada, D., Pascuta, M.S., Barta, G., Varvara, R.A., Stefanescu, B.E., Vodnar, D.C., 2023. Fruit and vegetable waste and by-products for pigments and color. In: Fruit and vegetable waste utilization and sustainability. Elsevier, Amsterdam, pp. 77–100. doi:10.1016/B978-0-323-91743-8.00011-3.

Darko, G., Azanu, D., Logo, N.K, Ng, C.A, 2016. Accumulation of toxic metals in fish raised from sewage-fed aquaculture and estimated health risks associated with their consumption. Cogent Environ. Sci. 2 (1), 1190116. doi:10.1080/23311843.2016.1190116.

D'Ascenzo, F., Vinci, G., Maddaloni, L., Ruggeri, M., Savastano, M., 2024. Application of life cycle assessment in beer production: systematic review. Beverages 10 (3), 86. doi:10.3390/beverages10030086.

Dawood, M.A.O., Habotta, O.A.E., Elsabagh, M., Azra, M.N., Doan, H.V., Kari, Z.A., Sewilam, H., 2022. Fruit processing by-products in the aquafeed industry: a feasible strategy for aquaculture sustainability. Rev. Aquac. 14 (4), 1945–1965. doi:10.1111/raq.12680.

Dawood, M.A.O., Magouz, F.I., Mansour, M., Saleh, A.A., Asely, A.M.E., Fadl, S.E., Ahmed, H.A., Al-Ghanim, K.A., Mahboob, S., Al-Misned, F., 2020. Evaluation of yeast fermented poultry by-product meal in Nile Tilapia (Oreochromis niloticus) feed: effects on growth performance, digestive enzymes activity, innate immunity, and antioxidant capacity. Front. Vet. Sci. 6. doi:10.3389/fvets.2019.00516.

Dong, F.M., Hardy, R.W., Haard, N.F., Barrows, F.T., Rasco, B.A., Fairgrieve, W.T., Forster, I.P., 1993. Chemical composition and protein digestibility of poultry by-product meals for salmonid diets. Aquaculture 116 (2-3), 149–158. doi:10.1016/0044-8486(93)90005-J.

Dongmeza, E., Steinbronn, S., Francis, G., Focken, U., Becker, K., 2009. Investigations on the nutrient and antinutrient content of typical plants used as fish feed in small scale aquaculture in the mountainous regions of Northern Vietnam. Anim. Feed Sci. Technol. 149 (1–2), 162–178. doi:10.1016/j.anifeedsci.2008.04.012.

Dorothy, M.S., Raman, S., Nautiyal, V., Singh, K., Yogananda, T., Kamei, M., 2018. Use of potential plant leaves as ingredient in fish feed-a review. Int. J. Curr. Microbiol. Appl. Sci. 7 (07), 112–125. doi:10.20546/ijcmas.2018.707.014.

FAO, 2020. The State of Food and Agriculture 2020 : Overcoming Water Challenges in Agriculture. FAO, Rome doi:10.4060/cb1447en.

Friedman, M., 1996. Nutritional value of proteins from different food sources: a review. J. Agric. Food Chem. 44 (1), 6–29. doi:10.1021/jf9400167.

Gaylord, T.G., Rawles, S.D., 2005. The modification of poultry by-product meal for use in hybrid striped bass Morone chrysops X M. saxatilis diets. J. World Aquac. Soc. 36 (3), 363–374. doi:10.1111/j.1749-7345.2005.tb00340.x.

Ghaly, A.E., Ramakrishnan, V.V., Brooks, M.S., Budge, S.M., Dave, D., 2013. Fish processing wastes as a potential source of proteins, amino acids and oils: a critical review. J. Microb. Biochem. Technol. 5 (4), 107–129. doi:10.4172/1948-5948.1000110.

Ghosh, A., Saha, S.K., Roy, A.K., Chakraborti, P.K., 1985. Package of practices for using domestic sewage in carp production. In: Aquaculture Extension Manual, 8. Central Inland Fisheries Research Institute (CIFRI), Barrackpore, India, pp. 1–19.

Gokulakrishnan, M., Kumar, R., Ferosekhan, S., Siddaiah, G.M., Nanda, S., Pillai, B.R., Swain, S.K., 2023. Bio-utilization of brewery waste (Brewer's spent yeast) in global aquafeed production and its efficiency in replacing fishmeal: From a sustainability viewpoint. Aquaculture 565, 739161. doi:10.1016/j.aquaculture.2022.739161.

Grainger, R., Yingliang, X., Shengha, L., Zhijie, G., 2005. Production and utilization of trash fish in selected Chinese ports. In: Regional workshop on low value and "trash fish' in the Asia-Pacific region". VietNam., Hanoi, pp. 7–9.

Hamam, M., Chinnici, G., Vita, G.Di, Pappalardo, G., Pecorino, B., Maesano, G., D'Amico, M., 2021. Circular economy models in agro-food systems: a review. Sustainability. 13 (6), 3453. doi:10.3390/su13063453.

Hosseini, S.M, Hoseinifar, S.H, Mazandarani, M., Paknejad, H., Doan, H.V, El-Haroun, E.R., 2020. The potential benefits of orange peels derived pectin on serum and skin mucus immune parameters, antioxidant defence and growth performance in common carp (Cyprinus carpio). Fish Shellfish Immunol. 103, 17–22. doi:10.1016/j.fsi.2020.04.019.

IFT, 2021. Facing the food waste crisis. https://www.ift.org/news-and-publications/food-technology-magazine/issues/2021/december/features/facing-the-food-waste-crisis (Accessed: 27 November 2024).

Irshad, A., Sharma, B.D., 2015. Abattoir by-product utilization for sustainable meat industry: a review. J. Anim. Prod. Adv. 5 (6), 681–696.

Jana, B.B., 1998. Sewage-fed aquaculture: the Calcutta model. Ecol. Eng. 11 (1–4), 73–85. doi:10.1016/s0925-8574(98)00024-x.

Jayant, M., Hassan, M.A., Srivastava, P.P., Meena, D.K., Kumar, P., Kumar, A., Wagde, M.S., 2018. Brewer's spent grains (BSGs) as feedstuff for striped catfish, Pangasianodon hypophthalmus fingerlings: an approach to transform waste into wealth. J. Cleaner Prod. 199, 716–722. doi:10.1016/j.jclepro.2018.07.213.

Jayathilakan, K., Sultana, K., Radhakrishna, K., Bawa, A.S., 2012. Utilization of byproducts and waste materials from meat, poultry and fish processing industries: a review. J. Food Sci. Technol. 49 (3), 278–293. doi:10.1007/s13197-011-0290-7.

Jhingran, V.G., 1991. Fish and Fisheries of India, 3rd rev. and enl. edn Hindustan Publishing Corporation (India), Delhi.

Joshi, H.C., 1999. Bio-energy potential of distillery effluents. Bioenergy News 3 (3), 10–15.

Kang, H.Y., Yang, P.Y., Dominy, W.G., Lee, C.S., 2010. Bioprocessing papaya processing waste for potential aquaculture feed supplement – economic and nutrient analysis with shrimp feeding trial. Bioresour. Technol. 101 (20), 7973–7979. doi:10.1016/j.biortech.2010.05.058.

Kapoor, R., Ghosh, P., Kumar, M., Sengupta, S., Gupta, A., Kumar, S.S., Vijay, V., Kumar, V., Kumar Vijay, V., Pant, D., 2020. Valorization of agricultural waste for biogas based circular economy in India: a research outlook. Bioresour. Technol. 304, 123036. doi:10.1016/j.biortech.2020.123036.

Karapanagiotidis, I.T., Psofakis, P., Mente, E., Malandrakis, E., Golomazou, E., 2019. Effect of fishmeal replacement by poultry by-product meal on growth performance, proximate composition, digestive enzyme activity, haematological parameters and gene expression of gilthead seabream (Sparus aurata). Aquacult. Nutr. 25 (1), 3–14. doi:10.1111/anu.12824.

Karlsen, F., Skov, P.V., 2022. Review–potentials and limitations of utilising brewer's spent grain as a protein source in aquaculture feeds. J. Cleaner Prod. 357, 131986.

Kaur, V.I., Saxena, P.K., 2004. Incorporation of brewery waste in supplementary feed and its impact on growth in some carps. Bioresour. Technol. 91 (1), 101–104. doi:10.1016/S0960-8524(03)00073-7.

Kesbiç, O.S., Acar, Ü., Mohammady, E.Y., Salem, S.M.R., Ragaza, J.A., El-Haroun, E., Hassaan, M.S., 2022. The beneficial effects of citrus peel waste and its extract on fish performance and health status: a review. Aquac. Res. 53 (12), 4217–4232. doi:10.1111/are.15945.

Keshavanath, P., Shivanna, B.G., 2006. Evaluation of sugarcane by-product pressmud as a manure in carp culture. Bioresour. Technol. 97 (4), 628–634. doi:10.1016/j.biortech.2005.03.019.

Kokou, F., Fountoulaki, E., 2018. Aquaculture waste production associated with antinutrient presence in common fish feed plant ingredients. Aquaculture 495, 295–310. doi:10.1016/j.aquaculture.2018.06.003.

Kumar, D., Chaturvedi, M.K.M., Sharma, S.K., Asolekar, S.R., 2015. Sewage-fed aquaculture: a sustainable approach for wastewater treatment and reuse. Environ. Monit. Assess. 187 (10). doi:10.1007/s10661-015-4883-x.

Kumar, D., Hiremath, A.M., Asolekar, S.R., 2014. Integrated management of wastewater through sewage fed aquaculture for resource recovery and reuse of treated effluent: a case study. APCBEE Proc. 10, 74–78. doi:10.1016/j.apcbee.2014.10.019.

Kumar, S., Sahu, N.P., Shamna, N., Ranjan, A., 2018. Feeding higher level of de-oiled rice bran causes stress to Labeo rohita fingerlings. Aquaculture 484, 184–190. doi:10.1016/j.aquaculture.2017.11.029.

Lee, Y.C., Chang, C.C., Lin, Y.H., Lin, Y.H., 2024. Effect of fermented lemon peel as a functional feed additive on growth, non-specific immune responses, and Vibrio alginolyticus resistance in whiteleg shrimp, Litopenaeus vannamei. Aquac. Rep. 34. doi:10.1016/j.aqrep.2024.101918.

Lu, B., Xu, Z., Li, J., Chai, X., 2018. Removal of water nutrients by different aquatic plant species: an alternative way to remediate polluted rural rivers. Ecol. Eng. 110, 18–26. doi:10.1016/j.ecoleng.2017.09.016.

Maizatul, A.Y., Radin Mohamed, R.M.S., Al-Gheethi, A.A., Hashim, M.K.A., 2017. An overview of the utilisation of microalgae biomass derived from nutrient recycling of wet market wastewater and slaughterhouse wastewater. Int. Aquat. Res. 9 (3), 177–193. doi:10.1007/s40071-017-0168-z.

Mazid, M.A., Zaher, M., Begum, N.N., Ali, M.Z., Nahar, F., 1997. Formulation of cost-effective feeds from locally available ingredients for carp polyculture system for increased production. Aquaculture 151 (1–4), 71–78. doi:10.1016/s0044-8486(96)01504-9.

Melanie, K., Nazlia, S., Saputra, J., Febri, S.P., Aprita, I.R., 2024. The effect of banana peel flour (Musa paradisiaca) as additional feed raw materials for common carp (Cyprinus carpio) BIO Web of Conferences, 87. EDP Sciences, Indonesia doi:10.1051/bioconf/20248703024 21174458. https://www.bio-conferences.org/87.

Mirzaei-Aghsaghali, A., Maheri-Sis, N., 2008. Nutritive value of some agro-industrial by-products for ruminants-a review. World J. Zool. 3 (2), 40–46.

Mishra, B., Ayyappan, S., Jena, J.K., Patri, P., Behera, P.K., Aravindakshan, P.K., Purushothaman, C.S., 2009. Carp polyculture in aquaculture sewage treatment plant and its impact on soil and water characteristics. Indian J. Fish. 56 (4), 271–276.

Mo, W.Y., Man, Y.B., Wong, M.H., 2018. Use of food waste, fish waste and food processing waste for China's aquaculture industry: needs and challenge. Sci. Total Environ. 613–614, 635–643. doi:10.1016/j.scitotenv.2017.08.321.

Mohamed, R.A., Yousef, Y.M., El-Tras, W.F., Khalafallaa, M.M., 2021. Dietary essential oil extract from sweet orange (Citrus sinensis) and bitter lemon (Citrus limon) peels improved Nile tilapia performance and health status. Aquac. Res. 52 (4), 1463–1479. doi:10.1111/are.15000.

Morais, D.R., Rotta, E.M., Sargi, S.C., Bonafe, E.G., Suzuki, R.M., Souza, N.E., Matsushita, M., Visentainer, J.V., 2017. Proximate composition, mineral contents and fatty acid composition of the different parts and dried peels of tropical fruits cultivated in Brazil. J. Braz. Chem. Soc. 28 (2), 308–318. doi:10.5935/0103-5053.20160178.

Obi, F.O, Ugwuishiwu, B.O, Nwakaire, J.N, 2016. Agricultural waste concept, generation, utilization, and management. Niger. J. Technol. 35 (4), 957. doi:10.4314/njt.v35i4.34.

Okino Delgado, C.H., Fleuri, L.F., 2016. Orange and mango by-products: agro-industrial waste as source of bioactive compounds and botanical versus commercial description—a review. Food Rev. Int. 32 (1), 1–14. doi:10.1080/87559129.2015.1041183.

Omojowo, T.M., Omojowo, F.S., Alatise, P.S., 2012. Growth response and nutritional evaluation of mango peel-based diets on tilapia (Oreochromis niloticus) fingerlings. Researcher 2 (6), 44–49.

Outama, P., Le Xuan, C., Wannavijit, S., Lumsangkul, C., Vu Linh, N., Montha, N., Tongsiri, S., Chitmanat, C., Van Doan, H., 2022. Modulation of growth, immune response, and immune-antioxidant related gene expression of Nile tilapia (Oreochromis niloticus) reared under biofloc system using mango peel powder. Fish Shellfish Immunol. 131, 1136–1143. doi:10.1016/j.fsi.2022.09.031.

Pardo, G., Martin-Garcia, I., Arco, A., Yañez-Ruiz, D.R., Moral, R., del Prado, A., 2016. Greenhouse-gas mitigation potential of agro-industrial by-products in the diet of dairy goats in Spain: a life-cycle perspective. Anim. Prod. Sci. 56 (3), 646–654. doi:10.1071/an15620.

Patel, B.M., Patel, C.A., Talpada, P.M., 1972. Evaluation of mango seed kernels and tomato waste in the ration of bullocks. Indian J. Nutr. Dietet. 9 (6), 347–350.

Pereira, A.G., Fraga-Corral, M., Garcia-Oliveira, P., Otero, P., Soria-Lopez, A., Cassani, L., Cao, H., Xiao, J., Prieto, M.A., Simal-Gandara, J., 2022. Single-cell proteins obtained by circular economy intended as a feed ingredient in aquaculture. Foods 11 (18). doi:10.3390/foods11182831.

Pillaiyar, P., Mohandoss, R., 1981. Cooking qualities of parboiled rices produced at low and high temperatures. J. Sci. Food Agric. 32 (5), 475–480. doi:10.1002/jsfa.2740320508.

Racioppo, A., Speranza, B., Campaniello, D., Sinigaglia, M., Corbo, M.R, Bevilacqua, A., 2021. Fish loss/waste and low-value fish challenges: state of art, advances, and perspectives. Foods 10 (11), 2725. doi:10.3390/foods10112725.

Rahman, M.M., Mat, K., Ishigaki, G., Akashi, R., 2021. A review of okara (soybean curd residue) utilization as animal feed: Nutritive value and animal performance aspects. Anim. Sci. J. 92 (1). doi:10.1111/asj.13594.

Ramachandra, T.V., 2006. Management of Municipal Solid Waste. Environmental Engineering Series. The Energy and Resources Institute (TERI), New Delhi, India.

Raman, S., Deo, A.D., Aklakur, M., Sahu, N.P., Jayant, M., Varghese, T., 2024. Comparative evaluation of dietary raw and solid-state fermented Sesbania leaf meal in Labeo rohita (Hamilton, 1822). Indian J. Anim. Res. 58 (3), 433–439. doi:10.18805/IJAR.B-4889.

Ranjan, A., Kumar, S., Sahu, N.P., Jain, K.K., Deo, A.D., 2022. Strategies for maximizing utilization of de-oiled rice bran (DORB) in the fish feed. Aquac. Int. 30 (1), 99–114. doi:10.1007/s10499-021-00791-6.

Ranjan, A., Sahu, N.P, Deo, A.D, Kumar, S., Kumar, S., Jain, K.K, 2017. Xylanase and phytase supplementation in the de-oiled rice bran (DORB) based diet improves the growth performance of Labeo rohita. Int. J. Curr. Microbiol. Appl. Sci. 6 (6), 1493–1503. doi:10.20546/ijcmas.2017.606.176.

Rawles, S.D., Thompson, K.R., Brady, Y.J., Metts, L.S., Aksoy, M.Y., Gannam, A.L., Twibell, R.G., Ostrand, S., Webster, C.D., 2011. Effects of replacing fish meal with poultry by-product meal and soybean meal and reduced protein level on the performance and immune status of pond-grown sunshine bass (Morone chrysops × M. saxatilis). Aquacult. Nutr. 17 (3), e708. doi:10.1111/j.1365-2095.2010.00831.x.

Reda, R.M., El-Rahim, M.I.A., Elkerdawy, D.A., Metwally, M.M.M., Said, N., 2024. Effects of mandarin peel powder on growth, biochemical, immune, and intestinal health in Oreochromis niloticus at suboptimal temperatures. BMC Vet. Res. 20 (1). doi:10.1186/s12917-024-04273-8.

Saeed, S.M., Abdel-Mageed, S.A., 2011. Impact of poultry slaughterhouse by-product on the aquatic environment of fish ponds. Environment, 6 (1), 1–18.

Saha, N.K., Balakrishnan, M., Batra, V.S., 2005. Improving industrial water use: case study for an Indian distillery. Resour. Conserv. Recycl. 43 (2), 163–174. doi:10.1016/j.resconrec.2004.04.016.

San Martin, D., Orive, M., Iñarra, B., Castelo, J., Estévez, A., Nazzaro, J., Iloro, I., Elortza, F., Zufía, J., 2020. Brewers' spent yeast and grain protein hydrolysates as second-generation feedstuff for aquaculture feed. Waste Biomass Valor. 11 (10), 5307–5320. doi:10.1007/s12649-020-01145-8.

Santos, C., Fonseca, J., Coutinho, J., Trindade, H., Jensen, L.S., 2021. Chemical properties of agro-waste compost affect greenhouse gas emission from soils through changed C and N mineralisation. Biol. Fertil. Soils 57 (6), 781–792. doi:10.1007/s00374-021-01560-6.

Schildbach, R., Ritter, W., Schmithals, K., Burbidge, M., 1992. New developments in the environmentally safe disposal of spent grains and waste kieselguhr from breweries. Proc. Conv.-Inst. Brew. (Asia Pac. Sect.) 22, 139–143.

Sepehrfar, D., Sudagar, M., Paknejad, H., Siahkalroodi, Y.S., Norouzitallab, P., 2023. Role of phytochemicals in farmed fish reproductive performance: a review. Iran. J. Fish. Sci. 22 (5), 1039–1068. doi:10.22092/ijfs.2023.130580.

Shapawi, R., Ng, W.K., Mustafa, S., 2007. Replacement of fish meal with poultry by-product meal in diets formulated for the humpback grouper, Cromileptes altivelis. Aquaculture 273 (1), 118–126. doi:10.1016/j.aquaculture.2007.09.014.

Sin, A.W., Chiu, M.T.L., 1987. The culture of silver carp, bighead, grass carp and common carp in secondary effluents of a pilot sewage treatment plant. Resour. Conserv. Recycl. 13 (2-4), 231–246. doi:10.1016/0166-3097(87)90065-4.

Srivastava, J., Gupta, A., Chandra, H., 2008. Managing water quality with aquatic macrophytes. Rev. Environ. Sci. Biotechnol. 7 (3), 255–266. doi:10.1007/s11157-008-9135-x.

Surnar, S.R, Ojha, M.L, Saini, V.P, Singh, E., Yadav, R., 2019. An overview on the potential of agro-industrial wastes as fertilizer in aquaculture. Int. J. Farm Sci. 9 (3), 101. doi:10.5958/2250-0499.2019.00082.x.

Swanepoel, J.C., Goosen, N.J., 2018. Evaluation of fish protein hydrolysates in juvenile African catfish (Clarias gariepinus) diets. Aquaculture 496, 262–269. doi:10.1016/j.aquaculture.2018.06.084.

Takagi, S., Hosokawa, H., Shimeno, S., Ukawa, M., 2000. Utilization of poultry by-product meal in a diet for red sea bream Pagrus major. Nippon Suisan Gakkaishi (Japanese Edition) 66 (3), 428–438. doi:10.2331/suisan.66.428.

Tawfik, W., Nassef, E., Bakr, A., Hegazi, E., Ismail, T.A., Abdelazim, A.M., El-Nagar, S.H., Sabike, I., Fadl, S.E., Sharoba, A.M., 2022. Orange pulp in Nile tilapia (Oreochromis niloticus) diets: growth performance, biochemical parameters and gene expression for growth and fat metabolism. Aquac. Rep. 22. doi:10.1016/j.aqrep.2021.100970.

United Nations, 2019. United Nations, Department of Economic and Social Affairs, Population Division (2019). World Population Prospects 2019, New York: United Nations. https://population.un.org/wpp/download/.

Woodard, F., 2001. Industrial Waste Treatment Handbook. Elsevier Butterworth-Heinemann, Boston, MA, USA, pp. 520–pp.

Zerai, D.B., Fitzsimmons, K.M., Collier, R.J., Duff, G.C., 2008. Evaluation of brewer's waste as partial replacement of fish meal protein in nile tilapia, oreochromis niloticus, diets. J. World Aquac. Soc. 39 (4), 556–564. doi:10.1111/j.1749-7345.2008.00186.x.

CHAPTER

15

Aquamimicry: Concept and application for sustainable aquaculture

S.M. Rafiquzzaman[a], *Md Shoebul Islam*[b], *Koushik Chakroborty*[c] *and Mohammad Hossein Khanjani*[d]

[a]Department of Fisheries Biology and Aquatic Environment, Gazipur Agricultural University, Gazipur, Bangladesh [b]Shrimp Research Station, Bangladesh Fisheries Research Institute, Bagerhat, Bangladesh [c]Department of Aquatic Environment and Resource Management, Sher-e-Bangla Agricultural University, Dhaka, Bangladesh [d]Department of Fisheries Sciences and Engineering, Faculty of Natural Resources, University of Jiroft, Jiroft, Iran

15.1 Introduction

Aquaculture continues to serve as a significant provider of food and nutrition for millions of individuals globally, playing a crucial role in addressing the growing demand for food. World aquaculture production reached a new record of 130.9 million tons (MT) in 2022, up by 6.6% from 2020 and comprising 94.4 MT of aquatic animals and 36.5 MT of algae. Asia contributed 91.4% of total aquaculture production, followed by Latin America and the Caribbean (3.3%), Europe (2.7 %), Africa (1.9%), North America (0.5%), and Oceania (0.2%). Ten leading countries (China, Indonesia, India, Viet Nam, Bangladesh, Philippines, Republic of Korea, Norway, Egypt, and Chile) produced 89.8% of the total. For the first time in history, in 2022, production of animal species from aquaculture (51%) surpassed that from capture fisheries. This production increased by 6.7 MT (FAO, 2024). This rapid expansion and intensification of aquaculture has also come at the cost of significant environmental and water quality degradation, the emergence of diseases, and inefficient resource allocation, which has raised concerns regarding the long-term growth and sustainability of the industry.

Zero-Waste Technologies in Aquaculture: Global Perspectives and Methods toward a Circular Economy
DOI: https://doi.org/10.1016/B978-0-443-34689-7.00019-7

To accomplish sustainable production with environmental considerations, it is imperative to develop green and organic technologies. Thus in recent years, technological advancements have been employed to mitigate harmful environmental effects, enhance biosecurity, and lower production expenses. Furthermore, there are various approaches to effectively manage aquaculture wastewater, including the utilization of nanomaterials, bioaugmentation technology, implementation of solid-state thermophilic aerobic fermentation, and in situ water bioremediation technologies such as biofloc technology (BFT) (Khanjani et al., 2022). Nevertheless, shellfish farming practices in resource-poor countries, particularly shrimp aquaculture, have typically been traditional (extensive or semiintensive). If good aquaculture practices are not adhered to, these farms could give rise to the aforementioned problems. To minimize these issues, it is vital for the industry to adopt technologies that support green aquaculture, even in semiintensive farms. Among various strategies, aquamimicry is considered a microbial-based approach that aims to achieve high yields while ensuring improved environmental practices (Khanjani et al., 2023). Aquamimicry is a versatile process that can be applied in different settings with adjustments, making it a potential transformative addition to environmentally friendly aquaculture, particularly for lower-income countries.

15.2 Principles of aquamimicry: Definition and principles

Aquamimicry acts as a link between conventional systems and technologically advanced systems. The introduction of this technique was a consequence of improving the fertilization process in shrimp ponds and reducing production expenses by implementing low stocking densities to avoid the appearance of diseases, such as early mortality syndrome. The "Aquamimicry System" was established in 2013 in Thailand by two experienced shrimp cultivators. This strategy bases itself on the idea that prawn farming practices can achieve more sustainability by replicating the natural aquatic environment within the confines of aquaculture conditions. In this system, the usage of chemicals is circumvented by implementing a synbiotic system, which is established through the utilization of prebiotics and probiotics that exhibit a beneficial influence on the host (Zeng et al., 2020).

The Aquamimicry Aquaculture Alliance defines aquamimicry as a simulator of the natural conditions that stimulates the growth of beneficial bacteria, phytoplankton blooms, and zooplankton populations, which serve as supplemental nutrition for cultivated shrimp and help maintain water quality (Catalani et al., 2023). It is a well-balanced and economical method that mimics natural circumstances by feeding shrimps with zooplankton and using beneficial bacteria to enhance water quality. The fundamental concept behind this approach is the conviction that shrimp or fish farming can attain enhanced sustainability by replicating the natural aquatic environment in aquaculture facilities. Thus the basic principle of this system is consistently using fermented carbon sources, such as rice bran, soybean meal, and wheat bran, in tandem with water probiotics such as *Bacillus* sp., which significantly boost the growth of zooplankton populations (dos Santos et al., 2022; Santos et al., 2022). This technology recreates the environmental conditions prevalent in pond culture and effectively produces food organisms in a manner that is similar to their natural habitat. This method primarily aims to promote the growth of copepods, one of the dominating species of zooplankton, which serve as a vital food source for cultured species. Planktons are regarded as a source of additional nourishment and contribute to enhancing the water quality in fish and prawn farming (Khanjani et al., 2023).

Microorganisms (bacteria and yeasts) are used to increase the solubility of the carbon source in water and accelerate the breakdown of nutrients. They are recognized for their ability to secrete hydrolytic enzymes that reduce fiber and carbohydrates, while also increasing the solubility of proteins in carbon sources (Romano, 2017). Water additives, such as sodium bicarbonate, can be utilized to regulate the pH levels in the water body. The presence of live food, particularly copepods that are abundant in long-chain polyunsaturated fatty acids (LC-PUFA), minerals, pigments, and a variety of free amino acids (such as taurine), has the potential to promote feed efficiency and boost the immune system of the shrimps. Also, the use of fermented carbon sources and probiotics can enhance water quality and improve the recycling of nitrogenous metabolites within the system. The synbiotic relationship between fermented carbon source derivatives (such as oligosaccharides) and probiotics (such as *Bacillus* species) in this technology helps to decrease the use of medicinal substances and promotes environmentally friendly aquaculture practices (Khanjani et al., 2022; Nisar et al., 2022). Moreover, aquamimicry addresses the constraints of its closest culture technique, the typical biofloc system, by integrating components of conventional approaches. This modification has demonstrated encouraging outcomes by upgrading the quality of water and improving the zootechnical performance in the farming environment (Catalani et al., 2023).

15.3 Processes involved in aquamimicry

15.3.1 Aquamimicry pond design

15.3.1.1 Key considerations

The design of aquamimicry ponds is fundamental to replicating natural aquatic ecosystems, creating conditions that support a healthy microbial community, efficient nutrient cycling, and a stable environment for cultured species such as fish and shrimp. Key design elements include:

(1) Layout: The shape and overall layout of aquamimicry ponds must promote uniform water movement to minimize the formation of stagnant areas or "dead zones," where water does not circulate properly (Hadiyanto et al., 2013; Huang et al., 2015). Typically, rectangular pond shapes are preferred, as they allow for better water flow dynamics (Naghizadeh et al., 2022). This design ensures that nutrients, microorganisms, and oxygen are evenly distributed, benefiting both the cultured species and microbial populations that contribute to the system's health.

(2) Water flow: Controlled and continuous water flow is a critical element in aquamimicry pond design, as it helps simulate natural tidal or riverine movements. Water flow ensures the removal of waste materials, promotes nutrient cycling, and sustains the microbial ecosystem essential for maintaining water quality (Zhou et al., 2024). Strategically placed inlets and outlets promote circular water currents that enhance microbial activity and maintain oxygen-rich conditions throughout the pond. This flow system helps ensure that the pond water remains nutrient-balanced and properly oxygenated, supporting both the cultured species and the microbial communities.

(3) Central pit for waste removal: One distinctive feature of aquamimicry pond design is the inclusion of a central pit at the pond's lowest point (Fig. 15.1).

FIGURE 15.1 Design of aquamimicry pond with central pit. *From Islam et al. (2024).*

FIGURE 15.2 Aerators in the Pond of aquamimicry culture system pond.

This pit is designed to collect waste, including uneaten feed, feces, and other organic debris that accumulate during the culture process. The central pit helps concentrate waste materials, allowing for easier removal without disturbing the pond's microbial balance. This design feature helps maintain water quality by reducing the buildup of harmful compounds, such as ammonia and hydrogen sulfide, which can be detrimental to both fish and shrimp if not properly managed. Periodic waste removal from the central pit is an essential practice for keeping the pond environment clean and supporting optimal growth conditions (Islam et al., 2024).

Aeration: Adequate aeration is essential to maintaining high dissolved oxygen levels, which are critical for both the cultured species and the aerobic microbial communities responsible for breaking down organic matter (Raza et al., 2024). Aeration systems such as paddlewheels, air diffusers, or airlift pumps are commonly used to ensure proper oxygen transfer throughout the water column (Fig. 15.2).

These systems help to maintain an oxygen-rich environment, which supports the activities of beneficial microbes that decompose organic waste and convert harmful compounds into less toxic forms (Zhang et al., 2023). Aeration also mimics natural processes, such as water flow and plankton photosynthesis, ensuring that oxygen levels remain sufficient for the pond ecosystem.

Aquamimicry pond design, with its focus on water movement, aeration, and efficient waste management, creates a stable and self-sustaining environment that mimics natural aquatic habitats. This design approach offers multiple benefits for the cultured species:

(1) **Efficient waste management**: The central pit, combined with controlled water flow and aeration, helps concentrate and remove organic waste, reducing the risk of harmful byproducts such as ammonia buildup. This leads to a cleaner and healthier environment for the cultured species (Nisar et al., 2022; Islam et al., 2024).
(2) **Enhanced water quality:** Continuous water circulation and aeration support microbial activity and nutrient cycling, helping to maintain water quality at optimal levels (Khanjani et al., 2022; Kumar et al., 2024). These systems reduce the accumulation of harmful substances, supporting the health and growth of the fish or shrimp.
(3) **Improved species health:** The carefully managed environment reduces stress on the cultured species by providing stable conditions that mimic their natural habitats (Anderson, 1996; Canosa and Bertucci, 2023). Reduced stress leads to better growth performance, lower disease susceptibility, and improved overall health.

15.3.2 Microbial dynamics in aquamimicry systems

15.3.2.1 Role of probiotics in aquamimicry

In aquamimicry, microbial dynamics play a critical role in maintaining pond health and ensuring optimal growth conditions for the cultured species. Probiotics, which are beneficial bacteria introduced into the system, form the cornerstone of microbial management. These bacteria promote the establishment of a stable and diverse microbial community, which is essential for nutrient cycling, water purification, and disease control (Abakari et al., 2022; Khanjani et al., 2022). Different bacteria and their function are described in Table 15.1.

These microorganisms work synergistically within the pond to maintain water quality, gut health, and help in feed utilization, enhance nutrient cycling, and provide a healthier environment for the cultured species. Their ability to break down organic waste and inhibit harmful pathogens is a cornerstone of the sustainability of aquamimicry systems.

One of the most significant roles of probiotics in aquamimicry is disease prevention (Islam et al., 2024; Khanjani et al., 2022). Probiotics help suppress the growth of pathogenic bacteria such as *Vibrio* species, which can cause vibriosis in shrimp (Tamilselvan and Raja, 2024; Wei et al., 2024). Probiotics achieve this through a process called competitive exclusion, where beneficial bacteria occupy surfaces and resources that would otherwise be used by pathogens.

By reducing the presence of harmful bacteria and boosting the immune response of shrimp and fish, probiotics help minimize the need for chemical treatments and antibiotics, which supports more sustainable and eco-friendly aquaculture practices.

15.3.3 Plankton-based mimicry: The natural environment

Aquamimicry systems are specifically designed to foster the growth of natural plankton populations, which play critical roles in maintaining a healthy ecosystem. Phytoplankton, through photosynthesis, contribute to oxygenation during daylight, while zooplankton, especially copepods, serve as essential natural food sources for fish and shrimp (Islam et al., 2024; Khanjani et al., 2022). In aquamimicry, by carefully managing light and nutrient levels, pond operators encourage a diverse plankton community, improving both water quality and overall pond health.

TABLE 15.1 Role of different probiotics for aquamimicry systems.

Type of Bacteria	Bacterial species	Functions	References
Lactic acid bacteria	*Lactobacillus* sp.	Known for enhancing gut microbiota, improving feed conversion, and boosting immunity	Chizhayeva et al. (2022); Rastogi and Singh (2022)
	Pediococcus sp.	Plays a role in promoting digestive health and inhibiting pathogenic bacteria	Porto et al. (2017); Eissa et al. (2022); Eissa et al. (2023)
	Bifidobacterium	Contribute to maintaining a balanced gut flora, enhancing nutrient absorption, and supporting immune responses	Huang et al. (2022); Gavzy et al. (2023)
	Enterococcus sp.	Contribute to maintaining a balanced gut flora, enhancing nutrient absorption, and supporting immune responses	Tachibana et al. (2020); Sun et al. (2024)
Gram-positive spore-forming bacteria	*Bacillus subtilis*	Improves digestive enzyme production, enhances gut health, and boosts immune function	Olmos et al. (2020); Monier et al. (2023); Islam et al. (2024)
	Bacillus licheniformis	Aids in the breakdown of organic matter, reducing ammonia levels, and promoting a healthier pond environment	Gopi et al. (2022); Monier et al. (2023)
	Bacillus pumilus	Known for their roles in reducing harmful bacteria and promoting overall water quality	Hlordzi et al. (2020); Ringø (2020)
	Bacillus megaterium	Known for their roles in reducing harmful bacteria and promoting overall water quality.	Hlordzi et al. (2020)
	Bacillus amyloliquefaciens	Helps in controlling pathogenic bacteria and supports the digestion of complex nutrients	Llario et al. (2019); Ngalimat et al. (2021)
	Bacillus clausii	Promotes immune responses in shrimp and aids in digestion	Ye et al. (2011); Wang et al. (2018)
Other important bacteria	*Rhodococcus*	Plays a critical role in degrading organic matter and enhancing the cycling of nutrients within the pond ecosystem	Bodhaguru et al. (2019); Islam et al. (2024)
	Rhodobacter	Help in managing the nitrogen cycle and improving water quality	Li et al. (2022); Islam et al. (2024)

15.3.3.1 *Nutrient cycling*

Plankton populations are central to nutrient cycling within aquamimicry ponds. Phytoplankton absorb inorganic nutrients, such as nitrogen and phosphorus, converting them into organic matter, which feeds zooplankton and other filter feeders (Khanjani et al., 2022; Romano et al., 2018) . This process helps prevent nutrient buildup, reducing the risk of water quality issues such as eutrophication. Copepod zooplankton, which are particularly promoted in aquamimicry systems, play an important role in this process by feeding on phytoplankton, thus supporting nutrient cycling and maintaining ecological balance (Islam et al., 2024; Khanjani et al., 2022).

15.3.3.2 *Natural food sources*

The production of copepod zooplankton in aquamimicry systems reduces reliance on external formulated feeds, making the aquaculture system more self-sustaining. This natural feeding mechanism significantly lowers feed costs and minimizes the ecological footprint of shrimp farming. Copepods are particularly valuable as a food source for shrimp, supporting their growth and health, while also contributing to more stable and efficient nutrient utilization in the pond environment (Khanjani et al., 2022).

By cultivating a rich planktonic ecosystem, aquamimicry promotes a sustainable and eco-friendly approach to aquaculture, with copepod zooplankton serving as a key factor for the success of shrimp farming.

15.3.4 Fermentation processes in aquamimicry

Fermentation is a key element in aquamimicry systems, especially in the preparation of feed and probiotics. The process of fermentation involves the breakdown of organic materials by beneficial microorganisms, which enhances the nutritional quality of feed and supports microbial communities in the pond environment (Abdel-Tawwab et al., 2022; Supriyati et al., 2015; Yang et al., 2021). Moreover, it has several functions:

15.3.4.1 *Nutrient enhancement*

Fermentation makes nutrients more bioavailable by breaking down complex molecules into simpler ones that are easier for fish and shrimp to digest (Siddiqui et al., 2025; Xiang et al., 2019). For example, proteins are partially hydrolyzed into peptides and amino acids, which are more readily absorbed by the digestive systems of the cultured species. This process also reduces antinutritional factors present in some plant-based feed ingredients, such as soybean meal (Islam et al., 2024).

15.3.4.2 *Probiotic enrichment*

Fermented feeds are rich in probiotics, which can further enhance the health of the cultured species by promoting a healthy gut microbiota. Probiotic bacteria such as *Lactobacillus* spp. and *Bacillus* spp. are often included in the fermentation process to enhance the feed's probiotic content. These bacteria continue to benefit the fish or shrimp after consumption by improving gut health and immunity (Lian et al., 2024).

15.3.4.3 *Reduced waste and improved water quality*

Fermentation pre-digests the feed, which results in less waste being excreted by the fish or shrimp. This reduction in waste output helps maintain cleaner pond water and reduces the need for frequent water exchange or additional water treatments (Islam et al., 2024; Khanjani et al., 2022) . The overall impact is a more efficient and sustainable system with fewer environmental impacts.

Besides, the source of carbohydrates used in the fermentation process plays a critical role in determining its effectiveness and nutritional outcomes. Rice bran, wheat bran, and soybean meal are commonly used carbohydrate sources, and each offers distinct benefits for fermentation and subsequent feed enrichment (Chen et al., 2024; Islam et al., 2024; Lim et al., 2024).

15.3.5 Water quality management in aquamimicry

Maintaining optimal water quality is critical to the success of aquamimicry systems, as it directly affects the health of both the cultured species and plankton, and the microbial communities upon which the system depends. Effective water quality management involves the integration of natural filtration systems and continuous monitoring of water parameters.

15.3.5.1 *Natural filtration systems*

In aquamimicry, natural filtration methods are often used to maintain water quality without relying on chemical treatments or artificial filtration systems (Islam et al., 2024). Constructed wetlands or biofilters, which utilize aquatic plants and microorganisms to remove excess nutrients, are commonly employed. These systems work by absorbing nitrogen and phosphorus, preventing eutrophication and maintaining water clarity (Khanjani et al., 2022).

15.3.5.2 *Biofilm formation*

Biofilms, which consist of microorganisms that attach to substrates such as pond walls and aerators. They play a key role in breaking down organic matter and processing waste (Sharma et al., 2023). These biofilms act as living filters, improving water quality by consuming organic pollutants and converting them into less harmful substances (Islam et al., 2024).

15.3.5.3 *Monitoring water parameters*

Continuous monitoring of critical water parameters is necessary to maintain the health of aquamimicry systems. Key parameters include dissolved oxygen, ammonia, nitrites, nitrates, and pH. Automated monitoring systems that provide real-time data are increasingly used in modern aquaculture operations. These systems enable farmers to make timely adjustments to prevent water quality problems that could lead to stress or disease outbreaks in the cultured species.

15.3.6 Aquamimicry culture techniques for fish and shrimp

The culture techniques employed in aquamimicry systems are designed to replicate the natural conditions in which fish and shrimp thrive. This involves careful consideration of species selection, stocking densities, feeding practices, and overall system management.

15.3.6.1 Species selection

Aquamimicry systems are most effective when used to culture species that are well-adapted to environments with fluctuating microbial and plankton dynamics. Species such as tilapia and shrimp (*Litopenaeus vannamei, Penaeus monodon*) are ideal for aquamimicry because they can tolerate variations in water quality and benefit from the natural food sources provided by plankton and bioflocs (Khanjani et al., 2022; Nisar et al., 2022).

15.3.6.2 Stocking densities

In aquamimicry, lower stocking densities are preferred to reduce stress, minimize disease risks, and maintain water quality by lowering the organic load in the system. However, farmers can stock at higher densities if they implement effective management strategies. By using fermented feed and probiotics, waste output is reduced, and gut health is enhanced, mitigating the impact of higher densities. In addition, proper aeration and BFT can help maintain water quality and nutrient recycling, allowing for sustainable, higher-density farming while still minimizing environmental stress. This balance ensures that aquamimicry can support both lower and higher stocking densities depending on the management approach.

15.3.6.3 Feeding practices

Natural food sources, such as plankton, form a significant part of the diet in aquamimicry systems. These are produced in an aquamimicry pond (Islam et al., 2024), which is nutritionally enhanced and contains probiotics to further support the health of the cultured species.

15.4 Understanding synbiosis in aquamimicry

The primary sources of organic carbon for aquamimicry typically consist of carbohydrates derived from monosaccharides and disaccharides (such as glucose, molasses, sucrose, and glycerol) to polysaccharides (such as rice bran, soy, wheat, and starch), with the specific choice depending on the geographical location. Polysaccharides, such as rice, wheat, and soybean bran, are usually fermented with *Bacillus* spp. and other probiotic microorganisms (such as *Lactobacillus* and *Saccharomyces*) to improve the levels of soluble sugars, crude protein, and lipid content, and reduce levels of crude fiber and antinutritional factors resulting from hydrolytic enzyme production and secondary metabolites such as liposaccharides and peptidoglycans (Santos et al., 2022). The utilization of rice bran in a synbiotic system has been found to be advantageous because of its cost-effectiveness and low protein content. This characteristic helps to prevent issues related to nitrogen in the system, resulting in positive impacts on fish, shrimps, and prawns (Romano et al., 2018).

This process, typically conducted in nearby bioreactors, particularly during the initial weeks of farming, enhances the influx of nutrients into the system and induces a substantial boost in the microbial population within the culture ponds. A synbiotic relationship is established when there is a balance between bacteria, phytoplankton, and zooplankton in the system, as shown by a microbial community color index (MCCI) of around 1–1.5 (Hargreaves, 2013). A biofertilizer is developed in this process by fermenting agroindustrial residue and probiotic bacteria using a semisolid and/or liquid fermentation technique that is enriched with nutrients and contains a high concentration of beneficial microbes for the environment.

Previously, it was discussed that synbiosis in aquamimicry leads to the production of zooplankton. In this system, copepods are considered the dominant zooplankton species. They are typically noticed from the second week of cultivation, suggesting a mature state of the system. Thus, this technology is referred to as "biomimicry" and "copefloc" in certain literature. The water's microbial ecology, primarily consisting of bacteria, may lack necessary amino acids and polyunsaturated fatty acids (ω-3), crucial for shrimp development (Castro et al., 2021). In this context, copepods serve as significant providers of proteins, lipids, amino acids, polyunsaturated fatty acids, and other nutrients. Moreover, they consume phytoplankton and serve as the primary source of nutrition for shrimps, enhancing the nutritional value and survivability of shrimp. Copepods can also contribute directly to the system by building microhabitats that promote bacterial development (Khanjani et al., 2023). A decreased C:N ratio can lead to a boom of phytoplankton, serving as a food source for zooplankton. The fermented carbon, probiotics, and supplements, and a balanced and diverse population of phytoplankton and zooplankton, can supply essential macro and micronutrients and establish a favorable environment for the growth and well-being of shrimp (dos Santos et al., 2022; Santos et al., 2022). Furthermore, aquamimicry systems maintain a suitable water quality for phytoplankton and enhance the level of dissolved oxygen, hence lowering the requirement for heavy aeration (Romano, 2017; Nisar et al., 2022).

The synbiosis relies on the principle of maintaining equilibrium in the interaction between phytoplankton and other microorganisms, leading to the creation of a natural habitat with optimal water quality conditions. In aquamimicry, the key concept is the regular use of fermented carbon/nitrogen sources. However, currently, there is a scarcity of scientific data regarding synbiotics and aquamimicry. Furthermore, there is no unanimous agreement regarding a singular optimal method (Khanjani et al., 2023).

15.5 Advantages and contributions toward sustainability

Aquamimicry, also called aquaculture mimicry, is a revolutionary idea that includes mimicking conducive natural conditions by live feed production, applying probiotics, and fermented feed. This concept is applicable in a wide range of conditions and can be a very useful tool to promote sustainable aquaculture.

15.5.1 Improved water quality and shrimp growth

Aquamimicry establishes an ecosystem that maintains optimal and consistent water quality, beyond that of the conventional culture system. The beneficial bacteria and natural production of plankton contribute to maintaining water quality and also serve as a food source. The shrimps in this system receive fermented carbohydrates, probiotics, and zooplankton as part of their diet. All of these components enhance the overall nutritional quality of farmed shrimps. By replicating natural conditions in this system, it is possible to stimulate enhanced growth performance in shrimps. The high availability of zooplankton, particularly copepods, reduces the feed conversion ratios.

15.5.2 Waste recycling

Furthermore, the aquamimicry process addresses the nutrient waste issue, allowing waste recycling within the system. Typically, the cycle of operation of nutrients depends solely on the soil microbes. However, the intensification of cultural practices demands an increased microbial population to effectively manage food residue and feces decomposition. The microorganisms responsible for the fermentation of carbon sources, in conjunction with water probiotics, improve nutrient recycling in aquamimicry, hence, diminishing waste production.

15.5.3 Immunity enhancement and biosecurity

Enhancing the ecological and welfare aspects of shrimp farming by replicating natural conditions decreases the probability of disease outbreaks. The cultured organism is farmed in an environment devoid of stress, since it does not foster conditions suitable for pathogenic and harmful organisms, and reduces the formation of black soil. In addition, probiotics that contain bacterial secondary metabolites, such as liposaccharides and peptidoglycans, enhance the immune system of shrimps, resulting in improved health (Deepak et al., 2020; Khanjani et al., 2023). Aquamimicry can be employed in extensive and semiintensive production systems, resulting in decreased feed consumption and water exchange rates. The restricted rate of water exchange mitigates stress conditions in cultivated species and improves biosecurity.

15.5.4 Low cost and feasible implementation

Implementing aquamimicry will lead to higher production, lower expenses, and improved profitability. Reducing reliance on commercial feeds minimizes the biological oxygen demand (BOD) and consequently, the requirement for vigorous aeration, resulting in enhanced energy consumption efficiency. Inexpensive carbon sources such as molasses and rice bran are utilized in this culture technique. This decreases the expenditure of feed, which is the largest contributor to costs in modern aquaculture. This enhances the practicality of aquamimicry in low-income communities. Aquaculture producers can easily adopt it with minimal technical experience. Because this system necessitates limited technology and infrastructure for its installation and operation (Deepak et al., 2020).

15.6 Challenges and technology gaps of aquamimicry

While aquamimicry offers a promising alternative for sustainable aquaculture, several challenges and gaps hinder its broader adoption and optimal functioning. Addressing these issues is crucial to ensure its success in promoting a zero waste, circular economy in aquaculture.

15.6.1 Lack of studies on the management and improvement of the system

One of the critical challenges facing aquamimicry is the lack of comprehensive research on how to effectively manage and improve the system over time. Aquamimicry relies on mimicking natural aquatic ecosystems (Islam et al., 2024), but the specific conditions that need to be replicated can vary widely between regions, climates, and species. Limited studies have been conducted to establish best practices for optimizing the performance of aquamimicry systems across diverse environments. This creates a gap in knowledge about how to adapt and fine-tune the technology for long-term efficiency and stability. Without robust data and case studies to guide management, practitioners may struggle to maintain system health, leading to suboptimal results.

15.6.2 No detailed understanding of synbiotics in aquamimicry

Synbiotics, which combine probiotics and prebiotics to support healthy microbial communities, are a critical component of aquamimicry (Khanjani et al., 2023). However, there is still no detailed understanding of how synbiotics function within the aquamimicry framework. The role of specific microbial interactions and their impacts on water quality, nutrient cycling, and fish health are underexplored, resulting in a lack of clear guidelines on how to optimize synbiotic usage. This gap in knowledge limits the potential of synbiotics to enhance aquamimicry's performance, particularly in terms of water purification and disease prevention. More research is needed to define the ideal microbial communities and understand how to maintain their balance in aquamimicry systems.

15.6.3 Commercialization and economics (Pricing)

Another challenge facing aquamimicry is the uncertainty surrounding its commercialization and the economics of implementing such systems on a large scale. While the concept has gained attention in small-scale and experimental settings, there is little clarity on how aquamimicry can be economically viable for larger commercial operations. The upfront costs of implementing the necessary infrastructure, along with the potential need for ongoing system monitoring and management, make it difficult to estimate the overall financial feasibility. Furthermore, without clear economic models, aquaculture operators may be hesitant to adopt aquamimicry, fearing it may not be cost-competitive with conventional farming techniques. This highlights the need for more detailed studies on pricing structures and cost-benefit analyses for aquamimicry at commercial scales.

15.6.4 Farming system

Aquamimicry introduces a unique approach to aquaculture farming by attempting to recreate natural ecosystem functions. However, integrating this concept into existing aquaculture farming systems presents significant challenges. The transition from traditional farming methods to aquamimicry-based systems requires a shift in both physical infrastructure and operational mindset. Many conventional aquaculture practices are based on intensive, high-input farming systems that may not be directly compatible with the ecological principles of aquamimicry. This misalignment creates a gap in knowledge about how to redesign farming systems to accommodate the ecological needs of aquamimicry, particularly in terms of pond design, feed management, and waste recycling. The lack of a standardized farming model also makes it difficult to scale aquamimicry consistently across different regions and species.

15.7 Aquamimicry and biofloc technology

Aquamimicry and BFT differ significantly in their basic concepts, ecosystem management approaches and implementation, specific methods for water quality management, and their goals and benefits. Both of them have similarities in creating more sustainable aquaculture, improving water quality, increasing productivity, disease management, and applicability to a variety of aquatic species.

15.7.1 The differences

(1) **Conceptual foundation:** In aquamimicry, the system simulates the natural aquatic ecosystem in aquaculture ponds, with the aim of creating a balanced environment where different aquatic organisms can grow together and promote the growth of the main cultured species. This approach focuses on mimicking the biological and physical properties of natural waters to enhance both sustainability and productivity. BFT, on the other hand, focuses on recycling nutrients and organic matter present in water through the cultivation of beneficial microbial communities, known as bioflocs. These microorganisms convert unused feed and waste materials into valuable biomass that can be consumed by the cultured species, improving water quality and reducing the need for water exchange.

(2) **Ecosystem management approaches and implementation:** In aquamimicry, the system involves the addition of various natural components and organisms, such as algae, zooplankton, and other beneficial microbes to establish a balanced ecosystem. It emphasizes the significance of biodiversity in maintaining water quality and supporting cultured species. In BFT, precise water quality management is essential for fostering the growth of advantageous microorganisms. This approach often includes the use of aeration and mixing systems to maintain high dissolved oxygen levels and keep flocs suspended, as well as careful management of the carbon/nitrogen ratio to encourage microbial growth.

(3) **Water quality management:** In aquamimicry, the system achieves water quality management through the balance of the ecosystem itself, relying on the natural processes of biofiltration, nutrient cycling, and predator-prey relationships. The role of zooplankton in the aquamimicry system is important, and a fermentable carbon source is required to create an aquamimicry system. The inclusion of a carbon source in the aquamimicry system is unrelated to nitrogen levels. BFT directly manages water quality through the biological conversion of wastes into microbial biomass, which requires active monitoring and management of water chemistry, especially carbon and nitrogen levels. Managing the C:N ratio requires experience and should be considered at least above 10 to stimulate the activity of heterotrophic bacteria (Khanjani et al., 2023).

(4) **The goals and benefits:** In aquamimicry, the objective of the designed system is to promote the health and well-being of cultivated species by establishing a more natural and stress-free environment. Such enhancements may lead to enhanced resistance to diseases, increased growth rates in animals, and the production of higher-quality products. Conversely, BFT emphasizes the recycling of waste and the reuse of water to enhance production efficiency, minimize environmental impacts, and reduce costs associated with water and feed exchange.

15.7.2 The similarities

(1) **Focus on sustainability:** The goal of both approaches is to make aquaculture more sustainable by reducing environmental impacts, improving water use efficiency, and decreasing reliance on external feed inputs.

(2) **Water quality improvement:** Although through different mechanisms, both aquamimicry and BFTs work toward improving water quality, reducing the need for frequent water changes, and minimizing effluents.

(3) **Increased productivity:** Both methods can enhance productivity and efficiency in aquaculture practices by optimizing conditions for cultivated species and, in the case of biofloc, generating additional feed resources.
(4) **Disease management:** By maintaining better water quality and a more balanced environment, both methods can contribute to reducing the prevalence of diseases and the need for chemical treatments.
(5) **Applicability to different species:** Although specific approaches may vary, both aquamimicry and biofloc can be adapted and applied to various aquaculture systems and species, including shrimp, fish, and other aquatic organisms.

15.8 Conclusion

The application of aquamimicry systems in the culture of aquatic animals, such as shrimp, offers greater sustainability compared to conventional systems. This approach avoids the use of detrimental chemicals and antibiotics while concurrently minimizing production costs through the cultivation of feed, thereby fostering both economic viability and environmental sustainability. This technology simulates natural aquatic ecosystems in aquaculture ponds, allowing different aquatic organisms to grow together and improve the overall performance of aquaculture. Overall, this system is based on comprehensive aquaculture management, environmentally friendly practices, sustainable biodiversity, and economic viability. Further studies in the future are necessary to explore aquamimicry as a modern feeding system, factors influencing its effectiveness, water consumption optimization, feed cost reduction, increased profitability, and organic fish production.

References

Abakari, G., Wu, X., He, X., Fan, L., Luo, G., 2022. Bacteria in biofloc technology aquaculture systems: Roles and mediating factors. Reviews in Aquaculture 14 (3), 1260–1284.

Abdel-Tawwab, M., Khalil, R. H., Nour, A. M., Elkhayat, B. K., Khalifa, E., Abdel-Latif, H. M., 2022. Effects of Bacillus subtilis-fermented rice bran on water quality, performance, antioxidants/oxidants, and immunity biomarkers of White leg shrimp (*Litopenaeus vannamei*) reared at different salinities with zero water exchange. Journal of Applied Aquaculture 34 (2), 332–357. doi:10.1080/10454438.2020.1844110.

Anderson, D.P., 1996. 7 Environmental factors in fish health: immunological aspects. Fish Physiol. 15, 289–310. doi:10.1016/S1546-5098(08)60277-0.

Bodhaguru, M., Prakash, S., Ramasubburayan, R., Ahila, N.K, Mariselvam, L., Immanuel, G., Palavesam, A., Kannapiran, E., 2019. Screening, partial purification of antivibriosis metabolite sterol-glycosides from Rhodococcus sp. against aquaculture associated pathogens. Microb. Pathog. 134, 103597. doi:10.1016/j.micpath.2019.103597.

Canosa, L.F., Bertucci, J.I., 2023. The effect of environmental stressors on growth in fish and its endocrine control. Front. Endocrinol. 14. doi:10.3389/fendo.2023.1109461.

Castro, L.F., Pinto, R.C.C., Nunes, A.J.P., 2021. Nutrient value and contribution of microbial floc to the growth performance of juvenile shrimp, Litopenaeus vannamei, fed fatty acid and amino acid-restrained diets under a zero-water exchange intensive system. Aquaculture 531. doi:10.1016/j.aquaculture.2020.735789.

Catalani, K., Wasielesky, W., Zuñiga, R., Silva de Souza, M.á, Foes, G., 2023. Comparison between biofloc technology system and aquamimicry in the cultivation of Litopenaeus vannamei in lined ponds in Southern Brazil. Bol. Inst. Pesca 49. doi:10.20950/1678-2305/bip.2023.49.e787.

Chen, S, Dai, J, Chen, Y, Chen, Q, Dong, F, Wang, C, Sun, Y, Wang, J, Han, T, 2024. Effects of Bacillus subtilis-fermented soybean meal replacing fish meal on antioxidant activity, immunity, endoplasmic reticulum stress and hepatopancreas histology in Pacific white shrimp (Litopenaeus vannamei). Front. Mar. Sci. 11, 1449066. doi:10.3389/fmars.2024.1449066.

Chizhayeva, A., Amangeldi, A., Oleinikova, Y., Alybaeva, A., Sadanov, A., 2022. Lactic acid bacteria as probiotics in sustainable development of aquaculture. Aquat. Living Resour. 35, 10. doi:10.1051/alr/2022011.

Deepak, A.P., Vasava, R.J., Elchelwar, V.R., Tandel, D.H., Vadher, K.H., Shrivastava, V., Prabhakar, P., 2020. Aquamimicry: new an innovative apporoach for sustainable development of aquaculture. J. Entomol. Zool. Stud. 8, 1029–1031.

Eissa, E.I.-S.H., Baghdady, E.S., Gaafar, A.Y., El-Badawi, A.A., Bazina, W.K., Al-Kareem, O.M.A., El-Hamed, N.N.B.A., Dawood, M., 2022. Assessing the influence of dietary Pediococcus acidilactici probiotic supplementation in the feed of European sea bass (Dicentrarchus labrax L.) (Linnaeus, 1758) on farm water quality, growth, feed utilization, survival rate, body composition, blood biochemical parameters, and intestinal histology. Aquacult. Nutr. 2022, 1–11. doi:10.1155/2022/5841220.

Eissa, M.E.H., Alaryani, F.S., Elbahnaswy, S., Khattab, M.S., Elfeky, A., AbouelFadl, K.Y., Eissa, E.S.H., Ahmed, R.A., Van Doan, H., El-Haroun, E., 2023. Dietary inclusion of Pediococcus acidilactici probiotic promoted the growth indices, hemato-biochemical indices, enzymatic profile, intestinal and liver histomorphology, and resistance of Nile Tilapia against Aspergillus flavus. Anim. Feed Sci. Technol. 306. doi:10.1016/j.anifeedsci.2023.115814.

FAO, In Brief to The State of World Fisheries and Aquaculture 2024: Blue Transformation in action. Food and Agriculture Organization of the United Nations. Rome (2024), doi: 10.4060/cd0690en.

Gavzy, S.J., Kensiski, A., Lee, Z.L., Mongodin, E.F., Ma, B., Bromberg, J.S., 2023. Bifidobacterium mechanisms of immune modulation and tolerance. Gut Microbes 15 (2). doi:10.1080/19490976.2023.2291164.

Gopi, N., Iswarya, A., Vijayakumar, S., Jayanthi, S., Nor, S.A.M., Velusamy, P., Vaseeharan, B., 2022. Protective effects of dietary supplementation of probiotic Bacillus licheniformis Dahb1 against ammonia induced immunotoxicity and oxidative stress in Oreochromis mossambicus. Comp. Biochem. Physiol. C Toxicol. Pharmacol. 259, 109379. doi:10.1016/j.cbpc.2022.109379.

Hadiyanto, H., Elmore, S., Van Gerven, T., Stankiewicz, A., 2013. Hydrodynamic evaluations in high rate algae pond (HRAP) design. Chem. Eng. J. 217, 231–239. doi:10.1016/j.cej.2012.12.015.

Hargreaves J.A., Biofloc production systems for aquaculture. (2013),

Hlordzi, V., Kuebutornye, F.K.A., Afriyie, G., Abarike, E.D., Lu, Y., Chi, S., Anokyewaa, M.A., 2020. The use of Bacillus species in maintenance of water quality in aquaculture: A review. Aquac. Rep. 18. doi:10.1016/j.aqrep.2020.100503.

Huang, H.T., Hu, Y.F., Lee, B.H., Huang, C.Y., Lin, Y.R., Huang, S.N., Chen, Y.Y., Chang, J.J., Nan, F.H., 2022. Dietary of Lactobacillus paracasei and Bifidobacterium longum improve nonspecific immune responses, growth performance, and resistance against Vibrio parahaemolyticus in Penaeus vannamei. Fish and Shellfish Immunol. 128, 307–315. doi:10.1016/j.fsi.2022.07.062.

Huang, J., Qu, X., Wan, M., Ying, J., Li, Y., Zhu, F., Wang, J., Shen, G., Chen, J., Li, W., 2015. Investigation on the performance of raceway ponds with internal structures by the means of CFD simulations and experiments. Algal Res. 10, 64–71. doi:10.1016/j.algal.2015.04.012.

Islam, M.D.S., Das, M., Chakroborty, K., Lee, J.M, Islam, M.D.R., Rafiquzzaman, S.M., 2024. Aquamimicry improves the growth performance and immunity of black tiger shrimp (Penaeus monodon) in low saline ponds. Aquac. Rep. 36, 102082. doi:10.1016/j.aqrep.2024.102082.

Khanjani, M.H., da Silva, L.O.B., Fóes, G.K., Vieira, F.d.N., Poli, M.A., Santos, M., Emerenciano, M.G.C., 2023. Synbiotics and aquamimicry as alternative microbial-based approaches in intensive shrimp farming and biofloc: novel disruptive techniques or complementary management tools? a scientific-based overview. Aquaculture 567. doi:10.1016/j.aquaculture.2023.739273.

Khanjani, M. H., Mozanzadeh, M. T., Fóes, G. K., 2022. Aquamimicry system: a sutiable strategy for shrimp aquaculture–a review. Annals of Animal Science 22 (4), 1201–1210. doi:10.2478/aoas-2022-0044.

Kumar, S., Srivastava, P. K., Kumar, V., Seth, C. S., 2024. Biofloc technology: A sustainable approach towards wastewater utilization and fish production. Lakes & Reservoirs: Research & Management 29, e12449. https://doi.org/10.1111/lre.12449.

Li, Z., Li, L., Sun, H., Wang, W., Yang, Y., Qi, Z., Liu, X., 2022. Ammonia assimilation: a double-edged sword influencing denitrification of Rhodobacter azotoformans and for nitrogen removal of aquaculture wastewater. Bioresour. Technol. 345, 126495. doi:10.1016/j.biortech.2021.126495.

Lian, X., Shi, M., Liang, Y., Lin, Q., Zhang, L., 2024. The effects of unconventional feed fermentation on intestinal oxidative stress in animals. Antioxidants 13 (3), 305. doi:10.3390/antiox13030305.

Lim, M.-J., Barathikannan, K., Jeong, Y.-J., Chelliah, R., Vijayalakshmi, S., Park, S.-J., Oh, D.-H., 2024. Exploring the Impact of Fermentation on Brown Rice: Health Benefits and Value-Added Foods—A Comprehensive Meta-Analysis. Fermentation 10 (1), 3. https://doi.org/10.3390/fermentation10010003.

Llario, F., Falco, S., Sebastiá-Frasquet, M.T., Escrivá, J., Rodilla, M., Poersch, L.H., 2019. The role of Bacillus amyloliquefaciens on Litopenaeus vannamei during the maturation of a biofloc system. J. Mar. Sci. Eng. 7 (7). doi:10.3390/jmse7070228.

Monier, M.N., Kabary, H., Elfeky, A., Saadony, S., El-Hamed, N.N.B.A., Eissa, M.E.H., Eissa, E.S.H., 2023. The effects of Bacillus species probiotics (Bacillus subtilis and B. licheniformis) on the water quality, immune responses, and resistance of whiteleg shrimp (Litopenaeus vannamei) against Fusarium solani infection. Aquac. Int. 31 (6), 3437–3455. doi:10.1007/s10499-023-01136-1.

Naghizadeh, M., Moghbeli, A., Zounemat-Kermani, M., 2022. Experimental study of flow pattern in rectangular fish breeding ponds. Res. Sq. doi:10.21203/rs.3.rs-2086176/v1.

Ngalimat, M.S., Yahaya, R.S.R., Baharudin, M.M.A.A., Yaminudin, S.M., Karim, M., Ahmad, S.A., Sabri, S., 2021. A review on the biotechnological applications of the operational group bacillus amyloliquefaciens. Microorganisms 9 (3), 1–18. doi:10.3390/microorganisms9030614.

Nisar, U., Peng, D., Mu, Y., Sun, Y.u, 2022. A solution for sustainable utilization of aquaculture waste: a comprehensive review of biofloc technology and aquamimicry. Front. Nutr. 8. doi:10.3389/fnut.2021.791738.

Olmos, J., Acosta, M., Mendoza, G., Pitones, V., 2020. Bacillus subtilis, an ideal probiotic bacterium to shrimp and fish aquaculture that increase feed digestibility, prevent microbial diseases, and avoid water pollution. Arch. Microbiol. 202 (3), 427–435. doi:10.1007/s00203-019-01757-2.

Porto, M.C.W., Kuniyoshi, T.M., Azevedo, P.O.S., Vitolo, M., Oliveira, R.P.S., 2017. Pediococcus spp.: an important genus of lactic acid bacteria and pediocin producers. Biotechnol. Adv. 35 (3), 361–374. doi:10.1016/j.biotechadv.2017.03.004.

Rastogi, S., Singh, A., 2022. Gut microbiome and human health: exploring how the probiotic genus Lactobacillus modulate immune responses. Front. Pharmacol. 13. doi:10.3389/fphar.2022.1042189.

Raza, B., Zheng, Z., Yang, W., 2024. A review on biofloc system technology, history, types, and future economical perceptions in aquaculture. Animals. 14 (10), 1489. doi:10.3390/ani14101489.

Ringø, E., 2020. Probiotics in shellfish aquaculture. Aquac. Fish. 5 (1), 1–27. doi:10.1016/j.aaf.2019.12.001.

Romano, N., 2017. Aquamimicry: a revolutionary concept for shrimp farming. Glob. Aquac. Advoc. 10.

Romano, N., Dauda, A.B., Ikhsan, N., Karim, M., Kamarudin, M.S., 2018. Fermenting rice bran as a carbon source for biofloc technology improved the water quality, growth, feeding efficiencies, and biochemical composition of African catfish Clarias gariepinus juveniles. Aquac. Res. 49 (12), 3691–3701. doi:10.1111/are.13837.

dos Santos, R.B., Coelho-Filho, P.A., Assunção, C.S., dos Santos, T.N., da Silva, J.H., Silva, G.C., Brito, L.O., 2022. The effect of different synbiotic preparation strategies on water fertilization and zootechnical performance of Macrobrachium rosenbergii reared in the nursery stage. Aquac. Int. 30 (6), 3159–3178. doi:10.1007/s10499-022-00955-y.

Santos, R.B, Filho, P.A.C, Gonçalves, A.P, Santos, R.A, Rodrigues, M.L, Correia, E.S, Oliveira, V.Q, Brito, L.O, 2022. Effects of organic carbon sources on water quality, microbial flocs protein and performance of Macrobrachium rosenbergii post-larvae reared in biofloc and synbiotic systems. Aquac. Res. 53 (2), 388–397. doi:10.1111/are.15580.

Sharma, S., Mohler, J., Mahajan, S.D., Schwartz, S.A., Bruggemann, L., Aalinkeel, R., 2023. Microbial biofilm: a review on formation, infection, antibiotic resistance, control measures, and innovative treatment. Microorganisms. 11 (6). doi:10.3390/microorganisms11061614.

Siddiqui, S.A., Lakshmikanth, D., Pradhan, C., Farajinejad, Z., Castro-Muñoz, R., Sasidharan, A., 2025. Implementing fermentation technology for comprehensive valorisation of seafood processing by-products: a critical review on recovering valuable nutrients and enhancing utilisation. Crit. Rev. Food Sci. Nutr. 65 (5), 964–991. doi:10.1080/10408398.2023.2286623.

Sun, X., Fang, Z., Yu, H., Zhao, H., Wang, Y., Zhou, F., Zhao, L., Sun, J., Tian, Y., 2024. Effects of Enterococcus faecium (R8a) on nonspecific immune gene expression, immunity and intestinal flora of giant tiger shrimp (Penaeus monodon). Sci. Rep. 14 (1). doi:10.1038/s41598-024-52496-4.

Supriyati, T. H., Susanti, T., Susana, I. W. R., 2015. Nutritional value of rice bran fermented by *Bacillus amyloliquefaciens* and humic substances and its utilization as a feed ingredient for broiler chickens. Asian-Australasian Journal of Animal Science 28 (2), 231–238. doi:10.5713/ajas.14.0039.

Tachibana, L., Telli, G.S., de Carla Dias, D., Gonçalves, G.S., Ishikawa, C.M., Cavalcante, R.B., Natori, M.M., Hamed, S.B., Ranzani-Paiva, M.J.T., 2020. Effect of feeding strategy of probiotic Enterococcus faecium on growth performance, hematologic, biochemical parameters and non-specific immune response of Nile tilapia. Aquac. Rep. 16. doi:10.1016/j.aqrep.2020.100277.

Tamilselvan, M., Raja, S., 2024. Exploring the role and mechanism of potential probiotics in mitigating the shrimp pathogens. Saudi J. Biol. Sci. 31 (3), 103938. doi:10.1016/j.sjbs.2024.103938.

Wang, J., Yang, H.-L., Xia, H.-Q., Ye, J.-D., Lu, K.-L., Hu, X., Feng, Y., Ruan, L., Sun, Y.-Z., 2018. Supplementation of heat-inactivated Bacillus clausii DE5 in diets for grouper, Epinephelus coioides, improves feed utilization, intestinal and systemic immune responses and not growth performance. Aquacult. Nutr. 24 (2), 821–831. doi:10.1111/anu.12611.

Wei, G., Tay, M.Y.F., Kamaja, V.K., Chan-Park, M.B., 2024. Synbiotic encapsulation against Vibrio parahaemolyticus infection in whiteleg shrimps. Aquaculture 590. doi:10.1016/j.aquaculture.2024.741051.

Xiang, H., Sun-Waterhouse, D., Waterhouse, G.I.N., Cui, C., Ruan, Z., 2019. Fermentation-enabled wellness foods: a fresh perspective. Food. Sci. Hum. Wellness 8 (3), 203–243. doi:10.1016/j.fshw.2019.08.003.

Yang, L., Zeng, X., Qiao, S., 2021. Advances in research on solid-state fermented feed and its utilization: The pioneer of private customization for intestinal microorganisms. Animal Nutrition 7 (4), 905–916. doi:10.1016/j.aninu.2021.06.002.

Ye, J.D., Wang, K., Li, F.D., Sun, Y.Z., 2011. Single or combined effects of fructo- and mannan oligosaccharide supplements and Bacillus clausii on the growth, feed utilization, body composition, digestive enzyme activity, innate immune response and lipid metabolism of the Japanese flounder Paralichthys olivaceus. Aquacult. Nutr. 17 (4), e902. doi:10.1111/j.1365-2095.2011.00863.x.

Zeng, S., Khoruamkid, S., Kongpakdee, W., Wei, D., Yu, L., Wang, H., Deng, Z., Weng, S., Huang, Z., He, J., Satapornvanit, K., 2020. Dissimilarity of microbial diversity of pond water, shrimp intestine and sediment in aquamimicry system. AMB Expr. 10 (1). doi:10.1186/s13568-020-01119-y.

Zhang, D., Wang, Y., Yang, H., Lan, S., Chen, C., Dai, B., Wang, C., Li, X., Xie, Y., 2023. Using intermittent moving aeration to repair hypereutrophic pond: nutrient removal efficiency and microbial diversity analysis. Environ. Sci. Pollut. Res. 30 (16), 46697–46710. doi:10.1007/s11356-023-25368-z.

Zhou, Q., Wu, K., Yao, L., Chen, R., Liu, S., Xing, H., Nie, L., Wu, Z., 2024. Bio-ecological remediation of freshwater aquaculture environments: a systematic review and bibliometric analysis. Water Biol. Secur. 3 (1), 100229. doi:10.1016/j.watbs.2023.100229.

CHAPTER

16

FLOCponics as a sustainable and environmentally friendly system

Otávio Augusto Lacerda Ferreira Pimentel[a], *Dariano Krummenauer*[b], *Wilson Wasielesky*[b] *and Mohammad Hossein Khanjani*[c]

[a]Department of Fisheries and Aquaculture, Federal Rural University of Pernambuco (UFRPE), Recife, Pernambuco, Brazil [b]Aquaculture Marine Station, Institute of Oceanography, Federal University of Rio Grande, Rio Grande, Brazil [c]Department of Fisheries Sciences and Engineering, Faculty of Natural Resources, University of Jiroft, Jiroft, Iran

16.1 Introduction

Over the years, aquaculture has demonstrated growth in the production of aquatic organisms, surpassing fishing production (The State of World Fisheries and Aquaculture 2026 – Blue Transformation: turning vision into impact, 2026). This growth is accompanied by the process of intensification of activity, which seeks to increase the productivity of the systems. However, the increase in animal biomass produced in tanks brings with it the accumulation of toxic nitrogenous compounds to animals, such as ammonia and nitrite, and organic matter in the system (Laktuka et al., 2023). This increases the number of discharges carried out during culture, releasing these wastes into natural systems after passing through settling tanks (Blancheton et al., 2007). These factors contributed strongly to the creation of closed systems that combined greater efficiency in water use with increased productivity.

Thus the biofloc technology system (BFT system) emerged to meet this need for sustainable and environmentally friendly systems. The BFT system is based on manipulating the carbon: nitrogen (C:N) ratio of water by the addition of organic carbon sources to immobilize the total ammonia produced by animals through the growth of heterotrophic bacteria (Avnimelech, 1999). In this system, chemoautotrophic bacteria also develop, being responsible for controlling nitrogenous compounds through the oxidation of ammonia to nitrite and finally nitrate, which is a less toxic compound for cultivated animals (Abakari et al., 2021). Stimulating

Zero-Waste Technologies in Aquaculture: Global Perspectives and Methods toward a Circular Economy
DOI: https://doi.org/10.1016/B978-0-443-34689-7.00008-2

the growth of microorganisms results in the microbial aggregates to be formed, consisting of bacteria, fungi, zooplankton, phytoplankton, and organic matter (Khanjani et al., 2022). These microbial aggregates are called bioflocs, and their function goes beyond controlling water quality. They also serve as supplementary food for animals, being constantly available in the system (Khanjani et al., 2022). The use of this system provides a reduction in water use, greater biosecurity, higher productivity, and less use of land (Emerenciano et al., 2013).

The fact that BFT is a system with reduced water exchange and exhaustive microbial activity, such as the nitrification process, causes the accumulation of compounds such as nitrate and orthophosphate over time (Da Silva et al., 2013). This raises concerns around the management of these nutrients, which must be improved to provide greater efficiency of this system in water reuse. These compounds can be removed from the water by treating these effluents with the denitrification process (Brandão et al., 2024), but in this process, only the nitrogen would be removed, leaving the phosphorus in the water. An alternative capable of removing nitrogen and phosphorus from the system is the integration of the culture with other species that use these inorganic nutrients for their growth, which is known as aquaponics (Pinho et al., 2021; Gonçalves et al., 2024). Vegetables and halophyte plants are considered for integration into aquatic animal culture, since, in addition to contributing to the removal of nutrients, they diversify production (Yep and Zheng, 2019). This tool, which combines the BFT system and aquaponics system, is called FLOCponics and has been tested in recent studies as a promising alternative that can contribute to the refinement of aquaculture techniques.

In this chapter, we will discuss several aspects related to FLOCponics, such as the main species used and stocking density, aspects of system configuration, management, and what has been studied in the main research recently carried out on this system. Furthermore, we demonstrate some perspectives on what can be done in the future in terms of research so that this system can become popular and consolidate itself as an environmentally friendly, safe, and productive tool.

16.2 Aquaponics and FLOCponics

Aquaponics is a cultivation method that integrates the production of aquatic organisms in a recirculating aquaculture system (RAS) and hydroponics system, which is the culture of plants in water instead of soil (Somerville et al., 2014) (Fig. 16.1A). In this system, water is treated by mechanical filtration to retain particulate matter and biological filtration (Fig. 16.1B), which oxidizes the ammonia produced by animals to nitrite and later to nitrate (Colt et al., 2022). The nitrate produced in the system meets the plants' growth needs and is consumed by them, purifying the water (Spradlin and Saha, 2022). The use of aquaponics allows the reduction of effluent emissions, reusing between 52% and 95% of the nitrogen and 85% of the phosphorus that would be discarded in cultures without the addition of plants (Gunning et al., 2016).

There are several fish species that are frequently used in commercial aquaponics, including Rainbow trout (*Oncorhynchus mykiss*) (Bordignon et al., 2022), common carp (*Cyprinus Carpio*) (Filep et al., 2016), largemouth bass (*Micropterus salmoides*) (Fischer et al., 2021), barramundi (*Lates calcarifer*) (Ardiansyah and Fotedar, 2016), tilapia (*Oreochromis niloticus*) (Pinho, et al., 2021), and Murray cod (*Maccullochella peelii*) (Lennard, 2021). For aquaponics to be successful, aquatic organisms must be able to tolerate elevated stocking densities as well as

FIGURE 16.1 Example of growing bed (A) and biological filtration compartment (B) in a traditional aquaponics system. *Photos: Dariano Krummenauer.*

substantial concentrations of total suspended solids, phosphorus, nitrogen, and potassium (Fischer et al., 2021). Carp and African catfish are frequently used and probably the most successful fish species in aquaponics (Abdel-Rahim et al., 2019).

Tilapia fish are microphagous low-trophic omnivores that feed on phytoplankton and other organic particles (Zappernick et al., 2022). The dissolved oxygen requirement of tilapia is low, since they do not require a lot of space to grow. They are therefore ideal for aquaponic systems that aim to meet plant nutrient needs (Abdel-Rahim et al., 2019; Zappernick et al., 2022). Current aquaponic literature does not address the effect of different species' excretion on nutrient concentrations and plant production (Pinho et al., 2017). Knaus and Palm (2017) found that common carp wastewater produced higher cucumber yields than tilapia wastewater. Tilapia effluent, however, increased tomato yields. Although tilapia have higher metabolic activity than carp, the exact reason for the superior growth of tomatoes with tilapia effluent compared with carp effluent is unknown (Knaus and Palm, 2017; Nadia et al., 2023). Probably because tilapia excretes more feces into the water. According to the authors, the use of various species could provide benefits when attempting to obtain a more comprehensive nutrient water profile (Abdel-Rahim et al., 2019). Several species of shrimp have been used in aquaponics, including *Penaeus vannamei* (Pinheiro et al., 2020; Alarcón-Silvas et al., 2021) and *Penaeus monodon* (Li et al., 2021). Polyculture, which involves the use of different aquatic species in aquaponics to enhance plant growth, has not been thoroughly explored yet (Pinho et al., 2022b).

With their short growing seasons, minimal nutritional requirements, and resistance to nitrogen-rich environments, leafy vegetables are traditionally grown in aquaponic systems (Barbosa et al., 2022). While flowering crops are more economically valuable than green vegetables, cultivating them in aquaponic systems presents greater challenges due to their increased phosphate and potassium fertilizer needs, susceptibility to pests and diseases, and slower growth cycles. Examples of crops used in commercial aquaponics are: *Ocimum basilicum* (basil) (Knaus et al., 2024), *Solanum lycopersicum* (tomatoes) (Nadia et al., 2023), *Lactuca sativa* (Khater et al., 2024), *Brassica oleracea* (Piñero et al., 2024), *Beta vulgaris* (Munekata et al., 2021), *Capsicum annuum* (pepper) (Romano et al., 2023), and *Cucumis sativus* (cucumbers) (Blanchard et al., 2020).

Aquaponic systems can also cultivate plants that thrive in saline environments (Gao et al., 2020). In aquaponics systems using salt or brackish water, Salicornia is an important plant species (Pinheiro et al., 2020). Castilla-Gavilán et al. (2024) described *Salicornia* as a halophyte that can tolerate elevated salinity levels and efficiently absorb phosphate and nitrate. As aquaponics has progressed, the variety of plant species that can be successfully grown has also increased (Nadia et al., 2023). There are, however, few scholarly studies that specifically address aquaponics' use for cultivating flowering plants (Sousa et al., 2024). The efficacy of aquaponics in the floriculture sector needs to be investigated further (Knaus et al., 2024).

A variation of aquaponics that has been expanding is FLOCponics. The term "FLOCponics" comes from the union of aquaponics and biofloc systems (Pinho et al., 2022a). This system replaces the traditional clear water RAS used in aquaponics with the biofloc system (Yu et al., 2023). Aquaponics + BFT system shares key characteristics of sustainable systems, such as water recirculation, nutrient cycling mediated by the system's microbial activity, and reduced use of water and land (Pinho et al., 2022b). The use of this system has strong potential to mitigate the accumulation of nitrate and phosphorus that occurs in the biofloc system through vegetable production (Yu et al., 2023).

Unlike aquaponics, which can be for personal use, as a hobby, on a family scale, and as an educational tool (Colt et al., 2022), FLOCponics can be more directed toward commercial production. This is possible due to the insertion of the biofloc system, which allows for an increase in the stocking density of both plants and animals and a consequent increase in productivity. One of the advantages brought by the growth of bioflocs in the system is the microbial communities grown in the environment, which allow the control of water quality and the provision of natural food for animals (Martinez-Cordova et al., 2023). Furthermore, these microorganisms colonize the intestinal tract of fish and shrimp, improving feed digestibility through the production of digestive enzymes, making the animals healthier and more resistant to disease outbreaks, as they also compete for nutrients and space with pathogenic microorganisms (Khanjani et al., 2022).

FLOCponics is a versatile system that can be adapted to different species of aquatic organisms and plants and different salinity conditions, such as seawater, brackish, and freshwater. This allows it to be taken to regions closer to consumer centers or even be inserted into urban farms producing aquatic organisms and vegetables, since they require less water and space (Martinez-Cordova et al., 2023).

16.3 Main species and stocking densities used

In FLOCponics, as well as in integrated multitrophic aquaculture systems (IMTA), the primary species is the one on which it feeds (i.e., fish and/or shrimp) while the extractive species are those that feed on organic or inorganic compounds produced by the species that are fed (i.e., vegetables or halophytic plants) (Buck et al., 2018). In this system, one of the aspects that can limit the development of animals and plants is their adaptation to an environment with bioflocs. Therefore the species that should have priority when choosing are those that are most resistant and that can grow in BFT systems.

The fish species most used in experimental cultures with FLOCponics is the Nile tilapia (*O. niloticus*), and one study used South American catfish (*Rhamdia quelen*) (Pinho et al., 2022b). Tilapia is one of the most used fish in biofloc system farming, in addition to being one of the most produced and consumed organisms in global aquaculture. This species can be used in aquaponics culture with BFT using a stocking density of 23–350 fish m^{-3} (Table 16.1). These crops can be carried out during the fish grow-out phase or during the nursery phase. Pinho et al. (2021) tested the integrated production of lettuce (*L. sativa*) with Nile tilapia in the nursery phase for 46 days. The authors used 1 g fingerlings and stocked them at a density of 300 fish m^{-3}. At the end of the experimental time, the animals reached a final weight that varied between 35 and 36 g and a survival rate above 90%.

In freshwater cultures, lettuce (*L. sativa*) is the species that is most used (Kotzen et al., 2019). Red lettuce, butter lettuce, and crispy lettuce are some lettuce variations that have already been introduced in FLOCponics (Pinho et al., 2017). Lettuce can be stocked at densities generally close to 20 plants m^{-2}, but this density can be increased to 210 plants m^{-2} (Table 16.1). da Rocha et al. (2017) tested the production of lettuce (*L. sativa*) and South American catfish (*R. quelen*) with a biofloc system. The authors cultivated the lettuces for a period of 46 days, using a stocking density of 20 plants m^{-2}. The plants reached an average fresh lettuce weight of 39.9 g.

At FLOCponics, the shrimp *P. vannamei* is the most widely cultured crustacean species, with the grow-out phase being the most adopted. In these cultivations, shrimp can grow from 1.4 to 12 g (Pinho et al., 2022b). In the most recent studies that reared shrimp integrated with plants in a BFT system, the stocking density varied between 250 and 600 shrimp m^{-3} (Table 16.1). *P. vannamei* is the shrimp species most cultivated in biofloc systems, and its growth and survival in FLOCponics systems is not limited by the presence of plants. Pinheiro et al. (2017) tested the production of halophytes (*Sarcocornia ambigua*) and the growth of *P. vannamei* in a culture integrated with a biofloc system. The authors used a stocking density of 250 shrimp m^{-3} and started culture at a weight of 1.4 g. At the end of the trial, the shrimp reached a mean weight of 11g and survival above 70%. The polyculture of *Macrobrachium rosenbergii* and Nile tilapia with lettuce and watercress in a BFT system was demonstrated by Barbosa et al. (2022). The authors reared the animals for a period of 45 days and used a stocking density of 5 shrimp m^{-3} and 23 fish m^{-3}. At the end of the trial, the animals reached a final weight of 565 g and 18.44 g for tilapia and shrimp, respectively.

In FLOCponics using *P. vannamei*, the plants used as an extractive species are generally halophytes of the genus *Sarcocornia*. This happens due to the characteristics of these plants being able to grow and survive in environments with high salinity. These plants are generally stocked at a density that can vary between 40 and 100 plants m^{-2} (Table 16.1). Pinheiro et al. (2020) tested the growth of *S. ambigua* and *P. vannamei* in a biofloc system at salinities of 8, 16, 24, and 32 g L^{-1}. The authors cultivated the plants for a period of 57 days and used a stocking density of 40 plants m^{-2}. At the end of the experiment, the mean final weight of the plants varied between 12.2 g at salinity 32 g L^{-1} and 21.4 g at salinity 16 g L^{-1}, and survival was above 60% in all treatments. Salinity reduction is possible in the culture of

TABLE 16.1 Key characteristics and main results from the latest research with FLOCponics.

Fish					Plant				
Species	**Stocking density**	**Final weight (g)**	**Survival (%)**	**Yield (kg m^{-3})**	**Species**	**Stocking density**	**Final weight (g)**	**Yield (kg m^{-2})**	**References**
Tilapia (*O. niloticus*)	3.2 kg $system^{-1}$	102.31	>95	—	Lettuce (*L. sativa* var. red, butter, and crispy)	20 plants m^{-2}	20.1 – 80.3	0.6 – 1.9	Pinho et al. (2017)
South American catfish (*R. quelen*)	504.64 g $tank^{-1}$	31.18	100	—	Lettuce (*L. sativa*)	20 plants m^{-2}	39.8	—	da Rocha et al. (2017)
Tilapia (*O. niloticus*)	40 fish m^{-3}	522.98	96	15	Cherry tomato (*Solanum lycopersicum*)	3.2 plants m^{-2}	—	7.4 – 11.4*	Pickens et al. (2020)
Tilapia (*O. niloticus*)	0.33 kg m^{-3}	36.69	96.48	8.05	Lettuce (*L. sativa*)	20 plants m^{-2}	100.02–164.70	1.80 – 2.96	Pinho et al. (2021)
Tilapia (*O. niloticus*)	150 – 350 fish m^{-3}	46.94– 75.33	100	11.03–16.43	Bell pepper	12 plants m^{-2}	26.40 – 31.60	286.89 – 338.10**	Saseendran et al. (2021)
Tilapia (*O. niloticus*)	184 fish m^{-3}	28.06 – 34.61	90.31 – 96.59	4.88 – 5.86	Lettuce (*L. sativa*)	19 plants m^{-2}	88.45 – 112.60	487.38 – 531.89 ***	Pinho et al. (2021)
Japanese eel (*Anguilla japonica*)	5 – 20 kg m^{-2}	132.9 – 168.2	91.6 – 99.4	—	Caipira lettuce (*L. sativa*)	210 plants $system^{-1}$	73.0 – 85.0	—	Hwang et al. (2023)
Tilapia (*O. niloticus*), *M. rosenbergii*	23 fish m^{-3}, 5 shrimp m^{-3}	565.00, 18.44	80.27, -	12.28, 0.08	Lettuce (*Lactuca sativa*), watercress (*Rorippa nasturtium-aquaticum*)	0.120 kg m^{-2}, 0.06 kg m^{-2}	0.120, -	0.94, 1.98	Barbosa et al. (2022)
Tilapia (*O. niloticus*)	98 fish m^{-3}	—	—	1.62 – 2.53	Asian rice (*Oryza sativa*)	250 plants m^{-2}	—	2.36 – 5.70	Rono et al. (2024)
Weather Loach (*Misgurnus anguillicaudatus*)	2.6 kg m^{-3}	1.99	91.06	—	Caipira Lettuce (*Lactuca sativa*)	100 plants bed^{-1}	163.60	—	Park et al. (2024)
Shrimp					**Plant**				
P. vannamei	250 shrimp m^{-3}	11.6	74.5	2.16	Halophyte (*Sarcocornia ambigua*)	100 plants m^{-2}	85.08	8.20	Pinheiro et al. (2017)
P. vannamei	300 shrimp m^{-3}	11.5 – 12.7	56.3–84	2.3 – 3.0	Halophyte (*Sarcocornia ambigua*)	40 plants m^{-2}	12.2 – 21.4	0.38 – 0.61	Pinheiro et al. (2020)
P. vannamei	600 shrimp m^{-3}	9.41 – 10.41	49.40 – 72.80	—	Red orache (*Atriplex hortensis*)	100 plants m^{-2}	14.3 – 16.5	—	Chu and Brown (2022)
					Okahijiki (*Salsola komarovii*)		9.70 – 11.30	—	
					Minutina (*Plantago coronopus*)		22.30 – 23.30	—	
P. vannamei	250 shrimp m^{-3}	14.57	18	—	New Zealand spinach (*Tetragonia tetragonioides*)	35 seedlings $system^{-1}$	4	519.41	Rodrigues et al. (2023)

- Data not available; *Yield from fruit production; **Yield in g $plant^{-1}$; ***Yield in g m^{-2}.

FIGURE 16.2 Example of aquaponics using NFT (A) and DWC (B) systems. *DWC*, Deep water culture; *NFT*, nutrient film technique. *Photos: Otávio Augusto L. F. Pimentel.*

P. vannamei, since this species is resistant to a wide range of salinity (Van Wyk et al., 1999). This allows greater flexibility in the cultivation conditions and allows extractive species to be inserted into the system.

16.4 System settings aspects

In traditional aquaponics, growing systems can be coupled or decoupled. Coupled systems are characterized by the recirculation of water in a continuous flow through the production compartments of aquatic organisms, water treatment (i.e., mechanical, biological filtration, and sterilization), and plant production (Palm et al., 2019). Decoupled systems are characterized by the physical separation of the water and nutrient circuits of both aquatic organisms and vegetable production units (Goddek et al., 2019). In these systems, there is a unidirectional flow from the aquaculture unit to the hydroponic unit, where the plants use the water provided by the aquaculture unit, which is topped up with clean water (Goddek et al., 2019). Water flow is controlled by demand. In FLOCponics systems, most studies use a coupled system, and only 30% of the studies that addressed this topic used a decoupled system (Pinho et al., 2022b).

Another facility aspect that must be observed in these systems is the configuration of the hydroponic compartment. The most common are nutrient film technique (NFT) (Fig. 16.2A) and deep water culture (DWC) (Fig. 16.2B). The NFT system is characterized by the arrangement of long channels with holes for inserting plants. In this technique, the roots touch the bottom of the channel where the water from the aquaculture compartment passes, the plants absorb the nutrients, and then the water returns to the aquaculture tank by gravity or by pumping (Colt et al., 2022). DWC is characterized by the suspension of plants in polystyrene sheets, where the plants float in water. This technique is the most common among large commercial

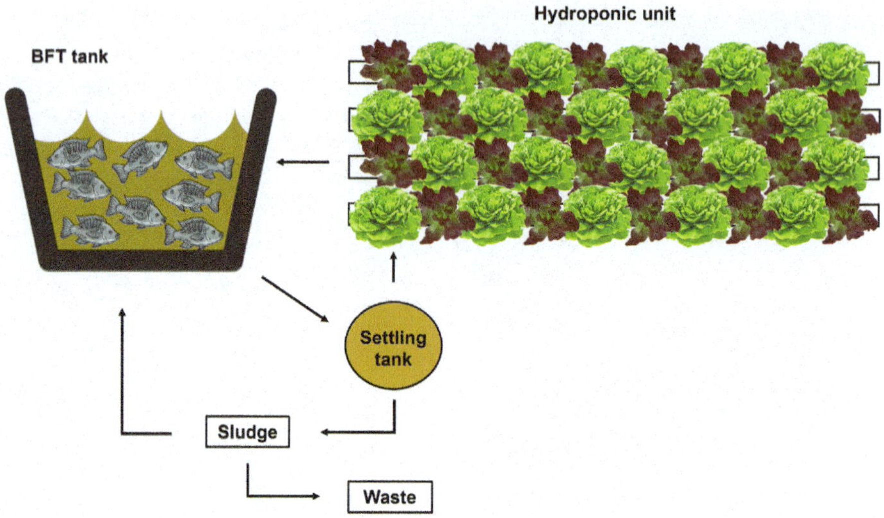

FIGURE 16.3 Scheme of a FLOCponics system using the NFT technique. Arrows indicate water/sludge flow. *NFT*, nutrient film technique.

aquaponics that use a single vegetable crop (Somerville et al., 2014). This demonstrates how versatile this system can be from a facility perspective, meeting different areas and investment conditions.

In FLOCponics, a compartment dedicated to solids control is inserted prior to the hydroponic compartment (Fig. 16.3). This is done to reduce the load of organic matter at the stage when water passes through the plants, so that the roots do not retain solids and hinder the absorption of nutrients and, consequently, their growth (Martinez-Cordova et al., 2023). In some studies, with the production of *P. vannamei* and halophyte plants, the water passes through a sedimentation chamber (settling tank or clarifier) and the sludge returns to the shrimp production tank, while the clean water is directed to the plant production compartment (Pinheiro et al., 2017; Pinheiro et al. 2020). The same can occur in FLOCponics with tilapia (*O. niloticus*) and lettuce (*L. sativa*), as demonstrated by Pinho et al. (2021). Another strategy for decoupled systems, as water does not flow continuously among compartments, water discharged from the aquaculture tank is directed to a sedimentation tank to separate excess solids, and sludge can be discarded, while water with a low concentration of solids is sent to the hydroponic unit (Martinez-Cordova et al., 2023; Pinho et al., 2023). In these systems, solids control is a key element, since the high concentration of solids, in addition to harming animals (Schveitzer et al., 2024), can affect plant roots and impact nutrient absorption, harming yield and the visual quality of plants (Kotzen et al., 2019).

The movement of water in the system is an important aspect in all hydroponic and aquaponic systems, as it can improve water oxygenation, also improving the release of nutrients into the system (Somerville et al., 2014). In FLOCponics systems, a large variation in the water flow rate through the hydroponic beds is reported, which can vary from 0.06 to 13.1 L min^{-1}(Pinho et al., 2022b). These aspects must be studied so that the production process can be optimized.

Aquaculture tanks in FLOCponics systems have different sizes and shapes, reaching volumes greater than 100 m^3, as reported by Pinho et al. (2022b). As it is a system that uses the BFT system, these tanks must follow the standards recommended for that system, with an aeration system that promotes higher mixing of water to keep the bioflocs in suspension (Samocha et al., 2019). Furthermore, the use of tools to improve the process of colonization of the system by nitrifying bacteria can be used. The use of artificial substrates, in addition to improving water quality by promoting a larger contact area for the fixation of nitrifiers, increases the relative stocking density and serves as a source of supplementary food for the animals, improving their growth (de Lara et al., 2021; Da Silveira et al., 2022).

16.5 Management and nutrient recycling

In BFT systems, microbes are crucial to recycling nutrients and maintaining ideal water quality for farmed animals (Khanjani et al., 2024a). Several organisms participate in the ammonia-nitrogen cycle, including phytoplankton, nitrifying bacteria, and heterotrophic bacteria (Khanjani et al., 2024b). By converting ammonia nitrogen into nitrate or adding it to bacteria's biomass, they do this. A number of nitrogen conversions occur simultaneously, and the prevalence of each form is dependent upon the nutrient management of the system (Arbour et al., 2024). The water's physicochemical properties must also meet the requirements of the microbes. It requires high levels of dissolved oxygen, alkalinity, and an elevated C:N ratio (Blanchard et al., 2020; Hou et al., 2024).

The management of a FLOCponics can be based mainly on what is done for a BFT system. In this system, the most used carbon source is liquid or powdered sugar cane molasses, following a C:N ratio that can vary between 12 and 20:1 (Pinheiro et al., 2020; Pinho et al., 2021; Barbosa et al., 2022). Wheat bran can also be used as an organic carbon source (Saseendran et al., 2021). However, it is important to highlight that vegetable brans are sources of slow carbon availability, making the control of nitrogenous compounds also slower. Maintaining a high C:N ratio is important in a biofloc system to sustain the dominance of the system by heterotrophic bacterial communities (Avnimelech, 1999).

The control of other variables, such as alkalinity and pH of the water in a BFT system, is an important factor for maintaining the activity of the community of nitrifying bacteria, which act strongly in the oxidation of toxic nitrogenous compounds to nitrate (Abakari et al., 2021), which is one of the essential nutrients for plant growth. Furthermore, alkalinity is responsible for increasing the buffering of water, which provides a lower variation of the water pH (Boyd, 2020). The alkalinity concentration of the water must respect the needs of the system and the species being cultivated. The optimal alkalinity and total hardness for aquaponics should vary between 60 and 140 mg L^{-1} (Table 16.2). A pH range of 5.5–7 is often recommended for hydroponics production (Vasdravanidis et al., 2022). FLOCponics systems, however, have typically operated at neutral pH values (Blanchard et al., 2020). The pH must be maintained between 7 and 8 for the system to function properly (Table 16.2). Table 16.2 demonstrates the recommended limits for an aquaponics system.

Nitrate is the least toxic form of nitrogen for fish and shrimp. However, this compound cannot reach concentrations above 250 mg L^{-1}, as it can have negative impacts on animals and plants being produced, leading to excessive growth and accumulation of nitrate in leaves

TABLE 16.2 Recommended limits of water quality variables for an aquaponics system.

Variable	Limit
Temperature (°C)	18.0–30.0
Dissolved oxygen (mg L^{-1})	> 5.0
pH	7.0–8.0
Ammonia (mg L^{-1})	< 1.0
Nitrite (mg L^{-1})	< 1.0
Nitrate (mg L^{-1})	5.0–150.0
Alkalinity and total hardness (mg L^{-1})	60–140

Modified from Somerville, C., Cohen, M., Pantanella, E., Stankus, A. and Lovatelli, A., 2014. Small-scale aquaponic food production: Integrated fish and plant farming. FAO Fisheries and Aquaculture Technical Paper No. 589. Rome: Food and Agriculture Organization of the United Nations, pp. 288

(Somerville et al., 2014). At a salinity of 11 g L^{-1}, nitrate levels above 220 mg L^{-1} can negatively affect the survival and growth of *P. vannamei* (Kuhn et al., 2010). For tilapia (*O. niloticus*), it is recommended that the nitrate concentration does not exceed 500 mg L^{-1}, so that it is possible to guarantee healthy growth of the fish (Monsees et al., 2017a). For *P. vannamei* grown in a BFT system, a concentration of up to 177 mg L^{-1} of nitrate for salinity 23 g L^{-1} is recommended (Furtado et al., 2015). For salinities 5 and 10 g L^{-1}, the safe levels are 60 and 127.6 mg L^{-1}, respectively (Alves Neto et al., 2019).

The concentration of total suspended solids is a key element in a BFT system. Total suspended solids management strategies must be adopted when the concentration of this variable is above 500 mg L^{-1} (Avnimelech, 2012). Hydroponic subsystems require a very low concentration of suspended particles in water to avoid the buildup of bioflocs in the plants' roots, which can hinder the plants' ability to breathe and absorb nutrients (Chandramenon et al., 2024). A strategy adopted in a study that tested the growth of the halophyte *Salicornia neei* in an integrated multitrophic aquaculture with shrimp *P. vannamei* and tilapia (*O. niloticus*) in a BFT system was washing weekly from the channels of the hydroponic compartment to remove excess solids retained in the roots of the plants, thus preventing their obstruction and water overflow (Gonçalves et al., 2024).

As a result of coupled FLOCponics, it is evident that one of the trade-offs associated with the concept is maintaining low solids levels in the hydroponics subsystems while maintaining biofloc levels in the fish tanks at optimal levels for animal production (Pinho et al., 2022b). There's a link between nutrient inflow and microbe transformation in all subsystems (Khanjani et al., 2024b) that makes the water quality optimal. It is believed that most nutrients used in traditional aquaponics are derived from recirculating aquaculture system (RAS) effluent; this is also expected to be the case in FLOCponics (Palm et al., 2024). Typically, feed is the primary source of nutrients in RAS's aquaculture subsystem. As a result of incorporating organic and inorganic carbon sources, FLOCponics can provide a higher concentration of nutrients (Emerenciano et al., 2021). Khanjani et al. (2024a) found that both processes are frequently necessary for BFT bacteria to multiply. There is not a lot of information on the specific features of FLOCponics systems, so it is hard to estimate how many nutrients plants will get (Pinheiro et al., 2020). Moreover, we do not know exactly how the BFT bacteria recycle nutrients and absorb nutrients, so projections are uncertain. By analyzing the nutritional composition of plant biomass, we are able to determine which nutrients are delivered at the lowest levels (Doncato and Costa, 2021).

Compared to hydroponic solutions, FLOCponic water has lower fertilizer concentrations (Pickens et al., 2020). However, unlike traditional aquaponics that utilize RAS, external carbon additives cause elevated levels of potassium (K), phosphorus (P), sulfur (S), calcium (Ca), and iron (Fe) in aquaponics (Pinho et al., 2021). The use of bioreactors to mineralize RAS sludge and produce liquid fertilizer has been demonstrated in multiloop aquaponics (Tetreault et al., 2023). In contrast, mineralized solids have received little attention as a potential nutrient source for plants in FLOCponics (Pickens et al., 2020). As part of nutrient recycling methods in FLOCponics research, plants have been used as filters to remove nutrients from water. Most of the research has focused on recovering nitrogen and phosphorus and converting them into plant biomass (Emerenciano et al., 2021). Halophyte plants were used by Pinheiro et al. (2020) to extract nitrogen and phosphorus from a FLOCponics with *P. vannamei* in different salinities. According to their findings, shrimp and plant production can eliminate 24.2%–28.7% of nitrogen (N) and 14.9%–19.4% P from feed. Both nutrients often accumulate in BFT water (Luo et al., 2020). Whenever these compounds are present in large amounts, they can be harmful to animals or contribute to water eutrophication if they are discharged into aquatic ecosystems (Raza et al., 2024).

16.6 Research

To date, little research has been done on the integrated culture of plants with the BFT system. Some studies that used fish as the primary species proposed comparing FLOCponics systems with traditional aquaponics and hydroponic systems. These studies demonstrated that the biofloc system, by providing a large amount of nutrients, improves the growth of fish and vegetables, being technically viable (da Rocha et al., 2017; Pinho et al., 2017; Pickens et al., 2020; Pinho et al., 2021; Park et al., 2024). Furthermore, FLOCponics has already been compared with the biofloc system without the presence of the hydroponic unit to demonstrate the efficiency of plants in absorbing nutrients. FLOCponics has been shown to be 10% and 27% more efficient in using water and nitrogen than the BFT system alone (Pinho et al., 2023).

From a feed point of view, the use of decoupled FLOCponics allows the production of lettuce and tilapia using feed with a lower crude protein content. Fish fed with 24% and 28% crude protein feed grew similarly to those fed with 32% crude protein feed (Pinho et al., 2021). Regarding the fertilization management of the FLOCponics system with tilapia and Asian rice, the sources lucerne hay, wheat bran, and Rhodes hay improved the yield of fish and rice crops (Rono et al., 2024). Furthermore, different densities of tilapia were tested in FLOCponics with bell pepper. The results indicated that a density of 150 fish m^{-3} and 12 plants m^{-2} can be used without compromising the growth and well-being of fish and plants (Saseendran et al., 2021).

FLOCponics using shrimp followed steps more focused on nutrient absorption, use of different salinities, and different protein contents of feed, using plants resistant to higher salinities. Pinheiro et al. (2017) demonstrated that the recovery of nitrogen contributed to the system by feed was higher in integrated systems of halophytes *S. ambigua* and *P. vannamei* than in systems without the presence of plants. Aiming at the integrated production of *P. vannamei* and *S. ambigua* in waters with low salinity, it was demonstrated that the salinity range between 16 and 24 g L^{-1} can be considered for this type of culture, without limiting shrimp growth, favoring plant growth, and the removal of nitrogen and phosphorus (Pinheiro et al., 2020). Furthermore, Chu and Brown (2022) found in FLOCponics with *P. vannamei* and the halophyte plants *Atriplex hortensis, Salsola komarovii*, and *Plantago coronopus*, that different crude protein contents in the feed do not affect shrimp growth. However, plants grew better in treatments that used feed with 35% crude protein in the initial and intermediate stages of culture. In the final stage of treatment, 30% had better growth.

From a sustainability perspective, FLOCponics, as well as the BFT and hydroponic systems, use a low amount of renewable natural resources and generally have a moderate environmental burden. However, they are systems sensitive to changes in the amount of electricity and equipment (Pinho et al., 2022c). Economically, Pinho et al. (2022a) suggested that it is more viable to produce tilapia and butter lettuce in conventional aquaponics systems than in FLOCponics. For FLOCponics to be profitable, at least 68.7% of the plants produced must have visual and marketable characteristics.

16.7 Sustainability

A number of emerging technologies have been developed to facilitate aquaculture's transition to more ecologically sound practices. Aquaculture sustainability refers to both technically feasible and economically viable systems. To meet the demands of current and future generations, safe and healthy food must be provided (Boyd et al., 2020). It can be beneficial to perform economic evaluations of different aquaculture operations to determine how to increase their resilience and longevity (Pickens et al., 2020; Soares et al., 2022). To develop appropriate public strategies that foster sustainable growth of an activity, it is essential to understand the social and ecological influences of a new production system, taking into account its interconnectedness.

Biological, technical, and economic issues are all included in this comprehension (David et al., 2018). According to Life Cycle Assessment research (Maucieri et al., 2018; Francisco et al., 2024), aquaponics production has the greatest environmental impact on infrastructure, power usage, and feed. In addition to low water consumption, aquaponics systems can improve the culture, recreation, education, and tourism of an area, as well as enhance its landscape (Pickens et al., 2020). Biofloc-based shrimp production does not have a significant carbon footprint as determined by life cycle assessment feed, but energy consumption does (Boyd et al., 2020). There are no sustainability assessments of FLOCponics systems in the literature. Most likely, these studies are lacking because FLOCponics systems do not have access to the large and comprehensive databases needed for these analyses (Pinho et al., 2022b). According to Emerenciano et al. (2021), FLOCponics has the potential to reduce certain unsustainable aspects of traditional aquaculture, despite the absence of sustainability assessment data. By replacing the RAS with BFT, both biofloc-based systems and regular aquaponics can be improved (Emerenciano et al., 2021). It is possible to incorporate this substitution into an environmentally friendly food production system, already recognized for its effectiveness (Pinho et al., 2022b).

As well as producing a variety of food items locally, FLOCponics systems also benefit the environment and are socially beneficial (Pickens et al., 2020). Also, FLOCponics produces pesticide-free, nutritious, and diverse foods in a variety of forms (fish/shrimp and vegetables) for consumers. The costs associated with FLOCponics' productivity can be offset by increased biomass production, addressing this economic concern when FLOCponics' productivity is confirmed (Pickens et al., 2020; Soares et al., 2022). For example, FLOCponics systems are expected to produce more food per unit of electricity consumed than biofloc-based monocultures (Knaus et al., 2024). FLOCponics systems could also be made more environmentally friendly by adding renewable energy sources, such as solar, wind, and biogas (Pinho et al., 2022). Inherently, food production systems have an impact on the environment. Our recommendation is to support systems that can achieve high levels of productivity while posing as few negative consequences as possible (David et al., 2021). Ultimately, FLOCponic systems need to be evaluated for their long-term viability, as well as their advantages and disadvantages. The feasibility of FLOCponics can be studied by developing a more comprehensive technical and economic database (Knaus et al., 2024) to achieve these objectives.

16.8 Challenges of FLOCponics

FLOCponics could be a viable alternative for investors seeking to establish an integrated agriaquaculture farm if technological obstacles are overcome. The successful operation of a FLOCponics system requires a comprehensive understanding of a variety of subjects (Pinheiro et al., 2017; Pinho et al., 2022b). In addition, the selection of a food production system must consider several factors, such as market demand, climatic conditions, producers' expertise, technical knowledge, input costs, and availability (David et al., 2021). Despite FLOCponics' potential benefits, it is imperative to conduct a comprehensive review of the entire production process before choosing the most appropriate method (Khater et al., 2024).

It is crucial to consider the system's configuration carefully to maximize the favorable conditions for the growth of aquatic plants and organisms, as well as the nourishment of BFT bacteria. To support fish and plant growth, ideal levels of suspended particles must be maintained in the water. FLOCponics systems are known to have a high concentration of solids, which hinders plant growth. With the removal of solid particles from the hydroponics subsystem, there is reduced access to food and bioflocs in the aquaculture compartment (Pinho et al., 2021; Khater et al., 2024). Mechanical separators can be developed to effectively separate solids and liquids from BFT effluent. As part of the aquaculture subsystem, nutrients and water are transferred from the biofloc into the hydroponics subsystem, while being reintegrated into the aquaculture subsystem (David et al., 2021).

In FLOCponics, a variety of filters should be considered, including bag filters with backwashing technology, drum filters, and sedimentation containers with meticulously engineered biofloc return flows (Pinho et al., 2021). Further, it is imperative to monitor their regular operation and control how much water is discharged into these systems. There is no doubt that each of these filters can be used in an interconnected FLOCponics system (David et al., 2021). However, the needs of plants and animals will always be compromised in interconnected systems (Monsees et al., 2017b; Khater et al., 2024). As FLOCponics systems' technical components are improved, these issues, mostly associated with solids management, should be effectively mitigated or perhaps resolved (Pinho et al., 2021). It is also possible to adjust the pH levels for each subsystem as well as to direct the addition of minerals to the hydroponic system using a decoupled design (Piñero et al., 2024).

In contrast to commercial hydroponics that use fully prepared fertilizers, FLOCponics may potentially reduce production costs by utilizing only specific nutrients (Pinheiro et al., 2020). Considering that BFT effluent contains an array of nutrients, this approach is reasonable. It is essential to have detailed information about the amount of nutrients and carbon sources present in feeds to achieve this goal (Monsees et al., 2017b; Khater et al., 2024). In addition, the micronutrient content of the BFT water needs to be evaluated, since they significantly affect plant physiological processes, such as photosynthesis (Monsees et al., 2017b; Piñero et al., 2024). By comparing FLOCponics systems with properly balanced hydroponic fertilizers, the quality and diversity of micronutrients may be able to shed light on the possibility of nutrient deficiencies (Nadia et al., 2023). Therefore, personalized supplementation regimens can be developed to maximize vegetable quality and yield (Sousa et al., 2024). Higher densities are also essential to achieve higher yields. Only a few animal species are capable of efficiently growing in BFT and, therefore, in FLOCponic systems.

16.9 Concluding remarks and perspectives

The use of the FLOCponics system has been studied and can be considered an environmentally sustainable alternative, diversifying production, recovering nutrients, and reverting them to plant growth, and increasing the productivity of the systems. However, this field is still new and needs to be further studied so that cultivation techniques become more refined and less expensive.

Studies based on the performance of plants and animals at different stocking densities should be carried out to better understand plant growth patterns based on nutrient availability in the system. The different types of bioflocs, predominantly heterotrophic or chemoautotrophic, have a direct influence on the production of nitrate in the system. This can have direct effects on the productivity of the plants in the hydroponic unit, improving the overall performance of the system. Furthermore, since total suspended solids can have a negative impact on plant growth, the use of predominantly chemoautotrophic bioflocs can benefit plants since these systems produce a lower microbial biomass (Ebeling et al., 2006). In heterotrophic systems, different levels of total suspended solids can be tested with different plant species to determine the concentration that provides better plant growth and the best water quality conditions in the system. Regarding the culture of marine shrimp (*P. vannamei*) integrated with plants, the lowest salinity can be tested, which does not limit the growth of plants, mostly cultivated in freshwater systems, such as lettuce. This can have positive impacts on the interiorization of *P. vannamei* culture, providing greater water reuse and less use of area.

Finally, the economic and sustainability aspects surrounding FLOCponics need to be better studied to better understand the viability of these systems and how profitable and sustainable they can be compared with traditional aquaponics systems.

References

Abakari, G., Luo, G., Kombat, E.O., 2021. Dynamics of nitrogenous compounds and their control in biofloc technology (BFT) systems: a review. Aquac. Fish. 6 (5), 441–447. doi:10.1016/j.aaf.2020.05.005.

Abdel-Rahim, M.M., Awad, Y.M., Abdallah, Y.A., Radwan, S.M., 2019. Effects of four medicinal plants on the bioeconomic analysis and water-use efficiency of nile tilapia, oreochromis niloticus fry nursed under a small-scale aquaponics system. AACL Bioflux 12 (3), 851–866. http://www.bioflux.com.ro/journal/.

Alarcón-Silvas, S.G., León-Cañedo, J.A., Fierro-Sañudo, J.F., Ramírez-Rochín, J., Fregoso-López, M.G., Frías-Espericueta, M.G., Osuna-Martínez, C.C., Páez-Osuna, F., 2021. Water quality, water usage, nutrient use efficiency and growth of shrimp Litopenaeus vannamei in an integrated aquaponic system with basil Ocimum basilicum. Aquaculture 543, 737023. doi:10.1016/j.aquaculture.2021.737023.

Alves Neto, I., Brandão, H., Furtado, P.S., Wasielesky Jr, W., 2019. Acute toxicity of nitrate in Litopenaeus vannamei juveniles at low salinity levels. Ciência Rural 49 (1). doi:10.1590/0103-8478cr20180439.

Arbour, A.J., Chu, Y.T., Brown, P.B., Huang, J.Y., 2024. Life cycle assessment on marine aquaponic production of shrimp, red orache, minutina and okahajiki. J. Environ. Manage. 353. doi:10.1016/j.jenvman.2024.120208.

Ardiansyah, R.F, 2016. Water quality, growth and stress responses of juvenile barramundi (Lates calcarifer Bloch), reared at four different densities in integrated recirculating aquaculture systems. Aquaculture 458, 113–120. doi:10.1016/j.aquaculture.2016.03.001.

Avnimelech, Y., 2012. Biofloc technology-a practical guide book, 2nd edition The World Aquaculture Society, Baton Rouge, LA.

Avnimelech, Y., 1999. Carbon/nitrogen ratio as a control element in aquaculture systems. Aquaculture 176 (3–4), 227–235. doi:10.1016/s0044-8486(99)00085-x.

Barbosa, P.T.L., Povh, J.A., Farias, K.N.N., da Silva, T.V., Teodoro, G.C., Ribeiro, J.S., Stringhetta, G.R., dos Santos Fernandes, C.E., Filho, R.A.C.C., 2022. Nile tilapia production in polyculture with freshwater shrimp using an aquaponic system and biofloc technology. Aquaculture 551. doi:10.1016/j.aquaculture.2022.737916.

Blanchard, C., Wells, D.E., Pickens, J.M., Blersch, D.M., 2020. Effect of pH on cucumber growth and nutrient availability in a decoupled aquaponic system with minimal solids removal. Horticulturae 6 (1). doi:10.3390/horticulturae6010010.

Blancheton, J.P., Piedrahita, R., Eding, E.H., OrbcasteL, D.E.R., Lemarié, G., Bergheim, A., Fivelstad, S., 2007. Intensification of landbased aquaculture production in single pass and reuse systems. Aquac. Eng. Environ. 27–47.

Bordignon, F., Sturaro, E., Trocino, A., Birolo, M., Xiccato, G., Berton, M., 2022. Comparative life cycle assessment of rainbow trout (Oncorhynchus mykiss) farming at two stocking densities in a low-tech aquaponic system. Aquaculture 556, 738264. doi:10.1016/j.aquaculture.2022.738264.

Boyd, C.E., D'Abramo, L.R., Glencross, B.D., Huyben, D.C., Juarez, L.M., Lockwood, G.S., McNevin, A.A., Tacon, A.G.J., Teletchea, F., Tomasso, J.R., Tucker, C.S., Valenti, W.C., 2020. Achieving sustainable aquaculture: historical and current perspectives and future needs and challenges. J. World Aquac. Soc. 51 (3), 578–633. doi:10.1111/jwas.12714.

Boyd, C.E., 2020. Water quality: an introduction, 3rd edition Springer, Cham doi:10.1007/978-3-030-23335-8.

Brandão, H., dos Reis, W.G., Krummenauer, D., Wasielesky, W., 2024. Growth performance of Litopenaeus vannamei under biofloc system using denitrified seawater. Aquac. Int. 32 (3), 3129–3145. doi:10.1007/s10499-023-01315-0.

Buck, B.H., Troell, M.F., Krause, G., Angel, D.L., Grote, B., Chopin, T., 2018. State of the art and challenges for offshore integrated multi-trophic aquaculture (IMTA). Front. Mar. Sci. 5. doi:10.3389/fmars.2018.00165.

Castilla-Gavilán, M., Muñoz-Martínez, M., Zuasti, E., Canoura-Baldonado, J., Mondoñedo, R., Hachero-Cruzado, I., 2024. Yield, nutrients uptake and lipid profile of the halophyte Salicornia ramosissima cultivated in two different integrated multi-trophic aquaculture systems (IMTA). Aquaculture 583, 740547. doi:10.1016/j.aquaculture.2024.740547.

Chandramenon, P., Aggoun, A., Tchuenbou-Magaia, F., 2024. Smart approaches to Aquaponics 4.0 with focus on water quality – comprehensive review. Comput. Electron. Agric. 225, 109256. doi:10.1016/j.compag.2024.109256.

Chu, Y.T., Brown, P.B., 2022. Optimal dietary crude protein in commercial feeds for shrimp and halophytes in marine aquaponic biofloc systems. Front. Mar. Sci. 9. doi:10.3389/fmars.2022.824973.

Colt, J., Schuur, A.M., Weaver, D., Semmens, K., 2022. Engineering design of aquaponics systems. Rev. Fish. Sci. Aquac. 30 (1), 33–80. doi:10.1080/23308249.2021.1886240.

da Rocha, A.F., Biazzetti Filho, M.L., Stech, M.R., da Silva, R.P., 2017. Lettuce production in aquaponic and biofloc systems with silver catfish Rhamdia quelen. Bol. Inst. Pesca 43, 64–73. doi:10.20950/1678-2305.2017.64.73.

de Lara, G.R, Poersch, L.H., Wasielesky, W., 2021. The quantity of artificial substrates influences the nitrogen cycle in the biofloc culture system of Litopenaeus vannamei. Aquacult. Eng. 94, 102171. doi:10.1016/j.aquaeng.2021.102171.

Da Silva, K.R., Wasielesky, W., Abreu, P.C., 2013. Nitrogen and phosphorus dynamics in the biofloc production of the Pacific white shrimp, Litopenaeus vannamei. J. World Aquac. Soc. 44 (1), 30–41. doi:10.1111/jwas.12009.

Da Silveira, L.G.P., Rosas, V.T., Krummenauer, D., Poersch, L.H., Wasielesky, W., 2022. Comparison between horizontal and vertical substrate in shrimp super-intensive culture in bioflocs system. Aquacult. Eng. 96. doi:10.1016/j.aquaeng.2021.102218.

David, L.H., Pinho, S.M., Agostinho, F., Kimpara, J.M., Keesman, K.J., Garcia, F., 2021. Emergy synthesis for aquaculture: a review on its constraints and potentials. Rev. Aquac. 13 (2), 1119–1138. doi:10.1111/raq.12519.

David, L.H.C., Pinho, S.M., Garcia, F., 2018. Improving the sustainability of tilapia cage farming in Brazil: an emergy approach. J. Cleaner Prod. 201, 1012–1018. doi:10.1016/j.jclepro.2018.08.124.

Doncato, K.B., Costa, C.S.B., 2021. Micronutrient supplementation needs for halophytes in saline aquaponics with BFT system water. Aquaculture 531. doi:10.1016/j.aquaculture.2020.735815.

Ebeling, J.M., Timmons, M.B., Bisogni, J.J., 2006. Engineering analysis of the stoichiometry of photoautotrophic, autotrophic, and heterotrophic removal of ammonia-nitrogen in aquaculture systems. Aquaculture 257 (1–4), 346–358. doi:10.1016/j.aquaculture.2006.03.019.

Emerenciano, M.G.C., Fitzsimmons, K., Rombenso, A.N., Miranda-Baeza, A., Martins, G.B., Lazzari, R., Fimbres-Acedo, Y.E., Pinho, S.M., 2021. Biofloc technology (BFT) in tilapia culture. In: Tilapia culture. Informa UK Limited, Boca Raton, FL, pp. 258–293. doi:10.1201/9781003004134-14.

Emerenciano, M., Gaxiola, G., Cuzon, G., 2013. Biofloc technology (BFT): a review for aquaculture application and animal food industry. InTech, Rijeka, Croatia doi:10.5772/53902.

Filep, R.M, Diaconescu, Ş, Costache, M., Stavrescu-Bedivan, M.-M., Bădulescu, L., Nicolae, C.G., 2016. Pilot aquaponic growing system of carp (Cyprinus Carpio) and basil (Ocimum Basilicum). Agric. Agric. Sci. Procedia 10, 255–260. doi:10.1016/j.aaspro.2016.09.062.

Fischer, H., Romano, N., Jones, J., Howe, J., Renukdas, N., Sinha, A.K, 2021. Comparing water quality/bacterial composition and productivity of largemouth bass Micropterus salmoides juveniles in a recirculating aquaculture system versus aquaponics as well as plant growth/mineral composition with or without media. Aquaculture 538, 736554. doi:10.1016/j.aquaculture.2021.736554.

Francisco, E.C., Freato, T.A., Piolli, A.L., Poz, M.E.S.D., 2024. Analysis of the aquaponic system sustainability via system dynamics modelling – FEW nexus approach. Circ. Econ. Sustain doi:10.1007/s43615-024-00408-z.

Furtado, P.S., Campos, B.R., Serra, F.P., Klosterhoff, M., Romano, L.A., Wasielesky, W., 2015. Effects of nitrate toxicity in the Pacific white shrimp, Litopenaeus vannamei, reared with biofloc technology (BFT). Aquac. Int. 23 (1), 315–327. doi:10.1007/s10499-014-9817-z.

Gao, Y., Wang, X., Li, J., Lee, C.T., Ong, P.Y., Zhang, Z., Li, C., 2020. Effect of aquaculture salinity on nitrification and microbial community in moving bed bioreactors with immobilized microbial granules. Bioresour. Technol. 297. doi:10.1016/j.biortech.2019.122427.

Goddek, S., Joyce, A., Wuertz, S., Körner, O., Bläser, I., Reuter, M., Keesman, K.J., 2019. Decoupled aquaponics systems. In: Aquaponics food production systems: Combined aquaculture and hydroponic production technologies for the future. Springer International Publishing, Cham, pp. 201–230. doi:10.1007/978-3-030-15943-6_8.

Gonçalves, M.S, Carvalho, A., Santos, J., Holanda, M., Poersch, L.H., Costa, C.S.B., 2024. Biofloc formation strategy effects on halophyte integration in IMTA with marine shrimp and tilapia. Aquac. J. 4 (4), 217–231. doi:10.3390/aquacj4040016.

Gunning, D., Maguire, J., Burnell, G., 2016. The development of sustainable saltwater-based food production systems: a review of established and novel concepts. Water 8 (12), 598. doi:10.3390/w8120598.

Hou, D., Li, H., Wang, S., Weng, S., He, J., 2024. Nitrite nitrogen stress disrupts the intestine bacterial community by altering host-community interactions in shrimp. Sci. Total Environ. 925, 171536. doi:10.1016/j.scitotenv.2024.171536.

Hwang, J.A., Park, J.S., Jeong, H.S., Kim, H., Oh, S.Y., 2023. Productivity of fish and crop growth and characteristics of bacterial communities in the FLOCponics system. Fishes. 8 (8). doi:10.3390/fishes8080422.

Khanjani, M.H., Mohammadi, A., Emerenciano, M.G.C., 2022. Microorganisms in biofloc aquaculture system. Aquac. Rep. 26. doi:10.1016/j.aqrep.2022.101300.

Khanjani, M.H., Mohammadi, A., Emerenciano, M.G.C., 2024a. Water quality in biofloc technology (BFT): an applied review for an evolving aquaculture. Aquac. Int. 32 (7), 9321–9374. doi:10.1007/s10499-024-01618-w.

Khanjani, M.H., Sharifinia, M., Emerenciano, M.G.C., 2024b. Biofloc technology (BFT) in aquaculture: what goes right, what goes wrong? A scientific-based snapshot. Aquacult. Nutr. 2024. doi:10.1155/2024/7496572.

Khater, E.S., Bahnasawy, A., Mosa, H., Abbas, W., Morsy, O., 2024. Nutrient supply systems and their effect on the performance of the Nile Tilapia (Oreochromis niloticus) and Lettuce (Lactuca sativa) plant integration system. Sci. Rep. 14 (1). doi:10.1038/s41598-024-54656-y.

Knaus, U., Hübner, D.H.D., Küchenmeister, C., Appelbaum, S., Iten, W., Palm, H.W., 2024. Aquaponic growth of basil (Ocimum basilicum) with African catfish (Clarias gariepinus) in standard substrate combined with a Humicacid Fiber-Substrate (HFS). Sci. Rep. 14 (1). doi:10.1038/s41598-024-68361-3.

Knaus, U., Palm, H.W., 2017. Effects of the fish species choice on vegetables in aquaponics under spring-summer conditions in northern Germany (Mecklenburg Western Pomerania). Aquaculture 473, 62–73. doi:10.1016/j.aquaculture.2017.01.020.

Kotzen, B., Emerenciano, M.G.C, Moheimani, N., Burnell, G.M., 2019. Aquaponics: alternative types and approaches. In: Aquaponics food production systems: combined aquaculture and hydroponic production technologies for the future. Springer Science and Business Media LLC, Cham, pp. 301–330. doi:10.1007/978-3-030-15943-6_12.

Kuhn, D.D., Smith, S.A., Boardman, G.D., Angier, M.W., Marsh, L., Flick, G.J., 2010. Chronic toxicity of nitrate to Pacific white shrimp, Litopenaeus vannamei: impacts on survival, growth, antennae length, and pathology. Aquaculture 309 (1–4), 109–114. doi:10.1016/j.aquaculture.2010.09.014.

Laktuka, K., Kalnbalkite, A., Sniega, L., Logins, K., Lauka, D., 2023. Towards the sustainable intensification of aquaculture: exploring possible ways forward. Sustainability 15 (24), 16952. doi:10.3390/su152416952.

Lennard, W.A., 2021. A comparison of buffering species and regimes applied within a research-scale, recirculating aquaponics system. Aquac. Fish. 6 (5), 495–505. doi:10.1016/j.aaf.2020.07.001.

Li, T., Zhang, B., Zhu, C., Su, J., Li, J., Chen, S., Qin, J., 2021. Effects of an ex situ shrimp-rice aquaponic system on the water quality of aquaculture ponds in the Pearl River estuary, China. Aquaculture 545, 737179. doi:10.1016/j.aquaculture.2021.737179.

Luo, G., Xu, J., Meng, H., 2020. Nitrate accumulation in biofloc aquaculture systems. Aquaculture 520, 734675. doi:10.1016/j.aquaculture.2019.734675.

Martinez-Cordova, L.R., Emerenciano, M.G.C., Miranda-Baeza, A., Pinho, S.M., Garibay-Valdez, E., Martínez-Porchas, M., 2023. Advancing toward a more integrated aquaculture with polyculture >aquaponics >biofloc technology >FLOCponics. Aquac. Int. 31 (2), 1057–1076. doi:10.1007/s10499-022-01016-0.

Maucieri, C., Forchino, A.A., Nicoletto, C., Junge, R., Pastres, R., Sambo, P., Borin, M., 2018. Life cycle assessment of a micro aquaponic system for educational purposes built using recovered material. J. Cleaner Prod. 172, 3119–3127. doi:10.1016/j.jclepro.2017.11.097.

Monsees, H., Klatt, L., Kloas, W., Wuertz, S., 2017a. Chronic exposure to nitrate significantly reduces growth and affects the health status of juvenile Nile tilapia (Oreochromis niloticus L.) in recirculating aquaculture systems. Aquac. Res. 48 (7), 3482–3492. doi:10.1111/are.13174.

Monsees, H., Kloas, W., Wuertz, S., Pardha-Saradhi, P., 2017b. Decoupled systems on trial: eliminating bottlenecks to improve aquaponic processes. PLoS One 12 (9), e0183056. doi:10.1371/journal.pone.0183056.

Munekata, P.E.S., Pateiro, M., Domínguez, R., Pollonio, M.A.R., Sepúlveda, N., Andres, S.C., Reyes, J., Santos, E.M., Lorenzo, J.M., 2021. Beta vulgaris as a natural nitrate source for meat products: a review. Foods 10 (9). doi:10.3390/foods10092094.

Nadia, Z.M., Akhi, A.R., Roy, P., Farhad, F.B., Hossain, M.M., Salam, M.A., 2023. Yielding of aquaponics using probiotics to grow tomatoes with tilapia. Aquac. Rep. 33. doi:10.1016/j.aqrep.2023.101799.

Palm, H.W., Knaus, U., Appelbaum, S., Strauch, S.M., Kotzen, B., 2019. Coupled aquaponics systems. In: Aquaponics food production systems: Combined aquaculture and hydroponic production technologies for the future. Springer, Cham, pp. 163–200. doi:10.1007/978-3-030-15943-6_7.

Palm, H.W., Knaus, U., Kotzen, B., 2024. Aquaponics nomenclature matters: it is about principles and technologies and not as much about coupling. Rev. Aquac. 16 (1), 473–490. doi:10.1111/raq.12847.

Park, J., Hwang, J.A., Choe, J., Lee, D., Kim, H., 2024. Enhancing indoor culture of weather Loach (Misgurnus anguillicaudatus) and Caipira Lettuce (Lactuca sativa) in a decoupled FLOCponics system. Fishes. 9 (5). doi:10.3390/fishes9050150.

Pickens, J.M., Danaher, J.J., Sibley, J.L., Chappell, J.A., Hanson, T.R., 2020. Integrating greenhouse cherry tomato production with biofloc tilapia production. Horticulturae 6 (3), 1–13. doi:10.3390/horticulturae6030044.

Piñero, M.C., Collado-González, J., Otálora, G., López-Marín, J., del Amor, F.M., 2024. Plant growth-promoting Rhizobacteria as tools to improve the growth of Kohlrabi (Brassica oleracea var. gongylodes) plants in an aquaponics system. Plants. 13 (5). doi:10.3390/plants13050595.

Pinheiro, I., Carneiro, R.F.S., Vieira, F.d.N., Gonzaga, L.V., Fett, R., Costa, A.C.d.O., Magallón-Barajas, F.J., Seiffert, W.Q., 2020. Aquaponic production of Sarcocornia ambigua and Pacific white shrimp in biofloc system at different salinities. Aquaculture 519. doi:10.1016/j.aquaculture.2019.734918.

Pinheiro, I., Arantes, R., do Espírito Santo, C.M., do Nascimento Vieira, F., Lapa, K.R., Gonzaga, L.V., Fett, R., Barcelos-Oliveira, J.L., Seiffert, W.Q., 2017. Production of the halophyte Sarcocornia ambigua and Pacific white shrimp in an aquaponic system with biofloc technology. Ecol. Eng. 100, 261–267. doi:10.1016/j.ecoleng.2016.12.024.

Pinho, S.M., Lima, J.P., David, L.H., Oliveira, M.S., Goddek, S., Carneiro, D.J., Keesman, K.J., Portella, M.C., 2021. Decoupled FLOCponics systems as an alternative approach to reduce the protein level of tilapia juveniles' diet in integrated agri-aquaculture production. Aquaculture 543. doi:10.1016/j.aquaculture.2021.736932.

Pinho, S.M., Flores, R.M.V., David, L.H., Emerenciano, M.G.C., Quagrainie, K.K., Portella, M.C., 2022a. Economic comparison between conventional aquaponics and FLOCponics systems. Aquaculture 552. doi:10.1016/j.aquaculture.2022.737987.

Pinho, S.M., Molinari, D., de Mello, G.L., Fitzsimmons, K.M., Coelho Emerenciano, M.G., 2017. Effluent from a biofloc technology (BFT) tilapia culture on the aquaponics production of different lettuce varieties. Ecol. Eng. 103, 146–153. doi:10.1016/j.ecoleng.2017.03.009.

Pinho, S.M., de Lima, J.P., David, L.H., Emerenciano, M.G.C., Goddek, S., Verdegem, M.C.J., Keesman, K.J., Portella, M.C., 2022b. FLOCponics: the integration of biofloc technology with plant production. Rev. Aquac. 14 (2), 647–675. doi:10.1111/raq.12617.

Pinho, S.M., David, L.H.C., Goddek, S., Emerenciano, M.G.C., Portella, M.C., 2021. Integrated production of Nile tilapia juveniles and lettuce using biofloc technology. Aquac. Int. 29 (1), 37–56. doi:10.1007/s10499-020-00608-y.

Pinho, S.M., de Lima, J.P., Tarigan, N.B., David, L.H., Portella, M.C., Keesman, K.J., 2023. Modelling FLOCponics systems: towards improved water and nitrogen use efficiency in biofloc-based fish culture. Biosyst. Eng. 229, 96–115. doi:10.1016/j.biosystemseng.2023.03.022.

Pinho, S.M., David, L.H., Garcia, F., Portella, M.C., Keesman, K.J., 2022c. Sustainability assessment of FLOCponics compared to stand-alone hydroponic and biofloc systems using emergy synthesis. Ecol. Indic. 141. doi:10.1016/j.ecolind.2022.109092.

Raza, B., Zheng, Z., Yang, W., 2024. A review on biofloc system technology, history, types, and future economical perceptions in aquaculture. Animals. 14 (10), 1489. doi:10.3390/ani14101489.

Rodrigues, A.B., Pinheiro, I.C., Nonato, T.C.M., Ferenhof, E.A., Sens, M.L., 2023. Valorization of reverse osmosis concentrate in the production of Litopenaeus vannamei and New Zealand spinach in an aquaponic system with biofloc. Desalination 549. doi:10.1016/j.desal.2022.116330.

Romano, N., Islam, S., Fischer, H., 2023. Ebb and flow versus constant water level on the sweet banana chili pepper (Capsicum annuum) production in an aquaponic system. Aquacult. Eng. 102, 102340. doi:10.1016/j.aquaeng.2023.102340.

Rono, K., Matolla, G., Manyala, J.O., Masese, F.O., 2024. Economic viability of agricultural carbon sources on Asian rice (Oryza sativa Lejeunia 1753) and Nile Tilapia (Oreochromis niloticus) production in a flocponic system. J. Aquat. Terr. Ecosyst. 2. doi:10.69897/jatems.v2i2.128.

Samocha, T.M., Prangnell, D.I., Castro, L.F., 2019. Site selection and production system requirements. In: Sustainable biofloc systems for marine shrimp. Elsevier, Amsterdam, pp. 59–117. doi:10.1016/B978-0-12-818040-2.00005-8.

Saseendran, S., Dube, K., Chandrakant, M.H., Babitha Rani, A.M., 2021. Enhanced growth response and stress mitigation of genetically improved farmed Tilapia in a biofloc integrated aquaponic system with bell pepper. Aquaculture 533, 736200. doi:10.1016/j.aquaculture.2020.736200.

Schveitzer, R., Baccarat, R.F.C., Gaona, C.A.P., Wasielesky, W., Arantes, R., 2024. Concentration of suspended solids in superintensive culture of the Pacific white shrimp Litopenaeus vannamei with biofloc technology (BFT): a review. Rev. Aquac. 16 (2), 785–795. doi:10.1111/raq.12867.

Soares, J., Martins, M.A., Castilho-Barros, L., do Espírito Santo, C.M., do Nascimento Vieira, F., Seiffert, W.Q., 2022. Reducing the feed input per unit of plant area as a means to improve the efficiency of sea asparagus and Pacific white shrimp biofloc technology-based aquaponics. Aquac. Res. 53 (18), 6536–6544. doi:10.1111/are.16122.

Somerville, C., Cohen, M., Pantanella, E., Stankus, A., Lovatelli, A., 2014. Small-scale aquaponic food production: integrated fish and plant farming. FAO Fisheries and Aquaculture Technical Paper No. 589, Rome, p. 288.

Sousa, R., Bragança, L., Silva, M.V., Oliveira, R.S., 2024. Challenges and solutions for sustainable food systems: the potential of home hydroponics. Sustainability 16 (2), 817. doi:10.3390/su16020817.

Spradlin, A., Saha, S., 2022. Saline aquaponics: a review of challenges, opportunities, components, and system design. Aquaculture 555, 738173. doi:10.1016/j.aquaculture.2022.738173.

Tetreault, J., Fogle, R.L., Ramos, A., Timmons, M.B., 2023. A predictive model of nutrient recovery from RAS drum-screen effluent for reuse in aquaponics. Horticulturae 9 (3). doi:10.3390/horticulturae9030403.

The State of World Fisheries and Aquaculture 2026 – Blue Transformation: turning vision into impact, 2026. Rome. https://doi.org/10.4060/cd8357en.

Van Wyk, P., Davis- Hodgkins, M., Laramore, R., Main, K.L., Mounta, J., Scarpa, J., 1999. Farming marine shrimp in recirculating freshwater systems. Florida Department of Agriculture and Consumer Services – Harbor Branch Oceanographic Institution, Fort Pierce, FL, p. 220.

Vasdravanidis, C., Alvanou, M.V., Lattos, A., Papadopoulos, D.K., Chatzigeorgiou, I., Ravani, M., Liantas, G., Georgoulis, I., Feidantsis, K., Ntinas, G.K., Giantsis, I.A., 2022. Aquaponics as a promising strategy to mitigate impacts of climate change on rainbow trout culture. Animals 12 (19). doi:10.3390/ani12192523.

Yep, B., Zheng, Y., 2019. Aquaponic trends and challenges – A review. J. Cleaner Prod. 228, 1586–1599. doi:10.1016/j.jclepro.2019.04.290.

Yu, Y.B., Choi, J.H., Lee, J.H., Jo, A.H., Han, S.W., Han, S.H., Choi, H.J., Choi, C.Y., Kang, J.C., Min, E.Y., Kim, J.H., 2023. Biofloc application using aquaponics and vertical aquaculture technology in aquaculture: review. Fishes 8 (11). doi:10.3390/fishes8110543.

Zappernick, N., Nedunuri, K.V., Islam, K.R., Khanal, S., Worley, T., Laki, S.L., Shah, A., 2022. Techno-economic analysis of a recirculating tilapia-lettuce aquaponics system. J. Cleaner Prod. 365, 132753. doi:10.1016/j.jclepro.2022.132753.

CHAPTER

17

Integrated multitrophic aquaculture: Advancing sustainability and environmental stewardship

Pronob Das[a], *Maibam Malemngamba Meitei*[b], *Muzammal Hoque*[a], *Syamala Karthireddy*[c], *Muralidhar P Ande*[c] *and Basanta Kumar Das*[d]

[a] ICAR-Central Inland Fisheries Research Institute, Regional Centre, Guwahati, Assam, India [b] ICAR-Central Institute of Fisheries Education, Mumbai, Maharashtra, India [c] ICAR-Central Institute of Fisheries Education, Kakinada Centre, Kakinada, Andhra Pradesh, India [d] ICAR-Central Inland Fisheries Research Institute, Barrackpore, Kolkata, West Bengal, India

17.1 Introduction

Aquaculture has emerged as the most rapidly expanding sector in global food production, currently contributing over 51% of the aquatic animal protein consumed worldwide (FAO, 2024). As capture fisheries approach their biological and ecological limits due to overexploitation, habitat degradation, and declining fish stocks, aquaculture is increasingly recognized as a vital component in ensuring global food and nutritional security (Ahmed et al., 2019; Naylor et al., 2021). However, the accelerated expansion of conventional monoculture systems characterized by high stocking densities, intensive feeding, and chemical inputs has precipitated a range of ecological concerns. These include nutrient overenrichment leading to eutrophication, habitat alteration, the spread of pathogens, and disruptions to local socioecological systems (Cao et al., 2015; Martínez-Porchas et al., 2023). Consequently, there is a growing imperative to transition toward more ecologically sound and resilient aquaculture practices.

Zero-Waste Technologies in Aquaculture: Global Perspectives and Methods toward a Circular Economy
DOI: https://doi.org/10.1016/B978-0-443-34689-7.00022-7

In response to these challenges, integrated multitrophic aquaculture (IMTA) has gained prominence as a holistic and systems-based approach that aligns productivity with environmental stewardship (Ridler et al., 2007). IMTA involves the cocultivation of species from different trophic levels within a single, interconnected system. Typically, it integrates fed species such as finfish or shrimp with extractive organisms, including filter feeders (e.g., bivalves) and primary producers (e.g., macroalgae or aquatic plants), which assimilate the organic and inorganic wastes generated by the fed species. This biomimetic system replicates the nutrient cycling processes found in natural ecosystems, thereby enhancing resource efficiency, reducing environmental externalities, and diversifying economic outputs. Beyond improving nutrient retention and waste mitigation, IMTA contributes to the provision of ecosystem services, promotes ecological stability, and fosters broader social acceptance of aquaculture by aligning production with sustainability goals (Barrington et al., 2009; Knowler et al., 2020).

The theoretical framework of IMTA is anchored in the interdisciplinary principles of ecological engineering, the circular economy, and biomimicry, all of which advocate for designing systems that replicate the functional complexity and nutrient dynamics of natural ecosystems (Troell et al., 2009). IMTA emphasizes the ecological interdependence among species, wherein the waste outputs of higher trophic-level organisms, such as finfish or shrimp, are transformed into inputs for lower trophic-level species, such as macroalgae and filter feeders (Khanjani et al., 2022). Nitrogen-rich effluents are efficiently absorbed by seaweeds, while suspended organic particulates are consumed by bivalves or detritivorous invertebrates, thereby reducing nutrient accumulation, improving water quality, and lowering the risk of eutrophication and harmful algal blooms (HABs).

From an economic perspective, IMTA enhances production efficiency and financial stability by enabling the simultaneous cultivation of complementary species within a single aquaculture system (Knowler et al., 2020). This integrated approach promotes income diversification, which is especially critical in buffering against market volatility and biological risks such as disease outbreaks. For smallholder and artisanal aquaculture enterprises, which are often more vulnerable to external shocks, IMTA offers a resilient production model that supports both livelihood sustainability and risk mitigation. Furthermore, the multioutput nature of IMTA can improve access to diverse aquatic food resources, thereby contributing to food security, employment generation, and nutritional outcomes in vulnerable rural and coastal communities. In this context, IMTA aligns strongly with the broader objectives of the United Nations Sustainable Development Goals (SDGs), particularly SDG 2 (zero hunger), SDG 12 (responsible consumption and production), and SDG 14 (life below water), positioning it as a viable strategy for inclusive, sustainable aquaculture development.

Moreover, the ecological footprint of aquaculture can be considerably reduced through the implementation of IMTA systems (Ghosh et al., 2025). By incorporating bioremediation within aquaculture operations, IMTA effectively mitigates nutrient accumulation in surrounding aquatic environments, decreases dependence on synthetic inputs such as fertilizers and chemical treatments, and contributes to both carbon and nitrogen sequestration (Nederlof et al., 2022a). In marine-based IMTA systems, macroalgae such as *Gracilaria* and *Saccharina latissima* have shown exceptional nutrient uptake efficiency, with studies indicating removal rates of up to 90% for dissolved inorganic nitrogen and 85% for phosphorus from aquaculture waste streams (Abreu et al., 2011; Zhu et al., 2025). These macroalgae also play a significant role in carbon capture; for instance, *Saccharina* species can sequester approximately 1.5 kg CO_2 m^{-2} annually, offering promising climate cobenefits through blue carbon storage (Duarte et al., 2017). In freshwater IMTA systems, fast-growing aquatic plants such as *Lemna minor* and *Azolla pinnata* are equally effective, with nutrient uptake capacities reaching 3.6 g N/m^2/day and 0.8 g P/m^2/day, while also contributing to improved oxygenation and water clarity (Cheng and Stomp, 2009). Filter-feeding bivalves, including mussels and freshwater pearl oysters, further enhance system performance by filtering 20–30 L of water per individual per day, thereby reducing turbidity, organic particulates, and phytoplankton density (Shumway et al., 2003). The integration of these extractive components fosters a more balanced ecosystem and supports the development of climate-resilient and environmentally responsible aquaculture.

In recent years, research and innovation in IMTA have increasingly focused on precision aquaculture technologies, sensor-based nutrient monitoring, and genetic improvement of extractive species, which hold promise for optimizing nutrient fluxes and increasing system productivity. Interdisciplinary collaboration among ecologists, aquaculturists, economists, and policymakers is essential to refine IMTA models and establish science-based best practices. Furthermore, robust regulatory frameworks and incentives are needed to encourage stakeholders to transition from linear to circular production paradigms in aquaculture.

IMTA thus represents a forward-looking strategy for achieving sustainable intensification in aquaculture. By addressing the core challenges of environmental degradation, production inefficiency, and socioeconomic vulnerability, IMTA positions itself as a cornerstone in the evolution of sustainable aquatic food systems. It embodies the principles of ecological integration, resource circularity, and resilience, hallmarks of sustainable development. As global demand for aquaculture products continues to rise, the successful adoption and mainstreaming of IMTA will rely on sustained investment in scientific research, effective stakeholder engagement, policy innovation, and targeted capacity-building initiatives. Through these efforts, IMTA can transition from a promising concept to a transformative practice that redefines the future of aquaculture.

17.2 Concepts and principles of integrated multitrophic aquaculture

Traditional fish culture is a long-established aquaculture practice. The sector has evolved to intensive aquaculture, particularly on monoculture species fed with a high-nutrient diet. However, all these systems generate a huge amount of waste, either uneaten feed or metabolic waste. This waste can be in the form of particulate organic matter (POM) and dissolved inorganic nutrients (DINs), which degrade both the culture environment as well as the surrounding ecosystems upon discharge, often leading to eutrophication. The POM can be assimilated by the filter-feeding organism, such as bivalves or echinoderms, and the DIN can be utilized by the seaweeds, halophytes, or other aquatic plants, and such organisms are called extractive species (as shown in Fig. 17.1A & B). Therefore to develop an integrated aquaculture system based on a balanced ecosystem concept, integrating these extractive species to utilize or extract the waste generated by the fed species. IMTA refers to the aquaculture practices that integrate fed species such as finfish and shellfish along with extractive species, either organic (bivalves, echinoderms, etc.) and/or inorganic extractive species (seaweeds, mangroves, etc.) having commercial value. This system develops a balanced ecosystem promoting environmental sustainability (biomitigation), economic stability (product diversification and risk reduction), and social acceptability (better management practices) (Khanjani et al., 2022). The system transforms the waste of the

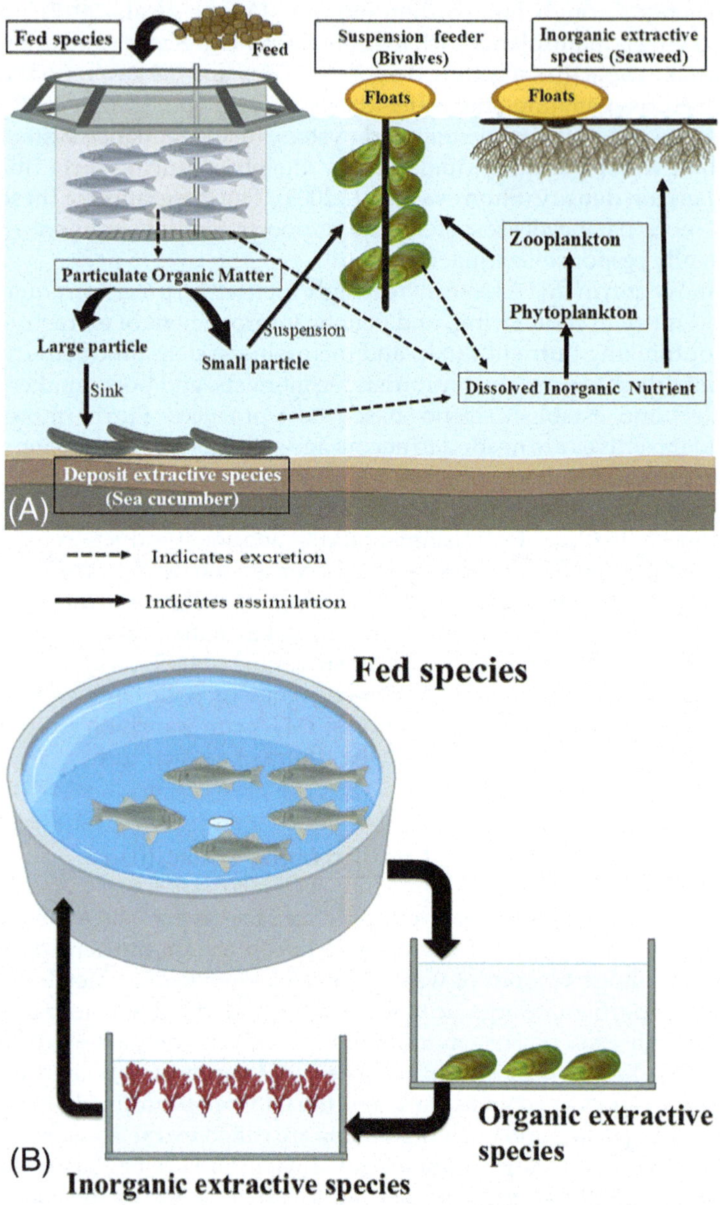

FIGURE 17.1 (A) Offshore IMTA model and (B) recirculating IMTA model. *IMTA*, Integrated multitrophic aquaculture.

feed species into additional biomass through the extractive species, thereby reducing nutrient load on the ecosystem(s) and promoting the circular economy.

This extractive species represents a different trophic level in the same ecosystem, having a different feeding habit to utilize waste (POM and DIN), thereby forming a multitrophic system. This makes the IMTA system different from the polyculture system, with which it is

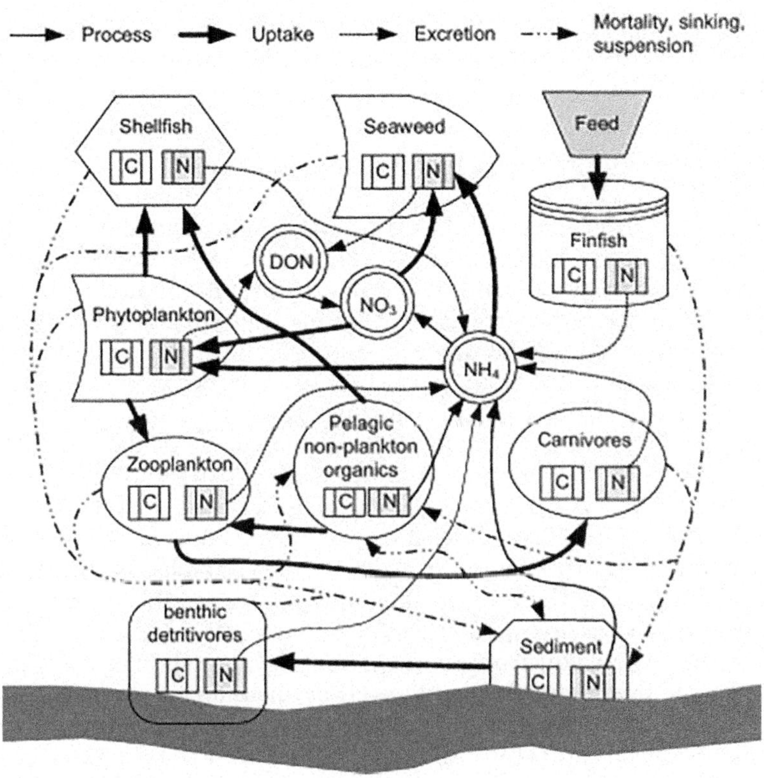

FIGURE 17.2 Ecophysiological and biogeochemical processes through carbon (C) and nitrogen (N) pathways between the various compartments.

often confused. Moreover, polyculture involves a combination of fish species having different ecological niches but sharing the same trophic level. For example, in an integrated culture system involving fed fish, bivalves, and seaweeds, the extractive species play complementary roles: Bivalves filter and consume POM, whereas seaweeds absorb DIN. Meanwhile, the fed species, such as finfish or shellfish, are provided with formulated diets, collectively forming a multitrophic aquaculture system within the same ecosystem.

Moreover, the IMTA system incorporates the concept of dynamic energy budgets (DEB) for each trophic group within the finfish–shellfish–detritivore–primary producer framework (Ren et al., 2012). As illustrated in Fig. 17.2, the system involves complex biogeochemical processes. Carbon and nitrogen released by the fed species are absorbed by phytoplankton, which serve as a link in transferring nutrients through the system. These plankton also contribute to sediment nutrient enrichment, which is then utilized by benthic organisms. In addition, suspension feeders filter both plankton and seston from the water column, while inorganic extractive species assimilate dissolved ammonia and carbon for their growth.

Fig. 17.2 shows the ecophysiological and biogeochemical processes through carbon (C) and nitrogen (N) pathways between the various compartments (Ren et al., 2012). The cultured trophic group comprises fed organisms (finfish), suspended filtering feeders (shellfish), nutrient extractive organisms (seaweed), and benthic detritivores (sea cucumber). The pelagic compartment includes phytoplankton, zooplankton, carnivores, dissolved inorganic nitrogen (DIN), dissolved organic nitrogen (DON), pelagic nonplankton organic carbon, and pelagic nonplankton organic nitrogen. DIN consists of ammonia nitrogen (NH_4) and nitrate nitrogen (NO_3). The benthic compartment comprises carbon sediment and nitrogen sediment.

17.3 History of integrated multitrophic aquaculture

IMTA is an ancient practice that has been believed to have originated along with the origin of aquaculture. During the Han dynasty (2200–2100 BC), You Hou Bin documented the integration of fish with aquatic plants and vegetable production, while the culture of paddies and fish was documented during 1977–1780 BC (Costa-Pierce, 2008). During 1550–1070 BC, there was evidence of tilapia culture in an integrated agriculture-aquaculture pond in Egypt (Chopin, 2018). Then, polyculture was developed during the Tang dynasty (1380–1100 BC), and the theory of mutualism in rice-fish culture was described in Lin Biao Lu Yi by Liu Chen in 1100 BC. In the same year, the integration of fish and fruit production was also documented. Further, in the book entitled "The Complete Book on Agriculture" by Xu Guangqi and the Treatise on Fish Culture by Huang Shenchen, published during 632–350 BC, described the rotation of fish with aquatic plant production (Costa-Pierce, 2008). The three decades of development were characterized by the emergence of ecologically engineered aquaculture, integrated mariculture or aquaculture, polyculture, and ecological aquaculture methods. However, the concept of modern IMTA evolved from the work "physical models of integrated waste-recycling marine polyculture systems" by Ryther et al. (1975), which approaches scientific integration of shellfish, microalgae, and seaweeds in sewage treatment. This is followed by the integration with their setup of macroalgivore, abalone, and carnivorous fish. Then, the work on the role of seaweeds in nutrient enrichment and uptake in aquaculture setups was started in 1975, and by the 1990s, the study on seaweed as a biofilter, including its cultivation with land-based salmon effluents, was started. The integration of oyster in the aquaculture has also started during the 1990s (Chopin, 2013). Then by 2004, the term "Integrated multitrophic aquaculture" system was born at a workshop in Saint John, New Brunswick, by combining the two terms "Integrated aquaculture" by Theiry Chopin and "Multi-trophic aquaculture" by Jack Taylor (Chopin, 2018). The evolution of the IMTA system is presented in a schematic way in Fig. 17.3.

17.4 Design of integrated multitrophic aquaculture

There is no rigid design for the IMTA system. However, an IMTA should have two components: A fed species and the extractive species, which can be either organic or inorganic. In general, organic extractive species should be positioned close to the fed species, with deposit feeders placed at the bottom to consume larger particles such as uneaten feed and faces, while filter feeders can be located either in the water column or near the bottom to capture suspended particles. In contrast, inorganic extractive species should be situated slightly farther from the fed species, aligned with the direction of water flow, to effectively absorb such as nitrogen and phosphorus. The combination of species in the IMTA system can be varied. For example, the IMTA system can be designed with the combination of different species in four trophic levels, such as fed species (finfish/shellfish), suspension feeders (bivalves/sea urchin), deposit feeders (sea cucumber/polychaete), and inorganic extractive species (seaweed/halophyte).

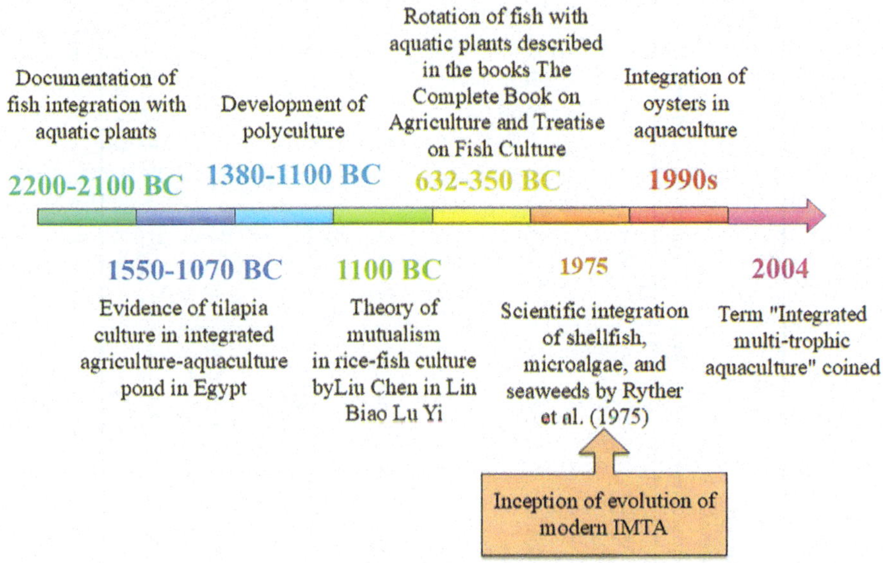

FIGURE 17.3 Evolution of the IMTA system. *IMTA*, Integrated multitrophic aquaculture.

Some IMTA systems can be fed species with either organic or inorganic or both for efficient nutrient utilization, as shown in Table 17.1. But, for an effective IMTA system, the selection, arrangement, and placement of the components are vital. The fed species (finfish/shellfish) occupy the upper or mid trophic level, releasing waste (uneaten feed or metabolic waste) while the extractive species occupy the lower trophic level that can be able to extract or assimilate the waste: POM (uneaten feed and metabolic waste) and DIN (ammonia, nitrate, phosphorous etc.) respectively released by the fed species (Nissar et al., 2023). The different components of the IMTA system will be discussed in depth in the following section.

The selection of the species for the IMTA is very necessary. The selected species of the IMTA system should be compatible and the extractive species should be able to utilize the nutrients released from the fed species, the species should be socially, commercially and ecologically acceptable, the species should be easily adaptable to the environment for which native species is more preferable, the primary (fed) species should provide sufficient nutrients for the extractive species so that they can enhance growth and survive, the size of the waste nutrient should be considered for selecting the organic extractive species.

In addition, the design of an IMTA system can be adapted and expanded based on ecological conditions. In freshwater environments, species such as rohu, tilapia, freshwater mussels, and aquatic plants, such as *Lemna* and *Wolffia*, can be integrated to form a freshwater IMTA (FIMTA) system. In brackish water conditions (salinity 0.5–30 ppt), euryhaline species such as mullet and milkfish can be cultured along with low-salinity bivalves (e.g., *Meretrix* and *Crassostrea*) and tolerant seaweeds or mangroves, forming a brackish water IMTA (BIMTA) system. Meanwhile, in offshore or high-salinity waters (salinity >30 ppt), integration of marine species such as salmon and pompano, with sea urchin and mussels, and seaweeds constitutes a marine IMTA (MIMTA) system. The components of the IMTA system can be placed either in the same system

TABLE 17.1 Diversification of IMTA system.

Fed species	Organic extractive species	Inorganic extractive species	Deposit extractive species	Mode of operation (Tank/ cage/ recirculating)	Water types (marine/ brackish/fresh)	IMTA types	Country	Reference
Colossoma macropomum	—	—	*Macrobrachium amazonicum*	Cages in earthen pond	Freshwater	Semiopen systems	Brazil	Flickinger et al. (2019)
Channos channos, *Oreochromis niloticus*, and *Pennaeus vanamae*	green mussel (*Perna viridis*)	*Gracilaria* seaweed	—	pond	Brackish water	Semiopen systems	Indonesia	Rejeki et al. (2016)
Sparus aurata	—	*Ulva lactuca*	—	Pond	Marine	Semiopen systems	Israel	Shpigel et al. (2018)
GIFT tilapia	*Lamellidens marginalis*	*Lemna minor*	—	Fiberglass Reinforced Plastic tanks	Freshwater	Semiopen systems	India	Shinde et al. (2024)
Labeo rohita	*Lamellidens marginalis*	*Wolffia globosa*	—	outdoor tank	Freshwater	Semiopen systems	India	Nath et al. (2021)
Trachinotus carolinus	—	*Ulva lactuca*	—	raceway	Brackish water	Semiopen systems	USA	Laramore et al. (2022)

(*continued on next page*)

TABLE 17.1 Diversification of IMTA system—cont'd

Fed species	Organic extractive species	Inorganic extractive species	Deposit extractive species	Mode of operation (Tank/ cage/ recirculating)	Water types (marine/ brackish/fresh)	IMTA types	Country	Reference
Penaeus monodon and *Chanos chanos*	*Meretrix casta*	*Bruguiera gymnorhiza* and *Avicennia officinalis*	—	Pond	Brackish water	Semiopen systems	India	Kumara et al. (2024)
Chanos chanos and *Penaeus vannamei*	*Crassostrea cuttackensis*	—	—	Tank	Brackish water	Semiopen systems	India	Naskar et al. (2022)
Salmon	—	*Saccharina latissima*	—	Circular cages	Marine	Open system	Norway	Fossberg et al. (2018)
Solea senegalensis	—	—	*Hediste diversicolor*	RAS tanks	Marine	Closed system	Portugal	Marques et al. (2018)
Litopenaeus vannamei	—	*Ulva lactuca*	—	RAS tank	Marine	Closed system	USA	Laramore et al. (2018)
Paracentrotus lividus	—	—	*Holothuria tubulosa*	30 L tanks	Marine	Closed system	Italy	Grosso et al. (2021)
Salmon	—	*Ulva lactuca*	Invertebrate grazer	RAS tank	Marine	Closed system	Australia	Qiu et al. (2022)
Dicentrarchus labrax and *Sparus aurata*	*Mytilus galloprovincialis*	—	—	Cages	Marine	Open system	Italy	Arduini et al. (2023a)

(cage, tank, or pond) or in a recirculating structure, such as aquaponics, as shown in Fig. 17.1A and B. The diverse IMTA systems that have been conducted globally are illustrated in Table 17.1.

Further, IMTA can be classified into three types based on the system openness, influenced by the surrounding environment (Nederlof et al., 2022), such as:

(1) *Closed systems*, where the environment can be controlled (e.g., recirculating aquaculture systems (RASs);
(2) *Open systems*, with minimal control over environmental influences (e.g., sea cages); and
(3) *Semiopen systems*, where environmental influences are partially controlled (e.g., land-based flow-through systems or ponds).

17.5 Components of integrated multitrophic aquaculture system

An IMTA system consists mainly of two components: The fed species and the extractive species.

17.5.1 Fed species

In an IMTA system, fed species occupy higher trophic levels and typically hold high market value, contributing significantly to the state's economy. These species may include finfish, shellfish, or a combination of both, with their selection based on socioeconomic preferences and market demand (Lal et al., 2023). Finfish contribute around 58.1% of the total aquaculture production, in which carps are the major farm species, followed by catfish and salmon (FAO, 2024). In addition, finfishes were the most traded species, forming 65% of the total value, and salmonids were the most valuable species (20% of the value). Crustacean aquaculture has also contributed 24.6% of the total aquaculture production, out of which penaeid shrimp contributes 62.2%, followed by red swamp crayfish (23.3%), Chinese mitten crab (6.4%), and river prawn (4.7%) (FAO, 2024). Recently, whiteleg shrimp (*Penaeus vannamei*) has gained more interest and became the top species produced in 2022 with 6.8 million tons (FAO, 2024).

These fed species are cultured with high-nutrient commercial diets, of which approximately 3%–5% is lost to the culture environment (Cromey et al., 2002; Reid et al., 2009), while about 65%–75% of the nutrients are assimilated by the organisms. This may contribute to the generation of aquaculture waste. Besides the uneaten feed, the fed fish also released fecal matter and other metabolic waste in the form of ammonia from the gills. This waste, particularly uneaten feed and fecal matter suspended in the water column or settled at the bottom of the water bodies, which is known as POM, while the other metabolic waste is dissolved into the water and is called dissolved inorganic nutrients (DIN) (Wang et al., 2012; Chopin, 2013). In addition, approximately 42%–57% phosphorus (P), 5%–45% nitrogen (N), and 6%–44% carbon (C) from diet become POM which are further broken down into smaller particles, called dissolved organic matter (DOM), accounting for around 2%–8% P, 1%–7% N, and 1%–6% C in feed (Reid et al., 2009). These POM and DIN are the potential pollutants that can cause eutrophication (Sanz-Lazaro and Sanchez-Jerez, 2020). However, these nutrients serve as nutrients for extractive organisms that can reduce pollution.

17.5.2 Extractive species

The extractive species play a major part in the IMTA system, which is dependent on the waste that is fed to aquaculture species and uses that waste for metabolism. The extractive species have the potential to be powerful biomitigators because of their capacity to absorb and assimilate the waste from fed aquaculture species (Chopin et al., 2001). Depending on the nutrient utilization, the extractive species can be classified into organic extractive species and inorganic extractive species. Organic extractive species consume POM, including suspended and deposited uneaten feed and fecal matter released from the fed organism, while inorganic extractive species that assimilate the DIN comprise dissolved inorganic nitrogen and dissolved inorganic phosphates (DIP). Inorganic extractive species include freshwater aquatic weeds such as ipomea, wolfia, lemna, whereas the marine or brackish water species include seaweeds and halophytes.

Further, the organic extractive species are often divided into two groups: deposit extractive species and suspension extractive species according to the water column (Marinho-Soriano et al., 2011). The suspension feeders are the filter-feeding organisms that feed on the small organic matter (planktons, bacteria, uneaten feed, and detritus) suspended in the water column through filtration (Hentschel and Shimeta, 2008). They are mostly kept in the column (mussel, oyster, etc.) at the bottom (clams and other jerking organisms). The suspension feeders include mussels, clams, oysters, and scallops. Deposits extractive species mainly feed on the larger organic particles and other organic matters that escape from the suspended organisms and occupy the bottom of the culture site. This includes sea cucumbers and polychaete.

The majority of aquaculture solid wastes are of fecal origin, and fecal losses from undigested feed nutrients comprise approximately 10%–30% of the feed (Kokou and Fountoulaki, 2018). However, there is no exact percentage of feed loss in the aquaculture industry, as it varies according to the operation of aquaculture. For example, in an Atlantic salmon farm with 93% feed consumption, 18% of the solid waste is produced, of which 3% is waste feed, and 15% is fecal waste (Reid et al., 2009; Wang et al., 2012), while less than 5% of the feed was lost for salmon reared in net pens (Cromey et al., 2002). In addition, untreated aquaculture waste from flow-through farms contains approximately 5–50 mg L^{-1} of solid waste (Turcios and Papenbrock, 2014). Overall, 5%–45% N, 42%–57% P, and 6%–44% C in feed are released as POM (Nederlof et al., 2022). In addition, fish excrete nitrogen primarily as ammonia (NH_3/NH_4^+), which is subsequently transformed into nitrite (NO_2^-) and nitrate (NO_3^-), and inorganic phosphorus (PO_4^{3-}) through both gill and urinary excretions as a result of metabolic processes, along with the dissolution of uneaten feed and feces into the surrounding environment. These nitrogen forms (NH_4^+, NO_2^-, and NO_3^-) are collectively referred to as dissolved inorganic nitrogen, while DIP denotes soluble phosphorus. Both dissolved inorganic nitrogen and DIP contribute to causing eutrophication in aquatic systems and can be harmful to fish health. According to mass balance models, approximately 39%–63% of nitrogen, 18%–30% of phosphorus, and 39%–70% of carbon from feed are released into the environment as inorganic waste (Nederlof et al., 2022). Both these POM and DIN can be utilized by the extractive species. Various potential extractive organism and their ecosystem services will be discussed.

17.5.3 Factors affecting nutrient uptake efficiency by the extractive species

There are factors that need to be considered for efficient nutrient assimilation for both organic and inorganic extractive species. The factors such as age, size, and species, particle size of the particulate matter, environmental factors such as temperature and salinity, velocity of particle sink, depth at which the bivalve will be placed, and density of the bivalve need to be considered for optimizing nutrient assimilation for organic extractive species, while the nutrient uptake rate of inorganic extractive species is influenced by environmental factors, water current, species, and stocking biomass.

17.5.3.1 Organic extractive species

17.5.3.1.1 Age, size, and species of bivalves

The filtration efficiency of bivalves is significantly influenced by their size. Larger bivalves generally demonstrate higher absolute filtration rates because of their greater gill surface area, which enables them to filter larger volumes of water. Under conditions of high algal density, they also tend to be more effective than smaller bivalves, as the gills of smaller individuals may become clogged with dense algal matter (Rajesh et al., 2001). However, aging in bivalves can lead to a decline in filtration efficiency (Sukhotin et al., 2003).

Despite their smaller size, young bivalves often exhibit a higher weight-specific filtration rate, which means they filter more efficiently relative to their body mass. This may be because of their increased activity levels and lower sensitivity compared with larger mussels (Tantanasarit et al., 2013; Kreeger et al., 2017). This suggests that while larger bivalves contribute more to overall system filtration, smaller ones may play a more effective role in nutrient cycling per unit of biomass. On the other hand, the types of bivalves and the environment also influenced the filtration capacity. For example, blood clam has relatively higher plankton filtration efficiency than the oyster and mussel in the microcosm experiment, but mussel has higher efficiency of filtration rate in the mesocosm experiment (Meitei et al., 2025). If comparing bivalve sizes, clams are generally shorter in length than other bivalves; however, their filtration efficiency depends more on morphological adaptations than size alone (Hopkins, 1933; Riisgård et al., 2003).

17.5.3.1.2 Particle size

Bivalves demonstrate species-specific particle retention efficiencies, largely influenced by particle size. Most can effectively retain particles ranging from 1–7 μm (Møhlenberg and Riisgård, 1978; Riisgård, 1988; Beecham, 2008), with nearly 100% retention for particles over 4 μm, but only about 50% for 1 μm particles (Shumway et al., 1985). Some species can even ingest particles as large as 400 μm (Beecham, 2008), although suspension feeders typically reject particles larger than 50 μm. Interestingly, retention efficiency in pearl oysters appears largely unaffected by particle size (Pouvreau et al., 1999). Marine aggregates may enhance nanoparticle (100 nm) uptake in suspension-feeding bivalves, which normally exhibit poor retention for particles <1 μm (Ward and Kach, 2009). While suspension feeders prefer small, lightweight particles, deposit feeders are better adapted to ingest larger, heavier particles (Ward and Shumway, 2004). Therefore incorporating deposit-feeding extractive species into IMTA systems may improve waste utilization efficiency.

Beyond particle size, the organic content of waste also influences nutrient absorption in extractive species. For example, sea cucumbers show higher absorption efficiency (>80%) when feeding on organic-rich particulate matter (more than 60% organic content), such as cultured microalgae or salmon waste, compared with their natural diet (Nelson et al., 2012). However, a direct correlation between the organic content of food and absorption efficiency is not always observed, indicating that additional factors may affect nutrient uptake.

In mussels, nutrient assimilation is influenced by the concentration of organic fish waste and surrounding seston levels. Their nutrient absorption is optimal in environments with high organic content and low seston concentrations (Cranford et al., 2013), which highlights the importance of environmental conditions in determining the efficiency of extractive species in IMTA systems.

17.5.3.1.3 Salinity and water temperature

Salinity and temperature play critical roles in regulating the filtration activity of bivalves, with both factors significantly affecting their physiological performance and ecological function.

Salinity exerts a notable influence on bivalve filtration rates, with most species exhibiting optimal performance at moderate to high salinity levels. For example, *Gomphina veneriformis* displayed peak filtration efficiency at 30–35 psu, whereas its activity declined substantially at lower salinities (Shin *et al.*, 2009). Similarly, *Mytilus edulis* from the Great Belt maintained stable filtration rates across a salinity range of 10–30 psu, but experienced a marked decline at 5 psu (Riisgård et al., 2013). Interestingly, a dwarfed form of *M. edulis* in the low-salinity Baltic Sea showed adaptation to its native salinity regime. Variations in salinity can influence several physiological responses in bivalves, including growth rate, oxygen consumption, heart rate, and immune function (Pourmozaffar et al., 2019). For instance, the Manila clam (*Ruditapes philippinarum*) tolerates salinity levels between 10 and 15 ppt, but experiences reduced oxygen consumption and mortality when salinity drops below 10 and 5 ppt, respectively (Kim et al., 2001). Such low-salinity conditions induce stress and ultimately reduce filtration activity.

Temperature is another key environmental driver affecting bivalve filtration. Generally, filtration rates increase with rising temperatures (Shin et al., 2009). At lower temperatures, bivalves tend to limit valve opening as a strategy to conserve metabolic energy, which in turn reduces their filtration of suspended particles (Kim et al., 2001). For example, *Mya arenaria* shows a linear increase in filtration rate between 4 and 22°C (Riisgård and Seerup, 2003), while *Dreissena polymorpha* exhibits optimal performance in the 14–26°C range. *G. veneriformis* is particularly sensitive to colder temperatures, showing significant reductions in filtration efficiency under such conditions (Shin et al., 2009).

Overall, both salinity and temperature need to be carefully considered when selecting and managing bivalves in aquaculture systems, particularly in IMTA, where their filtration capacity is vital for ecosystem functioning and water quality regulation.

17.5.3.1.4 Velocity of particle sink

Particles that sink slowly or at moderate rates tend to remain suspended in the water column for longer periods, which benefits suspension feeders such as bivalves by increasing their access to food. This prolonged suspension allows bivalves to filter particles more efficiently, reducing the energy required for filtration and promoting better growth. However, the sinking velocity of particles can vary based on their size and composition. Larger particles tend to sink rapidly and may settle to the bottom before suspension feeders can capture them, thereby reducing the food availability for bivalves. These fast-sinking particles, however, can serve as a food source for deposit feeders that inhabit the benthic zone. Therefore incorporating both suspension and deposit feeders in an IMTA system can improve waste utilization and enhance overall sustainability.

Water current speed also influences particle capture efficiency in bivalves. As current velocity increases, the ability of mussels to retain particles decreases. According to Cranford et al. (2013), mussels at a current speed of 2 cm/s were able to capture up to 3.5% of particles in horizontal flux, but this efficiency dropped to 1.7% and 0.9% at 4 cm/s and 8 cm/s, respectively. This demonstrates that higher current speeds can significantly limit the particle retention capacity of mussels. In addition, the direction of water flow plays a key role in the distribution of organic waste and nutrient absorption. In Gokasho Bay, Japan, for example, prevailing current directions indicated that aligning suspension feeders, such as mussels, with the flow enhances nutrient uptake efficiency (Zhang et al., 2016). While the deposit extractive species, such as polychaete, sea cucumber, should be kept under the fed fish as they prefer the large organic matter, such as uneaten feed and fecal matter derived from fed fish.

In summary, optimal waste reduction and nutrient absorption in IMTA systems require careful consideration of particle sinking rates, current speed, and flow direction. Integrating species with complementary feeding strategies, such as suspension and deposit feeders, can significantly improve system performance and ecological balance.

17.5.3.1.5 Depth of culture

The efficiency of waste nutrient assimilation by bivalves, such as mussels, is influenced by the depth at which they are suspended in aquaculture systems. Waste discharged from fed fish farms may remain suspended in the upper water column or sink toward the bottom, with larger particles settling more quickly. Mussels cultured at surface depths (0–10 m) have been shown to assimilate approximately 13% of waste materials, whereas those suspended at greater depths (18–28 m) can assimilate around 15%, indicating a slightly higher uptake at deeper levels (á Norði et al., 2023).

Moreover, slowly sinking particles tend to stay longer in the upper layers, allowing mussels positioned at shallower depths (1–3 m) to assimilate between 10% and 15% of the particulate waste (Brager et al., 2015). These findings highlight the importance of carefully selecting the suspension depth for mussels in IMTA systems. Positioning mussels within nutrient-rich water layers can significantly enhance their capacity to assimilate organic waste, thereby improving system efficiency and contributing to better waste management.

17.5.3.1.6 Stocking biomass

Optimizing the stocking density of organic extractive species is essential for maximizing nutrient uptake and maintaining environmental balance in aquaculture systems. This is because filtration rate is higher in low stocking densities (Tantanasarit et al., 2013). Increased densities of bivalves, such as oyster, impacted the water quality and growth of coculture organisms, such as shrimp (Lima et al., 2021). Moreover, higher stocking densities lead to a reduction in growth. Thus the effectiveness of nutrient assimilation and overall performance is closely linked to the biomass or population density of the extractive species. However, few works have been done in this regard in multitrophic studies. The stocking density of bivalves in various coculture and multitrophic analyses is presented in Table 17.2.

TABLE 17.2 Different stocking densities of bivalves in coculture studies.

Species	Stoking density or biomass	Objectives	Outcomes	References
Sydney rock oyster (*Saccostrea commercialis*)	18, 36, and 72 per 100 L	Nutrient reduction efficiency of oyster and macroalgae	The high stocking density (72 nos per 100 L) has high mortality, and thus the low stocking density is more preferable	Jones et al. (2002)
Mytilus galloprovincialis	220, 370, 500, 570, 700, 800, and 1150 individuals per meter of rope	Impact of mussel density on growth in suspended culture systems	Growth decreases with the increase in densities	Cubillo et al. (2012)
Green mussel (*Perna viridis*)	0.5, 1.0, 5.0, and 10 L ind^{-1}	Nutrient uptake by green mussel	Filtration rate was higher in the lower stocking densities	Tantanasarit et al. (2013)
Green mussel (*Perna viridis*)	0, 30, 45, 60, and 70 per container (1 m^{-2})	Effect of mussel on growth and survival rate of tiger shrimp with IMTA system	Shrimp performs good when cocultured with green mussels, with a stocking density of 30 green mussels per container (1 m^{-2})	Evania et al. (2018)
Sinanodonta woodiana	3.75; 7.5; 15; 30 and 60 kg m^{-3}	Bioremediation model using mussels in fresh water farm of rainbow trout	The bacterial densities were reduced to 72% at a stocking density of 7.5 kg m^{-3}	Sicuro et al. (2020)
Hyriopsis cumingii	0, 3, 6, 9, and 12 individuals per 100 L	Filtering capacity of *Hyriopsis cumingii* in pond water	Filtering capacity of TSS, FA, and POC reached 22.19%, 57.48%, and 21 % in the high-density group, and the filtering rate of chlorophyll ranges from 45.2% to 61.1%	Yu et al. (2021)

(*continued on next page*)

Crassostrea sp.	100, 200, and 300 oyster m^{-2}	Multitrophic biofloc system with *Litopenaeus vannamei* in nursery	Shrimp growth was unaffected by stocking densities of up to 200 oysters m^{-2}; however, water quality and shrimp performance were impacted by high oyster densities (300 oysters m^{-2})	Lima et al. (2021)
Aadara antiquata	3.5 inds. m^{-2}	Compatibility study with *Peneaus indicus*	Survival of shrimp, water quality, and economic return were improved without affecting the shrimp growth	Magondu et al. (2022)
Crassostrea cuttackensis	0.6, 1.2, and 1.8 kg m^{-3}	Effect of stocking densities on growth performances of milkfish (*Chanos chanos*) and Pacific white-leg shrimp (*Penaeus vannamei*) in the BIMTA system	Incorporating oysters in an IMTA system improved POM and PIM without affecting fish and shrimp growth, yielding a 1.6-fold higher benefit–cost ratio at a stocking density of 1.8 kg/m^3 compared to the control	Naskar et al. (2022)
Green mussels (*Perna viridis*)	20, 40, and 60 ind.m^{-2}	Tropical interaction between mussels and shrimp	Mussel stocking density of 40 ind. m^{-2} with shrimp performed better than the monoculture, with filtration rate ranging from 0.6 to 1.7 L day^{-1}	Nurfadillah et al. (2023)
Manila clam (*Ruditapes philippinarum*)	62.5, 125, 250 ind. m^{-2}	Effect of different densities and bivalves on the nitrogen cycle and microbial composition	*R. philippinarum*, at a stoking density of 125 ind. m^{-2}, has a higher activity of nitrogen cycling and reduces the nutrient load in the system	Yuan et al. (2023)

IMTA, integrated multitrophic aquaculture; *POM*, particulate organic matter; *PIM*, Particulate Inorganic Matter; *TSS*, Total Suspended Solids; *FA*, Fatty Acid; *POC*, Particulate Organic Carbon.

17.5.3.2 *Inorganic extractive species*

17.5.3.2.1 Temperature, salinity, pH, and light intensity

Environmental parameters such as temperature, salinity, pH, and light intensity play a critical role in regulating nutrient assimilation by macrophytes within an IMTA system.

(1) Temperature is a key factor affecting nutrient uptake in seaweeds, as it governs physiological processes such as enzymatic activity, metabolic reaction rates, membrane transport efficiency, and nutrient diffusion across boundary layers (Roleda and Hurd, 2019). For example, *Agarophyton vermiculophyllum* demonstrates higher nitrogen removal efficiency in land-based IMTA setups in northern Portugal during the warmer months (April to August), with diminished uptake in the colder seasons (Abreu et al., 2011). Similarly, nutrient uptake patterns in *Ulva lactuca* fluctuate seasonally—peaking during spring, increasing in autumn, and declining in summer (Macchiavello and Bulboa, 2014). However, the optimal temperature for nutrient assimilation is species-specific. For instance, *Chaetomorpha linum* thrives at 25–28°C, *Codium fragile* at 12–24°C, *Meristotheca* species at 18–28°C, and *Kappaphycus* performs best between 25 and 30°C. This indicates that while no universal temperature optimum exists, most species function efficiently across a relatively broad range.

(2) Salinity also significantly influences nutrient absorption in macrophytes. Most seaweed species exhibit optimal nutrient uptake at salinities between 20 and 30 ppt. For example, *Ulva* sp. and *Gracilaria lemaneiformis* show reduced growth and nutrient assimilation at salinities of 15–20 psu but perform effectively at 25 psu, making them suitable for wastewater treatment in aquaculture systems with moderate salinity (Glauco et al., 2024). However, stress from fluctuating salinity and elevated temperatures can impair physiological functions and lead to conditions such as ice-ice disease in seaweeds (Melendres, 2021).

(3) pH regulation is another crucial function of seaweeds in IMTA systems. Seaweeds contribute to pH stabilization through photosynthesis, which occurs during the day and consumes CO_2 and bicarbonate, leading to a rise in pH (De Prisco, 2020). At night, respiration releases CO_2, lowering the pH. This daily fluctuation alters the carbonate system and affects ammonia dynamics. Increased daytime pH raises the proportion of unionized ammonia (NH_3), a toxic form, whereas lower pH at night reduces it (Whitfield, 1974). A study by De Prisco (2020) found a positive linear correlation between total ammonia nitrogen (TAN) loading and TAN removal, but a negative correlation between pH and free ammonia nitrogen (FAN) removal, suggesting that elevated pH increases FAN concentration.

(4) Light intensity is another vital factor influencing macro- and microalgal growth and nutrient uptake, as photosynthesis is dependent on light availability. Nutrient assimilation is significantly higher during the light phase and may decrease by 20%–64% in darkness, depending on species. For example, *Gracilaria asiatica* achieves optimal growth under 90 ± 20 μmol photons m^{-2} s^{-1} (Wedchaparn et al., 2015). Furthermore, light intensity often interacts with temperature. In *Sargassum thunbergii*, the highest nitrogen uptake was observed at 140–180 μE m^{-2} s^{-1} under 15°C and at 60–100 μE m^{-2} s^{-1} under 20–25°C. Similarly, phosphorus uptake peaked at 60 μE m^{-2} s^{-1} and 25°C (Bao, 2008), indicating a combined effect of light and temperature on nutrient dynamics.

Therefore a comprehensive understanding of the interplay between temperature, salinity, pH, and light is essential for optimizing nutrient assimilation by macrophytes in IMTA systems, ensuring environmental stability and sustainable productivity.

17.5.3.2.2 Hydrodynamic factors

Water movement is a key environmental factor influencing nutrient uptake and the overall productivity of seaweeds (Roleda and Hurd, 2019). It plays a vital role in delivering essential nutrients and dissolved inorganic carbon (DIC) to the algal surface through advection (Hurd, 2017). Moreover, water flow affects the development and thickness of velocity and diffusion boundary layers (DBLs) around the seaweed thallus, which are critical in determining the efficiency of nutrient and DIC diffusion into the plant, thereby affecting its growth performance (Hurd, 2017).

However, increased water velocity does not always translate to higher nutrient absorption rates (Roleda and Hurd, 2019). In fact, Chopin et al. (2001) observed that optimal nutrient uptake may occur under conditions of limited water exchange. For instance, *U. lactuca* showed a decline in uptake efficiency for nutrients such as nitrate (NO_3^-), phosphate (PO_4^{3-}), and ammonium (NH_4^+) when exposed to higher water exchange rates. In contrast, *Gracilaria chilensis* exhibited reduced nutrient absorption under low exchange conditions, indicating that optimal flow rates are species-specific and must be tailored accordingly (Macchiavello and Bulboa, 2014).

17.5.3.2.3 Types of species and their nutrient assimilation

The capacity of macroalgae to assimilate inorganic nutrients is a key criterion in selecting appropriate species for integration into an IMTA system (Narvarte et al., 2024a). In recent years, a variety of microalgae, macroalgae, and aquatic macrophytes have been incorporated into IMTA systems as effective biofilters for dissolved nutrients. Notable macroalgal species that have demonstrated efficient nutrient uptake in aquaculture settings include *Gracilaria* spp., *Macrocystis pyrifera, S. latissima,* and *Enteromorpha intestinalis* (Buschmann et al., 2008; Abreu et al., 2009, 2011; Wang et al., 2014; Marinho et al., 2015; Naskar et al., 2023). In freshwater IMTA (FIMTA) systems, aquatic macrophytes such as *L. minor* have been widely utilized for their ability to absorb inorganic nutrients (Shinde et al., 2024a; Shinde et al., 2024b).

Narvarte et al. (2024a) evaluated the nutrient uptake efficiencies of five commercially valuable marine macroalgae *Caulerpa lentillifera, Ulva* sp. (blade), *Kappaphycus alvarezii, Kappaphycus striatus,* and *Eucheuma denticulatum*—in removing inorganic nutrients (NH_4^+, NO_3^-, NO_2^-, and PO_4^{3-}) from mariculture effluents. The results indicated that *C. lentillifera* and *K. alvarezii* exhibited higher NH_4^+ uptake rates than *K. striatus* and *E. denticulatum*. However, no significant difference in NH_4^+ uptake was observed between *C. lentillifera, Ulva* sp., and *K. alvarezii,* nor between *Ulva* sp., *K. striatus,* and *E. denticulatum*. For NO_3^- and NO_2^-, *Ulva* sp. showed the highest uptake rates, while other species demonstrated comparable and relatively lower absorption.

In terms of phosphate (PO_4^{3-}) uptake, both *C. lentillifera* and *K. alvarezii* had the highest rates, whereas *K. striatus* showed the lowest. The PO_4^{3-} uptake rates of *Ulva* sp., *K. alvarezii,* and *E. denticulatum* were statistically similar to those of both *C. lentillifera* and *K. striatus*. These findings highlight the nutrient-specific preferences and species-level variation in nutrient assimilation capabilities, which are essential for designing effective IMTA systems.

Furthermore, additional research by Narvarte et al. (2024b) assessed the response of three eucheumatoid species—*K. alvarezii, K. striatus,* and *E. denticulatum*—to varying phosphate concentrations in aquaculture effluents. The results revealed distinct species-specific responses: NH_4^+

uptake by *K. alvarezii* declined at high PO_4^{3-} levels, while *K. striatus* exhibited enhanced NH_4^+ uptake under the same conditions. *E. denticulatum*, in contrast, achieved peak NH_4^+ uptake at intermediate PO_4^{3-} levels. These observations emphasize the importance of selecting macroalgal species based on the specific nutrient dynamics of the target aquaculture environment to maximize the efficiency of nutrient removal in IMTA systems. However, the nutrient assimilation of the species depends on the site and environmental factors; thus a critical examination is required for the selection.

17.5.3.2.4 Stocking biomass

The ratio between macroalgae/ macrophytes and bivalves plays a critical role in optimizing nutrient removal within an IMTA system. Mao et al. (2009) investigated the efficiency of nutrient reduction by the red alga *G. lemaneiformis* when cocultured with the scallop *Chlamys farreri*, using various wet weight ratios of bivalve to alga: 1:0.16, 1:0.33, 1:0.61, and 1:0.80. Their findings revealed that nutrient removal improved with increasing algal biomass, particularly between the 1:0.33 and 1:0.80 ratios, with the highest specific growth rate (SGR) observed at the 1:0.33 ratio. In a related study, Han et al. (2013) reported that increased densities of macroalgae enhanced CO_2 absorption and contributed to a higher carbon sink, particularly when the bivalve-to-seaweed wet weight ratio was below 1:0.96 (ratios tested: 1:0.31, 1:0.63, and 1:0.96). Similarly, Chebil Ajjabi et al. (2018) found that a mussel-to-seaweed biomass ratio of 1:0.28 yielded optimal results for nutrient uptake efficiency and seaweed growth.

However, not all increases in algal density yield positive outcomes. According to Nakphet et al. (2017), excessive stocking densities can deteriorate water quality and negatively affect the growth and survival of the primary fed species. These studies underscore that determining the appropriate stocking biomass of macroalgae or macrophytes is essential for maximizing nutrient assimilation, ensuring balanced system functioning, and promoting sustainable growth within IMTA frameworks.

17.6 Ecosystem services provided by the various extractive species in an integrated multitrophic aquaculture system

17.6.1 Bivalves

Because of their exceptional filtering capabilities and ability to facilitate the removal of excess nutrients and N from sediments through bioturbation, bivalves can be employed as ideal species for IMTA models. In addition, by boosting microbial activity in sediments, bivalves can improve the denitrification process and bury N in sediments (Kellogg et al., 2014). Because they do not require pelleted feed created from culture environments, bivalves are considered environmentally beneficial because they can reduce the harmful effects of eutrophication and act as a carbon sink because of the creation of their shells. However, their role as a sink for carbon can vary depending on biotic and environmental factors, and in certain instances, they can even operate as a source of carbon (Filgueira et al., 2015). When cultivated in IMTA, filter-feeding species such as oysters and mussels can reduce the organic burden generated by species at higher trophic levels (Cranford et al., 2013). Notably, a sizable nitrogen mass influences the activity of bacteria in sediments and water bodies, underscoring the crucial function that bacteria play in the nitrogen cycle (Huang et al., 2020) and partially reflecting the quality of the water. Assimilation and oxidation of nitrogenous waste can occur in water and sediment due to the ability of certain heterotrophic bacteria to both nitrify and denitrify (Qi et al., 2009; Ma et al., 2015). Bivalves participate in the N cycle and have an impact on bacterial abundances through bioturbation (Kellogg et al., 2014). In the meantime, they can bury N compounds in sediments, increasing the relative abundance of denitrifying bacteria, and eliminate surplus nutrients from sediments via filter-feeding and modifying bacterial activity (Gallardi, 2014). Through the process of bioturbation, bivalves can release P, which will boost the number of *Cyanobacteria*. These bacteria then release dissolved oxygen into the surrounding water through photosynthesis, which improves nitrification and increases the system's ability to remove N from ammonia. Furthermore, by controlling the amount of dissolved oxygen, particularly heterotrophic bacteria, which are crucial for maintaining water quality, bivalves can boost the diversity and quantity of bacteria. This suggested that bivalves, an important aquatic organism in the IMTA system, play an effective role in controlling the quality of the water. In addition, the bivalves have a specific diet and favor plankton in water over nonliving objects. Furthermore, waste from fish farms is not a primary food source for mussels and contributes minimally to their overall diet (Sanz-Lazaro and Sanchez-Jerez, 2017). Moreover, it has been discovered that bivalves bioaccumulate human infections such as adenovirus, hepatitis A virus, human sapovirus, and *Vibrio spp*. A small number of studies indicate that shellfish may serve as reservoirs for several finfish diseases. As a result, the integration of IMTA for shellfish production in fish farms may change the dynamics of fish pathogen infection (Pietrak et al., 2012). According to research conducted by Lefebvre et al. (2000), the AE of *Crassostrea gigas* fed with a phytoplankton diet was higher, at 66%, than that of oysters fed feed-derived POM of 56%. When cultivated in the effluents of fish or shrimp cultures, bivalves removed up to 23% organic matter, 33% organic N, 96% chlorophyll a, 88% suspended particles, and 88% of bacterial biomass. In a fish–microalgae–bivalve system, the bivalves absorbed all of the microalgae, and the microalgae absorbed 47% of the P and 67% of the N (Hussenot et al., 1998). According to a recent investigation evaluated in laboratory fields as well as natural settings, it has been proven that commercially significant species such as fish and mussels are vulnerable to ingesting plastic (Solomando et al., 2020). Sedentary organisms, particularly mussels, are capable of filtering huge volumes of water and storing a variety of contaminants in their tissues (Abo-Al-Ela and Faggio, 2021). Because of this, mussels are frequently employed as bioindicators and sentinel species for pollution (Li et al., 2019). Several bivalves, including *Mytilus galloprovincialis* (Abidli et al., 2021), *C. gigas* (Corami et al., 2020), and *Anodonta anatina* (Berglund et al., 2019), have been observed to accumulate microplastics in their gastrointestinal tracts. Furthermore, prolonged exposure to polymers frequently causes physiological reactions such as a rise in the action of certain enzymes, particularly glutathione S-transferase (GST) (Capó et al., 2022). Similarly, it has been recorded that certain fish and mussel species exhibit altered behavior as a result of ingesting plastics (Capo et al., 2021). Due to their ecological and economic significance, the two most widely raised fish and mussel species that are regarded as bioindicator species of plastic ingestion globally, especially in Mediterranean Sea coastal areas, are *Sparus aurata* and *M. galloprovincialis* (Li et al., 2019). According to Alomar et al. (2022), 33% of *S. aurata*, 94% of *M. galloprovincialis*, and 100% of sea surface water samples had plastics detected at the IMTA system. Filter-feeding mussels consumed around twice as much plastic as fish did. Furthermore, the composition and the types of particles that were consumed varied by species, viz., fish might consume up to 70% of films and filaments, which consist of high-density polyethylene (HDPE) and low-density polyethylene (LDPE), whereas mussels consumed almost 97% of fibers, which consist of cellulose acetate.

17.6.2 Sponges

The incorporation of sponges into IMTA systems is gaining traction because of their capacity to produce collagen and contribute to filter-feeding while generating valuable bioproducts such as biosintering materials, biomedical compounds, and biosilica. The work of Pronzato et al. (1998) was among the earliest to highlight the potential of sponges as extractive species within IMTA frameworks. Owing to their rapid and efficient filter-feeding ability, sponges can significantly reduce organic waste suspended in the water column (Weisz et al., 2008). Being sessile suspension feeders, they can ingest organic particles ranging from 0.1 to 50 μm in size (Maldonado et al., 2010). Since sponges' rates of carbon uptake are similar to those of bivalves, they can be used to remediate organic pollutants in aquaculture and urban settings. According to Wang et al. (2013), seaweeds and fed aquaculture are the main sources of DOM. According to Benner (2002), DOM serves as sponges' primary source of organic food, with POM making up a very small percentage. The sponge loop route is used by sponges to transform DOM into POM (De Goeij et al., 2013). POM is then utilized as food by detritivores such as sea cucumbers (Maxwell et al., 2009) and small crustaceans (Rix et al., 2018). According to Stabili et al. (2006), Demospongiae can filter organic particles ranging in size from 0.1 to 50 μm, which consist of eukaryotes, bacteria, phytoplankton, and detritus. The bacteria can process a water column in 24 h with an 80% retention rate for suspended particles. Because of this, extensive sponge farming close to farms can significantly affect water quality through bioremediation (Stabili et al., 2006). Several in situ studies have documented the advantages sponges gain from nutrient enrichment originating from mariculture activities, cocultivation with mussels (Page et al., 2011), and nutrient inputs from urban wastewater runoff (Ledda et al., 2014). Multiple sponge species, such as *Dysidea avara, Chondrosia reniformis* (Osinga et al., 2010), *Chondrilla nucula* (Milanese et al., 2003), and *Spongia officinalis* var. *adriatica* (Stabili et al., 2006), have exhibited remarkable capacity for waste removal and compatibility with mariculture systems. Under natural rearing conditions, *D. avara* typically shows a 0%–50% annual growth decline; however, when cultivated beneath sea bass cages, explants doubled in size within four months and achieved a 100% survival rate (Osinga et al., 2010). Similarly, *C. reniformis* explants grown on PVC plates near a fish farm achieved annual survival and growth rates of 86% and 170%, respectively, compared with 39% and 79% under standard conditions. Evidence also suggests that some sponges thrive in polluted habitats; for instance, *Agelas oroides* and *Ircinia variabilis* exhibited enhanced particle retention rates in a contaminated site in the Ligurian Sea (Ledda et al., 2014). Furthermore, *Mycale hentscheli* cultured adjacent to mussel farms demonstrated an extraordinary growth increase of over 2000% per year compared to natural reference sites (Page et al., 2005).

17.6.3 Echinodermata

Because of their capacity to ingest larger particulate matter and their significant commercial value, echinoderms, particularly holothurians, are suitable candidates for use as detritus and deposit feeders in IMTA systems (Nelson et al., 2012). Incorporating holothurians to consume the particulate waste that settles after being processed by suspension feeders can effectively lower total organic carbon levels (Paltzat et al., 2008) and total organic matter concentrations (Neofitou et al., 2019) in the surrounding environment. This approach has been shown to work particularly well in oligotrophic waters, such as those found in the Mediterranean. In 2016, *Holothuria scabra* and *Holothuria lessoni* fetched market prices ranging from approximately US$ 369 to 1898 per kg in China (Purcell et al., 2018). When cultivated alongside mussels and fish, sea cucumbers have been reported to remove up to 70% of the organic material deposited on the seabed (Ren et al., 2012), resulting in reductions of 62.3% in total nitrogen and 60.3% in organic carbon from fish feces (Hannah et al., 2013). In addition, when reared either directly on the seabed or within suspended cages in association with bivalves (Yuan et al., 2006), holothurians have exhibited both lower mortality rates (Slater *et al.*, 2009) and notably high growth performance (Zhou et al., 2006). Global trade was facilitated by the IMTA models used to raise the common sea cucumber, *Apostichopus japonicus* (Yokoyama, 2013), sea urchins (Kang et al., 2003), abalones (Kang et al., 2003), and scallops (Zhou et al., 2006). When cocultured in IMTA, *A. japonicus* had a high growth rate of approximately 3.5 times with a 74% survival rate, in contrast to rockfish reared on natural sediments (Park et al., 2015). The same *A. japonicus* species showed a higher survival rate of 96% with SGR of 1.9% when kept in fish cages (Yokoyama, 2013). *Parastichopus californicus* cleans up oyster waste and can eat detritus to convert it into biomass, which sea cucumbers cannot use (Paltzat et al., 2008). *P. californicus* showed faster development and higher survival rates with 99.5% under sablefish net cages in an experimental IMTA setting, in contrast to fish raised outside of the fish farm (Hannah et al., 2013). After being raised for a year, the same *P. californicus* species also exhibited an increased mean weight of 42.9 g when cocultured with *Magallana gigas* (Paltzat et al., 2008). According to Nelson et al. (2012), *Cucumaria frondosa* exhibits an enhanced AE of more than 80% in response to high carbon particulate matter, indicating that it would be a good fit for the IMTA models. *Australostichopus mollis* fed green-lipped mussel wastes under mussel farms showed suitability for both an artificial diet (Slater et al., 2009) and a higher growth rate expectation (Zamora and Jeffs, 2011). Sea cucumbers have been shown to reduce the amounts of organic carbon and phytopigments associated with mussel farms, which may assist to counteract the pollution caused by bivalve farming (Slater and Carton, 2009). According to Zamora et al. (2014), when sea cucumbers and Pacific oysters are cocultured, the mortality rate is only 33%–52%. *Strongylocentrotus droebachiensis* absorbed sablefish waste at an ingestion rate of 0.43 g/individual/day by absorbing 40% of the ingested organic material (Orr et al., 2014). *U. lactuca* was fed with *Paracentrotus lividus* in an IMTA setting. Subsequently, the results demonstrated 13.86% absorbed N, good-quality gonads, an elevated gonadosomatic index, and an enhanced somatic growth rate of 1.4 mm/month (Shpigel et al., 2018). In the coculture of *S. droebachiensis* with *Mytilus spp.*, it was shown that this controls biofouling (Sterling et al., 2016). In the instance of *Holothuria leucospilota*, there was an increase in weight gain and SGR in IMTA models (Yu et al., 2012). According to the study, smothering by organic matter accumulation resulted in hypoxic circumstances that led to mass mortalities of species cultivated directly below the cage (Cutajar et al., 2022). Deposit feeders' bioturbation promotes the breakdown of organic matter in sediments, which improves the overall outcome of organic waste bioremediation (MacTavish et al., 2012). Sea cucumber AE in integrated systems has been documented, and they vary widely from 14% to 88%. From aquaculture waste that is fed directly to them or from sediments that have been enhanced with aquaculture waste, sea cucumbers can extract 0.1%–20% of organic matter, 3%–10% organic C, 7%–16% organic N, and 21%–25% organic P (Nissar et al., 2023).

17.6.4 Polychaetes

Marine polychaetes have been recognized for their important role in IMTA systems, particularly in nutrient recovery from POM, through the culture of species such as *Abarenicola pusilla, Alitta virens* (Brown et al., 2011), *Hediste diversicolor* (Wang et al., 2019), *Arenicola marina, Perinereis aibuhitensis* (Guerrero and Cremades, 2012), *Perinereis nuntia* (Palmer, 2010), *Perinereis helleri* (Palmer et al., 2014), *Capitella* sp., and *Ophryotrocha craigsmithi* (Nederlof et al., 2019). The Mediterranean polychaete *Sabella spallanzanii* has been shown to filter and remove various bacterial groups, including pathogenic human bacteria and vibrios (Stabili et al., 2010). Studies on its ability to filter *Vibrio alginolyticus* demonstrated that the species can effectively remove these bacteria from the water column (Licciano et al., 2005), highlighting its value as a biofilter. In addition, ragworms contribute to ecosystem engineering by acting as biodiffusors, engaging in bioturbation through burrow excavation (Shull, 2009), and performing bioirrigation (Quintana et al., 2015). The potential for IMTA is reflected by the polychaete-assisted sand filter (PASF), which helps retain POM found in effluent water and converts it to *H. diversicolor* biomass. When given elevated POM effluent water instead of declined POM water, the *H. diversicolor* in PASF demonstrated better reproductive success, with 70% POM bioremediation effectiveness recorded. In addition, marine invertebrates such as polychaetes *Diopatra neapolitana, Terebella lapidaria*, and *Sabella cf. pavonina* were shown to naturally colonize PASF. Palmer (2010) assessed the ability of two intertidal polychaetes, *P. helleri* and *P. nuntia*, grown in sand beds to produce harvestable polychaete biomass without further feeding and to treat wastewater from a shrimp farm using PASF. This study found that the polychaete filtering process and total suspended particles both significantly declined the amount of chlorophyll a. The settlement pond utilized in shrimp farms may have less area and retention time as a result of the treatment rates used in this investigation (Palmer, 2010). Nearly 50% of the N in the feces of *Paralichthys olivaceus* can be converted into body tissue by the polychaete *P. nuntia vallata* (Honda & Kikuchi, 2002). In addition to improving the fitness of the cocultured organism, research on the culture of *Nereis diversicolor* utilizing eel sludge suggests the former has the potential to reduce environmental wastes. Similar to sea cucumbers, AE recorded for polychaetes in the integrated model is increasingly varied between 24% and 71%. The efficiencies of waste extraction for polychaetes were raised compared to those of sea cucumbers, viz., 20%–85% of organic matter, 30%–91% of organic N, and 40%–91% of organic C of the aquaculture waste (Nederlof et al., 2022b).

17.6.5 Coelenterates

Aquaculture facilities frequently encounter biofouling agents, particularly hydrozoans and other coelenterates. According to Fitridge et al. (2012), benthic hydroids have been linked to a wide range of detrimental effects on finfish and shellfish, including competition, smothering of the shell, occlusion of the net, and more. According to Ferguson et al. (2010), these organisms can even damage fish gills and skin, leaving them vulnerable to illnesses. Conversely, the class Anthozoa, which includes sea anemones, has demonstrated IMTA potential. Research on *Anemonia viridis* along the Galician coast revealed that the species reared in unfiltered water from a turbot fish farm had a 50% elevated total biomass than the ones in filtered water, whose weight decreased by 33%. The study's findings, which show their potential application in IMTA, unequivocally show the benefits of fish feces and feed residues in promoting sea anemone growth (Guerrero and Cremades, 2012). *Rhopilemaesculenta* and *A. japonicus* were cocultured, and the results showed that the former increased the particle matter's settling velocity. In addition, the sea cucumbers from the coculture pond containing *Rhopilema esculenta* and *A. japonicus* had higher SGR than the monoculture specimens. In addition to providing a supplemental cash crop, coculturing jellyfish and sea cucumbers could reduce the benthic nutrient burden (Ren et al., 2012). China has reported producing several hundred kilograms of jellyfish per hectare in a very short amount of time. This illustrates its potential for enormous financial benefit, and in some regions of China, it has even supplanted the conventional aquaculture of shrimp and fish (You et al., 2007). The potential of coelenterates for IMTA has not been utilized since they are regarded as conventional biofouling agent inorganic extractive species.

17.6.6 Seaweed

Algae are a crucial part of the IMTA model because they can absorb nutrients by selecting excellent algae based on their capacity to eliminate N and P in addition to their ideal temperature range (Skriptsova and Miroshnikova, 2011). Along the Atlantic coast, it has been fairly common to integrate turbot and sea bass with *Gracilaria bursa-pastoris, Gracilaria gracilis, Chondrus crispus, Palmaria palmata, Porphyra dioica, Asparagopsis armata, Gracilariopsis longissima, Ulva rotundata*, and *Ulva intestinalis* as biofilters (Barrington et al., 2009). Algal growth models in the finfish, sea urchin, and seaweed system have shown that radiation, ambient temperature, and nutrients are typically the limiting factors for algal performance. Temperature and radiation should not be limiting considerations for algae farms; instead, nutrients may be the determining factor (Lamprianidou et al., 2015). According to Chopin et al. (1999), *Pyropia spp.* are a viable choice for rearing because they effectively absorb nutrients and have a respectable market value. Nori, on the other hand, costs about US$ 170/kg when it is roasted (Amazon, 2022, online price). Though *Pyropia* is not as good at storing nutrients as kelp, its rapid growth makes it a strong contender for nurturing in environmental surroundings (Chopin et al., 1999). Another viable option for the IMTA model is *Undaria pinnatifida*, and its dried wakame product sells for US$ 250/kg (Amazon, 2022, online price). Because of the presence of galactofucan sulfate, it has been traditionally consumed as food in China, Japan, and Korea, and also has a broad range of pharmacological benefits (Skriptsova et al., 2010). The productivity of *Gracilaria spp.* has increased over the past 20 years (FAO, 2018), and its refined agar product is priced at about US$50/kg (Amazon, 2022, online price), making it a viable contender for the IMTA models (Bird et al., 1982). Nitrogen is frequently considered a crucial factor that limits the growth performance of seaweed (Troell et al., 2003). When N loads are declined, the temperature is almost at its ideal level for rearing seaweed species such as *Ulva* and *Gracilaria*. According to Jones et al. (2002), inorganic N extraction efficiency up to 100% has been documented. According to some research, the inorganic P extraction efficiency varied from 3% to 95%, whereas Corey et al. (2014) stated that the C extraction efficiencies were between 2% and 5%. While the N extraction was 56%, the maximum N extraction efficiencies were up to 100% (Neori et al., 2003). This implies that nitrification and denitrification were frequently cited as insignificant processes, which may be responsible for the elevated efficiencies. According to Krom et al. (1995), denitrification eliminated about 8% of the inorganic N when sea bream and seaweed were reared in an IMTA system. Because the development of growth and tissue content may only be attained under high-nutrient loads (Buschmann et al., 2001). Investigation based on TAN/dissolved inorganic nitrogen levels in the rearing media may consequently overstate the N extraction efficiency of the seaweeds (Cohen and Neori, 1991). In 2020, 35.1 million tons of algae, mostly seaweed, valued at US$16.5 billion, were sold for human consumption, agrichemicals, pharmaceuticals, and phycocolloid feed supplements (FAO, 2022). Therefore it makes sense from an ecological and financial standpoint to diversify an IMTA system using extractive algae (Neori et al., 2004).

17.6.7 Halophytes

Halophytes are mangrove plants that thrive across a wide range of salinities and offer diverse applications in agriculture (Custódio et al., 2021). They can be cultivated in soil or hydroponic systems and demonstrate effective nutrient removal from wastewater sources such as sewage, aquaculture effluents, and others, making them promising extractive species for use in IMTA systems (Kumara et al., 2024). The halophyte species *Halimione portulacoides* exhibits a dissolved inorganic nitrogen (DIN-N) extraction rate of 1.5–2.8 mg L^{-1} day^{-1}, corresponding to 1.8–3.1 mg DIN-N per gram of biomass produced, and its dissolved inorganic phosphorus (DIP-P) extraction rate ranges from 0.1–0.2 mg L^{-1} day^{-1}, equivalent to 0.1–0.3 mg DIP-P per gram of biomass produced (Custódio et al., 2021). When halophytes such as *Avicennia officinalis* and *Bruguiera gymnorhiza* are integrated with *Meristic casta* as extractive species in a halophyte-based IMTA system, reductions of 68.5%–97.5% in TAN, 43.2%–56.1% in NO_2–N, 40.6%–43.3% in NO_3–N, and 71.7%–79.7% in PO_4–P can be achieved (Kumara *et al.*, 2024). Other halophytes such as *Salicornia dolichostachya, Tripolium pannonicum*, and *Plantago coronopus* are also capable of assimilating 167 mg of nitrogen and 23 mg of phosphorus, 61 mg of nitrogen and 9 mg of phosphorus, and 61 mg of nitrogen and 9 mg of phosphorus, respectively, in an RAS–IMTA system (Waller et al., 2015).

17.7 Integrated multitrophic aquaculture as a carbon sequester and carbon sink

IMTA offers a more versatile and cost-effective approach to the ecological management of aquaculture resulting from its intensification. The seaweed (macroalgae) component in the IMTA can alleviate ocean acidification by altering the marine ecosystem's ability to absorb CO_2 at the site of cultivation (Gattuso et al., 2018). Excess CO_2 is absorbed by the seaweeds via photosynthesis and raises the pH of the surrounding waters through physiological processes. They also generate a significant amount of dissolved organic carbon (DOC) throughout their growth stage. This DOC can either form refractory DOC (RDOC) or enter into the food web, eventually leading to carbon sequestration (Li et al., 2018). Seaweed farming on a large scale can mitigate ocean acidification and improve the carbon sink capacity of the ocean (Duarte et al., 2017). Besides, IMTA can restore the seagrass by placing it in the regions where seagrass is colonized, which may improve the sequestration of blue carbon (Greiner et al., 2013).

Atmospheric carbon dioxide, which is absorbed and stored in the marine and coastal ecosystem, is referred to as blue carbon. Three habitats, viz. mangrove swamps, salt marshes, and seagrass meadows, are generally referred to as the blue carbon ecosystem, which is a key component of the global carbon cycle. They are accountable for absorbing and retaining up to 71% of the global blue carbon (Nellemann and Corcoran, 2009). Mangroves have the largest blue carbon pool (6.5 billion tons), followed by seagrasses (3 billion tons) and salt marshes (2 billion tons). A total of about 53 million tons of blue carbon are sequestered globally each year, where seagrasses sequestered 26 million tons (49%), followed by mangroves with 16 million tons (30%), and salt marshes with 11 million tons (21%) (Siikamäki et al., 2012). However, the detrimental impacts of anthropogenic activity on coastal and marine ecosystems have significantly increased global emissions of blue carbon. The incorporation of mangroves or halophyte plants in the IMTA system is also an approach for conserving or promoting mangroves, which leads toward blue carbon.

Further, the culture of shellfish (bivalves) in IMTA is beneficial to the environment as they are filter feeders, removing up to 54% of nutrients (Reid et al., 2010). Shellfish are known to sequester blue carbon and act as a carbon sink (Papageorgiou et al., 2023). SARF (2012) reported the carbon sequestration rate of oyster and mussel shells to be 0.44 and 0.22 tons per harvest, respectively. Harvesting of shellfish from IMTA also removes the carbon trapped in the shells, mitigating ocean acidification, which is linked to climate change. There is mounting information that the $CaCO_3$ shells produced by the calcification during the growth of shellfish have the ability to remove carbon from coastal ecosystems, increasing the shelf-edge sea's capacity to absorb carbon and having a major impact on the carbon cycle of coastal ecosystems (Liu et al., 2022). Thus the most practical and economically viable method for nutrient recovery in open water is the shellfish–macroalgae IMTA model, which has shown great success (Fang et al., 2020), and it also sequesters carbon.

17.8 Profitability and social acceptability of IMTA

The waste generated by the fed species was utilized by the extractive species (bivalves/seaweed), which leads to the biomass gain. The extra biomass gain integrates the additional income to the cultivators, which gives a livelihood opportunity. For example, a net present value (NPV) for salmon and mussel monoculture systems is US$ 1.7 million and 0.650 million, respectively, which is lower than the integrated system of both, accounting for US$ 2.63 million, assuming mussel production rates were 20% higher in the integrated system (Whitmarsh et al., 2006).

Chopin et al. (2001) reported that implementing IMTA reduced the annual environmental cost of producing 250 tons of gross fish production to US$ 64,000, compared with US$ 201,441 for conventional practices. Similarly, Shi et al. (2013) observed greater economic and environmental sustainability in IMTA systems than in monocultures. In another case, Nobre et al. (2010) found that an abalone–seaweed IMTA operation increased profits by 1.4%–5% compared with abalone monoculture, with annual earnings estimated between US$ 1.1 and 3.0 million. The same study estimated the NPV of the integrated system at US$ 2.63 million, surpassing the combined NPVs of both salmon monoculture (US$ 1.7 million) and mussel monoculture (US$ 0.65 million), assuming mussel yields were 20% higher in the integrated configuration.

Seaweed cultivation not only supports environmental stability but also provides an additional source of income for livelihoods. For instance, Troell et al. (1997) evaluated the economic gains when *G. chilensis* was grown alongside salmon in an integrated salmon–seaweed farming model in southern Chile. Their analysis showed that seaweed production contributed an extra US$ 34,000 annually to farm revenue, equivalent to about US$ 0.28 per kg of fish. Similarly, Fonseca et al. (2017) reported that IMTA systems were more profitable than salmon monoculture, with the NPV increasing further if a 10% price premium was applied to IMTA products. In another example, Bunting and Shpigel (2009) assessed the potential of a land-based IMTA configuration involving sea urchins, shrimp, and seaweed using bioeconomic modeling. They observed that the system could yield one million sea urchins annually from the third year onward, with a 10-year internal rate of return (IRR) of 18%. Reducing annual sea urchin mortality to 9%–15% increased the IRR to 29.4%, while *Salicornia* spp. production reaching 33 kg m^{-2} y^{-1} in wetlands could raise the IRR to 133.4%. Likewise, a seahorse–shrimp–oyster IMTA system achieved an IRR of 131% and a payback period of under two years with the initial investment (Fonseca et al., 2017). These findings collectively confirm that IMTA systems offer higher profitability compared to monoculture operations.

Besides, low risk and profitability, IMTA should also gain social acceptability, and the local communities should benefit, and consumers must be willing to purchase the end product (Gempesaw et al., 1995). There are studies that have proved the IMTA system has social acceptability. A survey by Barrington et al. (2009) in Eastern Canada confirmed that IMTA products are safe for consumption, with 50% of respondents indicating a willingness to pay a 10% premium for IMTA-labeled products. Survey results indicated that 65% of respondents believed IMTA could help reduce the environmental impacts of salmon farming. In addition, all participants (100%) agreed that it could enhance waste management in aquaculture and food production. A large majority also felt that IMTA could improve employment opportunities (91%), strengthen community economies (95%), boost industry competitiveness (95%), and increase the overall sustainability of aquaculture (73%). Another survey for market analysis in New York City demonstrated that mussels produced through the IMTA system could attract a price premium. Although IMTA is not widely recognized, a survey conducted in Charlotte County involving 1,220 participants comprising both the general public and industry groups found that 90% of the public and 89% of industry respondents believe it has the potential to be a successful venture (Robinson et al., 2018).

17.9 Case studies

17.9.1 China

A study was carried out in Sanggou Bay, China, by Shi et al. (2013) to investigate the ecological and economic benefits of monocultures and IMTA in open-water systems. Three systems were compared, viz., kelp monoculture, scallop monoculture, and kelp-scallop IMTA model. The study resulted in the finding that the nutrient removal efficiency was significantly higher in the IMTA model, with nutrient removal of 202 Mt/km^2 nitrogen, while the nutrient removal of kelp and scallop monoculture was 193 and 49 Mt/km^2 nitrogen, respectively. This enhances the nutrient recycling capacity. However, the yield of kelp and scallop in the monoculture achieved 3750 and 4650 Mt/km^2, respectively, which was lower than the IMTA system, yielding 3000 Mt/km^2 for kelp and 4650 Mt/km^2 of scallops with a higher investment cost in the IMTA system. Though the IMTA system lowers individual production and has a higher investment cost, the IMTA system has a high total production of 7650 Mt/km^2 (3000 Mt/km^2 of kelps and 4650 Mt/km^2 of scallops) with minimal maintenance, resulting in the highest NPV, BCR, and RC. Furthermore, the IMTA approach demonstrated the potential to maximize economic returns, yielding 25.02 million yuan per km^2, which was substantially higher than the returns from monoculture systems of kelp (15 million yuan per km^2) or scallop (13.23 million yuan per km^2)

in Sanggou Bay. These results highlight that, from both economic and ecological perspectives, IMTA offers greater sustainability compared with monoculture practices.

17.9.2 India

Biswas et al. (2019) implemented a brackishwater IMTA model in six ponds, each measuring 600 m^2, in the Indian Sundarbans over a 150-day culture period. In the IMTA setup, mullets (*Mugil cephalus* and *Liza parsia*) and tiger shrimp (*Penaeus monodon*) served as fed species, while estuarine oysters (*Crassostrea cuttackensis*) and seaweed (*Enteromorpha* spp.) functioned as extractive species. Stocking densities were 10,000 and 30,000 individuals per hectare for the mullets and shrimp, respectively; oysters were stocked at 1,600 individuals per hectare, suspended in baskets within the water column; and *Enteromorpha* spp. was introduced at a biomass of 200 kg per hectare. The control treatment consisted of a polyculture of only mullets and tiger shrimp. Results showed that the IMTA system produced 19% higher yields, with notable growth improvements in both mullets and shrimp. In addition, the apparent feed conversion ratio decreased by 22%, while net income and benefit–cost ratio increased by 69% and 30%, respectively, compared with the control.

17.9.3 Italy

Borghese et al. (2025) carried out a comparative assessment between conventional fish monoculture and an IMTA setup located on the southwest side of the Mar Grande in Taranto, Southeast Italy (Ionian Sea). The farm consisted of 15 cages, each 22 m in diameter, positioned at depths ranging from 7 to 12 m, producing approximately 100 tons per year of European seabass (*Dicentrarchus labrax*) or gilthead sea bream (*S. aurata*). Of these, six cages were converted into an IMTA configuration incorporating the polychaete (*S. spallanzanii*), sponge (*Sarcotragus spinosulus*), mussel (*M. galloprovincialis*), and macroalga (*C. linum*). Compared with the IMTA site, the monoculture site exhibited higher lipid levels (151.73 $\mu g\ L^{-1}$) and organic matter content (189 $mg\ g^{-1}$). Conversely, the IMTA area supported a more abundant zooplankton community, attributed to the bioremediating activity of the cocultured species, which enhanced the quality and availability of seston. These findings highlight IMTA as an effective form of ecological engineering.

17.9.4 Peru

The Peruvian coast has high productivity of phytoplankton, leading to the suitable culture of *Argopecten purpuratus* (Loayza-Aguilar et al., 2023). It is mostly cultured through a long line of 100 m at a depth of 6 m and harvested at 106.5 g (Loayza-Aguilar et al., 2023). It also provides more than 10,162 new jobs in 2019, and incomes of US$ 420.7 million in 2021, from exportation (FAO, 2022), contributing foreign exchange to the country. However, the culture system causes a serious issue in the environment, such as eutrophication, high ammonia loading, biodeposition of feces and pseudofeces, and biofouling. Thus an IMTA model consisting of suspension organic extractive species, *A. purpuratus*, microalgal species (*Chondracanthus chemissei, Porphyra Columbina, Gracilariopsis lemaniformis,* and *U. Lactuca*), and deposit extractive species (holothurids: *Patallus mollis* and *Pseudocnus debiosus*) (Loayza-Aguilar et al., 2023). As a result, the organic matter (feces and pseudofeces) and the nutrient loadings (ammonia and phosphate) were biotransformed. This enhanced the biodiversity and water quality. In addition, it also reduced eutrophication and sequestered C, N, and P from the environment.

17.9.5 Canada

A pilot initiative in the Bay of Fundy, Atlantic Canada, integrated the cultivation of seaweeds (*Laminaria saccharina* and *Alaria esculenta*), mussels (*M. edulis*), and Atlantic salmon within an IMTA framework to assess both growth performance and food safety of the cultured species (Ridler et al., 2007). The findings demonstrated that implementing a three-species IMTA model on a Canadian salmon farm not only enhanced overall profitability but also mitigated economic risks linked to fluctuations in production and shifting market demands. Using a capital budgeting approach, the financial performance of the IMTA system was compared to that of a salmon monoculture, drawing on data from a pilot salmon–mussel–kelp farm in Eastern Canada. After more than a decade, and with a 5% discount rate, the IMTA system achieved an NPV of US$ 3,296,037—24% higher than the US$ 2,664,112 NPV from the monoculture system. These outcomes suggest that utilizing the waste from one species as a resource for another can improve profitability. Even when subjected to a sensitivity analysis in which all product prices were reduced by 12% for the entire 10-year period, the IMTA model maintained a profit margin of 3.2%, compared with only 0.3% for the monoculture. Furthermore, three simulated scenarios assessing variability in salmon harvest because of disease and adverse weather indicated that in every case, the IMTA system remained more profitable, underscoring the advantages of product diversification and the associated reduction in production risks.

17.9.6 Greece

The IMTA model was set in three different aquaculture farms in the Aegean, Mediterranean Sea (Chatzivasileiou et al., 2022). The first farm consists of three cages (1×104 m^2), which are 100–400 m from the nearest shore and at a depth of 45 m. It produced gilthead sea bream (*S. aurata)* and European sea bass (*D. labrax*) of 250 tons year^{-1}. The second farm consists of four cages (8×104 m^2) at a distance of 50–300 m and a depth of 42 m with a production capacity of 1600 tons yr^{-1}, and the third farm has only 1 cage (1×104 m^2) at a distance of 100 m from the shore and a depth of 34 m.

In the study, Mediterranean mussels (*M. galloprovincialis*) were suspended on longlines using ropes that connected fish cages to buoys, positioned 2.6 m away from the cages and spaced 80 cm apart within each fish farm. The stocking densities were 1,307 individuals per net in the first farm and 3,400 individuals per net in the second, while no mussels were stocked in the third farm. Pearl oysters (*Pinctada imbricata radiata*) were deployed centrally within the fish cages at a depth of 3 m across all farms. In addition, sea cucumber (*Holothuria poli*) was incorporated into the system using SEAPA baskets, with 13 individuals placed at the bottom of each farm.

After 9 months of culture, mussel condition index (CI) reached 42% and 33% in the first and second aquaculture farms, respectively, which is normally 35% in a typical Mediterranean mussel farm. On the other hand, pearl oysters have CI of 53%, 56%, and 19% in the first, second, and third farms, respectively, which range from 30% to 45% in the natural populations. However, holothurians did not gain weight under the fish cage regime despite the high survival rate that ranges from 8.3g to 18.3 g, which is very less compared with the 80 g of mean weight in the natural population.

17.10 Challenges

IMTA is a complex system where the nutrient removal efficiency by the extractive species cannot be quantified exactly, as it is the ratio of how much extractive species can utilize the waste generated by the fed species (Reid et al., 2008). Species in the trophic level depend on the spatial distribution for efficient nutrient recovery. For instance, organic extractive species consume particulate matter, but due to their metabolic activities, additional organic particulate matter may be added to the environment. Despite this, a portion of the nutrients remains unaccounted for, as some losses occur that are difficult to quantify. In addition, there is also controversy that the mussel can efficiently remove the fecal matter of the fed species, as it is not the preferred food for the mussel (Sanz-Lazaro and Sanchez-Jerez, 2017). Therefore the species selection and their interaction are very important to be considered for the IMTA design. Only scientific research at the commercial IMTA site will provide a real outcome of the aforementioned challenge. In addition, the open IMTA system requires high investment costs and sophisticated infrastructures to withstand the strong current velocities (Troell et al., 2009). There may be an increase in land conflicts between the aquaculture and other inland, coastal, or marine activities due to the IMTA expansion. The implementation of inland IMTA systems remains a significant challenge. In addition, regular monitoring and control of fouling organisms are essential to maintain system efficiency.

IMTA is an eco-friendly farming practice that promotes the SDGs. However, the system is in an infant stage. Numerous species remain underutilized, offering opportunities for both sustainable use and conservation, thereby supporting greater biodiversity and system diversification. A variety of system combinations came up in IMTA systems for efficient nutrient utilization, such as the multitrophic recirculating aquaculture system (MRAS) (Sri-uam et al., 2016), IMTA-RAS (Alamrousi et al., 2022; Marincola et al., 2025) system, biofloc–IMTA (BFT–IMTA) (Costa et al., 2021) system. Moreover, it is essential to raise awareness among farmers and the general public about the IMTA system and its associated benefits. Furthermore, advancing IMTA requires the establishment of suitable infrastructure, educational initiatives, awareness campaigns, applied research, and supportive government policies and legislation. Ultimately, the expansion and implementation of both land-based and offshore IMTA systems depend heavily on political will and effective policy decisions.

17.11 Conclusion

IMTA emerges as a viable solution for the adverse effects associated with the traditional aquaculture system, offering long-term sustainability for both open-water and land-based systems. The incorporation of extractive species such as filter feeders, deposit feeders, and seaweeds or other macrophytes, along with the fed species, serves as biofilters in the aquaculture system. Moreover, the various potential extractive species, such as bivalves, sponges, coelenterates, polychaete, sea cucumber, seaweeds, and halophytes, efficiently remove the waste in multitrophic layers. This approach enhances the eco-friendliness of the system, lowers risks, and improves profitability, making it appealing to a wide range of stakeholders, including researchers. However, large-scale implementation of IMTA requires well-defined regulations and effective coordination between governmental and nongovernmental organizations.

References

Abidli, S., Pinheiro, M., Lahbib, Y., Neuparth, T., Santos, M.M., Trigui El Menif, N., 2021. Effects of environmentally relevant levels of polyethylene microplastic on Mytilus galloprovincialis (Mollusca: Bivalvia): filtration rate and oxidative stress. Environ. Sci. Pollut. Res. 28 (21), 26643–26652. doi:10.1007/s11356-021-12506-8.

Abo-Al-Ela, H.G., Faggio, C., 2021. MicroRNA-mediated stress response in bivalve species. Ecotoxicol. Environ. Saf. 208, 111442. doi:10.1016/j.ecoenv.2020.111442.

Abreu, M.H., Pereira, R., Yarish, C., Buschmann, A.H., Sousa-Pinto, I., 2011. IMTA with Gracilaria vermiculophylla: productivity and nutrient removal performance of the seaweed in a land-based pilot scale system. Aquaculture 312 (1–4), 77–87. doi:10.1016/j.aquaculture.2010.12.036.

Abreu, M.H., Varela, D.A., Henriquez, L., Villarroel, A., Yarish, C., Sousa-Pinto, I., Buschmann, A.H., 2009. Traditional vs. integrated multi-trophic aquaculture of Gracilaria chilensis: productivity and physiological performance. Aquaculture 293 (3–4), 211–220. doi:10.1016/j.aquaculture.2009.03.043.

Ahmed, N., Thompson, S., Glaser, M., 2019. Global aquaculture productivity, environmental sustainability, and climate change adaptability. Environ. Manage. 63 (2), 159–172. doi:10.1007/s00267-018-1117-3.

Alamrousi, A., Casais, E., García-Cardesín, É, Masaló, I., Pintado, J., Cremades, J., 2022. Influence of pH, N, P, N: P ratio, and dissolved inorganic carbon on Ulva ohnoi growth and biomass quality: potential implications in IMTA-RAS. Aquac. J. 2 (4), 285–301. doi:10.3390/aquacj2040017.

Alomar, C., Compa, M., Sanz-Martín, M., Fagiano, V., Álvarez, E., Valencia, J.M., Deudero, S., 2022. A holistic approach to plastic pollution in integrated multi-trophic aquaculture facilities: plastic ingestion in Sparus aurata and Mytilus galloprovincialis. Aquaculture 561, 738666. doi:10.1016/j.aquaculture.2022.738666.

Arduini, D., Portacci, G., Giangrande, A., Immacolata Acquaviva, M., Borghese, J., Calabrese, C., Giandomenico, S., Quarta, E., Stabili, L., 2023. Growth performance of Mytilus galloprovincialis Lamarck, 1819 under an innovative integrated multi-trophic aquaculture system (IMTA) in the Mar Grande of Taranto (Mediterranean Sea, Italy). Water 15 (10), 1922. doi:10.3390/w15101922.

Bao, J., Tian, X.L., Dong, S.L., 2008. Effect of temperature, salinity and light intensity on nitrogen and phosphorus uptake by Sargassum thunbergii. J. Fish. Sci. China 15 (2), 293–300.

Barrington, K., Chopin, T., Robinson, S., 2009. Integrated multi-trophic aquaculture (IMTA) in marine temperate waters. In: Soto, D. (Ed.), Integrated mariculture: a global review. FAO Fisheries and Aquaculture Technical Paper No. 529. Food and Agriculture Organization of the United Nations, Rome, pp. 7–46.

Beecham, J., 2008. Literature review on particle assimilation by molluscs and crustaceans. In: The centre for environment, fisheries and aquaculture science. Lowestoft Laboratory, Lowestoft, UK, p. 15.

Benner, R., 2002. Chemical composition and reactivity. In: Hansell, D., Carlson, C. (Eds.), Biogeochemistry of marine dissolved organic matter. Academic Press, San Diego, CA, pp. 59–90. doi:10.1016/B978-012323841-2/50005-1.

Berglund, E., Fogelberg, V., Nilsson, P.A., Hollander, J., 2019. Microplastics in a freshwater mussel (Anodonta anatina) in Northern Europe. Sci. Total Environ. 697, 134192. doi:10.1016/j.scitotenv.2019.134192.

Bird, K.T., Habig, C., DeBusk, T., 1982. Nitrogen allocation and storage patterns in Gracilaria tikvahiae (Rhodophyta). J. Phycol. 18 (3), 344–348. doi:10.1111/j.1529-8817.1982.tb03194.x.

Biswas, G., Kumar, P., Kailasam, M., Ghoshal, T.K., Bera, A., Vijayan, K.K., 2019. Application of integrated multi trophic aquaculture (IMTA) concept in brackishwater ecosystem: the first exploratory trial in the Sundarban, India. J. Coastal Res. 86 (1), 49–55. doi:10.2112/SI86-007.1.

Borghese, J., Giangrande, A., Arduini, D., Trani, R., Doria, L., Anglano, M., Rizzo, L., Rossi, S., 2025. Influence of an innovative IMTA system (Mediterranean Sea, Italy) on environmental and biological parameters: Seasonal analysis. Aquaculture 596, 741726. doi:10.1016/j.aquaculture.2024.741726.

Brager, L.M., Cranford, P.J., Grant, J., Robinson, S.M.C., 2015. Spatial distribution of suspended particulate wastes at open-water Atlantic salmon and sablefish aquaculture farms in Canada. Aquacult. Env. Interact. 6 (2), 135–149. doi:10.3354/aei00120.

Brown, N., Eddy, S., Plaud, S., 2011. Utilization of waste from a marine recirculating fish culture system as a feed source for the polychaete worm, Nereis virens. Aquaculture 322-323, 177–183. doi:10.1016/j.aquaculture.2011.09.017.

Bunting, S.W., Shpigel, M., 2009. Evaluating the economic potential of horizontally integrated land-based marine aquaculture. Aquaculture 294 (1–2), 43–51. doi:10.1016/j.aquaculture.2009.04.017.

Buschmann, A.H., Varela, D.A., Hernández-González, M.C., Huovinen, P., 2008. Opportunities and challenges for the development of an integrated seaweed-based aquaculture activity in Chile: determining the physiological capabilities of Macrocystis and Gracilaria as biofilters. J. Appl. Phycol. 20 (5), 571–577. doi:10.1007/s10811-007-9297-x.

Buschmann, A.H., Troell, M., Kautsky, N., 2001. Integrated algal farming: a review. Cah. Biol. Mar. 42 (1–2), 83–90.

Cao, L., Naylor, R., Henriksson, P., Leadbitter, D., Metian, M., Troell, M., Zhang, W., 2015. China's aquaculture and the world's wild fisheries. Science 347 (6218), 133–135. doi:10.1126/science.1260149.

Capó, X., Alomar, C., Compa, M., Sole, M., Sanahuja, I., Rojas, D.LS., González, G.P., Martínez, R.M.G., Deudero, S., 2022. Quantification of differential tissue biomarker responses to microplastic ingestion and plasticizer bioaccumulation in aquaculture reared sea bream Sparus aurata. Environmental Research 211, 113063. https://doi.org/10.1016/j.envres.2022.113063.

Capo, X., Rubio, M., Solomando, A., Alomar, C., Compa, M., Sureda, A., Deudero, S., 2021. Microplastic intake and enzymatic responses in Mytilus galloprovincialis reared at the vicinities of an aquaculture station. Chemosphere 280, 130575. doi:10.1016/j.chemosphere.2021.130575.

Corami, F., Rosso, B., Roman, M., Picone, M., Gambaro, A., Barbante, C., 2020. Evidence of small microplastics (<100 μm) ingestion by Pacific oysters (*Crassostrea gigas*): a novel method of extraction, purification, and analysis using micro-FTIR. Mar. Pollut. Bull. 160, 111606, doi:10.1016/j.marpolbul.2020.111606.

Marincola, F.C., Palmas, C., Couto, M.L., Paz, I., Cremades, J., Pintado, J., Bruni, L., Picone, G., 2025. Metabolic profile of senegalese sole (Solea senegalensis) muscle: effect of fish–macroalgae IMTA-RAS aquaculture. Molecules 30 (12), 2518. doi:10.3390/molecules30122518.

Chatzivasileiou, D., Dimitriou, P.D., Theodorou, J., Kalantzi, I., Magiopoulos, I., Papageorgiou, N., Pitta, P., Tsapakis, M., Karakassis, I., 2022. An IMTA in Greece: co-culture of fish, bivalves, and holothurians. J. Mar. Sci. Eng. 10 (6), 776. doi:10.3390/jmse10060776.

Chebil Ajjabi, L., Abaab, M., Segni, R., 2018. The red macroalga Gracilaria verrucosa in co-culture with the Mediterranean mussels Mytilus galloprovincialis: productivity and nutrient removal performance. Aquac. Int. 26 (1), 253–266. doi:10.1007/s10499-017-0206-2.

Cheng, J.J., Stomp, A.M., 2009. Growing duckweed to recover nutrients from wastewaters and for production of fuel ethanol and animal feed. Clean Soil, Air, Water 37 (1), 17–26. doi:10.1002/clen.200800210.

Chopin, T., Buschmann, A.H., Halling, C., Troell, M., Kautsky, N., Neori, A., Kraemer, G.P., Zertuche-González, J.A., Yarish, C., Neefus, C., 2001. Integrating seaweeds into marine aquaculture systems: a key toward sustainability. J. Phycol. 37 (6), 975–986. doi:10.1046/j.1529-8817.2001.01137.x.

Chopin, T., Yarish, C., Wilkes, R., Belyea, E., Lu, S., Mathieson, A., 1999. Developing porphyra/salmon integrated aquaculture for bioremediation and diversification of the aquaculture industry. J. Appl. Phycol. 11 (5), 463–472. doi:10.1023/a:1008114112852.

Chopin, T., 2013. Integrated multi-trophic aquaculture–ancient, adaptable concept focuses on ecological integration. Glob. Aquac. Advocate 16 (2), 16–19.

Chopin, T., 2018. Integrated multi-trophic aquaculture (IMTA): a responsible approach to farming our waters. Aquaculturists. https://theaquaculturists.blogspot.com/2018/10/integrated-multi-trophic-aquaculture.html.

Cohen, I., Neori, A., 1991. Ulva lactuca biofilters for marine fishpond effluents I. ammonia uptake kinetics and nitrogen content. Bot. Mar. 34 (6), 475–482. doi:10.1515/botm.1991.34.6.475.

Corey, P., Kim, J.K., Duston, J., Garbary, D.J., 2014. Growth and nutrient uptake by Palmaria palmata integrated with Atlantic halibut in a land-based aquaculture system. Algae 29 (1), 35–45. doi:10.4490/algae.2014.29.1.035.

Costa, L.C.O, Poersch, L.H.S., Abreu, P.C, 2021. Biofloc removal by the oyster Crassostrea gasar as a candidate species to an integrated multi-trophic aquaculture (IMTA) system with the marine shrimp Litopenaeus vannamei. Aquaculture. 540, 736731. doi:10.1016/j.aquaculture.2021.736731.

, 2008. In: Costa-Pierce, B.A. (Ed.), Ecological aquaculture: The evolution of the blue revolution. John Wiley & Sons, Hoboken, NJ., p. 398.

Cranford, P.J., Reid, G.K., Robinson, S.M.C., 2013. Open water integrated multi-trophic aquaculture: constraints on the effectiveness of mussels as an organic extractive component. Aquacult. Env. Interact. 4 (2), 163–173. doi:10.3354/aei00081Canada.

Cromey, C.J., Nickell, T.D., Black, K.D., 2002. Depomodelling the deposition and biological effects of waste solids from marine cage farms. Aquaculture 214 (1–4), 211–239. doi:10.1016/S0044-8486(02)00368-X.

Cubillo, A.M., Peteiro, L.G., Fernández-Reiriz, M.J., Labarta, U., 2012. Influence of stocking density on growth of mussels (Mytilus galloprovincialis) in suspended culture. Aquaculture 342–343 (1), 103–111. doi:10.1016/j.aquaculture.2012.02.017.

Custódio, M., Villasante, S., Calado, R., Lillebø, A.I., 2021. Testing the hydroponic performance of the edible halophyte Halimione portulacoides, a potential extractive species for coastal integrated multi-trophic aquaculture. Sci. Total Environ. 766, 144378. doi:10.1016/j.scitotenv.2020.144378.

Cutajar, K., Falconer, L., Massa-Gallucci, A., Cox, R.E., Schenke, L., Bardócz, T., Sharman, A., Deguara, S., Telfer, T.C., 2022. Culturing the sea cucumber Holothuria poli in open-water integrated multi-trophic aquaculture at a coastal Mediterranean fish farm. Aquaculture 550. doi:10.1016/j.aquaculture.2021.737881.

De Goeij, J.M., Van Oevelen, D., Vermeij, M.J.A., Osinga, R., Middelburg, J.J., De Goeij, A.F.P.M., Admiraal, W., 2013. Surviving in a marine desert: the sponge loop retains resources within coral reefs. Science 342 (6154), 108–110. doi:10.1126/science.1241981.

De Prisco, J.A., 2020. An investigation of some key physico-chemical water quality parameters of an integrated multi-trophic aquaculture (IMTA) system operating recirculation methodology in the Western Cape of South Africa. University of Cape Town, Cape Town, South Africa.

Duarte, C.M., Wu, J., Xiao, X., Bruhn, A., Krause-Jensen, D., 2017. Can seaweed farming play a role in climate change mitigation and adaptation? Front. Mar. Sci. 4. doi:10.3389/fmars.2017.00100.

Evania, C., Rejeki, S., Aryati, R.W., 2018. Performa pertumbuhan udang windu (Penaeus monodon) yang dibudidayakan bersama kerang hijau (Perna viridis) dengan sistem IMTA. Sains Akuakultur Trop.: Indones. J. Trop. Aquac. 2 (2), 44–52. doi:10.14710/sat.v2i2.3945.

Fang, J., Fang, J., Chen, Q., Mao, Y., Jiang, Z., Du, M., Gao, Y., Lin, F., 2020. Assessing the effects of oyster/kelp weight ratio on water column properties: an experimental IMTA study at Sanggou Bay, China. J. Oceanol. Limnol. 38 (6), 1914–1924. doi:10.1007/s00343-019-9109-6.

FAO, 2018. The state of world fisheries and aquaculture 2018 – Meeting the sustainable development goals. Food and Agriculture Organization of the United Nations, Rome. http://www.fao.org/3/I9540EN/i9540en.pdf.

FAO, 2022. The state of world fisheries and aquaculture 2022: towards blue transformation. Food and Agriculture Organization of the United Nations, Rome doi:10.4060/cc0461en.

Ferguson, H.W., Christian, M.J.D., Hay, S., Nicolson, J., Sutherland, D., Crumlish, M., 2010. Jellyfish as vectors of bacterial disease for farmed Salmon (Salmo Salar). J. Vet. Diagn. Invest. 22 (3), 376–382. doi:10.1177/104063871002200305.

Filgueira, R., Comeau, L.A., Guyondet, T., McKindsey, C.W., Byron, C.J., 2015. Modelling carrying capacity of bivalve aquaculture: a review of definitions and methods. In: Smaal, A.F., Ferreira, J.G., Grant, J., Petersen, J.K., Strand, Ø. (Eds.), Goods and services of marine bivalves. Springer, Cham, pp. 1–33. doi:10.1007/978-1-4939-2493-6_945-1.

Fitridge, I., Dempster, T., Guenther, J., de Nys, R., 2012. The impact and control of biofouling in marine aquaculture: a review. Biofouling 28 (7), 649–669. doi:10.1080/08927014.2012.700478.

Flickinger, D.L., Costa, G.A., Dantas, D.P., Moraes-Valenti, P., Valenti, W.C., 2019. The budget of nitrogen in the grow-out of the Amazon River prawn (Macrobrachium amazonicum Heller) and tambaqui (Colossoma macropomum Cuvier) farmed in monoculture and in integrated multitrophic aquaculture systems. Aquac. Res. 50 (11), 3444–3461. doi:10.1111/are.14304.

Fonseca, T., David, F.S., Ribeiro, F.A.S., Wainberg, A.A., Valenti, W.C., 2017. Technical and economic feasibility of integrating seahorse culture in shrimp/oyster farms. Aquac. Res. 48 (2), 655–664. doi:10.1111/are.12912.

Food and Agriculture Organization of the United Nations, 2024. The State of World Fisheries and Aquaculture 2024: blue transformation in action. FAO, Rome doi:10.4060/cc5902en.

Fossberg, J., Forbord, S., Broch, O.J., Malzahn, A.M., Jansen, H., Handå, A., Førde, H., Bergvik, M., Fleddum, A.L., Skjermo, J., Olsen, Y., 2018. The potential for upscaling kelp (Saccharina latissima) cultivation in salmon-driven integrated multi-trophic aquaculture (IMTA). Front. Mar. Sci. 9. doi:10.3389/fmars.2018.00418.

Gallardi, D., 2014. Effects of bivalve aquaculture on the environment and their possible mitigation: a review. Fish. Aquac. J. 05 (03). doi:10.4172/2150-3508.1000105.

Gattuso, J.P., Magnan, A.K., Bopp, L., Cheung, W.W.L., Duarte, C.M., Hinkel, J., Mcleod, E., Micheli, F., Oschlies, A., Williamson, P., Billé, R., Chalastani, V.I., Gates, R.D., Irisson, J.O., Middelburg, J.J., Pörtner, H.O., Rau, G.H., 2018.

Ocean solutions to address climate change and its effects on marine ecosystems. Front. Mar. Sci. 5. doi:10.3389/fmars.2018.00337.

Gempesaw, C.M., Bacon, R., Wessells, C.R., Manalo, A., 1995. Consumer perceptions of aquaculture products. Am. J. Agric. Econ. 77 (5), 1306–1312. doi:10.2307/1243366.

Ghosh, A.K., Hasanuzzaman, A.F.M., Islam, S.S., Sarower, M.G., Mistry, S.K., Arafat, S.T., Huq, K.A., 2025. Integrated multi-trophic aquaculture (IMTA): enhancing growth, production, immunological responses, and environmental management in aquaculture. Aquac. Int. 33 (5). doi:10.1007/s10499-025-02021-9.

Glauco, F., He, P., Chen, Z., 2024. Combine effects of multiple environmental factors on growth and nutrient uptake of euryhaline seaweed growth in integrated multitrophic aquaculture systems. Algal Res. 77, 103347. doi:10.1016/j.algal.2023.103347.

Greiner, J.T., McGlathery, K.J., Gunnell, J., McKee, B.A., 2013. Seagrass restoration enhances "blue carbon" sequestration in coastal waters. PLoS One 8 (8). doi:10.1371/journal.pone.0072469.

Grosso, L., Rakaj, A., Fianchini, A., Morroni, L., Cataudella, S., Scardi, M., 2021. Integrated Multi-Trophic Aquaculture (IMTA) system combining the sea urchin Paracentrotus lividus, as primary species, and the sea cucumber Holothuria tubulosa as extractive species. Aquaculture. 534, 736268. doi:10.1016/j.aquaculture.2020.736268.

Guerrero, S., Cremades, J., 2012. Integrated multi-trophic aquaculture (IMTA): a sustainable, pioneering alternative for marine cultures in Galicia. J. Shellfish Res. 31 (1), 19–20. doi:10.2983/035.031.0112.

Han, T., Jiang, Z., Fang, J., Zhang, J., Mao, Y., Zou, J., Huang, Y., Wang, D., 2013. Carbon dioxide fixation by the seaweed Gracilaria lemaneiformis in integrated multi-trophic aquaculture with the scallop Chlamys farreri in Sanggou Bay. Aquac. Int. 21 (5), 1035–1043. doi:10.1007/s10499-012-9610-9.

Hannah, L., Pearce, C.M., Cross, S.F., 2013. Growth and survival of California sea cucumbers (Parastichopus californicus) cultivated with sablefish (Anoplopoma fimbria) at an integrated multi-trophic aquaculture site. Aquaculture 406-407, 34–42. doi:10.1016/j.aquaculture.2013.04.022.

Hentschel, B.T., Shimeta, J., 2008. Suspension feeders. In: Encyclopedia of ecology. Elsevier, Oxford, pp. 3437–3442. doi:10.1016/B978-008045405-4.00371-7.

Honda, H., Kikuchi, K., 2002. Nitrogen budget of polychaete Perinereis nuntia vallata fed on the feces of Japanese flounder. Fish. Sci. 68 (6), 1304–1308. doi:10.1046/j.1444-2906.2002.00568.x.

Hopkins, A.E., 1933. Experiments on the feeding behavior of the oyster, Ostrea gigas. J. Exp. Zool. 64 (3), 469–494. doi:10.1002/jez.1400640306.

Huang, Y., Ciais, P., Goll, D.S., Sardans, J., Peñuelas, J., Cresto-Aleina, F., Zhang, H., 2020. The shift of phosphorus transfers in global fisheries and aquaculture. Nat. Commun. 11 (1). doi:10.1038/s41467-019-14242-7.

Hurd, C.L., 2017. Shaken and stirred: the fundamental role of water motion in resource acquisition and seaweed productivity. Perspect. Phycol. 4 (2), 73–81. doi:10.1127/pip/2017/0072.

Hussenot, J., Lefebvre, S., BROSSARD, N., 1998. Open-air treatment of wastewater from land-based marine fish farms in extensive and intensive systems: current technology and future perspectives. Aquat. Living Resour. 11 (4), 297–304. doi:10.1016/s0990-7440(98)80015-6.

Jones, A.B, Preston, N.P, Dennison, W.C, 2002. The efficiency and condition of oysters and macroalgae used as biological filters of shrimp pond effluent. Aquac. Res. 33 (1), 1–19. doi:10.1046/j.1355-557x.2001.00637.x.

Kang, K.H., Kwon, J.Y., Kim, Y.M., 2003. A beneficial coculture: Charm abalone Haliotis discus hannai and sea cucumber Stichopus japonicus. Aquaculture 216 (1-4), 87–93. doi:10.1016/S0044-8486(02)00203-X.

Kellogg, M.L., Smyth, A.R., Luckenbach, M.W., Carmichael, R.H., Brown, B.L., Cornwell, J.C., Piehler, M.F., Owens, M.S., Dalrymple, D.J., Higgins, C.B., 2014. Use of oysters to mitigate eutrophication in coastal waters. Estuar. Coast. Shelf Sci. 151, 156–168. doi:10.1016/j.ecss.2014.09.025.

Khanjani, M.H., Zahedi, S., Mohammadi, A., 2022. Integrated multitrophic aquaculture (IMTA) as an environmentally friendly system for sustainable aquaculture: functionality, species, and application of biofloc technology (BFT). Environ. Sci. Pollut. Res. 29 (45), 67513–67531. doi:10.1007/s11356-022-22371-8.

Kim, W.S., Huh, H.T., Huh, S.H., Lee, T.W., 2001. Effects of salinity on endogenous rhythm of the Manila clam, *Ruditapes philippinarum* (Bivalvia: Veneridae). Mar. Biol. 138, 157–162. https://doi.org/10.1007/s002270000430.

Knowler, D., Chopin, T., Martínez-Espiñeira, R., Neori, A., Nobre, A., Noce, A., Reid, G., 2020. The economics of integrated multi-trophic aquaculture: where are we now and where do we need to go? Rev. Aquac. 12 (3), 1579–1594. doi:10.1111/raq.12399.

Kokou, F., Fountoulaki, E., 2018. Aquaculture waste production associated with antinutrient presence in common fish feed plant ingredients. Aquaculture 495, 295–310. doi:10.1016/j.aquaculture.2018.06.003.

Kreeger, D.A., Gatenby, C.M., Bergstrom, P.W., 2017. Comparative evaluation of native bivalve species restoration for water quality improvement in the Chesapeake Bay and other mid-Atlantic watersheds. Partnership for the Delaware Estuary, Wilmington, DE Report No. 17-05.

Krom, M.D., Ellner, S., van Rijn, J., Neori, A., 1995. Nitrogen and phosphorus cycling and transformations in a prototype non-polluting' integrated mariculture system, Eilat, Israel. Mar. Ecol. Prog. Ser. 118 (1–3), 25–36. doi:10.3354/meps118025.

Kumara, R., Syamala, K., Shyne Anand, P.S., Chadha, N.K., Sawant, P.B., Chithira, P., Muralidhar, A.P., 2024. Halophyte and bivalve-based integrated multi-trophic aquaculture (IMTA): effect on growth, water quality, digestive and antioxidant enzymes of Penaeus monodon and Chanos chanos reared in brackishwater ponds. Aquac. Int. 32 (2), 1927–1953. doi:10.1007/s10499-023-01251-z.

Lal, J., Singh, S.K., Pawar, L., Biswas, P., Meitei, M.M., Meena, D.K., 2023. Integrated multi-trophic aquaculture: A balanced ecosystem approach to blue revolution. In: Advances in resting-state functional MRI. Woodhead Publishing, Sawston, Cambridge, UK, pp. 513–535. doi:10.1016/B978-0-323-90759-8.00025-1.

Lamprianidou, F., Telfer, T., Ross, L.G., 2015. A model for optimization of the productivity and bioremediation efficiency of marine integrated multitrophic aquaculture. Estuar. Coast. Shelf Sci. 164, 253–264. doi:10.1016/j.ecss.2015.07.045.

Laramore, S., Baptiste, R., Wills, P.S., Hanisak, M.D., 2018. Utilization of IMTA-produced Ulva lactuca to supplement or partially replace pelleted diets in shrimp (Litopenaeus vannamei) reared in a clear water production system. J. Appl. Phycol. 30 (6), 3603–3610. doi:10.1007/s10811-018-1485-3.

Laramore, S.E., Wills, P.S., Hanisak, M.D., 2022. Seasonal variation in the nutritional profile of Ulva lactuca produced in a land-based IMTA system. Aquac. Int. 30 (6), 3067–3079. doi:10.1007/s10499-022-00950-3.

Ledda, F.D., Pronzato, R., Manconi, R., 2014. Mariculture for bacterial and organic waste removal: a field study of sponge filtering activity in experimental farming. Aquac. Res. 45 (8), 1389–1401. doi:10.1111/are.12084.

Lefebvre, S., Barillé, L., Clerc, M., 2000. Pacific oyster (Crassostrea gigas) feeding responses to a fish-farm effluent. Aquaculture 187 (1–2), 185–198. doi:10.1016/s0044-8486(99)00390-7.

Li, H., Zhang, Y., Liang, Y., Chen, J., Zhu, Y., Zhao, Y., Jiao, N., 2018. Impacts of maricultural activities on characteristics of dissolved organic carbon and nutrients in a typical raft-culture area of the Yellow Sea, North China. Mar. Pollut. Bull. 137, 456–464. doi:10.1016/j.marpolbul.2018.10.048.

Li, J., Lusher, A.L., Rotchell, J.M., Deudero, S., Turra, A., Bråte, I.L.N., Sun, C., Shahadat Hossain, M., Li, Q., Kolandhasamy, P., Shi, H., 2019. Using mussel as a global bioindicator of coastal microplastic pollution. Environ. Pollut. 244, 522–533. doi:10.1016/j.envpol.2018.10.032.

Licciano, M., Stabili, L., Giangrande, A., 2005. Clearance rates of Sabella spallanzanii and Branchiomma luctuosum (Annelida: Polychaeta) on a pure culture of Vibrio alginolyticus. Water Res. 39 (18), 4375–4384. doi:10.1016/j.watres.2005.09.003.

Lima, P.C.M., Silva, A.E.M., Silva, D.A., Silva, S.M.B.C., Brito, L.O., Gálvez, A.O., 2021. Effect of stocking density of Crassostrea sp. in a multitrophic biofloc system with Litopenaeus vannamei in nursery. Aquaculture 530. doi:10.1016/j.aquaculture.2020.735913.

Liu, Y.i, Zhang, J., Wu, W., Zhong, Y., Li, H., Wang, X., Yang, J., Zhang, Y., 2022. Effects of shellfish and macro-algae IMTA in North China on the environment, inorganic carbon system, organic carbon system, and sea–air CO_2 fluxes. Front. Mar. Sci. 9. doi:10.3389/fmars.2022.864306.

Loayza-Aguilar, R.E., Huamancondor-Paz, Y.P., Saldaña-Rojas, G.B., Olivos-Ramirez, G.E., 2023. Integrated multi-trophic aquaculture (IMTA): strategic model for sustainable mariculture in Samanco Bay. Front. Mar. Sci. 10. doi:10.3389/fmars.2023.1151810.

Ma, Y., Hu, A., Yu, C.P., Yan, Q., Yan, X., Wang, Y., Deng, F., Xiong, H., 2015. Response of microbial communities to bioturbation by artificially introducing macrobenthos to mudflat sediments for in situ bioremediation in a typical semi-enclosed bay, southeast China. Mar. Pollut. Bull. 94 (1-2), 114–122. doi:10.1016/j.marpolbul.2015.03.003.

Macchiavello, J., Bulboa, C., 2014. Nutrient uptake efficiency of Gracilaria chilensis and Ulva lactuca in an IMTA system with the red abalone Haliotis rufescens. Lat. Am. J. Aquat. Res. 42 (3), 523–533. doi:10.3856/vol42-issue3-fulltext-12.

MacTavish, T., Stenton-Dozey, J., Vopel, K., Savage, C., Solan, M., 2012. Deposit-feeding sea cucumbers enhance mineralization and nutrient cycling in organically-enriched coastal sediments. PLoS One 7 (11), e50031. doi:10.1371/journal.pone.0050031.

Magondu, E.W., Fulanda, B.M., Munguti, J.M., Mlewa, C.M., 2022. Toward integration of sea cucumber and cockles with culture of shrimps in earthen ponds in Kenya. J. World Aquac. Soc. 53 (5), 948–962. doi:10.1111/jwas.12861.

Maldonado, M., Zhang, X., Cao, X., Xue, L., Cao, H., Zhang, W., 2010. Selective feeding by sponges on pathogenic microbes: a reassessment of potential for abatement of microbial pollution. Mar. Ecol. Prog. Ser. 403, 75–89. doi:10.3354/meps08411.

Mao, Y., Yang, H., Zhou, Y., Ye, N., 2009. Jianguang Fang, Potential of the seaweed Gracilaria lemaneiformis for integrated multi-trophic aquaculture with scallop Chlamys farreri in North China. J. Appl. Phycol. 21 (6), 649–656. doi:10.1007/s10811-008-9398-1.

Marinho, G.S., Holdt, S.L., Birkeland, M.J., Angelidaki, I., 2015. Commercial cultivation and bioremediation potential of sugar kelp, Saccharina latissima, in Danish waters. J. Appl. Phycol. 27 (5), 1963–1973. doi:10.1007/s10811-014-0519-8.

Marinho-Soriano, E., Azevedo, C.A.A., Trigueiro, T.G., Pereira, D.C., Carneiro, M.A.A., Camara, M.R., 2011. Bioremediation of aquaculture wastewater using macroalgae and Artemia. Int. Biodeterior. Biodegrad. 65 (1), 253–257. doi:10.1016/j.ibiod.2010.10.001.

Marques, B., Lillebø, A.I., Ricardo, F., Nunes, C., Coimbra, M.A., Calado, R., 2018. Adding value to ragworms (Hediste diversicolor) through the bioremediation of a super-intensive marine fish farm. Aquacult. Env. Interact. 10, 79–88. doi:10.3354/aei00255.

Martínez-Porchas, M., Vargas-Albores, F., Garibay-Valdez, E., Martínez-Córdova, L.R., 2023. Advances and perspectives in integrated multi-trophic aquaculture. Front. Mar. Sci. 10, 1197343. doi:10.3389/fmars.2023.1197343.

Maxwell, K.H., Gardner, J.P.A., Heath, P.L., 2009. The effect of diet on the energy budget of the brown sea cucumber, stichopus mollis (Hutton). J. World Aquac. Soc. 40 (2), 157–170. doi:10.1111/j.1749-7345.2009.00239.x.

Meitei, M.M., Muralidhar, A.P., Syamala, K., Sureesh, S., Biswas, G., Megarajan, S., Munilkumar, S., 2025. Assessment of filtration capacity of different bivalve species suitable for integrated multi-trophic aquaculture (IMTA) systems vis-à-vis waste valorization for sustainable environment. Discov. Sustain. 6 (1). doi:10.1007/s43621-025-01104-0.

Melendres, A.R, 2021. Growth and absorption response of Eucheuma denticulatum(Burman) Collins & Hervey) to fish farm water quality condition. J. Aquac. Mar. Biol. 10 (4), 178–186. doi:10.15406/jamb.2021.10.00319.

Milanese, M., Chelossi, E., Manconi, R., Sarà, A., Sidri, M., Pronzato, R., 2003. The marine sponge Chondrilla nucula Schmidt, 1862 as an elective candidate for bioremediation in integrated aquaculture. Biomol. Eng. 20 (4–6), 363–368. doi:10.1016/S1389-0344(03)00052-2.

Møhlenberg, F., Riisgård, H.U., 1978. Efficiency of particle retention in 13 species of suspension feeding bivalves. Ophelia 17 (2), 239–246. doi:10.1080/00785326.1978.10425487.

Nakphet, S., Ritchie, R.J., Kiriratnikom, S., 2017. Aquatic plants for bioremediation in red hybrid tilapia (Oreochromis niloticus × Oreochromis mossambicus) recirculating aquaculture. Aquac. Int. 25 (2), 619–633. doi:10.1007/s10499-016-0060-7.

Narvarte, B.C.V., Hinaloc, L.A.R., Gonzaga, S.M.C., Roleda, M.Y., 2024a. Exploring the suitable extractive species in an Imta: inorganic nutrient removal from mariculture effluents by commercially important marine macroalgae. SSRN doi:10.2139/ssrn.4711471.

Narvarte, B.C.V., Hinaloc, L.A.R., Gonzaga, S.M.C., Roleda, M.Y., 2024b. Impacts of aquaculture nutrient sources: ammonium uptake of commercially important eucheumatoids depends on phosphate levels. J. Appl. Phycol. 36 (2), 557–565. doi:10.1007/s10811-023-03073-1.

Naskar, S., Biswas, G., Kumar, P., De, D., Das, S., Sawant, P.B., Chadha, N.K., Behera, P., 2023. The green seaweed, Enteromorpha intestinalis: an efficient inorganic extractive species for environmental remediation and improved performances of fed species in brackishwater integrated multi-trophic aquaculture (BIMTA) system. Aquaculture 569. doi:10.1016/j.aquaculture.2023.739359.

Naskar, S., Biswas, G., Kumar, P., De, D., Banerjee Sawant, P., Das, S., Roy, U., 2022. Effects of estuarine oyster, Crassostrea cuttackensis as the extractive species at varied densities on productivity and culture environment in brackishwater integrated multi-trophic aquaculture (BIMTA) system. Aquaculture 554, 738128. doi:10.1016/j.aquaculture.2022.738128.

Nath, K., Munilkumar, S., Patel, A.B., Kamilya, D., Pandey, P.K., Sawant, P.B., 2021. Lamellidens and Wolffia canopy improves growth, feed utilization and welfare of Labeo rohita (Hamilton,1822) in integrated multi-trophic freshwater aquaculture system. Aquaculture 534, 736207. doi:10.1016/j.aquaculture.2020.736207.

Naylor, R.L., Hardy, R.W., Buschmann, A.H., Bush, S.R., Cao, L., Klinger, D.H., Little, D.C., Lubchenco, J., Shumway, S.E., Troell, M., 2021. A 20-year retrospective review of global aquaculture. Nature 591 (7851), 551–563. doi:10.1038/s41586-021-03308-6.

Nederlof, M.A.J., Jansen, H.M., Dahlgren, T.G., Fang, J., Meier, S., Strand, Ø, Sveier, H., Verdegem, M.C.J., Smaal, A.C., 2019. Application of polychaetes in (de)coupled integrated aquaculture: production of a high-quality marine resource. Aquacult. Env. Interact. 11, 221–237. doi:10.3354/aei00309.

Nederlof, M.A.J., Verdegem, M.C.J., Smaal, A.C., Jansen, H.M., 2022a. Nutrient retention efficiencies in integrated multi-trophic aquaculture. Rev. Aquac. 14 (3), 1194–1212. doi:10.1111/raq.12645.

Nederlof, M.A.J., Neori, A., Verdegem, M.C.J., Smaal, A.C., Jansen, H.M., 2022b. Ulva spp. performance and biomitigation potential under high nutrient concentrations: implications for recirculating IMTA systems. J. Appl. Phycol. 34 (4), 2157–2171. doi:10.1007/s10811-022-02751-w.

, 2009. In: Nellemann, C., Corcoran, E. (Eds.), Blue carbon: the role of healthy oceans in binding carbon: a rapid response assessment. United Nations Environment Programme/Earthprint, Nairobi, Kenya, p. 78.

Nelson, E.J., MacDonald, B.A., Robinson, S.M.C., 2012. The absorption efficiency of the suspension-feeding sea cucumber, Cucumaria frondosa, and its potential as an extractive integrated multi-trophic aquaculture (IMTA) species. Aquaculture 370-371, 19–25. doi:10.1016/j.aquaculture.2012.09.029.

Neofitou, N., Lolas, A., Ballios, I., Skordas, K., Tziantziou, L., Vafidis, D., 2019. Contribution of sea cucumber Holothuria tubulosa on organic load reduction from fish farming operation. Aquaculture 501, 97–103. doi:10.1016/j.aquaculture.2018.10.071.

Neori, A., Msuya, F.E., Shauli, L., Schuenhoff, A., Kopel, F., Shpigel, M., 2003. A novel three-stage seaweed (Ulva lactuca) biofilter design for integrated mariculture. J. Appl. Phycol. 15 (6), 543–553. doi:10.1023/B:JAPH.0000004382.89142.2d.

Neori, A., Chopin, T., Troell, M., Buschmann, A.H., Kraemer, G.P., Halling, C., Shpigel, M., Yarish, C., 2004. Integrated aquaculture: rationale, evolution and state of the art emphasizing seaweed biofiltration in modern mariculture. Aquaculture 231 (1–4), 361–391. doi:10.1016/j.aquaculture.2003.11.015.

Nissar, S., Bakhtiyar, Y., Arafat, M.Y., Andrabi, S., Mir, Z.A., Khan, N.A., Langer, S., 2023. The evolution of integrated multi-trophic aquaculture in context of its design and components paving way to valorization via optimization and diversification. Aquaculture 565. doi:10.1016/j.aquaculture.2022.739074.

Nobre, A.M., Robertson-Andersson, D., Neori, A., Sankar, K., 2010. Ecological–economic assessment of aquaculture options: comparison between abalone monoculture and integrated multi-trophic aquaculture of abalone and seaweeds. Aquaculture 306 (1–4), 116–126. doi:10.1016/j.aquaculture.2010.06.002.

á Norði, G., Lund, I., Andreasen, B., Taylor, D., Johannesen, T.T., Jacobsen, B., Hughes, A.D., 2023. Modeling particulate waste assimilation by blue mussels within the spatial constraints of a commercial fish farm: implications for multitrophic aquaculture. Front. Mar. Sci. 10. doi:10.3389/fmars.2023.1236294.

Nurfadillah, N., Rusydi, I., Hasfiandi, H., 2023. Study of trophic interactions of green mussels (Perna viridis) as bioremediation of waste for cultivation of windu shrimp (Penaeus monodon) in Gampong Alue Naga, Banda Aceh: application of IMTA system (Integrated Multi-Trophic Aquaculture). IOP Conf. Ser.: Earth and Environ. Sci. 1221 (1), 012047. doi:10.1088/1755-1315/1221/1/012047.

Orr, L.C., Curtis, D.L., Cross, S.F., Gurney-Smith, H., Shanks, A., Pearce, C.M., 2014. Ingestion rate, absorption efficiency, oxygen consumption, and fecal production in green sea urchins (Strongylocentrotus droebachiensis) fed waste from sablefish (Anoplopoma fimbria) culture. Aquaculture 422–423, 184–192. doi:10.1016/j.aquaculture.2013.11.030.

Osinga, R., Sidri, M., Cerig, E., Gokalp, S.Z., Gokalp, M., 2010. Sponge aquaculture trials in the East-Mediterranean Sea: new approaches to earlier ideas. Open Mar. Biol. J. 4 (1), 74–81. doi:10.2174/1874450801004010074.

Page, M.J., Northcote, P.T., Webb, V.L., Mackey, S., Handley, S.J., 2005. Aquaculture trials for the production of biologically active metabolites in the New Zealand sponge Mycale hentscheli (Demospongiae: Poecilosclerida). Aquaculture 250 (1–2), 256–269. doi:10.1016/j.aquaculture.2005.04.069.

Page, M.J., Handley, S.J., Northcote, P.T., Cairney, D., Willan, R.C., 2011. Successes and pitfalls of the aquaculture of the sponge Mycale hentscheli. Aquaculture 312 (1–4), 52–61. doi:10.1016/j.aquaculture.2010.12.006.

Palmer, P.J., Wang, S., Houlihan, A., Brock, I., 2014. Nutritional status of a nereidid polychaete cultured in sand filters of mariculture wastewater. Aquacult. Nutr. 20 (6), 675–691. doi:10.1111/anu.12129.

Palmer, P.J., 2010. Polychaete-assisted sand filters. Aquaculture 306 (1–4), 369–377. doi:10.1016/j.aquaculture.2010.06.011.

Paltzat, D.L., Pearce, C.M., Barnes, P.A., McKinley, R.S., 2008. Growth and production of California sea cucumbers (Parastichopus californicus Stimpson) co-cultured with suspended Pacific oysters (Crassostrea gigas Thunberg). Aquaculture 275 (1–4), 124–137. doi:10.1016/j.aquaculture.2007.12.014.

Papageorgiou, N., Dimitriou, P.D., Chatzivasileiou, D., Tsapakis, M., Karakassis, I., 2023. Can IMTA provide added ecosystem value services in the fish farms of Greece? Front. Mar. Sci. 9. doi:10.3389/fmars.2022.1083099.

Park, H.J., Han, E., Lee, W.C., Kwak, J.H., Kim, H.C., Park, M.S., Kang, C.K., 2015. Trophic structure in a pilot system for the integrated multi-trophic aquaculture off the east coast of Korean peninsula as determined by stable isotopes. Mar. Pollut. Bull. 95 (1), 207–214. doi:10.1016/j.marpolbul.2015.04.019.

Pietrak, M.R., Molloy, S.D., Bouchard, D.A., Singer, J.T., Bricknell, I., 2012. Potential role of Mytilus edulis in modulating the infectious pressure of Vibrio anguillarum 02β on an integrated multi-trophic aquaculture farm. Aquaculture 326–329, 36–39. doi:10.1016/j.aquaculture.2011.11.024.

Pourmozaffar, S., Jahromi, S.T., Rameshi, H., Sadeghi, A., Bagheri, T., Behzadi, S., Gozari, M., Zahedi, M.R., Abrari Lazarjani, S., 2019. The role of salinity in physiological responses of bivalves. Rev. Aquac. 12 (3), 1548–1566. doi:10.1111/raq.12397.

Pouvreau, S.té, Jonquières, G.é, Buestel, D., 1999. Filtration by the pearl oyster, Pinctada margaritifera, under conditions of low seston load and small particle size in a tropical lagoon habitat. Aquaculture 176 (3–4), 295–314. doi:10.1016/s0044-8486(99)00102-7.

Amazon. (2022). Amazon (online price). http://www.amazon.com.

Pronzato, R., Cerrano, C., Cubeddu, T. (1998). In H. Grizel, P. Kesmont (Eds.), Aquaculture and water: Fish culture, shellfish culture and water usage (Special Publication No. 26, pp. 231–232). European Aquaculture Society. Oostende, Belgium.

Purcell, S.W., Williamson, D.H., Ngaluafe, P., 2018. Chinese market prices of beche-de-mer: implications for fisheries and aquaculture. Mar. Policy 91, 58–65. doi:10.1016/j.marpol.2018.02.005.

Qi, Z., Zhang, X.H., Boon, N., Bossier, P., 2009. Probiotics in aquaculture of China - Current state, problems and prospect. Aquaculture 290 (1–2), 15–21. doi:10.1016/j.aquaculture.2009.02.012.

Qiu, X., Carter, C.G., Hilder, P.E., Hadley, S., 2022. A dynamic nutrient mass balance model for optimizing waste treatment in RAS and associated IMTA system. Aquaculture 555. doi:10.1016/j.aquaculture.2022.738216.

Quintana, C.O., Shimabukuro, M., Pereira, C.O., Alves, B.G.R., Moraes, P.C., Valdemarsen, T., Kristensen, E., Sumida, P.Y.G., 2015. Carbon mineralization pathways and bioturbation in coastal Brazilian sediments. Sci. Rep. 5. doi:10.1038/srep16122.

Rajesh, K.V., Mohamed, K.S., Kripa, V., 2001. Influence of algal cell concentration, salinity and body size on the Filtration and Ingestion Rates of cultivable Indian bivalves. Indian J. Mar. Sci. 30 (2), 87–92.

Reid, G.K., Liutkus, M., Robinson, S.M.C., Chopin, T.R., Blair, T., Lander, T., Mullen, J., Page, F., Moccia, R.D., 2009. A review of the biophysical properties of salmonid faeces: implications for aquaculture waste dispersal models and integrated multi-trophic aquaculture. Aquac. Res. 40 (3), 257–273. doi:10.1111/j.1365-2109.2008.02065.x.

Reid, G.K., Liutkus, M., Bennett, A., Robinson, S.M.C., MacDonald, B., Page, F., 2010. Absorption efficiency of blue mussels (Mytilus edulis and M. trossulus) feeding on Atlantic salmon (Salmo salar) feed and fecal particulates: implications for integrated multi-trophic aquaculture. Aquaculture 299 (1–4), 165–169. doi:10.1016/j.aquaculture.2009.12.002.

Reid, G.K., Robinson, S.M., Chopin, T., Mullen, J., Lander, T., Sawhney, M., Powell, E., 2008. Recent developments and challenges for open-water, integrated multi-trophic aquaculture (IMTA) in the Bay of Fundy, Canada. Bull. Aquac. Assoc. Canada 108 (2), 28–35.

Rejeki, S., Ariyati, R.W., Widowati, L.L., 2016. Application of integrated multi tropic aquaculture concept in an abraded brackish water pond. J. Teknol. 78 (4–2), 227–232. doi:10.11113/jt.v78.8213.

Ren, J.S., Stenton-Dozey, J., Plew, D.R., Fang, J., Gall, M., 2012. An ecosystem model for optimising production in integrated multitrophic aquaculture systems. Ecol. Modell. 246 (C), 34–46. doi:10.1016/j.ecolmodel.2012.07.020.

Ridler, N., Wowchuk, M., Robinson, B., Barrington, K., Chopin, T., Robinson, S., Page, F., Reid, G., Szemerda, M., Sewuster, J., Boyne-Travis, S., 2007. Integrated multi−trophic aquaculture (IMTA): a potential strategic choice for farmers. Aquac. Econ. Manag. 11 (1), 99–110. doi:10.1080/13657300701202767.

Riisgård, H.U, 1988. Efficiency of particle retention and filtration rate in 6 species of Northeast American bivalves. Mar. Ecol. Prog. Ser. 45, 217–223. doi:10.3354/meps045217.

Riisgård, H.U., Kittner, C., Seerup, D.F., 2003. Regulation of opening state and filtration rate in filter-feeding bivalves (Cardium edule, Mytilus edulis, Mya arenaria) in response to low algal concentration. J. Exp. Mar. Biol. Ecol. 284 (1–2), 105–127. doi:10.1016/S0022-0981(02)00496-3.

Riisgård, H.U., Lu škow, F., Pleissner, D., Lundgreen, K., Lo ṕez, M.A., ‚P., 2013. Effect of salinity on filtration rates of mussels Mytilus edulis with special emphasis on dwarfed mussels from the low-saline Central Baltic Sea. Helgol Mar Res 67 (3), 591–598. https://doi.org/10.1007/s10152-013-0347-2.

Riisgård, H.U., Seerup, D.F., 2003. Filtration rates in the soft clam Mya arenaria: effects of temperature and body size. Sarsia 88 (6), 416–428. doi:10.1080/00364820310003208.

Rix, L., De Goeij, J.M., Van Oevelen, D., Struck, U., Al-Horani, F.A., Wild, C., Naumann, M.S., 2018. Reef sponges facilitate the transfer of coral-derived organic matter to their associated fauna via the sponge loop. Mar. Ecol. Prog. Ser. 589, 85–96. doi:10.3354/meps12443.

Robinson, B., Ridler, N., Chopin, T., Bastarache, S., Belyea, E., Kaya, K., Barrow, C., 2018. The socio-economics of integrated aquaculture: a case study of the Bay of Fundy. Ocean Manage. 1, 1–11.

Roleda, M.Y., Hurd, C.L., 2019. Seaweed nutrient physiology: application of concepts to aquaculture and bioremediation. Phycologia 58 (5), 552–562. doi:10.1080/00318884.2019.1622920.

Ryther, J.H., Goldman, J.C., Gifford, C.E., Huguenin, J.E., Wing, A.S., Clarner, J.P., Williams, L.D., Lapointe, B.E., 1975. Physical models of integrated waste recycling-marine polyculture systems. Aquaculture 5 (2), 163–177. doi:10.1016/0044-8486(75)90096-4.

Sanz-Lazaro, C., Sanchez-Jerez, P., 2017. Mussels do not directly assimilate fish farm wastes: Shifting the rationale of integrated multi-trophic aquaculture to a broader scale. J. Environ. Manage. 201, 82–88. doi:10.1016/j.jenvman.2017.06.029.

Sanz-Lazaro, C., Sanchez-Jerez, P., 2020. Regional integrated multi-trophic aquaculture (RIMTA): spatially separated, ecologically linked. J. Environ. Manage. 271, 110921. doi:10.1016/j.jenvman.2020.110921.

Shi, H., Zheng, W., Zhang, X., Zhu, M., Ding, D., 2013. Ecological–economic assessment of monoculture and integrated multi-trophic aquaculture in Sanggou Bay of China. Aquaculture 410–411, 172–178. doi:10.1016/j.aquaculture.2013.06.033.

Shin, H.C., Lee, J.H., Jeong, H.J., Lee, J.S., Park, J.J., Kim, B.H., 2009. The influence of water temperature and salinity on filtration rates of the hard clam, Gomphina veneriformis (Bivalvia). Korean J. Malacol. 25 (2), 161–171.

Shinde, S.V., Sukhdhane, K.S., Sawant, S.S., Krishnani, K.K., Munilkumar, S., Majeedkutty, B.R.A., Chanu, T.I., Pathak, M.S., 2024. Amelioration of water quality and physiological performance of GIFT fish through the incorporation of Lemna minor and Lamellidens marginalis for ecological bioremediation in freshwater integrated multi-trophic aquaculture system. Aquac. Int. 32 (6), 7151–7171. doi:10.1007/s10499-024-01507-2.

Shinde, S.V., Sukhdhane, K.S., Krishnani, K.K., Rani A. M., B., Pathak, M.S., Chanu, T.I., Munilkumar, S., 2024. Inclusion of organic and inorganic extractive species improves growth, survival and physiological responses of GIFT fish reared in freshwater integrated multi-trophic aquaculture system. Aquaculture 580, 740346. doi:10.1016/j.aquaculture.2023.740346.

Shpigel, M., Shauli, L., Odintsov, V., Ben-Ezra, D., Neori, A., Guttman, L., 2018. The sea urchin, Paracentrotus lividus, in an integrated multi-trophic aquaculture (IMTA) system with fish (Sparus aurata) and seaweed (Ulva lactuca): nitrogen partitioning and proportional configurations. Aquaculture. 490, 260–269. doi:10.1016/j.aquaculture.2018.02.051.

Shull, D.H., 2009. Bioturbation. In: Steele, J.H., Turekian, K.K., Thorpe, S.A. (Eds.), Encyclopedia of ocean sciences. Academic Press, pp. 395–400. doi:10.1016/B978-012374473-9.00649-4.

Shumway, S.E., Cucci, T.L., Newell, R.C., Yentsch, C.M., 1985. Particle selection, ingestion, and absorption in filter-feeding bivalves. J. Exp. Mar. Biol. Ecol. 91 (1-2), 77–92. doi:10.1016/0022-0981(85)90222-9.

Shumway, S.E., Davis, C., Downey, R., Karney, R., Kraeuter, J., Parsons, J., Rheault, R., Wikfors, G., 2003. Shellfish aquaculture – in praise of sustainable economies and environments. World Aquac. 34 (4), 8–10.

Sicuro, B., Castelar, B., Mugetti, D., Pastorino, P., Chiarandon, A., Menconi, V., Galloni, M., Prearo, M., 2020. Bioremediation with freshwater bivalves: A sustainable approach to reducing the environmental impact of inland trout farms. J. Environ. Manage. 276, 111327. doi:10.1016/j.jenvman.2020.111327.

Siikamäki, J., Sanchirico, J.N., Jardine, S., McLaughlin, D., Morris, D.F., 2012. Blue carbon: global options for reducing emissions from the degradation and development of coastal ecosystems. Resources for the Future (RFF), Washington, DC, p. 60.

Skriptsova, A.V., Shevchenko, N.M., Zvyagintseva, T.N., Imbs, T.I., 2010. Monthly changes in the content and monosaccharide composition of fucoidan from Undaria pinnatifida (Laminariales, Phaeophyta). J. Appl. Phycol. 22 (1), 79–86. doi:10.1007/s10811-009-9438-5.

Skriptsova, A.V., Miroshnikova, N.V., 2011. Laboratory experiment to determine the potential of two macroalgae from the Russian Far-East as biofilters for integrated multi-trophic aquaculture (IMTA). Bioresour. Technol. 102 (3), 3149–3154. doi:10.1016/j.biortech.2010.10.093.

Slater, M.J., Jeffs, A.G., Carton, A.G., 2009. The use of the waste from green-lipped mussels as a food source for juvenile sea cucumber, Australostichopus mollis. Aquaculture. 292 (3–4), 219–224. doi:10.1016/j.aquaculture.2009.04.027.

Solomando, A., Capó, X., Alomar, C., Álvarez, E., Compa, M., Valencia, J.M., Pinya, S., Deudero, S., Sureda, A., 2020. Long-term exposure to microplastics induces oxidative stress and a pro-inflammatory response in the gut of Sparus aurata Linnaeus, 1758. Environ. Pollut. 266. doi:10.1016/j.envpol.2020.115295.

Sri-uam, P., Donnuea, S., Powtongsook, S., Pavasant, P., 2016. Integrated Multi-Trophic Recirculating Aquaculture System for Nile Tilapia (*Oreochlomis niloticus*). Sustainability 8 (7), 592. https://doi.org/10.3390/su8070592.

Stabili, L., Licciano, M., Giangrande, A., Longo, C., Mercurio, M., Marzano, C.N., Corriero, G., 2006. Filtering activity of Spongia officinalis var. adriatica (Schmidt), (Porifera, Demospongiae) on bacterioplankton: Implications for bioremediation of polluted seawater. Water Res. 40 (16), 3083–3090. doi:10.1016/j.watres.2006.06.012.

Stabili, L., Schirosi, R., Licciano, M., Mola, E., Giangrande, A., 2010. Bioremediation of bacteria in aquaculture waste using the polychaete Sabella spallanzanii. New Biotechnol. 27 (6), 774–781. doi:10.1016/j.nbt.2010.06.018.

Sterling, A.M., Cross, S.F., Pearce, C.M., 2016. Co-culturing green sea urchins (Strongylocentrotus droebachiensis) with mussels (Mytilus spp.) to control biofouling at an integrated multi-trophic aquaculture site. Aquaculture 464, 253–261. doi:10.1016/j.aquaculture.2016.06.010.

Sukhotin, A.A., Lajus, D.L., Lesin, P.A., 2003. Influence of age and size on pumping activity and stress resistance in the marine bivalve Mytilus edulis L. J. Exp. Mar. Biol. Ecol. 284 (1–2), 129–144. doi:10.1016/s0022-0981(02)00497-5.

Tantanasarit, C., Babel, S., Englande, A.J., Meksumpun, S., 2013. Influence of size and density on filtration rate modeling and nutrient uptake by green mussel (Perna viridis). Mar. Pollut. Bull. 68 (1–2), 38–45. doi:10.1016/j.marpolbul.2012.12.027.

Troell, M., Joyce, A., Chopin, T., Neori, A., Buschmann, A.H., Fang, J.G., 2009. Ecological engineering in aquaculture - potential for integrated multi-trophic aquaculture (IMTA) in marine offshore systems. Aquaculture 297 (1–4), 1–9. doi:10.1016/j.aquaculture.2009.09.010.

Troell, M., Halling, C., Neori, A., Chopin, T., Buschmann, A.H., Kautsky, N., Yarish, C., 2003. Integrated mariculture: asking the right questions. Aquaculture 226 (1-4), 69–90. doi:10.1016/s0044-8486(03)00469-1.

Troell, M., Halling, C., Nilsson, A., Buschmann, A.H., Kautsky, N., Kautsky, L., 1997. Integrated marine cultivation of Gracilaria chilensis (Gracilariales, Rhodophyta) and salmon cages for reduced environmental impact and increased economic output. Aquaculture 156 (1–2), 45–61. doi:10.1016/s0044-8486(97)00080-x.

Turcios, A., Papenbrock, J., 2014. Sustainable treatment of aquaculture effluents—what can we learn from the past for the future? Sustainability 6 (2), 836–856. doi:10.3390/su6020836.

Waller, U., Buhmann, A.K., Ernst, A., Hanke, V., Kulakowski, A., Wecker, B., Orellana, J., Papenbrock, J., 2015. Integrated multi-trophic aquaculture in a zero-exchange recirculation aquaculture system for marine fish and hydroponic halophyte production. Aquac. Int. 23 (6), 1473–1489. doi:10.1007/s10499-015-9898-3.

Wang, H., Seekamp, I., Malzahn, A., Hagemann, A., Carvajal, A.K., Slizyte, R., Standal, I.B., Handå, A., Reitan, K.I., 2019. Growth and nutritional composition of the polychaete Hediste diversicolor (OF Müller, 1776) cultivated on waste from land-based salmon smolt aquaculture. Aquaculture 502, 232–241. doi:10.1016/j.aquaculture.2018.12.047.

Wang, X., Broch, O.J., Forbord, S., Handå, A., Skjermo, J., Reitan, K.I., Vadstein, O., Olsen, Y., 2014. Assimilation of inorganic nutrients from salmon (Salmo salar) farming by the macroalgae (Saccharina latissima) in an exposed coastal environment: Implications for integrated multi-trophic aquaculture. J. Appl. Phycol. 26 (4), 1869–1878. doi:10.1007/s10811-013-0230-1.

Wang, X., Andresen, K., Handå, A., Jensen, B., Reitan, K.I, Olsen, Y., 2013. Chemical composition and release rate of waste discharge from an Atlantic salmon farm with an evaluation of IMTA feasibility. Aquacult. Env. Interact. 4 (2), 147–162. doi:10.3354/aei00079.

Wang, X., Olsen, L.M., Reitan, K.I., Olsen, Y., 2012. Discharge of nutrient wastes from salmon farms: environmental effects, and potential for integrated multi-trophic aquaculture. Aquacult. Env. Interact. 2 (3), 267–283. doi:10.3354/aei00044Norway.

Ward, J.E., Kach, D.J., 2009. Marine aggregates facilitate ingestion of nanoparticles by suspension-feeding bivalves. Mar. Environ. Res. 68 (3), 137–142. doi:10.1016/j.marenvres.2009.05.002.

Ward, J.E., Shumway, S.E., 2004. Separating the grain from the chaff: Particle selection in suspension- and deposit-feeding bivalves. J. Exp. Mar. Biol. Ecol. 300, 83–130. https://doi.org/10.1016/j.jembe.2004.03.002.

Wedchaparn, O., Ayisi, C.L., Huo, Y., He, P., 2015. Effects of different light intensity fluctuations on growth rate, nutrient uptake and photosynthetic efficiency of Gracilaria asiatica. J. Fish. Aquat. Sci. 10 (6), 533–542. doi:10.3923/jfas.2015.533.542.

Weisz, J.B., Lindquist, N., Martens, C.S., 2008. Do associated microbial abundances impact marine demosponge pumping rates and tissue densities? Oecologia 155 (2), 367–376. doi:10.1007/s00442-007-0910-0.

Whitfield, M., 1974. The hydrolysis of ammonium ions in sea water-a theoretical study. J. Mar. Biol. Assoc. U. K. 54 (3), 565–580. doi:10.1017/s002531540002275x.

Whitmarsh, D.J., Cook, E.J., Black, K.D., 2006. Searching for sustainability in aquaculture: an investigation into the economic prospects for an integrated salmon-mussel production system. Mar. Policy 30 (3), 293–298. doi:10.1016/j.marpol.2005.01.004.

Yokoyama, H., 2013. Growth and food source of the sea cucumber Apostichopus japonicus cultured below fish cages —potential for integrated multi-trophic aquaculture. Aquaculture 372–375, 28–38. doi:10.1016/j.aquaculture.2012.10.022.

You, K., Ma, C., Gao, H., Li, F., Zhang, M., Qiu, Y., Wang, B.o, 2007. Research on the jellyfish (Rhopilema esculentum Kishinouye) and associated aquaculture techniques in China: current status. Aquac. Int. 15 (6), 479–488. doi:10.1007/s10499-007-9114-1.

Yu, X., Yang, Q., Zhao, Z., Tang, X., Xiong, B.o, Su, S., Wu, Z., Yao, W., 2021. Ecological efficiency of the Mussel Hyriopsis cumingii (Lea, 1852) on particulate organic matter filtering, algal controlling and water quality regulation. Water. 13 (3), 297. doi:10.3390/w13030297.

Yu, Z., Hu, C., Zhou, Y.i, Li, H., Peng, P., 2012. Survival and growth of the sea cucumber H olothuria leucospilota Brandt: a comparison between suspended and bottom cultures in a subtropical fish farm during summer. Aquac. Res. 44 (1), 114–124. doi:10.1111/j.1365-2109.2011.03016.x.

Yuan, S.Y., Zhu, W.J., Neori, A., Zhang, Y., Li, M., Li, J., Chang, Z.Q., 2023. Benthic suspension-feeding clams affect sedimentary microbial communities and nitrogen cycling in seawater pond IMTA. Aquaculture 563. doi:10.1016/j.aquaculture.2022.738907.

Yuan, X., Yang, H., Zhou, Y., Mao, Y., Zhang, T., Liu, Y., 2006. The influence of diets containing dried bivalve feces and/or powdered algae on growth and energy distribution in sea cucumber Apostichopus japonicus (Selenka), (Echinodermata: Holothuroidea). Aquaculture 256 (1–4), 457–467. doi:10.1016/j.aquaculture.2006.01.029.

Zamora, L.N, Dollimore, J., Jeffs, A.G, 2014. Feasibility of co-culture of the Australasian sea cucumber (Australostichopus mollis) with the Pacific oyster (Crassostrea gigas) in northern New Zealand. N.Z. J. Mar. Freshwater Res. 48 (3), 394–404. doi:10.1080/00288330.2014.901232.

Zamora, L.N., Jeffs, A.G., 2011. Feeding, selection, digestion and absorption of the organic matter from mussel waste by juveniles of the deposit-feeding sea cucumber, Australostichopus mollis. Aquaculture 317 (1–4), 223–228. doi:10.1016/j.aquaculture.2011.04.011.

Zhang, J., Kitazawa, D., Yang, C., 2016. A numerical modeling approach to support decision-making on design of integrated multitrophic aquaculture for efficiently mitigating aquatic waste. Mitig. Adapt. Strateg. Glob. Change 21 (8), 1247–1261. doi:10.1007/s11027-015-9652-1.

Zhou, Y.i, Yang, H., Liu, S., Yuan, X., Mao, Y., Liu, Y., Xu, X., Zhang, F., 2006. Feeding and growth on bivalve biodeposits by the deposit feeder Stichopus japonicus Selenka (Echinodermata: Holothuroidea) co-cultured in lantern nets. Aquaculture 256 (1–4), 510–520. doi:10.1016/j.aquaculture.2006.02.005.

Zhu, Y., Wang, N., Wu, Z., Chen, S., Luo, J.i, Christakos, G., Wu, J., 2025. The role of seaweed cultivation in integrated multi-trophic aquaculture (IMTA): the current status and challenges. Rev. Aquac. 17 (3). doi:10.1111/raq.70042.

CHAPTER

18

Novel nutraceuticals from waste resources: Looking beyond the aqua-industry

Soibam Ngasotter[a], Sandhiya Venkatesh[b], Tao Kara[b], Madhulika[b], Sushil Lukram[b], David Waikhom[c], Domendra Dhruve[a] and K. A. Martin Xavier[d,]*

[a]LSPN College of Fisheries, Dau Shri Vasudev Chandrakar Kamdhenu Vishwavidyalaya, Kawardha, Chhattisgarh, India [b]ICAR-Central Institute of Fisheries Education, Mumbai, Maharashtra, India [c]College of Fisheries, Central Agricultural University (Imphal), Lembucherra, Tripura, India [d]ICAR-Central Institute of Fisheries Technology, Cochin, Kerala, India

***Corresponding author: martinxavierkochery@gmail.com**

18.1 Introduction

The increasing prevalence of lifestyle-related diseases and the growing interest in preventive healthcare have fueled a surge in demand for nutraceuticals, products derived from natural sources that provide health benefits beyond basic nutrition (Vignesh et al., 2024). Nutraceuticals, often marketed as functional foods or dietary supplements, play a crucial role in addressing modern health challenges such as cardiovascular diseases, diabetes, obesity, and immune system dysfunction (Puri et al., 2022). While traditionally sourced from plants and terrestrial animals, the exploration of marine and aquatic resources has opened new avenues for the discovery of novel bioactive compounds. Among these, aquatic waste has emerged as a promising yet under-utilized resource for nutraceutical development, combining health benefits with sustainability (Xia et al., 2024).

Aquatic waste, generated during the processing of fish, shellfish, algae, and other marine organisms, is a significant byproduct of the global fisheries and aquaculture industries (Xia et al., 2024). More than 50% of the total biomass from fish and shellfish remains unused, often

Zero-Waste Technologies in Aquaculture: Global Perspectives and Methods toward a Circular Economy
DOI: https://doi.org/10.1016/B978-0-443-34689-7.00004-5

discarded as waste (Yusoff et al., 2024). This includes viscera, scales, skin, shells, fins, and other byproducts, which not only contribute to environmental pollution but also represent a loss of valuable resources. These materials are rich in bioactive compounds such as proteins, peptides, polyunsaturated fatty acids (PUFAs), antioxidants, and polysaccharides, such as chitin and chitosan, all of which have demonstrated potential health benefits (Awuchi et al., 2022).

The valorization of aquatic waste for nutraceutical development addresses two critical global challenges: environmental sustainability and resource efficiency. By utilizing what was previously considered waste, industries can reduce their environmental footprint, lower waste management costs, and create value-added products that contribute to human health (Shahbaz et al., 2023). This approach aligns with the principles of a circular economy, promoting the sustainable use of resources and reducing reliance on traditional, often finite, sources of nutraceutical ingredients (Alazaiza et al., 2024).

A wide variety of bioactive compounds can be derived from aquatic waste, each offering unique health-promoting properties. For instance, fish skin and bones are excellent sources of collagen and gelatin, which have applications in joint health, skin care, and wound healing (Jafari et al., 2020). Similarly, shellfish waste, such as shrimp and crab shells, contains chitin and chitosan (Ngasotter et al., 2022a; Ngasotter et al., 2023), polysaccharides known for their antimicrobial, antioxidant, and cholesterol-lowering effects. Marine lipids, particularly omega-3 fatty acids extracted from fish and algae, are widely recognized for their role in cardiovascular health and cognitive function. Moreover, carotenoids such as astaxanthin, derived from crustacean shells, are powerful antioxidants with anti-inflammatory properties.

Despite the potential of aquatic waste as a source of nutraceuticals, several challenges hinder its full utilization. Traditional waste management practices, such as landfilling and incineration, are not only environmentally unsustainable but also fail to capitalize on the valuable bioactive compounds present in these byproducts (Abubakar et al., 2022). In addition, the extraction and purification of bioactive compounds require advanced technologies that balance efficiency, cost, and environmental impact (Mungwari et al., 2025). Regulatory challenges, including safety and quality standards for waste-derived products, further complicate the path to commercialization (Srivastava et al., 2022). Addressing these challenges necessitates interdisciplinary research and collaboration between academia, industry, and policymakers.

The economic implications of utilizing aquatic waste for nutraceutical development are equally significant. The global nutraceutical market was estimated at US$ 591.1 billion in 2024 and is projected to reach US$ 919.1 billion by 2030 (Grand View Research, 2025), driven by increasing consumer awareness and demand for health-promoting products. Valorizing aquatic waste not only provides a cost-effective raw material for this burgeoning industry but also creates opportunities for economic growth (Xia et al., 2024), particularly in coastal regions where fisheries and aquaculture are major economic activities. Small and medium enterprises (SMEs) in these regions can benefit from the development of localized processing technologies, creating jobs and promoting sustainable development.

This chapter aims to explore the potential of aquatic waste as a sustainable source of nutraceuticals, highlighting the bioactive compounds present in various waste streams and their applications in functional foods, pharmaceuticals, and cosmeceuticals. By examining the economic, environmental, and regulatory aspects of aquatic waste valorization, this chapter seeks to provide a comprehensive understanding of the challenges and opportunities in this field.

18.2 Types of aquatic waste

18.2.1 Fish processing waste

Fish processing waste constitutes a significant portion of the biomass generated by the fishing and aquaculture industries (Rajput et al., 2024). It is estimated that 30%–70% of the total fish biomass, depending on the species and processing methods, is discarded as waste (Ahuja et al., 2020). This includes viscera, skin, scales, bones, cartilage, heads, fins, and other byproducts that are typically considered low-value materials. However, these waste components are rich in valuable bioactive compounds with potential applications in nutraceuticals, pharmaceuticals, and other industries.

18.2.2 Fish viscera

Fish viscera, which includes the stomach, intestines, and other internal organs, is a rich source of proteins, lipids, and enzymes (Estiasih et al., 2021). Bioactive peptides derived from viscera have demonstrated antioxidant, antimicrobial, and antihypertensive properties (Cunha and Pintado, 2022). In addition, viscera lipids often contain omega-3 fatty acids, such as eicosapentaenoic acid (EPA) and docosahexaenoic acid (DHA), which are well-known for their cardiovascular and cognitive health benefits (Swanson et al., 2012).

18.2.2.1 Fish skin

Fish skin is another valuable component of processing waste. It serves as a rich source of collagen and gelatin, which have numerous applications in the nutraceutical and cosmeceutical industries (Joy et al., 2024). Collagen derived from fish skin is increasingly used in antiaging products, wound healing formulations, and joint health supplements because of its bioavailability and functional properties (Gaikwad and Kim, 2024).

18.2.2.2 Fish scales

Fish scales, although often overlooked, are composed primarily of calcium salts and collagen (Qin et al., 2022). They can be processed to extract hydroxyapatite, a biomaterial widely used in bone and dental health applications (Hernández-Ruiz et al., 2022). In addition, the collagen present in scales can be hydrolyzed into peptides with bioactive properties that support health (Jena et al., 2025).

18.2.2.3 Fish bones and cartilage

Fish bones and cartilage are rich in minerals such as calcium and phosphorus, as well as collagen and bioactive peptides (Jo et al., 2025). These components are commonly used to produce dietary supplements for bone and joint health, including calcium-enriched powders and hydrolyzed collagen.

18.2.2.4 Heads and fins

Fish heads and fins contain significant amounts of bioactive lipids and proteins (Ozogul et al., 2021). Omega-3 fatty acids extracted from these parts are widely utilized in functional foods and dietary supplements (Swanson et al., 2012).

18.2.3 Shellfish waste

Shellfish waste, a significant byproduct of the seafood processing industry, is generated during the harvesting, processing, and consumption of crustaceans (e.g., shrimp, crab, and lobster) and mollusks (e.g., oysters, clams, and scallops). This waste constitutes a substantial proportion of the total biomass, often exceeding 50% of the harvested material, depending on the species and processing methods (Venugopal, 2021). Shellfish waste includes exoskeletons, soft tissues, and residual meat, which are often discarded despite being rich in bioactive compounds with immense potential for nutraceutical, pharmaceutical, and industrial applications (Azelee et al., 2023).

18.2.3.1 Exoskeletons (shells)

The exoskeletons of crustaceans, including shrimp, crab, and lobster shells, are composed of chitin, a polysaccharide that forms the structural backbone of these materials (Ngasotter et al., 2022a; Ngasotter et al., 2025a). Chitin can be processed into chitosan, a versatile derivative with antimicrobial, antioxidant, and cholesterol-lowering properties (Scarcelli et al., 2025). Chitosan has applications in weight management, gut health, and wound healing (Shagdarova et al., 2023; Mawazi et al., 2024). In addition to chitin, exoskeletons are also rich in minerals, such as calcium and magnesium, which can be used in dietary supplements (Xu et al., 2020).

Shellfish exoskeletons also contain carotenoids, particularly astaxanthin, a powerful antioxidant responsible for the characteristic pink or red color of crustaceans (Dayakar et al., 2022; Dayakar et al., 2024a; Dayakar et al., 2024b). Astaxanthin has shown potential in reducing inflammation, improving cardiovascular health, and protecting against oxidative stress (Fassett and Coombes, 2011). It is widely utilized in nutraceuticals and cosmeceuticals for its antiaging and skin-protective properties.

Shells of mollusks, such as oysters and clams, contain high levels of calcium carbonate, which can be processed into calcium supplements or used as a raw material for bone health applications (Topić Popović et al., 2023). Calcium carbonate is also used in pharmaceutical formulations as an antacid and excipient (Supriadi et al., 2023).

18.2.3.2 Soft tissues and residual meat

The soft tissues of shellfish, including the mantle of mollusks and the head regions of shrimp and crab, are rich in proteins, lipids, and enzymes (Venugopal and Sasidharan, 2022). Bioactive peptides derived from these proteins exhibit antioxidant, antimicrobial, and antihypertensive activities. In addition, the lipid fraction often contains omega-3 fatty acids, which are known for their cardioprotective and anti-inflammatory effects.

18.2.4 Algal and seaweed residues

Algal and seaweed residues represent an abundant and underutilized form of aquatic waste with immense potential for nutraceutical applications (Trigueros et al., 2024). As global industries increasingly rely on algae and seaweeds for food, pharmaceuticals, cosmetics, and biofuels, the byproducts generated during their processing provide an opportunity to recover valuable bioactive compounds. These residues, often discarded as waste, can be transformed into nutraceutical ingredients that promote health and wellness.

Algal and seaweed waste is primarily derived from industrial processes that extract high-value compounds such as agar, carrageenan, and alginates (Luo et al., 2024). These polysaccharides are widely used in food and pharmaceutical industries as thickeners, stabilizers, and gelling agents (Liao et al., 2021). The leftover biomass from such processes is a significant component of aquatic waste. Similarly, the production of biofuels from algae results in residual material rich in proteins, polysaccharides, and pigments, which hold great potential for nutraceutical development (Kammler et al., 2024).

Seaweed farming, a major aquaculture activity, also generates waste in the form of trimmings, rejected or spoiled biomass, and postharvest residues (Yun et al., 2023). These byproducts are often overlooked but contain a variety of bioactive compounds. In addition, small-scale and industrial uses of algae in dietary supplements, animal feed, and cosmetics contribute to the accumulation of algal and seaweed waste.

18.3 Bioactive compounds from aquatic waste

18.3.1 Proteins and peptides (collagen, gelatin, and protein hydrolysates)

Fish processing generates substantial quantities of byproducts such as heads, skin, bones, scales, fins, and blood, which are rich sources of proteins and other bioactive compounds. The extracellular matrix of connective tissues, such as skin, bone, tendon, ligament, cartilage, intervertebral discs, and blood vessels, primarily consists of collagen, the most abundant structural protein in animals (Alberts et al., 2002). Collagen is inherently biocompatible, biodegradable, and bioactive, making it valuable for various biomedical and industrial applications. A total of 28 different types of collagens have been identified based on their structure and tissue distribution (Gelse et al., 2003).

Collagen can be extracted from vertebrate sources such as fish skin, bones, scales, fins, and cartilage, as well as invertebrate sources including mollusks, crustaceans, echinoderms, and poriferans (Cadar et al., 2024). The extraction and processing method significantly influence the structure, composition, and properties of the collagen-derived products. These include: (1) Native collagen (soluble or insoluble, preserving the triple-helical structure), (2) gelatin (partially denatured collagen), and (3) collagen hydrolysates/peptides (completely hydrolyzed to low molecular weight peptides and amino acids).

Collagen hydrolysates are typically produced via acidic, alkaline, or enzymatic hydrolysis, with enzymatic methods preferred for their mild processing conditions and preservation of bioactivity (Al Hajj et al., 2024). Hydrolyzed collagen offers superior bioavailability, including improved digestibility, absorption, and systemic distribution compared to native collagen (León-López et al., 2019).

Gelatin, formed through the partial denaturation of collagen, results from the disruption of intra- and intermolecular hydrogen bonds within the collagen triple helix (Rather et al., 2022). The physicochemical properties of gelatin depend largely on the molecular weight distribution, amino acid composition, and degree of denaturation of collagen fragments. Gelatin is commonly categorized into Type A (obtained via acid hydrolysis) and Type B (produced through alkaline hydrolysis). It primarily consists of glycine (~21%), proline (~12%), and hydroxyproline (~12%) (Kariduraganavar et al., 2014).

Globally, mammalian sources, such as pig skin and cowhides, are the primary contributors to gelatin production, accounting for approximately 46% of the total supply, followed by bones (23%), hooves (29%), and only about 1% derived from marine sources such as fish (Drioli and Giorno, 2010; Rakhmanova et al., 2018). Gelatin-based nanospheres can be produced via methods such as desolvation, coacervation, and water-in-oil (W/O) emulsion gelation, and have shown significant potential in drug delivery and pharmaceutical applications (Narayanaswamy et al., 2016).

Protein hydrolysates can also be produced from fishery waste through acid, alkaline, or enzymatic hydrolysis, with enzymatic hydrolysis being the most effective and environmentally sustainable approach (Nemati et al., 2024). These hydrolysates consist of short-chain peptides (typically 2–20 amino acids) possessing diverse functional and bioactive properties. Commonly employed enzymes include trypsin, pepsin, α-chymotrypsin, papain, proteinase K, alcalase, protamex, flavourzyme, and neutrase (Chi et al., 2015). The functionality of the resulting protein hydrolysates, such as emulsifying, antioxidant, or antimicrobial activity, is strongly influenced by the degree of hydrolysis, which in turn determines the molecular weight distribution of the peptides.

18.3.2 Polysaccharides (chitin, chitosan, and derivatives)

Seafood processing waste, particularly from crustaceans such as shrimp, crab, and lobster, is a rich reservoir of structural polysaccharides such as chitin and its deacetylated form, chitosan (Santos et al., 2020). These biopolymers, often discarded as part of shell waste, are gaining prominence as valuable nutraceuticals because of their biofunctional properties, biocompatibility, and biodegradability. The conversion of chitin-rich shell waste into high-value derivatives not only reduces environmental burden but also aligns with circular bioeconomy goals by generating health-promoting compounds from underutilized resources (Xia et al., 2024).

Chitin, a linear polysaccharide composed of *N*-acetyl-D-glucosamine units, is the second most abundant natural polymer after cellulose (Izadi et al., 2025). It is predominantly present in the exoskeletons of marine arthropods and fungi. However, because of its insolubility in water and most solvents, its direct application is limited. To overcome this, chitosan, a partially deacetylated derivative of chitin, is commonly produced via chemical or enzymatic deacetylation processes. Chitosan exhibits enhanced solubility under mildly acidic conditions and possesses functional groups that can be readily modified, making it highly versatile in nutraceutical applications (Aranaz et al., 2021).

Chitosan and its derivatives have been extensively studied for their antioxidant, antimicrobial, anti-inflammatory, antihypertensive, and cholesterol-lowering effects (Ngo et al., 2015). Their ability to scavenge free radicals and chelate metal ions contributes to oxidative stress mitigation, an important factor in preventing chronic diseases such as cardiovascular disorders, neurodegeneration, and cancer (Ivanova and Yaneva, 2020). Chitosan's positive charge also enables it to interact with negatively charged microbial membranes, leading to membrane disruption and microbial inhibition, which is beneficial in gut health and food preservation (Yan et al., 2021).

One of the most notable nutraceutical uses of chitosan is its role in weight management and lipid reduction (Shagdarova et al., 2023). Chitosan binds dietary fats in the gastrointestinal tract, reducing their absorption and contributing to hypolipidemic effects. Several clinical studies have demonstrated the efficacy of chitosan supplementation in reducing LDL cholesterol and triglyceride levels, thereby supporting cardiovascular health (Ahn et al., 2021).

Recent advancements have also focused on nano-sized chitin and chitosan derivatives, such as chitin nanocrystals (ChNCs) and chitin nanofibers (ChNFs), which offer superior bioactivity because of their increased surface area and enhanced bioavailability (Ngasotter et al., 2024). These nanomaterials show potential as functional food additives, encapsulating agents for nutraceuticals, and delivery systems for bioactive compounds (Ngasotter et al., 2023; Ngasotter et al., 2025a). Moreover, chemical modifications such as carboxymethylation, sulfation, and quaternization have further improved the solubility, stability, and biological functionality of chitosan derivatives (Suryani et al., 2024).

The eco-friendly extraction of chitin and chitosan using enzymatic hydrolysis and microbial fermentation is gaining momentum as a sustainable alternative to conventional chemical methods, minimizing the use of harsh reagents and reducing environmental pollution (Casadidio et al., 2019). Utilizing aquaculture and seafood processing waste for chitosan production not only contributes to waste valorization but also ensures a steady supply of raw materials for the development of functional foods, dietary supplements, and therapeutic agents.

18.3.3 Lipids (omega-3 fatty acids and phospholipids)

Fish lipids are a valuable source of long-chain polyunsaturated fatty acids (LC-PUFAs), particularly EPA (20:5 n-3) and DHA (22:6 n-3), which are widely recognized for their nutritional and therapeutic benefits (Abedi and Sahari, 2014). These fatty acids are commonly incorporated into dietary supplements to support cardiovascular, neural, and visual health. Fish oils have both edible and non-edible applications. In the food and nutraceutical industries, they are utilized for human consumption (Das et al., 2024), while in industrial sectors, fish oils are used in the formulation of hydraulic fluids, lubricants, and soap production (Fineberg and Johanson, 1967).

Fish oil is broadly classified into two categories: (1) Fish body oil and (2) fish liver oil. Fish body oil is typically obtained through the wet rendering (or wet reduction) process, wherein oil is recovered from the stick water produced during fish meal processing (Dave et al., 2024). This oil is subsequently separated via centrifugation and undergoes refining to meet quality standards. In India, oil sardine (*Sardinella longiceps*) is one of the most prominent species used for fish oil production (Chakraborty and Joseph, 2015). Fish liver oil, on the other hand, is extracted from species such as shark, whale, tuna, and halibut, and is particularly valued for its high content of vitamins A and D, in addition to LC-PUFAs (Bimbo, 2011).

Chemically, fish oil primarily consists of triglycerides, which are molecules composed of a glycerol backbone esterified with three fatty acid chains, which may be identical or varied. In addition to triglycerides, fish oil contains minor but important components such as phospholipids, glycerol ethers, and wax esters (Rodrigues et al., 2024).

The lipid composition of fish differs significantly from that of terrestrial animals and plant oils because of its high concentration of omega-3 fatty acids, particularly EPA and DHA. Among the minor components, phospholipids play a crucial role. These amphipathic molecules consist of a hydrophilic phosphate-containing head group and two hydrophobic fatty acid tails, often enriched in omega-3 PUFAs, making them essential for cellular membrane integrity and function.

18.3.4 Antioxidants (carotenoids, astaxanthin, and polyphenols)

Crustaceans, such as shrimp, are a notable source of carotenoids (Sachindra et al., 2006). Among these, astaxanthin is the predominant pigment, classified as a red-colored xanthophyll and an oxygenated derivative of the carotenoid family (Yaqoob et al., 2022). Structurally, astaxanthin is a lipid-soluble tetraterpenoid, consisting of 40 carbon atoms and a series of conjugated double bonds, which contribute to its potent antioxidant activity (López et al., 2021).

In shrimp shells, astaxanthin primarily exists in the form of carotenoproteins or carotenolipoproteins, where it is tightly bound to proteins because of its specific binding affinities (Dayakar et al., 2022). The extraction of carotenoids from shrimp waste is typically carried out using organic solvents such as acetone, ethanol, methanol, isopropanol, and dichloromethane (Sachindra et al., 2006). However, with increasing emphasis on environmental sustainability, green extraction techniques are gaining attention. These include supercritical fluid extraction (SFE) with carbon dioxide, pulsed electric field (PEF), high-pressure processing (HPP), microwave-assisted extraction (MAE), enzyme-assisted extraction, and ultrasound-assisted extraction (UAE), among others (Morón-Ortiz et al., 2024).

In addition to carotenoids, marine polyphenols represent another important group of bioactive compounds found in various marine organisms, including algae, finfish, and crustaceans (Pereira and Cotas, 2023). These compounds are known for their strong antioxidant and anti-inflammatory activities, attributed to the presence of multiple hydroxyl groups in their molecular structure. Marine polyphenols encompass several chemical classes such as flavonoids, phenolic acids, tannins, lignans, and stilbenes. Among these, flavonoids such as catechins, quercetin, and rutin are commonly detected in fish species (Pereira and Cotas, 2023).

Extraction of marine polyphenols can be performed through both conventional methods, such as liquid–liquid and solid–liquid extraction, and modern techniques, including SFE, MAE, subcritical water extraction (SWE), and UAE (Brglez Mojzer et al., 2016).

18.3.5 Other compounds (minerals and enzymes)

Fish bone, which constitutes approximately 9%–15% of the total fish weight, also contains 30%–45% protein and 15%–60% ash, mostly calcium, with minimal lipid content, depending on the fish species (Pérez et al., 2024). Fish bone is a rich source of calcium and offers a promising alternative to dairy-derived calcium (Torres et al., 2024). It typically contains about 30% collagen and 60%–70% calcium phosphate, primarily in the form of hydroxyapatite (Mondal et al., 2019). After enzymatic hydrolysis of fish frames, calcium can be efficiently extracted and concentrated. Using nanotechnology, this calcium can be further processed into nanocalcium with enhanced bioavailability, making it a potential dietary calcium source for human consumption (Fatmawati et al., 2025).

Fish processing byproducts, such as viscera, liver, and head, are also rich in industrially relevant enzymes, including proteases, lipases, and transglutaminases (Venugopal, 2016). Notably, fish viscera contain enzymes such as pepsin, trypsin, and chymotrypsin, which exhibit superior catalytic efficiency, activity at low temperatures, broad pH stability, and low substrate sensitivity (Khiari, 2024).

18.4 Applications of aquatic waste-derived nutraceuticals

18.4.1 Functional foods and beverages

Collagen has both hydrophilic and hydrophobic groups in protein side chains, making it and its derivatives suitable as food additives, such as emulsifying agents and film-forming proteins (Tang et al., 2022). Gelatin, the denatured form of collagen, is widely used in the food packaging industry because of its excellent oxygen barrier properties and favorable rheological characteristics (Rather et al., 2022). In the food industry, gelatin serves multiple functions, as an emulsifier, foaming agent, and gelling agent (Rather et al., 2022).

Collagen hydrolysates or peptides enhance food quality and are used in developing functional foods because of their antioxidant and antimicrobial activities (Irastorza et al., 2021). They also improve storage stability because of their anticoagulant properties, protecting food tissues from low-temperature damage and helping preserve the physical, chemical, and sensory attributes of food. Collagen is commonly used in sausage production for its rheological benefits and to reduce fat layer formation in meat products (Al Hajj et al., 2024). In addition, it serves as an edible coating that limits oxygen and moisture permeability, thereby extending shelf life (Khin et al., 2024). Hydrolyzed collagen is also employed as a clarifying agent in beverages (Al Hajj et al., 2024).

Fish protein hydrolysates, known for their strong antioxidant properties, help minimize lipid peroxidation in food (Halldorsdottir et al., 2014). They possess unique functional properties such as water-holding capacity, oil absorption, solubility, gelling ability, foaming capacity, and emulsification, making them suitable for formulating functional foods (Wangkheirakpam et al., 2019). Collagen hydrolysate-enriched herbal soups are commercially available and offer health benefits such as enhanced collagen synthesis and cell proliferation.

Chitin and chitosan derived from shrimp waste are increasingly utilized in the food industry for their multifunctional properties, serving as natural preservatives, antimicrobial and antioxidant agents, and edible coating or film-forming materials to extend shelf life, improve food safety, and maintain quality (Chattopadhyay et al., 2023; Ngasotter et al., 2025b). Their film-forming ability is applied in packaging systems, while their bioactivity enhances functional food development. In addition, chitosan is used as a dietary fiber, clarifying agent in beverages, and emulsifier in various formulations, contributing to cleaner labels, reduced synthetic additives, and the valorization of seafood processing byproducts into sustainable food ingredients (Iñiguez-Moreno et al., 2024).

Fish oil is used in margarine and shortening formulations (Sharma et al., 2014). Phospholipids, because of their amphiphilic nature, are effective emulsifiers in food systems, improving the stability of oil-in-water emulsions by reducing droplet size (Ilza and Diharmi, 2022).

Astaxanthin, a natural pigment, can replace synthetic colorants in food. However, its low water solubility, susceptibility to oxidation (because of PUFA content), and poor bioavailability limit its use (Montero et al., 2016). Encapsulation via spray drying significantly improves its solubility and stability (Montero et al., 2016).

Enzymes derived from fish waste have diverse applications in food processing (Venugopal, 2016). Lipases are used to modify flavors and selectively hydrolyze PUFA ethyl esters. Proteases from fish waste are utilized to produce fish protein hydrolysates, although the resulting bitterness can be reduced by treating with acidic serine carboxypeptidase extracted from squid hepatopancreas (Komai et al., 2007). Transglutaminases are applied for texture improvement by cross-linking proteins, particularly in surimi production (Zhang and Chang, 2023). Fish sauce is a widely consumed product obtained through the enzymatic degradation of fish by naturally occurring proteolytic enzymes (Irianto, 2017).

18.4.2 Dietary supplements

The consumption of collagen hydrolysates or peptides contributes to skin rejuvenation and may help alleviate cardiovascular conditions. Collagen supplementation has been shown to reduce wrinkles, enhance skin elasticity and hydration, and promote collagen synthesis, density, and content, factors closely associated with mitigating age-related skin damage. In addition, collagen supplements improve joint mobility and function, reduce pain, and support increased bone mass, density, and strength (Campos et al., 2023).

Fish oil is an excellent source of PUFAs, particularly EPA and DHA, collectively known as omega-3 fatty acids (Coppola et al., 2021). Since omega-3 fatty acids cannot be synthesized by the human body, they are considered essential nutrients that play a vital role in the development and function of the central nervous system and retina. Regular consumption of omega-3s is associated with a reduced risk of Alzheimer's disease (Wood et al., 2022) and cardiovascular conditions. Fish liver oil, obtained by steam cooking of fish livers, is especially rich in vitamins A and D and is typically classified as medicinal-grade oil.

Calcium is an essential mineral involved in numerous physiological functions, including nerve transmission, muscle contraction, the formation and maintenance of strong bones and teeth, and acting as a cofactor in various enzymatic reactions (Razzaque and Wimalawansa, 2025). A deficiency in calcium can impair bone health and contribute to conditions such as hypertension, hypercholesterolemia, and certain types of cancer. Calcium derived from fish bone waste serves as an effective supplement in preventing and treating osteopenia and osteoporosis, and also plays a vital role in managing hypocalcemic conditions such as hypoparathyroidism and pseudohypoparathyroidism (Saetang et al., 2024). In addition, calcium helps inhibit bone resorption, thereby supporting skeletal integrity.

18.4.3 Pharmaceutical applications

In tissue engineering, collagen functions as a natural scaffold because of its abundance, biocompatibility, biodegradability, and structural similarity to the extracellular matrix. It provides support to various connective tissues, including skin, tendons, bones, cartilage, blood vessels, and ligaments (Raman and Gopakumar, 2018). However, collagen derived from fish exhibits relatively low mechanical strength because of reduced cross-linking. To enhance its performance, it is often cross-linked with bioceramics that mimic the inorganic components of natural bone, making it more suitable for scaffold applications (Leonard et al., 2024). Marine collagen is also widely used in dentistry as a membrane, bone graft material, and hemostatic agent (Lim et al., 2019). Moreover, it is effective in controlling bleeding, with several commercially available sponge-like products that are highly absorbent and capable of retaining multiple times their weight in fluids.

Hydrolyzed collagen contains bioactive peptides capable of penetrating joint tissues and exerting chondroprotective effects. Various collagen-derived peptides with demonstrated in vitro activities, such as antioxidant, ACE-I inhibitory, and DPP-IV inhibitory properties, have been documented (Tang et al., 2022). Moreover, hydrolyzed collagen exhibits enhanced bioavailability, allowing it to resist degradation by peptidases, enter the bloodstream, and stimulate the synthesis of cartilage extracellular matrix (Martínez-Puig et al., 2023).

Gelatin forms a rigid gel in aqueous environments, making it suitable as a protective coating for pharmaceuticals. Because of its biocompatibility and nonimmunogenic properties, gelatin also serves as a substrate for matrix metalloproteinases (MMPs). Furthermore, gelatin nanospheres have shown promise as effective gene delivery systems. In biomedical applications, gelatin, particularly in the form of electrospun nanofibers, is utilized as a wound healing biopolymer (Mohiti-Asli and Loboa, 2016).

Protein hydrolysates exhibit a range of biological activities, such as antioxidant, antihypertensive, antimicrobial, immunomodulatory, and anticancer effects, depending on their amino acid composition and sequence (Ishak and Sarbon, 2018). Hydroxyapatite, a calcium-phosphate compound, is commonly utilized as a bone graft material in both medical and dental applications (Wangkheirakpam et al., 2019).

Omega-3 fatty acids possess diverse biological and pharmacological properties, including anti-inflammatory, anticancer, antiapoptotic, antioxidant, and cardio-, hepato-, and neuroprotective effects in the context of chronic diseases (Lafuente et al., 2021). Research indicates that long-chain omega-3 fatty acids play a crucial role in neuroprotection, neurotransmission, brain development, and neuromodulation (Sambra et al., 2021). Animal studies have shown that a deficiency of DHA in the brain and retina can lead to impaired learning and visual function (Lozada et al., 2017).

Marine polyphenols help reduce oxidative stress and inflammation, thereby lowering the risk of cardiovascular diseases. They also exhibit antidiabetic properties by regulating the glycemic index and enhancing insulin sensitivity (Rathod et al., 2023). In addition, marine polyphenols inhibit tumor cell proliferation, highlighting their significant therapeutic potential in cancer treatment (Dyshlovoy, 2021).

Chondroitin sulfate extracted from shark cartilage possesses antioxidant, antiatherosclerotic, antithrombotic, and anticoagulant properties. It is widely used in the management of osteoarthritis (Nurunnabi et al., 2017).

18.4.4 Cosmeceuticals and skin care products

Exposure to ultraviolet A (UVA) radiation can lead to the breakdown of dermal collagen cross-links because of the overproduction of reactive oxygen species (ROS). This results in visible signs of skin aging, such as wrinkles and pigmentation. However, the use of antioxidants can help mitigate these effects. Collagen not only retains moisture but also exhibits antioxidant properties, contributing to skin hydration and firmness. Today, a wide variety of collagen-infused products, ranging from foods and beverages to cosmetics and health supplements, are commercially available. Fish-derived collagen has gained popularity over mammalian sources because of its lower molecular weight, which enhances absorption, and its safety from diseases such as bovine spongiform encephalopathy (BSE) (Kato et al., 2011). Collagen extracted from chum salmon and cod has demonstrated excellent moisturizing effects because of its strong moisture-retention capabilities (Alves et al., 2017).

Phospholipids are commonly used as stabilizing agents in cosmetic formulations (Van Hoogevest and Fahr, 2019). Marine collagen powder, often combined with vitamin C, is increasingly popular in skincare products aimed at preventing aging and enhancing skin firmness (Geahchan et al., 2022). Astaxanthin, known for its strong antioxidant properties, is widely incorporated into antiaging cosmeceuticals. It helps protect skin cells by neutralizing ROS, making it an effective skin protectant. Astaxanthin provides comprehensive protection against a range of UV radiation-induced skin issues, from immediate effects such as sunburn, pigmentation, and erythema to long-term conditions such as inflammation, photoaging, and skin cancer, positioning it as a promising natural sunscreen (Dutta et al., 2023).

18.4.5 Feed additives and aquafeed fortification

A significant portion of fish waste is utilized in the production of fish meal, which typically contains around 70% protein, 10% moisture, 10% fat, and 10% minerals (Geethanjali et al., 2020). Fish meal serves as a highly nutritious source of high-quality animal protein, offering enhanced digestibility, palatability, appealing flavor, growth-promoting benefits, and immune-boosting effects. It is rich in essential minerals such as calcium, phosphorus, and magnesium, along with trace elements such as copper, zinc, manganese, and molybdenum. Fish meal is the primary protein component in formulated feeds for marine finfish aquaculture, contributing to improved growth performance and feed efficiency in farmed fish.

Fish silage, the liquid counterpart of fish meal, is produced through the enzymatic breakdown of fish waste in the presence of acid (Raghunath and Gopakumar, 2002). It consists of predigested proteins, making it especially suitable for use in animal feeds. In aquaculture, fish silage is widely used in larval diets because of its hydrolyzed proteins that are easily absorbed by fish larvae (Maksimenko et al., 2024). In addition, fish silage-based feeds act as effective attractants. Calcium extracted from fish bones is commonly incorporated into poultry and duck feed, where it enhances eggshell hardness (Likittrakulwong et al., 2021).

Carotenoids are now recognized for their vitamin-like functions and are commonly added to feed to enhance the coloration of salmon flesh (Vílchez et al., 2011). As powerful natural biocolorants, carotenoids contribute to visually appealing hues that attract consumers. In aquaculture, fish oil is extensively used because of its high nutritional value and superior digestibility, making it one of the most important ingredients in feeds for farmed fish (Glencross et al., 2025).

Protein hydrolysates are widely used in animal feed formulations, particularly in fish feed, where they significantly improve growth performance, including specific growth rate and weight gain (Tejpal et al., 2021). Research also indicates that incorporating protein hydrolysates into fish diets enhances muscle properties such as chewiness and adhesiveness. Furthermore, protein hydrolysates contribute to improved immune responses and exhibit antioxidant, antimicrobial, and digestive enzyme-stimulating activities in fish (Espinales et al., 2023).

18.5 Economic and environmental perspectives

18.5.1 Economic feasibility of nutraceutical production from aquaculture waste

The valorization of aquaculture waste for nutraceutical production represents a promising approach toward achieving both economic and environmental sustainability. As concerns over seafood industry waste management intensify, converting byproducts into high-value nutraceuticals offers a viable means of enhancing profitability while reducing ecological impact (Ghaly et al., 2013; Kurnianto et al., 2024).

Because of their minimal or negligible acquisition cost, aquaculture byproducts present an economically attractive raw material for value-added applications. However, ensuring effective collection, proper handling, and appropriate storage is crucial to preserving the quality and preventing degradation. The seafood industry produces considerable waste, with byproducts constituting a substantial share of the total biomass. These discards are nutrient-rich and abundant in bioactive compounds, making them ideal candidates for nutraceutical extraction (Mutalipassi et al., 2021).

Various processing methods are employed for the recovery of nutraceuticals from aquaculture waste, including enzymatic hydrolysis, fermentation, and SFE (Venugopal, 2022). Among these, enzymatic hydrolysis is a preferred method for obtaining bioactive peptides and collagen. However, it demands precise control over enzyme selection, temperature, and reaction time, contributing to elevated production costs. Similarly, the extraction of omega-3 fatty acids from fish byproducts involves sophisticated techniques such as solvent extraction or supercritical CO_2 extraction, which require significant capital investment (Roy et al., 2022).

Recent advances in Industry 4.0 technologies, especially the integration of non-thermal techniques with enzymatic hydrolysis, have significantly improved extraction efficiency and product quality (Hassoun et al., 2023). These methods are gaining momentum because of their eco-friendly nature and potential for higher yields of functional compounds. In addition, emerging green extraction technologies, such as ultrasound-assisted and enzyme-assisted processes, are enhancing cost-effectiveness and driving sustainability in nutraceutical production (Kumar et al., 2023).

Production costs vary depending on geographic location and processing methods. For example, in Colombia, the cost of producing 1 kg of chitosan ranges between US$ 10.5 and 12, influenced by factors such as raw material supply, labor, and technological infrastructure (Gómez-Ríos et al., 2017). The global nutraceutical market continues to expand, fueled by rising consumer demand for natural health supplements. Marine-derived products, such as collagen, omega-3s, and chitosan, are widely used across sectors, including pharmaceuticals, cosmetics, and functional foods. Marine collagen, in particular, has gained prominence in anti-aging skincare and functional nutrition, with market value projected to reach US$ 2,320.6 million by 2033, up from US$ 1,076.8 million in 2023, growing at a compound annual growth rate (CAGR) of 8.0% (Fact.MR, 2023). Likewise, chitosan, extracted from shrimp and crab shells, finds applications in wound healing, agriculture, and water purification, further broadening its market appeal (Shahidi and Synowiecki, 1991).

Commercializing nutraceuticals from aquaculture waste requires adherence to stringent food safety and environmental regulations, including those set by agencies such as the Food and Drug Administration (FDA) and European Food Safety Authority (EFSA). Regulatory compliance involves addressing potential risks such as heavy metal contamination, microbial load, and allergenicity, all of which necessitate rigorous quality control systems that can elevate production expenses. Moreover, securing certifications for organic labeling and sustainable sourcing can boost consumer trust and marketability, though it adds further regulatory and operational costs.

Ultimately, the economic viability of nutraceutical production from aquaculture waste hinges on the integration of cost-efficient extraction methods, robust market demand, and compliance with regulatory frameworks. While the availability of abundant raw materials and a growing global market present substantial economic opportunities, high processing costs and regulatory barriers remain significant challenges. Strategic investments in sustainable technologies and collaborative industry efforts are essential to enhance profitability, maximize resource utilization, and ensure long-term success in the nutraceutical sector.

18.5.2 Role in reducing environmental pollution and promoting sustainability

While aquaculture has become a critical sector for ensuring global food security, its rapid growth has also raised significant environmental concerns, including water pollution, excessive feed waste, and disease outbreaks. However, aquaculture waste represents a valuable resource for nutraceutical production, supporting circular economy practices, sustainable resource management, and environmental conservation.

Functional feed additives derived from aquaculture byproducts play an essential role in improving fish health and feed efficiency, while decreasing dependence on chemical antibiotics. These biobased additives strengthen the immune systems of aquatic organisms, support gut health, and enhance nutrient absorption, ultimately improving growth performance. In addition, by optimizing feed utilization, they reduce feed wastage and contribute to more sustainable aquaculture practices (Van Doan et al., 2023).

Chitosan and its composites are extensively employed in environmental management applications such as wastewater treatment, soil remediation, sustainable agriculture, air quality control, and microplastic removal, mainly because of their excellent adsorption capabilities. Numerous studies have confirmed chitosan's efficiency in removing heavy metals from contaminated environments (Ngah et al., 2011; Hamza et al., 2021; Nayak et al., 2025; Patel et al., 2025). With antimicrobial resistance (AMR) emerging as a global threat (Ngasotter et al., 2022b; Ngasotter et al., 2025c), biochar–chitosan composites have shown potential for sustainably removing antibiotics from various systems (Ngo et al., 2023; Verma et al., 2025).

Aquaculture generates substantial waste, including uneaten feed, fish excreta, and processing byproducts. If unmanaged, these contribute to environmental issues such as water pollution, eutrophication, and habitat degradation (Liu et al., 2024). In the fishing industry, only around 25% of the fish's total biomass is typically used, while the remaining 75%, comprising skin, bones, and scales, can be a rich source of collagen (Srikanya et al., 2017). Disposing of this biomass contributes to pollution; however, utilizing it for nutraceutical extraction not only reduces environmental impact but also enables the sustainable recovery of high-value bioproducts.

Shellfish waste, in particular, is largely insoluble and highly resistant to natural biodegradation, posing potential ecological and health risks (Mutalipassi et al., 2021). Nonetheless, this waste contains approximately 30% chitin, which can be extracted and utilized instead of being discarded.

Although the environmental impact of aquaculture's rapid expansion remains a concern, its byproducts offer significant opportunities for the development of nutraceuticals, functional feed additives, and sustainable waste management solutions. Harnessing these resources not only reduces pollution but also reinforces environmental sustainability and promotes a more circular, resource-efficient aquaculture industry.

18.5.3 Contribution to a circular economy model and zero-waste goals

The extraction of nutraceuticals and bioactive compounds from aquaculture waste plays a crucial role in advancing the circular economy and supporting a zero-waste model. By maximizing the utilization of waste materials, this approach promotes sustainable resource management while generating high-value products that contribute to economic development. A prime example is the conversion of crustacean shells into valuable bioactives such as collagen, omega-3 fatty acids, chitosan, and astaxanthin. Similarly, fish processing byproducts, including scales, bones, and viscera, can be transformed into functional ingredients (Rossi et al., 2024). This valorization not only optimizes resource efficiency but also minimizes environmental impact, in line with circular economy principles. For instance, fish viscera, which are rich in proteins and enzymes, can be used to produce nutraceutical-grade enzymes (Villamil et al., 2017), while the remaining waste can be repurposed into animal feed or organic fertilizer. Chitosan extracted from shrimp shells is widely applied in the development of biodegradable antimicrobial plastics, offering a sustainable alternative to conventional food packaging materials (Ahmed et al., 2025). Integrating such extraction and conversion processes into the aquaculture supply chain transforms waste into valuable resources, thereby reducing ecological footprints and opening new economic opportunities.

Moreover, recovering high-value compounds from aquaculture waste reduces the reliance on virgin raw materials, helping to conserve natural resources. Advanced bioprocessing methods, such as enzymatic hydrolysis and microbial fermentation, enable efficient nutrient recovery and the production of functional ingredients suitable for use in food, cosmetics, and pharmaceuticals. The residual biomass, postextraction, can be utilized in multiple ways: As high-protein animal feed, reducing dependence on traditional feed sources, or as organic fertilizer to improve soil fertility and promote sustainable agriculture. By adopting such integrated practices, the aquaculture industry can establish a closed-loop system that minimizes waste generation and maximizes resource recovery, contributing to a more environmentally responsible and sustainable production model.

Hence, the extraction of nutraceuticals from aquaculture waste aligns seamlessly with the principles of the circular bioeconomy. It enhances resource efficiency, reduces environmental impact, and delivers economic benefits. Through sustainable production systems and circular strategies, this approach supports a zero-waste goal while catering to the growing global demand for functional foods and dietary supplements.

18.6 Case studies and success stories

18.6.1 Examples of commercially successful nutraceuticals from aquaculture waste

Aquaculture waste, once considered a disposal challenge, is now a valuable resource for extracting high-value nutraceuticals. Many companies have successfully transformed fish processing byproducts into commercially viable health supplements, contributing to both sustainability and economic growth (Table 18.1). Collagen and gelatin, derived from fish skin, scales, and bones, are widely used in skincare, joint health, and functional foods. Leading brands such as Vital Proteins, Naticol, and Peptan have commercialized marine collagen supplements. Glycosaminoglycans (GAGs), including chondroitin sulfate from fish cartilage and shrimp shells, support joint health and are found in products from Bioiberica and NOW. Omega-3 fatty acids from fish waste, known for their cardiovascular and cognitive health benefits, are marketed by Nordic Naturals, OmegaVia, and Wiley's Finest. Chitin and chitosan, extracted from shrimp and crab shells, are used in cholesterol-lowering and weight management supplements, with brands such as KitoZyme and LipoSan Ultra. Other notable nutraceuticals include fish peptides (Seagarden, Copalis), marine calcium (Aquamin), astaxanthin (AstaReal, BioAstin), and marine enzymes (Vitalzym). These successful products showcase how innovative technologies can convert low-value aquaculture waste into high-value, eco-friendly nutraceuticals, meeting growing consumer demand for sustainable health solutions.

TABLE 18.1 Examples of commercially successful nutraceuticals from aquatic waste.

Nutraceutical compound	Source (Aquatic waste)	Health benefits	Commercial applications	Notable brands
Collagen and gelatin	Fish skin, scales, bones	Joint health, skin elasticity, wound healing	Dietary supplements, functional foods, cosmetics	Vital Proteins, Naticol, Peptan
Glycosaminoglycans (GAGs) (e.g., Chondroitin sulfate)	Fish cartilage, shrimp/crab shells	Joint health, anti-inflammatory properties	Supplements for arthritis, pharmaceutical applications	Bioiberica, Pure Encapsulations, NOW
Omega-3 fatty acids (EPA and DHA)	Fish processing waste (heads, viscera, skin)	Cardiovascular health, brain function, and anti-inflammatory	Omega-3 capsules, fortified foods, infant formula	Nordic Naturals, OmegaVia, Wiley's Finest
Chitin and Chitosan	Shrimp and crab shells	Antimicrobial, cholesterol-lowering, weight management	Dietary supplements, wound healing products, and drug delivery systems	KitoZyme, Primex Iceland, LipoSan Ultra
Fish peptides and hydrolysates	Fish skin, bones, and muscle waste	Antioxidant, antihypertensive, and immune-boosting properties	Protein supplements, sports nutrition, functional foods	Seagarden, Nutripeptin, Copalis
Calcium supplements	Fish bones, crustacean shells	Bone health, osteoporosis prevention	Calcium-fortified food, dietary supplements	Marine Calcium (by Marine Bio-Resources), Aquamin
Astaxanthin	Shrimp shells, krill waste	Antioxidant, skin health, eye protection	Functional foods, nutraceuticals, and antiaging supplements	AstaReal, BioAstin, Algatech
Squalene and fish liver oils	Fish liver waste	Skin hydration, immune support, and cholesterol regulation	Skin care products, dietary supplements	Amway Nutrilite, Carlson Labs, Bluebonnet
Marine Enzymes (e.g., proteases and lipases)	Fish viscera, shrimp processing waste	Digestive health, metabolic enhancement	Digestive enzyme supplements, pharmaceutical applications	Vitalzym, Wobenzym, Enzymedica

18.7 Conclusion and future perspectives

The exploration of aquatic waste as a source of nutraceuticals represents a shift from conventional waste disposal to value creation, emphasizing resource efficiency and innovation. Advances in extraction and processing technologies have made it feasible to isolate high-purity bioactives from even low-value or complex waste matrices. However, maximizing the full potential of aquatic waste-derived nutraceuticals demands more than just technological advancements; it necessitates a supportive ecosystem encompassing strong policy backing, increased investment in research and development, and the establishment of stringent quality assurance frameworks. Future perspectives lie in integrating digital tools such as artificial intelligence (AI), machine learning, and blockchain for optimizing processing, tracing raw material origin, and ensuring product authenticity. Moreover, the combination of aquatic-derived compounds with other nutraceutical ingredients (e.g., plant polyphenols and probiotics) could lead to synergistic formulations with enhanced health benefits. Exploring novel delivery systems, such as nanoencapsulation, will improve the stability and bioavailability of these compounds in complex food and biological environments. There is also a growing need to assess life cycle impacts and sustainability metrics of nutraceutical production from aquatic waste, enabling better environmental benchmarking. As global interest in sustainable health solutions grows, the future of aquatic waste-derived nutraceuticals will depend on aligning scientific innovation with consumer needs, regulatory frameworks, and environmental ethics.

References

Abedi, E., Sahari, M.A., 2014. Long-chain polyunsaturated fatty acid sources and evaluation of their nutritional and functional properties. Food Sci. Nutr. 2 (5), 443–463. doi:10.1002/fsn3.121.

Abubakar, I.R., Maniruzzaman, K.M., Dano, U.L., AlShihri, F.S., AlShammari, M.S., Ahmed, S.M.S., Alrawaf, T.I., 2022. Environmental sustainability impacts of solid waste management practices in the global South. Int. J. Environ. Res. Public Health 19 (19), 12717. doi:10.3390/ijerph191912717.

Ahmed, H., Noyon, M.A.R., Uddin, M.E., Rafid, M.M., Hosen, M.S., Layek, R.K., 2025. Development and characterization of chitosan-based antimicrobial films: A sustainable alternative to plastic packaging. Clean. Chem. Eng. 11, 100157. doi:10.1016/j.clce.2025.100157.

Ahn, S.I., Cho, S., Choi, N.J., 2021. Effectiveness of chitosan as a dietary supplement in lowering cholesterol in murine models: a meta-analysis. Mar. Drugs 19 (1), 26. doi:10.3390/md19010026.

Ahuja, I., Dauksas, E., Remme, J.F., Richardsen, R., Løes, A.K., 2020. Fish and fish waste-based fertilizers in organic farming–with status in Norway: a review. Waste Manage. 115, 95–112. doi:10.1016/j.wasman.2020.07.025.

Al Hajj, W., Salla, M., Krayem, M., Khaled, S., Hassan, H.F., El Khatib, S, 2024. Hydrolyzed collagen: exploring its applications in the food and beverage industries and assessing its impact on human health–a comprehensive review. Heliyon 10 (16), e36433. doi:10.1016/j.heliyon.2024.e36433.

Alazaiza, M.Y., Mokaizh, A.A.B., Baarimah, A.O., Al-Zghoul, T., 2024. From agro-waste to bioactive wealth: analyzing nutraceutical extraction and applications. Case Stud. Chem. Environ. Eng. 11, 101066. doi:10.1016/j.cscee.2024.101066.

Alberts, B., Johnson, A., Lewis, J., Raff, M., Roberts, K., Walter, P., 2002. The extracellular matrix of animals. Molecular biology of the cell, 4th ed. Garland Science, New York.

Alves, A.L., Marques, A.L., Martins, E., Silva, T.H., Reis, R.L., 2017. Cosmetic potential of marine fish skin collagen. Cosmetics 4 (4), 39. doi:10.3390/cosmetics4040039.

Aranaz, I., Alcántara, A.R., Civera, M.C., Arias, C., Elorza, B., Heras Caballero, A., Acosta, N., 2021. Chitosan: an overview of its properties and applications. Polymers 13 (19), 3256. doi:10.3390/polym13193256.

Awuchi, C.G., Chukwu, C.N., Iyiola, A.O., Noreen, S., Morya, S., Adeleye, A.O., Twinomuhwezi, H., Leicht, K., Mitaki, N.B., Okpala, C.O.R., 2022. Bioactive compounds and therapeutics from fish: revisiting their suitability in functional foods to enhance human wellbeing. BioMed. Res. Int. 2022 (1), 3661866. doi:10.1155/2022/3661866.

Azelee, N.I.W., Dahiya, D., Ayothiraman, S., Noor, N.M., Abd Rasid, Z.I., Ramli, A.N.M., Selvasembian, R., 2023. Sustainable valorization approaches on crustacean wastes for the extraction of chitin, bioactive compounds and their applications-a review. Int. J. Biol. Macromol. 253, 126492. doi:10.1016/j.ijbiomac.2023.126492.

Bimbo, A.P., 2011. Production of marine oils. In: Omega-3 oils. AOCS Press, Urbana, IL, pp. 73–105. doi:10.1016/B978-1-893997-82-0.50007-4.

Brglez Mojzer, E., Knez Hrnčič, M., Škerget, M., Knez, Ž., Bren, U., 2016. Polyphenols: Extraction methods, antioxidative action, bioavailability and anticarcinogenic effects. Molecules 21 (7), 901. doi:10.3390/molecules21070901.

Cadar, E., Pesterau, A.M., Prasacu, I., Ionescu, A.M., Pascale, C., Dragan, A.M.L., Tomescu, C.L., 2024. Marine antioxidants from marine collagen and collagen peptides with nutraceuticals applications: a review. Antioxidants 13 (8), 919. doi:10.3390/antiox13080919.

Campos, L.D., Junior, V.D.A.S., Pimentel, J.D., Carregã, G.L.F., Cazarin, C.B.B., 2023. Collagen supplementation in skin and orthopedic diseases: a review of the literature. Heliyon 9 (4), e14961. doi:10.1016/j.heliyon.2023.e14961.

Casadidio, C., Peregrina, D.V., Gigliobianco, M.R., Deng, S., Censi, R., Di Martino, P., 2019. Chitin and chitosans: characteristics, eco-friendly processes, and applications in cosmetic science. Mar. Drugs 17 (6), 369. doi:10.3390/md17060369.

Chakraborty, K., Joseph, D., 2015. Production and characterization of refined oils obtained from Indian oil sardine (*Sardinella longiceps*). J. Agric. Food Chem. 63 (3), 998–1009. doi:10.1021/jf505127e.

Chattopadhyay, K., Xavier, K.M., Ngasotter, S., Karmakar, S., Balange, A., Nayak, B.B., 2023. Chitosan gel prepared with citric acid as the food acidulant: effect of the chitosan concentration and gel pH on physicochemical and functional properties of fish protein emulsion sausages. ACS Omega 8 (8), 7829–7837. doi:10.1021/acsomega.2c07538.

Chi, C.F., Wang, B., Hu, F.Y., Wang, Y.M., Zhang, B., Deng, S.G., Wu, C.W., 2015. Purification and identification of three novel antioxidant peptides from protein hydrolysate of bluefin leatherjacket (*Navodon septentrionalis*) skin. Food Res. Int. 73, 124–129. doi:10.1016/j.foodres.2014.08.038.

Coppola, D., Lauritano, C., Palma Esposito, F., Riccio, G., Rizzo, C., de Pascale, D., 2021. Fish waste: from problem to valuable resource. Mar. Drugs 19 (2), 116. doi:10.3390/md19020116.

Cunha, S.A., Pintado, M.E., 2022. Bioactive peptides derived from marine sources: biological and functional properties. Trends Food Sci. Technol. 119, 348–370. doi:10.1016/j.tifs.2021.08.017.

Das, P., Dutta, A., Panchali, T., Khatun, A., Kar, R., Das, T.K., Pradhan, S., 2024. Advances in therapeutic applications of fish oil: a review. Measurement: Food 13, 100142. doi:10.1016/j.meafoo.2024.100142.

Dave, J., Ali, A.M.M., Kumar, N., Nagarajan, M., Kieliszek, M., Bavisetty, S.C.B., 2024. Investigating the impact of wet rendering (solventless method) on PUFA-rich oil from catfish (*Clarias magur*) viscera. Open Life Sci. 19 (1), 20220903. doi:10.1515/biol-2022-0903.

Dayakar, B., Xavier, M., Ngasotter, S., Dhanabalan, V., Porayil, L., Balange, A.K., Nayak, B.B., 2024a. Extraction, optimization, and functional quality evaluation of carotenoproteins from shrimp processing side streams through enzymatic process. Environ. Sci. Pollut. Res. 31 (53), 62315–62328. doi:10.1007/s11356-023-30232-1.

Dayakar, B., Ngasotter, S., Layana, P., Balange, A.K., Nayak, B.B., Xavier, K.M., 2024b. Eco-friendly extraction, optimization, and characterization of carotenoprotein from shrimp waste biomass using a novel alkaline protease, Ecoenzyme (ECOENZYME–ALKP). Biomass Conv. Biorefin. 15, 12655–12671. doi:10.1007/s13399-024-06087-y.

Dayakar, B., Xavier, K.M., Ngasotter, S., Layana, P., Balange, A.K., Priyadarshini, B., Nayak, B.B., 2022. Characterization of spray-dried carotenoprotein powder from Pacific white shrimp (*Litopenaeus vannamei*) shells and head waste extracted using papain: antioxidant, spectroscopic, and microstructural properties. LWT 159, 113188. doi:10.1016/j.lwt.2022.113188.

Drioli, E., Giorno, L. (Eds.). (2010). Comprehensive membrane science and engineering (Vol. 1). Newnes. Oxford

Dutta, S., Kumar, S.J., Banerjee, R., 2023. A comprehensive review on astaxanthin sources, structure, biochemistry and applications in the cosmetic industry. Algal Res. 74, 103168. doi:10.1016/j.algal.2023.103168.

Dyshlovoy, S.A., 2021. Recent updates on marine cancer-preventive compounds. Mar. Drugs 19 (10), 558. doi:10.3390/md19100558.

Espinales, C., Romero-Peña, M., Calderón, G., Vergara, K., Cáceres, P.J., Castillo, P., 2023. Collagen, protein hydrolysates and chitin from byproducts of fish and shellfish: an overview. Heliyon 9 (4), e14937. doi:10.1016/j.heliyon.2023.e14937.

Estiasih, T., Ahmadi, K., Ali, D.Y., Nisa, F.C., Suseno, S.H., Lestari, L.A., 2021. Valorisation of viscera from fish processing for food industry utilizations. IOP Conf. Ser.: Earth Environ. Sci. 924 (1), 012024. doi:10.1088/1755-1315/924/1/012024.

Fact.MR, 2023. Marine collagen ingredient market outlook (2023 to 2033). Market research company, Rockville, MD. https://www.factmr.com/report/marine-collagen-ingredients-market.

Fassett, R.G., Coombes, J.S., 2011. Astaxanthin: a potential therapeutic agent in cardiovascular disease. Mar. Drugs 9 (3), 447–465. doi:10.3390/md9030447.

Fatmawati, S., Istiqomah, S.M., Hasanah, N., Helan, M.E.I.K., Santoso, M., Nugraheni, Z.V., Rachmawati, H., 2025. Physico-chemical characterization of natural nano calcium extracted from different fish bones in catfish (*Clarias gariepinus*) and snakehead fish (*Channa striata*). Case Stud. Chem. Environ. Eng. 11, 101080. doi:10.1016/j.cscee.2024.101080.

Fineberg, H., Johanson, A.G., 1967. Industrial use of fish oils, 278. US Bureau of Commercial Fisheries, Washington, DC.

Gaikwad, S., Kim, M.J., 2024. Fish byproduct collagen extraction using different methods and their application. Mar. Drugs 22 (2), 60. doi:10.3390/md22020060.

Geahchan, S., Baharlouei, P., Rahman, A., 2022. Marine collagen: a promising biomaterial for wound healing, skin anti-aging, and bone regeneration. Mar. Drugs 20 (1), 61. doi:10.3390/md20010061.

Geethanjali, S., Subash, A., Govindan, K., Pandiyan, M., Paramasivam, V., 2020. An overview of fish visceral waste pollution and its eco-friendly management practices. Int. J. Chem. Stud. 8 (5), 19–26.

Gelse, K., Pöschl, E., Aigner, T., 2003. Collagens—structure, function, and biosynthesis. Adv. Drug. Deliv. Rev. 55 (12), 1531–1546. doi:10.1016/j.addr.2003.08.002.

Ghaly, A.E., Ramakrishnan, V.V., Brooks, M.S., Budge, S.M., Dave, D., 2013. Fish processing wastes as a potential source of proteins. Amino acids and oils: a critical review. J. Microb. Biochem. Technol. 5 (4), 107–129. doi:10.4172/1948-5948.1000110.

Glencross, B.D., Bachis, E., Betancor, M.B., Calder, P., Liland, N., Newton, R., Ruyter, B., 2025. Omega-3 futures in aquaculture: exploring the supply and demands for long-chain omega-3 essential fatty acids by aquaculture species. Rev. Fish. Sci. Aquac. 33 (2), 167–216. doi:10.1080/23308249.2024.2388563.

Gómez-Ríos, D., Barrera-Zapata, R., Ríos-Estepa, R., 2017. Comparison of process technologies for chitosan production from shrimp shell waste: a techno-economic approach using Aspen Plus®. Food Bioprod. Process. 103, 49–57. doi:10.1016/j.fbp.2017.02.010.

Grand View Research. Nutraceuticals market size, share & trends analysis report by product (dietary supplements, functional food), by application (diabetes, eye health), by distribution channel, by region, and segment forecasts, 2025 –2030. Grand View Research.

Halldorsdottir, S.M., Sveinsdottir, H., Gudmundsdottir, A., Thorkelsson, G., Kristinsson, H.G., 2014. High quality fish protein hydrolysates prepared from byproduct material with Fucus vesiculosus extract. J. Funct. Foods 9, 10–17. doi:10.1016/j.jff.2014.04.009.

Hamza, M.F., Hamad, N.A., Hamad, D.M., Khalafalla, M.S., Abdel-Rahman, A.A.H., Zeid, I.F., Salem, W.M., 2021. Synthesis of eco-friendly biopolymer, alginate-chitosan composite to adsorb the heavy metals, Cd (II) and Pb (II) from contaminated effluents. Materials 14 (9), 2189. doi:10.3390/ma14092189.

Hassoun, A., Cropotova, J., Trollman, H., Jagtap, S., Garcia-Garcia, G., Parra-López, C., Bono, G., 2023. Use of industry 4.0 technologies to reduce and valorize seafood waste and byproducts: a narrative review on current knowledge. Curr. Res. Food Sci. 6, 100505. doi:10.1016/j.crfs.2023.100505.

Hernández-Ruiz, K.L., López-Cervantes, J., Sánchez-Machado, D.I., del Rosario Martínez-Macias, M., Correa-Murrieta, M.A., Sanches-Silva, A., 2022. Hydroxyapatite recovery from fish byproducts for biomedical applications. Sustain. Chem. Pharm. 28, 100726. doi:10.1016/j.scp.2022.100726.

Ilza, M., Diharmi, A., 2022. Processing and utilization of natural phospholipids from fish belly fat waste of jambal siam fish (*Pangasius hypophthalamus*). IOP Conference Series: Earth and Environmental Science, 1118. IOP Publishing, Bristol, UK.

Iñiguez-Moreno, M., Santiesteban-Romero, B., Melchor-Martínez, E.M., Parra-Saldívar, R., González-González, R.B., 2024. Valorization of fishery industry waste: Chitosan extraction and its application in the industry. MethodsX 13, 102892. doi:10.1016/j.mex.2024.102892.

Irastorza, A., Zarandona, I., Andonegi, M., Guerrero, P., de la Caba, K., 2021. The versatility of collagen and chitosan: from food to biomedical applications. Food Hydrocoll. 116, 106633. doi:10.1016/j.foodhyd.2021.106633.

Irianto, H.E., 2017. Enzymes in fermented fish. In: Advances in food and nutrition research, 80. Academic Press, London, pp. 199–216. doi:10.1016/bs.afnr.2016.10.004.

Ishak, N.H., Sarbon, N.M., 2018. A review of protein hydrolysates and bioactive peptides deriving from wastes generated by fish processing. Food Bioprocess Technol. 11 (1), 2–16. doi:10.1007/s11947-017-1940-1.

Ivanova, D.G., Yaneva, Z.L., 2020. Antioxidant properties and redox-modulating activity of chitosan and its derivatives: biomaterials with application in cancer therapy. BioRes. Open Access 9 (1), 64–72. doi:10.1089/biores.2019.0028.

Izadi, H., Asadi, H., Bemani, M., 2025. Chitin: a comparison between its main sources. Front. Mater. 12, 1537067. doi:10.3389/fmats.2025.1537067.

Jafari, H., Lista, A., Siekapen, M.M., Ghaffari-Bohlouli, P., Nie, L., Alimoradi, H., Shavandi, A., 2020. Fish collagen: extraction, characterization, and applications for biomaterials engineering. Polymers 12 (10), 2230. doi:10.3390/polym12102230.

Jena, A., Sivaraman, B., Ganesan, P., Shalini, R., Renuka, V., Arisekar, U., 2025. Extraction and antioxidative, antihypertensive, and antidiabetic properties of gelatin hydrolysates from Lethrinid fish scales. J. Food Process. Preserv. 2025 (1), 5577122. doi:10.1155/jfpp/5577122.

Jo, S., Hwangbo, H., Francis, N., Lee, J., Pei, M., Kim, G., 2025. Fish-derived biomaterials for tissue engineering: advances in scaffold fabrication and applications in regenerative medicine and cancer therapy. Theranostics 15 (12), 5666. doi:10.7150/thno.109186.

Joy, J.M., Padmaprakashan, A., Pradeep, A., Paul, P.T., Mannuthy, R.J., Mathew, S., 2024. A review on fish skin-derived gelatin: elucidating the gelatin peptides—preparation, bioactivity, mechanistic insights, and strategies for stability improvement. Foods 13 (17), 2793. doi:10.3390/foods13172793.

Kammler, S., Romero, A.M., Burkhardt, C., Baruth, L., Antranikian, G., Liese, A., Kaltschmitt, M., 2024. Macroalgae valorization for the production of polymers, chemicals, and energy. Biomass Bioenergy 183, 107105. doi:10.1016/j.biombioe.2024.107105.

Kariduraganavar, M.Y., Kittur, A.A., Kamble, R.R., 2014. Polymer synthesis and processing. In: Natural and synthetic biomedical polymers. Elsevier, pp. 12–31. doi:10.1016/B978-0-12-396983-5.00001-6.

Kato, S., Matsui, H., Saitoh, Y., Miwa, N., 2011. Fish collagen-containing drink is subcutaneously absorbed and attenuates the UVA-induced tissue-integrity destruction and DNA damages in 3D-human skin tissue model. J. Funct. Foods 3 (1), 50–55. doi:10.1016/j.jff.2010.11.005.

Khiari, Z., 2024. Enzymes from fishery and aquaculture waste: research trends in the era of artificial intelligence and circular bio-economy. Mar. Drugs 22 (9), 411. doi:10.3390/md22090411.

Khin, M.N., Ahammed, S., Kamal, M.M., Saqib, M.N., Liu, F., Zhong, F., 2024. Investigating next-generation edible packaging: protein-based films and coatings for delivering active compounds. Food Hydrocoll. Health 6, 100182. doi:10.1016/j.fhfh.2024.100182.

Komai, T., Kawabata, C., Tojo, H., Gocho, S., Ichishima, E., 2007. Purification of serine carboxypeptidase from the hepatopancreas of Japanese common squid Todarodes pacificus and its application for elimination of bitterness from bitter peptides. Fish. Sci. 73 (2), 404–411. doi:10.1111/j.1444-2906.2007.01348.x.

Kumar, L.R., Chatterjee, N.S., Mathew, S., 2023. Green technologies for developing high-value novel marine nutraceuticals: scope and innovations. In: Advances in fish processing technologies: preservation, waste utilization, and safety assurance. Apple Academic Press, New York, pp. 271–286.

Kurnianto, M.A., Defri, I., Syahbanu, F., Aulia, S.S., 2024. Fish-derived bioactive peptide: bioactivity potency, structural characteristics, and conventional and bioinformatics approaches for identification. Future Foods 9, 100386. doi:10.1016/j.fufo.2024.100386.

Lafuente, M., Rodríguez González-Herrero, M.E., Romeo Villadóniga, S., Domingo, J.C., 2021. Antioxidant activity and neuroprotective role of docosahexaenoic acid (DHA) supplementation in eye diseases that can lead to blindness: a narrative review. Antioxidants 10 (3), 386. doi:10.3390/antiox10030386.

Leonard, A.R., Cumming, M.H., Ali, M.A., Cabral, J.D., 2024. Fish collagen cross-linking strategies to improve mechanical and bioactive capabilities for tissue engineering and regenerative medicine. Adv. Funct. Mater. 34 (45), 2405335. doi:10.1002/adfm.202405335.

León-López, A., Morales-Peñaloza, A., Martínez-Juárez, V.M., Vargas-Torres, A., Zeugolis, D.I., Aguirre-Álvarez, G., 2019. Hydrolyzed collagen—sources and applications. Molecules 24 (22), 4031. doi:10.3390/molecules24224031.

Liao, Y.C., Chang, C.C., Nagarajan, D., Chen, C.Y., Chang, J.S., 2021. Algae-derived hydrocolloids in foods: applications and health-related issues. Bioengineered 12 (1), 3787–3801. doi:10.1080/21655979.2021.1946359.

Likittrakulwong, W., Moonsatan, S., Incharoen, T., 2021. Enhancement of tibia bone and eggshell hardness through the supplementation of bio-calcium derived from fish bone mixed with chelated trace minerals and vitamin D3 in laying duck diet. Vet. Anim. Sci. 14, 100204. doi:10.1016/j.vas.2021.100204.

Lim, Y.S., Ok, Y.J., Hwang, S.Y., Kwak, J.Y., Yoon, S., 2019. Marine collagen as a promising biomaterial for biomedical applications. Mar. Drugs 17 (8), 467. doi:10.3390/md18040214.

Liu, X., Wang, Y., Liu, H., Zhang, Y., Zhou, Q., Wen, X., Zhang, Z., 2024. A systematic review on aquaculture wastewater: pollutants, impacts, and treatment technology. Environ. Res. 262, 119793. doi:10.1016/j.envres.2024.119793.

López, A.M.Q., Santos, F.A.R.D., Martins, E.S., Silva, A.L.D.S., Santos, E.C.L.D., 2020. Pink and white shrimps from the Brazilian coast: pigment identification, antioxidant activity and microbial quality under different freezing-times. Food Sci. Technol. 41, 447–457.

Lozada, L.E., Desai, A., Kevala, K., Lee, J.W., Kim, H.Y., 2017. Perinatal brain docosahexaenoic acid concentration has a lasting impact on cognition in mice. J. Nutr. 147 (9), 1795–1805. doi:10.3945/jn.117.249953.

Luo, X., Wu, Y., Wang, S., Xiong, Z., Du, X., Zheng, M., Chen, Z., 2024. Frontier review of key reduction technologies and resource utilization of waste during the seaweed gel production process. Blue Biotechnol. 1 (1), 12. doi:10.1186/s44315-024-00013-7.

Maksimenko, A., Belyi, L., Podvolotskaya, A., Son, O., Tekutyeva, L., 2024. Exploring sustainable aquafeed alternatives with a specific focus on the ensilaging technology of fish waste. Fermentation 10 (5), 258. doi:10.3390/fermentation10050258.

Martínez-Puig, D., Costa-Larrión, E., Rubio-Rodríguez, N., Gálvez-Martín, P., 2023. Collagen supplementation for joint health: The link between composition and scientific knowledge. Nutrients 15 (6), 1332. doi:10.3390/nu15061332.

Mawazi, S.M., Kumar, M., Ahmad, N., Ge, Y., Mahmood, S., 2024. Recent applications of chitosan and its derivatives in antibacterial, anticancer, wound healing, and tissue engineering fields. Polymers 16 (10), 1351. doi:10.3390/polym16101351.

Mohiti-Asli, M., Loboa, E.G., 2016. Nanofibrous smart bandages for wound care. In: Wound healing biomaterials. Woodhead Publishing, Cambridge, pp. 483–499. doi:10.1016/B978-1-78242-456-7.00023-4.

Mondal, S., Hoang, G., Manivasagan, P., Moorthy, M.S., Kim, H.H., Phan, T.T.V., Oh, J., 2019. Comparative characterization of biogenic and chemical synthesized hydroxyapatite biomaterials for potential biomedical application. Mater. Chem. Phys. 228, 344–356. doi:10.1016/j.matchemphys.2019.02.021.

Montero, P., Calvo, M.M., Gómez-Guillén, M.C., Gómez-Estaca, J., 2016. Microcapsules containing astaxanthin from shrimp waste as potential food coloring and functional ingredient: characterization, stability, and bioaccessibility. LWT 70, 229–236. doi:10.1016/j.lwt.2016.02.040.

Morón-Ortiz, Á., Mapelli-Brahm, P., Meléndez-Martínez, A.J., 2024. Sustainable green extraction of carotenoid pigments: innovative technologies and bio-based solvents. Antioxidants 13 (2), 239. doi:10.3390/antiox13020239.

Mungwari, C.P., King'ondu, C.K., Sigauke, P., Obadele, B.A., 2025. Conventional and modern techniques for bioactive compounds recovery from plants. Sci. Afr. 27, e02509. doi:10.1016/j.sciaf.2024.e02509.

Mutalipassi, M., Esposito, R., Ruocco, N., Viel, T., Costantini, M., Zupo, V., 2021. Bioactive compounds of nutraceutical value from fishery and aquaculture discards. Foods 10 (7), 1495. doi:10.3390/foods10071495.

Narayanaswamy, R., Kanagesan, S., Pandurangan, A., Padmanabhan, P., 2016. Basics to different imaging techniques, different nanobiomaterials for image enhancement. In: Nanobiomaterials in medical imaging. William Andrew Publishing, Norwich, NY, pp. 101–129.

Nayak, M.A., Tripathi, A., Raj, A., Gauba, P., Bhatt, E., 2025. Chitosan: a novel approach and sustainable way to remove contaminants and treat wastewater. Starch-Stärke 77 (4), e202400171. doi:10.1002/star.202400171.

Nemati, M., Shahosseini, S.R., Ariaii, P., 2024. Review of fish protein hydrolysates: production methods, antioxidant and antimicrobial activity and nanoencapsulation. Food Sci. Biotechnol. 33 (8), 1789–1803. doi:10.1007/s10068-024-01554-8.

Ngah, W.W., Teong, L.C., Hanafiah, M.M., 2011. Adsorption of dyes and heavy metal ions by chitosan composites: a review. Carbohydr. Polym. 83 (4), 1446–1456. doi:10.1016/j.carbpol.2010.11.004.

Ngasotter, S., Meitei, M.M., Xavier, K.M., Sharma, S., Singh, S.K., Jakhar, J.K., Ninan, G., 2025a. Global trends and perspectives in nanochitin research: a comprehensive review of types, properties, applications, and scientometric analysis. Int. J. Biol. Macromol. 315, 144438. doi:10.1016/j.ijbiomac.2025.144438.

Ngasotter, S., Xavier, K.M., Sagarnaik, C., Sasikala, R., Mohan, C.O., Jaganath, B., Ninan, G., 2025b. Evaluating the reinforcing potential of steam-exploded chitin nanocrystals in chitosan-based biodegradable nanocomposite films for food packaging applications. Carbohydr. Polym. 348, 122841. doi:10.1016/j.carbpol.2024.122841.

Ngasotter, S., Xavier, K.M., Debbarma, S., Meinam, M., Madhulika, T., Muthukumar, A., Yirang, Y., 2025c. Prevalence and antimicrobial resistance of seafood-borne bacterial pathogens. In: Antibiotic residue and resistance in seafood safety and quality. Springer Nature Singapore, Singapore, pp. 41–72. doi:10.1007/978-981-96-4640-1_3.

Ngasotter, S., Sampath, L., Xavier, K.M., 2022a. Nanochitin: an update review on advances in preparation methods and food applications. Carbohydr. Polym. 291, 119627. doi:10.1016/j.carbpol.2022.119627.

Ngasotter, S., Mukherjee, S., Singh, S.K., Bharti, D., Haque, R., Varshney, S., Singh, A.S., 2022b. Prevalence, virulence, and antibiotic resistance of *Vibrio parahaemolyticus* from seafood and its environment: an updated review. Mediterranean J. Infect. Microb. Antimicrob. 11 (1), 1. doi:10.4274/mjima.galenos.2021.2021.1.

Ngasotter, S., Xavier, K.M., Meitei, M.M., Waikhom, D., Pathak, J., Singh, S.K., 2023. Crustacean shell waste derived chitin and chitin nanomaterials for application in agriculture, food, and health–a review. Carbohydr. Polym. Technol. Appl. 6, 100349. doi:10.1016/j.carpta.2023.100349.

Ngasotter, S., Xavier, K.M., Porayil, L., Balange, A.K., Nayak, B.B., Ninan, G., 2024. Facile green production of chitin nanomaterials from shrimp shell chitin using recyclable maleic acid and microwave irradiation. ACS Sustainable Chem. Eng. 12 (47), 17222–17235. doi:10.1021/acssuschemeng.4c06234.

Ngo, D.H., Vo, T.S., Ngo, D.N., Kang, K.H., Je, J.Y., Pham, H.N.D., Kim, S.K., 2015. Biological effects of chitosan and its derivatives. Food Hydrocoll. 51, 200–216. doi:10.1016/j.foodhyd.2015.05.023.

Ngo, H.H., Guo, W., Nguyen, T.H., Luong, T.M.L., Nguyen, X.H., Phan, T.L.A., Nguyen, M.K., 2023. New chitosan-biochar composite derived from agricultural waste for removing sulfamethoxazole antibiotics in water. Bioresour. Technol. 385, 129384. doi:10.1016/j.biortech.2023.129384.

Nurunnabi, M., Revuri, V., Huh, K.M., Lee, Y.K., 2017. Polysaccharide based nano/microformulation: an effective and versatile oral drug delivery system. In: Nanostructures for oral medicine. Elsevier, Oxford, pp. 409–433. doi:10.1016/B978-0-323-47720-8.00015-8.

Ozogul, F., Cagalj, M., Šimat, V., Ozogul, Y., Tkaczewska, J., Hassoun, A., Phadke, G.G., 2021. Recent developments in valorisation of bioactive ingredients in discard/seafood processing byproducts. Trends Food Sci. Technol. 116, 559–582. doi:10.1016/j.tifs.2021.08.007.

Patel, P.K., Pandey, L.M., Uppaluri, R.V., 2025. Conceptual techno-economic analysis of chitosan derivative resin and commercial sorbents for multi-heavy metal removal. Sep. Purif. Technol. 362, 131732. doi:10.1016/j.seppur.2025.131732.

Pereira, L., Cotas, J., 2023. Therapeutic potential of polyphenols and other micronutrients of marine origin. Mar. Drugs 21 (6), 323. doi:10.3390/md21060323.

Pérez, A., Ruz, M., García, P., Jiménez, P., Valencia, P., Ramírez, C., Almonacid, S., 2024. Nutritional properties of fish bones: potential applications in the food industry. Food Rev. Int. 40 (1), 79–91. doi:10.1080/87559129.2022.2153136.

Puri, V., Nagpal, M., Singh, I., Singh, M., Dhingra, G.A., Huanbutta, K., Dheer, D., Sharma, A., Sangnim, T., 2022. A comprehensive review on nutraceuticals: therapy support and formulation challenges. Nutrients 14 (21), 4637. doi:10.3390/nu14214637.

Qin, D., Bi, S., You, X., Wang, M., Cong, X., Yuan, C., Chen, X.G., 2022. Development and application of fish scale wastes as versatile natural biomaterials. Chem. Eng. J. 428, 131102. doi:10.1016/j.cej.2021.131102.

Raghunath, M.R., Gopakumar, K., 2002. Trends in production and utilization of fish silage. J. Food Sci. Technol. 39 (2), 103–110.

Rajput, V., Naik, B., Kumar, V., Kumar, V., Bhatt, S.C., Rustagi, S., 2024. Advancements in energy storage applications: harnessing the potential of fish industry waste. Discover Mater. 4 (1), 95. doi:10.1007/s43939-024-00161-y.

Rakhmanova, A., Khan, Z.A., Sharif, R., Lv, X., 2018. Meeting the requirements of halal gelatin: a mini review. MOJ Food Process. Technol. 6 (6), 477–482. doi:10.15406/mojfpt.2018.06.00209.

Raman, M., Gopakumar, K., 2018. Fish collagen and its applications in food and pharmaceutical industry: a review. EC Nutrition 13 (12), 752–767.

Rather, J.A., Akhter, N., Ashraf, Q.S., Mir, S.A., Makroo, H.A., Majid, D., Dar, B.N., 2022. A comprehensive review on gelatin: understanding impact of the sources, extraction methods, and modifications on potential packaging applications. Food Packag. Shelf Life 34, 100945. doi:10.1016/j.fpsl.2022.100945.

Rathod, N.B., Elabed, N., Punia, S., Ozogul, F., Kim, S.K., Rocha, J.M., 2023. Recent developments in polyphenol applications on human health: a review with current knowledge. Plants 12 (6), 1217. doi:10.3390/plants12061217.

Razzaque, M.S., Wimalawansa, S.J., 2025. Minerals and human health: from deficiency to toxicity. Nutrients 17 (3), 454. doi:10.3390/nu17030454.

Rodrigues, M., Rosa, A., Almeida, A., Martins, R., Ribeiro, T., Pintado, M., Caleja, C., 2024. Omega-3 fatty acids from fish byproducts: innovative extraction and application in food and feed. Food Bioprod. Process. 145, 32–41. doi:10.1016/j.fbp.2024.02.007.

Rossi, N., Grosso, C., Delerue-Matos, C., 2024. Shrimp waste upcycling: unveiling the potential of polysaccharides, proteins, carotenoids, and fatty acids with emphasis on extraction techniques and bioactive properties. Mar. Drugs 22 (4), 153. doi:10.3390/md22040153.

Roy, V.C., Park, J.S., Ho, T.C., Chun, B.S., 2022. Lipid indexes and quality evaluation of omega-3 rich oil from the waste of Japanese Spanish mackerel extracted by supercritical CO_2. Mar. Drugs 20 (1), 70. doi:10.3390/md20010070.

Sachindra, N.M., Bhaskar, N., Mahendrakar, N.S., 2006. Recovery of carotenoids from shrimp waste in organic solvents. Waste Manage. 26 (10), 1092–1098. doi:10.1016/j.wasman.2005.07.002.

Saetang, J., Issuriya, A., Suyapoh, W., Sornying, P., Nilsuwan, K., Benjakul, S., 2024. Bio-calcium from skipjack tuna frame attenuates bone loss in ovariectomy-induced osteoporosis rats. Mar. Drugs 22 (10), 472. doi:10.3390/md22100472.

Sambra, V., Echeverria, F., Valenzuela, A., Chouinard-Watkins, R., Valenzuela, R., 2021. Docosahexaenoic and arachidonic acids as neuroprotective nutrients throughout the life cycle. Nutrients 13 (3), 986. doi:10.3390/nu13030986.

Santos, V.P., Marques, N.S., Maia, P.C., Lima, M.A.B.D., Franco, L.D.O., Campos-Takaki, G.M.D., 2020. Seafood waste as attractive source of chitin and chitosan production and their applications. Int. J. Mol. Sci. 21 (12), 4290. doi:10.3390/ijms21124290.

Scarcelli, E., Catalano, A., Iacopetta, D., Ceramella, J., Sinicropi, M.S., Aiello, F., 2025. Chitin, chitosan and its derivatives: antimicrobials and/or mitigators of water. Macromolecules 5 (2), 15. doi:10.3390/macromol5020015.

Shagdarova, B., Konovalova, M., Varlamov, V., Svirshchevskaya, E., 2023. Anti-obesity effects of chitosan and its derivatives. Polymers 15 (19), 3967. doi:10.3390/polym15193967.

Shahbaz, M., Rashid, N., Saleem, J., Mackey, H., McKay, G., Al-Ansari, T., 2023. A review of waste management approaches to maximise sustainable value of waste from the oil and gas industry and potential for the State of Qatar. Fuel 332, 126220. doi:10.1016/j.fuel.2022.126220.

Shahidi, F., Synowiecki, J., 1991. Isolation and characterization of nutrients and value-added products from snow crab (*Chionoecetes opilio*) and shrimp (*Pandalus borealis*) processing discards. J. Agric. Food Chem. 39 (8), 1527–1532. doi:10.1021/jf00008a032.

Sharma, Y.C, Singh, B., Madhu, D., Liu, Y., Yaakob, Z., 2014. Fast synthesis of high quality biodiesel from 'waste fish oil' by single step transesterification. Biofuel Res. J. 1 (3), 78–80. doi:10.18331/BRJ2015.1.3.3.

Srikanya, A., Dhanapal, K., Sravani, K., Madhavi, K., Kumar, G.P., 2017. A study on optimization of fish protein hydrolysate preparation by enzymatic hydrolysis from tilapia fish waste mince. Int. J. Curr. Microbiol. Appl. Sci. 6 (12), 3220–3229. doi:10.20546/ijcmas.2017.612.375.

Srivastava, R.R., Rajak, D.K., Ilyas, S., Kim, H., Pathak, P., 2022. Challenges, regulations, and case studies on sustainable management of industrial waste. Minerals 13 (1), 51. doi:10.3390/min13010051.

Supriadi, D., Tiana Milanda, A.Y.C., Muctaridi, M.A., 2023. Calcium carbonate isolation from eggshell to meet pharmacopoeial standards and its effectiveness as an antacids. Int. J. Adv. Pharm. 15 (5), 204–209.

Suryani, S., Chaerunisaa, A.Y., Joni, I.M., Ruslin, R., Aspadiah, V., Anton, A., Ramadhan, L.O.A.N., 2024. The chemical modification to improve solubility of chitosan and its derivatives application, preparation method, toxicity as a nanoparticles. Nanotechnol. Sci. Appl. 17, 41–57. doi:10.2147/NSA.S450026.

Swanson, D., Block, R., Mousa, S.A., 2012. Omega-3 fatty acids EPA and DHA: health benefits throughout life. Adv. Nutr. 3 (1), 1–7. doi:10.3945/an.111.000893.

Tang, C., Zhou, K., Zhu, Y., Zhang, W., Xie, Y., Wang, Z., Xu, B., 2022. Collagen and its derivatives: from structure and properties to their applications in food industry. Food Hydrocoll. 131, 107748. doi:10.1016/j.foodhyd.2022.107748.

Tejpal, C.S., Vijayagopal, P., Elavarasan, K., Prabu, D.L., Lekshmi, R.G.K., Anandan, R., Ravishankar, C.N., 2021. Evaluation of pepsin derived tilapia fish waste protein hydrolysate as a feed ingredient for silver pompano (*Trachinotus blochii*) fingerlings: Influence on growth, metabolism, immune and disease resistance. Anim. Feed Sci. Technol. 272, 114748. doi:10.1016/j.anifeedsci.2020.114748.

Topić Popović, N., Lorencin, V., Strunjak-Perović, I., Čož-Rakovac, R., 2023. Shell waste management and utilization: mitigating organic pollution and enhancing sustainability. Appl. Sci. 13 (1), 623. doi:10.3390/app13010623.

Torres, B., Pérez, A., García, P., Jiménez, P., Abrigo, K., Valencia, P., Ruz, M., 2024. Fish bones as calcium source: bioavailability of micro and nano particles. Foods 13 (12), 1840. doi:10.3390/foods13121840.

Trigueros, E., Oliveira, A.P., Andrade, P.B., Videira, R.A., de Pinho, P.G., Sanz, M.T., Beltrán, S., 2024. Exploring the bioactive potential of algae residue extract via subcritical water extraction: insights into chemical composition and biological activity. Food Chem. 458, 140310. doi:10.1016/j.foodchem.2024.140310.

Van Doan, H., Prakash, P., Hoseinifar, S.H., Ringø, E., El-Haroun, E., Faggio, C., Dawood, M.A., 2023. Marine-derived products as functional feed additives in aquaculture: a review. Aquac. Rep. 31, 101679. doi:10.1016/j.aqrep.2023.101679.

Van Hoogevest, P., Fahr, A., 2019. Phospholipids in cosmetic carriers. In: Nanocosmetics: from ideas to products. Springer International Publishing, Cham, pp. 95–140. doi:10.1007/978-3-030-16573-4_6.

Venugopal, V., 2016. Enzymes from seafood processing waste and their applications in seafood processing. Adv. Food. Nutr. Res. 78, 47–69. doi:10.1016/bs.afnr.2016.06.004.

Venugopal, V., 2021. Valorization of seafood processing discards: bioconversion and bio-refinery approaches. Front. Sustain. Food Syst. 5, 611835. doi:10.3389/fsufs.2021.611835.

Venugopal, V., 2022. Green processing of seafood waste biomass towards blue economy. Curr. Res. Environ. Sustain. 4, 100164. doi:10.1016/j.crsust.2022.100164.

Venugopal, V., Sasidharan, A., 2022. Functional proteins through green refining of seafood side streams. Front. Nutr. 9, 974447. doi:10.3389/fnut.2022.974447.

Verma, A., Sharma, G., Wang, T., Kumar, A., Dhiman, P., Verma, Y., García-Penas, A., 2025. Graphene oxide/chitosan hydrogels for removal of antibiotics. Environ. Technol. 1–31. doi:10.1080/09593330.2025.2464267.

Vignesh, A., Amal, T.C., Sarvalingam, A., Vasanth, K., 2024. A review on the influence of nutraceuticals and functional foods on health. Food Chem. Adv. 5, 100749. doi:10.1016/j.focha.2024.100749.

Vílchez, C., Forján, E., Cuaresma, M., Bédmar, F., Garbayo, I., Vega, J.M., 2011. Marine carotenoids: biological functions and commercial applications. Mar. Drugs 9 (3), 319–333. doi:10.3390/md9030319.

Villamil, O., Váquiro, H., Solanilla, J.F., 2017. Fish viscera protein hydrolysates: production, potential applications and functional and bioactive properties. Food Chem. 224, 160–171. doi:10.1016/j.foodchem.2016.12.057.

Wangkheirakpam, M.R., Mahanand, S.S., Majumdar, R.K., Sharma, S., Hidangmayum, D.D., Netam, S., 2019. Fish waste utilization with reference to fish protein hydrolysate - a review. Fish. Technol. 56 (3), 169–178.

Wood, A.H., Chappell, H.F., Zulyniak, M.A., 2022. Dietary and supplemental long-chain omega-3 fatty acids as moderators of cognitive impairment and Alzheimer's disease. Eur. J. Nutr. 61 (2), 589–604. doi:10.1007/s00394-021-02655-4.

Xia, F.L.W., Supri, S., Djamaludin, H., Nurdiani, R., Seng, L.L., Yin, K.W., Rovina, K., 2024. Turning waste into value: extraction and effective valorization strategies of seafood byproducts. Waste Manag. Bull. 2 (3), 84–100. doi:10.1016/j.wmb.2024.06.008.

Xu, Y., Ye, J., Zhou, D., Su, L., 2020. Research progress on applications of calcium derived from marine organisms. Sci. Rep. 10 (1), 18425. doi:10.1038/s41598-020-75575-8.

Yan, D., Li, Y., Liu, Y., Li, N., Zhang, X., Yan, C., 2021. Antimicrobial properties of chitosan and chitosan derivatives in the treatment of enteric infections. Molecules 26 (23), 7136. doi:10.3390/molecules26237136.

Yaqoob, Z., Arshad, M.S., Imran, M., Munir, H., Qaisrani, T.B., Khalid, W., Suleria, H.A.R., 2022. Mechanistic role of astaxanthin derived from shrimp against certain metabolic disorders. Food Sci. Nutr. 10 (1), 12–20. doi:10.1002/fsn3.2623.

Yun, J.H., Archer, S.D., Price, N.N., 2023. Valorization of waste materials from seaweed industry: an industry survey based biorefinery approach. Rev. Aquac. 15 (3), 1020. doi:10.1111/raq.12748.

Yusoff, M.A., Mohammadi, P., Ahmad, F., Sanusi, N.A., Hosseinzadeh-Bandbafha, H., Vatanparast, H., Aghbashlo, M., Tabatabaei, M., 2024. Valorization of seafood waste: a review of life cycle assessment studies in biorefinery applications. Sci. Total Environ. 952, 175810. doi:10.1016/j.scitotenv.2024.175810.

Zhang, Y., Chang, S.K., 2023. Microbial transglutaminase cross-linking enhances the textural and rheological properties of the surimi-like gels made from alkali-extracted protein isolate from catfish byproducts and the role of disulfide bonds in gelling. Foods 12 (10), 2029. doi:10.3390/foods12102029.

CHAPTER

19

Zero-waste valorization of fibre crops: Towards a circular bioeconomy

Kanti Meena, A.K. Lavanya, Pandu Ranganath Diveti and Sonali Paul Mazumdar

ICAR-Central Research Institute for Jute and Allied Fibres, Kolkata, West Bengal, India

19.1 Introduction

"Zero waste from crops" is a way of farming in which the farmer tries to cut down on or get rid of all waste from their crops. Using all parts of the plant, composting organic materials back into the soil, and making the best use of resources are all ways to do this. The goal is to grow food with as little harm to the environment as possible by seeing crop waste as a useful resource instead of something to throw away. The zero-waste idea is a whole-system strategy that aims to cut down on the amount of trash we make every day by focusing on protecting resources, utilizing them responsibly, and making sure the ecosystem stays healthy for a long time. Zero-waste supporters do not only want to get rid of trash; they want to rethink how things are made, utilized, and thrown away to get rid of trash completely.

The zero-waste ideology is based on a few main ideas that assist people, businesses, and communities make less trash. The first step is to say no to things and packaging that are not required. You may do this by being careful about what you buy and staying away from things that have too many nonrecyclable materials. This helps the environment by stopping waste at the source. Next, it's important to cut back on what you buy by only getting what you need, picking products that will last, and not buying things on a whim. This helps lower the need for extracting resources and making things. Reusing things is a very important part of living with little waste. Instead of throwing things away after one use, supporters say you should fix, repurpose, and upcycle them. You may use jars, bags, and containers more than once, which cuts down on waste and saves you money. Recycling comes after reducing and reusing; it means gathering materials and turning them into new goods. But recycling takes energy and materials,

Zero-Waste Technologies in Aquaculture: Global Perspectives and Methods toward a Circular Economy
DOI: https://doi.org/10.1016/B978-0-443-34689-7.00023-9

so we should work on using fewer disposable products. Composting is another important part of zero waste. It turns organic waste, such as food scraps and yard trash, into compost that is full of nutrients. When you compost instead of throwing away biodegradable garbage, you keep it out of landfills, which stops methane from getting into the air and makes the soil richer.

Living a zero-waste lifestyle is not only about making less trash; it is also about changing to a more environmentally friendly and thoughtful way of life. People who live this way try to have as little of an impact on the environment as possible by following zero-waste rules in their daily lives. When they shop, they look for items that come in bulk, have less packaging, and are manufactured from natural or recycled materials. When it comes to food, they do not use plastic straws or takeaway containers that can only be used once. Instead, they plan their meals to cut down on food waste and compost their organic scraps. They choose to use cloth napkins, refillable containers, and eco-friendly cleaning supplies instead of disposable ones at home and at work. They also choose second-hand or repurposed furniture to cut down on waste. They embrace a minimalist wardrobe by buying second-hand clothes or investing in long-lasting, eco-friendly apparel that cuts down on textile waste and doesn't support fast fashion. People may have a big impact on the environment and help make the world more sustainable by making these choices.

Living a zero-waste lifestyle has a lot of benefits for the environment. It cuts down on trash in landfills, saves resources, and cuts down on pollution. All of these things help fight climate change and conserve ecosystems. Encouraging the smart use of materials, it lowers the need for raw materials, energy, and water, making sure that these resources will be available for future generations. Sustainable products may cost more at first, but following a zero-waste lifestyle can save you money in the long run since people buy less, reuse things, and value quality over quantity. Also, a zero-waste approach is better for both your health and the environment since it limits your exposure to dangerous chemicals, cuts down on your dependency on processed meals with too much packaging, and keeps you from coming into contact with plastics. By doing these things, people help make the world a better place and also improve their own health.

As more people choose materials that are good for the environment, research into natural fibers has sped up. Synthetic fibers such as Kevlar, nylon, glass, and aramid are manufactured by people and are bad for the environment because they release hazardous gases, such as CO, and poisonous smoke when burned. Natural fibers break down naturally, can be grown again, and are better for the environment. Because of growing environmental concerns, experts all around the world are looking at sustainable alternatives that lower pollution and protect biodiversity. Natural fibers are being looked at again as good alternatives to synthetic materials because they are good for the environment and easy to find. Many industries use these fibers, such as composites, construction, engineering, textiles, biomedical applications, biopolymers, biosensors, and smart packaging. Using them helps with problems caused by pollution, like too much trash piling up, too much use of landfills, hazardous emissions, and greenhouse gas generation. Natural fibers can have mechanical properties that are quite similar to those of synthetic fibers when they are treated in the right way (Chand and Fahim, 2008).

Plant fibers are grouped by where they come from and what section of the plant they come from. Plants that are farmed just for their fiber, such as flax, hemp, and jute, are used to make primary fibers. Coir, banana, and pineapple fibers are examples of secondary fibers that come from the waste of fruit-bearing plants. This is a long-lasting way to deal with agricultural waste. Also, fibers are grouped by the parts of the plant they come from: Bast, leaf, fruit/seed, and stalk fibers. Natural fibers are better than synthetic fibers in many ways, including being cheaper, lighter, stronger for their weight, more elastic, renewable, needing less energy, and not irritating. These advantages make them tough competitors to glass, carbon, wool, silk, and other materials in fields including textiles, paper, biocomposites, accessories, and crafts (Karimah et al., 2021). To have a future with no waste, fiber crops are very important. Hemp, flax, jute, and bamboo are all biodegradable crops that use fewer chemicals and less water than regular crops, such as cotton. Their fibers and byproducts are used in textiles, packaging, and building, and any leftovers are often utilized again or recycled, which is good for the environment. These crops help lower carbon emissions and reliance on nonrenewable resources by encouraging regenerative agriculture and supporting circular economies.

Using zero-waste methods in fiber crop production is good for the environment and the economy. These crops need fewer pesticides, save water, and improve the health of the soil, which is in line with sustainable farming. You may turn their byproducts, such as stalks and leaves, into bioenergy, compost, or eco-friendly building materials, which cuts down on trash in landfills. These methods not only reduce the environmental impact of farming, but they also open up new ways to make money, which helps rural economies stay strong. In fiber agriculture, a zero-waste approach means using every part of the crop in the best way possible. For example, fibers can be used to make textiles, stalks can be used to make bioplastics or buildings, and seeds can be used to make food or oil. This closed-loop technique turns trash from one step into a resource for the next, which saves natural resources and cuts down on pollution. This chapter looks at how fiber crops help zero-waste farming by being grown in a way that is good for the environment, processed in a way that is quick and easy, and used in new ways.

19.2 Unlocking the zero-waste potential of fiber crops

19.2.1 Revealing jute as a zero-waste crop

Jute (*Corchorus* spp.), also known as the "golden fiber," is India's second most important natural fiber crop after cotton. People have grown jute for hundreds of years because it is strong, long-lasting, and useful. It is a Kharif crop that grows in tropical and subtropical areas, especially in India, Bangladesh, Myanmar, China, Thailand, and parts of South America (Ghosh and Jethi, 2013). West Bengal, Assam, and Bihar are the key jute producers in India. West Bengal makes up over 75% of the country's jute production (Mondal and Bandyopadhyay, 2014). India has 0.76 million hectares of land and gets an average of 2.55 tons of fiber per hectare. This means that the country makes about 1.9 million tons of jute every year (Department of Agriculture & Farmers Welfare, 2023). Gunny bags, colorful fabrics, and carpets are some of the most common things made from jute. It makes jute sticks as a byproduct, but only 3.0–4.0 million tons of them are used each year. Jute is good for the environment because it does not need a lot of inputs. Jute grows a lot of biomass (40–60 tons per hectare), along with other crops such as kenaf and roselle. This makes it a good option for applications other than generating fiber. Only a limited amount of this biomass is used for business right now, mostly for fiber. The rest is burned in jute sticks, which is not very effective. Using all of the biomass might boost farmers' incomes by more than 40% and make jute more commercially viable.

- **Jute for bioethanol**: Jute fibers include a lot of cellulose (58%–63%), hemicellulose (20%–22%), and lignin (12%–14%) (Ramesh and Deepa, 2024). This makes them great for manufacturing bioethanol because they have a lot of cellulose and not a lot of lignin. Jute sticks have less cellulose (40.8%) and more lignin (23.5%), yet they might also be used to make ethanol. To make value-added products from jute in a way that lasts, it is important to use a biorefinery method that focuses on making the most of biomass. The lignin that is removed during this process can be used in many different ways in industry, such as in biodegradable polymers, adhesives, and textiles. Because jute has a lot of biomass and cellulose, it is similar to other commercial sources, including kenaf, hemp, and sugarcane bagasse (Satya and Maiti, 2013). Using biological approaches, research demonstrates that mesta gives 4.10 g/L of ethanol while jute sticks give 7.55 g/L of ethanol (Lavanya et al., 2019). There is now a set methodology for pretreatment and scarification of whole jute biomass, which is a key part in making bioethanol on a commercial scale (Sharma et al., 2023).

- **Biochar from jute biomass** India makes over 2.7 million tons of jute sticks every year, but most of them go to waste. Instead of open burning, this biomass can be turned into biochar, a fine-grained, porous, carbon-rich substance made by pyrolysis at 350–600°C in low-oxygen circumstances. Biochar has an aromatic composition that makes it very durable. It can stay in the soil for more than a hundred years, which means it can store carbon for a long time. Biochar not only reduces waste, but it also enhances the quality of soil by changing its physical, chemical, and biological qualities, such as how well it holds water and how easily nutrients are available. Biochar is a zero-waste product that helps fight climate change around the world. The Kon-Tiki-Kiln open-earth pyrolysis process has shown promise in turning jute and mesta sticks into biochar, with charcoal recovery rates of 28% and 34%, respectively (Ghorai et al., 2021). This technology could make roughly 750,000 tons of charcoal each year, which is worth about ₹4,914 crore on the market. Biochar directly boosts agricultural output by concentrating nutrients during pyrolysis. It also boosts agricultural productivity indirectly by increasing soil health and lowering greenhouse gas emissions. It also gives farmers a chance to make a living by giving them new ways to make money.

  Nanotechnology is looking into jute and its sticks more and more since they are mostly made up of cellulose, hemicellulose, and lignin, with very little ash and fat. These qualities make jute a great starting point for making nanocellulose, nanolignin, and nanocarbon compounds. People are starting to see the promise of jute-based nanomaterials in medicine, materials science, and protecting the environment. Jute-based nanocellulose has a lot of benefits, including being strong, stiff, biodegradable, light, and having a large surface-to-volume ratio. These qualities make it useful in many fields, including supercapacitors, lithium-ion batteries, biomedical devices (such as drug delivery systems and tissue scaffolds), water purification (getting rid of bacteria, viruses, and heavy metals), and lightweight composite materials (Jabbar et al., 2017).
- **Jute nanocarbons** (JC) are made by pyrolysis and have a lot of surface area and holes. They can be used to strengthen polymer composites and biomedical devices because they kill germs and clean up the environment (Aziz et al., 2020; Jabbar et al., 2020; Shafiul et al., 2020). Researchers are also looking at using jute nanocarbons in electronics, catalysis, and sensing technologies, as well as in energy storage systems, such as supercapacitors (Chen and Yang, 2017). So, jute-based nanomaterials could be used in many different fields, including solar energy, electronics, filtration, building, cosmetics, and making biofuels.
- **Other uses of jute** The jute industry makes about 40,000 tons of caddies every year. Caddies are short, unspinnable jute strands that are thrown away during processing (Ganguly et al., 2004). Making these caddies into goods that offer value can help the economy flourish, and the environment stay healthy (Nayak et al., 2012). Caddies are a prospective

source of cellulosic feedstock for biogas generation by anaerobic digestion, which makes renewable energy and nutrient-rich slurry. You may use this slurry, which is full of carbon and nutrients, as manure to help crops grow better and even grow more mushrooms. You may also turn jute caddies into nonwoven fabrics that can be used for a wide range of things, such as air filters, floor coverings, blankets, throwaway bags, automobile material, and wall and window treatments. These nonwoven fabrics can also be used instead of synthetic fibers, such as fiberglass, in composite materials, which is better for the environment. You can also turn jute caddies into briquettes or gasify them to make biomass energy. The calorific value of jute caddis briquettes is about 3000-3200 kcal/kg. This means that 1 kg of briquette is equal to 0.635 kg of coal or 0.33 kg of mineral oil (Nayak et al., 2011). Jute caddies can be used to make more than just electricity. They can also be used to make eco-friendly handmade paper and agrotextile mulches. These mulches keep weeds from growing, keep the soil moist, and make the soil more fertile, all of which help with sustainable farming.

19.3 Mesta

Mesta (*Hibiscus* sp.) is a herbaceous plant that grows every year or every other year. It is a member of the cotton family, Malvaceae. Two types of mesta are *Hibiscus sabdariffa* var. Wilson (2006) says that altissima (roselle) and *Hibiscus cannabinus* (kenaf) are two types of hibiscuses. In India, though, both species are usually called Mesta. After cotton and jute, it is the third most important fiber crop for business. Mesta can be produced on ground that is not good for food crops and does not need as much fertilizer and water. India grows mesta in 13 states: Andhra Pradesh, Assam, Bihar, Chhattisgarh, Jharkhand, Karnataka, Madhya Pradesh, Maharashtra, Meghalaya, Odisha, Tamil Nadu, Tripura, and West Bengal. Mesta can grow in a wider range of climates and soils than jute, and it is also particularly drought-resistant. The crop is grown on around 1.5 lakh ha of land, and the average yield is about 11 quintal ha^{-1}. Recent findings have expanded the use of different plant parts of mesta, such as its leaves, seeds, flowers, and stems, in the food, pharmaceutical, and medical industries beyond its traditional use as a fiber crop (Satya and Pandey, 2024). The extracts of these plant parts have a number of bio-activities, including antidiabetic, anticancer, antimicrobial, and antihypertensive properties, among many others (Islam, 2019). This demonstrates that mesta can also be safely regarded as a zero-waste crop by utilizing all plant parts in productive and sustainable ways to minimize or nullify the waste and make the world better, which is crucial for the sustainable development of any mesta-based products. Different plant parts of the mesta crop, used as a potential source for mankind, are represented in Fig. 19.1.

19.3.1 Harnessing plant parts of mesta

- **Fiber:** Mesta fiber comes from the stems of the plants; the fiber strands are up to 1.5 m long and rank next to jute in importance. Fiber is made up of cellulose (63.52%), lignin (7.6%), and other components, such as protein, fat, and dietary fiber. The bast fiber has a silky sheen, a long, stable length, high tensile strength, low extensibility, is biodegradable, and is resistant to heat, fire, and microbes. These qualities make it useful in industries such as textiles, paper, building, and automotive, as well as for saving soil, decorating, and furnishing. Studies have shown

FIGURE 19.1 Mesta plant parts images: (A) whole green plant, (B) seed, (C) flower, (D) leaf, (E) calyces, and (F) fiber.

that the new variety JBMP-5 (Vaibhav) makes 28.9 quintals ha^{-1} of fiber with good quality, low root content and defects, good strength (20.90 g tex^{-1}), and fineness (3.50 tex). This variety may be used to make high-quality fiber (Pandey et al., 2023). A novel approach to blending mesta fiber with light brown natural color cotton fibers in the proportion of 20:80, respectively, improved the suitability of yarn in fabric production (Vastrad et al., 2025). Mesta blended cotton fabric will be the rising sun shall provide better opportunities for the farmer in terms of income. Another useful product of fiber, which has industrial use, is nanocellulose, is one such product that can be extracted from residual fibers and used in bioplastics, electronics, and medical applications (Ilyas et al., 2021).

- **Stalks:** can be turned into briquettes, pellets, or biochar for energy production once the fiber is removed (Ghorai et al., 2021). The growing scarcity of hardwood is drawing more and more attention to mesta as a source of high-quality paper pulp, which can contribute to a great extent to reducing deforestation. Studies have shown that the kenaf variety, "MT 150," is the best paper pulp variety that produces 8 tons ha^{-1} of biomass, so it is a suitable variety for the paper pulp industry (Ghorai, 2015). Stalks that have been chopped can be composted or utilized as mulch to promote fertility and hold onto moisture. Chopped stalks can also be used as a substrate for growing mushrooms. Sometimes, stalks are used to make decorative items, basket weaving, and handicrafts. The retting (fiber separation from the woody core of the plant) process generates over 2.3 million tons of residues, which can be utilized to produce biomass-based fuel generation. Biofuel technology is a perfect example of highlighting zero-waste concepts, the conversion of waste into bioethanol, and the digested product can be used as organic manure. The chemical composition of mesta sticks consists of high holocellulose (64%–70%), lower lignin content (20%–22%), and ash content (3%–4%), with trace amounts of minerals, oils, soluble sugars, and other substances, because of these properties it is considered as one of the most potential lignocellulose feedstocks for bioethanol production (Lavanya et al., 2019). It has been shown that alkaline-pretreated mesta biomass yielded higher fermentable sugar and increased ethanol content (Sharma et al., 2025). Recently released kenaf variety JBMP-6 has shown special features such as fewer bristles on the stem, low lignin, and high cellulose content, which can be a good biofuel feedstock (Pandey et al., 2023).

- **Leaves:** Mesta leaves are a good source of vitamin C, its content is nine times higher than that of the citrus (*Citrus sinensis* L.) fruit and three times higher than the blackcurrant (*Ribes nigrum* L.) fruit. The leaf extract is also a good source of fiber, ash, calcium, phosphorus, iron, carbohydrate, protein, fat, thiamin, β-carotene, riboflavin (B2), and niacin (Islam et al., 2021). The tender leaves and stems of roselle have an acid, rhubarb-like flavor and are used for preparing pickles, chutney, eaten raw in salads, or cooked as greens alone or in combination with other vegetables and/or with meat. Because of high mucilage content of the leaves, it is used as a source of mucilage in the pharmacy and cosmetic industry. The leaf extracts have medicinal properties that can be processed into herbal products. Dried leaves and the tender portion of the plant are a rich source of nutrients for animal feed and can also be used to make compost or biofertilizers.
- **Roselle calyces:** Roselle calyces are found in three main color types: Dark red, red, and green. Among these, the red calyces are most commonly utilized because of their high content of anthocyanin pigments, specifically delphinidin 3-sambubioside and cyanidin 3-sambubioside. These calyces are rich in citric acid (approximately 4%) and pectin, making them ideal for producing jams, jellies, and similar food products. In addition, roselle calyx extracts are abundant in minerals, anthocyanins, and antioxidants, and contain about 5%–6% protein. As such, they are widely used in beverages such as teas and syrups, dietary supplements, and culinary items, including soups, cakes, and puddings. They also serve as natural coloring agents for syrups and alcoholic drinks (Ismail et al., 2008).

Moreover, roselle calyces can be a source of eco-friendly, natural dyes for textile applications. The extracts contain various bioactive compounds such as cardiac glycosides, flavonoids, saponins, and alkaloids, which contribute to their antibacterial properties (Sirag et al., 2014). In traditional medicine, these extracts have been used to treat ailments such as abscesses, bilious disorders, cancer, and coughs. Scientific studies have also demonstrated their antibacterial efficacy against numerous pathogenic microorganisms, including *Staphylococcus aureus, Bacillus stearothermophilus, Micrococcus luteus, Serratia mascences, Clostridium sporogenes, Escherichia coli, Klebsiella pneumoniae, Bacillus cereus, and Pseudomonas fluorescens* (Al-Hashimi, 2012). A recently introduced roselle variety, HSLC-1, yields 42.60 quintals per hectare of calyces and is notably rich in essential minerals such as potassium, calcium, magnesium, and iron, as well as vitamins C, B1, B3, and B5. Because of its superior nutritional profile and high yield, HSLC-1 is considered one of the most suitable varieties for calyx production (Pandey et al., 2023).

- **Seeds:** The mesta seeds are rich in protein (19%–22%), fat (19%–23%), and dietary fiber (40%–43%) content, is mainly used in cosmetics, lubricants, and biodiesel (Dhar et al., 2015). The seeds also yield a vegetable oil (22%–24%) that is edible with no toxins; these seeds are reported to be eaten in some parts of Africa. Besides, the seed oil contains a high percentage of omega polyunsaturated fatty acids (PUFAs), nearly 70%, which are essential fatty acids for normal growth and health. Furthermore, they are important for reducing cholesterol and

FIGURE 19.2 Banana plantation, leaf, fruit, blossom, stem, and many uses.

heart disease. Linoleic acid (C18:2) is the dominant PUFA, nearly 44%, followed by oleic acid (C18:1), nearly 28% (Islam, 2019). The seeds are also a good source of minerals such as phosphorus, magnesium, and calcium, amino acids such as lysine, tryptophan, arginine, leucine, phenylalanine, and glutamic acid. The roasted seeds can be used as a coffee substitute and as an aphrodisiac, and the ground powder is also used in soups and sauces. Mesta seed oil also has antioxidant properties, is used as a substitute for castor oil, so are used in cosmetics. Seed cake left after oil extraction contains approximately 32% of protein and can be used as high-protein livestock, fish, and poultry feed.

- **Roots:** The young roots of the mesta plant are edible, which are fibrous, mucilaginous in nature, and do not have much flavor. In some countries, such as the Philippines, it is consumed as a tonic because of its bitter taste, which has antibacterial and antifungal properties (Al-Hashimi, 2012). Furthermore, because of their fast growth, deep root systems, and high biomass production, plants hold great potential for phytoremediation because of their ability to absorb or degrade heavy metals and pollutants from soil and water (Uddin, 2016). It has been shown that mesta varieties, when grown in arsenic (As) contaminated soil, accumulate more and show the phytoremediation potential of the mesta plant.

19.4 Banana

The banana plant (*Musa* spp.) is well-known not just for its healthy fruit but also for the many ways its leaves, stem, trunk, and blossoms can be used. Because it is grown and eaten all over the world, it is now the second-largest fruit crop in the world. India, China, Uganda, Ecuador, the Philippines, and Nigeria are among the biggest producers in the world. Bananas are mostly cultivated for their fruit, but they also produce a lot of agricultural waste and byproducts that are not used very much. If they are not handled, they can be bad for the environment and represent a big loss of potentially valuable biomass. Fig. 19.2 shows how to grow a banana plant, its parts, and how to add value to it.

19.4.1 Uses of banana fiber and pseudo-stem

Banana fiber, which comes from the pseudostem, is known for being very strong, long-lasting, and able to break down naturally. The pseudostem waste is made up of about 40%–45% cellulose and other plant parts, such as leaves, peels, rachis, and peduncle. After going through certain physical and chemical procedures, this lignocellulosic biomass is used to make kraft paper. Mechanical decortication is still the best way to get fiber out of plants. The fiber that is taken out can be used in traditional crafts, the textile industry, and to make clothes. Around 500 factories in India make handmade paper and board from banana trash, which adds up to about 50,000 tons a year. This operation brings in ₹250,000 million in revenue and gives jobs to over 15,000 people, mostly in rural areas. Banana-based materials are becoming more popular as eco-friendly alternatives as government laws focus more and more on reducing single-use plastics.

19.4.2 Nutritional and agricultural value of banana peels

Banana peels are good for plants because they are high in cellulose (7.6%–9.6%), hemicellulose (6.4%–9.4%), pectin (10%–21%), lignin (6%–12%), and chlorophyll pigments (Yasin et al., 2025). Banana peels are high in important minerals, including potassium, calcium, and phosphorus, which help compost and make soil more fertile. The potassium in the soil helps plants grow, and the amino acids help seeds germinate, plants stay healthy, and chlorophyll production. Banana peels are becoming more common in smoothies, baked goods, and vegan foods since they are excellent for you. Banana compost makes the soil richer in nutrients, boosts microbial activity, and helps crops grow more (Yasin et al., 2025). Organic liquid fertilizers made from fruit and vegetable waste are valuable source for effective growth of short term plants (Ramamoorthy et al., 2024), this helps the environment and make people less reliant on synthetic fertilizers. In addition, mixing banana peel compost with biochar makes the soil hold more water and makes plants stronger (Islam et al., 2019). Banana biomass has also shown promise in making nanocellulose, which could be useful in high-value applications (Zaini et al., 2023).

19.4.3 Zero-waste innovations and sustainable enterprises

Innovations that do not waste anything and businesses that are good for the environment. Innovative businesses in India have adopted a zero-waste method for growing bananas. Jagat Kalyan, Satyam Kumar, and Nitish Verma started Taruwar Agro in 2021. The company uses all components of the banana plant—leaves, stems, and pulp—to make banana fiber for a wide range of eco-friendly products. These are things such as folders, yoga mats, baskets, coasters, sanitary pads, banana leather, and decorative goods. The leftover pulp is turned into high-quality vermicompost, while the liquid extract is an organic fertilizer that is full of micronutrients such as potassium and iron. The company processes about 600 kg of banana trash per month and sells its goods both online and in stores. It hopes to grow its business internationally, and Fig. 19.3 shows the zero-waste products of the start-up.

Ms. Varsha, who has an M.Tech degree from Oxford College of Engineering, is in charge of another important project. Her business, Aakruthi Eco-friendly Enterprises in Ummattur (Chamarajanagar district), turns banana plant waste into crafts and food. During the COVID-19 pandemic, she had the idea to turn banana trunks and stems into fiber-based handbags,

FIGURE 19.3 Banana trash and the items that can be made from it.

yoga mats, and dinnerware, as well as pickles, juices, and sweets from the inner pseudostem. This was because of problems with local farmers getting rid of their garbage. Composting solid waste and turning liquid extracts into potassium-rich organic fertilizers are two ways to use the waste. In the same way, Anusuya and Kasinath started the Jayadev Banana Farmers and Artisans Association in Balipatana (Bhubaneswar). Their project helps local farmers by turning banana biomass from after the harvest into both edible products, such as banana stem pickle, papad, jam, murabba, and juice, and useful things, such as ropes, mats, and coasters. The company follows a zero-waste concept, which means that all leftover materials from fiber processing are used for vermicomposting and organic liquid fertilizers.

19.4.4 Banana plant as food and nutrition

- **Fruit:** People eat bananas raw or cook them into traditional foods such as chips, cutlets, and curries. They are full of potassium, vitamin C, and dietary fiber, which are good for your heart, blood pressure, and digestion. Bananas are also used as natural sweeteners in smoothies, baked goods, and desserts. They are a better alternative to refined sugars (Kumari et al., 2023).
- **Leaves:** In South Indian areas such as Kerala, Karnataka, and Tamil Nadu, banana leaves are utilized a lot for both cooking and cultural purposes. They are natural and biodegradable alternatives to plastic plates that add flavor and smell when used to steam or grill food. Banana leaves are a good source of antioxidants since they have polyphenols and vitamins A and C.

- **Flower:** Banana flowers, or blossoms, are used in many different types of food, such as stir-fries, salads, and soups. They are high in nutrients and fiber, as well as vitamins A and C. Banana peels are also used for composting, polishing leather, and cleaning, which is in line with goals for sustainability.
- **Stem**: The banana stem (pith) is a component of the plant that may be eaten and has health benefits, such as helping to control blood sugar. It is high in fiber and low in calories. It is a good source of potassium, calcium, and vitamin B6. People use it in cooking to make stir-fries, soups, and pickles. People use its fibrous outer covering to make baskets, mats, and garland ties. Banana fibers can be woven into fabrics when mixed with cotton. Before eating, the outer layers of the stem are taken off, and the center is washed to get rid of the bitterness (Kumari et al., 2023).

19.5 Flax

Flax (*Linum usitatissimum*) is a great crop that is both useful and has been grown for thousands of years and has been cherished by many cultures for its dual purpose: it produces both high-quality fibers and seeds that are full of nutrients. Linseed is the name for the crop when it is grown mostly for its seeds, which are high in oil (Kaur et al., 2018). It can be used in textiles, food, medicine, and a number of other industrial settings. In the past, flax was very important to many societies. People grew it in Egypt as early as 5000 BCE. Linen, which is derived from flax fibers, was used for more than just clothes; it also had cultural meaning. It stood for purity and was utilized in religious ceremonies, such as mummification, when finely woven linen was wrapped around mummies to show their spiritual purity.

Similar uses were found in Mesopotamia and Phoenicia, which shows that flax has been important throughout history. Flax grows best in temperate climates with soils that are rich in organic content and drain well. It does not need a lot of chemicals, which makes it a good choice for the environment. Canada, Russia, China, and Kazakhstan are some of the biggest producers of flax. The crop is grown in more than 50 nations around the world (Stavropoulos et al., 2023). The fibers that come from the stem of the flax plant are what make it valuable. These natural cellulose fibers are well-known for being strong, breathable, biodegradable, and long-lasting. Linen, which comes from flax, is about 30% stronger than cotton and gets stronger when it gets wet. This makes it good for clothes, home furnishings, and many industrial uses. Flax has been a popular raw material for making biocomposites that are good for the environment in the last few years. Its strong but light fibers are great for strengthening both synthetic and biobased composite materials. They can be used in the automotive, construction, and sports industries (Amiri et al., 2018).

Flax fibers are also used in the paper industry to make high-quality items such as banknotes and cigarette paper since they are long and have a lot of tensile strength (Chand and Fahim, 2008). Growing flax is good for the environment because it helps keep it healthy. The plant takes in a lot of carbon dioxide and turns it into oxygen, which makes the air cleaner (Safilin). It has high pectin content and kills bacteria, which helps it hold onto water. This makes it useful for a wide range of ecological and industrial purposes. One of the best things about flax is that it does not squander anything. Long and short fibers are spun into textiles and technical materials. Woody shives are utilized to make mulch, animal bedding, insulation panels, or biofuel. Seeds are pressed to make oil. Even the dust is composted (original fibers). Flaxseeds are getting more and more attention for their rich composition and health benefits when it comes to nutrition. Flaxseeds are eaten a lot, especially in India, for their taste and health advantages. They are made up of 35%–45% oil, 20%–30% protein, and about 28% dietary fiber (Martinchik et al., 2012). Nutritionists and medical experts all around the world have been interested in this composition (Touré and Xueming, 2010), which makes flaxseed a very useful food (Oomah, 2001).

Flaxseed is recognized for having a lot of vital fatty acids, especially alpha-linolenic acid (ALA), which is an omega-3 fatty acid that is good for the heart and may help prevent cancer (Gonzalez et al., 1991). Flaxseeds are also rich in minerals such as calcium, magnesium, phosphorus, and sodium (Kajla et al., 2015). They have vitamins that dissolve in both fat and water. One of these is tocopherol (vitamin E), which helps fight free radicals and is good for heart health. Vitamin E helps keep sodium levels in check, stops lipids from oxidizing, and lowers the risk of heart disease, high blood pressure, and Alzheimer's disease (Kajla et al., 2015). Flaxseeds are also high in phenolic chemicals, which are known to fight cancer and protect cells from damage. Flaxseeds still have a lot of chlorogenic, gallic, and ferulic acids even after being defatted. They also have lower levels of 4-hydroxybenzoic acid, p-coumaric acid glucosides, and hydroxycinnamic acid glucosides (Beejmohun et al., 2007). Eating flaxseed on a regular basis has been related to a lower risk of long-term diseases, including type 2 diabetes (Adlercreutz, 2007). The crop has qualities that fight atherosclerosis, stop cell growth, and stop arrhythmias, which are good for both the heart and the cells (Dupasquier et al., 2007).

It is easily found in whole seeds, pulverized meal, oil, flour, capsules, and tablets. People use these goods a lot in diets to help with constipation, high cholesterol, diabetes, obesity, high blood pressure, and even some cancers (Goyal et al., 2018; Xu et al., 2022). Flaxseed has omega-3 fatty acids and dietary fiber that help the digestive system work better and keep metabolic processes in check. Its antioxidants fight oxidative stress and inflammation, which are two of the main causes of chronic illness. Because of this, flaxseed is often found in modern diets, supplements, and even skin care products. Flax has shown its worth in industry beyond only food and fiber. The woody component of the plant is used to make biodegradable things such as fiberboard, particle board, and plywood. Its fibers are necessary for making high-quality specialty papers, and its potential for biocomposite development is pushing green technology forward in fields including construction, automobile production, and packaging (More, 2021). This wide range of uses shows how important flax is for promoting environmentally friendly business operations.

In short, flax is a very versatile and environmentally friendly crop that may be used in many ways. Flax is an important part of living sustainably since its strong, breathable fibers are used in textiles and industrial products, and its seeds are rich in nutrients and used in foods and medicine. Growing it is good for the ecology, and its historical and present usage show how important it is. Flax will continue to be an important agricultural and industrial asset in the future as more and more people across the world become interested in eco-friendly products and meals that are good for their health (Fig. 19.4).

19.5.1 Flax: a crop with zero waste

Flax fiber is a natural cellulose bast fiber that is made up of many cells. When it is made into fabric, it becomes linen. It is one of the oldest and strongest textile fibers, noted for its incredible strength, long life, capacity to absorb water, and unique texture and shine. To get the fiber, the

FIGURE 19.4 Showing flax and its uses (A) plant; (B) flower; (C) seed; (D) fiber; (E) flaxseed oil; (F) composite; (G) snack (bar); (H) oil cake; (I) flax waste-based board; (J) flax-based product. *Source: www.google.com.*

woody sections of the stem are cut off, leaving long, flexible strands that can be spun into yarn or sewn into cloth. Flax is the best choice for apparel, linens, and upholstery since it is soft, absorbent, and shiny. This makes it perfect for bed linens, towels, tablecloths, and even sanitary napkins (Vastrad et al., 2025).

Flax is also good for ropes and twine since it has a high tensile strength. These materials have been used for load-bearing and seafaring purposes for a long time. It also serves a key part in generating high-quality paper that lasts, such as bookbinding paper and cigarette paper. Its structural qualities also make it a great material for reinforcing composites used in building and car parts. Flax is used in the automobile industry to make lightweight composite parts, such as door panels and trunk liners. This helps cars use less fuel and release fewer pollutants. Flax is also great for personal care goods, such as baby wipes, facial tissues, and sanitary napkins, since it is soft and absorbent. Flax is still a valuable crop for farming, even if it is used in industry. It is a resource of global importance since its seeds and oil are utilized in food, animal feed, and many other industrial uses.

- **Flax fiber processing:** By making good use of all the byproducts, flax fiber processing can be waste-free. In Western Europe, the production of fiber flax is done in such a way that there is no waste, and all of the leftovers are used by other industries. Retting is the first step in fiber extraction. It separates the fibers from the stem. There are three main types of retting: dew, water, and chemicals. The next stage is scutching, which cuts the woody stalks into pieces called "shives" without hurting the fibers. Shives are used again to make light building planks, potting soil aggregate, and bedding for animals. Hackling gets rid of the last shives and sorts the fibers by how good they are. This makes hackle tows for coarse yarns and flax line, which is the base for high-quality linen yarns. Classic linen spinning machines use high-quality tow, although it can be processed even more in staple fiber mills. Medium-grade tow is employed in nonwoven applications, while the lowest quality is used to make high-grade cellulose for specialist papers and filters. The way you spin depends on what you want to make. For example, delicate linen thread is wet-spun to make high-quality textiles, whereas coarse tow can be dry-spun to make thicker, more rustic yarns.

Flax fibers are still mostly used to make linen fabrics, which are known for being strong, biodegradable, and good for the environment. Linen is a more environmentally friendly option because it needs less water and chemicals to make than synthetics. Industrially, flax is used in ropes, twines, canvas, and ecocomposites. Construction and automotive industries utilize flax-based biocomposites for sustainable and lightweight materials. Byproducts are repurposed into insulation panels, biodegradable packaging, and reinforcements in bioplastics, cutting dependence on petroleum-based products. Flax is also a valuable raw material in papermaking, especially for producing banknotes, cigarette paper, and specialty stationery. Its inherent strength and durability offer a sustainable alternative to wood-based paper.

Nonwoven flax products and composites are biodegradable, reducing pollution and landfill waste. Flax residues can be converted into mulch, biofuel, or animal bedding. Research shows that flax straw, usually agricultural waste, can be used in insulation composites. Because of its porosity, flax straw offers high moisture absorption and low thermal conductivity, making it a viable material in eco-friendly construction, especially in rural regions (Cerny et al., 2024).

- **Flaxseed in food industry:** Flaxseed is widely used in the food industry, either as a whole ingredient or through its bioactive components to create functional foods and boost nutritional value (Edel et al., 2015). Flaxseed-enriched products-cheese, yogurt, snacks, milk, ice cream, biscuits, cookies, cakes, and bread-appeal to health-conscious consumers.
- **Dairy Industry:** Studies show that flaxseed mucilage combined with gum arabic enhances the viability of probiotics, such as *Lactobacillus acidophilus* and *Bifidobacterium lactis* in kefir, without affecting its composition (Sungatullina et al., 2023). Flaxseed oil and lignans have been added to ice cream, yogurt, cheese, butter, and whey beverages. One study revealed that 2% flaxseed oil could be incorporated into ice cream without affecting its functionality (Goh et al., 2006). Microencapsulated flaxseed powder added to curd provides omega-3 fatty acids and boosts cholesterol reduction, nutrient absorption, and antioxidant levels.
- **Baking industry:** Flaxseed is widely used in baked goods such as bread, biscuits, cakes, and noodles (Sharma et al., 2019). Research on gluten-free biscuits found that replacing baking powder with flaxseed powder improved protein, fiber, and antioxidant content (Nasiri et al., 2023). Because commercial cookies typically lack ALA and dietary fiber because of refined ingredients (Gökmen et al., 2011), flaxseed flour-enriched noodles offer improved nutritional quality (Yaver, 2023). Flaxseed is used in various forms, whole, roasted, ground, or as oil, in baked products such as muffins, pizza, cakes, bagels, and rolls (Kaur et al., 2018). Fermented cereal products containing live lactic acid bacteria can benefit from flaxseed-based fermentation, especially when combined with oat, buckwheat, or soy.
- **Snack products:** Flaxseed has a favorable lipid profile that modifies the omega-3 PUFA content in snacks, enhancing their nutritional quality and potentially reducing noncommunicable disease risks (Ansorena and Astiasarán, 2004). A corn snack fortified with flaxseed showed a sevenfold increase in dietary fiber, doubled protein content, and maintained consumer acceptability (Trevisan and Arêas, 2011).
- **Extruded products:** Flaxseed can be incorporated into extruded snacks or breakfast cereals, encouraging regular consumption and delivering health benefits (Wu et al., 2007). Gupta and Shivhare (2012) developed flaxseed-based pasta that retains its functional nutritional properties.

- **Other products:** Traditional foods such as chapatti, khakhra, chilla, and manchurian have been supplemented with flaxseed flour to enhance nutritional and functional value (Rathi and Mogra, 2012). Flaxseed has also been used in soups, vermicelli, and bhelpuri chaat. These low-fat, heart-healthy foods retain high nutraceutical potential (Jain et al., 2014). Flaxseed chutney powder, made by blending roasted flax seeds with spices, offers a palatable and functional food option (Rao et al., 2011).
- **Fodder:** The byproduct of oil extraction, linseed meal, is a protein-rich animal feed suitable for animal husbandry (ruminants, rabbits, and fish swine, poultry, horses, and dogs) (Xu et al., 2022).
- **Flaxseed oil:** Flaxseed oil is rich in PUFAs, especially ALA. Historically, it was used in the manufacture of drying oils, paints, and printing inks. In recent years, its value as an edible oil with nutraceutical properties has been recognized. Because of its sensitivity to heat, light, and oxygen, cold-pressing is used to extract edible oil (Choo et al., 2007). However, because of its high ALA content, flaxseed oil is prone to oxidation, rancidity, and reduced sensory quality (Wiesenborn et al., 2005), but continues to find wide-ranging applications across industries.

19.6 Ramie an ancient fiber crop, as zero waste

Ramie (*Boehmeria nivea*) is a perennial plant from the nettle family cultivated for thousands of years for its exceptionally strong and lustrous natural fiber. Historically, ramie played a significant role in ancient civilizations. In China, it was used over 2,000 years ago for burial shrouds, predating the use of cotton. In Egypt, it was part of mummification practices, and in India, its presence in traditional clothing dates back to at least 600 BC. As one of the world's earliest cultivated fibers, ramie has remained relevant because of its durability, strength, and wide range of applications. Today, China is the global leader in ramie production, responsible for more than 96% of the world's supply. The plant is believed to have originated and been domesticated in China, where wild varieties are still found. Other producing countries include Brazil, Taiwan, Korea, Japan, the Philippines, Indonesia, and Vietnam. In India, ramie is primarily cultivated in the northeastern states such as Assam, Arunachal Pradesh, Meghalaya, Manipur, and Sikkim. Its adaptability to diverse soils and climates offers strong potential for expanding cultivation in Northeast India, boosting regional economies, and supporting sustainable textile production. Ramie is notable for its fast growth and regenerative capacity. It can be harvested every 60 days without replanting, thanks to its perennial root system. This makes it highly productive and resource-efficient. The crop requires minimal pesticide and fertilizer input, and its deep roots help combat soil erosion and improve soil health by promoting nutrient recycling. Ramie's high biomass yield means it produces more usable fiber per acre than many other natural fibers, enhancing its economic and environmental value.

All parts of the plant can be utilized, making it a zero-waste crop. Its fibers are fully biodegradable, posing no long-term waste concerns. Naturally resistant to mold, mildew, bacteria, insects, and rot, ramie does not require chemical treatments for preservation. These qualities make it one of the most sustainable fibers available today. The fiber is extracted from the bark of the plant's stalk through a retting and degumming process, similar to that used for flax. Ramie fibers are fine, silky, naturally white, and lustrous, giving fabrics a luxurious appearance. Among natural fibers, ramie ranks among the strongest, twice as strong as linen and even more resilient when wet. However, pure ramie fabric tends to be brittle and lacks elasticity, making it prone to wrinkling and wear. For this reason, ramie is often blended with other fibers such as cotton, wool, silk, or viscose to improve its flexibility and durability. These blends result in textiles that resemble fine linen or durable canvas, suitable for a variety of uses.

Ramie is known for its excellent absorbency, ease of dyeing, and resistance to shrinkage even after repeated washes. These features make it well-suited for clothing, curtains, upholstery, towels, sewing threads, fishing nets, filter cloths, and marine packing materials. When blended with wool, ramie significantly reduces shrinkage compared to garments made from pure wool. Beyond textiles, ramie's applications extend into industrial and commercial fields. Short fibers and processing byproducts are used to make high-quality paper, including cigarette paper and currency notes. Studies conducted in the United States have identified ramie as a promising non-wood source of pulp for paper production. The woody core of the plant is also utilized in the manufacture of biodegradable materials, such as fiberboard, plywood, and particleboard. Based on the fiber's gum content they are used in textiles, apparel and cordage, and industrial uses such as hoses, bags, and canvas.

In agriculture, ramie also shows promise as a high-protein forage crop for pigs, because of its substantial biomass yield. In addition, the plant possesses anti-inflammatory and antioxidant properties, and its leaves are used in traditional medicine in various regions, supporting its role as a multipurpose crop. Although ramie has long been an integral part of Asian textile traditions, it remained relatively obscure in Western markets until the 1980s. Its popularity briefly surged when U.S. regulations allowed garments with more than 50% ramie content to enter without import quotas. This loophole spurred a wave of ramie-based apparel in the American market. However, in 1986, the policy was revised, removing its quota-free status and limiting its widespread adoption. Despite this, ramie has maintained a niche presence in the global textile industry.

With rising environmental awareness and the demand for sustainable materials, designers and manufacturers are once again embracing ramie as a viable green alternative to synthetic fibers. Its renewability, low environmental footprint, and strength align well with modern sustainability goals. Looking forward, ramie's ecological advantages and multifunctional properties make it a key player in sustainable development. Expanding its cultivation, particularly in underutilized regions such as Northeast India, could revitalize traditional farming, foster economic growth, and contribute to environmental conservation. As one of the oldest known fibers, ramie exemplifies how ancient resources can offer innovative solutions to today's sustainability challenges.

19.6.1 Ramie as a factory for zero waste

- **Fibers (Stem bark):** Ramie fiber, which comes from the plant's stem bark, is known for its natural shine, strong tensile strength, and long-lastingness. Because it works well with both natural and synthetic fibers, it is a popular choice in the textile sector. Ramie is great for clothes made for tropical areas since it absorbs moisture so well when mixed with synthetic fibers. The fiber's intrinsic resistance to microbes and its increased strength when wet make it even more useful. Because of this, it is used in a lot of different things, such as curtains, towels, threads, twines, pulley belts, ropes, cordages, fishing nets, fire hoses, water-carrying bags, industrial packing materials, canvas, upholstery, furnishing fabrics, and shoe stitching threads (Rehman et al., 2019).

- **Leaves:** Ramie leaves are safe to eat and are often used in East Asian cooking. They are a common part of rice dumplings and gelatinous rice cakes. People eat the young shoots and leaves as healthy vegetables by stir-frying, simmering, or adding them to salads, soups, and stews. Ramie leaves are a healthy part of your diet because they are full of vitamins, minerals, and dietary fiber. These leaves are very important in traditional medicine since they help with inflammation and healing (Lee et al., 2022). Ramie leaves are a great source of protein and nutrients for animals, especially cows, pigs, and rabbits, in addition to being good for people. They are extremely popular in cattle feed since they include about 24% protein. They turn dried leaves into fish meal, chicken feed, and bovine feed (Rehman et al., 2019).
- **Fertilizer:** You can compost or mix leftover leaves after harvesting into the soil to make it more fertile (cropshub.com). The liquid waste product from microbial degumming of ramie fibers, called wasted liquor, is also full of pectinolytic enzymes. It is possible to turn these enzymes into liquids or solids that are valuable in the textile and food processing industries. In addition, the semisolid waste from this procedure can be turned into useful biofertilizers (Mitra et al., 2014).
- **Roots:** Ramie's strong and deep root system helps the environment by holding soil in place and reducing erosion, especially on sloped or damaged land (Rehman et al., 2019). **Pulp for paper:** Leftover fibers from processing ramie fibers can be used to make high-quality paper without cutting down trees. This means that ramie is a good choice for making paper pulp due to their high cellulose content (Rowell, 2012) and is good for the environment. **Bioextracts:** Scientists are still looking at how ramie could be used to make bioplastics (Jamilah et al., 2026) and natural colors. These new ideas promise environmentally favorable options for application in a wide range of fields, from fashion to packaging.
- **Specialty papers:** You may also make high-quality specialty papers out of leftover ramie fibers. These papers are great for fine creative projects such as painting, calligraphy, and printmaking since they have a shiny look and a unique texture. They are especially preferred for archive records and conservation efforts because they are strong and do not go old. Their distinctive aesthetic properties also make them great for ornamental papers used in handicraft and interior design.
- **Stems:** The woody core of ramie stems could be used in a way that is good for the environment, in addition to fiber extraction. It can be turned into biobased composites that can be used for eco-friendly furniture, packaging, and building materials. In addition, the woody material is a great source of biomass for bioenergy by thermochemical processes (combustion, gasification, and pyrolysis) (Cheng et al., 2011). Fig 19.5 represents the utilization of ramie plant parts in various products.

19.7 Cotton

Cotton is a natural fiber that comes from the seeds of plants in the Gossypium genus, which is part of the Malvaceae family. For thousands of years, it has been very important to people. This delicate, fluffy fiber forms inside a protective boll that holds the plant's seeds. Cotton is one of the most frequently grown crops in the world. It is cheap and easy to grow, which makes cotton-based goods easy to find and buy. Cotton fabric is great for everyday use because it is strong and lasts a long time. It takes in dyes easily, is usually easy to clean, and can handle high heat when you iron it. Also, its capacity to absorb and release moisture well makes it more breathable and

FIGURE 19.5 Ramie is a crop that does not waste anything: Planting, extracting fiber, evaluating fiber, using it for medicine, and making special paper.

comfortable, which is why it is a popular choice for clothes (Weigmann, 2025). Cotton is grown mostly in subtropical areas and is a key part of the worldwide textile industry since it is soft, flexible, and strong. It is very important in history, notably during the Industrial Revolution, when it was a key part of trade and economic growth around the world (The Editors of Encyclopaedia Britannica, 2024). The cotton industry is now using zero-waste farming methods to reduce its impact on the environment by making full use of every part of the plant. Cotton seeds that used to be thrown out are now turned into cottonseed oil and high-protein feed for animals, which makes them more useful in terms of money. New ways of recycling textiles also make it possible to use cotton waste again, which helps the growth of a circular economy. Some imaginative ideas, such as turning cotton dust into packaging that can be composted, make attempts to be more sustainable even better (RecoverTM, 2024). These actions show a bigger commitment to ethical sourcing, production, and recycling that will help the fashion industry become more sustainable. When you harvest cotton, you get about 5–7 tons of byproducts per hectare. These include stalks, leaves, bolls, and seed-laden lint, which are all known as cotton straw or cotton wood. Most of the time, the irrigated types are dry, but some can still have up to 30% green leaves. These leftovers are generally burned, buried, or plowed back into the ground because they could attract pests (Huang et al., 2012). But when the levels of pesticides are safe, these things can be used again as organic fertilizer, firewood, briquettes, or animal feed (Saeidy, 2004). By using zero-waste methods in cotton farming, all byproducts are used in the best way possible. Cotton burrs are being turned into biodegradable packaging (Cotton, 2024), and gin waste with short fibers is being used to make biosynthetic sulfur pigments, which are a better choice than petroleum-based dyes. Zero-waste pattern-making and other methods that reduce waste help get rid of fabric scraps during garment manufacture (Zero Waste Clothing Production, 2024).

Composting cotton fibers also helps improve the soil. Recycling cotton textiles that have already been used also helps reduce the need for new cotton, even if it is currently hard to gather and sort. Cotton might become a zero-waste crop if we keep coming up with new ways to use

FIGURE 19.6 (A) cotton plant; (B) cotton; (C) cotton lint; (D) seeds; (E) oil; (F) cotton cake; (G) meal; and (H) appraisal.

its fibers, seeds, and agricultural waste in a way that does not waste them. But problems that keep coming up, such as pesticide contamination and a lack of recycling facilities, still need to be fixed. Cotton, as an emerging cash crop, has zero waste and its uses are represented in Fig. 19.6.

19.7.1 The range of cotton as a crop with no waste

Cotton is not usually thought of as a zero-waste crop such as flax is, but new technologies and eco-friendly methods are bringing it closer to that status. A zero-waste crop is one where every part of the plant is used in a useful way, which cuts down on or gets rid of waste. Almost all parts of the cotton plant may be reused or recycled, which makes it more environmentally friendly. Cotton seeds are small, oval-shaped, and found inside cotton bolls. They are covered in long, fuzzy threads called lint, which is the main raw material for cotton textiles, and shorter strands called linters. During the processing of cotton, a cotton gin is used to get rid of the lint from the seeds. The seeds are ground up after they are picked to make oil and meal.

The lint, which is the soft, white threads that are spun into yarn and fabric, is the most important portion of the cotton plant. It is regarded for its strength, comfort, and ability to adapt. It is also biodegradable, renewable, and recyclable. Because it is strong and absorbent, it is great for clothes, home textiles, and industrial purposes, including tarps, tents, hotel linens, military uniforms, and even spacesuits for astronauts. You can weave or knit cotton fibers into many different types of fabric, such as velvet, corduroy, chambray, velour, jersey, and flannel. Cotton also works well with both natural fibers such as wool and synthetic fibers such as polyester. People utilize cotton to make many things, such as fishnets, coffee filters, tents, bookbinding, and archival paper. Linters are the short fibers that stay on cottonseeds after they have been ginned. They are very important for making bandages, swabs, banknotes, cotton buds, and X-ray film. Cotton is one of the most valuable natural fibers in the world since it can be used for so many things in both everyday life and industry (Cotton Australia: Uses of Cotton).

Cottonseed oil is used a lot, especially in countries that grow cotton, such as the United States. Because it has a high smoke point, it is great for frying, sautéing, and baking. In addition to being used in cooking, it is also a major part of soap, margarine, emulsifiers, cosmetics, pharmaceuticals, rubber, and plastics. The oil has no cholesterol and a lot of polyunsaturated fats. It also has antioxidants such as vitamin E, which help protect cells from oxidative stress and keep skin healthy. Its phytosterols cut Low-Density Lipoprotein (LDL) cholesterol, which is good for heart health, and its anti-inflammatory actions reduce inflammation. People typically use cottonseed oil in skin care products because it keeps skin moist and safe. After harvest, cotton stalks and stems are left in the field. They are being looked into as a renewable energy source for biomass fuel, compost to improve soil health and fertility, particleboard and paper that can be made into low-grade boards and biodegradable paper products, and mushroom cultivation substrate in some areas. The short fibers, sticks, leaves, and dirt that come from ginning are no longer thought of as trash. Instead, they are used as animal bedding for livestock and poultry, as organic matter to improve soil structure, and as fillers in the automotive and construction sectors. Cottonseed meal is a byproduct of getting oil out of cotton. It is a high-protein feed that is good for cattle. Cottonseed hulls are also a good source of nourishment for cows since they add important nutrients to their diets (Sekhar et al., 2025). Cottonseed press cake has a lot of protein, up to 49% dry matter. This is 3–6 six times more than cereals and up to 20 times more than some fodder crops. It is also the plant-based feed with the most phosphorus. Presscake can also be used as fertilizer and as a substrate for growing mushrooms in agriculture. Presscake is sold all around the world, and most producers send it to other countries.

19.8 Pineapple

Pineapple is a short, tropical, herbaceous perennial plant that belongs to the Ananas genus and the Bromeliaceae family (Hazarika et al., 2017). It usually grows to be 1.5 meters tall and does best in tropical and coastal areas. India has about 2,250,000 acres of land where pineapples are farmed (Mulla, 1985). The plant does best when it gets 100–150 cm of rain each year and temperatures between 22 and 32°C. Pineapple comes from Central and South America, although it is now grown all around the world. Each kilogram of pineapple leaves (PLs) makes roughly 15–18 strands of fiber that look like silk (Textile Coach, 2020). The main goal of growing pineapples is to make juice, jam, concentrates, and candied fruit. This processing creates byproducts such as bark, peel, and bagasse, which make up 25%–35% of the fruit's total weight (Larrauri et al., 1997). These byproducts have a lot of potential for turning biomass into nanomaterials that can be used to strengthen composites (Pereria et al., 2021).

Pineapple is not only a fruit; it also has several uses in industry. The hull is turned into animal feed and organic fertilizer, while the leaf fibers are used to make paper. People are looking at turning the fiber waste from processing facilities into handmade paper that might be sold. Researchers have looked at many parts of plants, such as leaves, seeds, crowns, peels, stalks, straw, roots, bark, and branches, to see how they could be used in nanocellulose, paper making, textiles, and strengthening polymer matrices. Researchers are also looking at using pineapple byproducts in adsorbents, composites, biodiesel synthesis, hydrogels, aerogels, and nanocrystals (Pereira et al., 2021).

Pineapple is a great source of vitamin C, manganese, vitamin B6, copper, folate, potassium, and antioxidants. It also has bromelain, an enzyme that is known to help with inflammation. However, only around 52% of the fruit is eaten right away. The rest, which is high in lignin and cellulose, is usually thrown away. This waste can be used to make textile fibers, which can be mixed with polyester or silk to make fabrics that are good for the environment. The peel has lignocellulosic parts that can be turned into bioethanol. It also has bioactive chemicals such as

FIGURE 19.7 This picture shows a pineapple plant, fruit, leaves, fiber, and fiber-based products, including jam, jelly, and juice. *Source: www.google.com.*

salicylic acid, myricetin, and ferulic acid (Li et al., 2014), which could be used in the cosmetics and pharmaceutical industries (Rivera et al., 2023).

19.8.1 Pineapple farming with no waste

The pineapple plant takes its name from its fruit, which looks like a pine cone. This tropical fruit tastes great and is good for you since it has a lot of vitamin C, manganese, vitamin B6, copper, thiamin, folate, potassium, magnesium, niacin, riboflavin, and iron. Pineapples are the only food known to contain bromelain, a group of enzymes that break down protein and have anti-inflammatory and pain-relieving effects. Pineapples are also full of flavonoids and phenolic acids, which are strong antioxidants that protect cells from free radicals that can cause chronic diseases. The fruit has a lot of lignin and cellulose in it as well. Even though it has health benefits, only around 52% of the fruit is eaten or used to make things such as jam. The remainder is thrown away as organic waste. But this garbage has a lot of cellulose and lignin in it, which makes it good for getting fiber out of it. When mixed with polyester or silk, these fibers can make fabrics that are soft and silky and last a long time.

The peel has lignocellulosic components, including cellulose, hemicellulose, and lignin, that can be broken down into simple sugars for bioethanol and other products that are worth more. Scientists have shown that pineapple peel has antioxidant chemicals such as salicylic acid, myricetin, trans-cinnamic acid, tannic acid, and p-coumaric acid. Li et al. (2014) discovered phenolics such as gallic acid, epicatechin, catechin, and ferulic acid, whereas Sepúlveda et al. (2018) said that ferulic and syringic acids have antibacterial characteristics. According to Rivera et al. (2023), these chemicals are useful in the pharmaceutical, cosmetic, and food industries. Its uses are depicted in Fig. 19.7, which represents it as a zero-waste crop with every part of the plant as a source of raw material for the production of various products with nutritionally important both for humans and the soil.

PLs are a lot of agricultural waste in India, especially in the North East. They are high in cellulose and have great mechanical qualities, which makes them a cheap, renewable resource for the paper industry. PL are biodegradable, cheap, and grow back every year. They have a hydrophobic waxy layer and 2.5%–3.5% fiber inside. Fiber extraction alone isn't cost-effective, but the residual sludge can be utilized for vermicomposting and other things. PL trash is not being used to its full potential right now, which makes things hard for farmers. PL has a holocellulose content of 40%–87.7%, which makes it a very useful feedstock in the circular economy. Pineapple leaf fiber (PALF) is a long-lasting option for many sectors since it has a lot of cellulose (70%–82%), is strong, and is good for the environment. PALF, which is made up of 70%–82% cellulose, 5%–12% lignin, and 1.1% ash (Praveen, 2020), is a renewable supply of strong, environmentally friendly paper in the paper and pulp business, where cutting down trees leads to deforestation. PALF is useful in the automotive, packaging, and biological fields because of its structural and chemical properties (Saha et al., 2021).

Pineapple fiber: The long leaves of the pineapple plant are used to make pineapple fibers, which are a sustainable and eco-friendly option for many uses. People usually throw away these leaves, which are a byproduct of pineapple growing. Using them to make fiber is a renewable way to use waste. The staple fibers are between 0.5 and 1 m long, and they are long, shiny, and strong. They can be mixed with other fibers to replace jute and other bast fibers. PALFs are great for making fine yarns because they have a better chemical makeup and mechanical strength than jute. They are also utilized in industrial settings, such as tires and conveyor belts, as well as in heavier textiles for upholstery and furnishings. The fibers are also widely used in clothing and technical textiles, which helps reduce waste and provide jobs in rural areas.

The leaves have a waxy covering, and they go through a cleaning and decortication procedure to get rid of the outer coating. The fibrous residue is sticky and is soaked in water mixed with urea to speed up the retting process. After that, it is dried. Pineapple fiber is a natural and biodegradable substance that does not make microplastics. This means less trash in landfills and less harm to the environment. It gives the global fashion industry a clean and legal way to make things when it is handled properly. The fiber is white, silky, and shiny, like silk. It has a high tensile strength and is softer than other natural fibers. It has a lot of cellulose, which makes it very hydrophilic (Cherian et al., 2010).

PALF's natural whiteness, smooth texture, high tensile strength, stiffness, and hydrophilic nature similar to cotton (82.7% cellulose) make it a good choice for recyclable, nonfood-based composite materials. It is as strong as jute and has great flexural stiffness for usage in industry. The semicrystalline form of cellulose makes it possible to make nanocellulose by using chemo-mechanical bleaching to separate it, then removing lignin, grinding it, and sonication. This process produces nanocellulose that may be sold. PALF is used to make "Pinatex," a nonwoven fabric that looks like animal leather, for leather replacements. Pinatex is 85%–90% biodegradable and is made up of 80% pineapple fiber and 20% polylactic acid. It is soft, resilient, breathable, and resistant to heat, which makes it similar to leather made from animals (Mazotto, 2022). PALF is also a source of bioethanol. When COSLIF is used to treat pineapple waste, which is high in lignocellulose, it increases the conversion of glucose to 84%, resulting in 212 L of ethanol per ton of dry leaves (Chen et al., 2020; Mund et al., 2021; Saini et al., 2022). PALF is a more environmentally friendly option for fake eyelashes in cosmetics. Expert evaluations showed that Widowati and Amalia (2021) had an 88.75% chance of success.

FIGURE 19.8 Sun hemp and how it helps with a zero-waste approach: As a cover crop (flower, pod, and seed), as feed, as retting, as fiber, as byproducts, as garments, as paper, and as biocompost.

19.9 *Crotalaria juncea* or sunn hemp

Sunn hemp (*Crotalaria juncea*) is an annual leguminous plant from the Fabaceae family. It is one of the oldest fiber-yielding crops that grows naturally in the Indian subcontinent (Bhatt and Saha, 2014). It has been grown for a long time because it may be used in many ways, such as for green manure, making fiber, and improving soil. Sunn hemp is a bast fiber crop that grows quickly and is generally planted close together to keep the leaves from growing sideways. It does best in loamy, well-drained soils, although it can also grow in poor, dry soils. It grows to a height of 2.5–3 m and takes 60–90 days to mature. It is often grown with rice, maize, and cotton. The plant has pointed green leaves that are 5–7.5 cm long and bright yellow flowers that grow in bunches at the leaf axils. The crop is cut down when the pods are ready and then retted, stripped, washed, and dried to get the fiber (Bhandari et al., 2022). Sunn hemp is the second most important bast fiber after jute. It is stronger, lighter in color, and more resistant to mildew, dampness, and microbial deterioration (Bhatt and Saha, 2014). Its fibers are thick, coarse, stiff, and shiny. They can be white, gray, or yellow, and they are solid, such as hemp and stronger than jute. Sunn hemp is very helpful for sustainable farming in terms of agriculture. It can fix nitrogen from the air because it works with *Rhizobium* bacteria. This makes the soil more fertile. Its thick roots make the soil stronger and stop erosion. Also, its fast growth and thick canopy keep weeds from growing, which means you don't have to use as many chemical herbicides. It is a great cover crop that helps the environment and makes the soil more organic. Fig 19.8 represents its uses as a zero-waste crop for the benefit of humans, plants, and soil.

Sunn hemp is valuable for its high biomass output, ability to keep weeds from growing, resilience to pests and diseases, and ability to fix nitrogen in the soil (Maheshwari et al., 2020). The plant gives forth roughly 3% fiber when it is fresh and 13% when it is dry. Sunn hemp fibers come in different grades: White-grade I, Ganjam or green-grade II, and dew Huddy-grade III. They can be used to make high-quality rope or lower-quality things such as cigarette papers and cheap cordage (Chaudhary et al., 2010). Sunn hemp is a fast-growing legume that is used for both its green manure and its natural fiber. Sunn hemp is a plant that grows well in India and can even survive in bad weather, such as drought, saltiness, and alkalinity. It grows quickly, which makes it a great green manure crop. Sunn hemp helps the soil hold on to nutrients by stopping leaching and keeping moisture in the soil. The plant makes the soil healthier by changing its structure, storing more water, and stopping erosion. Sunn hemp can add up to 120 pounds of nitrogen per acre, which makes the soil much richer. It can also stop weeds from growing by up to 90% (Anne, 2018).

Because the leaves of sunn hemp are high in nitrogen and break down quickly, they are great for green manuring. Adding sunn hemp to the soil raises the levels of organic carbon and makes nutrients more available, which makes the soil more fertile. Researchers have shown that utilizing sunn hemp as a green manure improves the organic matter in the soil and helps with sustainable farming (Bhayal et al., 2018).

- **Fiber:** The bast fibers that come from the stems of the sunn hemp plant are also very useful. These fibers shine and come in shades of white, gray, or yellow. The fibers are between 1 and 1.5 m (3.3 and 5 ft) long and are made up of cylindrical cells with striations. Sunn hemp fibers are almost as strong as hemp and last longer than jute. Notably, they get stronger when wet and are quite resistant to mildew. These traits make sunn hemp great for making ropes, fishing nets, sacking textiles, canvas, and rug yarns. The paper industry also uses it, mostly to make tissue and cigarette papers. The high cellulose and low lignin content of sunn hemp fibers make them better for use in industry. They are a green option in many fields because they are strong, last a long time, and break down naturally (Bhayal et al., 2018). This is especially true for making rope, twine, and similar items.
- **Animal feed:** Sunn hemp is useful for more than just farming and industry; it can also be good for livestock. Cattle do not like the raw seeds, but they can be processed and mixed with other feed to make a healthy supplement for livestock. This is good for the economy and encourages people to use the whole harvest (Hemantha Kumar et al., 2019).
- **Biofuel:** Sunn hemp has a lot of fiber and nitrogen (N), which makes it a good base for digestion. Sinbuathong et al. (2019) say that mixing sunn hemp with cow manure for digestion could be a great way to make biogas. However, its lignocellulosic structure is hard to break down and does not break down easily by microbes, which limits biogas production. To get around this problem, we need to use pretreatment procedures to break down the hard lignocellulosic matrix and make it easier to digest (Rajendran et al., 2018). Sadhukhan and Sarkar (2015) also talked about how sunn hemp seed oil could be a good source for making biodiesel, but some changes need to be made to make it work better as a biofuel.

The **paper industry** uses the fiber to make high-quality, eco-friendly paper, especially for packaging and stationery, because it has long, strong fibers. In the past, traditional Islamic paper from North India was made from a mix of raw sunn hemp fiber and recycled sunn hemp material.

Sunn hemp is known for its strong, saltwater-resistant fibers, which make it great for manufacturing ropes and textiles. People often use it to manufacture fishing nets, threads, and ropes that are stronger than jute but not quite as strong as real hemp. The coarse fibers are also used in marine caulking to seal boat joints during the retting process (Chaudhury et al., 2012). Sunn hemp fibers are turned into strong, rough fabrics that are great for industrial uses, including hoses, belts, and canvas. In places such as the United Kingdom and Belgium, the fiber is used to make a lot of different things, such as twines, cords, mats, sacking, tarpaulins, and even shoes, such as sandals and shoe bottoms. Sunn hemp is also used in domestic textiles including rugs, carpets, webbing, and table and bed linens, which shows how versatile it is (Satya et al., 2016). **Composites and bioplastics:**Chaudhary et al. (2010) talked about how sunn hemp fibers could be used to strengthen waste polyethylene bags to make cheap thermoplastic composite materials. These composites are quite robust, therefore, they may be used in many different ways, such as for door panels, furniture, and packaging. Using sunn hemp this way helps reduce pollution from throwing away plastic bags. Raj et al. (2024) also showed that sunn hemp-reinforced poly(lactic acid) green composites are tougher when they break because they lock together, and they also break down in a regulated way. The main economic significance of sunn hemp lies in its bast fibers, which are extracted from the plant's stems. These fibers are widely used in the production of ropes, twines, fishing nets, and other similar products. The suitability of sunn hemp fibers for these applications is attributed to their chemical makeup—high cellulose content and low lignin levels—making them both strong and environmentally beneficial (Bhandari et al., 2022).

Sunn hemp serves as an effective cover crop, offering multiple agronomic benefits such as controlling soil erosion, enhancing soil structure, and improving fertility. Its ability to produce substantial biomass and fix large quantities of atmospheric nitrogen contributes significantly to overall soil health and productivity (Schomberg et al., 2007). Because of its fast growth and high biomass yield, sunn hemp is highly effective in weed suppression. When cultivated as a cover crop, it forms a dense canopy that limits sunlight penetration, thereby reducing the emergence and growth of weeds (Collins et al., 2007).

19.10 Sisal

Sisal (*Agave sisalana*), a hardy member of the Asparagaceae family, is renowned for its robust leaf fibers that have been utilized since pre-Columbian times. Native to Central America, the plant gained global economic importance in the late 19th century with the rise in demand for low-cost twine. This led to widespread cultivation in regions such as the Bahamas, Brazil, Tanzania, Kenya, and Mozambique. The plant forms a rosette of rigid, lance-shaped leaves measuring 60–180 cm (2–6 ft) long and grows up to 90 cm (3 ft) tall. It flowers once after 4–8 years, producing a tall stalk with yellow flowers, after which it dies. However, it regenerates naturally through bulbils, small plantlets that root and grow into new plants.

Sisal is highly valued for its strong, durable fibers, primarily used in products such as ropes, twines, mats, rugs, brushes, and industrial textiles. The fibers, extracted through a mechanical decortication process, are creamy white, glossy, coarse, and measure 100–125 cm (40–50 in.) in length. These qualities make them ideal for marine and industrial applications because of their tensile strength, stretchability, dye absorption, and saltwater resistance (Ashby, 2013). What makes sisal truly remarkable is its potential as a zero-waste crop. Despite only 4%–5% of the plant's biomass being used for fiber extraction, the remaining 95% can be repurposed across various sectors (Santos et al., 2015). The leftover biomass, once seen as waste, is now utilized to generate biogas, organic fertilizers, eco-friendly paper, biochar, and even animal feed. It also serves as a substrate for mushroom cultivation and as a source of valuable compounds for pharmaceuticals, biofuels, and cosmetic products. Cultivated in warm, well-drained soils, sisal is propagated using bulbils or rhizomes. Young plants spend 12–18 months in nurseries before transplantation to the field, where they mature in 3–5 years. They remain productive for about 7–8 years, yielding up to 300 leaves throughout their lifespan. The first harvest typically provides 70 leaves, while subsequent annual yields average around 25 leaves. Sisal's ability to grow in arid regions with minimal inputs enhances both its environmental and economic viability. As industries increasingly seek sustainable alternatives, sisal stands out as a model of the circular economy and resource efficiency. Its comprehensive utility—from fiber to energy and biochemicals—demonstrates how an underutilized crop can help mitigate environmental challenges while driving sustainable innovation.

19.10.1 Plant parts of the sisal for zero waste

Sisal (*Agave sisalana*) serves not only as a source of durable fiber but also as a multipurpose crop with extensive industrial and environmental applications. Traditionally, sisal is planted along earthen embankments and cattle-proof trench-cum-mound structures to protect cultivated or forested land from erosion and predatory animals.

- **Composites and industrial applications:** Sisal fiber composites show strong market potential in sectors such as automotive, aerospace, marine, and construction, offering a sustainable alternative to fiberglass and plastic reinforcements (Li et al., 2000). These composites are already used in the manufacture of ropes, mats, carpets, and other products, positioning sisal as a competitive material for future eco-friendly innovations.
- **Sisal pulp for paper**: The sisal spinning industry generates significant textile waste, often ending up in landfills. Mechanical (>90%) and semichemimechanical (>75%) pulping of sisal fibers yields high pulp volumes. Bosco et al. (2022) explored recycling this fiber waste into paper pulp, finding that it outperforms many other nonwood materials. This pulp is suitable for making products such as cigarette paper, banknotes, and specialty papers because of its strength, cost-effectiveness, and eco-friendliness. It also enhances the durability of recycled paper and grows faster than traditional wood sources.
- **Biogas and renewable energy:** Only about 2% of the sisal plant is used for fiber; the remaining 98%, including leaf residues, boles, roots, and poles—holds significant potential. With appropriate investment and technology, this biomass can fuel innovation and generate additional income. For example, Katani Ltd. in Hale operates the first commercial sisal biogas plant, developed by UNIDO and funded by the CFC. This facility converts 96% of the postdecorticated leaf into electricity. The resulting biogas residue serves as a high-quality organic fertilizer, enhancing crop yields and supporting organic farming. Kivaisi and Mshandete et al. (2017) found that codigesting sisal waste with cow dung in a continuous stirred tank reactor produced about 400 L of methane per cubic meter daily, along with 260 L of superior biogas manure.
- **Animal feed and fertilizer:** Sisal's year-round growth and arid-land adaptability make it suitable for commercial animal feed production. Its residual biomass is commonly used for feed and organic fertilizer. The biogas byproduct, or digestate, from anaerobic digestion has demonstrated greater efficacy than traditional NPK fertilizers, offering a sustainable option for agriculture (Kivaisi et al., 2017).
- **Chemical extraction and biobased products:** Sisal juice extracted from boles and leaves contains valuable chemicals such as inulin, sugars, acids, and saponins, which are used in biofuels, preservatives, and food products. The biomass also yields pharmaceuticals such as cortisones and waxes, and is a rich source of antioxidants. Advanced processes have isolated

FIGURE 19.9 Sisal and its value addition for zero waste approach: plant, extraction, fiber, byproducts, biocomposte, biochar, clothes, and sisal-based cosmetics. *Source: www.google.com.*

pectin, mannitol, succinic acid, and kaempferol, useful in chemical and pharmaceutical formulations (Santos et al., 2015).

- **Mushroom cultivation and cosmetics:** Sisal agroindustrial waste is a viable substrate for oyster mushroom cultivation. Biological efficiencies range from 26% to 86%, with commercial yields reaching 250 kg of mushrooms per ton of wet substrate (Kivaisi et al., 2017). Research also supports the use of sisal residues in producing high-protein mushrooms (do Carmo et al., 2021) and in developing antioxidant-rich cosmetic products (Daher et al., 2023).
- **Biochar and activated carbon:** With a lignocellulosic makeup of roughly 44% cellulose, 19% hemicellulose, and 19% lignin, sisal waste is an ideal feedstock for biochar and activated carbon. Utilization of sisal as a zero waste is represented in Fig. 19.9.

19.11 Conclusion

The widespread appreciation of fiber crops such as jute, ramie, banana, sisal, mesta, sunn hemp, flax, cotton, and pineapple marks a major shift in agriculture and industry, where sustainability, economic growth, and caring for the environment all come together. These natural fiber crops are a good example of a holistic zero-waste model, in which every part of the plant, from the main fibers to the leaves, roots, stems, and even the garbage left over after processing, is put to good use in many different areas. These crops have a lot of potential to power a regenerative, circular economy since they can be used in so many different ways, such as in textiles, bioplastics, bioenergy, food, pharmaceuticals, and building materials.

Industries can cut down on their usage of nonrenewable, petroleum-based resources and the damage that synthetic fibers and industrial waste do to the environment by making the most of the biomass and byproducts of these crops. Ramie, mesta, and sunn hemp are among the most useful crops since they can grow in poor soils, do not need a lot of inputs, and can be used in textiles, animal feed, soil improvement, and energy production. At the same time, bananas and pineapples show how crops that are usually grown for food can also be important for fiber innovation. They can be utilized as biodegradable, sustainable alternatives to synthetic materials, and they can help rural people make more money through value-added entrepreneurship.

This zero-waste method has a lot of good effects on the environment. Many of these crops are good for the soil, stop erosion, and need fewer chemicals, which is in line with sustainable farming practices. Using agricultural waste to make biochar, biogas, organic fertilizers, and sophisticated biomaterials not only reduces pollution and greenhouse gas emissions but also adds value to the economy and helps store carbon. Using fiber waste in eco-friendly packaging, paper, composites, and textiles is an example of how businesses can move toward alternatives that are less harmful to the environment and break down naturally.

Adopting zero-waste solutions diversifies income for farmers and processors, which promotes rural economic resilience. From entrepreneurial success stories in banana fiber invention in India to the increasing applications of jute and flax in nanotechnology and high-performance materials, these crops offer opportunities for both grassroots and high-tech innovation. Furthermore, developments in research and breeding, such as high-yielding mesta types or biorefined cotton byproducts, contribute to the practicality and scalability of zero-waste fiber systems.

To summarize, natural fiber crops provide an inspiring example for sustainable growth in which agriculture, the environment, and industry coexist in harmony. Embracing their zero-waste potential is not only an environmental imperative but also an economic opportunity, allowing for the development of climate-resilient systems, the promotion of circular economies, and the path to a greener, more egalitarian future. As global demand turns toward sustainable materials and ethical production, these fiber crops are poised to lead the way to a regenerative and waste-free bioeconomy.

References

Adlercreutz, H., 2007. Lignans and human health. Crit. Rev. Clin. Lab. Sci. 44 (5–6), 483–525. doi:10.1080/10408360701612942.

Al-Hashimi, A.G., 2012. Antioxidant and antibacterial activities of *Hibiscus sabdariffa* L. extracts. Afr. J. Food Sci. 6 (21), 506–511.

Amiri, A., Krosbakken, T., Schoen, W., Theisen, D., Ulven, C.A., 2018. Design and manufacturing of a hybrid flax/carbon fiber composite bicycle frame. Proc. Inst. Mech. Eng. Part P J. Sports Eng. Technol. 232 (1), 28–38.

Ansorena, D., Astiasarán, I., 2004. The use of linseed oil improves nutritional quality of the lipid fraction of dry-fermented sausages. Food Chem. 87 (1), 69–74.

Anne C. Randle (2018). Sunn Hemp: A Super Crop for Forage and Soil Health. Eco Friendly Daily. https://www.ecofriendlydaily.com/

Ashby, M.F., 2013. Materials and the Environment, 2nd ed. Butterworth-Heinemann, Elsevier, Oxford.

Aziz, M., Islam, S., Islam, N., Hasan, M.M., 2020. Jute as a sustainable and versatile bio-material for bio-medical applications. Int. J. Biol. Macromol. 163, 758–768.

Beejmohun, V., Fliniaux, O., Grand, E., Lamblin, F., Bensaddek, L., Christen, P., Kovensky, J., Fliniaux, M.A., Mesnard, F., 2007. Microwave-assisted extraction of the main phenolic compounds in flaxseed. Phytochem. Anal. 18 (4), 275–282. doi:10.1002/pca.973.

Bhandari, H.R., Shivakumar, K.V., Kar, C.S., Bera, A., Meena, J.K., 2022. Sunn hemp: a climate-smart crop. In: Jha, U.C., Nayyar, H., Agrawal, S.K., Siddique, K.H.M. (Eds.), Developing climate resilient grain and forage legumes. Springer, Singapore doi:10.1007/978-981-16-9848-4_13.

Bhatt, K.C., Saha, D., 2014. Indigenous knowledge on fibre extraction of sunnhemp in Bundelkhand region. Indian J. Nat. Prod. Resour. 5 (1), 92–96.

Bhayal, D., Khaddar, V.K., Bhayal, L., Yadav, T.C., Bangar, K.S., Singh, B., 2018. Effect of sunnhemp green manuring and intercropping on soil properties. Int. J. Curr. Microbiol. Appl. Sci. 7 (12), 371–384.

Bosco, M., Mbise, E., Minia, R.J.A., 2022. Production of paper pulp using sisal fiber waste from sisal spinning processes. Tanzan. J. Eng. Technol. 41 (2).

Cerny, P., Babenko, M., Bartos, P., Kononets, Y., Kriz, P., Rabenseifer, R., Spalek, F., 2024. Complex study of the composite building material based on flax straw organic waste: hygrothermal and physicochemical properties. Waste Biomass Valorization 15, 2231–2247. doi:10.1007/s12649-023-02273-7.

Chand, N., Fahim, M., 2008. Jute reinforced polymer composites. In: Chand, N., Fahim, M. (Eds.), Tribology of natural fiber polymer composites. Woodhead Publishing, pp. 108–128.

Chaudhary, S., Borkar, S., Mantha, S., 2010. Sunnhemp fiber-reinforced waste polyethylene bag composites. J. Reinf. Plast. Compos. 29 (15), 2241–2252. doi:10.1177/0731684409345615.

Chaudhury, J., D.P., Singh, S.K., Hazra, S., 2012. Sunnhemp (Crotalaria juncea L.). Barrackpore: CRIJAF (ICAR), 743, 24-Parganas (North), West Bengal. http://assamagribusiness.nic.in/Sunnhemp.pdf.

Chen, A., Guan, Y.J., Bustamante Roman, M., Uribe, L., Murillo-Roos, M., Liu, Y., 2020. Production of renewable fuel and value-added bio-products using pineapple leaves in Costa Rica. Biomass Bioenergy 141 (3), 105675.

Chen, X., Yang, X., 2017. Nanocellulose: A versatile material for energy storage, water treatment, and bio-medical applications. ACS Sustain. Chem. Eng. 5 (3), 1800–1827.

Cheng, G., Zhang, L., He, P., Yan, F., Xiao, B., Xu, T., Jiang, C., Zhang, Y., Guo, D., 2011. Pyrolysis of ramie residue: kinetic study and fuel gas produced in a cyclone furnace. Bioresour. Technol. 102 (3), 3451–3456.

Cherian, B.M., Leão, A.L., de Souza, S.F., Thomas, S., Pothan, L.A., Kottaisamy, M., 2010. Isolation of nanocellulose from pineapple leaf fibres by steam explosion. Carbohydr. Polym. 81 (3), 720–725.

Choo, W.S., Birch, J., Dufour, J.P., 2007. Physicochemical and quality characteristics of cold-pressed flaxseed oils. J. Food Compos. Anal. 20 (3–4), 202–211.

Collins, A.S., Chase, C.A., Stall, W.M., Hutchinson, C.M., 2007. Competitiveness of three leguminous cover crops with yellow nutsedge (Cyperus esculentus) and smooth pigweed (Amaranthus hybridus). Weed Sci. 55, 613–618.

Cotton Incorporated. (2024). Cotton Sustainability Basics. CottonWorks. Available at: https://cottonworks.com/ (Accessed 21 July 2026).

Daher, C.C., Barreto, S.M.A.G., de Brito Damasceno, G.A., de Santana Oliveira, A., Leite, P.I.P., Reginaldo, F.P.S., Escudeiro, C.C., Ostrosky, E.A., Giordani, R.B., Ferrari, M., 2023. Use of sisal industrial waste (Agave sisalana Perrine) in sustainable and multifunctional cosmetic products. Int. J. Cosmet. Sci. 45 (6), 815–833. doi:10.1111/ics.12890.

Department of Agriculture & Farmers Welfare. (2023). Agricultural Statistics at a Glance 2023. Directorate of Economics and Statistics, Ministry of Agriculture & Farmers Welfare, Government of India, New Delhi. Available at: https://nccd.gov.in/uploads/Agricultural_Statistics_at_a_Glance_2023_64c3ac54bf.pdf

Dhar, P., Kar, C.S., Ojha, D., Pandey, S.K., Mitra, J., 2015. Chemistry, phytotechnology, pharmacology and nutraceutical functions of kenaf (Hibiscus cannabinus L.) and roselle (Hibiscus sabdariffa L.) seed oil: an overview. Ind. Crops Prod. 77, 323–332.

do Carmo, C.O., da Silva, R.M., de Souza Rodrigues, M., Soares, A.C., 2021. Bio-conversion of sisal agro-industrial waste into high protein oyster mushrooms. Bioresour. Technol. Rep. 14, 100657.

Dupasquier, C.M., Dibrov, E., Kneesh, A.L., Cheung, P.K., Lee, K.G., Alexander, H.K., Yeganeh, B.K., Moghadasian, M.H., Pierce, G.N., 2007. Dietary flaxseed inhibits atherosclerosis in the LDL receptor-deficient mouse in part through antiproliferative and anti-inflammatory actions. Am. J. Physiol. Heart Circ. Physiol. 293 (4), H2394–H2402.

Edel, A.L., Rodriguez-Leyva, D., Maddaford, T.G., Caligiuri, S.P., Austria, J.A., Weighell, W., Guzman, R., Aliani, M., Pierce, G.N., 2015. Dietary flaxseed independently lowers circulating cholesterol and lowers it beyond the effects of cholesterol-lowering medications alone in patients with peripheral artery disease. J. Nutr. 145 (4), 749–757.

Ganguly, P.K., Bhaduri, S.K., Day, A., 2004. Jute caddies: a potential raw material for handmade paper. J. Sci. Ind. Res. 63, 417–419.

Ghorai, A.K., Mazumdar, S.P., Datta, D., 2021. Jute and mesta stick charcoal production using smokeless fire in Kon-Tiki-Kiln, an open earth pyrolysis process. Int. J. Curr. Microbiol. Appl. Sci. 10, 2319–7706. doi:10.20546/ijcmas.2021.1002.xx.

Ghorai, A.K., Mitra, S., Mahapatra, B.S., 2015. Improved production technology of jute and mesta in India: an overview. Indian Farming 60 (3).

Ghosh, B.K., Jethi, A., 2013. Growth and instability in world jute production: a disaggregated analysis. Int. J. Electron. Commun. Technol. 4, 191–195.

Goh, K.K.T., Ye, A., Dale, N., 2006. Characterisation of ice cream containing flaxseed oil. Int. Dairy J. 41 (8), 946–953.

Gökmen, V., Mogol, B.A., Lumaga, R.B., Fogliano, V., Kaplun, Z., Shimoni, E., 2011. Development of functional bread containing nanoencapsulated omega-3 fatty acids. J. Food Eng. 105 (4), 585–591.

Gonzalez-Esquerra, R., Leeson, S., 2000. Effects of menhaden oil and flaxseed in broiler diets on sensory quality and lipid composition of poultry meat. Br. Poult. Sci. 41 (4), 481–488.

Goyal, A., Patel, A., Sihag, M.K., Shah, N., Tanwar, B., 2018. Therapeutic potential of flaxseed. In: Therapeutic, probiotic, and unconventional foods. Academic Press, London, pp. 255–274.

Gupta, R., Shivhare, U.S., 2012. Optimization of process variables for the development of flaxseed and defatted flaxseed-based pasta. Journal of Nutrition & Food Sciences 2 (10), 1–9.

Hazarika, D., Gogoi, N., Jose, S., Das, R., Basu, G., 2017. Exploration of future prospects of Indian pineapple leaf, an agro waste for textile application. J. Cleaner Prod. 141, 580–586.

Hemantha Kumar, J., Ravi Kumar, K., Srinivas, D., Sri Ranjitha, P., Raghu Rami Reddy, P., Prasad, Y.G., Prasad, J.V., 2019. Paddy-sunhemp system as an alternative resilient technology to paddy–fallow system. Int. J. Curr. Microbiol. Appl. Sci. 8 (3), 1654–1658.

Huang, W., Bai, Z., Hoefel, D., Hu, Q., Lv, X., Zhuang, G., Xu, S., Qi, H., Zhang, H., 2012. Effects of cotton straw amendment on soil fertility and microbial communities. Front. Environ. Sci. Eng. (6) 336–349.

Ilyas, R.A., Sapuan, S.M., Kirubaanand, W., Zahfiq, Z.M., Atikah, M.S.N., Ibrahim, R., Asyraf, M.R.M., Ishak, M.R., Norrrahim, M.N.F., Nurazzi, N.M., Razman, M.R., Abral, H., Syafiq, R., Sabaruddin, F.A., 2021. Roselle: production, product, development, and composites. In: Sapuan, S.M. (Ed.), Roselle: production, processing, products and biocomposites. Academic Press, London, pp. 1–24.

Islam, A.A., Osman, M.B., Mohamad, M.B., Islam, A.M., 2021. Vegetable mesta (Hibiscus sabdariffa L. var. sabdariffa): a potential industrial crop for Southeast Asia. In: Sapuan, S.M. (Ed.), Roselle: production, processing, products and biocomposites. Academic Press, London, pp. 25–42.

Islam, M., 2019. Food and medicinal values of roselle (Hibiscus sabdariffa L. Linn e Malvaceae) plant parts. Rev. Fed. Am. Health Syst. 1, 14–20.

Ismail, A., Ikram, E.H.K., Nazri, H.S.M., 2008. Roselle (Hibiscus sabdariffa L.) seeds nutritional composition, protein quality and health benefits. Food 2 (1), 1–16.

Jabbar, M.A., Islam, M.N., Islam, M.S., 2020. Preparation, characterization and mechanical properties of nanocellulose/epoxy-resin composites. Compos. Struct. 238, 111711.

Jabbar, A., Militký, J., Wiener, J., Kale, B.M., Ali, U., Rwahwire, S., 2017. Nanocellulose coated woven jute/green epoxy composites: Characterization of mechanical and dynamic mechanical behavior. Composite Structures 161, 340–349. https://doi.org/10.1016/j.compstruct.2016.11.062.

Jain, M., Singh, C., Gupta, K., Jain, P., 2014. Formulation and sensory evaluation of healthy heart foods. Nutraceuticals 3, 1–5.

Jamilah, U., Hidayatillah, N., Hidayah, E., 2026. Physical and mechanical properties of chitosan bioplastics with ramie fiber concentration variations. J. Sains Mater. Indones. 27 (2), 136–143.

Kajla, P., Sharma, A., Sood, D.R., 2015. Flaxseed—a potential functional food source. J. Food Sci. Technol. 52 (4), 1857–1871.

Karimah, A., Ridho, M.R., Munawar, S.S., Ismadi, Amin, Y., Damayanti, R., Lubis, M.A.R., Wulandari, A.P., Nurindah, Iswanto, A.H., Fudholi, A., Asrofi, M., Saedah, E., Sari, N.H., Pratama, B.R., Fatriasari, W., Nawawi, D.S., Rangappa, S.M., & Siengchin, S. (2021). A comprehensive review on natural fibers: Technological and socio-economical aspects. Polymers, 13(24), 4280. https://doi.org/10.3390/polym13244280

Kaur, P., Waghmare, R., Kumar, V., Rasane, P., Kaur, S., Gat, Y., 2018. Recent advances in utilization of flaxseed as potential source for value addition. OCL 25 (3), A304.

Kivaisi, A., Mshandete, A., 2017. Exploring technological strategies for valorization of solid sisal waste: a research review. Tanz. J. Sci. 43 (1), 47–61.

Kumari, P., Gaur, S. S., Tiwari, R. K., 2023. Banana and its by-products: A comprehensive review on its nutritional composition and pharmacological benefits. eFood 4 (5), e110.

Larrauri, J.A., Rupérez, P., Saura Calixto, F., 1997. Pineapple shell as a source of dietary fiber with associated polyphenols. J. Agric. Food Chem. 45 (10), 4028–4031.

Lavanya, A.K., Sharma, A., Choudhary, S.B., Sharma, H.K., Nain, P.K.S., Singh, S., Nain, L., 2019. Mesta (Hibiscus spp.) – a potential feedstock for bio-ethanol production. Energy Sources Part A 42 (21), 2664–2677. doi:10.1080/15567036.2019.1618980.

Li, T., Shen, P., Liu, W., Liu, C., Liang, R., Yan, N., Chen, J., 2014. Major polyphenolics in pineapple peels and their antioxidant interactions. Int. J. Food Prop. 17 (8), 1805–1817.

Lee, S.K., Lee, Z.X., Lim, Y.Y., Liew, K.B., Akowuah, G.A., Chew, Y.L., 2022. Ramie (Boehmeria nivea): A functional food and herbal medicine for inflammatory bowel disease (IBD) and the promotion of optimal health. Nat. Prod. J. 12 (7), 53–61.

Li, Y., Mai, Y.W., Ye, L., 2000. Sisal fibre and its composites: a review of recent developments. Compos. Sci. Technol. 60 (11), 2037–2051.

Maheshwari, N.K., Singh, R.P., Manchanda, G., Dubey, R.C., Maheshwari, D.K., 2020. Sunn hemp (*Crotalaria juncea*) nodulating bacteria capable of high antagonistic potential and plant growth promotion attributes: sun hemp nodulating rhizobia. J. Microbiol. Biotechnol. Food Sci. 10 (3), 385–389.

Martinchik, A.N., Baturin, A.K., Zubtsov, V.V., VIu, M., 2012. Nutritional value and functional properties of flaxseed. Vopr. Pitan. 81 (3), 4–10.

Mazotto, Ana., 2022. Biopolymers in the Leather Industry doi:10.1007/978-3-030-98392-5_12.

Mitra, S., Saha, S., Guha, B., Chakrabarti, K., Satya, P., Sharma, A.K., Gawande, S., Kumar, M., Saha, M., 2014. Ramie: the strongest bast fibre of nature. Technical Bulletin No. 8/2013. Barrackpore, Kolkata: Central Research Institute for Jute and Allied Fibres (CRIJAF), Indian Council of Agricultural Research (ICAR). 10.13140/2.1.3519.5842.</bib>

Mondal, D., Bandyopadhyay, A.K., 2014. Adoption of jute production technology in West Bengal. Econ. Aff. 59, 701–709.

Mund, N.K., Dash, D., Mishra, P., Nayak, N.R., 2021. Cellulose solvent–based pretreatment and enzymatic hydrolysis of pineapple leaf waste bio-mass for efficient release of glucose towards bio-fuel production. Biomass Convers. Biorefinery doi:10.1007/s13399-021-01512-1.

Mulla, H. B. (1985). Problems and Prospects of Khadi and Village Industries in India (1951–1983). Ph.D. thesis, Department of Economics, Shivaji University, Kolhapur, Maharashtra, India.

Nasiri, V., Heidarlou, H.B., Alchin, A.A., Moradi, F., Rahmanian, S., Afshari, S., Maftei, C., Griess, V.C., 2023. How do conservation policies, climate and socioeconomic changes impact Hyrcanian forests of northern Iran? Ecological Informatics 78, 102351.

Nayak, L.K., Ammayappan, Ray, D.P., 2012. Conversion of jute caddies (Jute Mill Waste) into value added products: a review. Asian J. Text. 2, 1–5.

Nayak, L.K., Roy, A.K., Das, S., 2011. Some characteristics of jute caddies with reference to briquetting and gasification. J. Indian Chem. Soc. 88, 599–601.

Oomah, B.D., 2001. Flaxseed as a functional food source. J. Sci. Food Agric. 81 (9), 889–894.

Pandey, S.K., Satya, P., Mitra, S., Alam, N.M., 2023. Identification of jute and allied fibre crop varieties for central release. Crijaf Bull. 1 (1), 7.

Pandey, S.K., Satya, P., Mitra, S., Alam, N.M., 2023. Release and notification of jute and mesta varieties. Crijaf Bull. 1 (1), 6–7.

Pereira, P.H.F., Ornaghi, H.L., Arantes, V., Cioffi, M.O.H., 2021. Effect of chemical treatment of pineapple crown fiber in the production, chemical composition, crystalline structure, thermal stability and thermal degradation kinetic properties of cellulosic materials. Carbohydr. Res. 499, 108227.

Praveen Kumar, A., 2020. Pineapple leaf fibers: potential green resources for pulp and paper production. In: Jawaid, M., Asim, M., Tahir, P., Nasir, M. (Eds.), Pineapple leaf fibers. Springer, Singapore, pp. 297–308. doi:10.1007/978-981-15-1416-6_15.

Raj, H., Tripathi, S., Bauri, S., Mandal, S., Choudhury, A.M., Maiti, P., 2024. Sunn hemp reinforced poly(lactic acid) green composites: Controlled bio-degradation and improved fracture toughness through interlocking mechanism. Ind. Crops Prod. 223, 120245. doi:10.1016/j.indcrop.2024.120245.

Rajendran, K., Drielak, E., Sudarshan Varma, V., Muthusamy, S., Kumar, G., 2018. Updates on the pretreatment of lignocellulosic feedstocks for bio-energy production – a review. Biomass Convers. Biorefin. 8, 471–483.

Ramamoorthy, K., Dhanraj, R., Vijayakumar, N., Ma, Y., Al Obaid, S., Narayanan, M., 2024. Vegetable and fruit wastes: Valuable source for organic fertilizer for effective growth of short-term crops: Solanum lycopersicum and Capsicum annuum. Environmental Research 251, 118727.

Ramesh, M., Deepa, C., 2024. Processing and properties of jute (Corchorus olitorius L.) fibres and their sustainable composite materials: A review. Journal of Materials Chemistry A 12, 1923–1997.

Rao, P.P., Rao, G.N., Mala, K.S., Balaswamy, K., Satyanarayana, A., 2013. Preparation and storage stability of flaxseed chutney powder, a functional food adjunct. J. Food Sci. Technol. 50, 129–134.

Rathi, P., Mogra, R., 2012. Development and sensory evaluation of superior products with flaxseed. International Journal of Food Science and Nutrition 2, 12–15.

Recover™. (2024). Zero-Waste Innovation: Transforming Cotton Dust into Compostable Packaging with Ecovia. Available at: https://recoverfiber.com/newsroom/zero-waste-compostable-packaging-ecovia (Accessed 21 July 2026).

Rehman, M., Gang, D., Liu, Q., Chen, Y., Wang, B., Peng, D., Liu, L., 2019. Ramie, a multipurpose crop: Potential applications, constraints and improvement strategies. Ind. Crops Prod. 137, 300–307. doi:10.1016/j.indcrop.2019.05.029.

Rivera, A.M.P., Toro, C.R., Londoño, L., Bolivar, G., Ascacio, J.A., Aguilar, C.N., 2023. Bio-processing of pineapple waste bio-mass for sustainable production of bio-active compounds with high antioxidant activity. Food Measure 17, 586–606. doi:10.1007/s11694-022-01627-4.

Rowell, R.M. (Ed.), 2012. Handbook of Wood Chemistry and Wood Composites, 2nd ed. CRC Press, Boca Raton, FL.

Sadhukhan, S., Sarkar, U., 2016. Production of bio-diesel from *Crotalaria juncea* (sunn-hemp) oil using catalytic trans-esterification: process optimisation using a factorial and Box–Behnken design. Waste Biomass Valorization 7, 343–355.

Saha, A., Kumar, S., Kumar, A., 2021. Influence of pineapple leaf particulate on mechanical, thermal and bio-degradation characteristics of pineapple leaf fiber reinforced polymer composite. J. Polym. Res. 28.

Saini, R., Chen, C.W., Patel, A.K., Saini, J.K., Dong, C.D., Singhania, R.R., 2022. Valorization of pineapple leaves waste for the production of bio-ethanol. Bioengineering 9, 557.

Santos, J.D.G., Vieira, I.J.C., Braz-Filho, R., Branco, A., 2015. Chemicals from *Agave sisalana* bio-mass: isolation and identification. Int. J. Mol. Sci. 16, 8761–8771. doi:10.3390/ijms16048761.

Satya, P., Banerjee, R., Karan, M., Mukhopadhyay, E., Chaudhary, B., Bera, A., Maruthi, R.T., Sarkar, S.K., 2016. Insight into genetic relation and diversity of cultivated and semi-domesticated under-utilized *Crotalaria* species gained using start codon targeted (SCoT) markers. Biochem. Syst. Ecol. 66, 24–32.

Satya, P., Maiti, R., 2013. Bast and leaf fibre crops: Kenaf, hemp, jute, agave, etc. In: Singh, B.P. (Ed.), Bio-fuel crops: production, physiology and genetics. CABI, Wallingford, pp. 292–311.

Satya, P., Pandey, S., 2024. Breeding mesta (kenaf/roselle): from conventional to 21st century applications. In: Kar, G. (Ed.), A systematic overview of research developments in jute and allied fibre crops. Cambridge Scholars Publishing, Cambridge, pp. 72–102.

Saeidy, E.E., (2004). Technological fundamentals of briquetting cotton stalks as a bio-fuel. PhD thesis. Faculty of Agriculture and Horticulture, Humboldt University of Berlin, Germany.

Schomberg, H.H., Martini, N.L., Diaz-Perez, J.C., Phatak, S.C., Balkcom, K.S., Bhardwaj, H.L., 2007. Potential for using sunn-hemp as a source of bio-mass and nitrogen for the Piedmont and coastal plain regions of the southeastern USA. Agron. J. 99, 1448–1457.

Sekhar, S.C., Teja, A., Emmaniel, R., Basha, S.C., Sekhar, C.C., 2025. Cottonseed hulls as a source of roughage for animals: An overview. Indian J. Small Rumin. 31 (2), 123–135.

Sepúlveda, L., Romaní, A., Aguilar, C.N., Teixeira, J., 2018. Valorization of pineapple waste for the extraction of bioactive compounds and glycosides using autohydrolysis. Innovative Food Sci. Emerg. Technol. 47, 38–45.

Shafiul, I., Islam, M.N., Islam, M.S., 2020. Cellulose–gelatin nanocomposites reinforced with jute nanocarbon for bio-medical applications. Carbohydr. Polym. 228, 115943.

Sharma, L., Alam, N.M., Roy, S., Satya, P., Kar, G., Ghosh, S., Goswami, Tinku, Majumdar, Bijan, 2023. Optimization of alkali pretreatment and enzymatic saccharification of jute (*Corchorus olitorius* L.) biomass using response surface methodology. Bioresour. Technol. 368, 128318.

Sharma, L., Satya, P., Dev, B., Alam, N.M., Goswami, T., Roy, S., Pandey, S.K., 2024. Combined NaOH–Laccase mediator system assisted pretreatment and additive mediated saccharification enhances bio-ethanol production from kenaf (*Hibiscus cannabinus* L.) bio-mass. SSRN 1–24.

Sharma, L., Satya, P., Dev, B., Goswami, T., Alam, N.M., Roy, S., Singh, A.K., Pandey, S.K., Majumdar, B., Lavanya, A.K., Kar, G., 2025. Enhanced bioethanol production from sequential alkali-laccase-hydroxybenzotriazole pretreatment and additive-mediated saccharification in kenaf (Hibiscus cannabinus L.). Biomass and Bioenergy 201, 108123. https://doi.org/10.1016/j.biombioe.2025.108123.

Sinbuathong, N., Khun-Anake, R., Sawanon, S., 2019. Bio-gas production from sunn hemp. Int. J. Glob. Warm. 19 (1–2), 24–36.

Sirag, N., Elhadi, M.M., Algaili, A.M., Hassan, H.M., Ohaj, M., 2014. Determination of total phenolic content and antioxidant activity of *Roselle* (Hibiscus sabdariffa L.) calyx ethanolic extract. Stand. Res. J. Pharm. Pharmacol. 1 (2), 34–39.

Stavropoulos, P., Mavroeidis, A., Papadopoulos, G., Roussis, I., Bilalis, D., Kakabouki, I., 2023. On the path towards a "Greener" EU: a mini review on flax (*Linum usitatissimum* L.) as a case study. Plants 12, 1102.

Sungatullina, A., Petrova, T., Kharina, M., Mikshina, P., Nikitina, E., 2023. Effect of flaxseed mucilage on the probiotic, antioxidant, and structural-mechanical properties of different Lactobacillus cells. Fermentation 9 (5), 486.

Textile Coach, 2020. Pineapple Fibre/ Pineapple Leaf Fibre (PALEF). https://www.textilecoach.net/post/pineapple-fiber-pineapple-leaf-fibre-pale.

The Editors of Encyclopaedia Britannica. (2024). Cotton. Encyclopaedia Britannica. https://www.britannica.com/plant/cotton-fibre.

Touré, A., Xueming, X., 2010. Flaxseed lignans: Source, bio-synthesis, metabolism, antioxidant activity, bioactive components, and health benefits. Compr. Rev. Food Sci. Food Saf. 9 (3), 261–269. doi:10.1111/j.1541-4337.2009.00105.x.

Trevisan, A.J., Arêas, J.A., 2012. Development of corn and flaxseed snacks with high-fibre content using response surface methodology (RSM). Int. J. Food Sci. Nutr. 63 (3), 362–367.

Uddin Nizam, M., 2016. Phytoremediation potential of kenaf (*Hibiscus cannabinus* L.), mesta (*Hibiscus sabdariffa* L.) and jute (*Corchorus capsularis* L.) in arsenic-contaminated soil. Korean J. Environ. Agric. 35 (2), 111–120.

Vastrad, J.V., Pujari, M., Sannapapamma, K., Mahale, G., 2025. Blending mesta with natural colour cotton: a novel approach. Fibre2Fashion 1–5.

Weigmann, H.-D., 2025. Cotton. Encyclopaedia Britannica. https://www.britannica.com/topic/cotton-fibre-and-plant. [Accessed 4 March 2025].

Widowati, T., Amalia, N.S.A., 2021. Utilization of pineapple leaf fiber (*Ananas comosus*) as material for false eyelashes production. IOP Conf. Ser.: Earth Environ. Sci. 700, 012034.

Wiesenborn, D., Tostenson, K., Kangas, N., 2005. Processing flaxseed for food and feed uses. Food Sci. Biotechnol. 14 (3), 305–310.

Wilson, F.D., 2006. A distributional and cytological survey of the presently recognized taxa of *Hibiscus* section *Furcaria* (Malvaceae). Bonplandia 53–62.

Wu, W., Huff, H.E., Hsieh, F., 2007. Processing and properties of extruded flaxseed-corn puff. Journal of Food Processing and Preservation 31, 211–226.

Xu, L., Wei, Z., Guo, B., Bai, R., Liu, J., Li, Y., Sun, W., Jiang, X., Li, X., Pi, Y., 2022. Flaxseed meal and its application in animal husbandry: a review. Agriculture 12, 2027.

Xu, L., Wei, Z., Guo, B., Bai, R., Liu, J., Li, Y., Sun, W., Jiang, X., Li, X., Pi, Y., & others. (2022). Flaxseed meal and its application in animal husbandry: A review. Agriculture, 12(12), 2027.

Yasin, M., Gangan, S., Panchal, S.K., 2025. Banana peels: A genuine waste or a wonderful opportunity? Applied Sciences 15 (6), 3195. https://doi.org/10.3390/app15063195.

Yaver, E., 2023. Dephytinized flaxseed flours by phytase enzyme and fermentation: functional ingredients to enhance the nutritional quality of noodles. J. Sci. Food Agric. 103 (4), 1946–1953.

Zaini, H.M., Saallah, S., Roslan, J., Sulaiman, N.S., Munsu, E., Wahab, N.A., Pindi, W., 2023. Banana biomass waste: A prospective nanocellulose source and its potential application in food industry—A review. Heliyon 9 (8), e18734.

Zero Waste Clothing Production: Best Practices (2024). Sustainable Sourcing. Available at : https://www.oncatex.com/post/zero-waste-clothing-production-best-practices.

CHAPTER

20

Utilization of agroindustrial waste in ornamental fish culture

Mukesh Kumar Bairwa[a,*], *Dharmendra Kumar Meena*[b], *Biswajit Dash*[a] *and Saroj Kumar Swain*[a]

[a]ICAR-Central Institute of Freshwater Aquaculture, Bhubaneswar, Odisha, India
[b]ICAR-Central Inland Fisheries Research Institute, Barrackpore, Kolkata, West Bengal, India
*Corresponding author: mukeshbairwa2@gmail.com

20.1 Introduction

The aquaculture sector is one of the fastest-growing food production industries globally, and this growth extends beyond food fish to include the vibrant and economically significant ornamental fish trade. Currently, India's contribution to the global ornamental fish market is modest at just 0.01%, yet it possesses immense scope and demand in the domestic market, focused on captive breeding. The shift toward culture-based production systems, instead of wild collection, signifies a critical move toward sustainable trade in the ornamental fish industry. As a result, there is an increasing demand for controlled and reliable inputs, particularly in terms of feed.

In this context, the use of agroindustrial waste emerges as an important strategy, offering a sustainable and cost-effective alternative to conventional feed sources, which often impose significant economic burdens and environmental problems. A key challenge in aquaculture arises from the increasing costs of traditional fish feed, which frequently constitutes around 50% of the operational expense. Conventional feeds usually contain fishmeal and fish oil from small pelagic fishes, which can be used for human consumption. This practice raises serious concerns about resource competition and the long-term sustainability of harvesting small wild fish for the above purposes.

With this, aquaculture operations also produce waste, including uneaten food and fish fecal matter. The mishandling of aquaculture waste can lead to pollution, eutrophication, ecosystem disruption, and disease outbreaks. The collective pressures of high operational costs and the remarkable environmental impact associated with traditional fish feed and waste management

Zero-Waste Technologies in Aquaculture: Global Perspectives and Methods toward a Circular Economy
DOI: https://doi.org/10.1016/B978-0-443-34689-7.00010-0

create a convincing need to explore alternative ways out. This effort goes beyond simply finding an alternative; it pursues a sustainable option that addresses multiple serious issues: Reducing operational costs, decreasing dependence on wild-captured fish for feed, and mitigating the environmental effects of fish farming waste. Therefore adaptation of agroindustrial waste is not just a required choice but a fundamental necessity for the future of the fish farming industry. Adopting a shift from linear production models to more circular resource management may show a crucial path toward economic viability and environmental responsibility.

Agroindustrial waste consists of a diverse group of byproducts produced from agricultural production and processing activities. The agrowaste includes crop residues, various types of animal manure, and waste materials from food processing plants. Usually, these agro-wastes are discarded and often contribute to environmental pollution. These wastes are highly rich in valuable compounds, including essential macronutrients such as proteins, lipids, and starch, along with micronutrients, dietary fibers, and various bioactive compounds, such as phenolic compounds, terpenes, and β-glucans. Characterizing agrowaste as a source of "bioactive compounds" and "potentially functional compounds" elevates its value beyond meager caloric or protein content. This recommends that agrowaste can not only fulfil the basic nutritional requirements of fish but also improve their overall health, increase immune function, and support a healthy gut microbiota, which ultimately leads to greater resilience and vitality for farmed fish.

Utilizing agroindustrial waste in aquaculture is guided by an insightful "triple objective," i.e., (1) to add value to existing production chains, (2) significantly reduce environmental pollution, and (3) enhance the well-being of farmed organisms through improved nutrition. This multifaceted approach seamlessly supports the principles of a circular economy, which advises the "reduce, reuse, and recycle" of resources. The consistent emphasis on this "triple objective" will serve as a foundational principle for sustainable aquaculture. It signifies a paradigm shift, transforming the perception of waste from a liability into a valuable asset capable of generating economic, environmental, and biological returns simultaneously. This holistic perspective is crucial for guiding policy development and encouraging widespread industry adoption, as it frames agrowaste utilization as a systemic solution rather than an isolated practice. The success in integrating agrowaste will be measured not only by metrics such as fish growth rates or cost savings but also by its comprehensive contribution towards a more resilient and sustainable fish industry.

20.2 Types and characteristics of agroindustrial wastes for aquaculture

The potential of agroindustrial waste in aquaculture is underscored by the diverse range of waste streams available, each offering unique characteristics suitable for various applications. These can be broadly categorized as follows:

Crop residues: This category includes readily available materials such as rice bran, maize bran, wheat bran, and groundnut oil cake (GNOC), mustard oil cake (MOC). Various edible plant leaves, including cassava leaves and moringa leaves, also fall into this group. These materials are frequently used as feed inputs in small-scale aquaculture. In the case of ornamental fish, floating fish feed made from the above ingredients, including skin color enhancer ingredients such as spirulina, is used.

Animal manures and byproducts: Waste from livestock such as chicken, pig, and cattle manures serve as substrates for the cultivation of housefly larvae (Musca).[1] Animal waste and its extracts have also been successfully incorporated directly into the pond to enhance natural food (zooplankton).

Fish processing waste (FPW): This consists of discarded parts of fish from wholesalers. It is a locally abundant and high-quality protein source, comparable in nutritional value to conventional fishmeal.[7]

Organic food waste: The wastage of different fruits and vegetables is a good source of all the nutrients, with a significant amount of different carotenoids. The high concentration of carotenoids helps to enhance the skin color of ornamental fish, which is a prime factor in getting a higher price for ornamental fish.

Aquaculture sludge/wastewater solids: These are nutrient-rich solids generated as byproducts from aquaculture operations themselves, specifically retrieved from hatchery settling ponds and recirculating aquaculture systems (RAS) effluents. These waste materials can be effectively recycled within or across various aquaculture systems.

The wide range of agrowaste, including crop, animal, fish processing, food, and even aquaculture's own waste, highlights the broader applicability and scope of localized, context-specific solutions. This suggests that there is no single "agrowaste" solution that can be commonly applicable to all scenarios. Instead, a customized approach is often required, one that considers the specific type of waste available in a particular region and integrate them with the unique needs of the aquaculture operation. This adaptability raises the development of localized circular economies and opens avenues for intersectoral collaboration between agriculture and aquaculture (Table 20.1).

20.3 Nutritional profiles of agroindustrial wastes

Agroindustrial wastes are not only available in bulk, but also they are valuable reservoirs of essential nutrients and functional compounds. They are rich sources of proteins, lipids, starch, micronutrients, and dietary fibers, all crucial for fish growth and health (Renalda and Hieromin, 2021). Beyond these fundamental components, these wastes also contain a range of bioactive compounds, i.e., terpenes, carotenoids, and β-glucans, which have demonstrated properties such as antioxidant, antimicrobial, gut microbiota modulation, and immune system enhancement (Leyva et al., 2020). FPW, for instance, is particularly noted for its high-quality protein content, which is equivalent to that found in fishmeal. Similarly, organic food waste, which has a protein content of 10%–25%, can be significantly increased through fermentation, making it suitable as a feed ingredient.

The presence of both essential macronutrients (proteins, lipids, and starch) and specific bioactive compounds in agrowaste suggests a dual advantage for aquaculture. It indicates that these materials can fulfill basic nutritional requirements while simultaneously contributing to enhanced fish health and disease resistance. These properties make this agrowaste more suitable for ornamental fish. In this sector, most desirable traits such as vibrant coloration, intricate fin shape, and robust disease resistance are highly valued (Shinde and Sukhdhane, 2023). Therefore the use of agrowaste can contribute to more hardy and healthier ornamental fish stocks, potentially reducing the need for chemical treatments and ultimately improving overall farm productivity and profitability. The use of agrowastes also directly supports the objective of "improved well-being of organisms" within the broader sustainability framework.

TABLE 20.1 Major agroindustrial wastes and their potential nutritional value for ornamental fish feed.

Waste type	Primary nutritional components	Potential benefits to fish	References
Rice bran, maize bran, wheat bran, groundnut bran	Starch, lipids, fiber, protein	Growth, low-cost feed	Renalda and Hieromin (2021)
Cassava leaves, moringa leaf, *Brassica oleracea*	Protein, fiber, micronutrients	Growth performance, immune system, disease resistance	Renalda and Hieromin (2021)
Animal manure (chicken, pig, cattle)	Organic matter, minerals, nitrogen	Stimulates phytoplankton growth, substrate for live feeds	Renalda and Hieromin (2021)
Fish processing waste (FPW)	High-quality protein, lipids	Growth, feed cost reduction, water stability	Renalda and Hieromin (2021)
Organic food waste	Protein (10%–22%), lipids, carbohydrates	Growth, feed availability, improved digestibility (after fermentation)	Cahya et al. (2021)
Aquaculture sludge/wastewater solids	Nitrogen, phosphorus, organic matter	Nutrient recycling, soil amendment, substrate for vermicomposting	Kumar et al. (2025)
Algae	Protein, lipids, fatty acids	Feed for young hatchlings/fry, Omega-3 source	Cahya et al. (2021)
Earthworms, tubifex, mosquito larvae, daphnia, Black soldier fly (BSF) larvae, polychaete worms	High protein, lipids, omega-3	Direct live feed, growth, reproductive performance, bioremediation	Campanati et al. (2021)

20.4 Applications of agroindustrial waste in ornamental fish farming

The versatility of agroindustrial waste makes it suitable for its application in ornamental fish farming in multiple capacities, ranging from direct feed ingredients to enhancers of natural pond productivity and substrates for valuable live foods.

20.4.1 Agrowaste as direct feed ingredients

Agrowaste can be directly incorporated into fish diets, either as primary ingredients or as supplements, offering a cost-effective and nutritionally beneficial alternative to conventional feeds. Crop residues, for instance, are commonly utilized as pond inputs in small-scale aquaculture, with materials such as rice bran, maize bran, and seed cakes proving effective (Renalda and Hieromin, 2021). Studies involving wheat bran, rice bran, and groundnut bran have also explored their economic effectiveness in fish production, though growth rates can vary depending on the specific byproduct used. Beyond grains, edible plant leaves, such as cassava leaves and moringa leaves, have been experimentally shown to enhance fish growth performance, bolster their immune systems, and improve disease resistance. FPW stands out as a high-quality protein source, comparable to fishmeal. It can be transformed into a water-stable supplemental feed through the pelletizing method with the help of an extruder. Organic food waste, with its inherent protein content ranging from 10% to 22%, represents another viable basic material for fish feed (Cahya et al., 2021). Crucially, fermentation can significantly enhance its nutritional profile, reportedly increasing protein content by more than 50% while reducing crude fiber and fat, thereby improving overall feed quality and digestibility (Cahya et al., 2021).

The suitability of these agrowaste-derived feeds varies considerably depending on the specific ornamental fish species and their life stages. Fish have diverse nutritional needs; herbivorous species thrive on plant-protein–based mixtures (e.g., soy, corn, and vegetable oils), while carnivorous fish typically require fish oils and proteins, though plant proteins can also be included. This highlights the importance of tailoring agrowaste utilization to the dietary habits of the target ornamental species. For the most delicate life stages, such as fry, the smallest feed size is essential, and they begin seeking food shortly after yolk absorption (Adhikari and Fedler, 2020). Livebearer fry, being larger, can commence feeding immediately postbirth, whereas egg layer fry require a few days for yolk sac absorption before they start actively foraging (Adhikari and Fedler, 2020). Furthermore, the nutritional demands of broodstock are paramount for successful reproductive performance and the quality of gametes, often necessitating formulated diets supplemented with natural live foods (Shinde and Sukhdhane, 2023).

20.4.2 Use of agrowaste for culture of live fish food

Live food sources are highly essential in ornamental farming because these live foods not only provide superior nutrition but also stimulate natural feeding behaviors. The live food is often used to condition adult fish for breeding, as it signals a state of optimal health and nutrition. Live foods are also essential for many fish larvae; they are too small to consume commercial dry foods and require microscopic live organisms, such as rotifers, microalgae, or newly hatched brine shrimp as their first food. The small-scale farmers, who often face financial constraints and cannot afford commercial pellet feeds or brine shrimp larvae, heavily rely on low-cost alternative live foods (Shinde and Sukhdhane, 2023). Common options include green water, water fleas (daphnia, cyclops), tubifex (sludge worms), mosquito larvae, bloodworm, and chopped earthworms (Shinde and Sukhdhane, 2023).

These live feeds are often collected locally, with some individuals even making a livelihood from their collection and sale (Shinde and Sukhdhane, 2023). Earthworms can be cultured in soil (Adhikari and Fedler, 2020), while sludge worms and mosquito larvae are frequently gathered from sewage water channels (Shinde and Sukhdhane, 2023). The culture of tubifex can be done in running drain water from a fish tank, where these worms feed on fish excreta and unutilized feed. Tiny algae species, such as *Protococcus*, *Chlamydomonas*, and *Chlorella*, serve as crucial initial feeds for young hatchlings and fry after yolk absorption (Adhikari and Fedler, 2020). Furthermore, bioconversion methods leverage agrowaste as a medium for growing organisms such as housefly larvae or various worms, which are then harvested as high-protein components for animal feed (Renalda and Hieromin, 2021). Vermicomposting of aquaculture solids, for instance, produces nutrient-rich vermicompost, and the worms themselves can be sold as fishing bait or processed into supplemental feed. Marine polychaete worms, fed on aquaculture effluents, have been shown to provide a suitable nutritional balance for shrimp and fish, including essential omega-3 fatty acids (Kumar et al., 2025).

20.4.3 Agrowaste as pond fertilizers and water conditioners

Beyond direct feed, agroindustrial waste plays a crucial role in enhancing natural pond productivity and conditioning water quality. When animal waste and crop residues are introduced into fish ponds, they can be directly consumed by filter-feeding fish or decompose to form a significant portion of the detritus. More importantly, these organic resources stimulate the proliferation of phytoplankton, microscopic algae rich in protein that form the foundational basis of the pond's food web, thereby supporting the growth of a wide range of herbivorous and omnivorous fish (Renalda and Hieromin, 2021). In this context, wastes serve a triple function: As direct feed, as a source of minerals for autotrophic production (e.g., algae), and as a source of organic matter for heterotrophic production (e.g., bacteria and detritivores).

The extensive discussion of agrowaste as pond fertilizer and the reciprocal use of aquaculture waste as agricultural fertilizer underscores a powerful circular economy model. This signifies a symbiotic relationship where waste from one sector transforms into a valuable input for another, thereby reducing the overall environmental burden and enhancing resource efficiency across the entire agrifood system, not just within the confines of aquaculture. By effectively closing nutrient loops, this integrated system diminishes the need for external, often energy-intensive, synthetic fertilizers in agriculture, while concurrently managing aquaculture waste and mitigating its environmental impact. This creates a resilient, integrated farming system that optimizes resource flow and minimizes waste, offering a compelling blueprint for sustainable food production that extends beyond the boundaries of a single farm or sector, moving toward a regenerative agricultural paradigm.

20.4.4 Agrowaste as carotenoid supplements for color enhancement of ornamental fishes

Carotenoid supplements are used in ornamental fish feed to enhance their coloration, as fish cannot produce these pigments naturally. Various sources of carotenoids, both natural and synthetic, can be incorporated into fish diets to achieve desired color intensity and hues, such as red, orange, and yellow. Popular choices include astaxanthin, lutein, and zeaxanthin. The agrowaste is a good source of carotenoids, and scientific reports also support their use in fish feed as carotenoid supplements (Table 20.2). The primary function of these compounds in fish farming is to provide the precursors for the pigments that are deposited in the integument, scales, and muscle tissues. However, the role of carotenoids extends far beyond pigmentation. They are also vital for a range of physiological processes, including enhancing immune response, acting as powerful antioxidants, and supporting reproductive health

20.5 Benefits of utilizing agroindustrial waste in ornamental fish farming

The strategic integration of agroindustrial waste into ornamental fish farming offers a multitude of benefits, spanning economic advantages, environmental sustainability, improved fish health, and significant socioeconomic impacts.

TABLE 20.2 Agrowaste as carotenoid supplements in fish feed.

Agrowaste source	Key carotenoids (concentration)	Effects on fish	Scientific references
Carrot	Beta-carotene: Ranges from approximately 6.6 to 11.5 mg per 100 g fresh weight, but can be much higher in dried powders	Promotes growth, enhances red/orange pigmentation, and acts as an antioxidant. Studies show a direct correlation between dietary carrot inclusion and increased carotenoid deposition in fish tissues	Haldar et al. (2016), Kaur et al. (2016), Sarmah et al. (2019)
Marigold flowers	Lutein: A dominant carotenoid, often comprising up to 70% of the total carotenoids. Dried marigold flower meal can contain high concentrations, with some extracts reaching 27.22 mg/g. Also contains zeaxanthin	Excellent source of xanthophylls for yellow and orange pigmentation. Studies show improved growth, feed utilization, and enhanced skin coloration	Ezhil et al. (2008), Alma et al. (2013)
Mango Peels	Beta-carotene, beta-cryptoxanthin, lutein: Peels are a rich source of beta-carotene, with concentrations range of 3.1–6.1 mg/100 g	Acts as a natural antioxidant and provitamin A source. Research has shown improved growth and fry production in broodstock fed with diets containing mango peel silage	Aya et al. (2020)
Papaya	Lycopene: 1.44–3.39 mg/100 g in fresh pulp. It also contains some beta-carotene	Significant enhancement of red body pigmentation, improved growth performance, and higher survival rates. The effects are largely attributed to its high lycopene content	Das and Biswas (2020), Farina et al. (2020)
Crab waste Meal	Crab waste is a very rich natural source of astaxanthin, a highly valued pigment	As effective as synthetic astaxanthin in improving growth, total carotenoid content, and stress resistance. Provides the desirable pink/red coloration in fish flesh	Karadal et al. (2017)
Pumpkin	Beta-carotene: Varies by species, with concentrations of approximately 3.1–4.3 mg/100 g in fresh pulp	Used for flesh coloration and as a provitamin A source	Pailan et al. (2012)

20.5.1 Economic advantages: Cost reduction in feed inputs, added value to waste streams

A primary driver for the adoption of agrowaste in aquaculture is its profound economic advantage. Feed constitutes a major and often the most costliest input in fish farming. Agrowaste provides a significantly cheaper source of functional compounds, directly translating into reduced operational costs. The incorporation of animal manure and its extracts, for example, has been shown to increase fish production at a low cost (Renalda and Hieromin, 2021). Furthermore, the use of organic food waste as a fundamental material for fish feed directly addresses issues of feed availability and substantially lowers purchasing costs. FPW, which often represents a disposal cost for processors, can be valorized into a high-protein feed. This not only mitigates disposal expenses but also reduces overall feeding costs by lessening the dependency on more expensive, often imported, commercial feeds. Ultimately, the exploitation of agroindustrial waste inherently provides added value to existing production chains (Leyva-López et al., 2020). The consistent emphasis on "low cost" (Renalda and Hieromin, 2021) and "reducing feeding costs" clearly demonstrates that economic viability is a paramount factor driving the adoption of agrowaste, particularly for small-scale farmers who operate with limited capital (Shinde and Sukhdhane, 2023).

20.5.2 Environmental sustainability: Waste reduction, pollution mitigation, circular economy principles

The utilization of agroindustrial waste is a keystone of environmental sustainability in aquaculture. It directly contributes to reducing overall pollution by diverting waste materials from landfills or direct discharge into waterways. This approach perfectly symbolizes the principles of a circular economy, which fundamentally focuses on the "reduce, reuse, and recycle" of resources (Renalda and Hieromin, 2021). By valorizing waste from animal and food sources into valuable compounds and inputs, aquaculture actively participates in a regenerative system. For instance, recycling aquaculture wastewater solids (sludge) through land application or vermicomposting not only manages aquaculture-specific waste but also reduces the environmental impacts traditionally associated with mining and producing artificial fertilizers (Kumar et al., 2025). The integration of agrowaste into aquaculture thus represents a transformation from a linear "take-make-dispose" model to a regenerative one. This implies a systemic shift that not only mitigates pollution from existing waste streams but also significantly reduces the environmental footprint linked to the production of conventional feed ingredients and synthetic fertilizers, leading to a net positive environmental impact. This positions agrowaste utilization as a fundamental component of ecological aquaculture, contributing substantially to broader environmental objectives such as biodiversity conservation, improved water quality, and climate change mitigation through reduced waste emissions and decreased resource depletion.

20.5.3 Improved fish health and growth performance

Beyond economic and environmental benefits, agrowaste contributes directly to the health and growth performance of ornamental fish. These waste materials contain a spectrum of bioactive compounds that function as antioxidants, antimicrobials, and modulators of gut microbiota and the immune system (Leyva-López et al., 2020). This nutritional enhancement directly improves the overall well-being of farmed organisms. Specific examples include the use of plant leaves, which have been shown to contribute positively to fish growth performance, immune system function, skin color enhancement, and disease resistance.[1] Similarly, fermented food waste has demonstrated positive effects on fish growth rates and feed conversion ratios (Cahya et al., 2021). Furthermore, polychaete worms, which can be cultivated using aquaculture effluents, provide a suitable nutritional balance for fish, including vital omega-3 fatty acids (Kumar et al., 2025). The evidence suggesting improved fish health and disease resistance from agrowaste components extends beyond mere nutritional adequacy. This indicates that agrowaste can function as a natural prophylactic agent, potentially reducing the reliance on antibiotics and other chemical treatments. This aligns perfectly with sustainable aquaculture practices and directly addresses a major challenge within the industry, namely the responsible management of diseases and chemical use. This positions agrowaste as a valuable tool for proactive health management in ornamental fish farming, leading to more resilient fish stocks, lower operational risks, and ultimately, a more marketable product, such as "antibiotic-free" fish.

20.5.4 Socioeconomic impact: Livelihood generation, self-employment, and women's empowerment in rural areas

The utilization of agroindustrial waste also yields profound socioeconomic benefits, particularly in rural and marginalized communities. Backyard ornamental fish farming has proven to provide positive economic returns for many rural poor, fostering their social and financial empowerment (Bhattacharya et al., 2025). Initiatives such as those undertaken by the Indian Council of Agricultural Research (ICAR)-Central Institute of Freshwater Aquaculture and ICAR-Central Inland Fisheries Research Institute (ICAR-CIFRI) in India exemplify this, empowering tribal and Scheduled Caste women to create self-employment opportunities through ornamental fish farming.[2] These programs demonstrate that with minimal capital investment, a small backyard pond, and the use of homemade fish feed (which often incorporates agrowaste), marginalized women can achieve economic independence (Bhattacharya et al., 2025). Within a mere 5–6 months, women involved in these initiatives began earning significant income, with approximately 58% successfully establishing their own farming units (Bhattacharya et al., 2025). Furthermore, the collection and sale of low-cost live feeds, such as daphnia, tubifex, and mosquito larvae, which are often sourced from local ponds and sewage channels, can provide a viable livelihood for individuals (Shinde and Sukhdhane, 2023). The strong socioeconomic benefits, particularly the empowerment of women and the generation of livelihoods, reveal that agrowaste utilization is not merely a technical solution but a powerful instrument for community development. This implies that promoting the integration of agrowaste can serve as a strategic component of broader rural development programs, fostering sustainable livelihoods and contributing significantly to poverty alleviation, especially in regions abundant with agricultural byproducts and a growing demand for ornamental fish.

20.6 Challenges and mitigation strategies in agrowaste utilization

While the benefits of utilizing agroindustrial waste in ornamental fish farming are substantial, its successful implementation is not without challenges. These primarily revolve around maintaining optimal water quality, controlling disease, ensuring nutritional balance, and navigating regulatory and safety considerations.

20.6.1 Water quality management

The introduction of organic inputs, such as agrowaste, while beneficial, can pose significant risks to water quality if not properly managed. Overfeeding, for instance, leads to the deterioration of water quality because of excess uneaten feed (Bhattacharya et al., 2025). Aquaculture wastewater, laden with nutrients from fish feces and uneaten feed, is often discharged untreated, contributing to eutrophication, depletion of dissolved oxygen (hypoxia) in surface waters, groundwater contamination (e.g., increased nitrate levels), and adverse impacts on soil health, including increased salinity, altered microbial communities, and chemical residues (Ruben et al., 2025)

Mitigation strategies are, therefore, crucial for maintaining a healthy aquatic environment. The regular addition of beneficial bacteria is fundamental; these microorganisms colonize biological pond filters, breaking down harmful ammonia from fish waste and decomposing debris into less toxic nitrates, which can then serve as nutrients for pond plants. Sludge and muck removers, which enhance the activity of beneficial bacteria and enzymes, are used to reduce organic accumulation at the pond bottom. However, their use necessitates good aeration to prevent the depletion of dissolved oxygen, especially in warmer water where oxygen solubility is lower (Lackner and Besharati, 2025). Nitrifying bacteria specifically facilitate the nitrogen cycle, converting ammonia to nitrites and then to nitrates, a process that can take up to eight weeks to complete. Chemical treatments, such as algaecides, can control algae growth, but it is best practice to manually remove as much algae as possible beforehand to minimize nutrient release from dead plant matter (Lackner and Besharati, 2025). Flocculants are employed to bind and clump suspended particles, thereby clarifying cloudy pond water. For more controlled environments, RASs offer a land-based approach where water is continuously reused and treated, significantly reducing pollution and disease spread. The energy generated from byproducts can even support intensified production within RAS (Kumar et al., 2025). Intermediary settlement ponds or filters are also vital for treating wastewater before discharge or recirculation. This implies that while agrowaste offers benefits, it also introduces a higher risk of organic loading and nutrient imbalances, requiring sophisticated biological and mechanical filtration, careful monitoring, and potentially advanced systems such as RAS to maintain a healthy environment for ornamental fish.

20.6.2 Disease control and biosecurity

Disease outbreaks pose a significant threat in aquaculture, particularly in crowded conditions. The use of live feeds, such as water fleas, carries an inherent risk of introducing unwanted parasites or pathogens into the system (Adhikari and Fedler, 2020). Furthermore, untreated waste can harbor a substantial microbial load that contaminates water (Ruben et al., 2025). Effective strategies for disease control and biosecurity are therefore essential. Proper waste treatment, such as composting aquaculture wastewater, can significantly reduce biosolid pathogens (Kumar et al., 2025). Regular monitoring for disease symptoms is a fundamental best practice. Maintaining excellent water quality, as discussed previously, is foundational to minimizing disease risk. When treatments are necessary, they should be applied only for diagnosed diseases, using proper dosages, and with accurate record-keeping. For live feeds such as daphnia, it is advisable to source them from safe ponds because of the difficulty of effective sterilization (Adhikari and Fedler, 2020). Controlled environments, such as land-based systems and RAS, offer superior control over water quality and significantly reduce the spread of disease compared with open systems. The inherent risk of pathogen introduction from raw or improperly treated organic waste necessitates stringent biosecurity protocols. This implies that while agrowaste offers economic benefits, its use must be balanced with robust health management strategies, including careful sourcing, processing, and ongoing monitoring, to prevent disease outbreaks that could negate any cost savings.

20.6.3 Nutritional balance and palatability

Ensuring a complete and balanced diet for ornamental fish, particularly when incorporating diverse agrowaste materials, presents distinct challenges. Fish require approximately 40 essential nutrients, including vitamins, minerals, amino acids (protein building blocks), and certain fats. These nutritional needs vary considerably by species. A significant challenge lies in the limited research on the specific dietary nutritional needs of ornamental fish, especially for broodstock, where comprehensive information is scarce. This knowledge gap extends to optimal diet composition, specific feed ingredients, feeding frequency, and the precise nutritional content required for overall health, reproductive performance, and the quality of gametes and larvae. Ensuring the palatability of novel feed formulations is also crucial for feed uptake and efficacy (Shinde and Sukhdhane, 2023).

Strategies to address these challenges include the development of carefully formulated diets with known nutritional compositions, which are essential for consistent broodstock performance (Shinde and Sukhdhane, 2023). Agrowaste can be integrated into these formulations as a base. Even in the food fish industry, diets are often supplemented with raw or live natural foods (e.g., squid, fish, mussels, and *Artemia*), suggesting that agrowaste could serve a similar supplementary role. Processing methods, such as cooking and proper formulation (e.g., creating fish paste or pellets from FPW), can significantly improve the acceptance and palatability of feeds. Furthermore, the inherent variability in the composition of agro-byproducts can influence fish growth rates (Renalda and Hieromin, 2021), indicating the need for careful selection, blending, and potentially fortification to achieve consistent and optimal nutritional profiles. The significant knowledge gap regarding the specific nutritional requirements of ornamental fish presents a critical challenge for developing effective agrowaste-based feeds. This suggests that while agrowaste offers a cost-effective bulk, extensive research is still needed to develop species-specific formulations that ensure complete nutritional profiles, palatability, and optimal growth and reproduction, particularly for high-value ornamental species.

20.6.4 Regulatory and safety considerations

The utilization of agrowaste in aquaculture operates within a framework of regulatory and safety considerations. In many countries, aquaculture farmers are required to obtain permits and adhere to strict standards for the quality of water discharged from their farms. However, enforcing such regulations across numerous small-scale farms, particularly in regions such as Southeast Asia, can be challenging. There are also concerns regarding the improper or overuse of antibiotics, which can lead to the emergence of antibiotic-resistant bacteria, potentially impacting human health.

Mitigation strategies involve strict adherence to established water quality standards for discharged water. Responsible chemical use is paramount, meaning treatments should only be applied for diagnosed diseases, at proper dosages, and with posttreatment water management before discharge. Crucially, proper processing of agrowaste, such as composting to reduce pathogens (Kumar et al., 2025), is vital for ensuring safety, especially when the waste is in direct contact with fish or subsequently used for agricultural purposes. The mention of strict regulatory standards for aquaculture discharge and the potential for chemical and heavy metal contamination from waste underscores the importance of a robust regulatory framework and continuous monitoring. This implies that while agrowaste offers environmental benefits, its utilization must be carefully managed within a legal and safety compliance context to prevent unintended negative consequences on the environment and human health.

20.7 Future perspectives

The field of using farm and food waste in aquaculture is rapidly advancing day by day. Instead of just using raw waste, future innovations are focused on creating more valuable products and integrated, closed-loop systems. A key trend is enhancing waste's nutritional value through fermentation with probiotics, making it a better feed ingredient. Another promising method is bioconversion, where insects such as black soldier fly larvae or worms turn waste and animal manure into high-protein biomass for fish. Beyond direct feeding, new technologies are emerging for nutrient recycling, such as using biofilters to recover valuable nutrients from aquaculture wastewater. There is also growing interest in generating energy from byproducts to power fish farms, creating a self-sustaining model. The development of alternative feed ingredients, such as oils from algae, further reduces reliance on wild-caught fish. These advancements show a move beyond simple waste application toward sophisticated biotechnologies that maximize resource recovery and minimize environmental impact, potentially leading to new industries centered on waste processing.

20.8 Conclusion

Agroindustrial waste represents a significant, yet often underutilized, resource with transformative potential for the ornamental fish farming sector. Its strategic integration offers a compelling "triple objective" (1) providing added value to existing production chains, (2) substantially reducing environmental pollution, and (3) enhancing the well-being of farmed organisms through improved nutrition. These diverse waste streams are rich in essential nutrients and valuable bioactive compounds, making them highly suitable for various applications. They can serve as direct feed ingredients, effective supplements, or as substrates for culturing crucial live feeds. Beyond direct nutritional inputs, agrowaste and aquaculture byproducts can be effectively utilized as pond fertilizers, stimulating natural productivity and contributing significantly to nutrient recycling within a robust circular economy framework. The success of this utilization, however, hinges on the application of appropriate processing and pretreatment methods, such as fermentation, cooking, and bioconversion, which are vital for enhancing nutritional value, improving digestibility, and ensuring the safety of the final products. The valorization of agrowaste stands poised to transform ornamental fish farming into a more economically viable, environmentally sustainable, and socially equitable industry. It propels the sector toward a true circular economy model, systematically reducing its reliance on finite resources and mitigating its environmental footprint. Simultaneously, it actively fosters rural development and empowers marginalized communities by creating new economic opportunities. This implies a future where waste is systematically integrated into the production cycle, driven by both economic necessity and a profound commitment to environmental stewardship and social responsibility. Continued investment in research and development, coupled with supportive policies and targeted capacity-building initiatives, will be crucial to fully unlock the potential of agrowaste as a cornerstone of future sustainable aquaculture practices.

References

Adhikari, K., Fedler, C., 2020. Water sustainability using pond-in-pond wastewater treatment system: case studies. J. Water Process Eng. 36, 101281.

Alma, M.H., et al., 2013. Effect of dietary marigold flower meal on growth performance, body composition and pigmentation of red swordtail (*Xiphophorus helleri*). J. Fish. Aquat. Sci. 8 (3), 323–328.

Aya, F.A., et al., 2020. Agri-wastes eyed as cheap, eco-friendly source of fish feed. SEAFDEC/AQD. https://www.seafdec.org.ph/2020/agri-wastes-eyed-as-cheap-eco-friendly-source-of-fish-feed/.

Bhattacharya, S., Das, B., Ekka, A., Kumar, P., Roy, A., Lianthuamluaia, L., Chakraborty, S., Saha, A., 2025. Empowering women through alternative sustainable livelihood: a case study of ornamental fish farming in India. Front. Sustain. Food Syst 9. doi:10.3389/fsufs.2025.1507917.

Cahya, M.D., Andriani, Y., Junianto, Y., 2021. Utilization of food waste as raw material for fish feed: a review. Glob. Sci. J. 9 (1), 435–439.

Campanati, C., Willer, D., Schubert, J., Aldridge, D.C., 2021. Sustainable intensification of aquaculture through nutrient recycling and circular economies: more fish, less waste, blue growth. Rev. Fish. Sci. Aquac. 30 (2), 143–169.

Das, S., Biswas, T., 2020. Effect of dietary papaya powder on growth, survival and pigmentation of banded gourami (*Trichogaster fasciata*) fry. J. Aquac. Fish. 8 (3), 221–228.

Ezhil, J., Santhanam, R., Arulvel, S., Uthayakumar, V.S., 2008. Effect of dietary marigold flower meal on pigmentation and growth of tiger barb (*Puntius tetrazona*). Asian J. Anim. Sci. 3 (1), 17–21.

Farina, A., De Melo, M.T.C.P.C., De, B.N.R., et al., 2020. Bioactives from agri-food wastes: present insights and future challenges. Molecules 25 (10), 2390.

Haldar, A., Mukherjee, D., Mandal, A., Adhikari, S., 2016. Effect of different levels of dietary carrot meal on growth performance, body composition and pigmentation of goldfish (*Carassius auratus*). Int. J. Res. Fish. Aquac. 6 (3), 11–18.

Karadal, A., Guroy, D., Turkmen, G., 2017. Evaluation of natural and synthetic carotenoid supplementation on growth, survival, total carotenoid content, fatty acids profile and stress resistance of European Seabass (*Dicentrarchus labrax*) Fry. Aquac. Stud. 17 (3), 163–172.

Kaur, S., Sahu, N.P., Sarma, K., Sinha, P., 2016. Effect of dietary supplementation of dried carrot powder on growth and color enhancement in koi carp (*Cyprinus carpio*). J. Appl. Nat. Sci. 8 (2), 794–798.

Kumar, A., Kumar, D., Ananth, A., Mohammed, A., Singh, G., Kumar, S., Deepa, Madalageri, 2025. Ornamental fish culture: prospects, challenges, and economic significance. J. Surv. Fish. Sci. 10 (1), 17296–17303.

Lackner, M., Besharati, M., 2025. Agricultural waste: challenges and solutions, a review. Waste 3 (2), 18.

Leyva-López, N., Lizárraga-Velázquez, C.E., Hernández, C., Sánchez-Gutiérrez, E.Y., 2020. Exploitation of agro-industrial waste as potential source of bioactive compounds for aquaculture. Foods 9 (7), 843.

Pailan, G.H., Parhi, J., Debbarma, J., 2012. Effect of dietary supplementation of pumpkin pulp on skin colour and growth performance of goldfish. Indian J. Fish. 59 (1), 115–119.

Renalda, M., Hieromin, L., 2021. Animal waste and agro-by-products: Valuable resources for producing fish at low costs in Sub-Sahara. In: Novo de Barros, A., Gouvinhas, I. (Eds.), Innovation in the food sector through the valorization of food and agro-food by-products. IntechOpen. https://doi.org/10.5772/intechopen.95057.

Ruben, M.O., Akinsanola, A.B., Okon, M.E., Shitu, T., Jagunna, I.I., 2025. Emerging challenges in aquaculture: current perspectives and human health implications. Vet World 18 (1), 15–28.

Sarmah, P., Sarma, B., Sharma, A., Das, D.N., 2019. Effect of carotenoid in growth and colour enhancement in gold fish, *Carassius auratus* (L.). Int. J. Fish. Aquat. Stud. 7 (5), 769–775.

Shinde, S.V., Sukhdhane, K., 2023. Advances in ornamental fish food preparation. Just Agric. 3 (7), 25–29.

CHAPTER

21

Nutritional interventions for sustainable aquaculture in a biobased circular economy

Nazeemashahul Shamna[a] *and Femi John Fawole*[b]

[a]**Fish Nutrition, Biochemistry and Physiology Division, ICAR-Central Institute of Fisheries Education, Mumbai, Maharashtra, India** [b]**Agricultural Research and Development Program, Central State University, Wilberforce, OH, United States**

21.1 Introduction

In recent decades, environmental issues such as pollution, biodiversity loss, and resource overuse have been closely linked to climate change, as shown by several scientific reports (Almond et al., 2020). Climate change poses a global threat, particularly to resource-dependent sectors such as agriculture and forestry, with the potential to trigger substantial market and economic disruptions (Kardung et al., 2021). Growing concern over these impacts has prompted international agreements aimed at reducing the greenhouse gas footprint of human endeavor (Singh et al., 2024; Almond et al., 2020). Among the promising strategies to achieve this goal is the promotion of the bioeconomy principle, an economic system built on renewable biological materials such as food waste, agricultural residues, or algae. While the improper disposal of biowaste poses serious risks to terrestrial and aquatic ecosystems, the responsible management of bioresources offers opportunities to create new value chains and enhance ecosystem resilience (Kardung et al., 2021). Moreover, increasing global concern for environmental issues is reshaping production models, with companies increasingly seeking certifications that verify sustainable practices and product quality (Osmundsen et al., 2020). Consequently, there has been growing interest in sustainable production models, especially the circular economy, as a strategy to balance environmental and economic goals.

Aquaculture has grown rapidly in recent decades to meet global protein demands, but its current production model poses sustainability challenges. Feed is the largest cost input in fish farming, and most of the material used in aquafeed production is often processed to improve pelleting efficiency. As the aquaculture industry yearns for alternative ingredients as one of the

Zero-Waste Technologies in Aquaculture: Global Perspectives and Methods toward a Circular Economy
DOI: https://doi.org/10.1016/B978-0-443-34689-7.00006-9

antidotes for sustainable production, efforts must be directed at reducing waste along the feed production value chain. The traditional linear production process that transforms feed/food resources into final usable bioproducts often generates significant amounts of waste, reducing raw material utilization efficiency. Furthermore, about 10–12 million fish are lost annually from post-harvest loss, representing almost 10% of the global catch (Mavuru et al., 2022; Mramba and Mkude, 2022). Given this huge loss, the Food and Agriculture Organization emphasizes that waste emanating from food-fish systems must be repurposed by recycling the nutrients into other usable value-added products, thus limiting nutrient loss and wastage (FAO, 2024).

Transforming material waste into inputs for alternative use can decrease the costs associated with waste management and enhance the value obtained from the original raw materials. This approach is confined to the concept of a circular economy, which is essential for promoting sustainable aquaculture. The circular economy seeks to replace the traditional "end-of-life" paradigm by minimizing waste through reuse, recycling, and recovery, which in turn fosters environmental, economic, and social sustainability (Kirchherr et al., 2017). By optimizing raw material use, recycling, and producing additional value cost-effectively, the concept would inadvertently reduce waste generation and promote sustainable use of nature-derived resources. In addition to promoting climate-resilient aquaculture practices, the integration of a circular economy principle into aquafeed management would substantially contribute to an alternative ingredient portfolio (such as insect meal, algae, and bacterial protein).

Biocircular economy is an important component of the circular economy, playing a significant role in advancing both resource efficiency and environmental sustainability (Coppola et al., 2021). Carus and Dammer (2018) succinctly described it as the intersection of the bioeconomy and the circular economy. Although the bioeconomy itself is not circular, its intersection is most effective whenever biomass can be recycled between industries. Several biomaterial wastes with low economic value have significant potential to be transformed into valuable, protein- and oil-rich resources that could serve as input for aquaculture use. For example, brewery-spent grain (BSG), a byproduct of the brewing process, poses a major challenge for brewers because of its high water content. The high moisture level makes BSG susceptible to microbial degradation and spoilage (Mussatto et al., 2006; Robertson et al., 2010; Santos et al., 2023), which can lead to potential environmental issues. The use of this large stream of organic waste is further complicated by the high lignocellulose content, which renders it less suited for animal feed production, especially for fish. Hence, the low market demand relative to output by other sectors makes the method of recovering nutrients from this organic waste sacrosanct for effective waste management.

Hansen et al., 2023 stated that entomoremediation could be an alternative approach to brewery waste management, and several studies have shown that insects such as black soldier fly larvae (BSFL) and housefly worms can recycle brewery waste into high-quality, protein-rich larval biomass (Jucker et al., 2019; Fawole et al., 2023; Hansen et al., 2023; Resconi et al., 2024). The approach provides value-added waste management solutions for BSG by serving as inputs for insect meal production and indirectly reducing the land acreage required for growing conventional feed ingredients, such as soybeans, and lessening the pressure on fishery resources used in aquafeed production. Therefore the aquaculture nutrition concept of a biocircular economy is maximizing resource utilization and minimizing waste in aquaculture. This involves strategies such as reducing food waste, utilizing fish waste to create value-added products, and implementing integrated aquaculture systems, where multiple species are cultured together to utilize one another's waste. Closing nutrient loops and reducing reliance on synthetic inputs can render aquaculture more environmentally friendly and sustainable. Thus this chapter aims to highlight the principle of a biobased economy in aquaculture and its application in aquafeed development.

21.2 Waste valorization and its benefits in aquaculture: A nutritional viewpoint

Waste valorization is the transformation of waste into higher-value products, such as energy, materials, or chemicals, adding value while decreasing the environmental footprint. In recent decades, solid organic waste has increased significantly owing to population growth and urbanization. In the future, it is estimated that global solid waste will amount to around 249 million tons by 2030 and go up to 340 million tons by 2050, growing by 36.5% (Roy et al., 2022). Waste and byproducts from agriculture are ideal for valorization because they are abundant, renewable, and have high organic content for aquaculture. There are two levels of recycling and reuse of waste in aquaculture from a nutritional point of view. One is the output of aquaculture as an input in nutrient balancing of the system, and the second one is other industry byproducts as input in aquaculture ventures. In the first category, the nutrient recycling in the system is done by intervention in different trophic levels. Systems such as integrated multitrophic aquaculture (IMTA) systems and aquaponics are the best examples for this category. In integrated aquaculture systems, many fish species or algae are cultivated together, and a closed-loop system can be produced in which the waste from one species is used as fertilizer or food by another. Fish waste, be it solids or liquid effluent, can be processed and used as a nutrition source for crops or other aquatic animals. This is one way of waste reduction and nutrient recycling. Food waste can be minimized by the use of aquaculture byproducts (such as bones and offal of fish), formulation of optimal feed, and minimizing feed wastage during production and feeding.

21.2.1 Aquaculture waste valorization

In recent years, the European Union has encouraged the use of sustainable resources and appropriate waste management, stimulating the use of raw materials such as biomass and water (Geeter and Boute, 2006; European Parliament and Council, 2008). To fit in with this framework and continue to be an important source of animal protein, aquaculture is moving in the direction of sustainable and economically viable production systems. An innovative approach is IMTA, where waste, e.g., uneaten feed and metabolic byproducts, is used as a feed resource for other species, achieving maximum productivity and reducing environmental footprint (Lee et al., 2019; Hossain et al., 2022). IMTA imitates natural food webs by cultivating species from various trophic levels. Fishes, crustaceans, and cephalopods form the first level; filter feeders such as mollusks and sea cucumbers comprise the second, feeding on organic waste; and marine macroalgae comprise the third level, taking up dissolved nutrients and carbon dioxide (Guerrero and Creamades, 2012; Irisarri et al., 2013). This facilitates the coculture of high-value species with lower inputs and reduced environmental footprint (Singh et al., 2024).

Aquaculture waste and unused byproducts constitute a profitable substrate for the production of single-cell protein (SCP), which is obtained from algae, yeast, fungi, or bacteria. Microorganisms are able to grow on heterogeneous substrates and only need a carbon source with supplementary minerals and nitrogen. Aquaculture residues and effluents can therefore be a low-cost input for SCP production. This sustainable practice not only limits waste but also forms a circular process, as the resulting SCP can be employed as an alternative ingredient in fish feed production. It partially substitutes animal protein, which encourages a more environmentally friendly and resource-friendly aquaculture model (Gasco et al., 2018). In addition, aquaculture effluents can be used to culture micro- and macroalgae that can be further utilized as aquafeed ingredients. The utilization of unconsumed feed or fish waste as a fertilizer for vegetable production in aquaponic systems supports the circular economy concepts.

21.2.2 Valorization of agricultural waste and byproducts

The agriculture and food processing industries produce enormous volumes of organic waste and byproducts. In particular, crop residues, oilseed cakes, fruit pomace, vegetable trimmings, or fisheries waste are often discarded or underutilized. While these wastes may pose environmental problems when improperly managed, they represent a valuable resource because of their rich content of carbohydrates, proteins, fibers, and bioactive compounds. Integrating valorized waste streams as feed ingredients in aquaculture offers a viable strategy to lower feed costs, improve sustainability, and balance nutrient cycles in food production systems.

Food loss is an important global issue that is prevalent in both developed and developing countries (Skaf et al., 2021). Of all the forms of food loss, fruit and vegetables have the highest percentage (FAO, 2019), which includes seeds, skins, rinds, and pomace (Sagar et al., 2018). These byproducts, such as carrot pomace, gourd and baby corn residue, wheat and rice husks, mango and banana skins, pea skins, and citrus pulp, can be well reused in aquaculture as feed ingredients, providing an effective approach to food waste management (Wadhwa and Bakshi, 2013). Crop residues (e.g., straw, husks, and stalks), fruit and vegetable peels, animal manure, and agroindustrial waste are of great potential for the valorization of waste (Table 21.1). Valorization of waste is a process where waste material is transformed into valuable products such as energy, fertilizer, biobased chemicals, and materials. These methods involve different technologies, such as anaerobic digestion for the generation of biogas, combustion and gasification for energy production, and fermentation to generate bioethanol, biodiesel, and organic acids.

Composting and vermicomposting convert biodegradable waste into nutrient value organic manures, while pyrolysis of biomass generates biochar, which benefits soil health and carbon sequestration. Agricultural waste can also be utilized for the synthesis of nanomaterials and biodegradable plastics, or for animal feed following appropriate processing. These valorization processes help to minimize greenhouse gas emissions and pollution, and generate economic opportunities, particularly in rural communities.

TABLE 21.1 Categories of agricultural and agroindustrial waste for feed valorization.

Sl No.	Category	Components	Examples
1.	**Crop residues (Field-based agricultural waste):** These are left behind after the harvesting of crops and are rich in fiber and structural carbohydrates	**Straw**	Wheat straw, rice straw
		Husks and hulls	Rice husk, groundnut husk
		Stalks and stovers	Maize stover, corn stalks
		Leaves	Sugarcane leaves, legume foliage
2.	**Agricultural byproducts (Harvest and milling byproducts):** Generated during crop processing or milling, these materials are often high in nutrients	**Cereal byproducts**	Rice bran, wheat bran, corn cobs
		Oilseed cakes	Soybean meal, groundnut cake, mustard cake (after oil extraction)
		Legume residues	Pigeon pea husk, lentil bran
		Sugar industry byproducts	Molasses, bagasse
3.	**Fruit and vegetable (F&V) industry waste:** Derived from the processing of fruits and vegetables, this category contains high moisture content and is rich in carbohydrates, vitamins, and phytochemicals	**Fruit peels and pomace**	Banana peel, citrus peel, apple pomace, mango peel, grape pomace
		Vegetable trimmings	Carrot tops, cabbage leaves, tomato skins
		Seeds and cores	Papaya seeds, melon seeds, mango kernels
4.	**Animal waste (Livestock and slaughterhouse waste):** These materials, after proper treatment, can be reused as fertilizer or in feed	**Manure and litter**	Cow dung, poultry litter
		Processing waste	Blood meal, bone meal, offal from slaughterhouses
5.	**Agroindustrial waste (Byproducts from food processing industries):** Waste generated during food and beverage production, often nutrient-dense but variable	**Dairy waste**	Whey, spoiled milk, skimmed milk residues
		Sugar industry	Press mud, filter cake (in addition to molasses and bagasse)
		Oil industry	Palm kernel cake, oil sludge
		Brewery and distillery waste	Spent grain, yeast sludge
6.	**Biovalorization**: Utilization of the waste for producing other organisms	**Insect meal**	Black soldier fly larvae, house fly worm, superworm, yellow mealworm
		Live worms	Tubifex, polycheate
		Livefeed	Wheel animacules, methanotrophs, bacteria

21.2.3 Methods of waste valorization for aquafeed

Agricultural byproducts can be effectively valorized for aquafeed production through various processing methods. Fermentation, either solid-state or submerged, enhances the digestibility and palatability of feed ingredients while reducing antinutritional factors. This method utilizes microorganisms such as *Aspergillus, Rhizopus, Saccharomyces*, and *Bacillus* species to treat materials such as oilseed cakes, fruit and vegetable pomace, cassava peels, and rice bran, resulting in improved protein content, amino acid profiles, and increased shelf life. Enzymatic treatment further aids in breaking down complex polysaccharides and antinutrients found in materials such as soybean hulls, wheat bran, and corn stover, using enzymes such as cellulase, xylanase, protease, and phytase to increase nutrient bioavailability and energy value. Drying and grinding, either by sun or mechanical means, convert perishable byproducts such as banana peels, fruit pulps, and vegetable trimmings into stable, storable feed powders with extended shelf life. Oil obtained either by mechanical pressing or solvent extraction from oilseed byproducts, including groundnut cake or mustard seed meal, produces defatted meals for fish diets with improved storage stability. Silage making stabilizes high-moisture crop residues such as cassava pulp, BSG, fruit and vegetable waste, and aquatic weeds through controlled fermentation, so as to ensure nutritional quality retention over time. Also, mixing these byproducts with protein supplements such as soybean meal, groundnut meal, or fermented leaf meals balances the amino acid profile to form cost-efficient and nutritionally adequate aquafeeds. These feeds can be either pelletized or extruded to form stable, water-resistant pellets that enhance feed intake, minimize nutrient leaching, and ease handling. Incorporation of pro- and prebiotics isolated from sources such as fermented residues, onion or garlic peels enhances gut health, immunity, and feed conversion efficiency in aquatic animals. Thus bipolarization methods provide sustainable and economically feasible options for formulating aquafeed from agricultural byproducts.

21.3 Biovalorization of organic wastes for use as feed ingredients in aquaculture

Biovalorization is a promising strategy for organic waste management to reduce landfill waste and greenhouse gas emissions (Govorushko, 2019; Mannaa et al., 2024). This approach centers on transforming organic waste into valuable products, transitioning from traditional linear waste processing methods to a circular bioeconomy model (Fig. 21.1). Various biological agents, such as insects, yeasts, and microorganisms, have been considered in this transformation process. Furthermore, enzymes are touted as powerful tools in the bioconversion process, enabling the degradation of complex organic materials. This was demonstrated when protease enzyme, amyloglucosidase, and α-amylase were used to extract dietary fiber from mango peel. It was noted that the extracted fiber showed strong functional properties that could be employed in food and feed applications (Kaur et al., 2024).

Agricultural and food processing byproducts, especially those from legumes, present substantial potential for animal feed inputs because they have greater acceptability and digestibility (de Boer et al., 2006; De Vries and de Boer, 2010; Food and Agriculture Organization (FAO), 2017), as well as relatively small proportions of antinutritional factors (Mateos-Aparicio et al., 2010; Mateos-Aparicio et al., 2013; Tassoni et al., 2020). In the same way, fruit

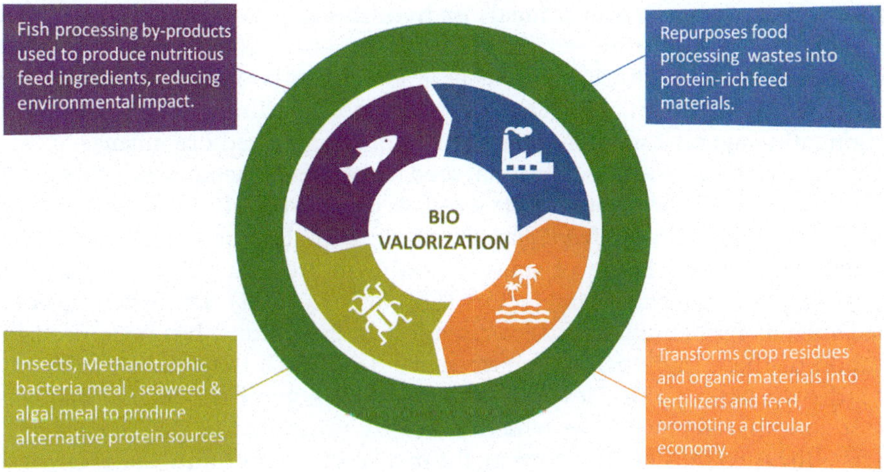

FIGURE 21.1 Diagram illustrating various sources of organic waste and their biovalorization within a circular bioeconomy model.

and vegetable processing wastes are nutrient-rich and are especially characterized as being high in polyphenol (Peschel et al., 2006; Amado et al., 2014). Various vegetable byproducts, e.g., baby corn, cabbage, carrot, cauliflower, cucumber, jackfruit, peas, potato, sweet corn, tomato, and radish leaves, have been effectively utilized both in ruminant and monogastric animals (Bakshi et al., 2016), contributing to their protein and energy supply.

21.3.1 Industrial waste

Global industrial growth has significantly contributed to food production systems, including food processing technologies and distribution, while enhancing the shelf life of food and, thus contributing to food security. Although this intervention improved the food production system, it unintentionally led to increased waste generation and greenhouse gas emissions. According to published data, global food and agricultural waste is projected to be between 1.05 and 1.3 billion tons and 659 and 998 million tons per year, respectively (WWF, 2021; Kolawole et al., 2024; Raut et al., 2023; UNEP, 2024). The majority of this waste is produced by industries such as agriculture, food processing, and manufacturing. Nonetheless, leveraging industrial waste in the animal feed sector could lead to new product development, maximize resource efficiency, and promote sustainability. Therefore integrating industrial waste streams into aquaculture and fish feed production would reduce reliance on traditional feed ingredients. For instance, studies have shown that alternative ingredients produced from food processing waste can effectively replace fishmeal in fish feed without obliterating the growth performance of aquatic animals (Dawood and Koshio, 2020). In the fermented soy pulp (FSP) trial, improved growth performance and immunity were reported when 50% FSP was substituted for fishmeal in the African catfish diet (Kari et al., 2021). The improvement in the fish's performance was associated with the enhanced production of lactic acid bacteria in the gut, thus maintaining a balanced microbial community within the fish gut. Waste from the shrimp industry processed into meals or hydrolysate has been utilized as a substitute for traditional protein source ingredients (Cavalheiro et al., 2007; Fall et al., 2012), thereby reducing pressure on wild stock. This waste is known to be rich in bioactive compounds, such as chitin and astaxanthin, which act as antioxidants, antiinflammatories, antimicrobials, and/or pigmenting agents for color enhancement of the cultured organisms.

Industrial byproducts such as jatropha seed cake (JSC), neem seed cake, rubber seed cake, and kernel meals of various seeds used in the nonedible oil industry are found to be potential ingredients in aquafeed (Fawole et al., 2016; Shamna et al., 2017; Phulia et al., 2018; Fawole et al., 2022). Some of these ingredients are rich in toxins, such as phorbol esters (PEs) and other antinutritional factors (Phulia et al., 2021). The application of different treatments, such as fermentation and/or chemical interventions, has been used to improve the utilization of these ingredients (Sanap et al., 2022). JSC, the primary byproduct of the biodiesel industry, has a protein level of 27%–33% and a good amino acid balance (Makkar et al., 2008). Jatropha seed toxicity, which is mostly attributed to the presence of PEs, has hindered its utilization thus far. In addition, the meal contains antinutritional elements such as lectin, phytate, and trypsin inhibitors (Shamna et al., 2017). Therefore as opposed to using raw seed cake directly, repurposing it into protein isolate or concentrate will guarantee a healthier animal and improved growth performance of fish. Fawole and his colleagues used protein isolation techniques to produce jatropha protein isolate (JPI) containing 87.5% crude protein (CP), helping to reengage the biofuel industry waste and repurpose it into a new product. Their research demonstrated that JPI, when supplemented with lysine, could replace soy protein isolate completely in the diet of *Labeo rohita* without adversely affecting fish performance (Fawole et al., 2018; Fawole et al., 2022). Thus it is worthy of note that waste generated from industrial processes could be repurposed and transformed into useful resources suitable for fish feed production, providing a more sustainable approach to waste management and resource efficiency. Table 21.2 summarizes the production performance of fish raised on industrial waste.

21.3.2 Fisheries waste

The global fishing industry generates substantial amounts of waste, which constitutes approximately two-thirds of total output, encompassing both fish discards and waste from the fish processing industry, raising significant socioeconomic and environmental management challenges (Nawaz et al., 2020; Coppola et al., 2021). The projection estimates show that fish production would reach 200 million tons by 2030, with more than 50% emanating from aquaculture (Suat Dikel and Çiğdem Dikel, 2024); however, the current approach to fish processing and handling loss will result in waste if efforts are not directed at repurposing them into a useful product. Recently, efforts to reduce this waste and transform it into a useful raw material for food/feed applications or to serve as an input in producing functional ingredients for fish health management are gaining traction. Most waste generated from fish processing plants is used in fishmeal and fish oil production or as fertilizer (Kagali et al., 2022); however, studies have further explored their potential as a source of biological compounds, including bioactive peptides (BPs), enzymes, collagen, and gelatin, Parma et al. (2023) studied the effect of BPs derived from Atlantic salmon byproducts on the growth and health of European sea bass (*Dicentrarchus labrax*) and found that the BPs improved lipid utilization

TABLE 21.2 Production performance of fish raised on industrial waste stream.

Organic waste stream	Biocircular intervention/ extraction method	Product obtained	Aquatic species	Feeding duration (days)	Suggested inclusion level in diet	Effects on fish performance and health	References
Jatropha curcas seed cake	Isoelectric precipitation and solid-state fermentation	Fermented protein concentrate	*Labeo rohita* (Rohu)	45	20%	Fermented jatropha protein concentrate can be utilized in the diet of rohu up to 20% without any detrimental effect on growth, hematoimmunological, and physiological response of rohu	Shamna et al. (2017)
Rubber seed cake	Alkali extraction and acid precipitation	Protein isolate (90.8% protein, DM)	*Labeo rohita*	60	18.7%	With no adverse effect on growth performance, feed efficiency, hematological parameters, total immunoglobulin, respiratory burst, and lysozyme activities, oxidative stress markers, and liver histology, rubber protein isolate was suggested as a potential substitute for soy protein isolate in rohu feed	Fawole et al. (2016), Fawole et al. (2017)
Agricultural wastes	Bacterial fermentation	Bacterial protein (77.48% protein, DM)	*Clarias gariepinus* (African catfish)	56	2.79%–9.30%	Bacteria protein (BP) improved growth performance, hematobiochemical parameters, intestinal histology, and liver functionality compared with the FM-fed group. The results indicate that BP can completely replace FM. However, a 30% replacement level (corresponding to 2.79% dietary BP inclusion) outperformed all other fed groups, including the FM group; therefore the author suggested this level as the most appropriate for farmed African catfish	Adeoye et al. (2021)

(*continued on next page*)

TABLE 21.2 Production performance of fish raised on industrial waste stream—cont'd

Organic waste stream	Biocircular intervention/ extraction method	Product obtained	Aquatic species	Feeding duration (days)	Suggested inclusion level in diet	Effects on fish performance and health	References
Wine industry waste (red grape pomace and wine lees)	Powdered wine byproducts	Phenolic-rich compound	*Liza aurata* (Gray mullet)	42	10%	The two phenolic-rich wine by-products improved fish growth and feed efficiency. Grape pomace boosted immunity, while wine lees reduced gut microbiota diversity and function. Both lowered cortisol levels and enhanced antioxidant enzyme activity were observed under hypoxic stress. The author concluded that these byproducts may benefit fish and support the circular economy	Martínez-Antequera et al. (2023)
Poultry slaughter house waste	—	Poultry viscera (60.67% protein, DM)	*Anabas testudineus* (Koi fish)	60	30%	Sun-dried poultry viscera completely replaced FM in Koi diets, improving final weight, SGR, and feed conversion. This supports poultry waste recycling for aquaculture	Bhaskar et al. (2015)
Agroindustrial waste	—	Brewer's Dried Grain (19.6% protein, DM)	*Clarias gariepinus* (African catfish)	74	11.69%	The African catfish tolerated the brewery spent grain (BSG) and performed relatively well as the control group. The weight gain and feed intake were unaffected, while the 40% replacement level recorded the lowest cost per kilogram of weight gain. Hence, BSG waste could be a good substitute for maize in the catfish diet	Omoike et al. (2019)

Olive oil industry waste	—	Olive waste cake (48% protein, DM)	*Oncorhynchus mykiss* (rainbow trout)	42	0.25%	Growth, feed conversion ratio, oxidative stress status, skin, and humoral immunity were improved in rainbow trout fed olive waste cake (OWC). The proinflammatory cytokines (IL-8 and TNF) and antiinflammatory cytokines (TGB) in the gut of the fish were positively influenced. OWC can be used as a functional ingredient for trout	Hoseinifar et al. (2020)
Industrial insect waste	Insect larva meal byproduct	Frass (21.6% protein, DM)	*Ictalurus punctatus* (Channel catfish)	70	30%	The use of frass in channel catfish diets proves beneficial by enhancing growth performance, feed palatability, hematological indices, serum complement activity, and resistance to *Flavobacterium covae* infection	Yıldırım-Aksoy et al. (2020, 2023)
Poultry slaughter-house waste	Enzymatic hydrolysis	Enzymatic feather meal	Hybrid tilapia (Oreochromis niloticus × O. aureus)	63	2.3%–4.6%	Enzymatic feather meals replaced FM with no adverse effect on growth, nutrient retention, and digestibility	Poolsawat et al. (2021)
Rainbow trout processing waste	Fish silage via the enzymatic method of protein hydrolysis	Fish silage oil	Mozambique tilapia (*Oreochromis mossambicus*)	52	10%	The silage oil had no negative effects on production performance, while it improved survival rates, PUFA content, and the immunity of the fish. The oil exhibited antimicrobial functions in the guts and feed. It could serve as a cost-effective alternative to fish oil	Goosen et al. (2014)

From Goosen, N.J., de Wet, L.F., Görgens, J.F., Jacobs, K. and de Bruyn, A., 2014. Fish silage oil from rainbow trout processing waste as alternative to conventional fish oil in formulated diets for Mozambique tilapia Oreochromis mossambicus. Anim. Feed Sci. Technol. 188, 74–84. https://doi.org/10.1016/j.anifeedsci.2013.10.019.

for energy production, especially after temperature stress challenge. In addition, the research indicated that BPs enhanced the immune response of the fish by upregulating the expression of the antiinflammatory cytokine interleukin 10 (IL-10) and downregulated proinflammatory cytokines IL-1β and IL-8. Fish viscera, which account for 5%–10% of the waste generated by the fish processing industry, have also been used as a source of natural enzymes for aquafeed applications. A study by García Martínez et al. (2024) demonstrated this by supplementing a high-plant–based diet fed to red claw crayfish (*Cherax quadricarinatus*) with a multienzyme extract derived from shrimp. The study indicated that adding the multienzyme extract significantly improved the endogenous digestive enzyme activities and enhanced the energy reserve by increasing muscle glycogen levels and hepatopancreatic lipid accumulation.

Furthermore, the utilization of value-enhanced products such as fish protein hydrolysate (FPH) produced via enzymatic hydrolysis of the fish muscle is another emerging concept of the biocircular economy that has been researched as a functional additives (Chaklader et al., 2020; Olaniran et al., 2024). Hydrolysate produced from fisheries waste has good biological and nutritional attributes that could potentially be used as ingredients for fish meal replacement and/or as supplementary additives for aquatic animal feed applications (Wei et al., 2020). For instance, Chaklader et al. (2020) supplemented 10% of three different FPHs with 90% poultry byproduct meal (PBM) as an additive in the diets of barramundi (*Lates calcarifer*). According to the study, feeding 100% PBM in place of Fishmeal (FM) negatively impacted the villus height and width, neutral mucin, and significantly increased the lamina propria area, thus resulting in enteritis. It was further reported that the PBM-based diet upregulated the expression of proinflammatory cytokines (TNF-α) while suppressing the antiinflammatory cytokines (IL-10). However, adding FPHs to PBM diets helped repair the intestinal lining and improved the damaged inflammatory and i-mucin c genes. The FPHs also enhanced intestinal homeostasis by improving microbial diversity and reversing the muscle atrophy mediated by PBM. Chaklader et al. (2020)reiterated that the concept of a biocircular economy could help in recovering functional bioactive peptides from fish waste and serve as a functional ingredient to improve fish performance and health. These findings and others show that waste from the fisheries industry can be used as an input in the production of functional ingredients for aquatic animal health management, thus strengthening the biobased circular economy concept for aquaculture. Table 21.3 summarizes the effect of some functional products derived from fisheries waste and the observed effects on aquatic species.

21.3.3 Insect meal

Biovalorization of organic waste using insects is growing at an unprecedented rate, and different species of insects have been raised on waste streams. Several species of insect have been evaluated for their ability to utilize organic waste streams in their diets; however, not all waste substrates are suitable for degradation by every species, and reduction/conversion efficiency varies across species (Mannaa et al., 2024; Siddiqui et al., 2024). For example, BSFL raised on soybean curd residues (SCR) showed a bioconversion rate (BR) of 5.0%, 49.0% dry mass reduction (DMR), and a feed conversion ratio (FCR) of 9.8. On the contrary, when the substrate was pretreated with *Lactobacillus buchneri*, the BC and DMR improved significantly to 6.9% and 55.7%, respectively, while the FCR decreased to 8.0% (Somroo et al., 2019). The codigestion with microbes improved the degradation of SCR and nutrient assimilation by

TABLE 21.3 Functional products derived from fisheries waste and their effects on aquatic animal performance.

Functional product/Source	Species	Feeding period (days)	Inclusion Level	Fish response	References
Fish protein hydrolysate	Largemouth bass *(Micropterus salmoides)*	56	0, 1.0%, 3.0%*, 5.0%.	• Enhanced WGR, SGR, PER, and a lower FCR value were recorded in the FPH group • Promoted protein synthesis and deposition in the muscle • Increase the activities of SOD, CAT, GSH-Px, POD, and T-AOAC • Higher expressions of GH1, IGF-1, TOR, and S6K genes • Upregulated the expressions of SOD, GSH-Px, IL1β, TNFβ, and downregulated the IL15, Caspase 3, 9, and 10 genes Conclusion: • FPH could substitute 16.3% of dietary fish meal without affecting the growth performance and intestinal health of the fish	Fan et al. (2022)
	Pabda catfish *(Ompok pabda)*	90	0, 0.5%, 1.0%, 2.0%	• Higher WG, SGR, TB, PER, and lower the FCR • SR, CF, and HSI were higher in 2% FPH • Increased neutrophil, monocyte, lymphocyte, red blood cell, and platelet levels were recorded in 2% FPH • Higher TP, globulin, and lower ALP were observed in FPH-fed fish • Histopathological observation of mid-intestinal tissues showed a well-anchored epithelial wall with well-arranged goblet cells, long villus structure, and tunica muscularis • Lower cumulative percent mortality (16.67%) against *Aeromonas hydrophila* infection was recorded in 2% FPH Conclusion: • A 2% inclusion of FPH significantly enhanced growth performance, improved health status, and increased disease resistance	Suma et al. (2023)

(*continued on next page*)

TABLE 21.3 Functional products derived from fisheries waste and their effects on aquatic animal performance—cont'd

Functional product/Source	Species	Feeding period (days)	Inclusion Level	Fish response	References
Tuna hydrolysate (Industrial residues)	Barramundi (*Lates calcarifer*)	70	10%	• Tuna hydrolysate (TH) improved the activity of the GPx enzyme in the fermented poultry byproduct meal supplemented group relative to the control • TH upregulated the expression of proinflammatory cytokines (IL-1β and TNF-α) and downregulated the antiinflammatory cytokine, IL-10, in the distal intestine • TH addition modulates the distal intestinal gut microbiome Conclusion • TH supplementation could enhance the antioxidant capacity and improve inflammatory responses in barramundi	Siddik et al. (2020)
Chitosan	Nile tilapia (*Oreochromis niloticus*)	60	0, 0.3%, 0.5%	• Improved the growth performance and body composition • Serum TP, albumin, and globulin levels increased in 0.5% chitosan • After bacterial infection, the serum AST and ALT levels were lower in the chitosan group compared to the control • Lysozyme and bactericidal activities increased significantly in the chitosan group before and after challenge with *Streptococcus agalactiae* • High survival rate was recorded postchallenge in chitosan-fed groups Conclusion • Feeding 0.3% or 0.5% chitosan enhanced growth, stimulated immune response, and reduced mortality	Fadl et al. (2020)

Bioactive peptides (BPs) (Atlantic salmon byproducts)	European sea bass (*Dicentrarchus labrax*)	58	0, 5.0, 10	• No significant differences in the FBW, SGR, FI, and FCR, but lipid efficiency was higher in BP 10% • Glucose levels decreased with increasing BP content alongside a reduction in total protein, cholesterol, high-density lipoprotein, and triglycerides following environmentally stressful conditions in the BP 5% fed group • A reduction in proinflammatory interleukin 1β (IL-1β), IL-8, and a significant increase in the expression of the antiinflammatory cytokine (IL-10) were observed in BP 5% Conclusion • Bioactive peptides derived from salmon byproducts hold potential as circular and functional ingredients for aquaculture, and they could be utilized to enhance mechanisms related to lipid metabolism and immune response	Parma et al. (2023)

ALP, alkaline phosphatase; **CF**, condition factor; **FBW**, final body weight; **FCR**, feed conversion rate; **GH**, growth hormone; **GSH-Px**, glutathione peroxidase; **HSI**, hepatosomatic index; **IGF-1**, insulin-like growth factor 1; **IL1β**, interleukin 1β; **IL15**, interleukin 15; **PER**, protein efficiency ratio; **S6K**, S6 kinase; **SGR**, specific growth rate; **SOD**, superoxide dismutase; **SR**, survival rate; **T-AOC**, total antioxidant activity; **TB**, total biomass; **TNF-β**, tumor necrosis factor β; **TOR**, target of rapamycin; **TP**, total protein; **WG**, weight gain; **WGR**, weight gain rate.

BSFL, resulting in increased protein and fat levels. This indicates that substrate types influence the degradation capacity and conversion efficiency of insects. Reviews by Koo et al., 2023 and Mannaa et al., 2024 provided more insights into the bioconversion efficiency of various insect species.

BSFL, a saprophytic insect that primarily feeds on organic wastes such as plant residues, animal manure, food waste, agricultural byproducts, or straw (Lalander et al., 2013; Banks et al., 2014; Nguyen et al., 2015), is one of the most researched insect species for biovalorization. BSFL has the ability to digest organic waste and assimilate nutrients rapidly in the process of degrading it. They convert organic waste into high biological value products in terms of protein and oils, BPs, and vitamins, with an innate ability to control certain harmful bacteria (such as *Salmonella* and *Escherichia coli*), pests, and flourish in a variety of environmental conditions (Lalander et al., 2013). The larvae and prepupae of BSF have drawn increased interest as an alternative protein to fishmeal (FM) among several insect species and are used in various animal feeds (mainly pets, pigs, and poultry) (Erickson et al., 2004; Liu et al., 2008). BSFL meals have a nutritional value that is comparable to FM and can be produced on an industrial scale (Henry et al., 2015; Fawole et al., 2020; Hossain et al., 2021). The nutrient composition of BSFL showed that it has between 37% and 63% CP, 7% and 39% crude lipid, and 9% and 28% ash (Makkar et al., 2014; Barragan-Fonseca et al., 2017).

Studies on other insects have also pointed out their suitability and potential to serve as a sustainable alternative protein source relative to traditional ingredients and are emerging as promising feed ingredients for aquatic animal feed production. Fawole et al. (2020) reported that African catfish fed 75% BSFL performed comparably to those fed FM-based diets in terms of growth, nutrient utilization, and physiological well-being. Similarly, replacement of 15% FM with defatted yellow mealworm (*Tenebrio molitor*) in large yellow croaker diet improved humoral immunity and intestinal health without negative effects on growth (Zhang et al., 2022).

21.3.4 Microbial protein meal

Agricultural waste (e.g., waste grain processing, fruit peels, and bagasse) and other lignocellulose materials can be converted into valuable protein through fermentation using microorganisms such as bacteria, fungi, yeast, and microalgae. This bioconversion process promotes a circular bioeconomy concept by providing a strategy for waste reduction and recycling, and also offers a sustainable means for producing ecologically friendly protein sources for aquafeed production. Microbial protein (MP), also referred to as SCP, has attracted attention as an alternative protein ingredient or additive in the animal feed and food supplement sector owing to its nutritional values, production efficiency, and reduced ecological footprint (Adeoye et al., 2021; Ayodele et al., 2024). MP is reported to be rich in protein (60%–82% DM) and has amino acid (AA) profiles similar to FM. It also serves as a source of nucleic acids, fatty acids, vitamins, and minerals (Adeoye et al., 2021; Sharif et al., 2021). However, the composition of MP depends on the type of microorganisms and the substrate used. This was emphasized by Yousufi (2012) and Sharif et al. (2021), who reported that the AA profile of yeast-derived protein is relatively similar to that of soy protein, whereas bacterial-derived proteins show profiles more comparable to fish protein. A wide range of waste materials, including fruit processing byproducts such as pineapple waste, citrus pulp, and apple pomace; animal waste and manure; as well as potato waste, wheat bran, and lignin waste, have been used as substrates for MP production. Moreover, MP production is considered environmentally friendly and is independent of season or varying climatic conditions, making it a sustainable alternative for animal feeding (Sharif et al., 2021).

Recent studies have indicated that microbial protein meals (MPM), when supplemented in aquaculture, meet the protein needs of fish and shrimp, with the potential to partially or completely substitute for animal- or plant-based protein ingredients. However, it is worth noting that MPs are deficient in certain essential AAs, especially sulfur-containing AAs such as methionine and cysteine, although they have relatively high levels of lysine, phenylalanine, threonine, and other essential AAs (Sharif et al., 2021; Vasilaki et al., 2023). Nevertheless, certain fungi grown on specific substrates (e.g., whey) can boost production of methionine and cysteine in MP (Ugalde and Castrillo, 2002; Ayodele et al., 2024). The use of MPs in feed formulation has been studied for decades with varying outcomes. Adeoye et al. (2021) investigated the replacement of fishmeal with bacterial protein (BP) in the diet of African catfish. Their findings indicated that incorporating BP at a 30% level (27.90 g/kg) resulted in significantly improved weight gain, weight gain percentage, specific growth rate, and enhanced whole-body protein and lipid content. Those fed 60% (55.70 g/kg) and 100% (93 g/kg) recorded similar growth responses relative to the control group. Irrespective of the diet fed, the hematological parameters, serum aspartate and alanine aminotransferase levels, alkaline phosphatase, and hepatocyte density were not influenced by the dietary BP. The author concluded that BP could be used to replace FM partially or completely in the African catfish diet, offering a sustainable alternative to FM dependency and improved waste utilization. In European sea bass, the inclusion of single-cell ingredients (SCI) up to 18% did not impact the growth, nutrient utilization, whole-body composition, fatty acid profile, or intestinal morphology, compared to the control group. The author indicated that the use of SCI could reduce the levels of fish meal trimming, plant protein ingredients, and fish oil by 7%, 24%, and 6%, respectively, without impacting the growth and welfare of the fish (Vasilaki et al., 2023). For rainbow trout, Rajesh et al. (2021) reported that methanotroph bacteria protein meal can be used in place of FM without compromising growth and feed efficiency, nutrient digestibility, tissue integrity, and reducing phosphorus excretion. It was further posited that higher inclusion resulted in poor growth performance and decreased whole-body protein and lipid digestibility. Several other studies in different fish species reported varying levels of BM inclusion, including 12.9% for largemouth bass (Zhang et al., 2022), 18.3% for juvenile turbot (Zheng et al., 2023), 8.5% in GIFT tilapia (Chama et al., 2021), and 17% in yellowtail kingfish (Biswas et al., 2020). The studies indicate that valuable protein with high biological value can be derived from waste streams and utilized in the aquaculture industry as an alternative protein source, thereby promoting both economic and environmental sustainability. Table 21.4 highlights some studies on insect and MP and the observed effects on fish/shrimp.

21.4 Feeding strategies to be adopted

To ensure the circular economy concept, feeding strategies also play an important role. The inclusion of microbial biomass developed from wastes, such as methanotrophic bacteria meal, purple sulfur bacteria meal, microalgae, and insect meal, in place of fishmeal thrusts on

TABLE 21.4 Nutrient composition of some insect species, microbial protein, and their effect on the performance and welfare of aquatic animals.

Insect species	Crude protein (% DM)	Crude lipid (% DM)	Crude fiber (% DM)	Aquatic animal species	Culture period (days)	Ingredient replaced and level in control diet	Suggested replacement level in diet	Results of main findings	References
BSFL meal	42.6	23.0	10.8	African catfish (*Clarias gariepinus*)	42	FM (15%)	50%	BSFL could be used to replace 50% of FM without negative effects on growth performance, survival, and welfare. However, 100% BSFL caused a reduction in growth, feed intake, and poor diet acceptability by the fish	Adeoye et al. (2020)
YM meal	74.8	12.6	4.7	Pacific White Shrimp (*Litopenaeus vannamei*)	60	FM (25%)	100%	Growth and feed intake across all YM-fed groups show significant improvement, with optimal performance recorded at 50% FM replacement. Enhanced immunity was observed in the YM groups, which contributed to improved resistance against *Vibrio parahaemolyticus*	Motte et al. (2019)
Defatted BSFL meal	62.5	4.03	7.0	Siberian sturgeon (*Acipenser baerii* Brandt)	118	FM (70%)	25%	Feed intake declined with higher BSFL inclusion, with complete rejection at 100%. BSFL incorporation raised whole-body lauric acid and saturated fatty acid levels, while 25% and 50% inclusion levels reduced crude protein digestibility. Up to 25% FM (18.5% BSFL) can be replaced without impacting growth, body indices, or composition in Siberian sturgeon	Caimi et al. (2020)
Defatted SPM	75.94	8.4%	—	Pacific White Shrimp (*Litopenaeus vannamei*)	56	FM (24%)	75%	Growth performance was similar across diets. SPM inclusion improved ADCs of DM, GE, and phosphorus, boosted antioxidant capacity, and lowered malondialdehyde levels. Complete FM replacement reduced molting time and increased chitin deacetylase activity, while β-N-acetyl-D-glucosaminidase activity and expression of ecdysteroid and retinoid X receptors were unaffected	Rahimnejad et al. (2019)
Methanotroph bacteria meal	75.14	8.31	—	Black sea bream (*Acanthopagrus schlegelii*)	70	FM (40%)	24.8%	Growth performance, feeding rate, whole-body protein, apparent digestibility of lipids, and mid-gut amylase activities were best at the 8.27% replacement level. However, dietary levels up to 24.8% did not negatively affect fish performance compared with the control group	Xu et al. (2021)

ADC, apparent digestibility coefficients; **BSFL**, black soldier fly (*Hermetia illucens*) larvae; **DM**, dry matter; **FM**, fishmeal; **GE**, gross energy; **SPM**, silkworm (*Bombyx mori* L.) pupae meal; **YM**, yellow mealworm (*Tenebrio molitor*).

sustainability and circular economy concepts. Aquaculture may maximize resource consumption and minimize waste by implementing circular economy methods. Enhancing feeding efficiency by utilizing digestible and bioavailable ingredients reduces feed wastage. Similarly, a reduction in certain ingredients, such as fishmeal, reduces the release of phosphorus and calcium into the water column that supports the circular economy concept. Moreover, removal of antinutritional factors using fermentation, exogenous enzyme addition, chemical hydrolysis, etc., can further reduce the nutrient loss to water. Feeding frequency and feeding rate also determine the feed wastage. Underfeeding or overfeeding affects not only the growth of the fish but also the water quality. Understanding the food and feeding habits and digestibility of the fish and formulating the feed accordingly can support the maximum feed utilization and minimize wastage.

21.5 Research/Studies

Agricultural waste valorization for animal and aquafeed offers a sustainable and economically rewarding solution to both food production and waste management problems. A good case study from Thailand, where pineapple peels and core waste from pineapple processing factories are utilized as livestock feed. Such byproducts have high fiber and sugar content, which can be used in diets for ruminants when suitably processed and supplemented. This practice not only minimizes waste disposal problems but also decreases farmers' feed costs, increasing their economic resilience. Another is the case of Vietnam, where rice bran, a byproduct of rice milling, is used extensively as pig and poultry feed, adding to Vietnam's food security and circular economy objectives.

In the aquaculture industry, India has progressed much in the use of fruit and vegetable processing waste, such as banana peels and papaya skins, in fish feed composition. From these case studies, it is shown that valorizing agricultural byproducts into aquafeeds and animal feeds can have a huge impact on the environment through organic waste reduction and emission lowering, while also developing cost-effective, locally available feed substitutes. This helps create a more circular and sustainable agrifood system, particularly in nations with big agriculture and high amounts of organic waste.

Substantial progress has been achieved in valorizing fruit processing waste for the production of aquafeed, especially in the research work of ICAR institutions, such as the Central Institute of Fisheries Technology (ICAR-CIFT) and the Central Institute of Fisheries Education (ICAR-CIFE). ICAR-CIFT led the production of low-cost fish feed using agroindustrial fruit wastes, such as mango seed kernel, papaya peel, and banana waste. These wastes, after being subjected to fermentation and drying treatment, were added to the carps' diets. The research indicated that the addition of the fruit wastes did not have any negative impact on the growth performance and health of the fish, and hence proved to be satisfactory substitutes for traditional feed ingredients. Studies by ICAR-CIFE have shown that processed agrowastes can be successfully integrated into aquafeeds without any adverse effect on fish growth or health (Patra et al., 2024; Suma et al., 2023; Tejaswini et al., 2024; Bhattacharya et al., 2025). These feeds are not only economical but also decrease dependence on traditional feed ingredients such as FM and SBM, which tend to be imported or environmentally expensive to produce.

ICAR-CIFE continued to expand studies on the applicability of fruit waste, including papaya peel meals, pineapple peel extract, and pomegranate peel meals as alternatives for conventional ingredients used in aquaculture. Sundararajan et al. (2024) noted enhanced feed efficiency and immune response in fish fed with papaya peel meal-based diets. Bhattacharya et al. (2025) proved the antioxidant effects and gut health improvement role played by pineapple peel extract in the tilapia diet. In a similar manner, Patra et al. (2024) illustrated that pomegranate peel meal can be a functional feed supplement, enhancing the growth of fish while delivering natural bioactive compounds with antimicrobial activity. Kaleeswaran et al. (2025) illustrated the application of cabbage and cauliflower waste in fermented and non-fermented form in aquafeed. The investigation established that deoiled ricebran was entirely substituted with the fermented mixed waste (1:1) ratio. A research study by Shalini (2023) on tomato pomace (TP) meal also established that raw and fermented TP meal substituted 100% deoiled ricebran, and the fermented product improved the weight gain percentage in rohu fingerlings.

Pomegranate peel, a major waste product of the pomegranate juice industry, accounts for 40%–50% of the fruit's weight. It is significantly high in polyphenolic compounds such as phenolic acids, tannins, flavonoids, and particularly anthocyanins, which are related to numerous biological activities, especially antimicrobial and antioxidant activities (Masci et al., 2016). Studies have proved that pomegranate peel extract possesses broad-spectrum antimicrobial activity, inhibiting Gram-positive as well as Gram-negative bacteria effectively. The antifungal potential is mainly because of the high polyphenol content, particularly punicalagin and ellagic acid (Singh et al., 2023). Such bioactive principles place pomegranate peel extract as a candidate phytomedicine, providing an alternative to traditional antibiotics and aiding cost savings in therapeutic use (Singh et al., 2019). Kanatt et al. (2010) investigated its application in the preservation of the shelf life of chicken products using its antimicrobial and antioxidant activities. Evidence for this comes from Safdar et al. (2021), who documented potent antioxidant activity of the extract against major foodborne pathogens *Staphylococcus aureus* and *Bacillus cereus*, indicating its applicability in the preservation and safety of foods.

Badrey et al. (2021) investigated the impact of incorporating varying levels of pomegranate peel meal (1%, 2%, 3%, 5%, 10%, 15%, and 20%) into a basal diet containing 25% CP for Nile tilapia (*Oreochromis niloticus*) reared under different salinity conditions (15‰, 25‰, and 37‰). The research showed that at a salinity of 15‰, mortality was zero in all the treatments, suggesting a high tolerance and survival in this case. In particular, the best growth performance and best FCR were recorded on a 20% inclusion of pomegranate peel meal under 15‰ salinity, implying that salinity and level of inclusion are important factors affecting optimal growth and health of Nile tilapia.

TP, a waste product of tomato processing, represents about 5% of the processed tomato volume (Palomo et al., 2019). The wet TP usually contains 33% seed, 27% skin, and 40% pulp, whereas the dried TP contains 44% seed and 56% pulp with skin (Sogi and Bawa, 1998). Nutritionally, TP is high in dietary fiber, sugars, proteins, lipids, vitamins, and minerals (Albanese et al., 2014), and also a good source of flavonoids, carotenoids, and nucleosides (Palomo et al., 2019). It particularly contains an array of powerful antioxidant compounds such as lycopene, folate, vitamin C, β-carotene, α-tocopherol, phenolics, and flavonoids (Sahin et al., 2008; Selim et al., 2013). The elevated concentration of bioactive constituents, particularly lycopene and β-carotene, has been reported to have high antioxidant activity, as reported in different studies (Szabo et al., 2019). Specifically, these characteristics render TP a potential functional ingredient for use in animal and aquaculture feeds.

Amirkolaie et al. (2015) explored the influence of dietary TP (10%, 20%, and 30%) on common carp (*Cyprinus carpio*) and reported that 10% inclusion increased final weight, specific growth rate, FCR, and weight gain significantly compared to the control. Furthermore, the 10% level improved apparent digestibility coefficients (ADCs) of dry matter and fat, but protein ADC dropped with increased inclusion levels. The research found that TP was suitable up to 30% of the inclusion without impairing growth.

These studies emphasize the value of fruit waste valorization not only in lowering fish feed costs and reliance on traditional resources such as FM and SBM, but also in advancing sustainability, enhancing fish health, and building a circular bioeconomy for aquaculture.

21.6 Conclusion and future perspectives

The global demand for food and animal feed is rising because of population growth, leading to an unprecedented increase in organic waste generation. Incorporating the circular economy principle into organic waste management and using the resulting outputs as ingredients in aquatic animal feed production is a promising way to improve the sustainability of the aquaculture industry. Although this approach offers innovative potential for sustainable aquaculture, it faces several challenges. These include inadequate waste collection systems, limited public awareness, and restricted access to processing technologies. Moreover, the variability in the composition and nature of organic waste affects the efficiency of bioconversion, causing inconsistency in product quality. Also, many biovalorization techniques remain in research and development stages, making industrial adoption difficult. Existing technologies may also not be easily adaptable to different types of organic waste streams. In addition, policy frameworks (e.g., regulatory approval) guiding the use of waste-derived feed ingredients need streamlining while ensuring safety and compliance from industry stakeholders. To address these issues, future research should focus on optimizing the conversion efficiency of waste streams within the circular economy to enhance the nutritional quality and consistency of waste-derived feed ingredients for sustainable aquatic animal feeds. Concerted efforts should promote the real-world application of circular principles to facilitate industry adoption, and the regulatory framework should be refined for greater efficiency, fostering a more sustainable and circular aquaculture industry.

References

Adeoye, A.A., Akegbejo-Samsons, Y., Fawole, F.J., Davies, S.J., 2020. Preliminary assessment of black soldier fly (Hermetia illucens) larval meal in the diet of African catfish (Clarias gariepinus): impact on growth, body index, and hematological parameters. J. World Aquac. Soc. 51 (4), 1024–1033. doi:10.1111/jwas.12691.

Adeoye, A.A., Akegbejo-Samsons, Y., Fawole, F.J., Olatunji, P.O., Muller, N., Wan, A.H., Davies, S.J., 2021. From waste to feed: dietary utilisation of bacterial protein from fermentation of agricultural wastes in African catfish (Clarias gariepinus) production and health. Aquaculture 531, 735850.

Albanese, D., Adiletta, G., D'Acunto, M., Cinquanta, L., Matteo, M.Di, 2014. Tomato peel drying and carotenoids stability of the extracts. Int. J. Food Sci. Technol. 49 (11), 2458–2463. doi:10.1111/ijfs.12602.

Almond, R.E.A., Monique G., Tom P. (2020). Living Planet Report 2020-bending the curve of biodiversity loss. World Wildlife Fund, Gland, Switzerland https://www.wwf.org.uk/sites/default/files/2020-09/LPR20_Full_report.pdf.

Amado, I.R., Franco, D., Sánchez, M., Zapata, C., Vázquez, J.A., 2014. Optimisation of antioxidant extraction from Solanum tuberosum potato peel waste by surface response methodology. Food Chem. 165, 290–299. doi:10.1016/j.foodchem.2014.05.103.

Amirkolaie, A.K., Dadashi, F., Ouraji, H., Khalili, K.J., 2015. The potential of tomato pomace as a feed ingredient in common carp (Cyprinus carpio L.) diet. J. Anim. Feed Sci. 24 (2), 153–159. doi:10.22358/jafs/65642/2015.

Ayodele, T., Tijani, A., Liadi, M., Alarape, K., Clementson, C., Hammed, A., 2024. Biomass-based microbial protein production: a review of processing and properties. Front. Biosci. (Elite Ed.) 16 (4). doi:10.31083/j.fbe1604040.

Badrey, A.E.A., Farrag, M.M.S., Toutou, M.M.M., Moustafa, M.A., Osman, A.G.M., 2021. Modulatory effect of pomegranate peel against cold and salinity stress in the monosex Nile tilapia, Oreochromis niloticus. Egypt. J. Aquat. Biol. Fish. 25 (1), 909–918. doi:10.21608/ejabf.2021.152628.

Bakshi, M.P.S., Wadhwa, M., Makkar, H.P., 2016. Waste to worth: vegetable wastes as animal feed. CABI Rev. 1–26.

Banks, I.J., Gibson, W.T., Cameron, M.M., 2014. Growth rates of black soldier fly larvae fed on fresh human faeces and their implication for improving sanitation. Trop. Med. Int. Health 19 (1), 14–22.

Barragan-Fonseca, K.B., Dicke, M., van Loon, J.J., 2017. Nutritional value of the black soldier fly (Hermetia illucens L.) and its suitability as animal feed–a review. J. Insects Food Feed 3 (2), 105–120.

Bhaskar, P., Pyne, S.K., Ray, A.K., 2015. Growth performance study of Koi fish, Anabas testudineus (Bloch) by utilization of poultry viscera, as a potential fish feed ingredient, replacing fishmeal. Int. J. Recycl. Org. Waste Agric. 4, 31–37.

Bhattacharya, S., Sardar, P., Sahu, N.P., Shamna, N., Jayant, M., Maiti, M.K., Sundararajan, A., 2025. Dietary pineapple peel ethanolic extract alleviates crowding stress in Labeo rohita fingerlings with enhancement of growth performance and IGF-1 expression, physio-biochemical profile, antioxidant status and immunological responses. Anim. Feed Sci. Technol. 320. doi:10.1016/j.anifeedsci.2024.116200.

Biswas, A., Takakuwa, F., Yamada, S., Matsuda, A., Saville, R.M., LeBlanc, A., Silverman, J.A., Sato, N., Tanaka, H., 2020. Methanotroph (Methylococcus capsulatus, Bath) bacteria meal as an alternative protein source for Japanese yellowtail, Seriola quinqueradiata. Aquaculture 529, 735700. doi:10.1016/j.aquaculture.2020.735700.

Caimi, C., Renna, M., Lussiana, C., Bonaldo, A., Gariglio, M., Meneguz, M., Dabbou, S., Schiavone, A., Gai, F., Elia, A.C, Prearo, M., Gasco, L., 2020. First insights on black soldier fly (Hermetia illucens L.) larvae meal dietary administration in Siberian sturgeon (Acipenser baerii Brandt) juveniles. Aquaculture 515, 734539. doi:10.1016/j.aquaculture.2019.734539.

Carus, M., Dammer, L., 2018. The circular bioeconomy—concepts, opportunities, and limitations. Ind. Biotechnol. 14 (2), 83–91. doi:10.1089/ind.2018.29121.mca.

Cavalheiro, J.M.O., de Souza, E.O., Bora, P.S., 2007. Utilization of shrimp industry waste in the formulation of tilapia (Oreochromis niloticus Linnaeus) feed. Bioresour. Technol. 98 (3), 602–606.

Chaklader, M.R., Fotedar, R., Howieson, J., Siddik, M.A.B., Foysal, M.J., 2020. The ameliorative effects of various fish protein hydrolysates in poultry by-product meal based diets on muscle quality, serum biochemistry and immunity in juvenile barramundi, Lates calcarifer. Fish Shellfish Immunol. 104, 567–578. doi:10.1016/j.fsi.2020.06.014.

Chama, M.K.H., Liang, H., Huang, D., Ge, X., Ren, M., Zhang, L., Wu, L., Ke, J., 2021. Methanotroph (Methylococcus capsulatus, Bath) as an alternative protein source for genetically improved farmed tilapia (GIFT: Oreochromis niloticus) and its effect on antioxidants and immune response. Aquac. Rep. 21. doi:10.1016/j.aqrep.2021.100872.

Coppola, D., Lauritano, C., Esposito, F.P, Riccio, G., Rizzo, C., Pascale, D., Waste, F., 2021. From problem to valuable resource. Mar. Drugs 19 (2), 116. doi:10.3390/md19020116.

Dawood, M.A.O., Koshio, S., 2020. Application of fermentation strategy in aquafeed for sustainable aquaculture. Rev. Aquac. 12 (2), 987–1002. doi:10.1111/raq.12368.

De Geeter, A., Boute, A., 2006. Directive 2006/32/EC on energy end-use efficiency and energy services: realising the transition to sustainable energy markets? J. Eur. Environ. Plan. Law 3 (5), 414–431.

De Vries, M., de Boer, I.J., 2010. Comparing environmental impacts for livestock products: a review of life cycle assessments. Livest. Sci. 128 (1–3), 1–11.

Dikel, S., Dikel, Ç., 2024. A gastronomic approach to industrial aquaculture waste utilization. Turk. J. Agric.-Food Sci. Technol. 12 (10), 1849–1857.

European Parliament and Council, 2008. Directive 2008/56/EC of the European Parliament and of the Council of 17 June 2008 establishing a framework for Community action in the field of marine environmental policy (Marine Strategy Framework Directive). Official Journal of the European Union L164, 19–40. https://eur-lex.europa.eu/eli/dir/2008/56/oj/eng.

Erickson, M.C., Islam, M., Sheppard, C., Liao, J., Doyle, M.P., 2004. Reduction of Escherichia coli O157:H7 and Salmonella enterica serovar enteritidis in chicken manure by larvae of the black soldier fly. J. Food Prot. 67 (4), 685–690. doi:10.4315/0362-028X-67.4.685.

Fadl, S.E., El-Gammal, G.A., Abdo, W.S., Barakat, M., Sakr, O.A., Nassef, E., Gad, D.M., El-Sheshtawy, H.S., 2020. Evaluation of dietary chitosan effects on growth performance, immunity, body composition and histopathology of Nile tilapia (Oreochromis niloticus) as well as the resistance to Streptococcus agalactiae infection. Aquac. Res. 51 (3), 1120–1132. doi:10.1111/are.14458.

Fall, J., Yi-Theng, T., Diegane, N., Shin-Shyn, S., 2012. The effects of replacement of soyabean meal by shrimp shell meal on the growth of hybrid tilapia (Oreochromis niloticus x Oreochromis aureus) reared under brackish water. Lut. J. Fish. Aquacult. 4 (5), 85–91.

Fan, Z., Wu, D., Li, J., Zhang, Y., Cui, Z., Li, T., Zheng, X., Liu, H., Wang, L., Li, H., 2022. Assessment of fish protein hydrolysates in juvenile Largemouth Bass (Micropterus salmoides) diets: effects on growth, intestinal antioxidant status, immunity, and microflora. Front. Nutrit. 9. doi:10.3389/fnut.2022.816341.

FAO, 2024. The State of World Fisheries and Aquaculture 2024: Blue Transformation in Action. Food and Agriculture Organization of the United Nations, Rome. https://openknowledge.fao.org/handle/20.500.14283/cd0683en.

Fawole, F.J., Adeoye, A.A., Tiamiyu, L.O., Ajala, K.I., Obadara, S.O., Ganiyu, I.O., 2020. Substituting fishmeal with Hermetia illucens in the diets of African catfish (Clarias gariepinus): effects on growth, nutrient utilization, haemato-physiological response, and oxidative stress biomarker. Aquaculture 518. doi:10.1016/j.aquaculture.2019.734849.

Fawole, F.J., Sahu, N.P., Jain, K.K., Gupta, S., Rajendran, K.V., Shamna, N., Poojary, N., 2017. Haemato-biochemical, non-specific immunity, antioxidant capacity and histopathological changes in Labeo rohita fingerlings fed rubber protein isolate. Fish Physiol. Biochem. 43 (3), 677–690.

Fawole, F.J., Sahu, N.P., Jain, K.K., Gupta, S., Shamna, N., Phulia, V., Prabu, D.L., 2016. Nutritional evaluation of protein isolate from rubber seed in the diet of Labeo rohita: effects on growth performance, nutrient utilization, whole body composition and metabolic enzymes activity. Anim. Feed Sci. Technol. 219, 189–199. doi:10.1016/j.anifeedsci.2016.06.014.

Fawole, F.J., Sahu, N.P., Shamna, N., Adeoye, A.A., Phulia, V., Emikpe, B.O., 2022. Effects of dietary detoxified jatropha curcas protein isolate on some physiological parameters, intestine, and liver morphology of Labeo rohita fingerlings. Turk. J. Fish. Aquat. Sci. 23 (1), 1–10.

Fawole, F.J., Sahu, N.P., Shamna, N., Phulia, V., Emikpe, B.O., Adeoye, A.A., Aderolu, A.Z., Popoola, O.M., 2018. Effects of detoxified Jatropha curcas protein isolate on growth performance, nutrient digestibility and physio-metabolic response of Labeo rohita fingerlings. Aquacult. Nutr. 24 (4), 1223–1233. doi:10.1111/anu.12660.

Fawole, F.J., Shamna, N., Memudu, H.A., Abdullahi, N., Hassaan, M.S., Gbadamosi, O.K., 2023. Housefly maggot meal complement soybean meal in a fish-free diet for hybrid catfish (Clarias gariepinus♀ x Heterobranchus longifilis♂): effect on growth, body composition, blood biochemistry and antioxidant enzyme activity. Anim. Feed Sci. Technol. 295, 115543.

Food and Agriculture Organization of the United Nations (FAO). (2019). Technical platform on the measurement and reduction of food loss and waste. https://www.fao.org/platform-food-loss-waste/flw-events/international-day-food-loss-and-waste/en.

Food and Agriculture Organization of the United Nations (FAO). Leave no one behind: Why investing in urban-rural areas is key to ending hunger and poverty.

García Martínez, E.S., Stumpf, L., Planas, M., Fernández Gimenez, A.V., López Greco, L.S., 2024. Fishery wastes as feed additive for the red claw crayfish Cherax quadricarinatus: impact on growth, biochemical composition, and digestive activity. Anim. Feed Sci. Technol. 318, 116116. doi:10.1016/j.anifeedsci.2024.116116.

Gasco, L., Gai, F., Maricchiolo, G., Genovese, L., Ragonese, S., Bottari, T., Caruso, G., 2018. Fishmeal alternative protein sources for aquaculture seeds. In: Feeding in Aquaculture. Springer Nature, Cham, pp. 1–28. doi:10.1007/978-3-319-77941-6_1.

Goosen, N.J., de Wet, L.F., Görgens, J.F., Jacobs, K., de Bruyn, A., 2014. Fish silage oil from rainbow trout processing waste as alternative to conventional fish oil in formulated diets for Mozambique tilapia Oreochromis mossambicus. Anim. Feed Sci. Technol. 188, 74–84. doi:10.1016/j.anifeedsci.2013.10.019.

Govorushko, S., 2019. Global status of insects as food and feed source: a review. Trends Food Sci. Technol. 91, 436–445. doi:10.1016/j.tifs.2019.07.032.

Guerrero, S., Cremades, J. (Eds.), 2012. Integrated Multi-Trophic Aquaculture (IMTA): A Sustainable, Pioneering Alternative for Marine Cultures in Galicia, 1st ed. Marine Research Centre, Regional Government of Galicia, Regional Council of the Rural and Regional Maritime Environment, Galicia, Spain, pp. 1–58.

Hansen, R.J., Nielsen, S.H.M., Johansen, M., Nielsen, F.K., Dragsbæk, F.B., Sørensen, O.S.B., Eriksen, N.T., 2023. Metabolic performance of black soldier fly larvae during entomoremediation of brewery waste. J. Appl. Entomol. 147 (6), 423–431. doi:10.1111/jen.13124.

Henry, M., Gasco, L., Piccolo, G., Fountoulaki, E., 2015. Review on the use of insects in the diet of farmed fish: past and future. Anim. Feed Sci. Technol. 203 (1), 1–22. doi:10.1016/j.anifeedsci.2015.03.001.

Hoseinifar, S.H, Shakouri, M., Yousefi, S., Doan, H.V, Shafiei, S., Yousefi, M., Mazandarani, M., Torfi Mozanzadeh, M., Tulino, M.G., Faggio, C., 2020. Humoral and skin mucosal immune parameters, intestinal immune related genes expression and antioxidant defense in rainbow trout (Oncorhynchus mykiss) fed olive (Olea europea L.) waste. Fish Shellfish Immunol. 100, 171–178. doi:10.1016/j.fsi.2020.02.067.

Hossain, A., Senff, P., Glaser, M., 2022. Lessons for coastal applications of IMTA as a way towards sustainable development: a review. Appl. Sci. 12 (23), 11920.

Hossain, M.S., Fawole, F.J., Labh, S.N., Small, B.C., Overturf, K., Kumar, V., 2021. Insect meal inclusion as a novel feed ingredient in soy-based diets improves performance of rainbow trout (Oncorhynchus mykiss). Aquaculture 544. doi:10.1016/j.aquaculture.2021.737096.

Irisarri, J., Cubillo, A.M., Fernández-Reiriz, M.J., Labarta, U., 2015. Growth variations within a farm of mussel (*Mytilus galloprovincialis*) held near fish cages: importance for the implementation of integrated aquaculture. Aquac. Res. 46 (8), 1988–2002.

Jucker, C., Leonardi, M.G., Rigamonti, I., Lupi, D., Savoldelli, S., 2019. Brewery's waste streams as a valuable substrate for black soldier fly Hermetia illucens (Diptera: Stratiomyidae). J. Entomol. Acarol. Res. 51 (3). doi:10.4081/jear. 2019.8876.

Kagali, R.N, Ogello, E.O, Kiama, C.W, Kim, H.-J., Wullur, S., Sakakura, Y., Hagiwara, A., 2022. Culturing live foods for fish larviculture using non-microalgal diet: the role of waste-generated bacteria and selected commercial probiotics—a review. Aquac. Fish Fish. 2 (2), 71–81. doi:10.1002/aff2.33.

Kaleeswaran V., (2023). Nutritional Evaluation of Cabbage and Cauliflower Waste Meal in the Diet of *Labeo rohita* (Hamilton, 1822) Fingerlings, MFSc Dissertation, ICAR-CIFE, Mumbai pp. 143.

Kanatt, S.R., Chander, R., Sharma, A., 2010. Antioxidant and antimicrobial activity of pomegranate peel extract improves the shelf life of chicken products. Int. J. Food Sci. Technol. 45 (2), 216–222.

Kardung, M., Cingiz, K., Costenoble, O., Delahaye, R., Heijman, W., Lovrić, M., Leeuwen, M., M'Barek, R., Meijl, H., Piotrowski, S., Ronzon, T., Sauer, J., Verhoog, D., Verkerk, P.J, Vrachioli, M., Wesseler, J.H.H., Zhu, B.X, 2021. Development of the circular bioeconomy: drivers and indicators. Sustainability 13 (1), 413. doi:10.3390/su13010413.

Kari, Z.A., Kabir, M.A., Mat, K., Rusli, N.D., Razab, M.K.A.A., Ariff, N.S.N.A., Wei, L.S., 2021. The possibility of replacing fish meal with fermented soy pulp on the growth performance, blood biochemistry, liver, and intestinal morphology of African catfish (Clarias gariepinus). Aquac. Rep. 21, 100815.

Kaur, B., Panesar, P.S., Thakur, A., 2024. Biovalorization of mango byproduct through enzymatic extraction of dietary fiber. Environ. Sci. Pollut. Res. 31 (46), 57455–57465.

Kirchherr, J., Reike, D., Hekkert, M., 2017. Conceptualizing the circular economy: an analysis of 114 definitions. Resour. Conserv. Recycl. 127, 221–232.

Kolawole, I.D., Kolawole, G.O., Sanni-Manuel, B.A., Kolawole, S.K., Ewansiha, J.U., Kolawole, V.A., Kolawole, F.O., 2024. Economic impact of waste from food, water, and agriculture in Nigeria: challenges, implications, and applications—a review. Discov. Environ. 2 (1). doi:10.1007/s44274-024-00086-6.

Koo, B., Park, J.Y., Kim, E., Kim, Y., Park, K., 2023. Bioconversion and growth performance of Hermetia illucens in single fruit by-products. Int. J. Indust. Entomol. Biomater. 46 (2), 34–40.

Lalander, C., Diener, S., Magri, M.E., Zurbrügg, C., Lindström, A., Vinnerås, B., 2013. Faecal sludge management with the larvae of the black soldier fly (Hermetia illucens) - from a hygiene aspect. Sci. Total Environ. 458–460, 312–318. doi:10.1016/j.scitotenv.2013.04.033.

Lee, C.S., 2019. Challenges and opportunities of IMTA in Hawaii and beyond. Bull. Jpn. Fish. Res. Educ. Agen 49, 129–134.

Liu, X., Chen, X., Wang, H., Yang, Q., ur Rehman, K., Li, W., Zheng, L., 2017. Dynamic changes of nutrient composition throughout the entire life cycle of black soldier fly. PLoS One 12 (8), e0182601.

Makkar, H.P.S., Tran, G., Heuzé, V.é, 2014. Philippe Ankers, State-of-the-art on use of insects as animal feed. Anim. Feed Sci. Technol. 197, 1–33. doi:10.1016/j.anifeedsci.2014.07.008.

Mannaa, M., Mansour, A., Park, I., Lee, D.W., Seo, Y.S., 2024. Insect-based agri-food waste valorization: agricultural applications and roles of insect gut microbiota. Environ. Sci. Ecotechnol. 17. doi:10.1016/j.ese.2023.100287.

Martínez-Antequera, F.P., Molina-Roque, L., de las Heras, V., Mancera, J.M., Martos-Sitcha, J.A., Moyano, F.J., 2023. Feed supplementation with winery by-products improves the physiological status of juvenile Liza aurata during a short-term feeding trial and hypoxic challenge. Aquac. Rep. 31, 101667. doi:10.1016/j.aqrep.2023.101667.

Masci, A., Coccia, A., Lendaro, E., Mosca, L., Paolicelli, P., Cesa, S., 2016. Evaluation of different extraction methods from pomegranate whole fruit or peels and the antioxidant and antiproliferative activity of the polyphenolic fraction. Food Chem. 202, 59–69. doi:10.1016/j.foodchem.2016.01.106.

Mateos-Aparicio, I., Redondo-Cuenca, A., Tenorio-Sanz, M.D., 2013. Soybean: overview of the nutritional profile and the implications for nutrition and health effects. In: Legumes, types, nutritional composition and health benefits. Nova Science Publishers, Nova Biomedical, New York, pp. 125–159.

Mateos-Aparicio, I., Redondo-Cuenca, A., Villanueva-Suárez, M.-J., Zapata-Revilla, M.-A., Tenorio-Sanz, M.-D., 2010. Pea pod, broad bean pod and okara, potential sources of functional compounds. LWT-Food Sci. Technol. 43 (9), 1467–1470. doi:10.1016/j.lwt.2010.05.008.

Mavuru, A., Mhlanga, L., Nhiwatiwa, T., 2022. An assessment of post-harvest fish losses (PHFLs) in the artisanal fishery of lake kariba. Sci. Afr. 16, e01124.

Motte, C., Rios, A., Lefebvre, T., Do, H., Henry, M., Jintasataporn, O., 2019. Replacing fish meal with defatted insect meal (yellow mealworm Tenebrio molitor) improves the growth and immunity of Pacific white shrimp (Litopenaeus vannamei). Animals 9 (5), 258. doi:10.3390/ani9050258.

Mramba, R.P., Mkude, K.E., 2022. Determinants of fish catch and post-harvest fish spoilage in small-scale marine fisheries in the Bagamoyo district. Heliyon. 8 (6). doi:10.1016/j.heliyon.2022.e09574.

Mussatto, S.I., Dragone, G., Roberto, I.C., 2006. Brewers' spent grain: generation, characteristics and potential applications. J. Cereal Sci. 43 (1), 1–14.

Nawaz, A., Li, E., Irshad, S., Xiong, Z., Xiong, H., Shahbaz, H.M., Siddique, F., 2020. Valorization of fisheries by-products: challenges and technical concerns to food industry. Trends Food Sci. Technol. 99, 34–43. doi:10.1016/j.tifs.2020.02.022.

Nguyễn, N.H., Thái, T.T.T., Lương, K.V., Nguyễn, P.C.T., 2024. Emissions of gases during bio-conversion of agro-waste by black soldier fly larvae. J. Agric. Dev. (Special Issue 2) 143–154.

Olaniran, A.F., Adeoye, O.E., Oyadeyi, O.M., Okonkwo, C.E., Erinle, O.C., Malomo, A.A., Iranloye, Y.M., Olaniran, O.D., Faloye, O.R., 2024. Utilization and application of bioactive compounds generated from fish waste and by product as functional food ingredient: a review. IOP Conf. Ser.: Earth Environ. Sci. 1342 (1), 012014. doi:10.1088/1755-1315/1342/1/012014.

Omoike, A., Falye, A.E., Ogunyemi, A., 2019. Use of agro-industrial waste (Brewer's dried grain) to substitute maize in the diet of African catfish Clarias gariepinus. Afr. J. Resour. Manag. Fish. Aquat. Sci. 4, 91–98.

Osmundsen, T.C., Amundsen, V.S., Alexander, K.A., Asche, F., Bailey, J., Finstad, B., Olsen, M.S., Hernández, K., Salgado, H., 2020. The operationalisation of sustainability: sustainable aquaculture production as defined by certification schemes. Glob. Environ. Chang. 60. doi:10.1016/j.gloenvcha.2019.102025.

Palomo, I., Concha-Meyer, A., Lutz, M., Said, M., Sáez, B., Vásquez, A., Fuentes, E., 2019. Chemical characterization and antiplatelet potential of bioactive extract from tomato pomace (byproduct of tomato paste). Nutrients 11 (2), 456. doi:10.3390/nu11020456.

Parma, L., Busti, S., Ciulli, S., Volpe, E., Errani, F., Oterhals, Å., Bonaldo, A., 2023. Growth, plasma biochemistry and immune-related gene expression of European sea bass (Dicentrarchus labrax) fed bioactive peptides from farmed salmon by-products. Aquaculture 563, 738982.

Patra, O., Shamna, N., Sardar, P., Jayant, M., Sahu, N.P., Deo, A.D., Bhusare, S., 2024. Dietary supplemented pomegranate peel ethanolic extract improves growth, IGF-1 expression and immune responses in Labeo Rohita fingerlings reared in varying stocking densities. Waste Biomass Valor. 15 (12), 6847–6863.

Peschel, W., Sánchez-Rabaneda, F., Diekmann, W., Plescher, A., Gartzía, I., Jiménez, D., Lamuela-Raventós, R., Buxaderas, S., Codina, C., 2006. An industrial approach in the search of natural antioxidants from vegetable and fruit wastes. Food Chem. 97 (1), 137–150. doi:10.1016/j.foodchem.2005.03.033.

Phulia, V., Sardar, P., Sahu, N.P., Fawole, F.J., Shamna, N., Gupta, S., 2018. Substitution of soybean meal with fermented Jatropha kernel meal: effect on growth performance, body composition, and metabolic enzyme activity of Labeo rohita. Fish Physiol. Biochem. 44 (2), 475–487. doi:10.1007/s10695-017-0447-z.

Phulia, V., Sardar, P., Jamwal, A., Kumar, V., Shamna, N., Fawole, F.J., Sanap, B.N., Sahu, N.P., Gupta, S., 2021. Identification of best detoxification strategies for sustainable valorization of waste from Jatropha-based biodiesel industry: compounding the benefits of plant-based vehicular fuel. Environ. Technol. Innov. 24, 101911. doi:10.1016/j.eti.2021.101911.

Poolsawat, L., Yang, H., Sun, Y.F., Li, X.Q., Liang, G.Y., Leng, X.J., 2021. Effect of replacing fish meal with enzymatic feather meal on growth and feed utilization of tilapia (Oreochromis niloticus × O. aureus). Anim. Feed Sci. Technol. 274. doi:10.1016/j.anifeedsci.2021.114895.

Rahimnejad, S., Hu, S., Song, K., Wang, L., Lu, K., Wu, R., Zhang, C., 2019. Replacement of fish meal with defatted silkworm (Bombyx mori L.) pupae meal in diets for Pacific white shrimp (Litopenaeus vannamei). Aquaculture 510, 150–159. doi:10.1016/j.aquaculture.2019.05.054.

Rajesh, M., Kamalam, B.S., Sharma, P., Verma, V.C., Pandey, A., Dubey, M.K., Kaushik, S.J., 2022. Evaluation of a novel methanotroph bacteria meal grown on natural gas as fish meal substitute in rainbow trout, Oncorhynchus mykiss. Aquac. Res. 53 (6), 2159–2174.

Raut, N.A., Kokare, D.M., Randive, K.R., Bhanvase, B.A., Dhoble, S.J., 2023. Introduction: fundamentals of waste removal technologies. In: 360-Degree Waste Management, Volume 1: Fundamentals, Agricultural and Domestic Waste, and Remediation. Elsevier, India, pp. 1–16. doi:10.1016/B978-0-323-90760-6.00009-6.

Resconi, A., Bellezza Oddon, S., Ferrocino, I., Loiotine, Z., Caimi, C., Gasco, L., Biasato, I., 2024. Effects of brewery by-products on growth performance, bioconversion efficiency, nutritional profile, and microbiota and mycobiota of black soldier fly larvae. Animal 18 (9), 101288. doi:10.1016/j.animal.2024.101288.

Robertson, J.A., I'Anson, K.J.A., Treimo, J., Faulds, C.B., Brocklehurst, T.F., Eijsink, V.G.H., Waldron, K.W., 2010. Profiling brewers' spent grain for composition and microbial ecology at the site of production. LWT-Food Sci. Technol. 43 (6), 890–896. doi:10.1016/j.lwt.2010.01.019.

Roy, H., Alam, S.R., Bin-Masud, R., Prantika, T.R., Pervez, M.N., Islam, M.S., Naddeo, V., 2022. A review on characteristics, techniques, and waste-to-energy aspects of municipal solid waste management: Bangladesh perspective. Sustainability 14 (16). doi:10.3390/su141610265.

Safdar, M., Naqvi, S.A., Anjum, F., Pasha, I., Shahid, M., Waliullah, M.J.J., Khan, I.A., Aadil, R.M., 2021. Microbial biofilm inhibition, antioxidants, and chemical fingerprints of Afghani pomegranate peel extract documented by gas chromatography–mass spectrometry and Fourier transformation infrared. J. Food Process. Preserv. 45 (7). doi:10.1111/jfpp.15657.

Sagar, N.A., Pareek, S., Sharma, S., Yahia, E.M., Lobo, M.G., 2018. Fruit and vegetable waste: bioactive compounds, their extraction, and possible utilization. Compr. Rev. Food Sci. Food Saf. 17 (3), 512–531. doi:10.1111/1541-4337.12330.

Sahin, N., Orhan, C., Tuzcu, M., Sahin, K., Kucuk, O., 2008. The effects of tomato powder supplementation on performance and lipid peroxidation in quail. Poult. Sci. 87 (2), 276–283. doi:10.3382/ps.2007-00207.

Sahlini S. (2023). Optimization of Tomato Pomace Meal in the Diet of Labeo rohita (Hamilton,1822) through Fermentation and Exogeneous Enzyme Supplementation, MFSc Dissertation, ICAR-CIFE, Mumbai pp. 125.

Sanap, B.N., Sardar, P., Sahu, N.P., Phulia, V., Shamna, N., Dasgupta, S., Datta, S., 2018. Utilization of electron beam irradiated Jatropha kernel meal in the diet of Labeo rohita (Hamilton, 1822) fingerlings. Aquacult. Nutr. 24 (4), 1304–1317.

Santos, A.D.S.S., Morais, N.W.S., Santos, A.B., 2023. Brewery Spent Grain as sustainable source for value-added bioproducts: opportunities and new insights in the integrated lignocellulosic biorefinery concept. Ind. Crops Prod. 206.

Selim, N.A., Youssef, S.F., Abdel-Salam, A.F., Nada, S.A., 2013. Evaluation of some natural antioxidant sources in broiler diets: effect on growth, physiological, microbiological and immunological performance of broiler chicks. Int. J. Poult. Sci. 12 (10), 561–571. doi:10.3923/ijps.2013.561.571.

Shamna, N., Sardar, P., Sahu, N.P., Phulia, V., Rajesh, M., Fawole, F.J., Pal, A.K., Angel, G., 2017. Heamato-immunological and physiological responses of Labeo rohita fingerlings to dietary fermented Jatropha curcas protein concentrate. Anim. Feed Sci. Technol. 232, 198–206. doi:10.1016/j.anifeedsci.2016.10.020.

Sharif, M., Zafar, M.H., Aqib, A.I., Saeed, M., Farag, M.R., Alagawany, M., 2021. Single cell protein: sources, mechanism of production, nutritional value and its uses in aquaculture nutrition. Aquaculture 531. doi:10.1016/j.aquaculture.2020.735885.

Siddik, M.A.B., Chaklader, M.dR., Foysal, M.dJ., Howieson, J., Fotedar, R., Gupta, S.K., 2020. Influence of fish protein hydrolysate produced from industrial residues on antioxidant activity, cytokine expression and gut microbial communities in juvenile barramundi Lates calcarifer. Fish Shellfish Immunol. 97, 465–473. doi:10.1016/j.fsi.2019.12.057.

Siddiqui, S.A., Harahap, I.A., Osei-Owusu, J., Saikia, T., Wu, Y.S., Fernando, I., Perestrelo, R., Câmara, J.S., 2024. Bioconversion of organic waste by insects – a comprehensive review. Process Saf. Environ. Prot. 187, 1–25. doi:10.1016/j.psep.2024.04.122.

Singh, B., Singh, J.P., Kaur, A., Singh, N., 2019. Antimicrobial potential of pomegranate peel: a review. Int. J. Food Sci. Technol. 54 (4), 959–965. doi:10.1111/ijfs.13964.

Singh, S., Kiran, B.R., Mohan, S.V., 2024. Carbon farming: a circular framework to augment CO2 sinks and to combat climate change. Environmental Science Advances,3 ((4):) 522–542. https://doi.org/10.1039/d3va00296a.

Singh, J., Prasad, R., Kaur, H.P., Jajoria, K., Chahal, A.S., Verma, A., Kara, M., Assouguem, A., Bahhou, J., 2023. Bioactive compounds, pharmacological properties, and utilization of pomegranate (Punica granatum L.): a comprehensive review. Trop. J. Nat. Prod. Res. 7 (9), 3856–3873. doi:10.26538/tjnpr/v7i9.2.

Skaf, L., Franzese, P.P., Capone, R., Buonocore, E., 2021. Unfolding hidden environmental impacts of food waste: an assessment for fifteen countries of the world. J. Cleaner Prod. 310, 127523. doi:10.1016/j.jclepro.2021.127523.

Sogi, D.S, Bawa, A.S., 1998. Dehydration of tomato processing waste. Indian Food Pack. 52, 26–29.

Somroo, A.A., ur Rehman, K., Zheng, L., Cai, M., Xiao, X., Hu, S., Mathys, A., Gold, M., Yu, Z., Zhang, J., 2019. Influence of Lactobacillus buchneri on soybean curd residue co-conversion by black soldier fly larvae (Hermetia illucens) for food and feedstock production. Waste Manage. 86, 114–122. doi:10.1016/j.wasman.2019.01.022.

Suma, A.Y., Nandi, S.K., Abdul Kari, Z., Goh, K.W., Wei, L.S., Tahiluddin, A.B., Seguin, P., Herault, M., Al Mamun, A., Téllez-Isaías, G., Anamul Kabir, M., 2023. Beneficial effects of graded levels of fish protein hydrolysate (FPH) on the growth performance, blood biochemistry, liver and intestinal health, economics efficiency, and disease resistance to Aeromonas hydrophila of Pabda (Ompok pabda) Fingerling. Fishes 8 (3). doi:10.3390/fishes8030147.

Sundararajan, A., Sahu, N.P., Shamna, N., Jayant, M., Sardar, P., Vasanthakumaran, Sankar, S., 2024. Dietary papaya peel extract ameliorates the crowding stress, enhances growth and immunity in Labeo rohita fingerlings. Fish Physiol. Biochem. 50 (3), 1047–1064.

Szabo, K., Varvara, R.A., Ciont, C., Macri, A.M., Vodnar, D.C., 2025. An updated overview on the revalorization of bioactive compounds derived from tomato production and processing by-products. J. Cleaner Prod., 145151.

Tassoni, A., Tedeschi, T., Zurlini, C., Cigognini, I.M., Petrusan, J.I., Rodríguez, Ó., Neri, S., Celli, A., Sisti, L., Cinelli, P., Signori, F., Tsatsos, G., Bondi, M., Verstringe, S., Bruggerman, G., Corvini, P.F.X., 2020. State-of-the-art production chains for peas, beans and chickpeas—valorization of agro-industrial residues and applications of derived extracts. Molecules 25 (6). doi:10.3390/molecules25061383.

Tejaswini, K., Deo, A.D., Shamna, N., Jayant, M., Aklakur, M.d., Annadurai, R., 2024. Effect of flavanone rich lemon peel extract on feed intake and growth of Labeo rohita (Hamilton, 1822) fingerlings reared at low temperature recirculatory aquaculture system. Aquaculture 584, 740450. doi:10.1016/j.aquaculture.2023.740450.

Ugalde, U.O., Castrillo, J.I., 2002. Single cell proteins from fungi and yeasts. Appl. Mycol. Biotechnol. 2 (C), 123–149. doi:10.1016/S1874-5334(02)80008-9.

United Nations Environment Programme (2024). Food Waste Index Report 2024. Think Eat Save: tracking Progress to Halve Global Food Waste. United Nations Environment Programme. Nairobi https://wedocs.unep.org/20.500.11822/45230.

Vasilaki, A., Mente, E., Fountoulaki, E., Henry, M., Nikoloudaki, C., Berillis, P., Kousoulaki, K., Nengas, I., 2023. Fishmeal, plant protein, and fish oil substitution with single-cell ingredients in organic feeds for European sea bass (Dicentrarchus labrax). Front. Physiol. 14. doi:10.3389/fphys.2023.1199497.

Wadhwa M., Bakshi M.P.S., Utilization of fruit and vegetable wastes as livestock feed and as substrates for generation of other value-added products. RAP Publication No. 2013/04. Food and Agriculture Organization of the United Nations, Regional Office for Asia and the Pacific. Bangkok **4** (2013),

Wei, Y., Liang, M., Xu, H., 2020. Fish protein hydrolysate affected amino acid absorption and related gene expressions of IGF-1/AKT pathways in turbot (Scophthalmus maximus). Aquacult. Nutr. 26 (1), 145–155.

, 2021. Driven to Waste: The Global Impact of Food Loss and Waste on Farms. World Wide Fund for Nature (WWF), Woking, UK, pp. 1–24.

Xu, B., Liu, Y., Chen, K., Wang, L., Sagada, G., Tegomo, A.F., Yang, Y., Sun, Y., Zheng, L., Ullah, S., Shao, Q., 2021. Evaluation of methanotroph (Methylococcus capsulatus, Bath) bacteria meal (FeedKind®) as an alternative protein source for juvenile black sea bream, Acanthopagrus schlegelii. Front. Mar. Sci. 8. doi:10.3389/fmars.2021.778301.

Yıldırım-Aksoy, M., Eljack, R., Aksoy, J., Beck, B.H., 2023. Frass from black soldier fly larvae, Hermetia illucens, as a possible functional dietary ingredient in channel catfish feed. Fishes 8 (11), 542.

Yildirim-Aksoy, M., Eljack, R., Beck, B.H., 2020. Nutritional value of frass from black soldier fly larvae, Hermetia illucens, in a channel catfish, Ictalurus punctatus, diet. Aquacult. Nutr. 26 (3), 812–819.

Yousufi, M.K., 2012. To determine protein content of single cell protein produced by using various combinations of fruit wastes and two standard food fungi. Int. J. Adv. Biotechnol. Res. 3 (1), 533–536.

Zhang, J., Dong, Y., Song, K., Wang, L., Li, X., Tan, B., Zhang, C., 2022a. Effects of the replacement of dietary fish meal with defatted yellow mealworm (Tenebrio molitor) on juvenile large yellow croakers (Larimichthys crocea) growth and gut health. Animals 12 (19), 2659.

Zhang, Q., Liang, H., Longshaw, M., Wang, J., Ge, X., Zhu, J., Ren, M., 2022b. Effects of replacing fishmeal with methanotroph (Methylococcus capsulatus, Bath) bacteria meal (FeedKind®) on growth and intestinal health status of juvenile largemouth bass (Micropterus salmoides). Fish Shellfish Immunol. 122, 298–305.

Zheng, J., Zhang, W., Dan, Z., Cao, X., Cui, K., Zhu, S., Zhuang, Y., Mai, K., Ai, Q., 2023. Effects of fish meal replaced by methanotroph bacteria meal (Methylococcus capsulatus) on growth, body composition, antioxidant capacity, amino acids transporters and protein metabolism of turbot juveniles (Scophthalmus maximus L.). Aquaculture 562, 738782. doi:10.1016/j.aquaculture.2022.738782.

CHAPTER

22

Introducing insects into a circular-based economy in aquaculture

Moslem Sharifinia[a], *Mohammad Hossein Khanjani*[b] *and Zahra Afshari Bahmanbeigloo*[c]

[a]Aquaculture Department, Shrimp Research Center, Iranian Fisheries Sciences Research Institute, Agricultural Research, Education and Extension Organization (AREEO), Bushehr, Iran [b]Department of Fisheries Sciences and Engineering, Faculty of Natural Resources, University of Jiroft, Jiroft, Iran [c]Bushehr Regional Water Authority, Iran Water Resources Management Company, Bushehr, Iran

22.1 Introduction

The global aquaculture sector has become an essential element of food security, supplying a substantial source of protein for billions of people across the globe (Dewali et al., 2023; Verdegem et al., 2023; Sharifinia, 2024; Sharifinia, 2025). However, its rapid growth has led to challenges such as reliance on conventional feed sources, environmental degradation, and unsustainable practices (Sharifinia, 2024). With the human population increasing at 1.6% annually (Gras et al., 2023) and wild fish availability declining, aquaculture is becoming essential for protein supply (Alfiko et al., 2022). The circular economy concept emphasizes resource efficiency and waste reduction, making it crucial for aquaculture to adopt these principles (Cooney et al., 2023).

Insects are recognized as a high-quality protein source due to their rapid growth and ability to convert organic waste into biomass (Chaalala et al., 2018; Hawkey et al., 2021; Sharifinia, 2024; Sharifinia et al., 2024). The overexploitation of oceans has led to unsustainable pressure on wild fish stocks (Pauly, 2010; Fantatto et al., 2024; Sharifinia, 2024), making reliance on wild fish unsustainable (Gasco et al., 2018). The aquaculture industry needs alternative protein sources to replace fishmeal due to rising demand and costs (Bordignon et al., 2024; Fantatto et al., 2024; Sharifinia, 2024). While fishmeal has often been replaced with plant proteins, such as soybean, high amounts can negatively affect fish health (Wang et al., 2017; Macusi et al., 2023). Therefore researchers are exploring innovative feed ingredients, including insects, which offer nutritional

Zero-Waste Technologies in Aquaculture: Global Perspectives and Methods toward a Circular Economy
DOI: https://doi.org/10.1016/B978-0-443-34689-7.00001-X

profiles similar to fishmeal (Dicke, 2018; Basto et al., 2020; Fabrikov et al., 2021; Bordignon et al., 2024; Sharifinia, 2024; Sharifinia et al., 2024).

Insects such as *Hermetia illucens, Tenebrio molitor,* and *Musca domestica* have high nutritional value, containing 42%–63.3% crude protein and essential amino acids (Alfiko et al., 2022). However, challenges remain, including standardizing insect feeding substrates and determining the replacement percentage for different fish species (Huis et al., 2021; Gasco et al., 2023; Jankauskienė et al., 2024). The price of insect meals can vary based on production systems and substrates (Macombe et al., 2019; Veldkamp et al., 2022). Integrating insects into aquaculture aligns with circular economy principles by transforming waste into valuable resources and promoting biodiversity (Fantatto et al., 2024; Sharifinia, 2024). Utilizing insect larvae reduces reliance on traditional feed components linked to environmental degradation (Sharifinia, 2024). Insects improve the nutritional quality of aquaculture feeds, often outperforming plant-based proteins in digestibility and growth rates (Agbohessou et al., 2021; Alfiko et al., 2022).

This chapter will explore the diverse benefits of integrating insects into aquaculture, focusing on their contributions to waste management, economic viability, and environmental sustainability. We will cover different insect farming techniques, the nutritional benefits of insect-based feeds, and the regulatory aspects involved in their use. Furthermore, we will present case studies showcasing successful implementations of insect farming in aquaculture systems, demonstrating the practical applications and potential for scaling these innovations.

22.2 Circular economy

The circular economy concept signifies a fundamental shift away from the conventional linear economic model, which is typically characterized by a "take-make-dispose" methodology. In contrast, a circular economy is conceptualized as an economic framework that aims to eradicate waste and encourage the continuous utilization of resources (Emerenciano et al., 2025). This model highlights the significance of designing products and systems that enable the reuse, recycling, and regeneration of materials. The foundational principles of a circular economy encompass longevity in design, the maintenance and extension of product lifespan, resource recovery, systems thinking, and regeneration. By prioritizing these principles, a circular economy aspires to establish a closed-loop system that minimizes resource input, waste, emissions, and energy loss, thereby promoting sustainability and resilience (Emerenciano et al., 2025; Kirchherr et al., 2023).

Sustainability plays a crucial role in aquaculture, especially in light of the rising global demand for seafood driven by population growth and evolving dietary habits (Sharifinia, 2024). The Food and Agriculture Organization (FAO) reports that fish consumption worldwide has seen substantial increases in recent decades, with aquaculture now representing more than half of the total fish consumed globally (FAO, 2024). Implementing sustainable aquaculture practices is vital for alleviating the environmental challenges linked to fish farming, including water pollution, habitat degradation, and the overexploitation of wild fish populations (Lal et al., 2024; Morales-Nin et al., 2024). By embracing sustainable approaches, aquaculture can bolster food security by ensuring a consistent supply of protein and essential nutrients, while also enhancing economic viability through more efficient utilization of resources. In addition, sustainable aquaculture contributes to community development by generating employment opportunities and supporting local economies, particularly in coastal and rural regions where livelihoods are often intricately connected to marine resources (Pratiwy et al., 2024).

Despite its potential, the aquaculture industry encounters several significant challenges that jeopardize its sustainability. Waste management emerges as a critical concern, as aquaculture practices produce considerable waste, including uneaten feed, fecal matter, and deceased fish (Cherian et al., 2023; Chary et al., 2024; Mahari et al., 2024; Tan et al., 2024). Effluents generated from aquaculture can substantially contribute to nutrient loading in aquatic environments, resulting in eutrophication and harmful consequences for aquatic ecosystems (Yeganeh et al., 2020; Luo, 2023; Lan et al., 2024). If not effectively managed, this waste can lead to water pollution and eutrophication, negatively impacting surrounding ecosystems. Furthermore, the dependence on wild-caught fish for feed, particularly in the form of fishmeal and fish oil, intensifies resource depletion and exerts pressure on marine ecosystems (Majluf et al., 2024; Sharifinia, 2024). Overfishing and habitat degradation pose significant threats to the sustainability of these resources, highlighting the urgent need for alternative feed sources that maintain environmental integrity.

Furthermore, the environmental repercussions of aquaculture extend beyond waste management and resource depletion. Aquaculture can lead to various ecological concerns, including the transmission of diseases to wild fish populations and modifications to local habitats (Fujita et al., 2023; Bøhn et al., 2024; Sanda et al., 2024). For example, the introduction of non-native species within aquaculture systems can disrupt local ecosystems and result in a loss of biodiversity (Kang et al., 2023). Tackling these environmental challenges is vital for ensuring the long-term sustainability of the industry. In addition, regulatory and market challenges, such as inconsistent standards for sustainability and consumer perceptions of farmed products, present further obstacles to growth and innovation in the aquaculture sector. Effectively navigating these complexities is crucial for promoting a sustainable aquaculture industry that aligns with the principles of a circular economy. In recent years, the exploration of alternative protein sources, such as insect-based feeds, has offered a promising avenue for boosting the sustainability of aquaculture. Insects are high in protein and can be cultivated on organic waste, which helps in reducing waste and promoting resource recovery. Research has shown that insect-based feeds can enhance growth rates and feed conversion efficiency in fish, making them a viable substitute for traditional fishmeal. This transition toward alternative feed sources not only eases the strain on wild fish populations but also aligns with the circular economy's focus on resource recovery and waste minimization (Sharifinia, 2024; Sharifinia et al., 2023a, 2023b). Embracing these innovative solutions can significantly contribute to a more sustainable and resilient aquaculture sector.

22.3 Role of insects in aquaculture

The role of insects in aquaculture is increasingly recognized as a transformative solution to the challenges faced by the industry, particularly in the context of sustainability, nutrition, and economic viability. As the global population continues to rise, the demand for protein sources is projected to increase significantly. Traditional aquaculture practices, which often rely heavily on fish meal and soy, are becoming increasingly unsustainable because of overfishing, habitat destruction, and the environmental impacts associated with large-scale agricultural practices. In this landscape, insects present a compelling alternative that not only meets the nutritional needs of aquatic species but also aligns with the principles of sustainable food production (Sharifinia et al., 2023a, 2023b; Sharifinia, 2024).

Insects, such as crickets, mealworms, and black soldier fly larvae, are particularly advantageous due to their rapid growth rates and high reproductive capacity. These characteristics enable them to be produced in large quantities with relatively low resource inputs. For instance, black soldier fly larvae can convert organic waste into high-quality protein in a matter of weeks, making them an efficient and sustainable feed source. This rapid production cycle is crucial for meeting the increasing demand for protein in aquaculture, particularly in regions where traditional feed sources may be scarce or prohibitively expensive (Lock et al., 2016; Agbohessou et al., 2021; Alvanou et al., 2023; Bordignon et al., 2024; Fantatto et al., 2024; Oliveira et al., 2024; Sharifinia, 2024).

The nutritional profile of insects is another significant factor contributing to their appeal in aquaculture. Insects are not only rich in protein but also provide a diverse range of essential nutrients that are vital for the growth and health of aquatic species. For example, many insects are excellent sources of essential amino acids, which are critical for protein synthesis and overall growth. In addition, insects contain bioactive compounds, such as chitin, which may enhance gut health and immune function in fish. The presence of these nutrients can lead to improved feed efficiency, growth rates, and overall health in aquaculture species, ultimately resulting in higher yields for producers (Barroso et al., 2014; Makkar et al., 2014; Henry et al., 2015; de Castro et al., 2018; Basto et al., 2020; Cadinu et al., 2020; Anankware et al., 2021; Hossain et al., 2023; Fantatto et al., 2024).

Moreover, the environmental benefits of using insects as a feed source cannot be overstated. Insect farming has a significantly lower carbon footprint compared to traditional livestock and fish farming. The production of insects generates fewer greenhouse gas emissions and requires less water and land. For instance, it is estimated that producing 1 kg of insect protein requires only a fraction of the water needed to produce the same amount of beef or pork. This resource efficiency is particularly important in the context of climate change and resource scarcity, as it allows for the production of high-quality protein with minimal environmental impact (Van Huis and Oonincx, 2017; Patel, 2019; Lin et al., 2023; Sharifinia, 2024).

Insects also play a crucial role in waste management within aquaculture systems. By utilizing organic waste materials, such as food scraps and agricultural byproducts, insects can help reduce the environmental burden associated with waste disposal. This waste-to-feed approach not only contributes to a more sustainable aquaculture system but also promotes a circular economy, where nutrients are recycled and reused rather than discarded. This integration of insect farming into aquaculture can lead to more resilient and self-sufficient production systems, reducing reliance on external feed sources and enhancing food security (van Huis, 2013; van Huis, 2015; Veldkamp et al., 2022).

The economic implications of incorporating insects into aquaculture are also noteworthy. As the demand for sustainable protein sources continues to rise, the market for insect-based feeds is expected to grow significantly. This presents opportunities for entrepreneurs and farmers to diversify their operations and tap into a burgeoning market. In addition, the relatively low startup costs associated with insect farming make it an accessible option for small-scale producers, particularly in developing regions where traditional feed sources may be limited. By investing in insect farming, aquaculture producers can enhance their profitability while contributing to sustainable food systems (Cappellozza et al., 2019; Aaqillah-Amr et al., 2024; Auzins et al., 2024).

Furthermore, the integration of insects into aquaculture aligns with the growing consumer demand for transparency and sustainability in food production. As consumers become more conscious of the environmental and ethical implications of their dietary choices, there is a shift toward products that are sourced sustainably. By incorporating insects into aquaculture practices, producers can appeal to this market trend, positioning themselves as leaders in sustainable food production. This shift not only benefits the environment but also enhances the marketability of aquaculture products, potentially leading to higher prices and increased consumer loyalty (Mulazzani et al., 2021; Roccatello et al., 2024).

The future of aquaculture may very well hinge on the successful incorporation of insects as a primary feed source. As research continues to explore the various applications of insects in aquaculture, including their potential to replace or supplement traditional feed sources, the industry is likely to see a paradigm shift toward more sustainable and efficient practices. Collaborative efforts between researchers, policymakers, and industry stakeholders will be essential in promoting the adoption of insect-based feeds and ensuring that the benefits are realized across the aquaculture sector.

Insects, such as crickets, mealworms, and black soldier fly larvae, are particularly advantageous due to their rapid growth rates and high reproductive capacity. These characteristics enable them to be produced in large quantities with relatively low resource inputs. For instance, black soldier fly larvae can convert organic waste into high-quality protein in a matter of weeks, making them an efficient and sustainable feed source. This rapid production cycle is crucial for meeting the increasing demand for protein in aquaculture, particularly in regions where traditional feed sources may be scarce or prohibitively expensive (Lock et al., 2016; Agbohessou et al., 2021; Alvanou et al., 2023; Bordignon et al., 2024; Fantatto et al., 2024; Oliveira et al., 2024; Sharifinia, 2024).

The nutritional profile of insects is another significant factor contributing to their appeal in aquaculture. Insects are not only rich in protein but also provide a diverse range of essential nutrients that are vital for the growth and health of aquatic species. For example, many insects are excellent sources of essential amino acids, which are critical for protein synthesis and overall growth. In addition, insects contain bioactive compounds, such as chitin, which may enhance gut health and immune function in fish. The presence of these nutrients can lead to improved feed efficiency, growth rates, and overall health in aquaculture species, ultimately resulting in higher yields for producers (Barroso et al., 2014; Makkar et al., 2014; Henry et al., 2015; de Castro et al., 2018; Basto et al., 2020; Cadinu et al., 2020; Anankware et al., 2021; Hossain et al., 2023; Fantatto et al., 2024).

Moreover, the environmental benefits of using insects as a feed source cannot be overstated. Insect farming has a significantly lower carbon footprint compared with traditional livestock and fish farming. The production of insects generates fewer greenhouse gas emissions and requires less water and land. For instance, it is estimated that producing 1 kg of insect protein requires only a fraction of the water needed to produce the same amount of beef or pork. This resource efficiency is particularly important in the context of climate change and resource scarcity, as it allows for the production of high-quality protein with minimal environmental impact (van Huis and Oonincx, 2017; Patel, 2019; Lin et al., 2023; Sharifinia, 2024).

Insects also play a crucial role in waste management within aquaculture systems. By utilizing organic waste materials, such as food scraps and agricultural byproducts, insects can help reduce the environmental burden associated with waste disposal. This waste-to-feed approach not only contributes to a more sustainable aquaculture system but also promotes a circular economy, where nutrients are recycled and reused rather than discarded. This integration of insect farming into aquaculture can lead to more resilient and self-sufficient production systems, reducing reliance on external feed sources and enhancing food security (van Huis, 2013; van Huis, 2015; Veldkamp et al., 2022).

The economic implications of incorporating insects into aquaculture are also noteworthy. As the demand for sustainable protein sources continues to rise, the market for insect-based feeds is expected to grow significantly. This presents opportunities for entrepreneurs and farmers to diversify their operations and tap into a burgeoning market. In addition, the relatively low startup costs associated with insect farming make it an accessible option for small-scale producers, particularly in developing regions where traditional feed sources may be limited. By investing in insect farming, aquaculture producers can enhance their profitability while contributing to sustainable food systems (Cappellozza et al., 2019; Aaqillah-Amr et al., 2024; Auzins et al., 2024).

Furthermore, the integration of insects into aquaculture aligns with the growing consumer demand for transparency and sustainability in food production. As consumers become more conscious of the environmental and ethical implications of their dietary choices, there is a shift toward products that are sourced sustainably. By incorporating insects into aquaculture practices, producers can appeal to this market trend, positioning themselves as leaders in sustainable food production. This shift not only benefits the environment but also enhances the marketability of aquaculture products, potentially leading to higher prices and increased consumer loyalty (Mulazzani et al., 2021; Roccatello et al., 2024).

The future of aquaculture may very well hinge on the successful incorporation of insects as a primary feed source. As research continues to explore the various applications of insects in aquaculture, including their potential to replace or supplement traditional feed sources, the industry is likely to see a paradigm shift toward more sustainable and efficient practices. Collaborative efforts between researchers, policymakers, and industry stakeholders will be essential in promoting the adoption of insect-based feeds and ensuring that the benefits are realized across the aquaculture sector.

22.4 Insect farming techniques

Insect farming is increasingly recognized as a sustainable alternative to traditional livestock production, offering a viable solution to global food security and environmental sustainability challenges. As the world population grows, the demand for protein is expected to rise significantly, necessitating environmentally friendly and economically viable nutritional sources. Insects, particularly species such as black soldier flies and mealworms, present a compelling option due to their high feed conversion efficiency, rapid growth rates, and ability to thrive on organic waste. This innovative approach not only addresses nutritional needs but also promotes waste reduction and resource efficiency, aligning with global efforts to create resilient food systems capable of withstanding climate change and resource depletion (Weinreis et al., 2023; Barragán-Fonseca et al., 2024).

The advantages of insect farming extend beyond protein production. Insects are highly efficient protein producers, requiring significantly less land, water, and feed compared to traditional livestock. For instance, black soldier flies can convert organic waste into protein at an impressive rate, showcasing their unique digestive systems that can break down complex organic materials. In contrast, traditional livestock requires much more feed to produce the same amount of protein, highlighting the stark differences in resource utilization. In addition, insects emit fewer greenhouse gases than conventional livestock, making their farming a crucial component in combating climate change. By integrating insect farming into food production systems, we can minimize the environmental impact of protein production and advance sustainable agricultural practices. Moreover, insects' ability to convert waste into high-quality protein supports a circular economy, where waste is minimized, and resources are reused effectively (Weinreis et al., 2023; Barragán-Fonseca et al., 2024).

Insect farming methods can be categorized into various approaches based on the species cultivated and their intended use. Two primary methods are vertical farming and batch farming. Vertical farming involves stacking insect-rearing units to maximize space utilization, particularly advantageous in urban settings. This method allows for controlled environmental conditions optimized for specific insect species, enhancing production efficiency and reducing the overall footprint of insect farming operations. Conversely, batch farming raises insects in discrete batches, allowing for controlled growth cycles and easier management. This method enables farmers to synchronize life cycles for consistent harvests and experiment with different feed formulations (Van Huis and Oonincx, 2017).

Creating optimal conditions for insect growth is essential for maximizing production efficiency and ensuring insect health. Key factors include temperature, humidity, and feed composition. Different insect species have varying temperature requirements, and maintaining appropriate ranges is crucial for promoting rapid growth and minimizing mortality rates. Humidity levels also significantly influence insect farming, as proper control prevents issues such as mold growth and dehydration. In addition, the choice of feed is critical for promoting optimal growth and nutritional quality. A balanced diet rich in protein and essential nutrients enhances growth performance and improves the nutritional value of harvested insects (Caruso et al., 2014; Morales-Ramos et al., 2024; Siddiqui et al., 2024).

Insect farming can be conducted at various scales, from small-scale operations to large industrial farms, each with its advantages and challenges. Small-scale insect farming often involves localized production for personal use or local markets, promoting sustainability by utilizing organic waste from local sources. However, small-scale operations may struggle with consistent production levels and market access. In contrast, industrial-scale insect farming involves large facilities designed for high-volume production, employing advanced technologies to optimize growth conditions. While industrial farms can achieve significant economies of scale, they also require substantial initial investments and may face regulatory challenges. Balancing the benefits of large-scale production with responsible environmental stewardship is crucial for the long-term viability of the insect farming industry (FAO, 2021; Barragán-Fonseca et al., 2024).

As insect farming continues to expand, regulatory frameworks are being developed to ensure food safety, animal welfare, and environmental sustainability. Different countries have varying regulations regarding the production and sale of edible insects, impacting market access and consumer acceptance. Establishing clear guidelines for insect farming practices is essential for building consumer trust and promoting the industry. Ethical considerations also play a role in the acceptance of insect farming; ensuring humane conditions and minimizing suffering during harvesting and processing are vital for addressing ethical concerns and gaining public acceptance (Pali-Schöll et al., 2019; Meijer et al., 2025).

The future of insect farming is promising, with ongoing research and innovation aimed at improving production techniques, enhancing nutritional profiles, and expanding market opportunities. As consumer awareness of sustainability issues increases, the demand for insect-based products is expected to rise, presenting significant opportunities for entrepreneurs and farmers. Research is also exploring the potential of insects in bioconversion processes, transforming organic waste into valuable protein and other byproducts. This approach addresses waste management challenges and contributes to a circular economy, where resources are reused and recycled. Furthermore, integrating insect farming into existing agricultural systems can enhance resilience and sustainability, as insects can serve as a feed source for livestock and aquaculture, providing a high-quality protein alternative to traditional feed ingredients. This shift reduces pressure on overfished marine resources and decreases the environmental footprint associated with conventional feed production.

22.5 Waste management and resource recovery

Waste management and resource recovery are increasingly recognized as essential components of sustainable agricultural practices, particularly in aquaculture, where organic waste generation poses significant environmental challenges. The aquaculture industry produces substantial amounts of organic waste, including fish byproducts such as offal, bones, and uneaten feed. Traditionally viewed as a disposal issue, this waste has led to pollution and inefficiencies in resource utilization. However, innovative strategies are emerging that transform organic waste into valuable resources, particularly through the use of insects for waste conversion. By employing insects as a feed source, aquaculture operations can effectively convert organic waste into high-quality protein, addressing both waste management and resource recovery in a synergistic manner (Boyd et al., 2020; Ingrao et al., 2021; Laktuka et al., 2023).

Insects, such as black soldier flies (*H. illucens*) and mealworms (*T. molitor*), are particularly suited for this purpose due to their remarkable ability to thrive on organic waste materials. Their efficient digestive systems enable them to decompose complex organic matter, converting it into protein-rich biomass. Utilizing organic waste from aquaculture as feed for insects not only reduces waste volume but also facilitates the recycling of essential nutrients back into the food production system. This nutrient recycling is vital for promoting sustainability, as it reduces reliance on synthetic fertilizers and mitigates the environmental impact of waste disposal. Furthermore, incorporating insects into aquaculture waste management strategies enhances resource use efficiency, contributing to a more circular economy within the agricultural sector. Numerous case studies demonstrate the effective implementation of waste-to-insect conversion in aquaculture. For example, fish processing byproducts have been successfully used as feed for black soldier fly larvae, which efficiently convert fish waste into high-quality protein for various aquaculture species. This innovative approach alleviates the environmental burden of fish waste disposal while providing a sustainable feed source for fish farming operations. In another significant case, insect larvae were employed to process organic waste from salmon farming, resulting in substantial reductions in waste volume and the production of nutrient-dense larvae suitable for both aquaculture and livestock (Govaerts, 2018; Lopes et al., 2020; Romero-Lorente et al., 2022; Abasubong et al., 2023; Arena et al., 2023; Rodrigues et al., 2025).

In addition to waste reduction, insects contribute to nutrient recycling by converting organic waste into valuable protein and other byproducts. The biomass produced from insect farming can serve as animal feed and a source of essential nutrients for human consumption. By integrating insect farming into aquaculture waste management practices, producers can create a closed-loop system that maximizes resource efficiency and minimizes environmental impact. This approach aligns with circular economy principles, transforming waste into valuable resources and enhancing the sustainability of food production systems (Siddiqui et al., 2024; Sanchez-Hernandez and Megharaj, 2025).

The environmental benefits of utilizing insects for waste management extend beyond waste reduction. Converting organic waste into insect biomass significantly lowers greenhouse gas emissions associated with traditional disposal methods, such as landfilling or incineration. Insects produce fewer greenhouse gases during growth compared to conventional livestock, making them an environmentally friendly alternative for protein production. In addition, using organic waste as feed for insects can help mitigate the environmental impact of aquaculture by reducing nutrient runoff and pollution associated with waste discharge (Aidoo et al., 2023; Fu et al., 2024).

Integrating insect farming into aquaculture also presents economic advantages. By converting waste into a valuable protein source, aquaculture operations can lower feed costs and enhance profitability. The cost of using organic waste as feed for insects is often significantly lower than purchasing conventional feed ingredients, providing economic relief to producers and encouraging sustainable practices. Furthermore, the production of insect protein creates new market opportunities, as demand for alternative protein sources in both animal feed and human consumption continues to grow (Madau et al., 2020; Barragán-Fonseca et al., 2022; Auzins et al., 2024). As consumer awareness of sustainability issues increases, the market for insect-based products is expected to expand, providing additional revenue streams for aquaculture operations that embrace this innovative approach.

22.6 Economic viability

The economic viability of insect farming as a feed source in aquaculture is essential for producers seeking sustainable alternatives to traditional feeds. Insect production offers several economic advantages, primarily because of its efficient conversion of organic waste into high-quality protein. For example, black soldier fly larvae can utilize agricultural byproducts and food waste, which are often cheaper than traditional feed components such as fishmeal or soybean meal. This efficiency not only reduces overall feed costs but also enhances economic resilience by providing a stable supply of feed (Alfiko et al., 2022; Barragán-Fonseca et al., 2024; Munguti et al., 2025).

Market demand for insect-based feed is increasing as consumers become more aware of sustainability issues and the environmental impacts of conventional feed production. Aquaculture producers are actively seeking alternative protein sources that align with these values, making insect-based feed an attractive option because of its lower ecological footprint and ability to recycle organic waste. Evolving regulatory frameworks are beginning to support the use of insect protein in animal feed, further driving demand.

Insect farming also has the potential to create jobs and stimulate economic growth in rural areas. It can be established with low capital investment, making it accessible for communities looking to diversify income sources. Employment opportunities arise in various sectors, including production, processing, distribution, and research, fostering local economies and entrepreneurship. The development of this industry can enhance collaboration among agricultural sectors, research institutions, and local governments, contributing to community resilience (Sokame et al., 2024).

Integrating insect farming into local economies can improve food security by providing a sustainable protein source for both livestock and human consumption. As global protein demand rises, insect farming minimizes the environmental impact associated with traditional livestock production. Local operations reduce reliance on imported feed, promoting a self-sufficient food system (Dickie et al., 2019; Sokame et al., 2024).

Educational and training opportunities related to insect farming can enhance the skills of rural populations. Institutions can develop training programs covering production techniques, waste management, and business development, empowering communities and fostering innovation. In addition, insect farming can stimulate ancillary industries, such as organic waste management and logistics, creating further job opportunities and economic activity.

The environmental benefits of insect farming also have positive economic implications. By reducing reliance on traditional feed sources, insect farming helps mitigate environmental degradation linked to conventional livestock production, such as deforestation and greenhouse gas emissions (Sharifinia, 2024). This transition to sustainable practices can enhance the reputation of aquaculture operations, attracting environmentally conscious consumers and investors. As sustainability drives market demand, aquaculture producers that adopt insect farming can position themselves favorably in a competitive marketplace.

22.7 Environmental impact

The environmental impact of insect farming highlights its potential as a sustainable alternative to traditional feed production methods. One of the primary benefits is the significant reduction in carbon emissions associated with animal feed production (Cadinu et al., 2020; Madau et al., 2020; Sharifinia, 2024). Insects, such as black soldier flies and mealworms, can efficiently convert organic waste into high-quality protein while emitting minimal greenhouse gases (Boakye-Yiadom et al., 2022; Siddiqui et al., 2024). Unlike conventional livestock, which require extensive land and resources, insects can be cultivated in controlled environments, significantly lowering the overall emissions related to feed production. This reduction not only aids in climate change mitigation but also aligns with global sustainability goals aimed at minimizing the environmental footprint of food systems (Boakye-Yiadom et al., 2022; Sharifinia, 2024; Siddiqui et al., 2024).

Water usage is another critical factor in assessing the environmental impact of insect farming compared to traditional feed production. The water footprint of insect farming is considerably lower than that of conventional livestock feed production, such as soybeans or fishmeal (Van Huis and Oonincx, 2017). Producing one kilogram of insect protein requires significantly less water than the same amount derived from traditional sources. This efficiency is particularly important in regions facing water scarcity, where sustainable food production methods are increasingly necessary. By utilizing organic waste streams and requiring less water, insect farming presents a viable solution to food security and environmental sustainability challenges (Van Huis and Oonincx, 2017; Sharifinia, 2024).

Insect farming also positively impacts biodiversity and alleviates pressures related to overfishing (Sharifinia, 2024; Sharifinia et al., 2023a). Traditional feed production often relies on fishmeal from wild-caught fish, leading to overfishing and depletion of marine resources, which threatens aquatic ecosystems. In contrast, insect farming offers an alternative protein source that does not depend on wild fish stocks. By converting organic waste into insect protein, we can reduce the demand for fishmeal and promote marine biodiversity conservation. In addition, integrating insect farming into existing agricultural systems can enhance biodiversity on farms by introducing a new layer of production that complements traditional crops and livestock.

Moreover, insect farming contributes to soil health and nutrient recycling. Insects can be fed organic waste materials, such as food scraps and agricultural byproducts, which not only reduces waste but also enriches the soil when insect frass (excrement) is used as fertilizer. This natural fertilizer is nutrient-rich and can improve soil fertility, promoting healthier crop growth. By closing the nutrient loop, insect farming supports sustainable agricultural practices that enhance soil health and reduce reliance on synthetic fertilizers, which can negatively impact the environment (Torgerson et al., 2021; Abajue and Gbarakoro, 2023).

The scalability of insect farming presents an opportunity to tackle global food security challenges. As the world's population grows, the demand for protein sources is expected to rise dramatically. Insect farming offers a scalable solution that can be implemented in various contexts, from small-scale operations in rural communities to large commercial facilities. This adaptability makes insect farming an attractive option for regions facing food insecurity, providing a reliable and sustainable protein source while utilizing local resources and waste streams (Glover and Sexton, 2015; Ameixa et al., 2020; Lisboa et al., 2024).

Furthermore, the educational and economic opportunities associated with insect farming can empower communities and foster sustainable development. Training programs focused on insect farming can equip individuals with the skills needed to engage in this innovative sector, promoting entrepreneurship and job creation. As communities adopt insect farming practices, they can experience increased income, improved food security, and enhanced resilience to economic and environmental challenges.

22.8 Regulatory and safety considerations

The regulatory landscape governing insect farming and its application in aquaculture is rapidly evolving, reflecting the industry's growth and increasing recognition of insects as a viable protein source (Hameed et al., 2022; Sharifinia et al., 2023b; Fantatto et al., 2024). Various countries are developing frameworks to regulate the production, processing, and use of insects in animal feed, ensuring adherence to safety, environmental, and ethical standards. In the European Union, the European Food Safety Authority (EFSA) has established guidelines that emphasize the necessity of thorough risk assessments and compliance with existing food safety legislation (Meijer et al., 2025). These guidelines are crucial for ensuring that insect farming practices meet stringent safety standards.

In addition to national regulations, international standards are being developed to facilitate trade and ensure the safety of insect-based products across borders (Lähteenmäki-Uutela et al., 2021; Delgado et al., 2023). Organizations such as the FAO and the World Health Organization (WHO) are working to create guidelines that address the safety and quality of insect production (Vreysen et al., 2021; WHO, 2021). These international standards aim to harmonize regulations globally, making compliance easier for producers and fostering consumer trust in insect-derived products. Food safety is a critical concern in the production of insect-based feed. To ensure the safety of these products, stringent measures must be implemented throughout the farming and processing stages. This includes monitoring the insects' diet, ensuring they are fed safe and approved organic materials to prevent contamination. Processing methods, such as heat treatment, can be employed to eliminate pathogens and parasites, enhancing the safety of the final product. Regular testing for contaminants, including heavy metals and microbial pathogens, is essential to ensure that insect-based feed meets safety standards before being introduced into aquaculture systems (Schlüter et al., 2017; van der Fels-Klerx et al., 2018; Imathiu, 2020).

The potential for bioaccumulation of harmful substances in insects is another significant concern (Liu et al., 2018; Hierlmeier et al., 2022). Insects can absorb toxins from their environment, including heavy metals and pesticides, which can pose risks to animal and human health. Therefore establishing monitoring protocols to assess the safety of the substrate used for insect farming is essential. By ensuring that the feedstock is free from contaminants, producers can mitigate the risks associated with bioaccumulation and enhance the safety of insect-based products. Consumer acceptance of insect-based products plays a crucial role in their successful integration into the food supply chain. While there is a growing recognition of the environmental and nutritional benefits of insect protein, consumer attitudes can vary widely based on cultural, social, and psychological factors. In many Western countries, insects are often viewed with skepticism or disgust, which can hinder their acceptance as a viable food source or feed ingredient. To address these concerns, educational campaigns and marketing strategies are essential to inform consumers about the benefits of insect-based products and to demystify their use in food systems (House, 2016; Sogari et al., 2023; Gambelli et al., 2024).

Cultural attitudes toward insects significantly influence consumer acceptance. In regions such as Asia, Africa, and Latin America, insects are already a traditional food source and are widely accepted as nutritious and sustainable protein options. In contrast, in Western societies, efforts must be made to reframe the narrative around insects as food. Engaging influential figures, such as chefs and nutritionists, can help promote insect-based products as gourmet and health-conscious choices. In addition, incorporating insects into familiar food products, such as protein bars or snacks, can ease consumers into accepting insect-based ingredients. Transparency in the production process and clear labeling of insect-based products can enhance consumer trust. Providing information about the sourcing, processing, and nutritional content of insect products empowers consumers to make informed choices. Engaging with stakeholders, including chefs, nutritionists, and environmental advocates, can also help promote insect-based products as sustainable alternatives to traditional protein sources (Tan et al., 2015; Tan and House, 2018; Wendin and Nyberg, 2021).

22.9 Future perspectives and innovations

The future of insect farming is set for significant advancements through innovative technologies that enhance farming and processing methods. Automated rearing systems, precision feeding techniques, and advanced monitoring tools are optimizing insect cultivation, improving feed conversion rates, and reducing labor costs. In addition, artificial intelligence and machine learning enable real-time data analysis, allowing farmers to make informed decisions about insect health and environmental conditions, which is crucial for scaling operations to meet global protein demands.

Genetic improvements in insect species also hold promise for enhancing yield and nutritional content. Selective breeding and genetic engineering can develop strains with superior

TABLE 22.1 Comprehensive overview of the role of insects in aquaculture: Functions, benefits, challenges, and economic impact.

Category	Details	Potential challenges	Economic impact
Natural protein source	Protein content: Ranges from 9.3% to 76% Fatty acids: Provides omega-3 and omega-6 fatty acids Amino acids: Rich in essential amino acids Optimal species: Particularly beneficial for omnivorous fish like catfish and carp	Variability in protein content among species	Reduces feed costs by providing a cheaper protein alternative
Waste management solution	- Organic waste conversion: Effective in recycling organic waste into valuable nutrients - Food waste processed - Environmental impact: Reduces landfill usage and greenhouse gas emissions through bioconversion	Public perception and acceptance of insect meal	Reduction in waste disposal costs; potential revenue from waste recycling
Feed innovation	- Sustainable alternatives: Acts as a sustainable replacement for traditional fishmeal - Product forms: Available as meal, oil, pulp, paste, and even live insects for feeding	Regulatory hurdles for insect products in feed	Growth in demand for sustainable aquafeeds
Key species used	- Black soldier fly (BSF) - Common house fly - Silkworm moth - Yellow mealworm	Limited geographical availability of all species	Economic value driven by high protein efficiency and low feed conversion ratio

(continued on next page)

TABLE 22.1 Comprehensive overview of the role of insects in aquaculture: Functions, benefits, challenges, and economic impact—cont'd

Category	Details	Potential challenges	Economic impact
Environmental impact	- Land usage: Requires significantly less land area compared to conventional feed crops - Water consumption: Much lower water use relative to traditional livestock and fishmeal production - Greenhouse gas emissions: Potentially significantly lower emissions compared to livestock	Environmental variability affecting insect farming	Long-term savings in resource use compared to traditional feeds
Waste management benefits	- Conversion of agricultural waste: Utilizes byproducts from agriculture, thereby reducing waste - Supports circular economy: Integrates into sustainable food systems - Pollution minimization: Reduces nutrient runoff and soil degradation risks	Need for infrastructure to support waste collection	Potential cost savings in waste management for local economies

TABLE 22.2 Overview of insect farming techniques: Sustainability, efficiency, and future prospects.

Category	Details	Advantages	Challenges
Farming techniques	Vertical farming: • Utilizes stacked rearing units for space efficiency • Allows controlled environmental conditions 2. Batch farming: • Raises insects in discrete batches for easier management and synchronized harvests	Optimizes space, reduces land use, and enhances production efficiency	Requires precise environmental controls for specific species and consistent monitoring
Environmental parameters	- Temperature control - Humidity levels	Enhances growth and minimizes mortality rates	Maintaining proper humidity to prevent mold growth and dehydration is critical
Feed composition	• Variety of organic materials, including food scraps, grains, and agricultural byproducts • Balanced diets promote growth and high nutritional quality	Reduces food waste while providing nutrient-rich feed for insects	Sourcing consistent high-quality feed can be challenging

(*continued on next page*)

TABLE 22.2 Overview of insect farming techniques: Sustainability, efficiency, and future prospects—cont'd

Category	Details	Advantages	Challenges
Scale of operations	1. Small-scale farms: • Local production utilizing organic waste; less investment • Promotes sustainability 2. Industrial-scale farms: • Advanced technologies for high-volume production; requires significant investment	Flexibility and lower startup costs for small-scale; economies of scale for industrial	Small-scale operations face market access issues; industrial farms require careful environmental management
Regulatory considerations	Different countries have varying regulations affecting production and market access	Establishes food safety and promotes consumer trust in insect products	Navigating diverse regulations can be complex and time-consuming
Future prospects	Growth in demand for insect-based products and innovations in cultivation techniques	Significant opportunities for entrepreneurs and farmers; advances in waste bioconversion	Continuous research and development needed to enhance production methods

growth rates, disease resilience, and improved nutritional profiles. Researchers are exploring modifications to increase protein content and enhance amino acid composition, leading to more efficient and nutritionally superior insect-based feeds for aquaculture. Integrating insects into aquaculture practices offers a sustainable feed source, reducing reliance on traditional ingredients such as fishmeal and soybean meal while improving feed quality. Utilizing organic byproducts from food processing as insect feed promotes waste reduction and supports a circular economy.

Symbiotic relationships between insect farming and aquaculture can further enhance sustainability. Byproducts from insect farming can serve as organic fertilizers in aquaculture, while aquaculture effluents can nourish insect larvae, creating a mutually beneficial cycle that maximizes resource utilization. Collaboration among researchers, industry stakeholders, and policymakers is essential for driving innovation and addressing challenges in the insect farming sector. Investment in research and development will explore new insect species and production techniques, expanding the applications of insect farming.

22.10 Conclusion

Integrating insects into aquaculture presents significant opportunities for sustainability, resource efficiency, and reduced environmental impact. By utilizing organic waste as insect feed, aquaculture can minimize waste and generate high-quality protein, decreasing reliance on traditional feed ingredients and alleviating pressure on marine ecosystems. Stakeholders in the aquaculture industry must embrace insect-based solutions and collaborate on research initiatives, best practices, and regulatory frameworks to support integration. As global protein demand rises, adopting insect farming practices is critical for long-term sustainability, paving the way for a resilient and environmentally responsible aquaculture industry that contributes to food security and ecological balance.

References

Aaqillah-Amr, M.A., Hidir, A., Ikhwanuddin, M., Fatan, N.A., Muhammad-Rahimi, R., Ning-Shahira, S., Siankhe, R.G.N., Vandenberg, G.W., Tanga, C.M., Yossa, R., 2024. Effects of insect meals on fish digestibility, blood parameters, and economic performance: a meta-analysis. J. Insects Food Feed 11 (5), 833–859. doi:10.1163/23524588-00001096.

Abajue, M.C., Gbarakoro, T.N., 2023. Prospects of insect farming for food security, environmental sustainability, and as an alternative to agrochemical use. Insects for sustainable food production and environmental management. Springer Science and Business Media LLC, Singapore, pp. 565–600. doi:10.1007/978-981-99-3439-3_21.

Abasubong, K.P., Gabriel, N.N., Adjoumani, J.-J.Y., Okon, A.O., 2023. Waste to feed: an emerging technology to improve aquaculture in Africa. Sustainable agriculture and food security. Springer Science and Business Media LLC, Singapore, pp. 103–130. doi:10.1007/978-981-19-7451-9_4.

Agbohessou, P.S., Mandiki, S.N.M., Gougbédji, A., Megido, R.C., Lima, L.M.W., Cornet, V., Lambert, J., Purcaro, G., Francis, F., Lalèyè, P.A., Kestemont, P., 2021. Efficiency of fatty acid-enriched dipteran-based meal on husbandry, digestive activity and immunological responses of Nile tilapia Oreochromis niloticus juveniles. Aquaculture 545. doi:10.1016/j.aquaculture.2021.737193.

Aidoo, O.F., Osei-Owusu, J., Asante, K., Dofuor, A.K., Boateng, B.O., Debrah, S.K., Ninsin, K.D., Siddiqui, S.A., Chia, S.Y., 2023. Insects as food and medicine: a sustainable solution for global health and environmental challenges. Front. Nutr. 10. doi:10.3389/fnut.2023.1113219.

Alfiko, Y., Xie, D., Astuti, R.T., Wong, J., Wang, L., 2022. Insects as a feed ingredient for fish culture: status and trends. Aquac. Fish. 7 (2), 166–178. doi:10.1016/j.aaf.2021.10.004.

Alvanou, M.V., Kyriakoudi, A., Makri, V., Lattos, A., Feidantsis, K., Papadopoulos, D.K., Georgoulis, I., Apostolidis, A.P., Michaelidis, B., Mourtzinos, I., Asimaki, A., Karapanagiotidis, I.T., Giantsis, I.A., 2023. Effects of dietary substitution of fishmeal by black soldier fly (Hermetia illucens) meal on growth performance, whole-body chemical composition, and fatty acid profile of Pontastacus leptodactylus juveniles. Front. Physiol. 14. doi:10.3389/fphys.2023.1156394.

Ameixa, O.M.C.C., Duarte, P.M., Rodrigues, D.P., 2020. Insects, food security, and sustainable aquaculture. Encyclopedia of the UN sustainable development goals: zero hunger. Springer Science and Business Media LLC, Cham, pp. 425–435. doi:10.1007/978-3-319-95675-6_111.

Anankware, J.P., Roberts, B.J., Cheseto, X., Osuga, I., Savolainen, V., Collins, C.M., 2021. The nutritional profiles of five important edible insect species from West Africa—an analytical and literature synthesis. Front. Nutr. 8. doi:10.3389/fnut.2021.792941.

Arena, R., Manuguerra, S., Curcuraci, E., Cusimano, M., Monaco, D.L.o, Di Bella, C., Santulli, A., Messina, C.M., 2023. Fisheries and aquaculture by-products modulate growth, body composition, and omega-3 polyunsaturated fatty acid content in black soldier fly (Hermetia illucens) larvae. Front. Anim. Sci. 4. doi:10.3389/fanim.2023.1204767.

Auzins, A., Leimane, I., Reissaar, R., Brobakk, J., Sakelaite, I., Grivins, M., Zihare, L., 2024. Assessing the socio-economic benefits and costs of insect meal as a fishmeal substitute in livestock and aquaculture. Animals 14 (10), 1461. doi:10.3390/ani14101461.

Barragán-Fonseca, K.B., Muñoz-Ramírez, A.P., Cune, J.nP., Dicke, M., Cortés, J., 2022. Fighting rural poverty in Colombia: circular agriculture by using insects as feed in aquaculture. Wageningen Livestock Research, Wageningen Wageningen Livestock Research Report 1353. https://doi.org/10.18174/561878 .

Barragán-Fonseca, K.B., Gómez, D., Lalander, C.H., Dzepe, D., Chia, S.Y., 2024. Review – insect farming for food and feed in the global south: focus on black soldier fly production. Animal, 101397 doi:10.1016/j.animal.2024.101397.

Barroso, F.G., de Haro, C., Sánchez-Muros, M.J., Venegas, E., Martínez-Sánchez, A., Pérez-Bañón, C., 2014. The potential of various insect species for use as food for fish. Aquaculture 422–423, 193–201. doi:10.1016/j.aquaculture.2013.12.024.

Basto, A., Matos, E., Valente, L.M.P., 2020. Nutritional value of different insect larvae meals as protein sources for European sea bass (Dicentrarchus labrax) juveniles. Aquaculture 521, 735085. doi:10.1016/j.aquaculture.2020.735085.

Armah Boakye-Yiadom, K., Ilari, A., Duca, D., 2022. Greenhouse gas emissions and life cycle assessment on the black soldier fly (Hermetia illucens L.). Sustainability 14 (16), 10456. doi:10.3390/su141610456.

Bøhn, T., Strøm, J.F., Sanchez-Jerez, P., Keeley, N.B., Johansen, T., Gjelland, K., Sandlund, N., Sæther, B.S., Sætra, I., Olsen, E.M., Skjæraasen, J.E., Meier, S., van der Meeren, T., Bjørn, P.A., 2024. Ecological interactions between farmed Atlantic salmon and wild Atlantic cod populations in Norway: a review of risk sources and knowledge gaps. Rev. Aquac. 16 (3), 1333–1350. doi:10.1111/raq.12899.

Bordignon, F., Trocino, A., Gasco, L., Bellezza Oddon, S., Xiccato, G., Berton, M., 2024. Towards circularity in aquaculture systems: environmental impact of Hermetia illucens meal inclusion in diets for rainbow trout reared in aquaponics. J. Clean. Prod. 466, 142901. doi:10.1016/j.jclepro.2024.142901.

Boyd, C.E., D'Abramo, L.R., Glencross, B.D., Huyben, D.C., Juarez, L.M., Lockwood, G.S., McNevin, A.A., Tacon, A.G.J., Teletchea, F., Tomasso, J.R., Tucker, C.S., Valenti, W.C., 2020. Achieving sustainable aquaculture: historical and current perspectives and future needs and challenges. J. World Aquac. Soc. 51 (3), 578–633. doi:10.1111/jwas.12714.

Cadinu, L.A., Barra, P., Torre, F., Delogu, F., Madau, F.A., 2020. Insect rearing: potential, challenges, and circularity. Sustainability. 12 (11), 4567. doi:10.3390/su12114567.

Cappellozza, S., Leonardi, M.G, Savoldelli, S., Carminati, D., Rizzolo, A., Cortellino, G., Terova, G., Moretto, E., Badaile, A., Concheri, G., Saviane, A., Bruno, D., Bonelli, M., Caccia, S., Casartelli, M., Tettamanti, G., 2019. A first attempt to produce proteins from insects by means of a circular economy. Animals 9 (5), 278. doi:10.3390/ani9050278.

Caruso, D., Devic, I.W.S., Talamond, P., Baras, E., 2014. Technical handbook of domestication and production of diptera Black Soldier Fly (BSF) Hermetia illucens (Stratiomyidae). IRD Éditions, Marseille.

de Castro, R.J.S., Ohara, A., Aguilar, J.G.S., Domingues, M.A.F., 2018. Nutritional, functional and biological properties of insect proteins: processes for obtaining, consumption and future challenges. Trends Food Sci. Technol. 76, 82–89. doi:10.1016/j.tifs.2018.04.006.

Chaalala, S., Leplat, A., Makkar, H., 2018. Importance of insects for use as animal feed in low-income countries. Edible insects in sustainable food systems. Springer International Publishing, France, pp. 303–319. doi:10.1007/978-3-319-74011-9_18.

Chary, K., van Riel, A.J., Muscat, A., Wilfart, A., Harchaoui, S., Verdegem, M., Filgueira, R., Troell, M., Henriksson, P.J.G., de Boer, I.J.M., Wiegertjes, G.F., 2024. Transforming sustainable aquaculture by applying circularity principles. Rev. Aquac. 16 (2), 656–673. doi:10.1111/raq.12860.

Cherian, T., Ragavendran, C., Vijayan, S., Kurien, S., Peijnenburg, W.J.G.M., 2023. A review on the fate, human health and environmental impacts, as well as regulation of antibiotics used in aquaculture. Environ. Adv. 13, 100411. doi:10.1016/j.envadv.2023.100411.

Cooney, R., de Sousa, D.B., Fernández-Ríos, A., Mellett, S., Rowan, N., Morse, A.P., Hayes, M., Laso, J., Regueiro, L., Wan, A.H., Clifford, E., 2023. A circular economy framework for seafood waste valorisation to meet challenges and opportunities for intensive production and sustainability. J. Clean. Prod. 392. doi:10.1016/j.jclepro.2023.136283.

Delgado, L., Garino, C., Moreno, F.J., Zagon, J., Broll, H., 2023. Sustainable food systems: EU regulatory framework and contribution of insects to the farm-to-fork strategy. Food Rev. Int. 39 (9), 6955–6976. doi:10.1080/87559129.2022.2130354.

Dewali, S., Sharma, N., Melkani, D., Arya, M., Kathayat, N., Panda, A.K., Bisht, S.S., 2023. Aquaculture: contributions to global food security emerging solutions in sustainable food and nutrition security. Springer International Publishing, India, pp. 123–139. doi:10.1007/978-3-031-40908-0_6.

Dicke, M., 2018. Insects as feed and the sustainable development goals. J. Insects Food Feed 4 (3), 147–156. doi:10.3920/jiff2018.0003.

Dickie, F., Miyamoto, M., Matilda (Tilly) Collins, C., 2019. The potential of insect farming to increase food security. IntechOpen, London doi:10.5772/intechopen.88106.

Emerenciano, M.G.C., Khanjani, M.H., Sharifinia, M., Miranda-Baeza, A., 2025. Could biofloc technology (BFT) pave the way toward a more sustainable aquaculture in line with the circular economy? doi:10.1155/are/1020045.

Fabrikov, D., Barroso, F.G., Sánchez-Muros, M.J., Hidalgo, M.C., Cardenete, G., Tomás-Almenar, C., Melenchón, F., Guil-Guerrero, J.L., 2021. Effect of feeding with insect meal diet on the fatty acid compositions of sea bream (Sparus aurata), tench (Tinca tinca) and rainbow trout (Oncorhynchus mykiss) fillets doi:10.1016/j.aquaculture.2021.737170.

Fantatto, R.R., Mota, J., Ligeiro, C., Vieira, I., Guilgur, L.G., Santos, M., Murta, D., 2024. Exploring sustainable alternatives in aquaculture feeding: the role of insects doi:10.1016/j.aqrep.2024.102228.

FAO, 2021. Looking at edible insects from a food safety perspective. challenges and opportunities for the sector. Angew. Chem., Int. Ed. doi:10.4060/cb4094en.

van der Fels-Klerx, H.J., Camenzuli, L., Belluco, S., Meijer, N., Ricci, A., 2018. Food safety issues related to uses of insects for feeds and foods. Compr. Rev. Food Sci. Food Saf. 17 (5), 1172–1183. doi:10.1111/1541-4337.12385.

Fu, C., Cheema, W.A., Mobashar, M., Shah, A.A., Alqahtani, M.M., 2025. Insects as sustainable feed: enhancing animal nutrition and reducing livestock environmental impression. J. Anim. Physiol. Anim. Nutr. 109 (2), 280–290. doi:10.1111/jpn.14055.

Fujita, R., Brittingham, P., Cao, L., Froehlich, H., Thompson, M., Voorhees, T., 2023. Toward an environmentally responsible offshore aquaculture industry in the United States: ecological risks, remedies, and knowledge gaps. Mar. Policy 147, 105351. doi:10.1016/j.marpol.2022.105351.

Gambelli, D., Vairo, D., Zanoli, R., Alleweldt, F., 2024. Future scenarios of the sector of insect as food and feed in Europe. J. Insects Food Feed 11 (17), 191–210. doi:10.1163/23524588-00001108.

Gasco, L., Bellezza Oddon, S., Vandenberg, G.W., Veldkamp, T., Biasato, I., 2023. Factors affecting the decision-making process of using insect-based products in animal feed formulations. J. Insects Food Feed 10 (10), 1–12. doi:10.3920/jiff2022.0164.

Gasco, L., Gai, F., Maricchiolo, G., Genovese, L., Ragonese, S., Bottari, T., Caruso, G., 2018. Fishmeal alternative protein sources for aquaculture feeds. Feeding in Aquaculture. Springer Nature, Cham, pp. 1–28. doi:10.1007/978-3-319-77941-6_1.

FAO, 2024. Brief to The State of World Fisheries and Aquaculture 2024. Towards Blue Transformation.FAO, Rome.

Glover D., Sexton A., Edible insects and the future of food: A foresight scenario exercise on entomophagy and global food security. (2015),

Govaerts, F., 2018. Introducing Insect-Based Salmon Feed. From a Nutritional, Economic, Legal and Marketing Perspective thesis. Wageningen University, Wageningen, The Netherlands.

Gras, M., Winker, H., Vasilakopoulos, P., Pinto, C., Konrad, C., Mannini, A., 2023. Scientific, Technical and Economic Committee for Fisheries (STECF): Monitoring the Performance of the Common Fisheries Policy (STECF-Adhoc-23-01). Publications Office of the European Union, Luxembourg.

Hameed, A., Majeed, W., Naveed, M., Ramzan, U., Bordiga, M., Hameed, M., Ur Rehman, S., Rana, N., 2022. Success of aquaculture industry with new insights of using insects as feed: a review. Fishes. 7 (6), 395. doi:10.3390/fishes7060395.

Hawkey, K.J., Lopez-Viso, C., Brameld, J.M., Parr, T., Salter, A.M., 2021. Insects: a potential source of protein and other nutrients for feed and food. Ann. Rev. Anim. Biosci. 9, 333–354. doi:10.1146/annurev-animal-021419-083930.

Henry, M., Gasco, L., Piccolo, G., Fountoulaki, E., 2015. Review on the use of insects in the diet of farmed fish: past and future. Anim. Feed Sci. Technol. 203 (1), 1–22. doi:10.1016/j.anifeedsci.2015.03.001.

Hierlmeier, V.R., Gurten, S., Freier, K.P., Schlick-Steiner, B.C., Steiner, F.M., 2022. Persistent, bioaccumulative, and toxic chemicals in insects: current state of research and where to from here? Sci. Total Environ. 825. doi:10.1016/j.scitotenv.2022.153830.

Hossain, M.S., Small, B.C., Hardy, R., 2023. Insect lipid in fish nutrition: recent knowledge and future application in aquaculture. Rev. Aquac. 15 (4), 1664–1685. doi:10.1111/raq.12810.

House, J., 2016. Consumer acceptance of insect-based foods in the Netherlands: academic and commercial implications. Appetite 107, 47–58. doi:10.1016/j.appet.2016.07.023.

Huis, 2015. Edible insects contributing to food security? Agric. Food Secur. 4, 1–9.

Imathiu, S., 2020. Benefits and food safety concerns associated with consumption of edible insects. NFS J. 18, 1–11. doi:10.1016/j.nfs.2019.11.002.

Ingrao, C., Arcidiacono, C., Siracusa, V., Niero, M., Traverso, M., 2021. Life cycle sustainability analysis of resource recovery from waste management systems in a circular economy perspective key findings from this special issue. Resources 10 (4), 32. doi:10.3390/resources10040032.

Jankauskienė, A., Aleknavičius, D., Kiselioviené, S., Antanaitis, Š., Falkauskas, R., Šumskienė, M., Juknienė, I., Kabašinskienė, A., 2024. The influence of different sustainable substrates on the nutritional value of Tenebrio molitor larvae. Foods 13 (3), 365. doi:10.3390/foods13030365.

Kang, B., Vitule, J.R.S., Li, S., Shuai, F., Huang, L., Huang, X., Fang, J., Shi, X., Zhu, Y., Xu, D., Yan, Y., Lou, F., 2023. Introduction of non-native fish for aquaculture in China: a systematic review. Rev. Aquac. 15 (2), 676–703. doi:10.1111/raq.12751.

Kirchherr, J., Yang, N.H.N., Schulze-Spüntrup, F., Heerink, M.J., Hartley, K., 2023. Conceptualizing the circular economy (Revisited): an analysis of 221 definitions. Resour. Conserv. Recycl. 194. doi:10.1016/j.resconrec.2023.107001.

Lähteenmäki-Uutela, A., Marimuthu, S.B., Meijer, N., 2021. Regulations on insects as food and feed: a global comparison. J. Insects Food Feed 7 (5), 849–856. doi:10.3920/jiff2020.0066.

Laktuka, K., Kalnbalkite, A., Sniega, L., Logins, K., Lauka, D., 2023. Towards the sustainable intensification of aquaculture: exploring possible ways forward. Sustainability 15 (24), 16952. doi:10.3390/su152416952.

Lal, S., Bala Ahirwar, S., Kanojia, S., Rai, S., Vidhya, C.S., Gupta, S., Nautiyal, C.T, Nautiyal, P., 2024. Robot-assisted aquaculture and sustainable seafood production for enhanced food security. Int. J. Environ. Clim. Change 14 (2), 215–220. doi:10.9734/ijecc/2024/v14i23938.

Lan, J., Liu, P., Hu, X., Zhu, S., 2024. Harmful algal blooms in eutrophic marine environments: causes, monitoring, and treatment. Water 16 (17), 2525. doi:10.3390/w16172525.

Lin, X., Wang, F., Lu, Y., Wang, J., Chen, J., Yu, Y., Tao, X., Xiao, Y., Peng, Y., 2023. A review on edible insects in China: nutritional supply, environmental benefits, and potential applications. Curr. Res. Food Sci. 7, 100596. doi:10.1016/j.crfs.2023.100596.

Lisboa, H.M., Nascimento, A., Arruda, A., Sarinho, A., Lima, J., Batista, L., Dantas, M.F., Andrade, R., 2024. Unlocking the potential of insect-based proteins: sustainable solutions for global food security and mutrition. Foods 13 (12). doi:10.3390/foods13121846.

Liu, Y., Luo, X.J., Huang, L.Q., Yu, L.H., Mai, B.X., 2018. Bioaccumulation of persistent halogenated organic pollutants in insects: common alterations to the pollutant pattern for different insects during metamorphosis. Environ. Sci. Technol. 52 (9), 5145–5153. doi:10.1021/acs.est.8b00616.

Lock, E.R., Arsiwalla, T., Waagbø, R., 2016. Insect larvae meal as an alternative source of nutrients in the diet of Atlantic salmon (Salmo salar) postsmolt. Aquacult. Nutr. 22 (6), 1202–1213. doi:10.1111/anu.12343.

Lopes, I.G., Lalander, C., Vidotti, R.M., Vinnerås, B., 2020. Using Hermetia illucens larvae to process biowaste from aquaculture production. J. Cleaner Prod. 251. doi:10.1016/j.jclepro.2019.119753.

Luo, G., 2023. Review of waste phosphorus from aquaculture: source, removal and recovery. Rev. Aquac. 15 (3), 1058–1082. doi:10.1111/raq.12727.

Macombe, C., Le Feon, S., Aubin, J., Maillard, F., 2019. Marketing and social effects of industrial scale insect value chains in Europe: case of mealworm for feed in France. J. Insects Food Feed 5 (3), 215–224. doi:10.3920/jiff2018.0047.

Macusi, E.D., Cayacay, M.A., Borazon, E.Q., Sales, A.C., Habib, A., Fadli, N., Santos, M.D., 2023. Protein fishmeal replacement in aquaculture: a systematic review and implications on growth and adoption viability. Sustainability 15 (16). doi:10.3390/su151612500.

Madau, F.A., Arru, B., Furesi, R., Pulina, P., 2020. Insect farming for feed and food production from a circular business model perspective. Sustainability 12 (13), 5418. doi:10.3390/su12135418.

Majluf, P., Matthews, K., Pauly, D., Skerritt, D.J., Palomares, M.L.D., 2024. A review of the global use of fishmeal and fish oil and the fish In:Fish Out metric. Sci. Adv. 10 (42). doi:10.1126/sciadv.adn5650.

Makkar, H.P.S., Tran, G., Heuzé, V., Ankers, P., 2014. State-of-the-art on use of insects as animal feed. Anim. Feed Sci. Technol. 197, 1–33. doi:10.1016/j.anifeedsci.2014.07.008.

Meijer, N., Safitri, R.A., Tao, W., Hoek-Van den Hil, E.F., 2025. Review: European Union legislation and regulatory framework for edible insect production – safety issues. Animal, 101468 doi:10.1016/j.animal.2025.101468.

Morales-Nin, B., Zimmermann, F., Merino, G., Antelo, L.T., Murillas-Maza, A., Moksness, E., 2024. Challenges on blue food provision. Front. Ocean Sustain. 2. doi:10.3389/focsu.2024.1271783.

Morales-Ramos, J.A., Tomberlin, J.K., Miranda, C., Rojas, M.G., 2024. Rearing methods of four insect species intended as feed, food, and food ingredients: a review. J. Econ. Entomol. 117 (4), 1210–1224. doi:10.1093/jee/toae040.

Mulazzani, L., Madau, F.A., Pulina, P., Malorgio, G., 2021. Acceptance of insect meal in aquaculture feeding: a stakeholder analysis for the Italian supply chains of trout and seabass. J. World Aquac. Soc. 52 (2), 378–394. doi:10.1111/jwas.12766.

Munguti, J., Muthoka, M., Mboya, J.B., Kyule, D., Meenakshisundaram, M., Tanga, C.M., 2025. Unraveling the potential of black soldier fly larvae as a sustainable protein source for Nile tilapia production in diverse aquaculture systems. Aquacult. Nutr. 2025 (1). doi:10.1155/anu/3598843.

Oliveira, L.A., Pereira, S.M.S., Dias, K.A., Paes, S.S., Grancieri, M., Jimenez, L.G.S., de Carvalho, C.W.P., de Oliveira, E.E., Martino, H.S.D., Della Lucia, C.M., 2024. Nutritional content, amino acid profile, and protein properties of edible insects (Tenebrio molitor and Gryllus assimilis) powders at different stages of development. J. Food Compos. Anal. 125. doi:10.1016/j.jfca.2023.105804.

Pali-Schöll, I., Binder, R., Moens, Y., Polesny, F., Monsó, S., 2019. Edible insects – defining knowledge gaps in biological and ethical considerations of entomophagy. Crit. Rev. Food Sci. Nutr. 59 (17), 2760–2771. doi:10.1080/10408398.2018.1468731.

Patel, S., 2019. Insects as a source of sustainable proteins: sustainable source, processing and applications. Sustainable protein sources. Elsevier, London, pp. 41–61. doi:10.1016/B978-0-12-816695-6.00002-7.

Pauly, D., 2010. Unsustainable marine fisheries Sustain. Dev. Law Policy 7, 10–12.

Pratiwy, F.M., Suryana, A.A.H., Othman, R., 2024. Exploring the depths: social-economic mapping for sustainable aquaculture communities and resilience. Essentials of aquaculture practices. Springer Science+Business Media, Singapore, pp. 183–195. doi:10.1007/978-981-97-6699-4_9.

Roccatello, R., Endrizzi, I., Aprea, E., Dabbou, S., 2024. Insect-based feed in aquaculture: a consumer attitudes study. Aquaculture 582, 740512. doi:10.1016/j.aquaculture.2023.740512.

Rodrigues, D.P., Calado, R., Pinho, M., Domingues, M.d.R., Vázquez, J.A., Ameixa, O.M.C.C., 2025. Growth rate prediction, performance, and biochemical enhancement of black soldier fly (Hermetia illucens) fed with marine by-products and co-products: a potential value-added resource for marine aquafeeds. Insects 16 (2). doi:10.3390/insects16020113.

Romero-Lorente, M.Á., Fabrikov, D., Montes, J., Morote, E., Barroso, F.G., Vargas-García, M.D.C., Varga, Á.T., Sánchez-Muros, M.J., 2022. Pre-treatment of fish by-products to optimize feeding of Tenebrio molitor L. larvae. Insects. 13 (2). doi:10.3390/insects13020125.

Sanchez-Hernandez, J.C., Megharaj, M., 2025. Insect farming: a bioeconomy-based opportunity to revalorize plastic wastes. Environ. Sci. Ecotechnol. 23, 100521. doi:10.1016/j.ese.2024.100521.

Sanda, M.K., Metcalfe, N.B., Mable, B.K., 2024. The potential impact of aquaculture on the genetic diversity and conservation of wild fish in sub-Saharan Africa. Aquat. Conserv. Mar. Freshw. Ecosyst. 34 (2). doi:10.1002/aqc.4105.

Schlüter, O., Rumpold, B., Holzhauser, T., Roth, A., Vogel, R.F., Quasigroch, W., Vogel, S., Heinz, V., Jäger, H., Bandick, N., Kulling, S., Knorr, D., Steinberg, P., Engel, K.H., 2017. Safety aspects of the production of foods and food ingredients from insects. Mol. Nutr. Food Res. 61 (6). doi:10.1002/mnfr.201600520.

Sharifinia, M., Bahmanbeigloo, Z.A., Keshavarzifard, M., Khanjani, M.H., Daliri, M., Koochaknejad, E., Jasour, M.S., 2023b. Fishmeal replacement by mealworm (Tenebrio molitor) in diet of farmed Pacific white shrimp (Litopenaeus vannamei): effects on growth performance, serum biochemistry, and immune response. Aquat. Living Resour. 36.

Sharifinia, M., Tayemeh, M.B., Dashtiannasab, A., Kaviani, A., Faghih, S., Pazir, M.K., Keshavarzifard, M., Aeinjamshid, K., Kalanaky, S., Fakharzade, S., Jand, Y., Nazaran, M.H., 2025. Enhancement of growth performance, meat quality, digestive system, and immunity of Litopenaeus vannamei fed with advanced chelate compounds-based minerals: a new perspective on chelation technology. Ann. Anim. Sci. 25 (2), 733–745. doi:10.2478/aoas-2024-0104.

Sharifinia, M., Bahmanbeigloo, Z.A., Keshavarzifard, M., Khanjani, M.H., Daliri, M., Koochaknejad, E., Jasour, M.S., 2023. The effects of replacing fishmeal by mealworm (Tenebrio molitor) on digestive enzymes activity and hepatopancreatic biochemical indices of Litopenaeus vannamei. Ann. Anim. Sci. doi:10.2478/aoas-2022-0098.

Sharifinia, M., 2025. From nutrient bioavailability to disease resistance: the comprehensive benefits of chelated minerals in aquaculture. Fish Shellfish Immunol. 160, 110218. doi:10.1016/j.fsi.2025.110218.

Sharifinia, M., 2024. Improve aquaculture with insect meal. Science 383 (6685), 838. doi:10.1126/science.ado0380.

Siddiqui, S.A., Harahap, I.A., Osei-Owusu, J., Saikia, T., Wu, Y.S., Fernando, I., Perestrelo, R., Câmara, J.S., 2024. Bioconversion of organic waste by insects – a comprehensive review. Process Saf. Environ. Prot. 187, 1–25. doi:10.1016/j.psep.2024.04.122.

Sogari, G., Bellezza Oddon, S., Gasco, L., van Huis, A., Spranghers, T., Mancini, S., 2023. Review: Recent advances in insect-based feeds: From animal farming to the acceptance of consumers and stakeholders. Animal 17 (2), 100904.

Sokame, B.M., Runyu, J.C., Tonnang, H.E.Z., 2024. Integrating edible insect into circular agriculture for sustainable production. Sustain. Prod. Consum. 52, 80–94. doi:10.1016/j.spc.2024.10.015.

Tan, H.S.G., Fischer, A.R.H., Tinchan, P., Stieger, M., Steenbekkers, L.P.A., van Trijp, H.C.M., 2015. Insects as food: exploring cultural exposure and individual experience as determinants of acceptance. Food Qual. Preference 42, 78–89. doi:10.1016/j.foodqual.2015.01.013.

Tan, H.S.G., House, J., 2018. Consumer acceptance of insects as food: Integrating psychological and socio-cultural perspectives. Edible insects in sustainable food systems. Springer International Publishing, Netherlands, pp. 375–386. doi:10.1007/978-3-319-74011-9_23.

Tan, S.Y., Sethupathi, S., Leong, K.H., Ahmad, T., 2024. Challenges and opportunities in sustaining aquaculture industry in Malaysia Aquac. Int. 32, 489–519 doi:10.1007/s10499-023-01173-w.

Torgerson, K.L., Meijering, J.V., Sok, J., Dicke, M., Oude Lansink, A.G.J.M., 2021. Towards circular agriculture – exploring insect waste streams as a crop and soil health promoter. J. Insects Food Feed 7 (3), 357–368. doi:10.3920/jiff2020.0095.

van Huis, A., Rumpold, B., Maya, C., Roos, N., 2021. Nutritional qualities and enhancement of edible insects. Annu. Rev. Nutr. 41 (1), 551–576. doi:10.1146/annurev-nutr-041520-010856.

van Huis, A., 2013. Potential of insects as food and feed in assuring food security. Annu. Rev. Entomol. 58 (1), 563–583. doi:10.1146/annurev-ento-120811-153704.

van Huis, A., Oonincx, D.G.A.B., 2017. The environmental sustainability of insects as food and feed. a review. Agron. Sustainable Dev. 37 (5). doi:10.1007/s13593-017-0452-8.

Veldkamp, T., Meijer, N., Alleweldt, F., Deruytter, D., Van Campenhout, L., Gasco, L., Roos, N., Smetana, S., Fernandes, A., van der Fels-Klerx, H.J., 2022. Overcoming technical and market barriers to enable sustainable large-scale production and consumption of insect proteins in Europe: a SUSINCHAIN perspective. Insects 13 (3), 281. doi:10.3390/insects13030281.

Verdegem, M., Buschmann, A.H., Latt, U.W., Dalsgaard, A.J.T., Lovatelli, A., 2023. The contribution of aquaculture systems to global aquaculture production. J. World Aquac. Soc. 54 (2), 206–250. doi:10.1111/jwas.12963.

Vreysen, M.J.B., Abd-Alla, A.M.M., Bourtzis, K., Bouyer, J., Caceres, C., Beer, C., Carvalho, D.O, Maiga, H., Mamai, W., Nikolouli, K., Yamada, H., Pereira, R., 2021. The insect pest control laboratory of the joint FAO/IAEA programme: ten years (2010–2020) of research and development, achievements and challenges in support of the sterile insect technique. Insects 12 (4), 346. doi:10.3390/insects12040346.

Wan Mahari, W.A., Waiho, K., Fazhan, H., Azwar, E., Shu-Chien, A.C., Hersi, M.A., Kasan, N.A., Foo, S.S., Wong, K.Y., Draman, A.S., Ma, N.L., Chang, J.S., Dong, C.D., Lam, S.S., 2024. Emerging paradigms in sustainable shellfish aquaculture: microalgae and biofloc technologies for wastewater treatment. Aquaculture 587. doi:10.1016/j.aquaculture.2024.740835.

Wang, Y., Wang, L., Zhang, C., Song, K., 2017. Effects of substituting fishmeal with soybean meal on growth performance and intestinal morphology in orange-spotted grouper (Epinephelus coioides). Aquac. Rep. 5, 52–57. doi:10.1016/j.aqrep.2016.12.005.

Weinreis, Y., Baum, C.M., Smetana, S., 2023. Insect production as a novel alternative to livestock farming: exploring interest and willingness to adopt among German farmers. Sustain. Prod. Consum. 35, 28–39. doi:10.1016/j.spc.2022.10.004.

Wendin, K.M., Nyberg, M.E., 2021. Factors influencing consumer perception and acceptability of insect-based foods. Curr. Opin. Food Sci. 40, 67–71. doi:10.1016/j.cofs.2021.01.007.

WHO, 2021. Managing pesticides in agriculture and public health: a compendium of FAO and WHO guidelines and other resources, 2nd edn World Health Organization, Geneva.

Yeganeh, V., Sharifinia, M., Mobaraki, S., Dashtiannasab, A., Aeinjamshid, K., Borazjani, J.M, Maghsoudloo, T., 2020. Survey of survival rate and histological alterations of gills and hepatopancreas of the Litopenaeus vannamei juveniles caused by exposure of Margalefidinium /Cochlodinium polykrikoides isolated from the Persian Gulf. Harmful Algae 97, 101856. doi:10.1016/j.hal.2020.101856.

CHAPTER

23

Novel flocculants from waste derivatives: A way to improve biofloc systems functioning

Reshmi Debbarma[a], *Soibam Khogen Singh*[a,b], *Pronob Das*[c], *Sourabh Debbarma*[d] and *Shrinath Vaijanath Gavhane*[e]

[a]College of Fisheries, Central Agricultural University, Agartala, Tripura, India [b]ICAR Research Complex for NEH Region, Krishi Vigyan Kendra, Imphal, Manipur, India [c]ICAR-Central Inland Fisheries Research Institute, Regional Centre, Guwahati, Assam, India [d]Fisheries College and Research Institute, Thoothukudi, Tamil Nadu Dr. J. Jayalalithaa Fisheries University, Tamil Nadu, India [e]Dr. M.G.R. Fisheries College and Research Institute, Ponneri, Tamil Nadu Dr. J. Jayalalithaa Fisheries University, Tamil Nadu, India

23.1 Introduction

Biofloc technology (BFT) has revolutionized the aquaculture industry by providing a sustainable and efficient method for managing water quality and enhancing the growth of aquatic organisms. BFT presents a viable and sustainable solution to tackle the aforementioned challenges. This technology allows for high-density production while minimizing environmental impact (Debbarma et al., 2022). This technology represents a significant shift from traditional water exchange methods to in situ waste recycling by fostering the development of heterotrophic microbial flocs via the addition of an external carbon source (Majhi et al., 2023). Biofloc consists of a diverse microbial community, including bacteria, protozoa, fungi, and microalgae, which aggregates with organic waste particles (Avnimelech, 2012). This system provides in situ protein production for farmed species, improving growth and immune function, while further enhancing feed efficiency by reducing reliance on supplementary feeding (Majhi et al., 2023). The development of floc is crucial for the establishment and stabilization of

Zero-Waste Technologies in Aquaculture: Global Perspectives and Methods toward a Circular Economy
DOI: https://doi.org/10.1016/B978-0-443-34689-7.00024-0

biofloc systems, involving a complex interplay of physical, chemical, and biological processes (Khanjani and Sharifinia, 2020). The development of robust floc is crucial for enhancing water quality in the system, given that nitrogen removal is directly associated with floc formation (Krummenauer et al., 2014). The development of a stable biofloc system usually takes several weeks, owing to the complex interactions between physical, chemical, and biological processes that contribute to microbial floc formation (Avnimelech, 2009). It is still difficult to keep up constant floc quality and avoid system crashes even after the system is first established. To address these limitations, in recent years, there has been a growing interest in the development and application of novel flocculants derived from waste materials (Chutia et al., 2022; Walunj et al., 2024). These waste-derived flocculants offer a promising alternative to conventional synthetic flocculants, as they are often more environmentally friendly, cost-effective, and sustainable (Singh et al., 2023). By utilizing waste derivatives, such as agricultural byproducts, industrial residues, insect waste, and organic waste, researchers aim to improve the flocculation efficiency while contributing to waste reduction and resource recovery (Zhang et al., 2023; Zhang et al., 2024). For instance, flocculants derived from chitosan extracted from crustacean shells, cellulose from plant residues, and proteins from agricultural waste have shown remarkable potential in enhancing flocculation efficiency (Fauzani et al., 2021; Jiang et al., 2021; Yin et al., 2021). These materials are often abundant and readily available, reducing the need for costly and environmentally harmful synthetic flocculants. Moreover, the biodegradability of these natural flocculants minimizes the risk of secondary pollution in aquatic environments.

23.1.1 Importance of biofloc in sustainable aquaculture

Aquaculture is the most rapidly expanding food-production method, currently surpassing catch fisheries in world fish biomass when nonedible quantities are considered (Edwards et al., 2019). The majority of this growth has transpired over the past 50 years, leading to an increasing concern for sustainability, especially environmental sustainability. Aquaculture produces large volumes of effluent, which, like other agricultural operations, can harm the environment if improperly treated. In addition to causing eutrophication, algal blooms, and other ecological disruptions, the discharge of untreated or insufficiently treated wastewater can also transfer illnesses to populations of wild fish. There has been a growing focus on raising knowledge of environmental issues and implementing practices aimed at reducing the ecological impact of aquaculture. While environmental deterioration was not regarded as the foremost issue for the industry 50 years ago, it has now become a critical focus across academic, governmental, industrial, and market sectors. The necessity for environmentally responsible and sustainable management of aquaculture output is no longer a subject of debate (Boyd et al., 2020). Thus to guarantee aquaculture's long-term sustainability and reduce its environmental impact, effective and sustainable wastewater treatment techniques are crucial.

An alternate way to reduce the hazards to human health and the environment posed by the use of chemical coagulants and flocculants is to utilize biocoagulants and bioflocculants (Kurniawan et al., 2020). Aggregates (flocs) of bacteria, protozoa, algae, and other types of particulate organic matter, including excrement and leftover feed, are known as bioflocs (Hargreaves, 2013). It is a closed-loop BFT that provides a superior approach to managing water quality parameters, biosecurity, and feed management compared to conventional pond or cage systems, and may be implemented in freshwater, brackish, and saltwater environments (Emerenciano et al., 2017). In addition, biofloc devices were created to stop illness from entering a farm through incoming water. In the past, water exchange (typically 10% daily) was a routine practice for shrimp ponds as a way to regulate the quality of the water (Hargreaves, 2013). BFT may be used in raceways, tanks, and ponds of different sizes. Shrimp are produced intensively using biofloc in lined ponds ranging in size from 500 to 20,000 m^2, with a seasonal yield of 10–20 tons/ha (Browdy et al., 2012). The technology utilizes the microbial community in the water column to ingest nitrogenous and phosphorus waste, transforming these substances into microbial biomass. This subsequently serves as an alternate diet for farmed aquatic animals, namely fish and shrimp (McCusker et al., 2023).

23.2 The functioning of biofloc systems

Essential functions that bioflocs offer are the treatment of feeding waste and the provision of nutrients through floc intake. Low water exchange rates are sufficient for biofloc systems to function. The formation of a thick and active biofloc community to improve the treatment of waste organic matter and nutrients is made possible by this extended water residence period. Internal waste treatment procedures are prioritized and promoted in biofloc systems, while the use of water exchange to control water quality is reduced (Hargreaves, 2013). At the same time, this process turns organic waste into microbial biomass, which can serve as an additional food source. The addition of external carbon is required to maintain an optimum C/N ratio of 10–20:1 to keep continuous heterotrophic bacterial growth. Continuous aeration keeps bioflocs suspended and provides oxygen for both microorganisms and cultured species. This system maintains optimal water quality and reduces feed conversion ratios, making aquaculture more efficient and environmentally friendly.

23.2.1 Floc formation in biofloc systems

Floc formation is a fundamental aspect of BFT, playing a crucial role in maintaining water quality and promoting the health of cultured aquatic organisms (Debbarma et al., 2021, 2022). In biofloc systems, flocs are aggregates of microorganisms, including bacteria, algae, protozoa, and other organic matter, which are bound together by extracellular polymeric substances (EPSs) (Tansel, 2018). These flocs are responsible for various biogeochemical processes that enhance the efficiency of the system.

23.2.2 Floc formation mechanism in biofloc system

Floc formation in biofloc systems occurs through several mechanisms, including:

(1) **Microbial aggregation**: Microorganisms in the culture water naturally aggregate because of interactions between their cell surfaces and EPSs. EPSs, produced by microorganisms, act as a glue, binding cells and organic matter together (Khanjani et al., 2020), leading to an increase in the size of bioflocs. Further, the abundance of other microorganisms,

such as flagellates, ciliates, and amoeboid forms, may increase because most of these microorganisms graze on the bacteria (Biddanda and Pomeroy, 1988). In addition, the cultured organisms, such as crustaceans and fish, can affect the dynamics of aggregation by grazing or movement (Dilling and Alldredge, 2000).

(2) Charge interaction and gravity: Balance between the force of the gravity (hydrogen bond, molecular, and dipole) and electrostatic repulsion forces plays an important role in floc formation. Most organisms are negatively charged and cause counterelectrostatic repulsion. If this repulsion is reduced, then strong gravity forces can occur; this is the case when the salt concentration is high and multivalent ions are present in the environment (Avnimelech, 2009). Calcium and aluminum ions stimulate the formation of stable flocs, and algae, fungi, or bacteria can help to bond between the components of different flocs (De Schryver et al., 2008; Avnimelech, 2009).

(3) Physical forces: Turbulence and mixing within the culture water, often induced by aeration or mechanical agitation, promote the collision and aggregation of microbial cells and particles, leading to the formation of flocs, while excessive turbulence would also promote the breakage of aggregation (Jarvis et al., 2005; Mietta et al., 2009). Under certain hydrodynamic conditions, weaker flocs break and reflocculate until they are strong enough to withstand shear stress. Increasing turbulence homogenizes floc size to smaller values (Tang and Maggi, 2016).

(4) Presence of flocculant: The presence of flocculants, whether naturally occurring or added to the system, facilitates the aggregation of particles. Flocculants neutralize the charges on particle surfaces, reducing repulsion and allowing them to come together (Alnawajha et al., 2022). Addition of flocculants leads to more stable and larger flocs in various hydrodynamic conditions (Fan et al., 2020). Flocculants facilitate the formation of flocs by bridging particles together, which helps in reducing the shear stress impact on individual particles. Flocculants also help in the production of a more homogeneous floc size and improved overall stability and resistance of flocs to shear force (Milligan and Hill, 1998; Tang and Maggi, 2016).

23.2.3 Benefits of floc formation

Floc formation offers several benefits to biofloc systems:

(1) Improves water quality by aggregating suspended particles, making water clearer and cleaner.
(2) Enhances nutrient recycling as flocs contain microbial biomass that fish can consume, reducing the need for additional feed.
(3) Helps in preventing disease outbreaks by maintaining better water quality and reducing harmful substances.
(4) Minimizes the discharge of pollutants into the environment, promoting sustainable aquaculture practices.
(5) Reduces the need for frequent water exchanges and external feed, lowering operational costs and increasing profitability.

23.3 Current challenges in biofloc waste management

The most crucial factor in BFT is effective water quality management, which significantly impacts the health and growth of aquatic organisms. Maintaining optimal water conditions is essential for system efficiency (Debbarma et al., 2022; Majhi et al., 2023). One major issue in BFT systems is the lack of sustainable floc formation, leading to unstable system performance. Several factors contribute to low floc formation, including inadequate carbon sources, improper carbon-to-nitrogen (C:N) ratios, insufficient flocculants, and low microbial diversity (Minaz and Kubilay, 2021). Poor floc formation results in suboptimal water quality, reduced nutrient recycling, and leads to the accumulation of toxic ammonia. Another significant problem is excessive sludge production in BFT systems. Sludge consists of settled organic matter, uneaten feed, and microbial biomass. If not managed properly, excessive sludge can lead to reduced water quality, increased biochemical oxygen demand (BOD), and the proliferation of harmful microorganisms. Sludge accumulation can also clog filters and aeration systems, causing operational challenges.

23.4 Role of flocculants in biofloc systems

Flocculants serve a significant role in boosting the efficacy of biofloc systems through enhanced floc formation in aquaculture. These additives help in the production and stability of bioflocs, which are vital for preserving water quality and giving nutritional advantages to cultured organisms. In the biofloc system, flocculant additions can affect the composition and organization of the microbial community (Lal et al., 2024). Larger, more stable flocs are created when suspended particles, microbes, and organic matter aggregate with the help of flocculants (Xavier et al., 2022). This procedure is essential to the biofloc system because it contributes to the formation of the distinctive suspended aggregates that have several uses in the aquaculture setting . Flocculants improve the settling characteristics of the biofloc by encouraging the production of bigger flocs. This is especially crucial in greenwater biofloc systems, where flocs settle readily in still water and usually have sizes between 50 and 200 μm (Hargreaves, 2013).

23.5 Flocculants in biofloc systems

Floc, also known as tiny flocculent material, is a loosely aggregating mass component that can be precipitated from a liquid or suspended (Nethaji et al., 2021). Flocculants are utilized in diverse wastewater treatments, drinking water purification, and dredge or downstream operations across multiple industrial sectors. Flocculants are often classified into inorganic, organic, and naturally occurring categories (Okaiyeto et al., 2016).

23.5.1 Flocculation mechanism of flocculants

Flocculation is a crucial step in water treatment, where flocculants are used to agglomerate dispersed particles into larger aggregates, known as flocs, that can be easily removed. The mechanism can be broken down into several distinct phases:

(1) **Dispersion:** When flocculants are introduced into the water, they must first disperse uniformly. This dispersion is critical to ensure that the flocculant interacts effectively with the suspended particles. Proper mixing is necessary to achieve an even distribution.
(2) **Charge neutralization:** In many cases, the suspended particles in water carry similar electric charges, often negative. These like charges cause mutual repulsion, preventing the particles from coming together. Flocculants, particularly cationic (positively charged) ones, neutralize these charges by attaching to the particle surfaces, thereby reducing the electrostatic repulsion and allowing the particles to come closer together. This process is known as coagulation.
(3) **Bridging:** Some flocculants are long-chain polymers that can attach themselves to multiple particles simultaneously, forming bridges between them. These polymer chains can either be natural (e.g., starch and alginates) or synthetic (e.g., polyacrylamides). Bridging occurs when the polymer chains adsorb onto the surface of the particles, creating a network that links them together into larger aggregates. This is particularly effective in forming dense and easily removable flocs.
(4) **Aggregation:** Following charge neutralization and bridging, particles start to aggregate, forming larger flocs. This aggregation is a gradual process, where smaller flocs collide and adhere to each other, growing in size. The presence of a flocculant enhances the rate and efficiency of this aggregation process.
(5) **Sedimentation/filtration:** Once the flocs have formed, they must be separated from the water. This is typically done through sedimentation or filtration. In sedimentation, the denser flocs settle to the bottom of the treatment tank because of gravity. In filtration, the water is passed through a medium (such as sand or cloth) that traps the flocs. The choice of separation method depends on the characteristics of the flocs and the specific requirements of the treatment process.

23.6 Types of flocculants

Experts from all over the world have recently become interested in bioflocculants, which are regarded as a low-cost, straightforward method that can also function as an agent that causes microorganisms to aggregate to produce bioflocs. Bioflocculation is the most popular method for treating wastewater from aquaculture in biofloc systems (Zaki et al., 2013). Three types of flocculants are generally distinguished as synthetic inorganic flocculants, such as polyethyleneimine and polyacrylamide, and synthetic organic flocculants, such as polyaluminum chloride and aluminum sulfate. In addition, there are microbial flocculants such as *Aspergillus niger* and *Scenedesmus*, as well as natural flocculants (bioflocculants) such as chitosan and other bioflocculants, including chitosan, eggshell, and *Moringa oleifera* (Jusoh et al., 2023).

The scope for using these flocculants in industrial settings is quite high. However, there are restrictions that must be addressed, such as limited flocculation efficiency and high dose requirements (Czemierska et al., 2017).

23.7 Problems with the synthetic flocculants

Many synthetic flocculants do not degrade naturally and can persist in the environment, leading to potential long-term pollution and adverse effects on aquatic ecosystems. For example, polyacrylamides can persist in water bodies, harming fish and aquatic plants. The use of synthetic flocculants can also result in residual chemicals being discharged into water bodies, posing risks to aquatic life and potentially entering the human food chain. Certain synthetic flocculants, such as polyacrylamides, can degrade into toxic monomers, such as acrylamide, a known neurotoxin and carcinogen. For instance, acrylamide contamination in drinking water has been linked to increased cancer risk. In addition, workers in wastewater treatment plants often report skin rashes and respiratory problems because of direct exposure to these chemicals.

Synthetic flocculants also come with economic and operational challenges. Their production is often complex and energy-intensive, leading to high costs and reliance on nonrenewable petrochemicals. For example, producing polyacrylamides involves multiple energy-intensive steps. Moreover, synthetic flocculants may require specific pH conditions to function optimally, necessitating costly pH adjustments. They also generate large volumes of sludge that require proper treatment and disposal. In addition, the application of synthetic flocculants is subject to stringent regulations because of their potential environmental and health risks, limiting their use and increasing compliance costs. Some synthetic flocculants have even been banned in various regions, leading to reliance on safer alternatives. Lastly, synthetic flocculants may struggle to effectively aggregate organic-rich wastewater, limiting their efficacy in systems with high levels of organic matter, such as those found in food processing plants.

23.8 Novel organic flocculants from waste derivatives

Waste-derived flocculants are gaining significant attention in aquaculture because of their eco-friendly and cost-effective properties. These natural alternatives to synthetic chemicals are derived from various waste materials, offering a sustainable solution for improving water quality and reducing environmental impact. Here are some notable examples of waste-derived flocculants used in aquaculture in the following sections.

23.8.1 Chitosan as novel flocculating agent

23.8.1.1 Mechanism

Similar to classical flocculation, chitosan-based biofocculation binds insoluble particles with dissolved organic matter to form larger aggregates that can be separated by sedimentation and filtration (Morin-Crini et al., 2019). First, during the coagulation process, chitosan chains destabilize the suspended and colloidal particles in wastewater by forming microflocs. The next process, called flocculation, aggregates the microflocs through agitation, causing the particles to clump together for easier subsequent solid removal. Charge neutralization and bridging mechanisms are the primary forces behind the mechanism. Bioflocculation may also be influenced by processes such as electrostatic patch and sweeping, adsorption, entrapment, complexation, chelation, and precipitation. Particles engage in charge neutralization when they come into contact with ions of the opposite charge, and particles engage in sweep flocculation when they become entangled in an expanding precipitate. Particles become agglomerates because of the bridging effects of polymer chains. Particles are adsorbed onto polymer chains by forming particle–polymer–particle complexes. Adsorbed polymer chains extend loops and tails that can bind to and bridge gaps between particles in an aqueous solution.

23.8.1.2 Properties of chitosan as flocculant

Chitosan's excellent flocculation ability makes it a promising biocoagulant that might eventually replace chemical coagulants. In addition, it is biodegradable, safe for both humans and animals, needs little dose, and settles quickly after harvest (Rashid et al., 2013). Microalgae cells with a negative charge can be coagulated using chitosan, which is known to have positive charges under acidic conditions. In addition, it creates repulsive interactions with the negatively charged microalgae cell to facilitate flocculation (Mohd Yunos et al., 2017).

As microbial floc formation is a complex process involving many physicochemical and biological phenomena, the initial development of biofloc takes a few weeks (Avnimelech, 2009). Many times, even when the biofloc is established, the system will eventually fail because of a lack of sustainable floc formation. However, it has been observed that various flocculating agents can be used to accelerate floc formation and improve floc quality. Biodegradability, biocompatibility, nontoxicity, and the ability to aggressively bind microbial cells, act as immunoadjuvants, and speed up bone formation are just some of the many beneficial properties of chitosan, a naturally abundant polymer that is a deacetylated product of chitin (Debbarma et al., 2026a; Dutta et al., 2004). It has a high hydrophobicity, which means it is insoluble in water and most organic solvents but soluble in acidic solutions (Debbarma et al., 2025; Dutta et al., 2004). It has the unusual property of selectively removing harmful bacteria such as *Edwarsiella ictaluri* from water while leaving beneficial nitrifying bacteria unaffected at doses ranging from 28 to 64 ppm (Chung, 2006). Yunos et al. (2017) investigated the use of chitosan as a natural biocoagulant to harvest *Chlorella* sp., focusing on the efficiency of coagulation–flocculation. The highest flocculation activity (30 ppm) was seen in treatments supplemented with the highest amount of chitosan, as determined by the work of Chutia et al. (2022). The authors found that 30 ppm chitosan-supplemented groups had higher flocculation activity than lower doses. Further, Walunj et al. (2024) demonstrated that biofloc developed with 30 ppm nanochitosan could shorten the production time of biofloc, maintain the water quality of the biofloc system, and improve the growth of GIFT tilapia reared in an in situ biofloc system with inland saline groundwater. Chitosan hydrogel is a gum-like substance that improves the attachment of other microbes, suspended particles, and dead organisms, as well as increasing the size (Debbarma et al., 2025), resulting in better floc utilization by fish. The surface morphology comparison of nonchitosan-supplemented biofloc and chitosan-supplemented biofloc shows that the density of organisms in the chitosan-supplemented groups is higher than in the control group. The chitosan-supplemented biofloc showed higher total suspended solids (TSSs) at 30 ppm, followed by 20 ppm in the inoculum, which can be attributed to the increased flocculating activity in the chitosan-supplemented groups, concluding that chitosan acts as a bioflocculant when supplemented at a moderate level in biofloc systems.

23.8.2 Chicken eggshell

Because of their nontoxicity, noncorrosive nature, and ease of handling, eggshells are highly suggested as a natural coagulating material. They have encouraging flocculation and adsorption capabilities in addition to being biodegradable and biocompatible. In addition, they are regenerable and reusable several times (Choi, 2015). The properties of eggshells that consist of an amorphous calcium carbonate matrix and the mineralized material contribute to the alkalinity replenishment, as the alkalinity level becomes critical in maintaining the aquaculture system. In addition, eggshells may aggressively adsorb and destabilize negative particles, including negatively charged microbial cells, particularly microalgae cells, because of their high cationic (Ca^{2+}) charge density. Granule development and expansion are significantly influenced by calcium, the primary precipitate. On the other hand, it seemed that the shear strength was unaffected by the microbial ecology of the granule or the position of calcium precipitates (Jusoh et al., 2023).

Jusoh et al. (2023) reported that eggshells demonstrated effective performance in biofloc harvesting and aquaculture wastewater treatment, making them a sustainable solution for promoting environmentally friendly practices in the aquaculture industry.

23.8.3 Moringa seed powder

M. oleifera (MO) is recognized as one of the most important natural flocculants available for water and wastewater treatment. The coagulation properties of *M. oleifera* are attributed to an active protein component that acts as a cationic polyelectrolyte. This protein attaches to soluble particles, forming bindings between them, which leads to the creation of large flocs in the water (Magalhães et al., 2021). The application of moringa seed powder significantly improves turbidity removal, as demonstrated by Muyibi et al. (2002), where the clear water obtained is comparable to conventional treatment methods. In addition, Salim et al. (2026) found that *M. oleifera* enhances the removal of various dyes from wastewater, showcasing its versatility in treating different pollutants. In terms of microbial contamination, Bichi et al. (2012) reported that Moringa seed powder effectively removes *E. coli* bacteria, contributing to safer and cleaner water.

More recently, research has focused on the application of Moringa seed extract in aquaculture, particularly in treating fish farm wastewater. Studies such as those by Tong et al. (2022) and Olusegun et al. (2016) have shown that the culture of African catfish (*Clarias gariepinus*) in moringa-treated water results in significant improvements in fish growth, survivability, and immune response. The positive effects on fish health are attributed to the bioactive compounds in *M. oleifera*, which enhance water quality and provide additional nutrients. Amin et al. (2024) and Debbarma et al. (2026b) further corroborate these findings by demonstrating the improved overall health and productivity of cultured fish in moringa-treated environments.

In conclusion, *M. oleifera* has proven to be an effective natural flocculant with diverse applications in water and wastewater treatment. Its ability to improve turbidity, dye, and microbial removal, combined with its benefits in aquaculture, underscores its potential in promoting sustainable practices and enhancing environmental management.

23.8.4 Other potential plant-based coagulants

(1) Tamarind kernel polysaccharide (TKP): Extracted from tamarind seeds, TKP is an effective natural flocculant known for its ability to reduce turbidity, TSS, and chemical oxygen demand (COD) in aquaculture effluents. Its polysaccharide content facilitates the aggregation of suspended particles, enhancing water clarity and quality (Al Juboury et al., 2022).

(2) Tannin-based substances (TBS): Derived from plant materials rich in tannins, TBS are utilized for their strong coagulation properties. These substances effectively reduce turbidity and improve water quality, making them a viable alternative to synthetic flocculants (Al Juboury et al., 2022).

(3) Plant gums and mucilage: Natural substances, such as guar gum and xanthan gum, derived from plants, serve as efficient flocculants in water treatment. Their polysaccharide content helps aggregate suspended particles, resulting in improved water clarity and quality (Javed et al., 2024)

23.8.5 Several factors can influence the flocculation properties of novel flocculants

(1) pH Levels: The pH of the solution can significantly impact the flocculation process. Different flocculants perform optimally at specific pH ranges.

(2) Flocculant dose: The amount of flocculant added to the solution is crucial. An optimal dose is necessary to achieve effective flocculation without causing overdosing issues.

(3) Salinity: The presence of salts in the solution can affect the flocculation efficiency. Higher salinity levels can enhance the flocculation process.

(4) Organic matter: Organic matter in the solution can inhibit the adsorption of flocculants onto particles, reducing flocculation efficiency.

23.9 Strategies for enhancement and factors on which floc formation depends

The primary idea for biofloc came from external biofilter systems, where the flocculation process happens in stages as a result of the electrostatic attraction of suspended particles, filamentous microorganisms, and loose mucus or EPS secreted by bacteria that produce flocculants during their growth (Hargreaves, 2013). Where flocculants were used, tiny particles of uneaten feed, waste, and sludge suspended in the water column will drop to the bottom (Ugbenyen and Okoh, 2014). The use of coagulants aids in the flocculation process by stabilizing charges and clumping the suspended particles to create floc by electrostatic attraction (Prakash et al., 2014). The development of biofloc is also aided by filamentous bacteria, whose cilia entrap suspended particles and other materials. The process of flocculation is further aided by the attachment of filamentous bacteria and other microbes to unconsumed feed, excrement, and sludge (Crab, 2010). Despite the use of coagulants and filamentous microorganisms, it was shown that adding bacteria that produce bioflocculants as an inoculum increased the development of biofloc. The following are the criteria on which floc formation is dependent and could be enhanced based on alterations in the following parameters:

23.9.1 Mixing intensity

The steady-state floc size, which is the equilibrium between the rate of aggregation and the rate of breaking, and the floc size distribution are determined by the mixing intensity imposed by a specified aeration device at a specific power input (Chaignon et al., 2002). The average floc size reduces with increasing mixing intensities and correspondingly higher shear rates as a result of more floc breaking. The steady-state floc size is a critical component of BFT since it has been demonstrated that the size of the food also affects the food quality for many aquaculture species (Garatun-Tjeldstø et al., 2006).

23.9.2 Organic carbon source

Because of aerobic microbial metabolism, the dosing of an organic carbon source to the culture water in biofloc ponds causes a reduction in dissolved oxygen levels. This could have a lethal impact on sensitive cultural species (Landman et al., 2005). The amount of organic matter required for an intensive pond may be calculated based on the amount of nitrogen emitted by the aquaculture species and should be added based on the requirement, as it could hamper the oxygen supply to the cultured fish.

23.9.3 Dissolved oxygen

In addition to being crucial for the metabolic activity of cells inside aerobic flocs, the Dissolved oxygen (DO) level is also believed to have an impact on floc shape. Wilen and Balmer (1999) observed a tendency toward bigger and more compact flocs at greater DO concentrations, but no direct correlation with average floc diameter was discovered. At low DO levels, filamentous bacteria dominate the community. Because of their increased affinity for oxygen, filaments can outcompete their zoogloeal counterparts during times of oxygen scarcity and take control of the microbial flocs. Thus alter the floc volume index (FVI) and the capability of fish to consume the floc in suspension. In addition, the amount of dissolved oxygen present in the water has a significant impact on the development characteristics and stress tolerance of aquaculture species (Colt, 2006).

23.9.4 Organic loading rate

The organic carbon can be given in larger quantities but at regular intervals (like once per day), or it can be added in smaller doses and thus nearly continuously. The availability and dosing of organic carbon determine a bacterial colony's dominance.

23.10 Conclusion

The use of new flocculants from waste products is a cost-efficient and sustainable way to enhance the performance of biofloc systems. In this chapter, we discussed the viability of using chicken eggshells, chitosan, *M. oleifera*, and other derivatives from plants as flocculants for improving water quality, microbial community stability, and system performance. These bioderived flocculants have important benefits, such as being biodegradable, nontoxic, and having the capability to enhance nutrient retention and bioavailability in aquaculture systems.

Future studies need to aim at maximizing the application rates of these flocculants, determining their long-term impacts on microbial communities, and evaluating their scalability for commercial aquaculture. Through furthering our understanding in this area, we can continue to advance eco-friendly and effective methods for sustainable aquaculture practices.

References

Al Juboury, M.F., Abdulredha, M., Nile, B.K., 2022. Photocatalysis and flocculation processes for recycling aquaculture effluent into nutrient-rich irrigation water. Water Supply 22 (3), 3103–3113. doi:10.2166/ws.2022.123.

Alnawajha, M.M., Kurniawan, S.B., Imron, M.F., Abdullah, S.R.S., Hasan, H.A., Othman, A.R., 2022. Plant-based coagulants/flocculants: characteristics, mechanisms, and possible utilization in treating aquaculture effluent and benefiting from the recovered nutrients. Environ. Sci. Pollut. Res. 29 (39), 58430–58453. doi:10.1007/s11356-022-21631-x.

Amin, A.A., Soliman, H.A., Hashem, K.S., 2024. Survival, growth, haematological and biochemical responses of *Clarias gariepinus* grown in *Moringa oleifera* seeds treated wastewater. Egypt. J. Chem. 67 (3), 1–11. doi:10.21608/ejchem.2024.123456.

Avnimelech, Y., 2009. Biofloc technology—a practical guide book, 1st edn. The World Aquaculture Society, Baton Rouge, LA.

Avnimelech, Y., 2012. Biofloc Technology – A Practical Guide Book, second ed. The World Aquaculture Society, Baton Rouge, Louisiana, United States.

Bichi, H.M, Agunwamba, C.J, Muyibi, A.S, 2012. Optimization of operating conditions for the application of Moringa oleifera (Zogale) seeds extract in water disinfection using response surface methodology. Afr. J. Biotechnol. 11 (92), 15875–15887. doi:10.5897/AJB12.1341.

Biddanda, B.A, Pomeroy, L.R, 1988. Microbial aggregation and degradation of phytoplankton-derived detritus in seawater. I. Microbial succession. Mar. Ecol. Prog. Ser. 42, 79–88. doi:10.3354/meps042079.

Boyd, C.E., D'Abramo, L.R., Glencross, B.D., Huyben, D.C., Juarez, L.M., Lockwood, G.S., McNevin, A.A., Tacon, A.G.J., Teletchea, F., Tomasso, J.R., Tucker, C.S., Valenti, W.C., 2020. Achieving sustainable aquaculture: historical and current perspectives and future needs and challenges. J. World Aquac. Soc. 51 (3), 578–633. doi:10.1111/jwas.12714.

Browdy, C.L., Ray, A.J., Leffler, J.W., Avnimelech, Y., 2012. Biofloc-based aquaculture systems aquaculture production systems. Wiley-Blackwell, Ames, IA, United States, pp. 278–307. doi:10.1002/9781118250105.ch12.

Chaignon, V., Lartiges, B.S., El Samrani, A., Mustin, C., 2002. Evolution of size distribution and transfer of mineral particles between flocs in activated sludges: an insight into floc exchange dynamics. Water Res 36 (3), 676–684. doi:10.1016/S0043-1354(01)00274-6.

Choi, H.J., 2015. Effect of eggshells for the harvesting of microalgae species. Biotechnol. Biotechnol. Equip. 29 (4), 666–672. doi:10.1080/13102818.2015.1031177.

Chung, Y.-C., 2006. Improvement of aquaculture wastewater using chitosan of different degrees of deacetylation. Environ. Technol. 27 (11), 1199–1208. doi:10.1080/09593332708618734.

Chutia, A., Martin Xavier, K.A., Shamna, N., Babitha Rani, A.M., 2022. Application of bioflocculating agent in inoculum enhances quality of biofloc and influences growth, feed utilization and stress responses of GIFT tilapia reared in-situ. Aquaculture 553, 738050. doi:10.1016/j.aquaculture.2022.738050.

Colt, J., 2006. Water quality requirements for reuse systems. Aquacult. Eng. 34 (3), 143–156. doi:10.1016/j.aquaeng.2005.08.011.

Crab, R., 2010. Bioflocs technology: an integrated system for the removal of nutrients and simultaneous production of feed in aquaculture. Ghent University, Faculty of Bioscience Engineering, Ghent, Belgium.

Czemierska, M., Szcześ, A., Hołysz, L., Wiater, A., Jarosz-Wilkołazka, A., 2017. Characterisation of exopolymer R-202 isolated from Rhodococcus rhodochrous and its flocculating properties. Eur. Polym. J. 88, 21–33. doi:10.1016/j.eurpolymj.2017.01.008.

De Schryver, P., Crab, R., Defoirdt, T., Boon, N., Verstraete, W., 2008. The basics of bio-flocs technology: the added value for aquaculture. Aquaculture 277 (3-4), 125–137. doi:10.1016/j.aquaculture.2008.02.019.

Debbarma, R., Biswas, P., Singh, S.K., 2021. An integrated biomarker approach to assess the welfare status of *Ompok bimaculatus* (Pabda) in biofloc system with altered C/N ratio and subjected to acute ammonia stress. Aquaculture 545, 737184. doi:10.1016/j.aquaculture.2021.737184.

Debbarma, R., Meena, D.K., Biswas, P., Meitei, M.M., Singh, S.K., 2022. Portioning of microbial waste into fish nutrition via frugal biofloc production: a sustainable paradigm for greening of environment. J. Cleaner Prod. 334. doi:10.1016/j.jclepro.2021.130246.

Debbarma, R., Singh, S.K., Ngasotter, S., Biswas, P., Waikhom, G., Chouhan, N., Priyadarshini, M.B., Choudhury, T.G., Mehta, N.K., 2025. Insect shell chitosan nanoparticle as novel flocculants in biofloc system: an innovative resource recovery approach for sustainable aquaculture. Bioresour. Technol. Rep. 102465. doi:10.1016/j.biteb.2025.102465.

Debbarma, R., Singh, S.K., Ngasotter, S., Choudhury, T.G., Biswas, P., Waikhom, G., 2026b. Characterization and evaluation of *Moringa oleifera* seed powder as a flocculating agent in biofloc culture of *Clarias magur* catfish. Sci. Rep. 16, Article 12345. doi:10.1038/s41598-026-12345-y.

Debbarma, R., Singh, S.K., Ngasotter, S., Waikhom, G., Biswas, P., Chouhan, N., Xavier, K.M., Patel, A.B., 2026a. Extraction and characterization of chitin, chitosan, and nanochitosan from black soldier fly (*Hermetia illucens*) pupal exoskeletons: Insights into structural, thermal, microstructural, color, and antimicrobial properties. Int. J. Biol. Macromol. 262, 152497. doi:10.1016/j.ijbiomac.2026.152497.

Dilling, L., Alldredge, A.L., 2000. Fragmentation of marine snow by swimming macrozooplankton: a new process impacting carbon cycling in the sea. Deep-Sea Res. Part I: Oceanogr. Res. Pap. 47 (7), 1227–1245. doi:10.1016/S0967-0637(99)00105-3.

Dutta, P.K., Dutta, J., Tripathi, V.S., 2004. Chitin and chitosan: chemistry, properties and applications. J. Sci. Ind. Res. 63 (1), 20–31.

Edwards, P., Zhang, W., Belton, B., Little, D.C., 2019. Misunderstandings, myths and mantras in aquaculture: Its contribution to world food supplies has been systematically over reported. Mar. Policy 106, 103547. doi:10.1016/j.marpol.2019.103547.

Emerenciano, M.GC., Martínez-Córdova, L.R, Martínez-Porchas, M., Miranda-Baeza, A., 2017. Biofloc technology (BFT): a tool for water quality management in aquaculture. InTech, Rijeka, Croatia., pp. 91–109. doi:10.5772/66416.

Fan, Y., Ma, X., Dong, X., Feng, Z., Dong, Y., 2020. Characterisation of floc size, effective density and sedimentation under various flocculation mechanisms. Water Sci. Technol. 82 (7), 1261–1271. doi:10.2166/wst.2020.153.

Fauzani, D., Notodarmojo, S., Handajani, M., Helmy, Q., Kardiansyah, T., 2021. Cellulose in natural flocculant applications: a review. J. Phys. Conf. Ser. 2047 (1), 012030. doi:10.1088/1742-6596/2047/1/012030.

Hargreaves J.A., Biofloc production systems for aquaculture. SRAC Publication No. 4503. Southern Regional Aquaculture Center. Stoneville, MS (2013), pp. 1–12

Garatun-Tjeldstø, O., Otterå, H., Julshamn, K., Austreng, E., 2006. effects of food particle size. ICES J. Mar. Sci. 63 (2), 311–319. doi:10.1016/j.icesjms.2005.11.004.

Jarvis, P., Jefferson, B., Gregory, J., Parsons, S.A., 2005. A review of floc strength and breakage. Water Res. 39 (14), 3121–3137. doi:10.1016/j.watres.2005.05.022.

Javed, S., Zulfiqar, Z., Fatima, Z., Muhammad, G., Hussain, M.A., Mushtaq, M., Haseeb, M.T., 2024. A comprehensive review of plant-based mucilages as promising candidates for water remediation. J. Environ. Chem. Eng. 12 (5). doi:10.1016/j.jece.2024.114035.

Jiang, X., Li, Y., Tang, X., Jiang, J., He, Q., Xiong, Z., Zheng, H., 2021. Biopolymer-based flocculants: a review of recent technologies. Environ. Sci. Pollut. Res. 28 (34), 46934–46963. doi:10.1007/s11356-021-15299-y.

Jusoh, H.H.W., Kasan, N.A., Manan, H., Nasir, N.M., Yunos, F.H.M., Hamzah, S., Jusoh, A., 2023. Chicken eggshell as an innovative bioflocculant in harvesting biofloc for aquaculture wastewater treatment. J. Renew. Mater. 11 (5), 2321–2332. doi:10.32604/jrm.2023.026086.

Khanjani, M.H., Sharifinia, M., 2020. Biofloc technology as a promising tool to improve aquaculture production. Rev. Aquac. 12 (3), 1836–1850. doi:10.1111/raq.12412.

Krummenauer, D., Poersch, L., Romano, L.A., Lara, G.R., Encarnação, P., Wasielesky, W., 2014. The effect of probiotics in a Litopenaeus vannamei biofloc culture system infected with Vibrio parahaemolyticus. J. Appl. Aquac. 26 (4), 370–379. doi:10.1080/10454438.2014.965575.

Kurniawan, S.B., Abdullah, S.R.S., Imron, M.F., Said, N.S.M., Ismail, N.'., Hasan, H.A., Othman, A.R., Purwanti, I.F., 2020. Challenges and opportunities of biocoagulant/bioflocculant application for drinking water and wastewater treatment and its potential for sludge recovery. Int. J. Environ. Res. Public Health 17 (24), 1–33. doi:10.3390/ijerph17249312.

Lal, J., Vaishnav, A., Brar, K.S, Sahil, Verma, D.K, Jayaswal, R., Lavkush, Debbarma, S., Devati, Kumar, S., 2024. Biofloc technology: optimizing aquaculture through microbial innovation. Adv. Microbiol. 24 (7), 11–24. doi:10.9734/jamb/2024/v24i7835.

Landman, M.J., Van Den Heuvel, M.R., Ling, N., 2005. Relative sensitivities of common freshwater fish and invertebrates to acute hypoxia. N.Z. J. Mar. Freshw. Res. 39 (5), 1061–1067. doi:10.1080/00288330.2005.9517369.

Magalhães, E.R.B., Filho, J.D.B.C., Padilha, C.E.A., Silva, F.L., Sousa, M.A.S.B., Santos, E.S., 2021. Activated sludge treatment for promoting the reuse of a synthetic produced water in irrigation. J. Environ. Sci. Health B 56 (2), 132–141. doi:10.1080/03601234.2020.1852855.

Majhi, S.S., Singh, S.K., Biswas, P., Debbarma, R., Parhi, J., Ngasotter, S., Waikhom, G., Meena, D.K., Devi, A.G., Mahanand, S.S., Xavier, K.A.M., Patel, A.B., 2023. Effect of stocking density on growth, water quality changes and cost efficiency of butter catfish (Ompok bimaculatus) during seed rearing in a biofloc system. Fishes 8 (2). doi:10.3390/fishes8020061.

McCusker, S., Warberg, M.B, Davies, S.J., Valente, C.S, Johnson, M.P., Cooney, R., Wan, A.H.L., 2023. Biofloc technology as part of a sustainable aquaculture system: a review on the status and innovations for its expansion. Aquac. Fish Fish. 3 (4), 331–352. doi:10.1002/aff2.108.

Mietta, F., Chassagne, C., Manning, A.J., Winterwerp, J.C., 2009. Influence of shear rate, organic matter content, pH and salinity on mud flocculation. Ocean Dyn. 59 (5), 751–763. doi:10.1007/s10236-009-0231-4.

Milligan, T.G., Hill, P.S., 1998. A laboratory assessment of the relative importance of turbulence, particle composition, and concentration in limiting maximal floc size and settling behaviour. J. Sea Res. 39 (3–4), 227–241. doi:10.1016/s1385-1101(97)00062-2.

Minaz, M., Kubilay, A., 2021. Operating parameters affecting biofloc technology: carbon source, carbon/nitrogen ratio, feeding regime, stocking density, salinity, aeration, and microbial community manipulation. Aquac. Int. 29 (3), 1121–1140. doi:10.1007/s10499-021-00681-x.

Mohd Yunos, F.H., Nasir, N.M., Wan Jusoh, H.H., Khatoon, H., Lam, S.S., Jusoh, A., 2017. Harvesting of microalgae (Chlorella sp.) from aquaculture bioflocs using an environmental-friendly chitosan-based bio-coagulant. Int. Biodeterior. Biodegrad. 124, 243–249. doi:10.1016/j.ibiod.2017.07.016.

Morin-Crini, N., Lichtfouse, E., Torri, G., Crini, G., 2019. Fundamentals and applications of chitosan. In: Sustainable agriculture reviews 35. Springer Science and Business Media LLC, Cham, Switzerland, pp. 49–123. doi:10.1007/978-3-030-16538-3_2.

Muyibi, S.A., Megat Johari, M.J.M., Leong, T.K., Loon, L.H., 2002. Effects of oil extraction from Moringa oleifera seeds on coagulation of turbid water. Int. J. Environ. Stud. 59 (2), 243–254. doi:10.1080/00207230210924.

Nethaji, M., Ahilan, B., Ravaneswaran, K., Tamilarasu, A., Mahadevi, N., Lingam, R.S.S., 2021. Biofloc: floc types and their importance in aquaculture. Biotechnol. Res. Today 3, 581–583.

Okaiyeto, K., Nwodo, U.U., Okoli, S.A., Mabinya, L.V., Okoh, A.I., 2016. Implications for public health demands alternatives to inorganic and synthetic flocculants: bioflocculants as important candidates. MicrobiologyOpen 5 (2), 177–211. doi:10.1002/mbo3.334.

Olusegun, A.A., Babatunde, D.A., Abiodun, O.O., 2016. Haematological response of *Clarias gariepinus* juveniles reared in treated wastewater after waste solids removal using alum or *Moringa oleifera* seed powder. Int. J. Aquacult. 6. doi:10.5376/ija.2016.06.0004.

Prakash, N.B., Sockan, V., Jayakaran, P., 2014. Waste water treatment by coagulation and flocculation. Int. J. Eng. Sci. Innov. Technol. 3 (2), 479–484.

Rashid, N., Rehman, S.U., Han, J.I., 2013. Rapid harvesting of freshwater microalgae using chitosan. Process Biochem 48 (7), 1107–1110. doi:10.1016/j.procbio.2013.05.010.

Salim, A., Nour, H., El Handaoui, E.M., Adel, M., El Bouari, A., Lghazi, Y., Jrifi, A., Tanane, O., 2026. Integrated experimental, thermodynamic and molecular-level investigation of anionic and cationic dye removal by raw *Moringa oleifera* seed powder: mechanistic and structural insights. React. Kinet. Mech. Catal. 1–23. doi:10.1007/s11144-026-02234-5.

Singh, H.M., Sharma, M., Tyagi, V.V., et al., 2023. Potential of biogenic and non-biogenic waste materials as flocculant for algal biomass harvesting: mechanism, parameters, challenges and future prospects. J. Environ. Manage. 337, 117591. doi:10.1016/j.jenvman.2023.117591.

Tang, F.H.M., Maggi, F., 2016. A mesocosm experiment of suspended particulate matter dynamics in nutrient- and biomass-affected waters. Water Res. 89, 76–86. doi:10.1016/j.watres.2015.11.033.

Tansel, B., 2018. Morphology, composition and aggregation mechanisms of soft bioflocs in marine snow and activated sludge: a comparative review. J. Environ. Manage. 205, 231–243. doi:10.1016/j.jenvman.2017.09.082.

Tong, C.Y., Yusuf, F.H.B.C., Derek, C.J.C., 2022. Optimization of *Moringa oleifera* seed extract and chitosan as natural coagulant in treatment of fish farm wastewater. Desalination Water Treat 256, 99–113. doi:10.5004/dwt.2022.28019.

Ugbenyen, A.M, Okoh, A.I, 2014. Characteristics of a bioflocculant produced by a consortium of *Cobetia*and *Bacillus* species and its application in the treatment of wastewaters. Water SA 40 (1), 140. doi:10.4314/wsa.v40i1.17.

Walunj, T., Kala, A.S., Susitharan, V., Rani, A.M.B., 2024. Natural and nano-conjugated polymers as bioflocculating agents influences biofloc development, water quality dynamics and growth performance in in-situ biofloc system with genetically improved farmed tilapia. Aquac. Int. 32 (3), 3563–3585. doi:10.1007/s10499-023-01338-7.

Wilén, B.M., Balmer, P., 1999. The effect of dissolved oxygen concentration on the structure, size and size distribution of activated sludge flocs. Water Res 33 (2), 391–400. doi:10.1016/S0043-1354(98)00219-5.

Xavier, M., Júnior, W.W, Hostins, B., Bequé, E., Krummenauer, D., 2022. The use of a flocculant additive and its effect on biofloc formation, nitrification, and zootechnical performance during the culture of Pacific white shrimp Penaeus vannamei (Boone, 1931) in a BFT system. Lat. Am. J. Aquat. Res. 50 (2), 181–196. doi:10.3856/vol50-issue2-fulltext-2777.

Yin, Z., Hu, D., Li, X., Zhong, Y., Zhu, L., 2021. Shell-derived chitosan as a green flocculant to harvest microalgae for biofuel production. Biofuels Bioprod. Biorefin. 15 (3), 637–645. doi:10.1002/bbb.2197.

Zaki, S.A., Elkady, M.F., Farag, S., Abd-El-Haleem, D., 2013. Characterization and flocculation properties of a carbohydrate bioflocculant from a newly isolated Bacillus velezensis 40B. J. Environ. Biol. 34 (1), 51–58. http://www.jeb.co.in/journal_issues/201301_jan13/paper_08.pdf.

Zhang, H., Sun, T., Tan, H., Lv, X., Liu, W., Luo, G., 2023. Using black soldier fly larvae (Hermetiaillucens) converted the bioflocs produced with shrimp solid waste. Aquaculture 568, 739329. doi:10.1016/j.aquaculture.2023.739329.

Zhang, S., Yi, X., He, D., Tang, X., Chen, Y., Zheng, H., 2024. Recent progress and perspectives of typical renewable bio-based flocculants: characteristics and application in wastewater treatment. Environ. Sci. Pollut. Res. 31 (34), 46877–46897. doi:10.1007/s11356-024-34199-5.

Index

Page numbers followed by "*f*" and "*t*" indicate, figures and tables respectively.

B

Z

Printed and bound by CPI Group (UK) Ltd, Croydon, CR0 4YY
29/09/2026
02224672-0004